Human-Computer Interaction

Design Issues, Solutions, and Applications

Human Factors and Ergonomics

Series Editor
Gavriel Salvendy

Published Titles

Conceptual Foundations of Human Factors Measurement, *D. Meister*

Designing for Accessibility: A Business Guide to Countering Design Exclusion, *S. Keates*

Handbook of Cognitive Task Design, *E. Hollnagel*

Handbook of Digital Human Modeling: Research for Applied Ergonomics and Human Factors Engineering, *V. G. Duffy*

Handbook of Human Factors and Ergonomics in Health Care and Patient Safety, *P. Carayon*

Handbook of Human Factors in Web Design, *R. Proctor and K. Vu*

Handbook of Standards and Guidelines in Ergonomics and Human Factors, *W. Karwowski*

Handbook of Virtual Environments: Design, Implementation, and Applications, *K. Stanney*

Handbook of Warnings, *M. Wogalter*

Human-Computer Interaction: Designing for Diverse Users and Domains, *A. Sears and J. A. Jacko*

Human-Computer Interaction: Design Issues, Solutions, and Applications, *A. Sears and J. A. Jacko*

Human-Computer Interaction: Development Process, *A. Sears and J. A. Jacko*

Human-Computer Interaction: Fundamentals, *A. Sears and J. A. Jacko*

Human Factors in System Design, Development, and Testing, *D. Meister and T. Enderwick*

Introduction to Human Factors and Ergonomics for Engineers, *M. R. Lehto and J. R. Buck*

Macroergonomics: Theory, Methods and Applications, *H. Hendrick and B. Kleiner*

The Handbook of Data Mining, *N. Ye*

The Human-Computer Interaction Handbook: Fundamentals, Evolving Technologies, and Emerging Applications, Second Edition, *A. Sears and J. A. Jacko*

Theories and Practice in Interaction Design, *S. Bagnara and G. Crampton-Smith*

Usability and Internationalization of Information Technology, *N. Aykin*

User Interfaces for All: Concepts, Methods, and Tools, *C. Stephanidis*

Forthcoming Titles

Computer-Aided Anthropometry for Research and Design, *K. M. Robinette*

Content Preparation Guidelines for the Web and Information Appliances: Cross-Cultural Comparisons, *Y. Guo, H. Liao, A. Savoy, and G. Salvendy*

Foundations of Human-Computer and Human-Machine Systems, *G. Johannsen*

Handbook of Healthcare Delivery Systems, *Y. Yih*

Human Performance Modeling: Design for Applications in Human Factors and Ergonomics, *D. L. Fisher, R. Schweickert, and C. G. Drury*

Smart Clothing: Technology and Applications, *G. Cho*

The Universal Access Handbook, *C. Stephanidis*

Human-Computer Interaction

Design Issues, Solutions, and Applications

Edited by

Andrew Sears
Julie A. Jacko

CRC Press
Taylor & Francis Group
Boca Raton London New York

CRC Press is an imprint of the
Taylor & Francis Group, an **informa** business

This material was previously published in *The Human-Computer Interaction Handbook: Fundamentals, Evolving Technologies and Emerging Applications, Second Edition,* © Taylor & Francis, 2007.

CRC Press
Taylor & Francis Group
6000 Broken Sound Parkway NW, Suite 300
Boca Raton, FL 33487-2742

First issued in paperback 2017

© 2009 by Taylor & Francis Group, LLC
CRC Press is an imprint of Taylor & Francis Group, an Informa business

No claim to original U.S. Government works

ISBN 13: 978-1-138-11425-8 (pbk)
ISBN 13: 978-1-4200-8885-4 (hbk)

Library of Congress Cataloging-in-Publication Data

Human-computer interaction. Design issues, solutions, and applications / editors, Andrew Sears, Julie A. Jacko.
    p. cm. -- (Human factors and ergonomics)
"Select set of chapters from the second edition of The Human computer interaction handbook"--Pref.
Includes bibliographical references and index.
ISBN 978-1-4200-8885-4 (hardcover : alk. paper)
    1. Human-computer interaction. I. Sears, Andrew. II. Jacko, Julie A. III. Human-computer interaction handbook. IV. Title. V. Series.

QA76.9.H85H85656 2008
004.01'9--dc22

2008049135

Visit the Taylor & Francis Web site at
http://www.taylorandfrancis.com

and the CRC Press Web site at
http://www.crcpress.com

For Beth, Nicole, Kristen, François, and Nicolas.

CONTENTS

Contributors ix

Advisory Board xi

Preface xiii

About the Editors xv

1   Visual Design Principles for Usable Interfaces: Everything Is Designed: Why We Should Think Before Doing 1
     Suzanne Watzman and Margaret Re

2   Global/Intercultural User Interface Design 27
     Aaron Marcus

3   Conversational Speech Interfaces and Technologies 53
     Jennifer Lai, Clare-Marie Karat, and Nicole Yankelovich

4   Multimedia User Interface Design 65
     Alistair Sutcliffe

5   Multimodal Interfaces 85
     Sharon Oviatt

6   Adaptive Interfaces and Agents 105
     Anthony Jameson

7   Mobile Interaction Design in the Age of Experience Ecosystems 131
     Marco Susani

8   Tangible User Interfaces 141
     Hiroshi Ishii

9   Achieving Psychological Simplicity: Measures and Methods to Reduce Cognitive Complexity .161
     John C. Thomas and John T. Richards

10   Information Visualization 181
     Stuart Card

11   Groupware and Computer-Supported Cooperative Work 217
     Gary M. Olson and Judith S. Olson

12   HCI and the Web 231
     Helen Ashman, Tim Brailsford, Gary Burnett, Jim Goulding,
     Adam Moore, Craig Stewart, and Mark Truran

13   Human-Centered Design of Decision-Support Systems .245
     Philip J. Smith, Norman D. Geddes and Roger Beatty

14   Online Communities 275
     Panayiotis Zaphiris, Chee Siang Ang, and Andrew Laghos

15   Virtual Environments 293
     Kay M. Stanney and Joseph V. Cohn

16   Human-Computer Interaction Viewed from the Intersection of Privacy, Security, and Trust 311
     John Karat, Clare-Marie Karat, and Carolyn Brodie

Author Index 331

Subject Index 351

CONTRIBUTORS

Chee Siang Ang
Centre for HCI Design, City University London, UK

Helen Ashman
School of Computer Science and Information Technology, University of Nottingham, UK

Roger Beatty
American Airlines, USA

Tim Brailsford
School of Computer Science and Information Technology, University of Nottingham, UK

Carolyn Brodie
IBM T.J. Watson Research Center, USA

Gary Burnett
School of Computer Science and Information Technology, University of Nottingham, UK

Stuart Card
User Interface Research Group, Palo Alto Research Center (PARC), USA

Joseph V. Cohn
Naval Research Laboratory, USA

Norman D. Geddes
Applied Systems Intelligence, Inc., USA

Jim Goulding
School of Computer Science and Information Technology, University of Nottingham, UK

Hiroshi Ishii
MIT Media Laboratory, USA

Anthony Jameson
German Research Center for Artificial Intelligence, and International University in Germany, Germany

Clare-Marie Karat
IBM T.J. Watson Research Center, USA

John Karat
IBM T.J. Watson Research Center, USA

Andrew Laghos
Centre for HCI Design, City University London, UK

Jennifer Lai
IBM T.J. Watson Research Center, USA

Aaron Marcus
Aaron Marcus and Associates, Inc., USA

Adam Moore
School of Computer Science and Information Technology, University of Nottingham, UK

Gary M. Olson
School of Information, The University of Michigan, USA

Judith S. Olson
School of Information, The University of Michigan, USA

Sharon Oviatt
Department of Computer Science and Engineering, Oregon Health and Science University, USA

Margaret Re
Visual Arts Department, UMBC, USA

John T. Richards
IBM T. J. Watson Research Center, USA

Philip J. Smith
Institute for Ergonomics, Ohio State University, USA

Kay M. Stanney
Industrial Engineering and Management Systems, University of Central Florida, USA

Craig Stewart
School of Computer Science and Information Technology, University of Nottingham, and Department of Electronic Engineering, Queen Mary, University of London, UK

Marco Susani
Design for Seamless Mobility, Motorola, USA

Alistair Sutcliffe
School of Informatics, University of Manchester, UK

John C. Thomas
IBM T. J. Watson Research Center, USA

Mark Truran
University of Teesside, UK

Suzanne Watzman
Watzman Information Design, USA

Nicole Yankelovich
Sun Microsystems, USA

Panayiotis Zaphiris
Centre for HCI Design, City University London, UK

ADVISORY BOARD

PREFACE

We are pleased to offer access to a select set of chapters from the second edition of *The Human–Computer Interaction Handbook*. Each of the four books in the set comprises select chapters that focus on specific issues including fundamentals that serve as the foundation for human–computer interactions, design issues, issues involved in designing solutions for diverse users, and the development process.

While human–computer interaction (HCI) may have emerged from within computing, significant contributions have come from a variety of fields including industrial engineering, psychology, education, and graphic design. The resulting interdisciplinary research has produced important outcomes including an improved understanding of the relationship between people and technology as well as more effective processes for utilizing this knowledge in the design and development of solutions that can increase productivity, quality of life, and competitiveness. HCI now has a home in every application, environment, and device, and is routinely used as a tool for inclusion. HCI is no longer just an area of specialization within more traditional academic disciplines, but has developed such that both undergraduate and graduate degrees are available that focus explicitly on the subject.

The HCI Handbook provides practitioners, researchers, students, and academicians with access to 67 chapters and nearly 2000 pages covering a vast array of issues that are important to the HCI community. Through four smaller books, readers can access select chapters from the Handbook. The first book, *Human–Computer Interaction: Fundamentals,* comprises 16 chapters that discuss fundamental issues about the technology involved in human–computer interactions as well as the users themselves. Examples include human information processing, motivation, emotion in HCI, sensor-based input solutions, and wearable computing. The second book, *Human–Computer Interaction: Design Issues*, also includes 16 chapters that address a variety of issues involved when designing the interactions between users and computing technologies. Example topics include adaptive interfaces, tangible interfaces, information visualization, designing for the web, and computer-supported cooperative work. The third book, *Human–Computer Interaction: Designing for Diverse Users and Domains,* includes eight chapters that address issues involved in designing solutions for diverse users including children, older adults, and individuals with physical, cognitive, visual, or hearing impairments. Five additional chapters discuss HCI in the context of specific domains including health care, games, and the aerospace industry. The final book, *Human–Computer Interaction: The Development Process,* includes fifteen chapters that address requirements specification, design and development, and testing and evaluation activities. Sample chapters address task analysis, contextual design, personas, scenario-based design, participatory design, and a variety of evaluation techniques including usability testing, inspection-based techniques, and survey design.

Andrew Sears and Julie A. Jacko

March 2008

ABOUT THE EDITORS

Andrew Sears is a Professor of Information Systems and the Chair of the Information Systems Department at UMBC. He is also the director of UMBC's Interactive Systems Research Center. Dr. Sears' research explores issues related to human-centered computing with an emphasis on accessibility. His current projects focus on accessibility, broadly defined, including the needs of individuals with physical disabilities and older users of information technologies as well as mobile computing, speech recognition, and te difficulties information technology users experience as a result of the environment in which they are working or the tasks in which they are engaged. His research projects have been supported by numerous corporations (e.g., IBM Corporation, Intel Corporation, Microsoft Corporation, Motorola), foundations (e.g., the Verizon Foundation), and government agencies (e.g., NASA, the National Institute on Disability and Rehabilitation Research, the National Science Foundation, and the State of Maryland). Dr. Sears is the author or co-author of numerous research publications including journal articles, books, book chapters, and conference proceedings. He is the Founding Co-Editor-in-Chief of the *ACM Transactions on Accessible Computing,* and serves on the editorial boards of the *International, Journal of Human–Computer Studies,* the *International Journal of Human–Computer Interaction,* the *International Journal of Mobil Human–Computer Interaction,* and *Universal Access in the Information Society,* and the advisory board of the upcoming *Universal Access Handbook.* He has served on a variety of conference committees including as Conference and Technical Program Co-Chair of the Association for Computing Machinery's Conference on Human Factors in Computing Systems (CHI 2001), Conference Chair of the ACM Conference on Accessible Computing (Assets 2005), and Program Chair for Asset 2004. He is currently Vice Chair of the ACM Special Interest Group on Accessible Computing. He earned his BS in Computer Science from Rensselaer Polytechnic Institute and his Ph.D. in Computer Science with an emphasis on Human–Computer Interaction from the University of Maryland—College Park.

Julie A. Jacko is Director of the Institute for Health Informatics at the University of Minnesota as well as a Professor in the School of Public Health and the School of Nursing. She is the author or co-author of over 120 research publications including journal articles, books, book chapters, and conference proceedings. Dr. Jacko's research activities focus on human–computer interaction, human aspects of computing, universal access to electronic information technologies, and health informatics. Her externally funded research has been supported by the Intel Corporation, Microsoft Corporation, the National Science Foundation, NASA, the Agency for Health Care Research and Quality (AHRQ), and the National Institute on Disability and Rehabilitation Research. Dr. Jacko received a National Science Foundation CAREER Award for her research titled, "Universal Access to the Graphical User Interface: Design For The Partially Sighted," and the National Science Foundation's Presidential Early Career Award for Scientists and Engineers, which is the highest honor bestowed on young scientists and engineers by the US government. She is Editor-in-Chief of the *International Journal of Human–Computer Interaction* and she is Associate Editor for the *International Journal of Human Computer Studies.* In 2001 she served as Conference and Technical Program Co-Chair for the ACM Conference on Human Factors in Computing Systems (CHI 2001). She also served as Program Chair for the Fifth ACM SIGCAPH Conference on Assistive Technologies (ASSETS 2002), and as General Conference Chair of ASSETS 2004. In 2006, Dr. Jacko was elected to serve a three-year term as President of SIGCHI. Dr. Jacko routinely provides expert consultancy for organizations and corporations on systems usability and accessibility, emphasizing human aspects of interactive systems design. She earned her Ph.D. in Industrial Engineering from Purdue University.

Human-Computer Interaction

Design Issues, Solutions, and Applications

· 1 ·

VISUAL DESIGN PRINCIPLES
FOR USABLE INTERFACES

Everything Is Designed:
Why We Should Think Before Doing

Suzanne Watzman
Watzman Information Design

Margaret Re
UMBC

Making Things Easier to Use and Understand:
Thinking about the User's Experience 3
Defining Visual Design 3
The Design Process 3
The Role of the Designer 4
The Process of Good Design—How Do We Get
There from Here? 4
An Information-Design Process Is an Informed
Design Process 4
    Phase 1: The Audit 5
        Audit Questions A 5
        Audit Questions B 5
    Phase 2: Design Development 5
    Phase 3: Implementation and Monitoring 5
Visual Design Principles 6
Universal Principles of Visual Communication
and Organization 6
Visual Design Tools and Techniques 6
    The Five Criteria for Good Design 7

Visual Design Principles at Work 7
    Typography 7
How the Human Eye Sees, and then Reads 9
Typeface Size and Selection 9
    Serif and sans serif 9
    Families of type 10
Variations in Letterforms 11
    Variations in Stress 11
    Variations in Thick and Thin 11
    Variations in Serifs 11
Typographic Guidelines 11
    Combining Typefaces 11
    Contrast in Weight (Boldness) 11
    Output Device and Viewing Environment 11
    Letter Spacing and Word Spacing 12
    Line Spacing/Leading 12
    Line Length/Column Width 12
    Justified vs. Ragged Right (Flush Left) 12
    Highlighting with Type 12

Decorative Typefaces 12
Black on White vs. White on Black and Dark
on Light Background vs. Light on
Dark Background 12
  Positive and negative type 12
Design Principles: Page Design **13**
Building the Design of a Page 13
  Gray page or screen 13
  Chunking 13
  Queuing 13
  Filtering 13
  Mixing modes 13
  Abstracting 15
Other Page Design Techniques 15
  White space 15
  The grid 15
  Field of vision 15
  Proximity 15
  The illusion of depth 16
Charts, Diagrams, Graphics, and Icons **16**
Tables, Charts, Diagrams 16
Icons and Visual Cues 16
Illustrations and Photographs 17
Guidelines 18
  Visuals should reinforce the message 18
  Create a consistent visual language 18
  Consider both function and style 18
  Focus on quality vs. quantity 18
  Work with a professional 18
  Build a library to create visual consistency,
  organizational identity, and a streamlined
  process 19
  Reinforce shared meaning (common
  visual language) 19
Color **19**
Basic Principles of Color 19
  Additive primaries 19
  Subtractive primaries 19
How to Use Color 19
  Less is more . . . useful and understandable 19
  Create a color logic; use color coding 19
  Create a palette of compatible colors 20

Use complementary colors with
extreme caution 20
Decisions regarding color in typography
are critical 20
Consider the viewing medium 20
Context is everything 20
Contrast is critical when making
color choices 20
Quantity affects perception 20
Use color as a redundant cue when
possible 20
We live in a global world, so when in Rome 20
Creating a System: Graphic Standards **20**
What does a system cover? 22
Developing the system 22
Audit 22
Development 22
Implementation 22
Designing the Experience **22**
Effective and appropriate use of the medium 23
The element of time 23
Consistent and Appropriate Visual Language **23**
Navigational Aids 23
Graphics/Icons 23
Metaphor 23
Color 23
Legibility 24
Readability 24
Guidelines 24
  Use the analogy of a poster as a guide
  to design 24
  Design for the most difficult common
  denominator 24
  Avoid overuse of saturated colors 24
  Consider different users' levels of skill 24
  Be aware of the fatigue factor 24
  Other differences to consider 24
  Use the "squint test" to check the design 24
Challenges and Opportunities 24
Creating Your Own Guidelines 24
References 25

Take a moment and visualize Las Vegas at night. What kind of image does this conjure up for you? Flashing lights from all directions, a hotel's lighting display designed to outdo its neighbor as well as that of the casino signage down the street. At first glance, everything is exciting, colorful, and beautiful. Now add a fireworks display to your picture. More color, more excitement. Where do you look first? There is so much going on; it is hard to see it all, but you don't want to miss a thing. Your head turns in all directions. You look there; then, out of the corner of your eye, you see something else. Look over there! Now the fireworks are at their peak, and the noise gets even louder. Any conversation with companions is impossible, yet it is also impossible to focus on any one thing for more than a split second. You are overwhelmed and overloaded. Everything is screaming for your attention. Can you manage to pay attention? For how long? Do you begin to shake your head in despair, and give up? Do you wish you were somewhere else—NOW?

MAKING THINGS EASIER TO USE AND UNDERSTAND: THINKING ABOUT THE USER'S EXPERIENCE

The previous description is unfortunately an accurate analogy of many users' experiences as they attempt to learn, work, play, and relax. New products, new services, and new technology with which you are unfamiliar can create confusion. Users of these new products, services, and technology are customers, electricians, grandparents, clerks, pilots, and students—you and me. And for most of us, it's a jungle out there! Las Vegas at night with fireworks, or monitors that are winking, blinking, distracting, disturbing, overwhelming—and, after a short time, visually deafening. Now add voices coming from boxes. . .! Although this may seem like an exaggeration, for many this situation is exactly their experience. User interface design focuses on designing flexible environments that have a positive impact on a user's ability to experience and interact with a product, whether that product is a mobile communication device, website, information kiosk, or appliance. It involves creating environments that include strong navigational devices that can be understood intuitively and used effortlessly. Designers have a responsibility to create user experiences that are simple and transparent. To do their job well, they must advocate on behalf of the user, ensuring that the interfaces they design are not just merely exercises in technology but that they truly assist and guide the user from task to task, enabling work to be done, and ultimately improving quality of life. When designers succeed, their products can be used effortlessly and are even pleasurable to use. Good design does not needlessly draw attention to itself. It just works. This is the role of good design.

DEFINING VISUAL DESIGN

The nautilus shell is an example of the synthesis between form and function found in nature (Fig. 1.1). Its form is the result of evolution, which is both transparent and beautiful. The nautilus shell is a perfect analogy for design and the design process because it creates valuable user experiences and usable interfaces.

The word *design* functions as both a noun and a verb. Many people use it to refer to the outward appearance or style of a product. However, design also refers to a process—that of intentionally establishing a plan or system by which a task can be accomplished or a goal reached. It includes tangible and intangible systems in which objects or processes are coherently organized to include the environments in which these objects or processes function. Design affects all people in every aspect of what they do. Good design performs for people. It is concerned with economics and the transmission of ideas. The challenge presented to a design team is to plan a

FIGURE 1.1. Nautilus Shell.

prototype with a clear purpose that is easy to use, meets user needs, addresses commercial considerations, and can be mass-produced. Its visual form, whether two- or three-dimensional, digital or analog, logically explains its purpose and efficiently leads the user through its function. Design is not a series of subjective choices based on personal preference, at best a cosmetic afterthought considered if time and money are leftover. Good design is the tangible representation of product goals. An iterative and interactive process that requires active learning, design unifies a wide range of disciplines. Good design is a significant activity that reveals multiple solutions to each problem. Design equally values different ways of thinking. It allows people with a variety of skills and learning abilities to work cooperatively to bring insights and expertise to problems and opportunities in order to better develop new and innovative solutions. Problems can be analyzed using a multitude of viewpoints and methods. Writing, drawing, statistical analysis, graphing, discussion, interviewing, personal observation, model-making, and diagramming are all legitimate methods for examination as the physical, social, and cultural contexts of possible answers are considered (Davis, Hawley, McMullan, & Spilka, 1997)

THE DESIGN PROCESS

Design as a Catalyst for Learning, a publication funded in part by the National Endowment for the Arts, argued that effective design that responds to human problems uses the following steps (Fig. 1.2):

- *Problem identification and definition:* A need or problem is identified, researched, and defined.

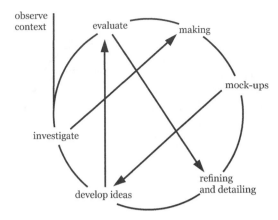

FIGURE 1.2. Interacting Design Loop. The interacting design loop developed by Richard Kimbell, founder of the Technology Education Research Unit at Goldsmiths College, and presented in *Design as a Catalyst for Learning*, captures the divergent, iterative and cyclical nature of the design process.

- *Gathering and analyzing information:* The focus is on learning what is not known. Assumptions are questioned. Wide and broad research is used to locate information and generate ideas.

- *Determining performance criteria for a successful solution:* Research continues as imagery is selected. Rules are declared and what is known is specified.

- *Generating alternative solutions and building prototypes:* Multiple solutions are generated. A variety of methods for analysis, such as drawing, interviewing, modeling or evaluating statistics, are used.

- *Implementing choices:* Project content, scope, and intent are formally established. Initial possibilities are represented and presented as prototypes.

- *Evaluating outcomes:* Prototypes are assessed, tested, evaluated, and judged. The knowledge gained is incorporated into further studies and refinements.

- *Production:* A prototype, which is a synthesis of the initial solutions made using this process, and specifications are released for making multiples to a manufacturer.

THE ROLE OF THE DESIGNER

Visual design decisions are based on project goals, user perspective, and informed decision making. While many aspects of design are quantifiable, there are visual principles that are less measurable but equally important. Even though the necessary skills to become visually literate and make competent design decisions can be learned, design involves a highly specialized knowledge base. A unique combination of creativity and skill differentiates and makes one design more attractive and desirable than another. Both education and talent are necessary to apply the principles required to present information in its most accessible, useful, and pleasing form. The role of the designer in the development of interfaces for interactive products is to understand the product goals and ensure that information is approachable, useful, and desirable. In an environment in which the interface is the only tangible representation of a product and user perception determines product success, appropriate information presentation and visual design are key. Designers understand visual principles in context, and know how to apply them appropriately to create innovative, functional and aesthetically pleasing solutions.

THE PROCESS OF GOOD DESIGN— HOW DO WE GET THERE FROM HERE?

Interface designers are responsible for defining what the experience will be like when a product is used. While print media dictates that users encounter content in a largely predetermined sequence, an interface offers the user greater flexibility over how content can be accessed based on users' needs and wants. A successful interface can be easily navigated. Interface designers define, decide, and then create the experience for users, so that an experience with a product is useful, meaningful, even pleasant and empowering. The designer must maintain an attitude of unbiased discovery and empathy for the user. The designer must develop clearly defined goals in order to create a good design that includes an evaluation process that supports and enhances these goals, and includes the flexibility to respond to changes as the process continues and products evolve.

AN INFORMATION-DESIGN PROCESS IS AN INFORMED DESIGN PROCESS

An information-design process (IDP) is a method of visually structuring and organizing information to develop effective communication. Information design is not superficial or decorative, but is rather a merging of functional, performance-based requirements with the most appropriate form to present these requirements. A thoughtful, well-designed solution will,

- *Motivate users:* It psychologically entices an audience, convincing members that information and tasks can be successfully handled.

- *Increase ease of use and accessibility:* The effort needed to comprehend information is decreased. A clear path that aids in skimming and referencing text and gives easy access is provided.

- *Increase the accuracy and retention of the information:* Users learn and retain information better when it is visually mapped and structured in obvious and intuitive ways.

- *Focus on the needs of its users:* Multiple audiences have different requirements and styles of learning. Solutions should be developed that provide alternative means of accessing information for different types of users. An information-design approach is part of a process that incorporates research, design, testing, and training to produce useful, cost-effective solutions.

Phase 1: The Audit

The goal of the audit is to create a blueprint for the project, much like architectural drawings are developed before constructing a building (Fig. 1.3). The audit process begins by asking and answering a number of questions and acknowledging ongoing change and an ever-increasing palette of products and services. Questions are asked throughout the entire product life cycle, since the answers/design solutions reflect the user/use environment and affect the ongoing usefulness and value of the product. To create an eloquent design, continually ask and answer the following questions:

Audit Questions A

- *Who* are the product users?
- *How* will this product be used?
- *When* will this product be used?
- *Why* will this product be used?
- *Where* will this product be used?
- *How* will the process evolve to support this product as it evolves?

After the first set of questions are asked and answered, a second set of questions must be asked and answered:

Audit Questions B

- What is the most efficient, effective way for a user to accomplish a set of tasks and move on to the next set of tasks?
- How can the information required for product ease of use be presented most efficiently and effectively?
- How can the design of this product be done to support ease of use and transition from task to task as a seamless, transparent, and even pleasurable experience?
- What are the technical and organizational limits and constraints?

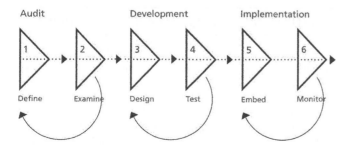

FIGURE 1.3. Information Design Process (IDP). IDP is phased to insure user and organizational needs are met. It is ongoing and iterative, throughout the lifecycle of a product. Any change can trigger a recycling of this process, to insure solutions remain appropriate and useful.

An audit focuses on discovery. Many disciplines and organizational resources must be consulted. Change is a given, since designers begin with assumptions and don't know all they need to know yet. The answers to their questions and their analysis in the context of organizational objectives provide the basis for the audit report, which serves as the guide in design development. The audit report can be as simple as a two-page list or as complex as a comprehensive hundred-page report. Since the goal is discovery, it includes every aspect of the organization concerned with the product-development cycle: project management, usability engineering, technical development, user support and documentation, visual communication and design, and content management. With these goals, the result is unbiased, accurate, comprehensive information that serves as the basis for design.

Phase 2: Design Development

The design-development phase uses the audit report as a guideline. This is an ongoing, iterative process with each iteration incorporating user test results to make the product appropriate to the particular set of needs. In reality, the length of this process is often defined and limited by real-world deadlines such as product release dates. The design-development phase includes design and testing. The designer or design team creates a number of solutions based on results and objectives determined by the audit report as well as other project specifics. Initially, design ideas should be very broad, incorporating many ideas and options no matter how unrealistic or unusual. As ideas are tested, user feedback incorporated, and other parameters defined, solutions naturally become more defined. Surviving design ideas are then based on solid information derived from user feedback, providing a strong basis for final design decisions. In the beginning, the focus is on high-level concepts and navigation. How will the product work? What will it feel like to use? As initial concepts are refined, design details become more specific. When the conceptual model and organizational framework are approved, the design of the look or product package begins. By the end of this phase, a prototype design to be carried out in implementation and monitoring is tested, approved, and specified.

Phase 3: Implementation and Monitoring

The implementation phase focuses on delivering what has been defined, designed, and documented in the preceding phases. It is the final part of a *holistic* process that defines everything necessary to make a product succeed on an ongoing basis. This includes not only the implementation of the design within the technology, but also any additional support such as the creation of training materials and other reinforcements that enhance use and productivity. Continuous monitoring is key to sustained success, because a successful product responds to evolving technology and user needs. This last phase is mostly consultative and ongoing throughout the product lifecycle in order to ensure that changes such as new technology and product developments are reflected in the product itself. These may in fact trigger another audit/design/testing cycle, although usually less extensive than the initial process. Though the implementation phase is called "the last phase," it reveals the evolutionary process of design and development. The goal of ongoing monitoring of solutions is to be aware of changes in user needs, technology, and competition that impact user acceptance and satisfaction. Changes here often result in the need to reevaluate and redesign to incorporate this new knowledge gained.

VISUAL DESIGN PRINCIPLES

Interaction design bridges many worlds: that of visual design, information presentation, and usability with aesthetics. Donis A. Donis (1973a), in *A Primer of Visual Literacy,* argued that art and its meaning have dramatically changed in contemporary times from one that involved a concern with function to one that views the process of creating art as that of making emotional maps that spring from the province of the intuitive and subjective. This argument extends to design. To someone unskilled in creating effective communications, visual design is often understood as personal preference limited to style or appearance. However, any form of effective design is a result of rigorous study, a concern for organization and usability combined with knowledge of the basic design principles of harmony, balance, and simplicity. Visual design is in fact a form of literacy.

UNIVERSAL PRINCIPLES OF VISUAL COMMUNICATION AND ORGANIZATION

The principles of harmony, balance, and simplicity are related yet distinct in meaning and application. *Harmony* is the grouping of related parts, so that all elements combine logically to make a unified whole. In interface design, as with other categories of design, this is achieved when all design elements work in unity. Transitions from place to place are effortless and the techniques used to achieve this harmony are unnoticed by the user. Visual harmony achieves the same goal as musical harmony in which notes combine to create a chord. The golden section, also known as the "golden mean" or "golden rectangle," is one of the most widely used methods for creating harmony. Architects, artists, musicians, mathematicians, and designers have used the golden section extensively for centuries to create proportional relationships (Fig. 1.4).

Balance offers equilibrium or rest. Donis stated that equilibrium is the strongest visual reference (Donis, 1973b). It provides the equivalent of a center of gravity that grounds the page. Without balance, the page collapses, all elements are seen as dispersed, and content is lost. Balance requires continual modification from page to page because while each page is part of a greater system, elements can vary and all have visual weight. In the same way that a clown balancing on a ball while juggling objects of different weights must continually make adjustments for actions that are occurring, visual balance requires the same concerns and adjustments as in the physical world. Regardless of how a design is organized, it must achieve stability and unity in order for a user to feel comfortable with the solution. Balance can be

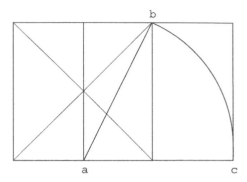

FIGURE 1.4. The Golden Rectangle. Divide a square to find the center point (a) from which length (ab) is found. From point (a) the length (ab) is swung as an arc to point (c) to create a rectangle that uses the proportions of the golden section.

achieved a number of ways. One obvious method uses *symmetry,* such as found on a page with text and image aligned on a centered axis. Deceptively simple, symmetry form is often considered easy to make; however, unless handled carefully a symmetrical composition can be predictable, boring, and static. *Asymmetry* employs nonaxial balance and uses contrast between elements such as weight, form, and color to create visual tension and drama. Both are valid approaches and require skill and knowledge of complex visual interaction to achieve.

Simplicity is the embodiment of clarity, elegance, and economy. A solution that offers simplicity is unambiguous and easily understood. It offers clarity working effortlessly devoid of unnecessary decoration. It appears deceivingly easy, accessible, and approachable, even though it may be conceptually rich. Simplicity involves distillation—every element is indispensable, if an element is removed, the composition falls apart. Achieving simplicity is no easy task. Two guidelines for creating simple design solutions are (a) "Less is more!" (attributed to Mies van der Rohe) and (b) "When in doubt, leave it out!" (Anonymous). The most refined design is direct and includes only essential elements. Removing any element breaks the composition rendering it unintelligible or radically different.

VISUAL DESIGN TOOLS AND TECHNIQUES

New technologies are rapidly being created that extend past the simple automation of tasks and communication; they are revolutionizing processes and the resulting products. Before the revolution brought about by electronic publishing technology, many disciplines such as writing, editing, design, publishing, and programming were discrete units that handled a defined step within a larger process. Today's new publishing environments encourage the possibility of a *renaissance* publisher—a person who can create, design, publish, and distribute. Yet the process used to arrive at successful solutions is very complex and extends past technical knowledge to include a mastery of visual and verbal language in order to build effective communication. Focus must be placed on the factors that determine success with constant evaluation and adjustment of these factors in light of new developments.

The Five Criteria for Good Design

Before any work begins, all participants in the process should have a clear understanding of the criteria for good design. The following questions are guidelines for evaluation of design solutions before, during, and after the process to ensure that all solutions remain valid as products, technology, and user needs evolve.

- **Is it appropriate?** Is the solution appropriate for the particular audience, environment, technology, and/or culture?
- **Is it durable?** Will the solution be useful over time? Can it be refined and transitioned as the product evolves and is redefined?
- **Is it verifiable?** Has the design been tested by typical users in the environment that it will be used in? Has feedback been properly evaluated and used to improve the product?
- **Does it have impact?** Does the design solution not only solve the problem, but also impact the look and feel, so that the user finds the product experience comfortable, useful, and desirable?
- **Is it cost effective?** Can the solution be implemented and maintained? Are individuals with the necessary skills and understanding to create, refine, and maintain the design available throughout the product's life? The cost of any design begins with the audit and design phases, but continues after the implementation phase to insure that it remains advantageous and cost-effective. The hard and soft costs of delivering the solution plus ongoing maintenance add up to the real design costs.

Visual Design Principles at Work

The following sections outline the disciplines and principles used to create quality design solutions. Each topic is worthy of extended study, because there is much to understand when evaluating how to effectively present information. As the design process evolves, insights and information are discovered that impact a solution. It is optimistic to base solutions on an initial exercise because the very nature of process means discovering what is unknown yet critical. Therefore, all members involved in the design process must remain open and ready to incorporate new information, which may change or delay results, but more accurately reflect user needs.

For example, if a new feature is developed that changes a product's target audience from mid-level managerial to executive users, most methods for critical interaction and content delivery should be reconsidered. Executives have less time and need different information. The result might be a simpler interface with streamlined content that uses a larger typeface and a more conservative visual language. The most important principle to remember when thinking about design is that there are no rules, only guidelines. Everything is context sensitive. Always consider and respect the users.

Typography

In *The Elements of Typographic Style*, the poet and typographer Robert Bringhurst (2005) described typography as frozen language. *Typography* is the visual representation of spoken and unspoken thought that allows an idea to be shared across time and distance independent of its creator. A functional and expressive art that shares many of the same concerns as writing and editing, typography involves organizing text so that its meaning is communicated according to an author's intent. In design, a literacy that provides an understanding of typography and how text can be structured in space is as important as a literacy that understands how to structure grammar in order to explicate content.

Typography is made from type, individual characters organized by visual characteristics into typefaces. Type is the smallest definable part of a design, much like a pixel is to a screen display. James Felici (2003), who has worked through evolutions in typesetting technologies in *The Complete Manual of Typography,* defined a font as the electronic file that contains the programming code needed to make the characters found in a typeface. Historically, a typeface consists of all the individual characters or glyphs at a given size: letterforms, punctuation, numbers, mathematical symbols, diacritical marks, and other accessory characters needed to fully compose a text. This definition serves as a reminder to read a text carefully and consider all needs before selecting a typeface and developing a presentation form (Fig. 1.5).

Effective typography is rational. It is concerned with clarity and comprehension; the ease in which characters and word shapes are recognized in reading environments and the ease in which they can be used. It extends past the shapes of individual letters and their potential combinations to include the relationships found between word and interword shapes, functional groupings that ultimately progress into issues of type weight, slope, width and scale, characteristics that act as interpretative devices in order to create influential and

Roman	Roman
Italic	*Oblique*
Bold	Condensed
Bold Italic	*Condensed Oblique*
Black	Extended
Black Italic	*Extended Oblique*
Ultra	
Ultra Italic	

FIGURE 1.5.   Type Family. A type family is built around four core members: roman, bold, italic, and bold italic. Additional members may include typefaces whose weight and width are variants of the core group. A family can also contain expert sets that offer additional or alternate characters such as small caps, fractions and non-aligning numbers.

persuasive form. If the principles of good typography can be understood and applied, then these same principles can be extended to more complex issues that follow such as page and product design. Typographic choice affects legibility and readability, the ability to easily see and understand what is on the page, in all media. Walter Tracy (1986), in *Letters of Credit: A View of Type Design,* offered the most useful definitions for legibility and readability. *Legibility,* the speed at which letters and the words built from them can be recognized, refers to perception. *Readability,* the facility and ease with which text can be read, refers to comprehension. Legibility and readability are related. Regardless of media, legibility and readability are determined by variables such as point size, letter pairing, word spacing, line length and leading, resolution, color, and organizational strategies such as text clustering. Together, legibility and readability comprise typography's functional aspects. Good typography, like good design, is invisible to the user—it just works (Fig. 1.6).

Selecting an appropriate typeface for a specific purpose and context requires experience and understanding (Fig. 1.7). With thousands of typefaces to choose from and numerous ways to manipulate them, finding the typeface best suited for an audience is not easy. With its lack of control, multiple media, and varied viewing contexts the current publishing environment makes this a complex task.

Typeface choice impacts whether and how a communication is read. Distinct typefaces and typographic styles create environments that influence a user's perception of text. The physical nature of the presentation itself helps determine content and acceptance. A typeface with extremely thick and thin strokes may appear sophisticated and readable in print but may look naïve and render text unreadable in a digital environment. Typefaces are frequently designed to solve issues of legibility and readability created by a technology. A typeface made for online use can increase page legibility, as well as the overall perception of approachability, quality of an interface, and ultimately product acceptance (Fig. 1.8, 1.9, & 1.10).

An informed selection can make reading enjoyable and effortless rather than frustrating and fatiguing. Though typography

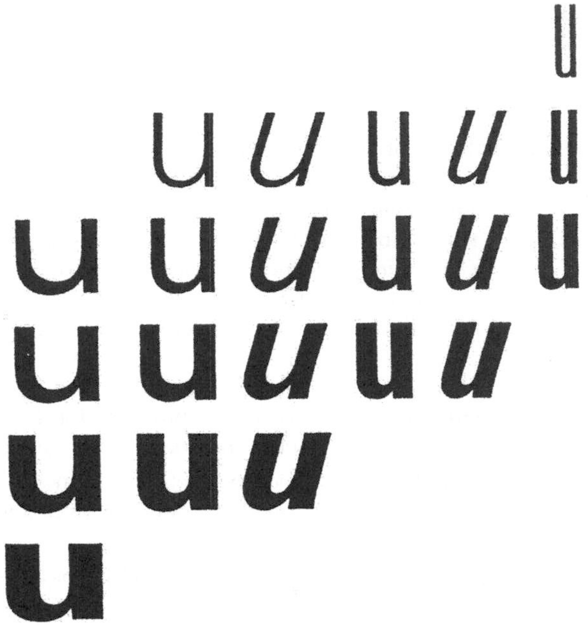

FIGURE 1.7.   Univers "U". Univers, a type family designed by Adrian Frutiger and released for commercial use in 1954, is composed of twenty-one fonts that together offer a wide range of weights, widths and slopes that allows a text to be organized so that its form is visually coherent and easily read.

Bell Centennial, 6 point

Address
ABCDEFGHIJKLMNOPQRSTUVWXYZ
abcdefghijklmnopqrstuvwxyz
1234567890 ([.,;:"-"/—)

Name & Number
ABCDEFGHIJKLMNOPQRSTUVWXYZ
abcdefghijklmnopqrstuvwxyz
1234567890 ([.,;:"-"/—)

BOLD LISTING
ABCDEFGHIJKLMNOPQRSTUVWXYZ
ABCDEFGHIJKLMNOPQRSTUVWXYZ
1234567890 ([.,;:"-"/—)

Sub-caption
ABCDEFGHIJKLMNOPQRSTUVWXYZ
abcdefghijklmnopqrstuvwxyz

FIGURE 1.8.   Bell Centennial: Technology-specific typefaces. AT&T commissioned Bell Centennial, a typeface designed at a very small size, for telephone directory use, in order to solve an industrial problem created by changing typesetting and printing technologies. The resulting type family designed for maximum legibility, readability and spatial efficiency provided the user with a clear information hierarchy. It reduced paper use and directory assistance calls. Here, Bell Centennial is shown at six point, the size at which it was intended to function.

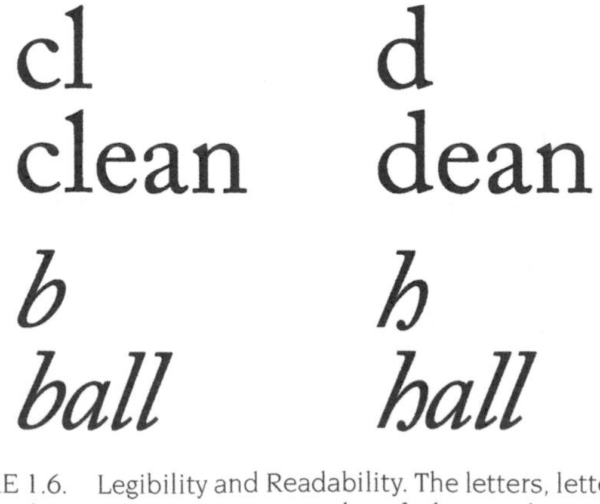

FIGURE 1.6.   Legibility and Readability. The letters, letter pairs and words shown above are examples of what can happen if the designer is not sensitive to issues of legibility and readability.

might seem to be an insignificant issue to a non-designer, it affects overall usability. A clear understanding of the concepts and principles that affect legibility and readability is crucial to determining effective typography.

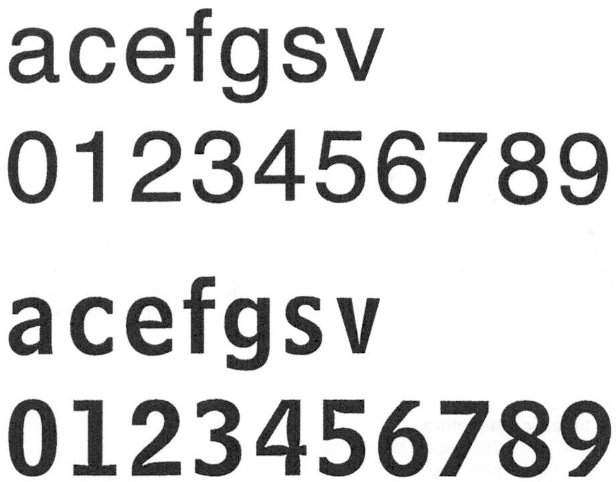

acefgsv
0123456789

acefgsv
0123456789

FIGURE 1.9.   Bell Centennial: Technology-specific typefaces. Bell Centennial's forms were opened to increase legibility and readability. Curved strokes were straightened and the horizontal and vertical juncture were notched so that they did not clog with ink when printed. Select letterforms shown at 32 point for Bell Centennial (bottom) and Helvetica (top) illustrate this.

FIGURE 1.10.   Pixel Font. Pixel fonts are screen fonts designed specifically for use with or as on-screen navigational elements. Their block-like forms are the result of working with the actual pixels themselves.

HOW THE HUMAN EYE SEES, AND THEN READS

Herbert Spencer (1969) in *The Visible Word*, a publication with an objective of introducing and uniting those who research legibility with those who work with typography, presented that the eye uses both outline word shapes and their internal patterns to move along a text line and steps and jumps as it groups text to form comprehensible phrases of information. This motion of the eye during reading is known as "saccadic movement." Sometimes during this process, the eye regresses and returns to what has been read. Optimal typography allows for fewer backward movements. Spencer argues that possessing a mastery of reading mechanics is important to a mastery of content. Typeface selection directly affects this skill, making it easier or more difficult for the eye

Now Read This

Now Read This

Now Read This

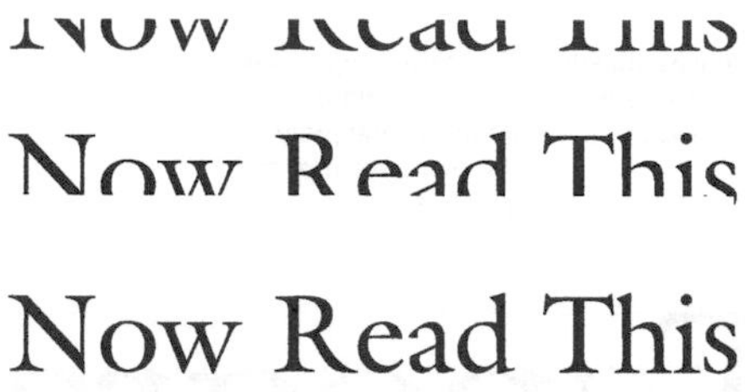

FIGURE 1.11.   Now Read This. The phrase, Now Read This, is shown in full, cropped at the bottom, and from the top suggesting the importance of outline word shape and internal pattern.

to group, read, and understand information (Fig. 1.11, 1.12, & 1.13).

TYPEFACE SIZE AND SELECTION

Type size is given in points, a measuring system unique to typography. In digital typesetting systems, a point equals 1/72 of an inch. Type size cannot be determined by physically measuring a letterform because when type existed solely in metal, the technology in which it was first used, size was measured by the height of the metal block on which a letter sat. This is one of the reasons why the same letter repeatedly set in differing typefaces at the same point size appears dissimilar when compared. Lowercase letters set in the same point size with differing typefaces can appear larger or smaller in comparison to each other due to variations in x-height. Other variables such as stroke weight, letter width, and ascender and descender length influence size perception and help make some typefaces more or less readable and legible than others. Type size is also dependent on the resolution offered by output and viewing devices, color usage, context, and other design issues. It is crucial to understand not only the characteristics of a typeface but also usage context and application environment when selecting a typeface.

Serif and sans serif.   Serif and sans serifs are general categories used for classifying type. *Serif* refers to a typeface with serifs, the short strokes that project off the end of letter strokes, as opposed to *sans serif*, a typeface without serifs. It is debatable whether serif typefaces, conventionally used for setting text in print, are more readable than sans serif typefaces, which have an even stroke weight and more open counters that have proven useful for setting text for on-screen reading. And, while sans serif is considered easier to read on screen, serifs can be made equally legible if the appropriate typeface, size, and color is specified. Some designers think that, in print, serifs aid in character recognition and readability; they help differentiate individual letters creating horizontal lines for the eye to follow. This has not been proven conclusively. Other designers hold the view that "we read best what we read most." It's likely that this discussion will continue. Recent technological developments that subdivide a pixel into red, green, and blue elements on LCD screens have resulted in new technologies that create a better immersive environment

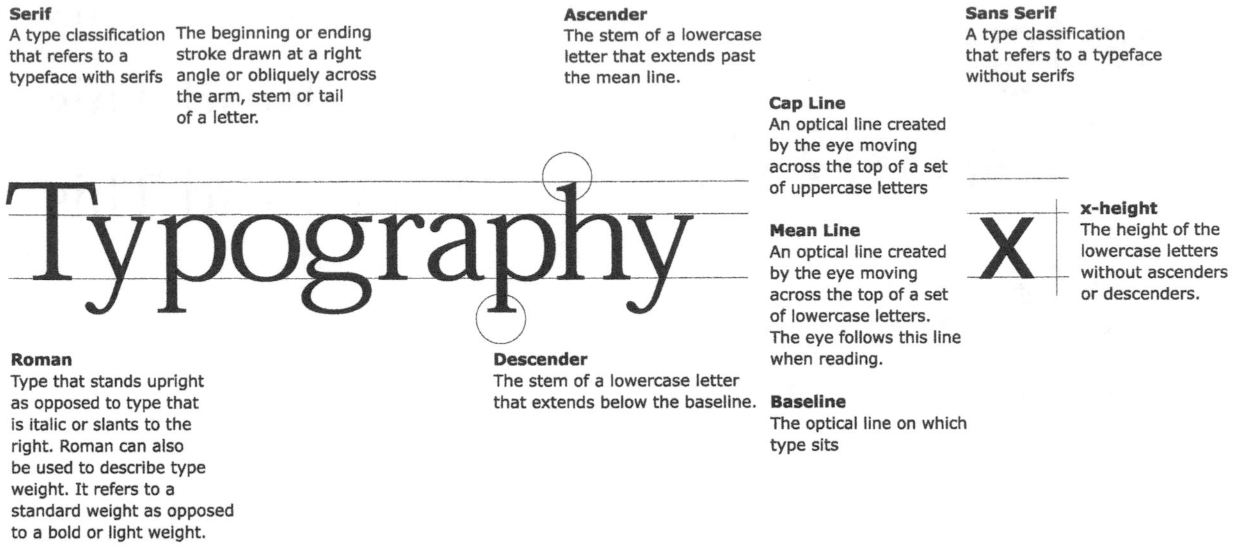

Serif
A type classification that refers to a typeface with serifs

The beginning or ending stroke drawn at a right angle or obliquely across the arm, stem or tail of a letter.

Ascender
The stem of a lowercase letter that extends past the mean line.

Sans Serif
A type classification that refers to a typeface without serifs

Cap Line
An optical line created by the eye moving across the top of a set of uppercase letters

Mean Line
An optical line created by the eye moving across the top of a set of lowercase letters. The eye follows this line when reading.

x-height
The height of the lowercase letters without ascenders or descenders.

Roman
Type that stands upright as opposed to type that is italic or slants to the right. Roman can also be used to describe type weight. It refers to a standard weight as opposed to a bold or light weight.

Descender
The stem of a lowercase letter that extends below the baseline.

Baseline
The optical line on which type sits

FIGURE 1.12. Anatomy of a letter. The typographic terms defined and illustrated above are used by designers in discussing principles that affect the legibility of type and overall quality of the communication.

PALATINO   BASKERVILLE   BODONI   SERIFA   HELVETICA

FIGURE 1.13. Hqx Illustration. The x-height of a typeface (based on the actual height of a lower-case x) is a key characteristic when deciding the visual size of a typeface, particularly when readability is critical. While the above typefaces are the same point size, some seem larger (e.g., Helvetica) and easier to read than others (e.g., Serifa).

for online reading. This, in turn, will spark new explorations in typographic form and its presentation.

Families of type. Many design students first learn to classify typefaces into five chronological families or organizational groupings popularized by James Craig et al., (2006) in *Designing with Type*: (1) old style, (b) transitional, (c) modern, (d) Egyptian, and (e) contemporary. Classifying typefaces into these families makes it easier to understand the differences and similarities in structure and fitness. Like anything else, type design does not happen in isolation. These categories, as with those of many other classification systems, are formed around historical junctures in which the overall design of letterforms shifted dramatically in response to technological, political, cultural, aesthetic, and economic concerns. Typefaces within each set have passed through the tenures of metal, photocomposition, and been adapted for digital technologies. Many of the problems solved through these older forms have proven inspirational to contemporary type designers who, as their predecessors did, continue to explore new forms for new purposes.

Verdana, a sans serif screen font designed by Matthew Carter, whose roots lie in the Industrial Revolution, was commissioned by Microsoft and released in 1996. Its form, which uses a visu-

ally even stroke with wide counters, helped signal a new software release. Verdana's members consist of Roman, italic, bold, and bold italic. It has two peers that use nonLatin alphabets: Verdana Greek and Verdana Cyrillic. Verdana is related to Tahoma, a condensed variation of Verdana designed for use in situations that require more information to fit in less space—such as with dialog boxes and menus; and Nina, a spatially efficient sans serif designed for situations that require more information to fit in even less space—such as with small handheld devices. John Berry (2004), who writes and consults extensively on typographic matters, said in *Now Read This* that Meiryo, Verdana's daughter, evolved partially as a response to technologies that enhance photometric resolution permitting more complex writing systems such as scripts to be optimally read on screen, and a demand for a Japanese type that weaves the character sets of Kanji, Kana, Katakana and Romanji together, fashioning a favorable reading environment for screen and print. Meiryo, too, has Greek and Cyrillic companions. That one family has siblings and offspring, manufactured around a variety of alphabets and writing systems, speaks in part to the economic and political concerns of an international corporation that must respond to the demands of different cultural markets as much as it does to the need for multilingual communication (Fig. 1.14).

1617 Old Style	1757 Transitional	1788 Modern
Garamond	Baskerville	Bodoni
Bembo	**Bell**	Didot
Galliard	Caslon	Waldbaum

1894 Egyptian	1957 Contemporary
Serifa	**Helvetica**
Rockwell	Frutiger
Georgia	**Verdana**

FIGURE 1.14. Type Classification. These five A's are representative of typographic style from the 1600s to present day, reflecting changes in tools, fashion and current events. Ultimately, choice of output media should determine typeface selection, given details such as stress, the thick and thin parts of letterforms, negative space, viewing environment, output resolution, etc.

VARIATIONS IN LETTERFORMS

Variations in Stress

Early type designers mimicked scribal letterforms because they knew and understood these forms. Old-style typefaces have a diagonal stress, a backwards slant—a visual remnant of the pen—that distributes weight through the thin part of the letterform. Over time, the stress traveled several degrees to the right as seen in transitional typefaces. With modern typefaces, the stress becomes rigidly vertical. Egyptians such as Century Schoolbook have a slight diagonal stress; other Egyptians with a stronger slab serif such as Serifa have no stress. Contemporary typefaces have no noticeable stress.

Variation in Thick and Thin

The degree of contrast between the thick and thin strokes of the letters can vary. Old style typefaces have little contrast in strokes. This contrast increases in transitional faces. Extreme stroke contrast is a dominant characteristic of modern typefaces. Egyptian typefaces return to less contrast. Contemporary typefaces have no perceptible thick and thin strokes.

Variations in Serifs

Serifs differ in weight and bracket, which is the shape created by the serif joining the vertical stroke of the letter. Old-style typefaces have heavy concave serifs with thickset brackets. The meticulous serifs of modern typefaces are refined and thin and without brackets. Many Egyptians have heavy, straight serifs with little or no bracket. Sans serifs are considered contemporary typefaces (Fig. 1.15).

TYPOGRAPHIC GUIDELINES

Combining Typefaces

Sans serif and serif typefaces can be effectively combined if changes are limited to prevent visual chaos. The key is to ensure that the result respects the content and reinforces the information hierarchy and overall design goals. When combining typefaces, decide whether harmony or contrast is important. Generally, do not use more than two different type families in a document. Remember, at a minimum, a type family offers a Roman, italic, bold, and bold italic. Consider the pattern and texture that the x-heights and stroke weights weave when combined. Excellent typography does not impede the user and the information. Too many typefaces jar and confuse the reader, create visual intrusions, and slow the reading process.

Contrast in Weight (Boldness)

Combining two classic typefaces with a strong differential factor such as Helvetica Extra Bold with Times New Roman can add useful contrast. Be wary of combining intricate typefaces such as Gill Sans Bold and Souvenir, which have structures that may not create compatible reading environments. Too much contrast and visual complexity can be detrimental.

Output Device and Viewing Environment

The quality of publishing technologies and viewing environment vary greatly—laser printer versus video versus electronic

FIGURE 1.15. Serif Versus Sans Serif. The serif typeface Century above versus the san serif typeface Univers, below. Understanding a typeface's physical characteristics and how it performs in different environments is important. Set a paragraph of text in both a serif and sans serif with exactly the same line length, size and spacing and compare the differences on screen and on paper.

media, etc. In choosing a typeface, its style, size, spacing, and leading, think about the final output medium, and examine this technology's effect on legibility. Low-quality monitors and poor lighting have a major impact: serifs sometimes disappear, letters in small bold type fill in and colored type may disappear altogether.

Letter Spacing and Word Spacing

While the spaces within and between letterforms and words are determined by a type designer in order to set a rhythm that reads well, this spacing can be altered or kerned. However, be careful! When letter spacing is too tight, the letters are hard to distinguish from each other and legibility decreases. When letter spacing is too wide, letter groups are not easily recognized. Spencer argues that optimal letter spacing is inconspicuous, the user can read quickly and easily and understand content. Tight word spacing makes distinguishing individual words difficult. When word spacing is too wide, word groups fall apart. When there is greater space between words than there is between lines, the reader's eye naturally falls to the closest word, which may be below instead of across the line. This frequently occurs with low-resolution or low-cost products.

Line Spacing/Leading

Leading is the distance measured in points between the baseline of one line of text and the baseline of the text line below it. Ascender and descender length influences how closely lines of type can be stacked. The space between lines of text, or leading, should increase in relation to type size. This adjustment is visual not mathematical. Overall legibility may be improved by increasing the leading in relation to column width.

Line Length/Column Width

The correct line length is just long enough for the eye to sweep across without losing its place and easily drop down to continue reading the following lines. A good rule of thumb is that a line of average length contain between 39 and 52 characters.

Justified vs. Ragged Right (Flush Left)

A justified text column can leave uneven word spacing, creating rivers, or vertical white spaces, within the paragraph. Rivers cause the eye to move vertically down the page, naturally connecting with what is closest in proximity, instead of moving easily across the line. It is very difficult to prevent rivers in justified text columns without spending considerable effort. Unless the type is manually set or adjusted, which is a time-consuming activity, it's better to use type that is set flush left, ragged right (Fig. 1.16).

A justified column can leave uneven word spacing, creating rivers, or vertical white spaces within the paragraph. These rivers cause the eye to move vertically down the page, to naturally connect visually what is closest in proximity, instead of easily across the line of type. It is very difficult to prevent rivers in justified columns, unless much time and effort are applied. For this reason, unless the type is manually set or adjusted, it is better to use a column set flush left, ragged right.

A justified column can leave uneven word spacing, creating rivers, or vertical white spaces within the paragraph. These rivers cause the eye to move vertically down the page, to naturally connect visually what is closest in proximity, instead of easily across the line of type. It is very difficult to prevent rivers in justified columns, unless much time and effort are applied. For this reason, unless the type is manually set or adjusted, it is better to use a column set flush left, ragged right.

FIGURE 1.16. With current technology, the difference between a justified text column and ragged right text column can make a huge difference in readability. In a poorly justified column, spaces within a justified line connect vertically down the page, distracting the eye from easily reading across a line of text.

Highlighting with Type

Content can be highlighted by modifying type weight, slope, or case. Weight can be shifted from Roman to bold or extra bold. Slope can be altered from Roman to italic. Be mindful that italics are appropriate for short phrases and not long text passages. The italic appears lighter and smaller on the page when compared with its companion Roman and its complex forms are more difficult to read. Case can change from upper and lowercase to all capitals or small caps. Using all caps for extended text passages impedes readability since word outlines are rectangular and harder for the eye to differentiate. Limited shape and size cues are available to help differentiate between letters, words, and sentences to create meaning. Use only one highlighting technique for emphasis.

Decorative Typefaces

Decorative typefaces are of limited use for body text, because their irregular design lessens legibility and should be used in headlines with caution. Because they are essentially typographic fashion statements, decorative typefaces can either reinforce or distract from the overall message or brand of a particular product or organization.

Black on White vs. White on Black and Dark on Light Background vs. Light on Dark Background

Positive and negative type. White on black (or light on a dark background) is generally regarded as less legible and much more difficult to read over large areas. To the eye, white letters on a black background appear smaller than their reversed equivalent. The amount of contrast between the color of type and the background is an especially important factor for online communication. Color adds exponential levels of complexity to these considerations since displays are inconsistent from one situation to another (Fig. 1.17 & Fig. 1.18). (Dair, 1967)

DESIGN PRINCIPLES: PAGE DESIGN

Typography deals with legibility and *page design* focuses on readability, the ability to read and comprehend information. Can the user find what is needed on the page? The two important functions of page design are motivation and accessibility. A well-designed page is inviting, drawing the eye into the information. Users are motivated to accept the invitation. An effective page design ensures that the reader continues by increasing the ease of understanding and accessibility of the information. (For purposes of simplicity, the term *page design* is used interchangeably to mean page, screen, and document design.) Motivation and accessibility are accomplished by providing the reader with ways to quickly understand the information hierarchy. At a glance, the page design should reveal easy navigation and clear, intuitive paths to discovering additional details and information. This is called "visual mapping."

A page, site, or product visually mapped for easy navigation has,

- an underlying visual structure or grid, organizational landmarks, graphic cues and other reader aids (Fig. 1.19)
- distinctly differentiated information types
- clearly structured examples, procedures, and reference tools
- well-captioned and annotated diagrams, matrices, charts, and graphics.

This kind of visual structuring helps the reader and provides an obvious path through the materials, aids in skimming, gives a conceptual framework, and prevents a feeling of information overload.

A table of contents is a simple visual map. It quickly provides a general overview of the order and some details about the structure and content. What it does not reveal, however, are priorities. Site maps or other diagrams provide this type of information as well.

Building the Design of a Page

Effective visual mapping is apparent in the sequence shown below that demonstrates the evolution of an accessible page from plain text. As design elements are added, the page becomes inviting to read and content becomes attainable. The final example organizes the content into units of information using line spacing and vertical thresholds or queues. Differentiation in typeface, weight, and scale reinforce structure. Information design techniques, drawn from cognitive science, can be used to improve communication effectiveness and performance.

Gray page or screen. Raw text interests few readers. When information is presented as a uniform, undifferentiated mass, it is difficult and irritating to use and easy to ignore (Fig. 1.20).

Chunking. Structure the visual field by breaking like kinds of information into manageable groups according to subject matter. Chunks in close proximity are read as related. Graphic devices such as rules and line spaces are used to reinforce a grouping and separate chunks (Fig. 1.21).

Queuing. Order information chunks visually to reflect the content hierarchy by addressing the user's requirements of subject matter, order, and importance (Fig. 1.22).

Filtering. Simplify linguistic and visual order by filtering out unnecessary background noise, which interferes with the information being transmitted. Filtering builds a sense of layers of information by using color, visual cues and symbols, and bulleted lists and headers to make a page effective for a range of users and uses (Fig. 1.23).

Mixing modes. People learn through different cognitive modes or styles. Some users favor text, others may prefer illustrations, photos, diagrams, or formulas. To suit these varied learning preferences, information must be translated into several different modes that are then carefully presented to reinforce content and organization (Fig. 1.24).

| PALATINO | BASKERVILLE | BODONI | SERIFA | HELVETICA |

FIGURE 1.17.   Black on White vs. White on Black. The x-height of a typeface (based on the actual height of a lowercase x) is a key characteristic when deciding the visual size of a typeface, particularly when readability is critical. While the above typefaces are the same point size, some seem larger (e.g., Helvetica) and easier to read than others (e.g., Serifa). Letterforms often appear as black/dark shapes on white/light backgrounds. The eye also reads the reverse, or negative shape around a letterform, which can create a shape that visually distracts and makes text difficult to read. Try setting a paragraph of type black on white background, then set the same exact paragraph with white type on a black background. You can also try this same exercise with a dark color on a light background, and then try the reverse. You will notice that the greater the contrast (e.g. white type on black) the harder it is to read in large amounts.

Size

Structure

Weight

Form

Size & Weight

Size & Structure

Weight & Structure

Structure & Form

FIGURE 1.18. Typographic Contrasts. This figure, which builds on the relationships of contrasts discussed and illustrated in *Design with Type* by Carl Dair (1967), shows typographic contrasts that can be easily used to build information heirarchies.

FIGURE 1.19. Grid. The lines that appear in this rectangle provide an understanding of what the grid, or underlying structure, is of this page. The grid is used as a guide to create more pages that use similar relationships for placement and alignments. It is a point of departure for one who understands the system, to create variations, be a bit more playful, yet still provide a consistent "feel" to the user.

Word Spacing The effect of too-tight word spacing is that words are more difficult to distinguish from eac h other. When word spacing is too wide, gaps betwe en words don't allow the eye to take inword groups

as easily. When there is greater space between wor ds than there is between lines, the reader's eye trav els down the page instead of across the line. This often occurs with low-resoluti\on or low-cost product s.Line Spacing/LeadingThe space between lines of text, or leading, should increase in relation totype

size. You can also improve legibility by increasing th e leading in relation to column widtLine Length/Colu mn WidthThe correct line length is just long enough for the eye to easily move across the line without los ing its place, and drop down to continue reading the following lines.Justified vs. Ragged Right (flush left)

FIGURE 1.21. Chunking.

mfdhjrkj mgjr kji hfhejw fmdg

bedfhgfen Word Spacing The effect of too-tight word spacing is that words are more difficult to distinguish from each ote When word spacing is too wide, gap s between words don't allow the eye

dfervjt gyth as easily. When there is greater spa ce between words than there is bet ween lines, the reader's eye travels

This often occurs with low-resolutio n or low-cost Line Spacing/Leading The space between lines of text, or leading, should increase in relation

FIGURE 1.22. Queuing.

Word Spacing The effect of too-tight word spacing is that words are more difficult to distinguish from eac h other. When word spacing is too wide, gaps betwe en words don't allow the eye to take inword groups as easily. When there is greater space between wor ds than there is between lines, the reader's eye trav els down the page instead of across the line. This often occurs with low-resoluti\on or low-cost product s.Line Spacing/LeadingThe space between lines of text, or leading, should increase in relation totype size. You can also improve legibility by increasing th e leading in relation to column widtLine Length/Colu mn WidthThe correct line length is just long enough for the eye to easily move across the line without los ing its place, and drop down to continue reading the following lines.Justified vs. Ragged Right (flush left) A justified column can leave uneven word spacing, creatig "rivers", or vertical white spaces, within in the

FIGURE 1.20. Gray page or screen.

mfdhjrkj mgjr kji hfhejw fmdg

bedfhgfen Word Spacing The effect of too-tight word spacing is that words are more difficult to distinguish from each ote When word spacing is too wide, gap

dfervjt gyth as easily. When there is greater spa ce between words than there is bet

ghytmn xds This often occurs with low-resolutio n or low-cost Line Spacing/Leading The space between lines of text, or leading, should increase in relation

FIGURE 1.23. Filtering.

FIGURE 1.24. Mixing Modes.

FIGURE 1.25. Abstracting.

Abstracting. The individual page or screen is a microcosm of the complete book, site, or product. The result is a complete codified system of graphic standards, which is effective for both the reader and the producer. Abstracting builds a system of standards that simplifies text organization, creates consistent approaches to preprocessing information, and establishes a customized look for an organization's products (Fig. 1.25).

Other Page Design Techniques

White space. White space (or empty space) is an underutilized but extremely effective design tool. It visually opens up a page, provides focus, helps group like kinds of information, provides resting points for the reader's eye, and creates the perception of simplicity and ease of use.

The grid. A grid is a controlled system of organization that allows for the distribution of visual elements in an intelligible order. A grid, as part of an overall design system, provides an underlying structure that determines the horizontal placement of columns and the vertical placement of headlines, text, graphics, and other artwork.

A grid is built on a series of consistent relationships, alignments, and spatial organizations. It acts as a blueprint that can be repeatedly used to create sequential pages, which are related but respond to different content. When the grid system is understood, it forms the basis for consistent application and extension of the design by others who also understand the intention of the system. Every strong design uses an underlying structure or grid to create a consistent look and feel to any form of visual communication. An analogy can be created between the horizontal and vertical lines that compose a grid used on a page and the metal beams that systematically make up the overall supporting structure of a high-rise building. While the struc-

tural supports in a building are consistent from floor to floor, the configuration of the space within each individual floor is based on each occupant's needs. The same holds true with the grid on a page. An important tool that improves usability, a grid enables a user to navigate a page quickly and easily. A grid specifies placement for all visual elements. The user anticipates where a button will appear or how help is accessed. Product or program usefulness and ultimately success are greatly increased through the consistency offered by a grid.

Field of vision. Field of vision refers to what a user can see on a page with little or no eye movement; it is the main area where the eye rests to view most of the page. A good design places key elements in the primary field of vision, reflecting and reinforcing the information hierarchy. Size, contrast, grouping, relationships, and movement are tools that create and reinforce field of vision. The user first sees what is visually strongest, not necessarily what is largest or highest. This is particularly true for online information due to the limitations of page real estate and dense information environments.

These concepts, as well as the strength of peripheral vision, can be experienced when viewing a page that has a banner advertisement or moving graphic.

It is virtually impossible to ignore or focus attention on the primary field of vision when there is winking and blinking elsewhere. In fact, superfluous visual devices reduce information's value by distracting and disturbing the user's desire and ability to focus, read and understand.

Proximity. This concept applies to the placement of visual elements physically close to each other so that it is understood that these are related elements.

The illusion of depth. Though the online world exists in two-dimensions, contrast can be used to create the illusion of depth. Contrasts created through size, weight, structure, direction, color, texture, and layering can form cues that reinforce hierarchy by giving the illusion that an element is on top of or in front of another.

CHARTS, DIAGRAMS, GRAPHICS, AND ICONS

The goal of any visual device is to provide the fastest, most efficient path to understanding ideas and to make these ideas clear and compelling. Useful, effective graphics act like visual shorthand, particularly important when the real estate of the page is limited (Fig. 1.26). A good visual eliminates the need for text and communicates across cultures. A bad graphic with an unclear meaning that must be reinforced by a long caption can be worse than none at all. The old cliché, a picture is worth a thousand words, is true only if the picture is efficient and effective. In stressful or difficult situations, people do *not* have time to read or the ability to focus on text and/or complex visuals. Though more difficult to achieve, brevity and simplicity in such cases have greater value. Product users prefer well-designed charts, diagrams, and illustrations that quickly and clearly communicate complex ideas and information. Studies show that visual images are retained long after the reader is finished. Designed correctly, visual images can make information memorable and effective. At a minimum, a powerful illustration or graphic can often improve performance simply because it increases user motivation. Visuals are robust communication tools used to (a) visualize and analyze data; (b) present new or abstract concepts; (c) make physical and technical concepts invisible to the eye; and (d) summarize information efficiently and effectively. Visuals explain and reinforce concepts, relationships and data, making them tangible. Photographs, charts, illustrations, icons, or diagrams become thinking tools. The information is clarified, made easier to evaluate, and has greater impact. Visuals are a very effective way to communicate a message, but choosing the appropriate presentation for a concept is critical to the user's ability to effectively comprehend a message. Understanding the limitations of the display medium is crucial to creating a successful visual.

Tables, Charts, Diagrams

These three types of graphics are discussed in order of complexity. Tables are the least difficult to create, charts the second most difficult, and diagrams the third. Illustrations, graphics, and other images and visuals are the most complex, require more conceptual and visual sophistication, and may require a consultant to create. When is one more appropriate than the other? Determining which format is the most effective is illustrated in Fig. 1.27. In addition to this list, it is important to remember that visual cues such as color, shading, texture, lines, and boxes should be considered redundant cues and only used to provide additional emphasis to support the concept.

Icons and Visual Cues

Icons and other visual cues are a form of visual shorthand, which helps users locate and remember information. Developing an easily understood style that is consistent with the overall program style is not easy. Choose a style that is simple and consistently reinforced throughout a product. More complex and unique symbols and icons can be used if usage takes place over a longer period, allowing product familiarity and learning to take place. The MasterCard logo consists of two intersecting circles. After many years of reinforcement, most people immediately recognize it without accompanying text or other explanation. It is very difficult to create an icon that, without explanation, communicates a concept across cultures. For example, the use of a freestanding rectangular box with an open door flap indicates mailbox or in-box. This kind of mailbox is rarely used today and was never used in Europe where mail is placed in slots or upright boxes. Even the concept of mail delivery can be considered strange. This is a case where meaning had to be learned. Although simple ideas presented as icons are appropriate, a program with many complex concepts using colloquial images can make using a program agonizing for users from other cultures. There is an important difference between an icon and an illustration, though the two concepts are often confused. If an icon must be labeled, it is really an illustration. The icon's value as visual shorthand is lost. Better to use a word or short phrase rather than word and image when screen space is at a minimum.

A successful icon is memorable with minimal reinforcement. If after viewing an icon several times a user cannot remember its meaning, then the icon is valueless and should be eliminated. Icon sets should share a similarity of style (businesslike or playful) and possess formal presentation properties consistent with the overall program or product to which they belong (Fig. 1.28).

FIGURE 1.26. Zen calligraphy is an example of the historically close relationship between word and image. The Zen master Hakuin (1768–1865) created this symbol to mean "dead", with additional notes saying "Whenever anyone understands this, then he is out of danger".

When to Use What Graphic

	If you want to show...	Use a...
Groups	Group of related items, with a specific order	numbered list
Relationships	Relationships and steps involved in a process	flow chart
	Relationships between categories of ideas	Table
	Relationships of tasks taking place over time	project plan table
Evaluate/Compare	Evaluate items against several criteria	rating table
	Evaluate items against one criteria	comparison table
	Compare more than one item to more than one variable	matrix diagram
	Compare several things in relation to one variable	bar chart
	Compare the relative parts that make up a whole	pie chart
Hierarchy	Hierarchical structure of an organization	organizational chart
Concepts	Concept	illustration and/or text icons, other graphics
	Abstract concept	complex images, interactive components

FIGURE 1.27.

FIGURE 1.28.   AIGA/DOT Icons. These six icons, all from the same set, were developed by the American Institute of Graphic Arts and the U.S. Department of Transportation for use in signage. From upper left to lower right: restaurant, no smoking, trash, stairway leading up, information and taxi.

Illustrations and Photographs

As technology improves, the only limit placed on the use of complex images will be by the designer. The most important consideration is appropriateness of the image for the intended audience. Do not use cartoons for a company brochure, or a low-resolution photograph of a control panel when a line illustration is more effective.

While understanding the meaning and implications of illustrations and photographs is no easy task, there are guidelines for making choices. A photograph can easily represent an existing object, but issues related to resolution and cross-media publishing can make it unintelligible. If a photograph can be reproduced with proper resolution, cropping and contrast, and emphasize a required detail, then photography is a good choice. A photograph can provide orientation and contextual cues that are more difficult to achieve in an illustration. No matter how simplified or cropped to focus attention, a photograph's reproduction quality is often unpredictable. In this situation, a technical illustration such as a line drawing is more effective. An obvious advantage of illustration is that it can present abstract concepts or objects that do not yet exist, or that may never exist. Another benefit is the ability to focus the viewer's attention on detail. For example, a line drawing can place attention on a specific machine part by changing line weight. To achieve a similar result in a photograph adds time, complicates the image, and possibly never simplifies the explanation.

Regardless of the method used to create a visual explanation, it must clarify and reinforce content. If the purpose of the image is to explain where to locate a piece of equipment, then an overview of the equipment in the environment is appropriate. If the goal is to show an aspect, such as a button location, then the illustration should only focus attention on that aspect. An image can be cropped to focus attention on what is being explained; it depends on the goal of the photograph or illustration.

Situations exist in which it is more effective to use a combination of photography and illustration than either alone. For example, a photograph of an object in its usage environment conveys more information than that of the object itself. If the objective is to show the location of a part of that object, then a line drawing in close proximity to or inset in the photograph is more useful than a photograph or illustration alone.

Guidelines

Visuals should reinforce the message.   Don't assume that the audience understands how a visual reinforces the argument (Fig. 1.29). A clear and concise argument must still be made that helps shorten the process of comprehension and learning and causes the user to say, "Aha, that's how it works together!" Visuals should,

- clarify complex ideas
- reinforce concepts
- help the user understand relationships

Create a consistent visual language.   Create a consistent visual language that works within the entire communication system. Graphics attract attention. When the user sees a screen, the eye automatically jumps to a visual, regardless of the

fact that it may interrupt reading. Graphics should conform to all elements on a page. Unharmonious graphics impede comprehension by increasing the effort needed to understand the relationship between the text and the visual.

Consider both function and style.   It is important to consider function versus decoration. Albeit wonderful to see an artistically illustrated tax form, is it appropriate to the content or image of both the message and organization it represents? The best graphic is appropriate to the context of the communication and reinforces and validates the message (Fig. 1.30).

Focus on quality vs. quantity.   Graphics are only effective if they are carefully planned, executed, and used sparingly. A well-considered diagram with a concise caption is more effective than several poorly thought-out diagrams that require long explanations.

Work with a professional.   Many individuals within an organization can write an internal report but a writer or public relations firm is usually commissioned to find the most effective, relevant, and interesting way to communicate a public message such as presented in an annual report or company brochure. Similarly, a designer or visual communications firm should be retained to oversee the development of user interfaces, graphics,

FIGURE 1.30.   It is obvious which of the above examples communicates an important message most quickly. The goal for the designer is to communicate the message in the most direct way, so that the user can understand and make decisions based on that information. Obviously some situations are more critical than others, but it is no less important to begin design with consideration of the needs of users.

FIGURE 1.29.   Thirty centuries of development separate the Chinese ancient characters on the left from the modern writing on the right. The meaning of the characters is (from top to bottom): sun, mountain, tree, middle, field, frontier, door.

and other visual elements that impact the look and feel and ultimately the overall success of a program.

Build a library to create visual consistency, organizational identity, and a streamlined process. Because graphics require a professional, they can be very time consuming and expensive to create. Once a visual language and style are established, start building a graphics library. If concepts are repeatedly illustrated, streamline the development process by collecting the supporting illustrations and making them available for reuse. An organizational style can be created for these visual explanations. Through repetition, users can learn to associate a style and method of explanation with an organization, which aids in understanding and reinforces product brand and identity.

Reinforce shared meaning (common visual language). A serious issue to consider when creating graphics, particularly conceptual diagrams, is shared meaning, whether it be across an organization or the globe. Individuals can interpret the same diagram in a variety of ways based on backgrounds and experiences.

Truly effective graphics require extra time and effort, but the payoff is tremendous. Graphics are invaluable tools for promoting additional learning and action because they (a) reinforce the message, (b) increase information retention, and (c) shorten comprehension time.

COLOR

Though color should be considered a reinforcing, or redundant, visual cue, it is by far the most strongly emotional element in visual communication. Color evokes immediate and forceful responses, both emotional and informational. Because color is a shared human experience, it is symbolic. And like fashion, the perception of color changes over time. In all communication, color can be used to trigger certain reactions or define a style. For example, in Western business culture, dark colors such as navy are generally considered conservative, while paler colors such as pink are regarded as feminine. In other cultures, these color choices have an entirely different meaning.

The appropriate use of color can make it easier for users to absorb large amounts of information and differentiate information types and hierarchies. Research on the effects of color in advertising show that ads using one spot of color are noticed 200% more often than black-and-white ads, while full-color ads produce a 500% increase in interest. Color is often used to:

- show qualitative differences
- act as a guide through information
- attract attention/highlight key data
- indicate quantitative changes
- depict physical objects accurately

All in all, color is an immensely powerful tool. Like the tools of typography and page design, it can easily be misused. Research shows that while one color, well used, can increase communication effectiveness, speed, accuracy, and retention, mul-

tiple colors when poorly used actually decrease effectiveness. Because it is readily available, it is very tempting to apply color in superficial ways. For color to be effective, it should be used as an integral part of the design program, to reinforce meaning and not simply as decoration. The choice of color—while ultimately based on individual choice—should follow and reinforce content as well as function.

Basic Principles of Color

Additive primaries. The entire spectrum of light is made up of red, green, and blue light, each representing a third of the spectrum. These three colors are known as "additive primaries," and all colors are made up of varying amounts of them. When all three are combined, they produce white light.

Subtractive primaries. If you add and subtract the three primaries, cyan, yellow, and magenta are produced. These are called "subtractive primaries."

Green + Blue – Red = Cyan
Red + Blue – Green = Magenta   Red + Green – Blue = Yellow

Color on a computer display is created by using different combinations of red, green, and blue light. In print, colors are created with pigments rather than light. All pigments are made up of varying amounts of the subtractive primaries. The three attributes of color are,

1. ***hue***—the actual color
2. ***saturation***—the intensity of the color
3. ***value***—includes lightness and brightness

"Lightness" refers to how light or dark a color appears. "Brightness" is often used interchangeably with lightness; however, lightness depends on the color of the object itself, and brightness depends on the amount of light illuminating the object.

How to Use Color

Less is more . . . useful and understandable. Just as you can overload a page or screen with too many typefaces, you can have too many colors. Given the unpredictability of color displays, users, and viewing situations, the choice can get very complicated. Color is often best used to highlight key information. As a general rule, use no more than three colors for primary information. An example is the use of black, red, and gray—black and red for contrasting information, gray for secondary. When thinking about color online, one must remember that each display will output color in a different way. Add to that the lighting situation and a variety of users. All these factors affect color choice.

Create a color logic; use color coding. Use a color scheme that reinforces the hierarchy of information. Don't miscue the audience by using different colors for the same elements. Whenever possible, try to use colors that work with the project identity or established visual language. Create a color code that is easily understood by the user and reinforces the information.

Create a palette of compatible colors. Harmonious color is created by using a monochromatic color scheme or by using differing intensities of the same hue. However, make them different enough to be easily recognized and simple enough to be easily reproduced, no matter what medium you are using.

Use complementary colors with extreme caution. These are colors that lie opposite each other on the color wheel. Let one dominate and use the other for accents. Never place them next to each other because the edges where they meet will vibrate. Though this was the goal of pop art in the 1960s, it makes pages impossible to read. One must check each particular display, as the calibration of monitors can unexpectedly cause this to happen.

Decisions regarding color in typography are critical. Colored type appears smaller to the human eye than the same type in black. This is important to consider when designing user interfaces. One must also consider the "smear" effect on typography in displays, based on the color chosen and interaction with colors around it. Additionally, quality and calibration of displays impact characteristics of color online.

Consider the viewing medium. The same color looks different when viewed on different viewing media such as a computer display, an LCD projector, color laser printer versus dot-matrix output, glossy versus dull paper.

Context is everything. Though printed color is very familiar and more controllable, projected color is inconsistent and varies depending on such things as lighting, size of the color area, size and quantity of colored elements, lighting, and output device. One must check all output/viewing possibilities to insure that a color is readable as well as legible, and not depend on cross-media specification for insuring consistency. What might look good on a laptop may not be readable when projected in a room for hundreds of people to view, and may look completely different when printed in a corporate brochure. The amount of color will affect how it is viewed as well as the best background choice. A blue headline is very readable on a white background, but if that background becomes a color, then readability can be reduced dramatically, depending on how it gets presented on each particular display.

Contrast is critical when making color choices. Contrast is the range of tones between the darkest and the lightest elements, whether one is considering black and white or color. The desired contrast between what is being "read" (this includes graphics, photographs, etc.) must be clearly and easily differentiated from the background it is presented against. If there is not enough contrast (of color, size, resolution, etc.), it will be difficult or impossible to read. This is particularly a problem with online displays, as the designer has no control of quality of the output display.

Quantity affects perception. A small amount of color will be perceived differently than the same color used in a large quantity. In the smaller area, the color will appear darker; in the larger area, the color will appear lighter and brighter.

Use color as a redundant cue when possible. At least 9% of the population, mostly male, is color-deficient to some degree, so it is generally not a good idea to call out warning points only through color. With a combination of color and a different typeface, etc., you won't leave anyone in the dark.

We live in a global world, so when in Rome. . . Remember that different colors have different connotations within various cultures, religions, professions, etc. For example, in the United States on February 14th, red means love, but in Korea, red means death, and in China, red is used in weddings and symbolizes good luck and fortune. In many other countries, red means revolution. To a competitor, red means first place, and to an accountant, red means a negative balance. To a motorist, red means stop, and in emergencies, a red cross means medical help (Fig. 1.31).

CREATING A SYSTEM: GRAPHIC STANDARDS

With the explosion of new publishing media in a global marketplace, the need for guidelines for developing and producing consistent, quality communication has taken on a new urgency. New technologies make it easy to generate images, offering a wealth of options for experienced and inexperienced publishers alike. The danger lies in creating visual chaos, with every element demanding attention beyond the point of sensory overload. With new tools, chaos can happen faster, at a lower cost, and with greater distribution. A graphic standards manual prevents this confusion.

A *graphic standards manual* is the physical manifestation of an identity system. The design historian and educator Philip Meggs (1998) writes in *A History of Graphic Design* that identity systems arose in the 1950s with the rise of multinational corporations that began to recognize the value and power found in presenting a cohesive visual image globally. A quality-control agent, a standards manual allows an organization to document guidelines and provide tools for organizing and structuring communications and reinforcing brand identity to diverse internal and external audiences. It explains the methodology behind the design, specifies written and visual language, production materials, and methods, and gives examples of how to and not to use the identity system so that standards can be implemented by different people, in different places, at different times. A standards manual supports expansion by explaining how to maintain a consistent brand and organizational look and feel as new products, features, and technology are introduced.

A graphic standards system provides

- Built-in quality: The system ensures that the correct organization/product image is communicated to all audiences. Standards promote consistency in handling information across product lines, divisions, projects, etc.

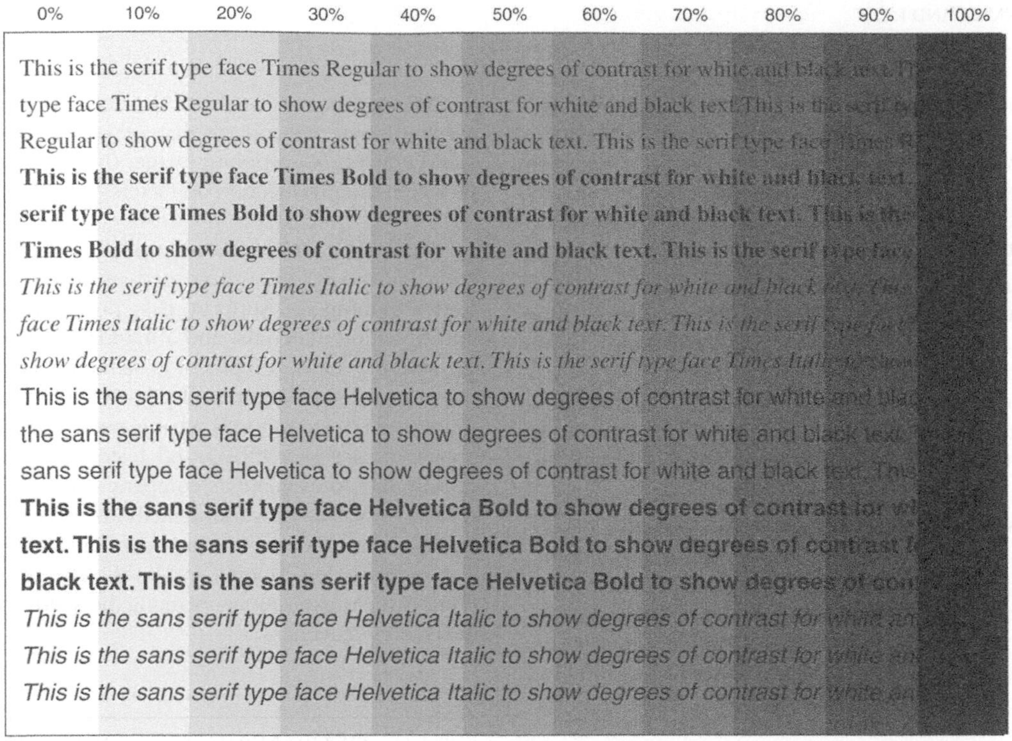

0% 10% 20% 30% 40% 50% 60% 70% 80% 90% 100%

This is the serif type face Times Regular to show degrees of contrast for white and black text. This is the serif type face Times Regular to show degrees of contrast for white and black text. This is the serif type face Times Regular to show degrees of contrast for white and black text. This is the serif type face Times R

This is the serif type face Times Bold to show degrees of contrast for white and black text. This is the serif type face Times Bold to show degrees of contrast for white and black text. This is the serif type face Times Bold to show degrees of contrast for white and black text. This is the serif type face

This is the serif type face Times Italic to show degrees of contrast for white and black text. This is the serif type face Times Italic to show degrees of contrast for white and black text. This is the serif type face Times Italic to show degrees of contrast for white and black text. This is the serif type face Times Italic to show

This is the sans serif type face Helvetica to show degrees of contrast for white and black text. This is the sans serif type face Helvetica to show degrees of contrast for white and black text. This is the sans serif type face Helvetica to show degrees of contrast for white and black text. This

This is the sans serif type face Helvetica Bold to show degrees of contrast for white and black text. This is the sans serif type face Helvetica Bold to show degrees of contrast for white and black text. This is the sans serif type face Helvetica Bold to show degrees of con

This is the sans serif type face Helvetica Italic to show degrees of contrast for white and black text. This is the sans serif type face Helvetica Italic to show degrees of contrast for white and black text. This is the sans serif type face Helvetica Italic to show degrees of contrast for white and black text.

(a)

0% 10% 20% 30% 40% 50% 60% 70% 80% 90% 100%

the serif type face Times Regular to show degrees of contrast for white and black text. This is the s Times Regular to show degrees of contrast for white and black text. This is the serif type face Ti to show degrees of contrast for white and black text. This is the serif type face Times Regular to

the serif type face Times Bold to show degrees of contrast for white and black text. This is th e face Times Bold to show degrees of contrast for white and black text. This is the serif type old to show degrees of contrast for white and black text. This is the serif type face Times Bo

the serif type face Times Italic to show degrees of contrast for white and black text. This is the seri es Italic to show degrees of contrast for white and black text. This is the serif type face Times Ita grees of contrast for white and black text. This is the serif type face Times Italic to show degrees

the sans serif type face Helvetica to show degrees of contrast for white and black text. T s serif type face Helvetica to show degrees of contrast for white and black text. This is t h rif type face Helvetica to show degrees of contrast for white and black text. This is the s a

the sans serif type face Helvetica Bold to show degrees of contrast for white and is is the sans serif type face Helvetica Bold to show degrees of contrast for white xt. This is the sans serif type face Helvetica Bold to show degrees of contrast for

the sans serif type face Helvetica Italic to show degrees of contrast for white and black t e sans serif type face Helvetica Italic to show degrees of contrast for white and black t e sans serif type face Helvetica Italic to show degrees of contrast for white and black t

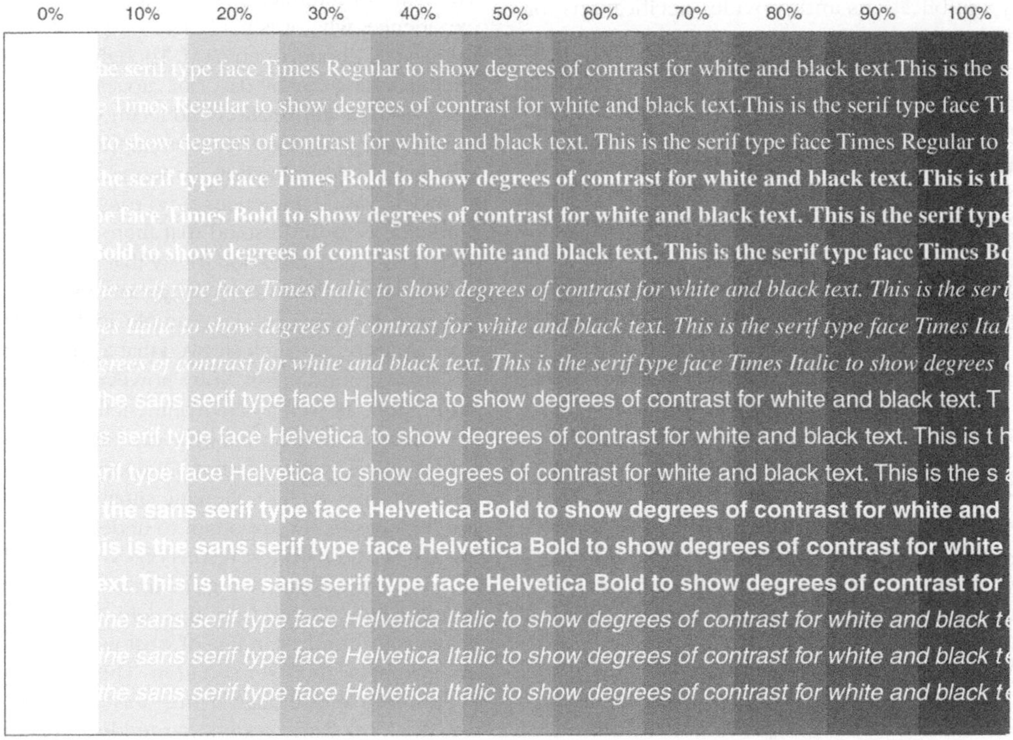

(b)

FIGURE 1.31.   Trying Examples in Your Context. Since color is not available in this particular edition, try your own experiment. Take a look at this illustration, and recreate a paragraph of text, with the background graded from 100% to 0%, choosing one color for the background. Then set lines of type in a variety of typefaces and sizes, to see where it becomes legible or totally impossible to read. The important thing to remember is to test out whatever choices you make within the particular context and parameters, including viewing/projecting devices. Such things as lighting, projection distance, and users' physiological constraints can make all the difference as to whether something can be read or not.

- Control over resources: A system provides dramatic managerial control over resources that use time, money, and materials. Well-developed standards build in flexibility. New communications are easily developed without the original designer and in many instances are developed within the organization itself.
- Streamlined development process: A graphic standards system helps structure thinking for content, design, and production by providing a guideline of predetermined solutions for communication problems. Typical problems are solved in advance or the first time they occur. Most importantly, a graphics standards system encourages the organization as a whole to progress to higher-level issues of communication effectiveness.

What does a system cover? Graphic standards historically have been applied to an organization's logo, stationery, business cards, and other printed materials. As the online portion of an organization's identity dominates, providing for cross-media guidelines is even more critical. Graphic standards are generally communicated to the organization in print and electronic form. Documentation often includes:

- Corporate identity manuals: Style guides available in print and online illustrate the application of the standards across the company's publications and provide specifications for production and expansion.
- Templates and guidelines: Templates and guidelines are available in paper and electronic form and are used to develop pages for both environments.
- Editorial style guides: Editorial guides determine the use of product/service names, punctuation, spelling, and writing styles.

Developing the system. When developing a corporate graphics standards system, consider the global publishing needs of the company, the resources available for producing documents, and the skill level of those in charge of production. To responsibly determine overall needs, a team effort is required. Personnel from areas such as information systems, graphic design, usability, and marketing along with engineers, writers, and users should be involved in the process. This team approach helps build support for, and commitment to, the corporate standards. The development of a comprehensive system follows the information design process of audit, development and implementation.

Audit. The audit is a critical step in determining the scope and parameters of an organization's corporate graphic standards. Specific questions for the audit phase include:

- What is the purpose?
- Who are the audiences?
- What are the differences and similarities between audiences?
- Who will do the work?
- How long will it take?
- What tools will be used?
- What is the desired company or product image?

Development. Goals for the development phase include: (a) the design of standards that are easy to read, use, and project a consistent corporate image; (b) design of products that fit within the production parameters of the company.

Implementation. The implementation phase must ensure that the system is accepted and used properly. This requires training and support, easy procedures for distributing and updating materials, and a manual explaining how to use the system. The development of standards is in itself an educational process. It requires all participants to be aware of communication objectives and what is needed to meet them. As alternatives are developed and tested, management has the opportunity to evaluate its company's purpose, nature, and direction as well as its working methods and communication procedures. The process requires commitment and involvement across many departments and levels. The result is an empowering of the organization—planting the seeds for growth and increased effectiveness.

DESIGNING THE EXPERIENCE

The heart of interface design is to define and create the user's experience; what it is really like for people facing the monitor, using a cell phone, or an ATM. Though presentation possibilities are expanding day by day, our capacity to understand, use, and integrate new information and technology has not grown at the same rate. Making the most appropriate media choices, whether image, animation, or sound, to explain complex ideas to widely varied audiences is no easy task. The most important guideline is to understand that there are no rules, only guidelines. It is a generalization to say that a visual principle works a certain way because any change in context changes the application of the principle. For example, in the early days of the software industry, research showed that a specific blue worked well as a background color. Now, however, depending on monitor calibration, as well as environmental lighting, that blue could be a disaster. In fact, that blue can often vibrate if type in particular colors is placed on it. Of course, it depends on the type quantity, size, weight, viewing situation, etc. Sound complex? For this reason, it's important to understand the principles, test the ideas, and then test results on every output device that will be used. Putting known guidelines together with experience continually gathered from the field allows the designer to develop a clear understanding of what works well in a well-defined environment and user situation. The next key guideline is to keep it simple. Although many tools are available, there is only one goal: to clearly communicate ideas. The designer must always ask, "What is the most efficient and effective way to communicate this idea?" A good illustration might work better (and take less bandwidth) than an animated sequence. Text set in a simple bold headline might allow the user to read the page more efficiently than text placed in a banner moving across the top of a page that constantly draws the eye upward. Animated icons are entertaining, but are they appropriate or necessary for serious financial information? It's tempting to use new tools. The best tip is to use a tool only if it can explain an idea better than any other tool, enhance an explanation, or illustrate a point that otherwise could not be made as effectively or efficiently. The best design is not noticed; it just works. Products are used to accomplish tasks, *not to draw attention to the design*. The best test of product success is the ease with which a user can understand and complete a task and move on to the next task. Real estate and online real estate are alike in that they both stress location, location, location! With such a premium on space, and so much to accomplish in so little time, be considerate and efficient with online real estate.

Use the elements found in a product's graphic standards appropriately. Constantly consider choices and context and review design principles. The following are issues and considerations to continually evaluate when presenting interactive information.

Effective and appropriate use of the medium. Transitioning a print document to an online environment requires rethinking how the document is presented. Viewing and navigating through online information requires radically different design considerations and methods. Users do not necessarily view the information in a linear way, in a specific order, or timeframe. Interactive media viewed on computer screens have quite different characteristics and potential, particularly as information crosses platforms, resolutions, and environments. The rich medium of print allows a book—a product—to be held, viewed, and read in a sequence determined by the user. The physicality of a book provides sensory cues that are not present on a two-dimensional monitor. Interface designers must find ways to provide equivalent cues that encourage people to handle products comfortably and with confidence.

The element of time. The element of time is the critical difference between static and interactive media. The sense of interaction with a product impacts the user's perception of usefulness and quality. Animated cues such as blinking cursors and other implied structural elements like handles around selected areas become powerful navigational tools if intuitively understood and predictably applied. Consider how the product will be used. Will the user sit down and calmly use the product or will he or she panic and fumble with a keypad? Will the task be completed at one time or at intervals over hours, days, months, and years? The element of time contributes to the design criteria and choices.

CONSISTENT AND APPROPRIATE VISUAL LANGUAGE

A major issue is the unpredictability and vastness of products. Providing way-finding devices that are easy to recognize, understand, and remember, include:

- clear and obvious metaphors.
- interface elements consistent with the visual style of other program parts, including consistent style for illustrations, icons, graphic elements, and dingbats.
- guidelines for navigational aids such as color, typography, and page/screen structure that are consistent with other parts of product support.

Navigational Aids

Progress through a book is seen and marked in many ways. Bookmarks and turned corners serve as placeholders. Pens act as mnemonic devices highlighting or underlining text. Table of contents and indices reveal content location. A finger marks a passage to be shared with a colleague as a book is cradled.

Unlike a book, a digital document or program cannot be seen or touched in its entirety. If a document cannot be held, how is specific location known in relation to overall location? How do users return to or move forward through content? How do users travel through unfamiliar space?

Navigational aids provide users with highways, maps, road signs, and landmarks as they move through the online landscape. They enhance discovering and communicate the underlying structure; thereby providing a sense of place so that users know where they are, where they have been, and how to move elsewhere or return to the beginning. Using or building on already familiar visual elements, such as those found in other products and earlier releases, leverages existing knowledge. Graphic standards support this as well. Consistent use of page layout and grid structure makes it easier to remember how information is organized and where it is placed or zoned. This ensures that whatever visual cues are applied can take advantage of the user's experience, ultimately saving time for both the designer and user.

Graphics/Icons

Visual representations such as site maps, graphics, and icons are effective devices for orienting users within a program. A site map offers an overall product view and shows sections or units of information and how these units are related. In the digital environment, graphics and icons assume the role of contents, indexes, and page numbers and can be more effective guides through and around a program than their counterparts in print, because tools such as roll-overs highlight functionality. Creating effective graphics and icons requires that intent and action are defined and designed.

Metaphor

Prior knowledge makes it easier to learn because it provides a conceptual framework on which information can be associated and expanded. When it was first introduced, the metaphor of a desktop with a filing system for a software interface that organized data in a program was easy to grasp because it built on a known experience. Using familiar visual analogies helps users easily understand and organize new information.

Color

Color is a free and very seductive design tool once a monitor is purchased. Use it intelligently. A monitor offers limited workspace. Color can replace or reinforce written explanations when meaning is assigned to it and it is applied methodically. A blue background is always utilized in a testing section and a yellow background in a section overview.

Legibility

Legibility is the ability to read information on the page. The page can be a screen, and as such, has special considerations. Color, size, background, movement, viewing environment, lighting, and resolution play a critical part in legibility.

Readability

Readable screens demand intelligent visual representations and concise, unambiguous text. Meaning can be implied or inferred by the placement of elements in designated areas or zones reserved for distinct information. This makes optimal use of a limited space and increases comprehension and accessibility.

Guidelines

Use the analogy of a poster as a guide to design. A home page is the equivalent of an attention-grabbing poster unpredictably placed in uncontrollable locations. Unlike a home page were a mouse-click on a speaker's name can give biographical information and a click on location can find directions, a poster's static format restricts the amount and depth of information that it can offer.

In print, information is presented in a fixed order. On the web, information is organized hierarchically in a manner that is radically different. Individual users can access the information offered on a website in a sequence that suits their intents and purposes. Online environments offer little regulation over how and in what order the product is accessed. While designers can make suggestions and guesses, this lack of control requires fundamental differences in information presentation. A well-designed home page, like a well-designed poster, should hint at all topics contained in the site, provide high-level information about these topics, and suggest easy paths to access this information. If information goes beyond a single screen, its design must visually communicate location through strong visual hints, so that the user investigates beyond what is immediately visible. Imagine the design considerations required for smaller, hand-held, voice-activated devices.

Design for the most difficult common denominator. Design the interface in anticipation of a worst-case scenario. If a manager will use a product in a quiet office with a fast connection, perfect lighting, and a large monitor, then the problem is different from that of a contractor accessing critical information on a laptop in the field. User profile is often unknown because *new technologies define new categories as new opportunities are recognized.* Consider the breadth of possibilities. Design from the user's perspective. Testing, viewing, and questioning can make the difference of product acceptance or not.

Avoid overuse of saturated colors. Saturated colors such as red tend to jump out at the viewer and can be distracting and irritating. Red is usually not a good choice for large areas of on-screen color. High impact is dependent on the contrast between background and foreground colors. On a black background, both yellow and white have a higher impact than red. Consider variations in every viewing situation including how these variations affect contrast among page elements and overall legibility and readability.

Consider different users' levels of skill. Navigational tools should be simple enough for a novice to use, but should not impede an expert. Detailed visual maps and other graphics should be available for those who need them, without hindering an experienced user who wants to bypass an explanation.

Be aware of the fatigue factor. Although there is no definitive answer on fatigue caused by looking at a computer screen for long periods of time, it is a central factor to consider. According to *Color and the Computer,* by H. John Durrett, looking at a well-designed computer screen should not cause any more fatigue than reading a book or writing a report. Though some would disagree with this statement, many people spend more time with their computer than a book and no doubt could offer additional opinions on this subject. As interactive media becomes a commodity, the focus will not be on what a program does, but on how it does it. This will make the difference between product acceptance and product failure. Success or failure will be judged by the ease with which a product is used and how easily users perceive its interface.

Other differences to consider. There are many differences that impact how and why interfaces are designed, and many of these differences are discussed in more detail in other chapters. A designer should never forget that physical and mental impairments impact an ability to read, comprehend, and use interfaces.

Use the "squint test" to check the design. The squint test is a very simple self-test that checks visual hierarchy. Simply squint at the page so that details are out of focus. What is the first, most dominant element on the page? Is this what should be seen first? What has secondary importance? Cognitive psychology calls this "visual queuing." Successful interaction design creates a visual order that the user can easily follow.

CHALLENGES AND OPPORTUNITIES

Creating Your Own Guidelines

Interactive communication designers face great challenges. How can products that are seen, read, understood, and acted upon be created? Given increasing variety and complexity, how can the power of new technologies be harnessed? How can informed visual choices be made? **WARNING:** No book, seminar, or technology will turn someone into a professional designer! Design is not a craft dependent upon aesthetic ability. Design requires education, training, and experience in a variety of related disciplines equivalent to that of an architect, engineer, surgeon, or cabinetmaker. The following guidelines are offered as starting points, first steps in understanding how to make informed design decisions—design that provides the best, most thoughtful, and appropriate integration of both form and function.

- **There are no universal rules, only guidelines.** If there were rules, everything would look the same and work perfectly according to these rules. Each situation is different with its own context and parameters.

- **Remember the audience: be a user advocate.** Think about audience needs first throughout the development process. Who is in the audience? What are their requirements? How and where will the audience use the product? The evaluation criteria used in the design-development process springs from the answers to these and other questions. Designers must understand and advocate for the user.

- **Structure the messages.** Analyze content to create a clear visual hierarchy of major and minor elements that reflects the information hierarchy. This visual layering of information helps the user focus on context and priorities.

- *Test the reading sequence.* Apply the squint test. How does the eye travel across the page, screen, or publishing medium? What is seen first, second, and third? Does this sequence support the objectives and priorities as defined in the audit?

- **Form follows function.** Be clear about the user and use environment first. An effective interface design represents and reinforces these goals.

- *Keep things simple.* Remember the objective is to communicate a message efficiently and effectively, so that users can perform a task. Fewer words, type styles, and graphic elements mean less visual noise and greater comprehension. An obvious metaphor enhances intuitive understanding and use. The goal is to transfer information, not show off features or graphics.

- **People don't have time to read.** Write clearly and concisely. Design information in an economical, accessible, intuitive format that is enhanced by a combination of graphics and typography. Graphics, if well thought out and designed as an integral part of the page, are very powerful and can efficiently and effectively provide explanations while saving space on a page.

- **Be consistent.** Consistent use of type, page structure and graphic and navigational elements creates a visual language that decreases the amount of effort it takes to read and understand a communication piece. The goal is to create a user experience that seems effortless and enjoyable throughout.

- **Start the design process early.** Don't wait. Assemble the development team of designers, usability professionals, engineers, researchers, writers, and user advocates at the beginning of the process. With interactive media, the traditional review and production process will change. The process is less of a handoff and more of a team effort; it's more like making a film than writing a book. Successfully applying the principles of good design enables an organization to communicate more effectively with its audiences and customers, improving the worth of its products and services and adding value to its brand and identity.

- **Good design is not about good luck.** Good design for usable interfaces appropriately applies the fundamentals of visual design to interactive products. Creating the most useful, successful design for an interactive product is difficult. The design process is iterative and experiential. There are usually several possible ways to solve a problem, and the final design decision is dictated by the best choices that work within the parameters at any particular time. Advocate on behalf of the user. Users are why designers are here and have this work to do. Users are everywhere, often in places not yet imagined. As the world grows smaller and becomes even more connected, the opportunity lies in where and what has not been discovered.

References

Berry, J. D. (Ed.). (2004). *Now read this: The Microsoft clear type font collection.* (pp. 15–17, 61–69). Microsoft Corporation.

Bringhurst, R. (2005). *The elements of typographics style* (pp. 17–24). Point Roberts, WA: Hartley & Marks Publishers.

Craig, J. et al. (2006). *Designing with type: The essential guide to typography* (pp. 23–26). New York: Watson-Guptill Publication.

Dair, C. (1967). *Design with type* (pp. 49–70). Toronto: University of Toronto Press.

Davis M., Hawley, P., McMullan, B., Spilka, G. (1997). *Design as a catalyst for learning* (pp. 3–12). Alexandria, VA: Association for Supervision and Curriculum Development.

Donis, D. A. (1973a). *A primer of visual literacy* (p. ix). Cambridge, MA: MIT Press.

Donis, D. A. (1973b). *A primer of visual literacy* (pp. 22–23). Cambridge, MA: MIT Press.

Felici, J. (2003). *The complete manual of typography: A guide to setting perfect type* (p. 29). Berkeley, CA: Peachpit Press.

Meggs, P. B. (1998). *A history of graphic design* (p. 363). New York: John Wiley & Son.

Re, M. (Ed.) (2002a). Reading Matthew Carter's Letters. *Typographical speaking: The art of Matthew Carter* (p. 20). Baltimore: Albin O. Kuhn Library & Gallery.

Re, M. (Ed.) (2002b). Reading Matthew Carter's Letters. *Typographical speaking: The art of Matthew Carter* (p. 23). Baltimore: Albin O. Kuhn Library & Gallery.

Re, M. (Ed.) (2002c). Reading Matthew Carter's Letters. "Type Vocabulary." *Typographical speaking: The art of Matthew Carter* (p. 82). Baltimore: Albin O. Kuhn Library & Gallery.

Spencer, H. (1969). *The visible word.* New York: Hastings House.

Tracy, W. (1986). *Letters of credit: A view of type design* (pp. 30–31). Boston, MA: David R. Godine Publisher.

·2·

GLOBAL/INTERCULTURAL USER INTERFACE DESIGN

Aaron Marcus

Aaron Marcus and Associates, Inc.

Globalization 29
Definitions of Globalization,
Internationalization, Localization 29
Advantages and Disadvantages of Globalization 31
Globalization Development Process 31
Critical Aspects for Globalization:
General Guidelines 32
   User demographics 32
   Technology 32
   Metaphors 32
   Mental models 33
   Navigation 33
   Interaction 33
   Appearance 33
An Example of Specific Guidelines: Appearance 33
   Layout and orientation 33
   Icons, symbols, and graphics 33
   Typography 34
   Color 35

   Aesthetics 35
   Language and verbal style 35
Globalization Case Study 35
   Planet Sabre 35
Culture Dimensions 36
   Introduction to Culture Dimensions 36
   Culture: An Additional Issue for Globally
   Oriented UI Designers 37
   Hofstede's Dimensions of Culture 38
   Power Distance 38
   Individualism versus Collectivism 38
   Masculinity versus Femininity (MAS) 41
   Uncertainty Avoidance (UA) 43
   Long- versus Short-Term Time Orientation 45
   Design Issues 46
Conclusions and Future Research
Issues 48
Acknowledgments 49
References 49

User interfaces (UIs) for successful products and services enable users around the world to access complex data and functions. Solutions to global user interface design consist of partially universal and partially local solutions to the design of metaphors, mental models, navigation, appearance, and interaction. The UI development process must account for many complex localization issues, in particular, the user group's culture. Culture dimensions, *i.e.*, those proposed by anthropologists and other analysts of culture, can provide insight into how designers can/should adjust UIs to better serve users. By managing users' experiences of familiar structures and processes, the UI designer can achieve compelling solutions that enable the UI to be more useful and appealing. Users across the globe will be more productive and engaged with the products and services.

The concept of *User Interfaces for All,* as set forth by Stephanidis (2000), implied the availability of and easy access to computer-based products and services among all peoples in all countries worldwide. Successful computer-based products and services developed for users in different countries and among different cultures (even within one country) consist of partially universal, general solutions and partially unique, local solutions to the design of UIs. Global enterprises seek to mass-distribute products and services with minimal changes to achieve cost-efficient production, maintenance, distribution, and user support. Nevertheless, it is becoming increasingly important, technically viable, and economically necessary to produce localized versions for certain markets. UIs must be designed for specific user groups, not merely translated and given a superficial "local" appearance for quick export to different markets.

Insufficient attention to localization can lead to embarrassing and sometimes critical miscommunication. For example, in Chinese, Coca-Cola means "bite the wax tadpole" or "female horse stuffed with wax," depending on the dialect, which caused the company to change its name to a phonetic equivalent, which means "happiness in the mouth." Similarly, Pepsi's slogan "Come alive with Pepsi" becomes "Pepsi brings your ancestors back from the grave." (Hendrix, 2001). Differences of culture can lead to significant business implications. For example, Saudi Arabia's Higher Committee for Scientific Research and Islamic Law banned Pokemon video games because they "possessed the minds" of Saudi children, thus closing off one of the Middle East's largest markets to the Japanese Nintendo's multibillion-dollar enterprise (Associated Press).

By contrast, attention to localization of language leads to greater comprehension, which can lead to a drop in customer-service costs when instructions are displayed in a user's native language. Moreover, localization can lead to greater attention and retention on the part of viewers/customers. This implication is especially significant for Web-based communication, where Forrester (1998) reported that visitors remain twice as long reviewing local-language sites as they do English-only sites, and business users are three times more likely to buy when communication is in their own language. Developers in the European Union are faced with the daunting but unavoidable challenge of providing websites in more than a dozen languages to succeed in national markets.

By managing the user's experience with familiar structures and processes, surprise at novel approaches, and preferences/expectations, the UI designer can achieve compelling forms that enable the UI to be more usable, useful, and appealing. *Usable* is defined by the International Standards Organization in Switzerland (ISO) as effective, efficient, and satisfying. Globalization of product distribution requires a strategy and tactics for the design process that enable efficient product development, marketing, distribution, and maintenance. Globalization of UI design, whose content and form is so much dependent upon visible languages and effective communication, improves the likelihood that users will be more productive and engaged with computer-based products and services in many different locations globally.

From the designer's perspectives, two primary objectives are (a) provide a consistent UI, and, more generally, user experience across all appropriate products and services, and (b) design products and services with their necessary support systems that are appropriately internationalized (prepared for localization) and localized, i.e., designed for specific markets. Before discussing globalization, localization, and culture issues, we review briefly essential concepts of UI design.

Demographics, experience, education, and roles in organizations of work or leisure characterize users. Their individual needs and wants, hence their goals, as well as their group roles define their tasks. User-centered, task-oriented design methods account for these aspects and facilitate the attainment of effective UI designs that acknowledge and respect users.

UIs conceptually consist of metaphors, mental models, navigation, interaction, and appearance, which may be defined as follows (Marcus, 1995, 1998):

- *Metaphors*: essential concepts conveyed through words and images, or through acoustic or tactile means. Metaphors concern both overarching concepts that characterize interaction, as well as individual items, like the "trash can" standing for "deletion" within the "desktop" metaphor.

- *Mental models*: organization of data, functions, tasks, roles, and people in groups at work or play. The term, similar to, but distinct from cognitive models, task models, user models, etc., is intended to convey the organization observed in the UI itself, which is presumably learned and understood by users and which reflects the content to be conveyed, as well as the user tasks.

- *Navigation*: movement through mental models, afforded by windows, menus, dialogue areas, control panels, etc. The term implies dialogue and process, as opposed to structure, i.e., sequences of content potentially accessed by users, as opposed to the static structure of that content.

- *Interaction*: the means by which users communicate input to the system and the feedback supplied by the system. The term implies all aspects of command-control devices (e.g., keyboards, mice, joysticks, microphones), as well as sensory feedback (e.g., changes of the state of virtual graphical buttons, auditory displays, and tactile surfaces).

- *Appearance*: verbal, visual, acoustic, and tactile perceptual characteristics of displays. The term implies all aspects of visible, acoustic, and haptic languages (e.g., typography or color; musical timbre or cultural accent within a spoken language; and surface texture or resistance to force).

It is important to bear in mind that localization concerns go well beyond only language translation. They may affect each component of a UI: from choices of metaphorical references, hierarchies in the mental model, and navigation complexity, to choices of input techniques, graphics, colors, sounds/voice/music, and use of vibration.

This chapter discusses the development of UIs intended for users in many different countries with different cultures and languages. The text presents a survey of important issues, as well as recommended steps in the development of UIs for an international and/or intercultural user population and products/services intended for most platforms, including desktop, Web, mobile devices, and vehicles. With the rise of the Internet and application-oriented websites, the challenge of designing good UIs becomes an immediate, practical matter, not only a theoretical issue. This topic is discussed from a user perspective, not a technology or code perspective. The chapter will (a) introduce fundamental definitions of globalization in UI design; (b) demonstrate why globalization is vital to the success of computer-based communication products; (c) introduce culture dimensions; (d) show their effect on one particular platform of desktop Web designs, suggest alternate culture dimensions, and recommend other issues that might relate to culture and user experience.

GLOBALIZATION

Definitions of Globalization, Internationalization, Localization

"Globalization" refers to the entire process of preparing products or services for worldwide production and consumption and includes issues at international, intercultural, and local scales. In an information-oriented society, globalization affects most computer-mediated communication, which, in turn, affects UI design. The discussion that follows refers particularly to UI design.

"Internationalization" refers to the process of preparing code that separates the localizable data and resources (that is, items that pertain to language and culture needed for input and output) from the primary functionality of the software. This separation may include the ability for the UI to work on different platforms in one or more geographic regions where technical, financial, political, and legal matters may affect the designs. Software created in this way does not need to be rewritten or recompiled for each local market. International issues refer to geographic, political, linguistic, and typographic issues of nations or groups of nations. The UI is the International Standards Organization's (ISO) draft of human factors standards in Europe for color legibility standards of cathode-ray tube (CRT) devices (International Organization for Standardization, 1989) is an example of an effort to establish international standards for some parts of a UI. Other examples are the legal requirement for bilingual English and French displays in Canada, and the quasi-legal denominations for currency, time, and physical measurements, which differ from country to country (Table 2.1).

Intercultural issues refer to the religious, historical, linguistic, aesthetic, and other, more humanistic issues of particular groups or peoples, sometimes crossing national boundaries. Examples (Table 2.2) include calendars that acknowledge various religious time cycles; terminology for color, type, and signs reflecting various popular cultures; and organization of content in Web search criteria reflecting cultural preferences. Sometimes these issues reach national political importance, such as the debate in the redesigned society of Iraq during 2005 over whether Thursday and Friday (the Moslem Sabbath) should constitute the "weekend" rather than Saturday (the Jewish Sabbath) or Sunday (the Christian Sabbath) (Kuhn, 2005).

"Localization" refers to the process of customizing (including language translation but potentially other changes, also, such as content hierarchies, graphics, colors, and icons/symbols) the data and resources of code needed for a specific market. Translation is accomplished manually by in-house staff or one or more outside contractors performing that service. Translation can also be accomplished semi-automatically using software provided by third-party firms, such as Systran (www.Systran.com). Localization can take place for specific, small-scale communities, often with unified language and culture, and usually at a scale smaller than countries, or significant cross-national ethnic "regions." Examples include affinity groups (e.g., French "twenty-somethings," or U.S. Saturn automobile owners), business or social organizations (e.g., German staff of DaimlerChrysler or Japanese golf club members), and specific intranational groups (e.g., India's untouchables, Swedish househusbands, or young Japanese professional women who are rejecting marriage as a life pattern.). Used informally, localization may apply to "corporate cultures" or other groups that may be geographically dispersed. With the spread of Web access, the term "localization" may be used more often to refer to groups of shared interests that may be geographically dispersed. *Note:* this broad definition of "culture" is not accepted by all theorists. For example, see Clausen (2000), as reported by Yardley (2000). However, for the purposes of this chapter, this broad definition is used.

TABLE 2.1. Examples of Differing Displays for Currency, Time, and Physical Measurements

Item	USA Examples	European Examples	Asian Examples
Currency	$1,234.00 (US Dollars)	DM1.234 (German marks)	¥1,234 (Japanese yen)
Time Measures	8:00 P.M., August 24, 1999	20:00, 24 August 1999 (England)	20:00, 1999.08.24, or Imperial Heisei 11,
	8:00 P.M., 8/24/99	20:00, 24.08.99 (Germany, traditional)	or H11 (Japan)
		20:00, 1999-08-24 (ISO 8601 Euro standard)	
Physical Measures chars.	3 lb, 14 oz	3.54 kg, 8.32 m (England)	3.54 kg, 8.32 m in Roman or Katakana
	3' 10", 3 feet and 10 inches	3,54 kg, 8,32 m (Euro standard)	

TABLE 2.2. Examples of Differing Displays for Other Data Formats (Partly from Aykin)

Item	USA Examples	European Examples	Asian Examples
Numerics	1,234.56 (also Can., China, UK)	1 234,56 (Finland, Fr, Lux., Portugal., Sweden) 1.234,56 (Albania, Arg,, Denmark, Greece, Neth.) 1'234.56 (Switz: Ger., Ital.) 1'234,56 (Switz: French)	1,234.56
Telephone Numbers	1-234-567-8901, ext. 23 1.234.567.8901 (123) 456-7890	1234 56 78 90 (Austria) (123) 4 5 6 78 90 (Germany) (12) 3456 789 (Italy) +46(0)12 345 67 +49 (1234) 5678-9 (Switz)	−1-53-478-1481 (Japan) 82 2 3142 1100 (Korea) −2-(0)2-535-3893 (Korea) 86 12 34567890 (China)
Address Formats	Title, First Name, MI, Last Name Department Company Number, Street, City, State, Zip Code Country	Paternal Name, Maternal Name, First Name Company, Department Street, Number City, District/Region Zip Code, Country (Order may vary from country to country)	Family Name, First Name Department Company Number, Street, Neighborhood, District Zip Code, City (Japan)

TABLE 2.3. Examples of Differing Cultural References

Item	N. America/Europe Example	Middle-Eastern Example	Asian Example
Sacred colors	White, blue, gold, scarlet (Judeo-Christian)	Green, light blue (Islam)	Saffron yellow (Buddhism)
Reading direction	Left to right	Right to left	Top to bottom
Item	USA	France, Germany	Japan
Web search	"Culture" doesn't imply political discussions	"Culture" implies political Discussions	"Culture" implies tea ceremony Discussions
Sports references	Baseball, football, basketball, golf is a sport	Soccer	Sumo wrestling, baseball, golf is a religion

Localization changes may need to consider any or all of the following:

Address formats
Alphabetic sequence and nomenclature
Arithmetic operations symbolism
Business standards (quotes, tariffs, contracts, agreement terms, etc.)
Calendar
Character handling
Colors
Content categories
Date and time formats
Documentation nomenclature and formats
Electrical and electronic plug formats and nomenclature
Energy formats
Environmental standards ("green"-compliancy, low energy, low pollution, etc.)
File formats
Font nomenclature, sizes, faces, and byte formats
Frequency (i.e., gigahertz)
Hyphenation and syllabification
Icons and symbols

Intellectual property (protection via patents, copyrights, trademarks)
Keyboard formats
Language differences
Legal processes
Licensing standards
Measurement units (length, volume, weight, electricity, energy, temperature, etc.)
Monetary or currency formats
Multilingual usage
Name formats
Negative formats
Numeric formats and number symbols:
Packaging
Paper formats
Punctuation symbols and usage
Reading/writing direction
Sorting sequences
Style formats
Telephone/fax, temperature formats
Text length
Video recording and playback formats
Voltage/amperage units and formats
Weight formats

Table 2.4 demonstrates the complexities, even within English-language users. Preparing texts in local languages may require use of additional or different characters. The ASCII system, which uses seven or eight bits to represent characters, supports English, and the ISO 8859-1 character set supports Western European languages, including Spanish, French, and German. Other character encoding systems include EBCDIC, Shift-JIS, UTF-8, and UTF-16. ISO has established specific character sets for languages such as Cyrillic, Modern Greek, Hebrew, Japanese, etc. The new Unicode system (ISO 10646) (Graham, 2000) uses 16 bits to represent 65,536 characters, which is sufficient to display Asian languages, like Japanese and Korean, and permits easier translation and presentation of character sets.

Advantages and Disadvantages of Globalization

The business justification for globalization of UIs is complex, but compelling. Clear business reasons can drive decisions to localize content on websites or for Web/mobile applications: Research shows that Web-site visitors stay twice as long at local language sites as they do at English-only sites, business users are three-times more likely to buy when the site is in their own language, and contacts with customer service decline when instructions are shown in the users' native languages (Forrester, 1998).

If the content (functions and data) is likely to be of value to target populations outside of the original market, it is usually worthwhile to plan for international and intercultural factors in developing a product or service, so that it may be customized efficiently, e.g., having separate text files that can be translated more easily. Rarely can a product achieve global acceptance with a "one-size-fits-all" solution. Developing a product for international, intercultural audiences usually involves more than merely a translation of verbal language, however. Visible (or otherwise perceptual, e.g., auditory) language must also be revised, and other UI characteristics may need to be altered.

While increasing initial development costs, developing products or services ready for global use gives rise to potential for increased international sales. However, for some countries, monolithic domestic markets may inhibit awareness of, and incentives for, globalization; e.g., because the United States has in the past been such a large producer and consumer of software, it is not surprising that some U.S. manufacturers have targeted only domestic needs. However, as many U.S. industries, such as movie making and games development, have discovered, foreign sales may be a significant portion of total sales or even larger than domestic sales. In order to penetrate some markets, the local language may be a nearly absolute requirement as in France. Recent reports show that European Union (EU) countries must provide local variations in order to gain user acceptance. As an example: English-only portals are a barrier to EU business success (Vickers, 2000).

Some software products are initiated with international versions (e.g., typically five to seven languages for global products originating in the United States), but they are usually released in sequence because of limited development resources. Other products are "retrofitted" to suit the needs of a particular country, language, or culture, as needs or opportunities arise. In some cases, the later, ad-hoc solution may suffer because of the lack of original planning for globalization.

Globalization Development Process

The "globalized" UI development process is a sequence of partially overlapping steps, some of which are partially, or completely iterative:

- *Plan:* Define the strategy, including the challenges or opportunities for globalization; establish objectives and tactics; determine budget, schedule, tasks, development team, and other resources. Globalization must be specifically accounted for in each item of project planning; otherwise, cost overruns, delays in schedule, and lack of resources are likely to occur. In most cases, business managers will expect to see a return-on-investment (ROI) analysis of the expected benefits, basis for benchmarking, likely tools and process, and metrics to be used in "proving" the results are better.

- *Research:* Investigate dimensions of global variables and techniques for all subsequent steps, e.g., techniques for analysis, criteria for evaluation, media for documentation, etc. In particular, identify items among data and functions that should be targets for change and identify sources of national/cultural/local reference. Globalized user-centered design stresses the need to adequately research users' wants and needs according to a sufficiently varied spectrum of potential users, across specific dimensions of differentiation. Recently, *Fortune*, a major business publication, noted that Microsoft hired a significant number of anthropologists to undertake ethnographic analyses and evaluations in the course of product development (Murphy, 2005). During the past few years, many companies have had anthropologists and ethnographers attend user interface conferences and anthropology conferences to report on their findings. One Internet-based discussion group, Anthropologists in Design (see URL references) regularly tracks inquiries about best information resources, tools, educational institutions, and any case studies related to current job tasks. Local informal gatherings within some urban environments enable designers and anthropologists/ethnographers to get together to exchange experiences.

TABLE 2.4. Comparison of English–Language User Community Conventions

	United States	United Kingdom
Dates	Month/Day/Year: March 17, 2001, 3/17/01	Day/Month/Year: 17 March 2001, 17/03/01
Time	12-hour clock, A.M./P.M. No leading zero (8:32 A.M.)	24-hour clock Leading zero (08:32)
Currency	$29.56, 56¢	GB£29.56, £29.56, 56p
Spelling	Center Color	Centre Colour
Terminology	Truck Bathroom	Lorrie Toilet
Book spine title	Top-down	Bottom up

- *Analyze:* Examine results of challenges or opportunities in the prospective markets, refine criteria for success in solving problems or exploiting opportunities (write marketing or technical requirements), determine key criteria for usability, usefulness, and appeal (i.e., the user experience), and define the design brief, or primary statement of the design's goals. At this stage, globalization targets should be itemized.

- *Design:* Visualize alternative ways to satisfy criteria using alternative prototypes; based on prior or current evaluations, select the design that best satisfies criteria for both general good UI design and globalization requirements; prepare documents that enable consistent, efficient, precise, and accurate implementation.

- *Implement:* Build the design to complete the final product, i.e., write code using appropriate, effective, efficient tools identified in planning and research steps.

- *Evaluate:* At any stage, review or test results in the marketplace against defined criteria for success, i.e., conduct focus groups, test usability on specific functions, gather sales, and user feedback. Identify and evaluate matches and mismatches, and then revise the designs. Test prototypes or final products with international, intercultural, or specific localized user groups to achieve globalized UI designs. As noted in recent proceedings of UI conferences such as the Association for Computing Machinery's Special Interest Group on Human-Computer Interaction (ACM/SIGCHI), Usability Professionals Association (UPA), Human-Computer Interaction International (HCII), and the International Conference on the Internationalization of Products and Services (IWIPS), the techniques of evaluation may need to be adjusted to be successful in different cultures and geographic circumstances (Chavan, 2005). Recent publications have focused specifically on studies of mobile phone users (Lee, Ryu, Smith-Jackson, Shin, Nussbaum, & Tamioka, 2005), as opposed to Web-site visitors.

- *Document:* Record the development history, issues, and decisions in specifications, guidelines, and recommendation documents. Honold (1999) for example, notes that German and Chinese cell-phone users require different strategies for documentation, as well as training, which are related to cultural differences predicted by classical models. Even designing the "ideal" documentation for a particular culture of developers may require user-centered design processes, e.g., those that prefer paper and text, and those that prefer visual, interactive means with minimal text reading.

- *Maintain:* Determine which documents, customer-response services, and other processes will require specialized multiple languages and changes in media or delivery techniques. Prepare appropriate guidelines and templates.

- *Train:* Determine which documents and processes will require multiple languages, different graphics, pacing, media, or distribution methods. Prepare appropriate guidelines and templates.

Critical Aspects for Globalization: General Guidelines

Beyond the UI development process steps identified in the previous section, the following guidelines can assist developers in preparing a "checklist" for specific tasks. The following recommendations are grouped under the UI design terms referred to earlier:

User Demographics

- Identify national and cultural target user populations and segments within those populations, and then identify possible needs for differentiation of UI components and the probable cost of delivering them.

- Identify potential savings in development time through the reuse of UI components based on common attributes among user groups. For example, certain primary (or top-level) controls in a Web-based, data-retrieval application might be designed for specific user groups to aid comprehension and to improve appeal. Lower-level controls, on the other hand, might be designed with standardized, unvarying form-like elements.

- Consider legal issues in target communities, which may involve issues of privacy, intellectual property, spamming, defamation, pornography and obscenity, vandalism (e.g., viruses), hate speech, fraud, theft, exploitation and abuse (children, environment, elderly, etc.), legal jurisdiction, seller/buyer protection, etc.

Technology

- Determine the appropriate media for the appropriate target user categories, e.g., emphasis of sound, visual, or three-dimensional (3D) tactile media; verbal versus visual content; etc.

- Account for international differences to support platform, population and software needs, including languages, scripts, fonts, colors, file formats, etc.

- Research and provide appropriate software for code development and content management systems.

Metaphors

- Determine optimum minimum number of concepts, terms, and primary images to meet target user needs.

- Check for hidden miscommunication and misunderstanding.

- Adjust metaphorical images or text to account for national or cultural differences. Publications and projects from/for China and India have suggested metaphors that differ considerably from Western stereotypes. For example, Chavan (1994) stated that Indians relate more easily to bookshelves, books, chapters, sections, and pages, rather than the desktop, file folders, and files. The Wukong prototype PDA, developed by Ericsson for Chinese users in 2002, used metaphors based on the Chinese business-social concept of Guang-xi, or relationship maintenance. This approach meant that people, knowledge, and relationships were more fundamental and pervasive than folders, files, and applications (Marcus, 2003).

Mental Models

- Determine optimum minimum varieties of content organization.
- Consider how hierarchies may need to change in detail and overall in terms of breadth and depth. In Choong & Salvendy (1999), the authors noted that Chinese and North-American users tended to organize the contents of a house in different ways, and in Carroll (1999), the author noted that if one group was given the hierarchies of the other, the group had more difficulty navigating the hierarchy.

Navigation

- Determine need for navigation variations to meet target user requirements, determine cost-benefit, and revise as feasible.

Interaction

- Determine optimum minimum variations of input and feedback. For example, because of Web-access speed differences for users in countries with very slow access, it is important to provide text-only versions, without extensive graphics, as well as alternative text labels to avoid graphics that take considerable time to appear.

Appearance

- Determine optimum minimum variations of visual and verbal attributes. Visual attributes include layout, icons and symbols, choice of graphics, typography, color, and general aesthetics. Verbal attributes include language, formats, and ordering sequences. For example, many Asian written languages, such as Chinese and Japanese, contain symbols with many small strokes. This factor seems to lead to an acceptance of higher visual density of marks in complex public-information displays than is typical for Western countries.

An Example of Specific Guidelines: Appearance

Guidelines for visual and verbal appearance follow. Details can be found in (Aykin, 2004; DelGaldo & Neilson, 1996; Fernandes, 1995; Marcus, 1999; Nielsen, 1990).

Layout and Orientation

- Adjust layout of menus, tables, dialogue boxes, and windows to account for varying reading directions and size of text. Roman languages read only left to right, but Asian languages may read in several directions. For example, in Japan, people read from the right and down, or down and to the right, and Arabic/Hebrew may include right-reading Roman text within left-reading lines of text.
- If areas where dialogue will appear use sentence-like structure with embedded data fields and/or controls, these areas will need special restructuring to account for language changes that significantly alter sentence format. For example, German sentences often have verbs at the ends of sentences, while in English and French verbs are placed in the middle of sentences.
- As appropriate, change layout or imagery that implies or requires a specific reading direction. Left-to-right sequencing may be inappropriate or confusing for use with right-to-left reading scripts and languages. A Web-site design for Arabia On-Line of the late 1990s featured left-to-right English text on a home page intended for Western business people and tourists, but the page layout was still right-to-left, as in Arabic, with primary links at the far right, secondary and tertiary links in the center and left sections of the screen, and small directional arrows pointing right to left. The designers had succumbed to Arabic-language influence to the detriment of English-language readers.
- Check for misleading arrangements of images that lead the viewer's eye in directions inconsistent with language reading directions.
- For references to paper and printing, use appropriate printing formats and sizes. For example, in the United States, standard office letterhead size is 8.5 × 11 inches, but in Europe it is 210mm × 297 mm.

Icons, Symbols, and Graphics

- Avoid the use of text elements and punctuation within icons and symbols to minimize the need for versions to account for varying languages and scripts.
- Adjust the appearance and orientation to account for national or cultural differences. For example, using a mailbox as an icon for e-mail may require different images for different countries.
- Consider using signs or derivatives from international signage systems developed for safety, mass transit, and communication (American Institute of Graphic Arts, 1981; Olgyay, 1995; Pierce, 1996). These signs require little or no translation and may require minimal culture-specific support information.
- Avoid puns and local, unique references that will not transfer well. Note: many "universal" signs may be covered by international trademark and copyright use, e.g., Mickey Mouse and the "Smiley" smiling face.
- Check for appropriateness and use with caution the following:

Animals	People
Body parts and positions	Puns, or plays on words
Colors	National emblems, signs
Hand gestures	Religious, mythological signs

- Consider whether selection symbols, such as the X or check marks, convey correct distinctions of selected versus not-selected items. Some users may interpret an X as crossing out what is not desired, not selection.

- Be aware that office equipment such as telephones, mailboxes, folders, and storage devices differ significantly from nation to nation.

Typography

- Consider character-coding schemes, which differ dramatically for different languages. ASCII is used primarily for English, but single-byte schemes are used for European languages, and double-byte schemes are used for Asian languages. These differences, as well as bidirectional fonts for Hebrew and Arabic display, make it more challenging to support multilingual UIs. Without accounting for character-coding schemes, it is more difficult for users to access content easily.
- Use fonts available for the range of languages required.
- Consider whether special font characters are required for currency, physical measurements, etc.
- Use appropriate alphabetic sequence and nomenclature (e.g., U.S. "zee" versus Canadian/English "zed")
- Ensure appropriate decimal, ordinal, and currency number usage. Formats and positioning of special symbols vary from language to language.
- Consider appropriate numeric formats for decimal numbers and their separators (Aykin, 2000):

1,234.56	Canada, China, United Kingdom, United States
1.234,56	Albania, Argentina, Denmark, Greece, Netherlands
1 234,56	Finland, France, Luxembourg, Portugal, Sweden
1'234.56	Switzerland (German, Italian)
1'234,56	Switzerland (French)

Other numeric issues include the following:

Names of characters
Standards for display of negative and positive numbers
Percent indication
Use of leading zeros for decimal values (e.g., 0.1 or .1)
List separators
Lucky and unlucky numbers (i.e., lucky telephone numbers in Asian countries sometimes sell for higher prices. Note that Chinese prefer even numbers, but Japanese prefer odd numbers.)

- Use appropriate temperature formats, i.e., Fahrenheit, Centigrade, Kelvin.
- Use appropriate typography and language for calendar, time zone, and telephone/fax references.
- Consider these date and time issues, among others:

Calendars (Gregorian, Moslem, Jewish, Indian, Chinese, Japanese, etc. Note that even in India there are three major Hindu religious calendars, one used by northern Indians, one by southern Indians, and one shared by both, separate from a secular calendar.)
Character representation (Hindu-Arabic, Arabic, Chinese, Roman, etc.)
Clock of 12 or 24 hours

Capitalization rules (e.g., main words in U.S. book titles versus British first only, and the use of capitals for all nouns in German)
Days considered for start of week and for weekend
Format field separators
Maximum and minimum lengths of date and time
Names and abbreviations for days of week and months (two-, three- and multicharacter standards)
Short and long date formats for dates and times
Time zone(s) appropriate for a country and their names
Use of AM and PM character strings
Use of daylight savings time
Use of leading zeros

- Consider monetary format issues:

Conventions for use with Euros (e.g., double listings)
Credit/debit-card formats, usage conventions
Currency names, denominations
Currency symbols (local versus international versions)
Currency conversion rates
Monetary formats, symbols, and names
Rules for combining different monetary formats: e.g., required multiple currency postings of Euro plus local national currency
Validating monetary input

- Consider name and address formats:

Address elements and number of lines
Address line order
Address: street numbers first or last punctuation:

e.g., 4, route de Monastère; 1504 South 58th Street; Motza Illit 11)
Address: zip/postal codes: alphanumerics
Address: zip/postal codes: sequence with city, province/state, country
Character sets
Datafield labels (family name versus last name versus surname; first name versus given name versus Christian name)
Field labels (last name versus surname, city/town/district/ province, etc.)
Location and location order: neighborhood, district, city/ town, state/province)
Name formats:

e.g., family name first for Asian names, family names in all caps in Asia): number of names, number of last names, or surnames (maternal versus paternal), prefixes/titles (e.g., German double title: Dr. Eng.), suffixes (e.g., Jr.)
Name formats:, double family names:

Note, for example, that even within Spanish-speaking countries, some list double family names with maternal first, and others list family names with paternal first. Number of last names/surnames (maternal, paternal)
Number of names (first, middle, initials, etc.)
Prefixes and suffixes

Zip and postal codes (numeric versus alphanumeric, typography, order in relation to city, state/province, or country)

- Consider telephone, fax, and mobile phone-number formats: Grouping of digits varies from country to country.

Internal dialing (initial area-code zeros) versus external (without)

Numeric versus alphanumeric (i.e., +1-510-767-2676 versus 510-POP-CORN) Number grouping, separators (e.g., (,), +, −, period, space, etc.)

Use of plus sign for country codes in telephone numbers

Use of parentheses for area codes

Format for multiple sequential numbers of businesses (slash, commas, etc.)

Color

- Follow perceptual guidelines for good color usage. For example, use warm colors for advancing elements and cool colors for receding elements; avoid requiring users to recall in short-term memory more than 5 ± 2 different coded colors.
- Respect national and cultural variations in colors, where feasible, for the target users.
- Follow appropriate professional/popular usage of colors, color names, denotation, and connotation.

Aesthetics

- Respect, where feasible, different aesthetic values among target users. For example, some cultures have significant attachment to wooded natural scenes, cartoon characters, textures, patterns, and imagery (e.g., the Finnish and the Japanese), which might be viewed as exotic or inappropriate by other cultures. For example, the Finnish and the Japanese enjoy natural-scenery imagery more than some other countries. The Chinese, Korean, and Japanese seem to enjoy cartoon characters appearing in their products and services more than would seem appropriate for other cultures.
- Consider specific culture-dependent attitudes. For example, Japanese viewers find disembodied body parts, such as eyes and mouths, unappealing in visual imagery.

Language and Verbal Style

- Consider which languages are appropriate for the target users, including the possibility of multiple national languages within one country. For example, English and French within Canada; French, German, and Italian within Switzerland, French or Dutch in Belgium, and in Israel: Hebrew, Arabic, French (official), and English (unofficial). India has more than 20; South Africa has seven, and English is third among native languages. Note also that some languages have different dialects, i.e., Mexican, Argentinian, and Castillian (Spain) Spanish; Parisian, Swiss, and Canadian French.

- Consider which dialects are appropriate within language groupings and check vocabulary carefully, e.g., for British versus American terms in English, Mexican versus Spanish terms in Spanish, or Mainland China versus Taiwanese terms in Mandarin Chinese.
- Consider the impact of varying languages on the length and layout of text. For example, German, French, and English versions of text generally have increasingly shorter lengths. Some Asian texts are 50%–80% shorter than English; some non-English Roman-character prose texts can be 50% to 200% longer. Some labels can be even longer.

Example (Aykin, 2000):

English	Undo	Dutch	Ongedaan maken
English	Autoscroll	Swedish	Automatisk rullning
English	Preferences	German	Bildschirmeinstellungen

- Consider the different alphabetic sorting or ordering sequences for the varied languages and scripts that may be necessary and prepare variations that correspond to the alphabets. Note that different languages may place the same letters in different locations, for example, Å comes after A in French but after Z in Finnish.
- Consider differences of hyphenation, insertion-point location, and emphasis, e.g., use of bold, italic, quotes, double quotes, brackets, etc.
- Use appropriate abbreviations for such typical items as dates, time, and physical measurements.
- Remember that different countries have different periods of time for "weekends" and the date on which the week begins.

GLOBALIZATION CASE STUDY

Planet Sabre

An early example of globalization is the UI design for Sabre's Planet Sabre™, one of the world's largest private extranets used by one-third of the all travel agents. Sabre contained approximately 42 terabytes of data about airline flights, hotels, and automobile rentals that enabled almost $2 billion of bookings annually, receiving up to one billion "hits" per day. The author's firm worked closely with Sabre over a period of five years to develop the Planet Sabre UI (Marcus, 2001).

The UI development process emphasized achieving global solutions from the beginning of the project. For example, requirements mentioned allowing for the space needs of multiple languages for labels in windows and dialogue boxes. Besides supporting English, Spanish, German, French, Italian, and Portuguese, the UI design proposed switching icons for primary application modules, so they would be more gender-, culture-, and nation-appropriate (see Fig. 2.1 & 2.2).

Figure 2.1 shows the initial screen of Planet SABRE, with icons representing the primary applications or modules within the system conveyed through the metaphor of objects on the surface of a planet. The mailbox representing the electronic

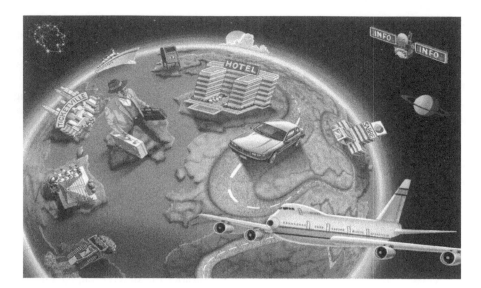

FIGURE 2.1.   Example of Planet SABRE home screen showing typical icons for Passenger, Airline Booking, Hotel Rental, Car Rental, and Email (mail box).

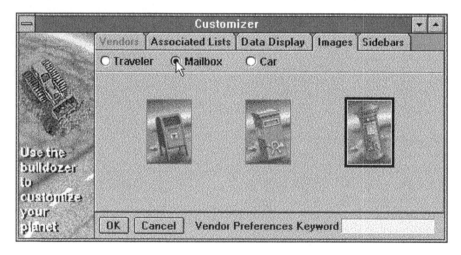

FIGURE 2.2.   Example of dialogue box in the Customizer application, by which users can change the icons to become more locally relevant.

mail application depicts an object that users in the USA would recognize immediately; however, users in many other countries might not recognize this object, because mailboxes come in very different physical forms.

Figure 2.2 shows a prototype version of a Customizer dialogue box, in which the user can change preferences; i.e., icons, so they appear throughout the UI with more recognizable images such as the depiction of the passenger.

At every major stage of prototyping, designs developed in the United States were taken to users in international locations for evaluation. User feedback was relayed to the development team and affected later decisions about all aspects of the UI design.

CULTURE DIMENSIONS

Introduction to Culture Dimensions

Localization includes considerations of the target-market cultures. Although relevant for client-server personal-computer software applications and mobile devices, the Web, in particular, enables global distribution of products and services through Internet websites, intranets, and extranets. Professional Web-oriented analysts and designers generally agree that well-designed UIs improve the performance and appeal of websites, helping to convert "tourists" or "browsers" to "residents" and "customers." In a global economy, user differences may reflect worldwide culture differences. This section analyzes some of the needs, wants, preferences, and expectations of different cultures. This analysis, as an example, uses a cross-cultural theory developed by Geert Hofstede; however, other competing theories exist, a discussion of which follows.

As an example of culture differences, consider the order in which one prefers to find information. If one is planning a trip by train, is it preferable to see the schedule information first or read about the information provider and assess its credibility? Different cultures look for different data to make decisions.

Culture: An Additional Issue for Globally Oriented UI Designers

Cultures, even within some individual countries, may be very different, e.g., French-Canadians within Canada, or Moslem groups within European countries. As previously noted, sacred colors in the Judeo-Christian West (e.g., red, blue, white, gold) are different from Buddhist saffron yellow or Islamic green. Subdued north-European, Finnish designs for background screen patterns (see Fig. 2.3) might not be equally suitable in south-European, Mediterranean climates. These differences reflect strong cultural values. Batchelor (2000), as reported by Boxer (2001), analyzes the history of Western culture's "fear" of color in a book appropriately titled *Chromophobia*. How might these cultural differences be understood?

Many analysts in organizational communication have studied cultures thoroughly and published classic theories; other authors have applied these theories to analyze the impact of culture on business relations and commerce (e.g., see Elashmawi & Harris 1998; Harris & Moran 1993; Lewis 1991). Culture-oriented works are becoming better known within the UI design community. Anthropological theorists include Geert Hofstede (1997), Edward T. Hall (1969), David Victor (1992),

Fons Trompenaars and C. H. Turner (1998), and Shalom Schwarts (2004), each of whom offers valuable insights into the the challenges of cross-cultural communication. Three recent books mention "cultural intelligence" or "cultural awareness" in their titles (Earley, Soon, & Soon, 2003; Kohls, Robert, & Knight, 1994; Thomas & Inkson, 2004).

Other recent publications not only add to the depth of this research, but also generate controversy regarding the permanence and identity of some characterizations of cultures. For example, Nisbett (2003) asserted that essential thinking patterns seem to be culturally influenced and are not universal. He described Western culture's devotion objects, logical reasoning, categorization, and desire to understand situations and events as linear cause-and-effect sequences; while Eastern minds emphasize relationships and seem to accept contradiction more readily.

Clausen (Clausen, 2000), as reported by Yardley (2000), argued that the anthropological term "culture" refers to "the [essentially inescapable] total structure of life of a particular society" while many today use culture to refer to (optional) "shared values." As decribed by Yardley (Yardley, 2000 B3), Clausen argued that ". . . the 'culture' of the Internet has none of the characteristics of a real culture. It is not a total way of life; it did not evolve among a distinct people; nobody inherited it or was raised in it; it makes no moral demands, has no religion at its center, and produces no art." Although complex, its rules are procedural. From the perspective of Internet "dwellers," some of his statements are likely to stir debate, as will his assertions that in terms of the strict definition, culture no longer exists in the United States.

The permanence of cultural attributes is questioned further in articles by Ono and Spindle (2000) and Herskovitz (2000), who note the rise of individualism in classically collectivist Japan, and the acceptance, as well as the influence, of Japanese pop-cultural artifacts (i.e., music, movies, and television) in Asian nations that were recently mortal enemies of Japan.

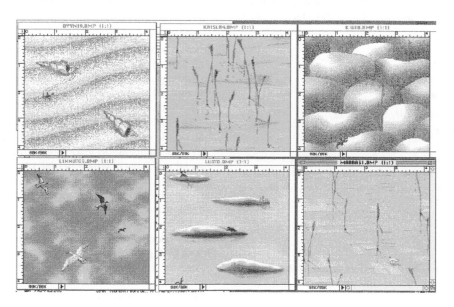

FIGURE 2.3. TeamWare Finnish screen patterns.

Some theorists, including Hofstede (1997), have discussed the relation of culture to economic success and have made differing assertions. Hofstede felt it is a complex mixture of culture plus geography and idiosyncratic drifts in technology, i.e., both culture and creativity. Other theorists argue for solely cultural determinism, as reported by Stille (2001). Still other students of media, sociology, and politics link culture to political and social constructs and discuss the ability of specific groups to manipulate people's cultural self images through media (Lamont, 1995; Brain, 2005).

Although many cultural anthropology theorists are potentially valuable for UI designers, the application of Hofstede's theories will demonstrate the value of this body of research for the fields of UI design, using examples from Web-site design. The following sections introduce Hofstede's concept of cultural dimensions, and apply these dimensions to analyzing Web-site UIs.

Hofstede's Dimensions of Culture

During 1978 to 1983, the Dutch cultural anthropologist Geert Hofstede conducted detailed interviews with hundreds of IBM employees in 53 countries. Through standard statistical analysis of large data sets, he was able to determine patterns of similarities and differences among their replies. From this analysis, he formulated his theory that world cultures vary along consistent, fundamental dimensions. Because his subjects were constrained to one multinational corporation's worldwide employees, and thus to one company culture, he ascribed their differences to the effects of their national cultures. (One debated characteristic of his approach is that he maintained that each country has just one dominant culture.) In 1997, Hofstede published a version of his research in *Cultures and Organizations: Software of the Mind.* His focus was not on defining culture as refinement of the mind but rather on essential patterns of thinking, feeling, and acting that are well established during childhood. These cultural differences manifest themselves in a culture's choices of symbols, heroes/heroines, rituals, and values.

Hofstede rated 53 countries on indices including five dimensions normalized to values (usually) of 0 to 100. His five dimensions (indices) of culture are,

- Power distance index (PDI)
- Collectivism/individualism (IDV)
- Femininity/masculinity (MAS)
- Uncertainty avoidance (UAI)
- Long/short-term time orientation (LTO)

Each of Hofstede's dimensions follows with an explanation of implications for UI (especially Web UI) design, and illustrations of characteristic websites. The complete data for all countries appears in Table 2.5.

Power Distance

"Power distance" (PD) refers to the extent to which less powerful members expect and accept unequal power distribution within a culture.

High PD countries tend to have centralized political power and exhibit tall hierarchies in organizations with large differences in salary and status. Subordinates may view the boss as a benevolent dictator who expects them to do as they are told. Parents teach children to be obedient and expect respect. Teachers are considered wise and are esteemed. Inequalities are expected, and even may be desired.

Low PD countries tend to view subordinates and supervisors more as equals and more interchangeable, with flatter hierarchies in organizations and less difference in salaries and status. Parents and children, as well as teachers and students, may view themselves more as equals. Equality is expected and generally desired.

Hofstede noted that these differences are hundreds or even thousands of years old. He does not believe they will disappear quickly from traditional cultures, even with powerful global telecommunication systems. Based on this definition, (high versus low) power distance may influence the following aspects of UI design:

- Access to information: highly structured versus not as highly structured.
- Hierarchies in mental models: tall versus shallow.
- Emphasis on the social and moral order (e.g., nationalism or religion) and its symbols: significant/frequent versus minor/infrequent use.
- Focus on expertise, authority, certifications, official logos: strong versus weak.
- Prominence given to leaders versus citizens, customers, or employees.
- Importance of security, restrictions, or barriers to access: explicit, enforced, frequent restrictions on users versus transparent, integrated, implicit freedom to roam.
- Social roles used to organize information (e.g., a managers' section that is obvious to all but sealed off from nonmanagers): frequent versus infrequent.

These PD differences are illustrated by university websites from two countries with very different PD indices (Fig. 2.4 & Fig. 2.5): the Universiti Utara Malaysia (www.uum.edu.my) in Malaysia, with a PD index of 104, the highest in Hofstede's analysis; and and the Technische Universiteit Eindhoven (www.tue.nl) in the Netherlands, with a PD index of 38.

The Malaysian website features strong axial symmetry, a focus on the official seal of the university, photographs of faculty or administration leaders conferring degrees, and monumental buildings in which people play a small role. A top-level menu selection provides a detailed explanation of the symbolism of the official seal and information about the leaders of the university.

The Dutch website, on the other hand, features an emphasis on students (not leaders), an asymmetric layout, and photos of both genders in illustrations. This website emphasizes the power of students as consumers and equals. In similar Dutch university websites, students even have the opportunity to operate a WebCam to take their own tours.

TABLE 2.5. Indices from Hofstede, Geert, *Cultures and Organizations: Software of the Mind*, (Hofstede).

Score	PDI Rank	Score	IDV Rank	Score	MAS Rank	Score	UAI Rank	Score	LTO Rank	Score
Arab Countries	7	80	26/27	38	23	53	27	68		
Argentina	35/36	49	22/23	46	20/21	56	10/15	86		
Australia	41	36	2	90	16	61	37	51	15	31
Austria	53	11	2	55	2	79	24/25	70		
Bangladesh									11	40
Belgium	20	65	8	75	22	54	5/6	94		
Brazil	14	69	26/27	38	27	49	21/22	76	6	65
Canada	39	39	4/5	80	24	52	41/42	48	20	23
Chile	24/25	63	38	23	46	28	10/15	86		
China									1	12
Columbia	17	67	49	13	11/12	64	20	80		
Costa Rica	42/44	35	46	15	48/49	21	10/15	86		
Denmark	51	2	9	74	50	16	51	23		
East Africa	21/23	64	33/35	27	39	41	36	52		
Equador	8/9	78	52	8	13/14	63	28	67		
Finland	46	33	17	63	47	26	31/32	59		
France	15/16	68	10/11	71	35/36	43	10/15	86		
Germany FR	42/44	35	15	67	9/10	66	29	65	14	31
Great Britain	42/44	35	3	89	9/10	66	47/48	35	2	25
Greece	27/28	60	30	35	2/19	57	1	112		
Guatemala	2/3	95	53	6	43	37	3	101		
Hong Kong	15/16	68	37	25	2/19	57	49/50	29	2	96
India	10/11	77	21	48	20/21	56	45	40	7	61
Indonesia	8/9	78	47/48	14	30/31	46	41/42	48		
Iran	29/30	58	24	41	35/36	43	31/32	59		
Ireland (Rep of)	49	28	12	70	7/8	68	47/48	35		
Israel	52	13	19	54	29	47	19	81		
Italy	34	50	7	76	4/5	70	23	75		
Jamaica	37	45	25	39	7/8	68	52	13		
Japan	33	54	22/23	46	1	95	7	92	4	80
Malaysia	1	104	36	26	25/26	50	46	36		
Mexico	5/6	81	32	30	6	69	2	82		
Netherlands	40	38	4/5	80	51	14	35	53	10	44
New Zealand	50	22	6	79	17	58	39/40	49	16	30
Nigeria									22	16
Norway	47/48	31	13	69	52	8	38	50		
Pakistan	32	55	47/48	14	25/26	50	24/25	70	23	0
Panama	2/3	95	51	11	34	44	10/15	86		
Peru	21/23	64	45	16	37/38	42	9	87		
Philippines	4	94	31	32	11/12	64	44	44	21	19
Poland									13	32
Portugal	24/25	63	33/35	27	45	31	2	104		
Salvador	2/19	66	42	19	40	40	5/6	94		
Singapore	13	74	39/41	20	28	48	53	8	9	48
South Africa	35/36	49	16	65	13/14	63	39/40	49		
South Korea	27/28	60	43	2	41	39	16/17	85	5	75
Spain	31	57	20	51	37/38	42	10/15	86		
Sweden	47/48	31	10/11	71	53	5	49/50	29	12	33
Switzerland	45	34	14	68	4/5	70	33	58		
Taiwan	29/30	58	44	17	32/33	45	26	69	3	87
Thailand	21/23	64	39/41	20	44	34	30	64	8	56
Turkey	2/19	66	28	37	32/3	45	16/17	85		
Uruguay	26	61	29	36	42	38	4	100		
USA	38	40	1	91	15	62	43	46	17	29
Venezuela	5/6	81	50	12	3	73	21/22	76		
West Africa	10/11	77	39/41	20	30/31	46	34	54		
Yugoslavia	12	76	33/35	27	48/49	21	8	88		
Zimbabwe									19	25

FIGURE 2.4. High-power distance: Malaysian University website.

Individualism versus Collectivism

Individualism in cultures implies loose ties; everyone is expected to look after one's self or immediate family but no one else. Collectivism implies that people are integrated from birth into strong, cohesive groups that protect them in exchange for unquestioning loyalty.

Hofstede found individualistic cultures value personal time, freedom, challenge, and such extrinsic motivators as material rewards at work. In family relations, they value being honest, talking things out, using guilt to achieve behavioral goals, and maintaining self-respect. Their societies and governments place individual social-economic interests over the group, maintain strong rights to privacy, nurture strong private opinions (expected from everyone), restrain the power of the state in the economy, emphasize the political power of voters, maintain strong freedom of the press, and profess the ideologies of self-actualization, self-realization, self-government, and freedom.

At work, collectivist cultures value training, physical conditions, skills, and the intrinsic rewards of mastery. In family relations, they value harmony more than honesty/truth (and silence more than speech), use shame to achieve behavioral goals, and strive to maintain face. Their societies and governments place collective social-economic interests over the individual, may invade private life and regulate opinions, favor laws and rights for groups over individuals, dominate the economy, control the press, and profess the ideologies of harmony, consensus, and equality.

Individualism and collectivism may influence, respectively, the following Web UI aspects:

- Motivation based on personal achievement: maximized (expect the extraordinary) for individualist cultures versus underplayed (in favor of group achievement) for collectivist cultures.

- Images of success: demonstrated through materialism and consumerism versus achievement of sociopolitical agendas.

- Rhetorical style: controversial/argumentative speech and tolerance or encouragement of extreme claims versus official slogans and subdued hyperbole and controversy.

- Prominence given to youth and action versus to aged, experienced, wise leaders and states of being.

- Importance of individuals versus products shown by themselves or with groups.

- Underlying sense of social morality: emphasis on truth versus relationships.

- Emphasis on change: what is new and unique versus tradition and history.

- Willingness to provide personal information versus protection of personal data differentiating the individual from the group.

The effects of these differences can be illustrated on the Web by examining national park websites from two countries with very different IC indices (Fig. 2.6 & Fig. 2.7). The Glacier Bay

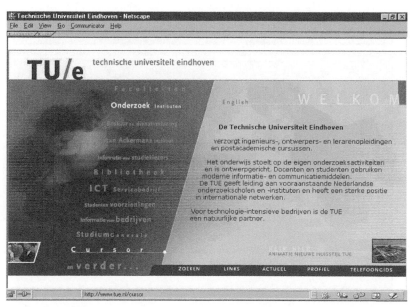

FIGURE 2.5.   Low-power distance: Dutch Educational website.

National Park website (www.nps.gov/glba/evc.htm is located in the United States, which has the highest IC index rating (91). The website from the national parks of Costa Rica (www.tourism-costarica.com) is located in a country with an IC index rating of 15. The third image (Fig. 2.8) shows a lower level of the Costa Rican website.

Note the differences in the two groups of websites. The U.S. website features an emphasis on the visitor, his or her goals, and possible actions in coming to the park. The Costa Rican website features an emphasis on nature, downplays the individual tourist, and uses a slogan to emphasize a national agenda. An even more significant difference lies below the What's Cool menu. Instead of a typical Western display of new technology or a new experience, the screen is filled with a major political announcement that the Costa Rican government has signed an international agreement against the exploitation of children and adolescents.

Masculinity versus Femininity (MAS)

"Masculinity" and "femininity" refer to gender roles, not physical characteristics. Hofstede focused on the traditional assignment to masculine roles of assertiveness, competition, and

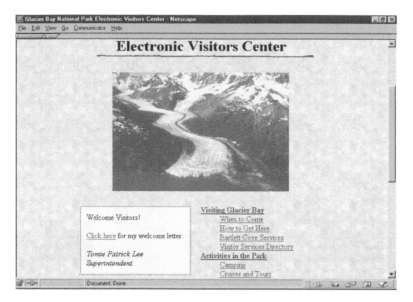

FIGURE 2.6.   High individualist (low collectivist) value: US National Park website.

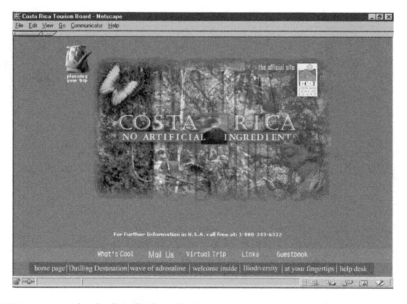

FIGURE 2.7.   Low individualist (high collectivist) value: Costa Rican National Park website.

toughness, and the assignment of feminine roles to home and children, people, and tenderness. He acknowledged that, in different cultures, different professions are dominated by different genders. (For example, women dominate the medical profession in the Soviet Union, while men dominate in the United States). However, in masculine cultures, the traditional distinctions are strongly maintained, while feminine cultures tend to collapse the distinctions and overlap gender roles (both men and women can exhibit modesty, tenderness, and a concern with both quality of life and material success.) Traditional masculine work goals include earnings, recognition, advancement, and challenge. Traditional feminine work goals include good

relations with supervisors, peers, and subordinates; good living and working conditions; and employment security.

The following list shows some typical masculinity (MAS) index values, where a high value implies a strongly masculine culture:

95	Japan
79	Austria
63	South Africa
62	United States
53	Arab countries
47	Israel
43	France

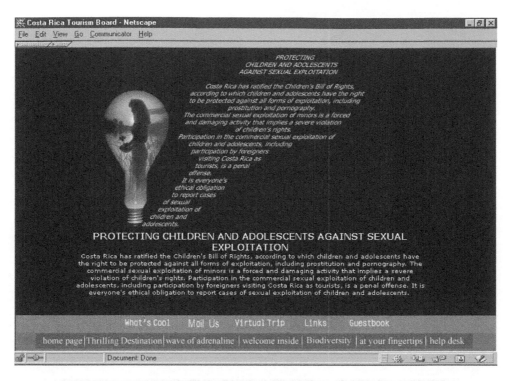

FIGURE 2.8. What's Cool: Political message about exploitation of children.

39 South Korea
05 Sweden

Since Hofstede's definition focused on the balance between roles and relationships, masculinity and femininity may be expressed on the Web through different emphases. High-masculinity cultures would focus on the following UI design elements:

- Traditional gender/family/age distinctions
- Work tasks, roles, and mastery, with quick results for limited tasks
- Navigation oriented to exploration and control
- Attention gained through games and competitions
- Graphics, sound, and animation used for utilitarian purposes

Feminine cultures would emphasize the following UI design elements:

- Blurring of gender roles
- Mutual cooperation, exchange, and support (versus mastery and winning)
- Attention gained through poetry, visual aesthetics, appeals to unifying values

Examples of MAS differences on the Web can be illustrated by examining websites from countries with very different MAS indices (Fig. 2.9, Fig. 2.10, & Fig. 2.11). The Woman.Excite website (woman.excite.co.jp) is located in Japan, which has the highest MAS value 95). The website narrowly orients its search portal toward a specific gender, which this company does not do in other countries.

The ChickClick (www.chickclick.com) U.S. website (MAS = 52) consciously promotes the autonomy of young women (leaving out later stages in a woman's life.)

The Excite website (www.excite.com.se) from Sweden, with the lowest MF value 5, makes no distinction in gender or age. (With the exception of the Netherlands, another low MAS country, all other European websites provide more preselected information.)

Uncertainty Avoidance (UA)

People vary in the extent to which they feel anxiety about uncertain or unknown matters, as opposed to the more specific feeling of fear caused by known or understood threats. Cultures vary in their avoidance of uncertainty, creating different rituals and having different values regarding formality, punctuality, legal-religious-social requirements, and tolerance for ambiguity.

Hofstede noted that cultures with high uncertainty avoidance tend to have high rates of suicide, alcoholism, and accidental deaths, and high numbers of prisoners per capita. Businesses may have more formal rules, require longer career commitments, and focus on tactical operations rather than strategy. These cultures tend to be expressive; people talk with their hands, raise their voices, and show emotions. People seem active, emotional, even aggressive. They shun ambiguous situations and expect structure in organizations, institutions, and

FIGURE 2.9. High masculinity website: Excite.com especially for women in Japan.

FIGURE 2.10. Medium masculinity website: ChickClick.com in the USA.

relationships to help make events easy to interpret interpret and predictable. Teachers are expected to be experts who know the answers and may speak in cryptic, academic language that excludes novices. In high UA cultures, what is different may be viewed as a threat, and what is "dirty" (unconventional) is often equated with what is dangerous.

By contrast, low UA cultures tend to be less expressive and less openly anxious; people behave quietly without showing aggression or strong emotions (though their excessive caffeine consumption may be to combat depression from their inability to express their feelings.) People seem easygoing, even relaxed. Teachers

may not know all the answers (or there may be more than one correct answer), conduct more open-ended classes, and are expected to speak in plain language. In these cultures, what is different may be viewed as simply curious, or perhaps ridiculous.

Based on this definition, uncertainty avoidance may influence contrary aspects of UI and Web design. High-UA cultures would emphasize the following:

• Simplicity, with clear metaphors, limited choices, and restricted data

• Attempts to reveal or forecast results of actions before users act

FIGURE 2.11. Low masculinity website: Swedish Excite.com.

- Navigation schemes intended to prevent users from becoming lost
- Mental models and help systems that focus on reducing "user errors"
- Redundant cues (color, typography, sound, etc.) to reduce ambiguity.

  Low-UA cultures would emphasize the reverse:

- Complexity with maximal content and choices
- Acceptance (even encouragement) of wandering and risk, with a stigma on "over-protection"
- Less control of navigation; for example, links might open new windows leading away from the original location
- Mental models and help systems might focus on understanding underlying concepts rather than narrow tasks
- Coding of color, typography, and sound maximize information

Online examples of UA differences can be seen on airline websites from two countries with very different UA indices (Fig. 2.12 & Fig. 2.13). The Sabena Airlines website (www.sabena.com) is located in Belgium, a country with a UA index of 94, the highest of the cultures studied. This website shows a home page with very simple, clear imagery and limited choices.

The British Airways website (www.britishairways.com) from the United Kingdom (UA = 35) shows more complexity of content and choices with pop-up windows, multiple types of controls, and "hidden" content that must be displayed by scrolling.

Long- versus Short-Term Time Orientation

In the early 1980s, shortly after Hofstede first formulated his cultural dimensions, he became convinced that a fifth dimension needed to be defined. Long-term time orientation (LTO) seemed to play an important role in Asian countries (influenced by

Confucian philosophy over thousands of years) that shared these beliefs:

- A stable society requires unequal relations.
- The family is the prototype of all social organizations; consequently, older people (parents) have more authority than younger (and men more than women).
- Virtuous behavior to others means not treating them as one would not like to be treated.
- Virtuous behavior in work means trying to acquire skills and education, working hard, and being frugal, patient, and persevering.

Western countries, by contrast, were more likely to promote equal relationships, emphasize individualism, focus on treating others as you would like to be treated, and find fulfillment through creativity and self-actualization. When Hofstede and Bond developed a survey specifically for Asia and reevaluated earlier data, they found that long-term orientation cancelled out some of the effects of masculinity/femininity and uncertainty avoidance. They concluded that Asian countries are oriented to practice and the search for virtuous behavior while Western countries are oriented to belief and the search for truth. Of the 23 countries compared, the following showed the most extreme values (ranked in parentheses):

118	China (ranked 1)
80	Japan (4)
29	United States (17)
0	Pakistan (23)

High-LTO countries would emphasize the following aspects of UI design:

- Content focused on practice and practical value
- Relationships as a source of information and credibility
- Patience in achieving results and goals

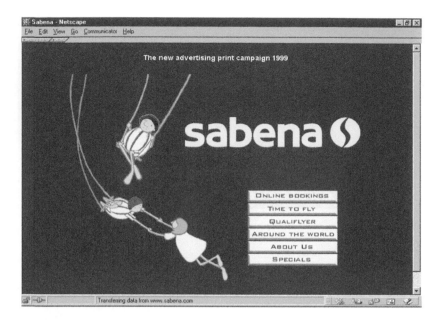

FIGURE 2.12.   High uncertainty avoidance: Sabena Airlines website from Belgium.

FIGURE. 2.13.   Low uncertainty avoidance: UK's British Airways website.

Low-LTO (i.e., short-term time-orientation) countries would emphasize the contrary:

• Content focused on truth and certainty of beliefs
• Rules as a source of information and credibility
• Desire for immediate results and achievement of goals

Online examples of LTO differences can be seen in two versions of the same company's website from two countries with different LTO values (Fig. 2.14 & Fig. 2.15). The Siemens website (www.siemens.co.de) from Germany (LTO of 31) shows a typ-

ical Western corporate layout that emphasizes crisp, clean functional design aimed at achieving goals quickly.

The Chinese version from Beijing requires more patience to achieve navigational and functional goals. Typically, such websites also have more emphasis on photos of people and more tolerance of informal, even cartoon-like imagery.

Design Issues

Hofstede noted that some cultural relativism is necessary: it is difficult to establish absolute criteria for what is noble and what

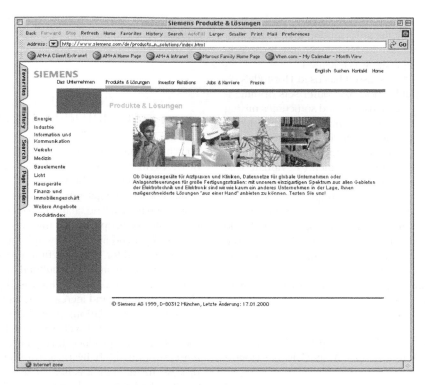

FIGURE 2.14. Low long-term orientation: website form Siemens Germany.

FIGURE 2.15. High long-term orientation: website from Siemens in China.

is disgusting. There is no escaping bias; all people develop cultural values based on their environment and early training as children. Not everyone in a society fits the cultural pattern precisely, but there is enough statistical regularity to identify trends and tendencies. These trends and tendencies should not be treated as defective or used to create negative stereotypes but recognized as different patterns of values and thought. In a mul-

ticultural world, it is necessary to cooperate to achieve practical goals without requiring everyone to think, act, and believe identically.

The cultural diffferences just identified seem to survive in websites over periods of time. A study by Marcus and Baumgartner (2003) showed that over a three-year period, the Sabena and British Airways websites cited earlier changed their designs dramatically. The main air-booking areas had approximately the same number of links, but the areas around this central area differed remarkably. The UK site had twice as many links, indicating again that more complexity was tolerated.

Although this analysis just discussed has used Hofstede's dimensions, many recent publications use other sets of dimensions, often numbering 7 ± 2 dimensions, and sometimes mixing several theorists. Marcus and Baumgartner (2004) published a summary of Baumgartner's thesis, which surveyed approximately 60 experts worldwide regarding their views of the best dimensions from 11 theories through which to evaluate UIs. Twenty-nine dimensions emerged, from which five dimensions seemed to be the highest esteemed: (a) high versus low context, (b) technology attitudes and status, (c) uncertainty avoidance, (d) time perception, and (e) authority perception. While the optimum set of dimensions continues to be debated, more publications within the UI community extol the value of using culture dimensions to analyze and design UIs.

This review of cultural dimensions raises many issues about UI design, especially for the Web:

- How formal or rewarding should interaction be?
- What will motivate different people? Money? Fame? Honor? Achievement?
- How much conflict can people tolerate in content or style of argumentation?
- Should sincerity, harmony, or honesty be used to make appeals?
- What role exists for personal opinion versus group opinion?
- How well are ambiguity and uncertainty avoidance received?
- Will shame or guilt constrain negative behavior?
- What role do community values play in individualist versus collectivist cultures?
- Does the objective of distance learning change what can be learned in individualist versus collectivist cultures? Should these sites focus on tradition? Skills? Expertise? Earning power?
- How should online teachers or trainers act: friends or gurus?
- Would job sites differ for individualist versus collectivist cultures?
- Should there be different sites for men and women in different cultures?
- Would personal Webcams be okay? Not okay?
- How much advertising hyperbole is likely to be tolerated?
- Would an emphasis on truth as opposed to practice and virtue require different types of procedural websites for Western or Asian audiences?

If cross-cultural theory becomes an accepted element of UI design, then we may need to change our current practices and tools (e.g., use appropriate testing and focus-group techniques or design-team collaboration techniques suitable for the cultures involved) or develop new ones. We need to make it feasible to develop multiple culturally distinguished versions of websites in a cost-effective manner, perhaps through templates or through specific versioning tools. As personal computer, Web, and mobile products/services continue to develop globally, answering these questions and exploring, then exploiting, these dimensions of culture will become a necessity, not an option, for successful theory and practice.

CONCLUSIONS AND FUTURE RESEARCH ISSUES

To achieve culturally sensitive, successful global access to UIs provides many design challenges in the UI-development process. Progress in technology increases the number and kinds of functions, data, platforms, and users of computer-based communication media. The challenge of enabling more people and more kinds of people to use this content and these tools effectively will depend increasingly upon global solutions. By recognizing the need for, and benefit to users of UI designs intended for international and intercultural markets, developers will achieve greater success and increased profitability through the global distribution and increased acceptance of their products.

The recommendations provided in this chapter are an initial set of heuristics that will assist developers in achieving global solutions to their product development. Design methodologies must support globalization throughout the development process. In addition, it is likely that some international and intercultural references will change rapidly, requiring frequent updating of designs. Future work on global UI design may address the following issues:

1. How might global UIs be designed to account for different kinds of intelligence? Gardner (1985) identified the following dimensions of intelligence. These dimensions suggest users might have varying strengths of conceptual competence with regard to using UIs on an individual basis, but these might also vary internationally, or interculturally, due to influences of language, history, or other factors:

   - Verbal/image comprehension
   - Word/image fluency
   - Numerical/graphical fluency
   - Spatial visualization
   - Associative memory
   - Perceptual speed
   - Reasoning
   - Interpersonal awareness
   - Self-awareness

2. How might metaphors, mental models, and navigation be designed precisely for different cultures that differ by such dimensions as age, gender, national or regional group, or profession? The author has posed this issue earlier as a question to the UI analysis/design community (Marcus, 1993, April). The topic is discussed broadly in DelGaldo and Nielson (1996). Further, what means can be developed to enable these variations to be produced in a cost-effective manner using templates?

The taxonomic analyses of global issues for UIs, the theoretical basis for their component selection, the criteria for their evaluation, and their design methodology have emerged in the UI-development field. Articles about the impact of culture differences on UI design and techniques of ethnographic analysis appear ever more frequently in primary industry publications and in professional conferences such as ACM/SIGCHI, the Usability Professionals Association, the American Anthropologists' Association, IWIPS, and others. The lively exchanges to be found on Anthropologists in Design's Internet discussions likewise attest to the growing numbers of professionals involved in this cross-disciplinary practice (see URL references for resources). Designers should be aware of the scope of the activity, know sources of insight, and incorporate professional techniques in their development process in order to improve the value and success of their international and intercultural computer-mediated products and services.

ACKNOWLEDGMENTS

The author acknowledges Marcus et al. (1999) and Marcus and Gould (2000), on which this chapter is based. Mr. Marcus thanks his staff at Aaron Marcus and Associates, Inc., Sabre, Dr. Emile Gould, Acadia University, the editors of this book, and Dr. Constantine Stephanidis, the editor of *User Interfaces for All*, for their assistance in preparing this chapter. The author also acknowledges DelGaldo and Nielsen (1996), Fernandes (1995), and Nielsen (1990), which provided a basis for some points raised in this chapter; and the advice of Peter Siemlinger, Dipl.-Ing, Dir., Internat. Inst. for Info. Design, Vienna, Austria, Europe; Prof. Andreas Schneider, Info. Design Dept, Tama Art Univ. Tokyo, Japan; and Dr. Nuray Aykin, New School, New York.

References

American Institute of Graphic Arts (AIGA). (1981). *Symbol Signs*. New York: Visual Communication Books, Hastings House.

Aykin, N. (2000). (personal communication, 12 March 2000).

Aykin, N. (Ed.). (2004). *Usability and Internationalization of Information Technology*. New York: Lawrence Erlbaum Publishers.

Batchelor, D. (2000). *Chromophobia*. London: Reaktion Books.

Boxer, S. (2001, April 28). Vivid color in a world of black and white. *New York Times*, p. A15ff.

Brain, D. (2005). Syllabus for course in sociology of culture at New College, Florida. Retrieved from http://www.ncf.edu/brain/courses/culture/culture_syl05.htm (31 Dec 2005).

Carroll, John M. (1999). "Using Design Rationale to Manage Culture-Bound Metaphors for International User Interfaces, *Proceedings, Internat. Conf. on Internationalization of Products and Services*, Rochester, NY, 20–22 May 1999, pp. 125–131.

Chavan, A. L. (1994). *A design solution project on alternative interface for MS Windows*. Unpublished master's thesis, Royal College of Ave, London, UK.

Chavan, A. L. (2005). Another culture, another method. *Proceedings of the Human Computer Interaction International Conference*, Las Vegas, NV, CD-ROM.

Choong Y., & Salvendy, G. (1999). Implications for design of computer interfaces for Chinese users in mainland China. *International Journal of Human-Computer Interaction, 11*(1), 29–46.

Clausen, Christopher. (2000). *Faded Mosaic: The Emergence of Post-Cultural America*. Chicago, IL: Ivan R. Dee Publisher.

Coriolis Group. (1998). *How to build a successful international website*. Scottsdale, AZ: Author.

Cox, T., Jr. (1994). *Cultural diversity in organizations*. San Francisco: Berrett-Koehler.

Crystal, D. (1987). *The Cambridge encyclopedia of language*. Cambridge, UK: Cambridge University Press.

Daniels, P. T., & Bright, W. (Eds.). (2000). *The world's writing systems*. Sandpoint, ID: MultiLingual Computing.

Day, D. L., del Galdo, E. M., & Prabhu, G. V. (Eds.). (2000, July). Designing for global markets 2. *Proceedings of the 2nd International Workshop on Internationalisation of Products and Systems*. Rochester, NY: Backhouse Press.

Day, D. (2000, July 13–15). Gauging the extent of internationalization activities. *Proceedings of the 2nd International Workshop on Internationalization of Products and Systems* (pp. 124–136). Baltimore, MD: Backhouse Press.

DelGado, E., & Nielsen, J. (Eds.) (1996). *International User Interface*. New York: Wiley.

Earley, P., Soon, C., & Soon, A. (2003). *Cultural intelligence: Individual interactions across cultures*. Stanford, CA: Stanford University Press.

Elashmawi, F., & Harris, P. R. (1998). *Multicultural management 2000: Essential cultural insights for global business success*. Houston, TX: Gulf.

Evers, V., et al. (Eds.) (2003). *Proceedings of the Fifth International Workshop on Internationalisation of Products and Systems (IWIPS 2003)*. Berlin, Germany: Products and Systems Internationalization.

Fernandes, T. (1995). *Global interface design: A guide to designing international user interfaces*. Boston: AP Professional.

Ferraro, G. (2006). *The cultural dimension of international business* (5th ed.). Upper Saddle River, NJ: Pearson Prentice Hall.

Forrester Research, Inc. (2001, March). *The global user experience*. Cambridge, MA: Forrester Research, Inc.

Gardner, H. (1985). *Frames of mind, the theory of multiple intelligences*. New York: Basic Books.

Goode, E. (2000, August 8). How culture molds habits of thought. *New York Times*, p. D1ff.

Gould, E. W., Zakaria, N., & Shafiz Affendi, M. Y. (2000, 25 September). Applying Culture to website design: A comparison of Malaysian and U.S. websites," *Proceedings of the International Professional Communication Conference, Special interest group on Documentation*, Boston.

Gudykunst, W. B. (2003). *Cross-cultural and intercultural communication*. Thousand Oaks, CA: Sage.

Gudykunst, W. B. (2005). *Theorizing about intercultural communication.*. Thousand Oaks, CA: Sage Publications.

Harris, J., & McCormack, R. (2000). *Translation is not enough*. San Francisco, CA: Sapient.

Harris, P. R., & Moran, R. T. (1993). *Managing cultural differences*. Houston, TX: Gulf.

Hendrix, A. (2001, April 15). The nuance of language. *San Francisco Chronicle*, p. A10.

Herskovitz, J. (2000, December 26). J-Pop takes off: Japanese music, movies, TV shows enthrall Asian nations. *San Francisco Chronicle*, p. C2.

Hofstede, G. (1997). *Cultures and organizations: Software of the mind, intercultural cooperation and its importance for survival.* New York: McGraw-Hill.

Honold, P. (1999). Learning how to use a cellular phone: Comparison between German and Chinese users. *Journal of Society for Technical Communication, 46*, 196–205.

International Organization for Standardization. (1989). *Computer display color* (Draft Standard Document 9241-8). Geneva, Switzerland: Author.

International Standards Organization. (1990). *ISO 7001: Public information symbols.* Geneva, Switzerland: The American National Standards Institute (ANSI).

Kimura, D. (1992). Sex differences in the brain, *Scientific American, 267*(3), September 1992, 12–125.

Konkka, K., & Koppinen, A. (2000). Mobile devices: Exploring cultural differences in separating professional and personal time, *Proceedings of the Second International Workshop on Internationalization of Products and Systems* (pp. 89–104). 13–15 July 2000 Rochester, NY: Backhouse Press.

Lamont, M. (1995). Syllabus for course in contemporary sociological theory. Princeton University. Retrieved from http://www.princeton.edu/~sociolog/grad/courses/fall1995/soc502.html (Retrieved: January 2005).

Lee, Y. S., Ryu, Y. S., Smith-Jackson, T. L., Shin, D. J., Nussbaum, M. A., & Tomioka, K. (2005, July). Usability testing with cultural groups in developing a cell phone navigation system. *Proceedings of the 11th International Conference on Human-Computer Interaction (HCII '05),* Las Vegas, NV. (CD-ROM)

Leventhal, L. (1996). Assessing user interfaces for diverse user groups: Evaluation strategies and defining characteristics. *Behaviour & Information Technology, 15*(3), 127–138.

Lingo Systems. (1999). *The guide to translation and localization,* Los Alamitos, CA IEEE Computer Society, ISBN 0-7695-0022-6.

LISA, The Localization Industry Primer. The Localization Industry Standards Association, (LISA) 7, rute du Monastère, 1173 Féchy, Switz., www.lisa.org, 1999, 35 pp.

Marcus, A. (1992). *Graphic design for electronic documents and user interfaces,* Reading MA: Addison-Wesley.

Marcus, A. (1993, April). Human communication issues in advanced UIs, *Communications of the ACM, 36*(4), 101–109.

Marcus, A. (1993, October). Designing for diversity. *Proceedings of the 37th Human Factors and Ergonomics Society,* Seattle, WA, 258–261.

Marcus, A. (1995). Principles of effective visual communication for graphical user interface design. In Baecker, R.; Grudin, J.; Buxton, W.; & Greenberg, S. (Eds.), *Readings in human–computer interaction* (2nd ed., pp. 425–441). Palo Alto, CA: Morgan Kaufman.

Marcus, A. (1998). Metaphor design in user interfaces. *Journal of Computer Documentation, 22*, 43–57.

Marcus, A., Gould, & Chen, E. (1999, June). Globalization of user interface design for the Web. *Proceedings of the 5th Human Factors and the Web Conference,* Gathersburg, MD. (CD-ROM)

Marcus, A. (2001). User interface design for air-travel booking: A case study of Sabre. *Information Design Journal, 10*(2), 26–206.

Marcus, A. (2003). 12 myths of mobile UI design. *Software Development Magazine, May,* 38–40.

Marcus, A. (2005). User interface design and culture. In Aykin, N. (Ed.), *Usability and Internationalization of Information Technology* (pp. 51–78). New York: Lawrence Erlbaum.

Marcus, A., & Baumgartner, V. J. (2004). Mapping user interface design components vs. culture dimensions in corporate websites," *Visible Language, 38*(1), 1–65.

Marcus, A., & Baumgartner, V. J. (2004). A practical set of culture dimension for evaluating user interface designs. *Proceedings of the Sixth Asia-Pacific Conference on Computer-Human Interaction (APCHI 2004),* Rotorua, New Zealand, 252–261.

Marcus, A., & Baumgartner, V. J. (2005). A practical set of culture dimensions for global user interface development. In M. Masoodian, S. Jones, B. Rogers, (Eds.), *Proceedings of the Computer Human Interaction: 6th Asia Pacific Conference, APCHI 2004,* Rotorua, New Zealand, 252–261.

Matsumoto, D., Weissmann, M. D., Preston, K., Brown, B. R., & Kupperburd, C. (1997). Context-specific measurement of individualism-collectivism on the individual level: The individualism-collectivism interpersonal assessment inventory. *Journal of Cross-Cultural Psychology, (28)*6, 743–767.

Matsumoto, D., & LeRoux, J. (2003). Measuring the psychological engine of intercultural adjustment: The intercultural adjustment potential scale (ICAPS). *Journal of Intercultural Communications,* 6, pp. 27–52.

Neustupny, J. V. (1987). *Communicating with the Japanese.* The Japan Times, Tokyo, Japan.

Nielsen, J. (Ed.). (1990). *Designing user interfaces for international use: Vol. 13. Advances in human factors/ergonomics.* Amsterdam: Elsevier Science.

Nisbett, R. E. (2003). *The geography of thought: How Asians and Westerners think differently . . . and why.* New York: Free Press.

Olgyay, N. (1995). *Safety symbols art.* New York: Van Nostrand Reinhold.

Ona, Y., & Spindle, B. (2000, December 30). Japan's long decline makes one thing rise: Individualism. *Wall Street Journal,* p. A1.

Ota, Y. (1973). *Locos: Lovers communications system* (in Japanese). Tokyo, Japan: Pictorial Institute.

Ota, Y. (1987). *Pictogram design.* Kashiwashobo Publishers, Tokyo, Japan.

Peng, K. (2000). *Readings in cultural psychology: Theoretical, methodological and empirical developments during the past decade* (1989–1999). New York, NY: John Wiley and Sons.

Perlman, G. (13–15 July). ACM SIGCHI Intercultural Issues. *Proceedings of the Second International Workshop on Internationalization of Products and Systems* (pp. 23–195). Rochester, NY: Backhouse Press.

Prabhu, G. V., & delGaldo, E. M. (Eds.). (1999, 20–22 May). Designing for global markets 1. *Proceedings of the First International Workshop on Internationalization of Products and Systems* (p. 226). Rochester, NY: Backhouse Press.

Prabhu, G., & Harel, D. (1999, August 22–26). GUI design preference validation for Japan and China: A case for KANSEI engineering? *Proceedings of the 8th International Conference on Human-Computer Interaction (HCI International '99),* Munich, Germany.

Saudi Arabia issues edict against Pokemon. (2001, March 27). *San Francisco Chronicle,* p. F2.

Shahar, L., & Kurz, D. (1995). *Border crossings: American interactions with Israelis.* Yarmouth, Maine: Intercultural Press.

Singh, N., & Matsuo, H. (2002). Measuring cultural adaptation on the Web: A content analytic study of U.S. and Japanese websites. *Journal of Business Research, 57*(8), 864–872.

Stille, A. (2001, January 13). An old key to why countries get rich: It's the culture that matters, some argue anew. *New York Times,* 81.

Storti, C. (1994). *Cross-cultural dialogues: 74 brief encounters with cultural difference.* Yarmouth, ME: Intercultural Press.

Tannen, D. (1990). *You just don't understand: Women and men in conversation.* New York: William Morrow and Company, Inc.

Thomas, D., & Inkson, K. (2004). *Cultural intelligence: People skills for global business.* San Francisco: Berrett-Koehler Publishers.

Vickers, B. (2000, November 22). Firms push to get multilingual on the Web. *Wall Street Journal,* p. B11A.

Victor, D. A. (1992). *International business communication.* New York: HarperCollins.

Würtz, E. (2005). A cross-cultural analysis of websites from high-context cultures and low-context cultures. *Journal of Computer-Mediated Communication, 11*(1), article 13.

Yardley, J. (2000, August 7). Faded Mosaic nixes idea of "cultures" in U.S. *San Francisco Examiner*, p. B3.

URLs and Other Information Resources
(Latest version available from author or email to www.AMandA.com)

ACM/SIGCHI Intercultural Issues database: www.acm.org/sigchi/intercultural/

ACM/SIGCHI Intercultural listserve: chi-intercultural@acm.org.

American National Standards Institute (ANSI): www.ansi.org

Anthropologists in Design: http://groups.yahoo.com/group/anthrodesign/

Bibliography of Intercultural publications: www.HCIBib.org//SIGCHI/Intercultural

China National Standards: China Commission for Conformity of Elect. Equip. (CCEE) Secretariat; 2 Shoudu Tiyuguan, NanLu, 100044, P.R. China; Tel: −6-1-8320088, ext. 2659, Fax: −6-1-832-0825

Cultural comparisons: www.culturebank.com, www.webofculture.com, www.iir-ny.com

Digital divide: www.digitaldivide.gov, www.digitaldivide.org, www.digitaldividenetwork.org

Globalization and Internet language statistics: language: www.euromktg.com/globstats/, www.sapient.com, www.worldready.com/biblio.htm

Glossary, six languages: www.bowneglobal.com/bowne.asp?page=9&language=1

International Standards Organization (ISO): http://www.iso.ch/

Internationalization providers: www.basistech.com, www.cij.com, www.Logisoft.com

Internationalization resources: www.world-ready.com/r_intl.htm, www.worldready.com/biblio.htm

Internet users survey, Nua: www.nua.ie/surveys/how_many_online

Japan Info. Processing Society; Kikai Shinko Bldg., No. 3-5-8 Shiba-Koen, Minato-ku, Tokyo 105, Japan; Tel: −1-3-3431-2808, Fax: −1-3-3431-6493

Japanese Industrial Standards Committee (JISC); Min. of Internat. Trade and Industry; 1-3-1, Kasumigaseki, Chiyoda-ku, Tokyo 100, Japan; Tel: −1-3-3501-9295/6, Fax: −1-3-3580-142

Java Internationalization: http://java.sun.com/docs/books/tutori

Localization Industry Standards Organization (LISA): www.lisa.org

Localization providers: www.Alpnet.com, www.Berlitz.com, www.globalsight.com, www.lhsl.com, www.Lionbridge.com, www.Logisoft.com, www.Logos-usa.com, www.translations.com, www.Uniscape.com

Machine translation providers: www.babelfish.altavista.com, www.IDC.com, www.e-Lingo.com, Lernout & Hauspie <www.lhsl.com>, www.Systransoft.com

Microsoft global development: www.eu.microsoft.com/globaldev/

Simplified English: userlab.com/SE.html

Unicode: www.unicode.org/, www−4ibm.com/software/developer/library/glossaries/unicode.html

World-Wide Web Consortium: www.w3.org/International, www.w3.org/WAI

Recommended Readings

Alvarez, G. M., Kasday, L. R., & Todd, S. (1998). How we made the website international and accessible: A case study. *Proceedings of the 4th Human Factors and the Web Conference*, 5 June 1998, Holmdel, NJ. (CD-ROM)

Day, D., Eves, V., F del Galdo, E. (2005). *Designing for global markets 7: Bridging cultural differences. Proceedings of the International Workshop on Internationalization of Products and Services 2005*, Amsterdam: Grafisch Centrum Amsterdam.

Day, D. L., & Dunckley, L. M. (2001, 12–14 July). *Proceedings of the Third International Workshop on Internationalisation of Products and Systems (IWIPS 2001)*. Milton Keynes, UK.

Doi, T. (1973). *The anatomy of dependence*. New York: Kodansha International.

Doi, T. (1986). *The anatomy of self: The individual versus society*. New York: Kodansha International.

Dreyfuss, H. (1966). *Symbol sourcebook*. New York: Van Nostrand Rhinehold.

Fetterman, D. M. (1998). *Ethnography: Step by step* (2nd ed.). Thousand Oaks, CA: Sage.

Forrester Research, Inc. (1998). *JIT Web localization*. Retrieved 4 July 1998, from www.Forrester.com.

French, T., & Smith, A. (2000, 13–15 July). Semiotically enhanced Web interfaces: Can semiotics help meet the challenge of cross- cultural design? Proceedings of the 2nd International Workshop on Internationalization of Products and Systems (pp. 23–38). Baltimore, MD. Rochester, NY: Backhouse Press.

Graham, T. (2000). *Unicode: A primer*. Sandpoint, ID: MultiLingual Computing and Technology.

Hall, E. T. (1969). *The hidden dimension*. New York: Doubleday.

Harel, D., & Girish, P. (1999, 22–23 May). Global User Experience (GLUE), Design for cultural diversity: Japan, China, and India. *Designing for Global Markets, Proceedings of the First International Workshop on Internationalization of Products and Systems (IWIPS-99)* (pp. 205–216). Rochester, NY: Backhouse Press.

Hoft, N. L. (1995). *International technical communication: How to export information about high technology*. New York: John Wiley and Sons, Inc.

Inglehart, R. F., Basanez, M., & Moreno, A. (Eds.) (1998). *Human values and beliefs: A cross-cultural sourcebook*. Ann Arbor, MI: University of Michigan Press.

International Standards Organization. (1993). *ISO 7001: Public information symbols: Amendment 1*. Geneva, Switzerland: The American National Standards Institute (ANSI).

Kohls, L. R., & Knight, J. M. (1994). *Developing intercultural awareness: A cross-cultural training handbook* (2nd ed.). Yarmouth: ME: Intercultural Press.

Kuhn, A. (2005, March 12). Mulling weekends, workdays in Iraq. *National Public Radio Web Archive*. Retrieved January 12, 2006, from http://www.npr.org/templates/story/story.php?storyId=4540715.

Kurosu, M. (1997, August 24–29). Dilemma of usability engineering. In G. Salvendy, M. Smith, & R. Koubek (Eds.), Design of computing systems: Social and ergonomics considerations, Volume 2. *Proceedings of the 7th International Conference on Human-Computer Interaction HCI International '97* (pp. 555–558). Amsterdam: Elsevier.

Lewis, R. (1991). *When cultures collide*. London: Nicholas Brealey.

Marcus, A. (2000). International and intercultural user interfaces. In Dr. C. Stephanidis (Ed.), *User Interfaces for All*. (pp. 47–63). New York: Lawrence Erlbaum Associates.

Marcus, A., & Gould, E. W. (2000). Crosscurrents: Cultural dimensions and global web user interface design. *Interactions, 7*, 32–46.

Murphy, R. (2005). Getting to know you. *Fortune Small Business*. Retrieved November 6, 2005, from http://www.fortune.com/fortune/smallbusiness/technology/articles/0,15114,1062892-1,00.html.

Nisbett, R. E., Kaipeng, P., Incheol, C., & Norenzayan, A. (2001). Culture and systems of thought: Holistic versus analytical cognition. *Psychological Review, 108*, 291–310.

Pierce, T. (1996). *The international pictograms standard*. Cincinnati, OH: S. T. Publications.

Prabhu, G. V., Chen, B., Bubie, W., & Koch, C. (1997, 24–29 August). Internationalization and localization for cultural diversity. In Salvendy G. (Ed.), *Design of Computing Systems: Cognitive Considerations. Vol. 1: Proceedings of the 7th International Conference on Human-Computer Interaction* (HCI International '97) (pp. 149–152). Amsterdam: Elsevier.

Schwartz, S. H. (2004). Mapping and interpreting cultural differences around the world. In H. Vinken, J. Soeters, & P. Ester (Eds.), *Comparing cultures, dimensions of culture in a comparative perspective* (pp. 43–73). Leiden, Netherlands: Brill.

Singh, N. (2004). From cultural models to cultural categories: A framework for cultural analysis. *The Journal of American Academy of Business*, 5(1/2), 1–8. [Vol 5, Nos. 1 & 2].

Singh, N., & Baack, D. W. (2004). website adaptation: A cross-cultural comparison of U. S. and Mexican websites. *Journal of Computer-Mediated Communication*, 9(4), Retrieved 16 September 2006 from http://JCMC.Indiana.edu/vol9/issue4/singh_back.html.

Spradley, J., & McCurdy, D. (1998). *The cultural experience: Ethnography in complex society*. Long Grove, IL: Waveland Press.

Stephanidis, C. (Ed.). (2000). *User interfaces for all*. New York: Lawrence Erlbaum Associates.

Trompenaars, F., & Turner, C. H. (1998). *Riding the waves of culture*. New York: McGraw-Hill.

Yeo, A. W. (2001). Global-software development lifecycle: An exploratory study. *Proceedings Computer-Human Interaction Conference 2001*, Seattle, WA, 104–111.

·3·

CONVERSATIONAL SPEECH INTERFACES AND TECHNOLOGIES

Jennifer Lai and Clare-Marie Karat
IBM T.J. *Watson Research Center*

Nicole Yankelovich
Sun Microsystems

Spoken Interface Life Cycle **55**
Selecting the Speech Technology **55**
   How Does ASR Work? 55
   How Does Text-to-Speech (TTS) Work? 56
   Current Capabilities and Limitations
   of Speech-Synthesis Software 57
   How Do Natural Language Processing (NLP) and
   Natural Language Understanding (NLU) Work? 57
   Current Capabilities and Limitations
   of NLU Systems 58
Crafting the Spoken Interaction **58**
   Dialogue Styles 59
      Directed dialogue (system initiated) 59

      User-initiated 59
      Conversational systems (mixed initiative) 59
   Inherent Challenges of Speech 59
   Prompt Design 59
   Providing Help 60
   Handling Errors 60
      Rejection errors 61
      Substitution errors 61
      Insertion errors 61
   Correction Strategies 61
Testing and Iterating **61**
Conclusion **62**
References **62**

A conversational speech interface for a computer emulates human-to-human interaction by calling on our inherent ability as humans to speak and listen. While human speech is a skill we acquire early and practice frequently, getting computers to map sounds to actions and to respond appropriately with either synthesized or recorded speech is a massive programming undertaking. Because we all speak a little differently from each other, and because the accuracy of the recognition is dependent on an audio signal that can be distorted by many factors, speech technology, like the other recognition technologies, lacks 100% accuracy. When designing a spoken interface, one must design to the strengths and weaknesses of the technology to optimize the overall user experience.

The goal for a spoken user interface is to emulate a conversation convincingly enough that the person interacting with the computer can use what he or she has learned in a lifetime of conversations. Successful communication occurs when the sender and receiver of the message achieve a shared understanding. Human-to-human conversations are characterized by turn taking, or shifts in initiative, as well as verbal and nonverbal feedback to indicate understanding. Herb Clark (393) said, "Speaking and listening are two parts of a collective activity." Because language use is deeply ingrained in human behavior, successful speech interfaces should be based on an understanding of the different ways that people use language to communicate. Speech applications should adopt language conventions that help people know what they should say next and avoid conversational patterns that violate standards of polite, cooperative behavior.

Many excellent examples of speech user interfaces emulate effective conversational partners. In these systems, the computer "speaker" appears to remember contextual information and gives the impression of understanding what the user is saying. For example, several airlines use speech applications to handle telephone reservations or lost baggage tracking (Cohen, Giangola, & Balogh, 2004). These systems are a substantial improvement over previous interactive voice response (IVR) systems that relied solely on telephone keypad input. Instead of pressing "1" for this and "2" for that, users can speak natural language phrases such as "I'd like to travel from Boston to New York."

A crucial factor in determining the success of a spoken application is whether the use of speech technology presents a clear benefit. Speech is best used when it enables something that cannot otherwise be done, such as accessing electronic mail or an online calendar over the telephone when a computer is not available. In general, it is effective to use speech applications for situations when speech can enable a task to be done more efficiently, such as when a user's hands and eyes are busy doing another task. Likewise, speech input is useful when no keyboard is available for text entry, or when a user has a physical disability that limits the use of his or her hands, or if a user is just not comfortable typing. Speech output is particularly liberating for people with visual impairments. In addition, it provides a way of communicating information if the user is in a divided attention state such as driving, and it can be used to grab users' attention or embody a particular personality for a computer system or character.

While speech seems like it might be the ideal way to communicate with a computer anytime, there are situations when it is best not to use a speech user interface. For example, speech output is ineffective for delivering large amounts of information. Not only is it difficult for users to maintain the information in short-term memory, but people can read much faster than they can listen. Speech input can also be problematic when the speaker is not in a private environment, or when other voices in the background might interfere with the speech recognition.

The success of a spoken interaction with a computer depends on not only the task and the motivation of the user, but also on the physical device(s) being used. Speech input and output capabilities of devices greatly vary. For example, personal computers (PCs) provide good quality audio subsystems and speakers, and a variety of microphones that perform very well under quiet conditions can be used with them. The audio channels for other devices are not on par with PCs. While most personal digital assistants (PDAs) have started to offer speech input capabilities, the quality of their microphones and audio subsystems degrade the speech signal, resulting in poor recognition performance for many applications. In addition, PDAs have insufficient computing resources to enable large vocabulary speech recognition processing. For telephony applications, when the input device is either a cell phone or a landline, the speech recognition engines are usually deployed on large servers. The accuracy of telephony systems for a given task is normally lower than for a similar PC configuration with a headset microphone in a quiet environment. Background noise, signal degradation, poor cellular connection and the application of compression techniques can substantially reduce the performance of speech recognition systems. Nevertheless, careful design can compensate for many of these problems and lead to successful interaction with a telephone-based speech application.

Speech technology can be used for the following types of tasks:

- **Composition.** Composition tasks have as their primary goal the creation of a document such as word processing documents, e-mails, or instant-messaging text. Composition includes dictating the text and fixing any recognition errors.
- **Transcription.** Transcription is similar to composition in that a document is created from speech, but it differs by virtue of the fact that the primary user task is parallel to the creation of a document. Broadcast news, business meetings, and business calls are examples of situations where having a permanent, textual record of the speech is valuable. A digital (textual) record of the conversation is searchable, readable by the deaf, and supports advanced business intelligence applications such as data mining. These are all examples of transcription tasks.
- **Transaction.** The third type of interaction, and a major focus of this chapter, is one where users have as their goal the completion of one or more transactions, rather than the creation of a document or a permanent record of a conversation. Examples of transactional applications include financial account management such as trading stocks, e-commerce applications such as the purchase of computer equipment, searching for information on the Internet, and controlling the environment.
- **Collaboration.** The final type of conversational task is collaboration. Collaborative conversational applications are

characterized by tasks that result in human-to-human communication. One example of this is when speech recognition is used as an input modality to an instant messaging application.

Designing a speech user interface is similar to designing any other interface for human-computer interaction (HCI). A good design relies on applying principles of user-centered design, and many of these same principles and techniques can be used with speech interfaces. In this chapter, we discuss the lifecycle of a speech application from the starting point of understanding the requirements for speech, through knowing which technology to use, selecting a dialog style, and designing the prompts. All these steps culminate with an iterative testing process.

SPOKEN INTERFACE LIFE CYCLE

Once a designer has clearly established that including speech technology for input, output, or both is an appropriate design decision, the designer must begin to define how speech best fits into the accomplishment of the task. In order to do this, the designer must first have a clear understanding of the task. With the task well understood, the designer needs to overlay speech onto that model by listening to how people speak in the domain. Observing humans interacting and speaking to accomplish the task may result in additional refinements to the model. The next step is to select the appropriate speech technology, or combination of technologies, for the task, in the contexts of use and user group. The heart of the work for a speech interface designer is to craft the prompts that the system will "speak" and to prepare the system to recognize what the users will most likely say in response. Lastly, the designer should create a series of tests for the system, initially in the lab and ultimately out in the real world with users accomplishing real tasks. In between the various forms of testing, a designer needs to allocate time to refine the design and make changes based on what he or she learns from the results of the testing. In this chapter, we will cover certain steps in greater detail than others. These steps include (a) listening to users speak when accomplishing the task, (b) selecting the technology, and (c) designing the system prompts.

Thus, the first step in the life cycle of a conversational interface is to model the task that users will be accomplishing. This requires the designer to sketch out the logical steps that make up the task, along with the pieces of information that must be exchanged in order to complete each step. With a speech application, each interaction between the system and a user is often referred to as a "turn." Thus, the *task model* tries to capture the expected number of turns, as well as the vocabulary that is required to support the information exchange during those turns. It is important for the designer to consider alternate flows, since not all users will necessarily approach the task in the same way. For example, when making a flight reservation, some users will select a flight based exclusively on time schedules, while others will opt for airline loyalty and price points, accepting any flight time that meets their criteria for loyalty and cost.

The model must operate within the set of constraints for the application. Constraints usually include the business goals for creating the speech application, as well as the requirements resulting from the environment in which users may find themselves (e.g., noisy environment) and the users themselves (e.g., the majority of users will be over 65 years old and uncomfortable with technology). For example, if a business goal is to cross-sell related items when a customer makes a purchase by calling into a speech-enabled call center application, the designer will want to incorporate a turn that involves the system suggesting related items that are on sale that day.

SELECTING THE SPEECH TECHNOLOGY

While good interaction design in a speech application can compensate for some shortcomings in speech technology, if a certain baseline level of accuracy is not achieved, the application probably will not succeed. Accuracy depends on the choice of the underlying speech technology, and making the best match between the technology, the task, the users, and the context of use. Automatic Speech Recognition (ASR) can have explicitly defined, rule-based grammars, or use statistical grammars such as language models. Usually, a transactional system uses explicitly defined grammars, while dictation systems or natural language understanding systems use statistical models. The designer will also have to decide whether to use synthetic speech in the system or not. Synthetic speech, also known as text-to-speech (TTS), is speech produced by a computer. Given that today's synthesizers still do not sound entirely natural, the choice of whether to use synthesized output, recorded output, or no speech output can be a difficult one. The next section discusses the various technologies in more detail.

How Does ASR Work?

ASR systems work by analyzing the acoustic signal received through a microphone connected to the computer (see Figure 3.1). The user speaks and the microphone captures the acoustic signal as digital data, which is then analyzed by an acoustic model and a language model. The different speech recognition systems on the market differ in the building blocks (e.g., phonemes) that they employ to analyze the acoustic signal data. The analysis employs algorithms based on Hidden Markov Models, a type of algorithm that uses stochastic modeling to decode a sequence of symbols, to complete the computations (Rabiner, 389; Roe & Wilpon, 393). After the acoustic analysis

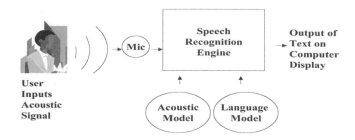

FIGURE 3.1. Overview of human-computer interaction model for speech recognition.

is complete, the system analyzes the resulting strings of building-block data using a language model that contains a base vocabulary and any specific domain topics (e.g., computer, medical) that may have been added. When the analysis is completed, the text appears on the computer screen.

Speech recognition systems require computers with an approximate minimum of 200MHz processor, 32MB of RAM, 300 MB of available hard disk, and a 16-bit sound card with a microphone input jack and good recording. These requirements enable local decoding of the recognized speech. It is possible to have a system, such as a small, pervasive device, capture a user's speech and then decode it on a remote server (e.g., Price & Sears, 2005) and return the decoded text to the user, albeit with a short time delay. Multiple users can work with one installation of a speech recognition system. In some systems, users create personal voice models and log on with their individual user names, and the system uses their personalized speech files for the recognition processing. Each user can also create several different user voice models in order to achieve the best recognition rates in environments with different levels of background noise (e.g., home, office, and mobile work locations). For results on the effectiveness of creating and using different user voice models, see Price, Ling, Feng, Goldman, Sears, & Jacko (2005).

Karat, Lai, Danis, and Wolf (399) reported ASR system accuracy rates in the mid-90s for "in vocabulary" words (e.g., words in the 20,000-word vocabulary included with the software) in people's initial use of continuous speech recognition software. Error rates of 2–5% are common with more frequent use. Karat, Horn, Halverson, and J. Karat (399) tested three commercially available speech recognition systems in 398 with users in initial and extended use and found initial use data of 13.6 corrected words per minute with an average of one incorrect, missing, or extra word for every 20 words and one formatting error every 75 words. Improved ASR system accuracy and higher user productivity were measured when users employed more recent versions of the speech recognition software (Sears, Karat, Oseitutu, Kaimullah, & Feng, 2000; Feng, Sears, & Karat, 2006), since users are able to correct errors more easily and quickly. The product also provided a better quality microphone to reduce the number of recognition errors that occurred in the first place.

In general, there are three types of errors that occur in ASR systems (Halverson, Horn, C. Karat, & J. Karat, 399). The first is that users can make direct errors where they misspeak, stutter, or press the wrong key. Second, users can make errors of intent, where they decide to restate something. The third type of error is an indirect error, where the speech recognition system mis-recognizes what the user says. The indirect errors are difficult to detect during proofreading. All three types of errors can lead to cascading errors where, in the process of correcting one error, others occur.

Most telephony systems are speaker-independent (e.g., no personalized training of the voice models required) speech-recognition systems. They are also usually server based and must handle the signal degradation that occurs across the telephone lines. Telephony systems can be created from a combination of speech recognition and natural language processing (NLP) technologies. In telephony systems, a dialog manager component works with the speech recognition software to handle the course of the conversation with the user. The system provides feedback to the user through the dialog manager using recordings of either human voice or text-to-speech. Telephony systems have capabilities such as "barge in" and "talk ahead" that enable the user to redirect the action of the system and complete multiple requests before being prompted for additional information necessary to complete the task. Conversational telephony systems include interactive voice response (IVR) systems and new systems built using voice XML (see section on NLP for description of the technology).

How Does Text-to-Speech (TTS) Work?

Text-to-speech (TTS) synthesis enables computers or other electronic systems such as telephones to output simulated human speech. Synthetic speech is based on the fields of (a) text analysis, (b) phonetics, (c) phonology, (d) syntax, (e) acoustic phonetics, and (f) signal processing. The quality and effectiveness of speech synthesis is based on a hierarchy. The base level of achievement is to produce speech synthesis that is intelligible by human beings. The second level is to produce speech synthesis that simulates the natural qualities of human speech. The third level of speech synthesis is to produce synthesized speech that is personalized to the person it is representing—that is, the intonation of the particular person's speech is represented. The fourth and highest level of achievement in synthesized speech is to produce speech based on a person's own voice recordings, so that the speech sounds just like the actual person being represented. Currently, speech synthesis technology has achieved the base level of quality and effectiveness, and concatenated synthesis can simulate the natural quality of human speech, although at great expense. The third and fourth levels of speech synthesis technology are the focus of research in laboratories around the world.

Two types of speech synthesis are commercially available today: (a) concatenated synthesis and (b) formant synthesis, which is the most prevalent type. Concatenated synthesis employs computers to assemble recorded voice sounds into speech output. It sounds natural, but can be prohibitively expensive for many applications, as it requires large disk storage space for the units of recorded speech and significant computational power to assemble the speech units on demand. Concatenated synthesizers rely on databases of diphones and demi-syllables to create the natural-sounding synthesized speech. Diphones are the transitions between phonemes. Demi-syllables are the half-syllables recorded from the beginning of a sound to the center point, or from the center point to the end of a sound (Weinschenk & Barker, 2000). After the voice units are recorded, then the database of units is coded for changes in frequency, pitch, and prosody (intonation and duration). The coding process enables the database of voice units to be as efficient as possible.

Formant synthesis is a rule-based process that creates machine-generated speech (see Figure 3.2). A set of phonological rules is applied to an audio waveform that simulates human speech. Formant synthesis involves two complex steps: the first covers the conversion of the input text into a phonetic representation, while the second encompasses the production of sound based on that phonetic representation. In the first step, the text is input from a database or file and is normalized so that any symbols or abbreviations are resolved as full alphabetic words. To convert the words into phonemes, a pronunciation dictionary is used for most words and a set of letter-to-sound

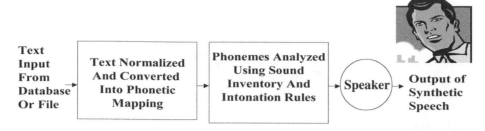

FIGURE 3.2. Model of text to speech synthesis.

rules is used for word exceptions not available in the dictionary. In the second step, the phonemes are analyzed using a sound inventory and intonation rules about pitch and duration. The speech synthesis is the resulting output heard by users through a speaker or headphone.

The quality of the synthetic speech should be evaluated along the lines of its acceptability, naturalness, and intelligibility. It is important to ask users to evaluate different speech against each other, because these qualities are always relative (Francis & Nusbaum, 399). Although these subjective differences in opinion will probably always exist, a study by Lai, Wood, and Considine (2000) showed that there were no significant differences in comprehension levels for longer messages (e.g., with a word length ranging from 100 to 500 words) among five major commercial text-to-speech engines.

Current Capabilities and Limitations of Speech-Synthesis Software

Formant synthesis produces speech that is highly intelligible, but sounds unnatural; however, it has the power to inexpensively produce nearly unlimited speech, from a resource point of view. The limitation of using formant synthesis is the complexity of the required linguistic rules to produce accurate speech output. Utilizing domain-specific information and assumptions produces a substantial improvement in the synthesizer's prosody. (*Prosody* refers to speech elements such as intonation, duration, pitch, rate, loudness, or rhythm.) Users will be able to comprehend the speech at a higher rate and will perceive the voice to be more natural.

Some applications of concatenated synthesis attempt to reduce costs by basing the voice recordings on whole words. Interactive voice systems (IVR) and voice mail systems that use concatenated speech synthesis based on voice recordings of whole words sound unnatural and unevenly paced, and this makes the synthetic voice hard to understand or remember. An application of concatenated synthesis needs to be done correctly or not at all.

How Do Natural Language Processing (NLP) and Natural Language Understanding (NLU) Work?

"Natural language processing (NLP)" refers to a wide range of processing techniques aimed at extracting, representing, responding to, and ultimately understanding the semantics of text.

Natural Language Understanding (NLU) is an area of NLP focused on the understanding of natural language text. It is the process of analyzing text and taking some action based on the meaning of the text. We include any technology that allows a user to communicate with a system using a language that is not rigidly structured (e.g, a "formal" language). The focus of this chapter is on systems in which communication between the user and the system has constructs similar in grammar and dialog to the language of everyday human-to-human communication.

As a technology, NLU is independent from speech recognition, although the combination of the two yields a powerful human-computer interaction (HCI) paradigm (see Figure 3.3). When combined with NLU, speech recognition transcribes an acoustic signal into text, which is then interpreted by an understanding component to extract meaning. In a conversational system, a dialog manager will then determine the appropriate response to give the user. Communication with the user can take place via a variety of modalities, including (a) speech input and output, (b) text input and output, (c) handwriting, or (d) some combination of these modalities. Figure 3.3 shows a block diagram of a prototypical multimodal conversational system that allows speech and keyboard natural language input, and speech and GUI text output.

NLU has been an active area of research for many decades. The promise of NLU lies in the "naturalness" of the interaction. Since humans have deep expertise in interacting with one another through language, the idea that leveraging a user's ability to interact using language will result in systems with greater usability has been an implicit and explicit hypothesis in a wide array of research studies and technology development efforts. Designing systems that use natural interaction techniques means the user is freed from learning the formal language of a system. Thus, instead of using formalisms (e.g., UNIX commands), scripting languages, or graphical menus and buttons, users can engage in a dialog with the system. Less user training, more rapid development of expertise and better error recovery are all promising aspects of systems that use NLU.

Dialog can be described as a series of related conversational interactions. It adds the richness of context and the knowledge of multiple interactions over time to the user interface, transforming natural language interfaces into conversational interfaces. Such interactions require the system to maintain a history of the interaction as well as the state of the interaction at all times (Chai, 2001). Two important components, which must be represented in any dialog system, are user goals and the current context of the interaction. The history of the interaction must be evaluated against the user goal, with prompts to the user designed to acquire the necessary information required to satisfy a goal. Dialog

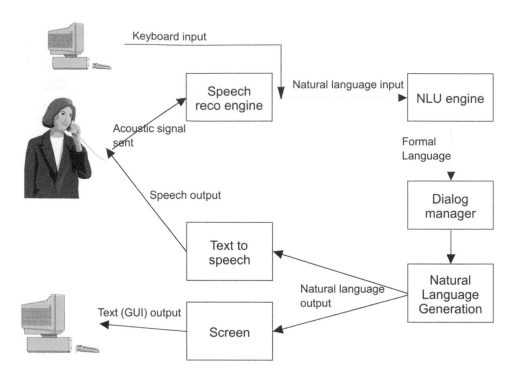

FIGURE 3.3. Diagram of a prototypical multimodal conversational system.

technology can be embedded in an application to enable either user-initiated, system-initiated, or mixed-initiative conversations (see descriptions of these different dialog styles following). The conversations are guided by dialog management technology.

Current Capabilities and Limitations of NLU Systems

Most NLU applications will easily run on the desktop and note-book computers commonly available in the year 2007. In addition, research prototype systems that move NLU functionality onto handheld systems are beginning to appear. Cars that accept natural voice commands to control a limited range of features are now available. NLU on the web is becoming more common (e.g., www.askjeeves.com, www.neuromedia.com), and illustrates client-server architecture for NLU. As computing power increases and digital technology pervades our everyday life, we can expect to see a dramatic increase in the use of NLU in new environments, supporting more everyday tasks.

Capturing and representing knowledge of a domain is a complex and labor-intensive process. As the scope of the application domain increases, it becomes increasingly difficult to build and maintain NLU applications. As a result, all successful examples of rich NLU interfaces have relatively narrow application domains. This has direct impact on the definition of user profiles for any given application. Simply stated, the narrower the conceptual, functional, syntactic, and lexical domains of the target user population, the greater the chance of building an NLU application that satisfies its users. There is currently no way to quantify and measure these characteristics, and determining whether a particular application represents a tractable problem in NLU is largely an issue of experience and instinct on the part of designers and

engineers. An NLU engine and speech input and output technologies are the building blocks for a conversational application. Now, let's turn our attention to the design of the application itself.

CRAFTING THE SPOKEN INTERACTION

The first step in designing a conversational interface is to educate yourself about the ways people speak in the domain of the task. The best approach is to find users who are doing an activity that is as close as possible to the target task. Keep in mind that the vocabulary and sentence structure used in graphical computer applications or in printed material may be quite different from the way people actually speak about a task. For example, imagine that you want to provide a speech interface to a calendar. Printed calendars and online calendars typically show day, week, and month views of calendar appointments. They always show days of the week and numbered dates. As soon as you listen to two people talk about appointments and scheduling, however, you learn that they only use numbered dates occasionally. The more common way to discuss dates with other people is to use relative descriptions: "What do you have a week from Monday? Are you busy this Thursday? What about the next day?" The concept of relative dates is completely absent from paper-based and online calendars; however, users will expect to be able to use them when speaking.

With a little creativity, you can find or create situations in which two people can talk about the target task. Although you might want to plan out the interactions a little bit, care must be taken not to use scripts or memorized phrases. The idea with this type of listening is to allow people's natural social instincts to drive the dialog. If possible, record the natural dialogs. They will serve as the basis for the speech interface design. Listening in this way is called a "natural dialogue study" (Yankelovich, in press). Using natural dialogue studies, you can learn (a) about concepts used when talking about the task that are absent in other media, (b) common vocabulary used when talking about the task, (c) tone of voice that is considered polite in the context of the task, (d) sentence structure typical for the domain patterns of interaction, and (e) methods people use to give one another feedback.

Dialogue Styles

Once you have an understanding of your task and the way people speak in this domain, you will need to decide if your application would benefit most from a directed, user-initiated, or conversational dialogue style.

Directed dialogue (system initiated). This is the most commonly used style of interaction in speech-based telephony systems on the market today. With a directed dialogue, the user is instructed or "directed" what to say at each prompt (Kamm, 394). Systems that use this style can often be recognized by the use of the word "say" in the prompts. "Welcome to ABC Bank. You can check an account balance, transfer funds, or pay a bill. Say, 'Balance,' 'Transfer,' or 'Pay.'" The reason most systems use this dialogue style is to increase the accuracy of the speech recognition. By telling the user what to say, the chances of the user speaking an out-of-vocabulary utterance are much lower. This style is sometimes referred to as "system initiated" because the user plays a more passive role, waiting for the system to prompt the user for what is needed.

User-initiated. Unlike a directed dialogue, which works well for novice users, user-initiated systems are intended for use by expert users. With this type of interaction style, the system is passive and the user is not prompted for specific information. For example,

Computer: How can I help you?
User: I would like to buy 100 shares of Amazon.com at the market price with funds from my current account number 34538.

These types of dialogues do not work well with new users who do not know what to say, and tend to have lower recognition rates.

Conversational systems (mixed initiative). The type of interaction that is easiest for users and feels the most natural is a mixed initiative system. In this style of interaction, the system and the user take turns initiating the communication depending on the flow of the conversation and the status of the task. With a mixed initiative system, the prompts can fluctuate between open-ended prompts and more directed prompts, depending on the context. Here's a portion of dialogue:

Computer: What would you like to do? (*Open-ended*)
User: What is the price of Fidelity Magellan? (*User-initiated*)

Computer: Price of Fidelity Magellan is $36.05 per share. What else?
User: I want to buy 500 shares. (*Which shares to buy is implied in the previous interaction.*)
Computer: Confirming purchase of 500 shares of Fidelity Magellan fund, please say, "Yes" or "No." (*Directed*)
User: No, skip the purchase. I'd like to transfer $1,000 from my index fund (*User-initiated*).
Computer: Purchase canceled. To which fund would you like to transfer? (*Prompt for missing information*)

In this example, we see the system functioning as a cooperative partner in the conversation.

Inherent Challenges of Speech

Designing conversational systems poses substantial challenges. Understanding these challenges and assessing the various trade-offs that must be made during the design process will help to produce the most effective interface.

By its nature, speech is *transient*. Once you hear it or say it, it's gone. By contrast, graphics are *persistent*. A graphical interface typically stays on the screen until the user performs some action. Listening to speech taxes users' short-term memory. Because speech is transient, users can remember only a limited number of items in a list, and they may forget important information provided at the beginning of a long sentence. Users' limited ability to remember transient information has substantial implications for the speech interface design. In general, transience means that speech is not a good medium for delivering large amounts of information. The transient nature of speech can also provide benefits. Because people can look and listen at the same time, speech is ideal for grabbing attention or for providing an alternate mechanism for feedback. Imagine receiving a notification about the arrival of an e-mail message while working on a spreadsheet. Speech might give the user the opportunity to ask for the sender or the subject of the message. The information can be delivered without forcing the user to switch contexts.

Speech is also *invisible*. The lack of visibility makes it challenging to communicate the functional boundaries of an application to the user (Yankelovich, 396). In a graphical application, menus and other screen elements make most or all of the functionality of an application visible to a user. By contrast, in a speech application, it is much more difficult to indicate to the user what actions they may perform, and what words and phrases they must say to perform those actions.

Prompt Design

The problems of transience and invisibility can often be mitigated by well-designed prompts. Many factors must be considered when designing prompts, but the most important is assessing the trade-off between flexibility and performance. The more you constrain what the user can say to an application, the less likely they are to encounter recognition errors. On the other hand, allowing users to speak information flexibly can often speed the interaction (if recognition succeeds), feel more natural, and avoid forcing users to memorize commands. Here are some tips for creating useful prompts.

Use *explicit prompts* when the user input must be tightly constrained. For example, after recording a message, the prompt might be "Say, 'Cancel,' 'Send,' or 'Review.'" This sort of prompt directs the user to say just one of those three keywords. Even conversational systems should fall back on this type of prompt when it is critical that the input be correctly recognized, as with transactions that involve transferring money or booking travel. "Are you sure you want to book this flight from JFK to LAX? Please say, 'Yes' or 'No.'"

Use *implicit prompts* when the application is able to accept input that is more flexible. These prompts rely on conversational conventions to constrain the user input. For example, if the user says, "Send mail to Bill," and "Bill" is ambiguous, the system prompt might be "Did you mean Bill Smith or Bill Jones?" Users are likely to respond with input such as "Smith" or "I meant Bill Jones." While possible, conversational convention makes it less likely that they would say, "Bill Jones is the one I want."

Using *variable prompts* is a good way to try to simulate a human-human conversation. Given a certain condition or state of the system (e.g., the ready state, or a system response to silence), it is preferable not to play the exact same system prompt every time. Subtle variations in the wording impart a much more natural feel to the interaction. Note the following possibilities for the ready state in a natural language understanding system: "What now?", "I'm ready to help," "What's next?"

Another interaction that we can model on human speech is the use of *tapered prompts*. Tapering can be accomplished in one of two ways. If an application is presenting a set of data such as current quotes for a stock portfolio, drop out unnecessary words once a pattern is established. For example, "As of 15 minutes ago, Acme Industries was trading at 45 up 1/2, BioStartup was at 83 up 1/2, and TechGiant was at 106 down 1/4." Tapering can also happen over time—that is, if you need to tell the user the same information more than once, make it shorter each time the function is used.

One way to speed interaction for expert users and provide help for less experienced users is to use *incremental prompts*. This technique involves starting with a short prompt. If the user does not respond within a time-out period, the application prompts again with instructions that are more detailed. For example, the initial prompt might be, "Which service?" If the user says nothing, then the prompt could be expanded to "Say 'Banking,' 'Address book,' or 'Yellow pages.'" Incremental prompts have been used successfully in a number of systems, but they suffer from several problems. A first-time user is just as likely to say something to the first prompt as they are to say nothing. This often results in a recognition error. Another common pitfall with incremental prompts is that they tend to cause collision errors where both the system and the user speak at the same time. It is common for timing problems to happen repeatedly once started, increasing the user's frustration along the way.

Providing Help

There are two basic types of help. In the first case, the user initiates by requesting help. Common queries for help in speech systems are, "Help", "What can I say?", or "What are my choices?" It is always a good idea to support several variations of this request in the grammar. In the second type of help, the system detects that the user is in trouble and automatically presents help. Although not all forms of user problems are detectable by the system, one can presume if a series of errors or long silences occur that the user probably needs some assistance.

Help, when given, needs to be specific to the user's current task. Few things are as frustrating as being stuck and asking for help only to find that the information presented, while long and detailed, has no bearing on the current situation. A good approach for help is to provide the user with an example of a successful utterance just as the user starts the downward spiral of repeated errors. If several errors have occurred in a row, the system should present the user with sample phrasing that would be expected to work, based on the context that he is in. For example, a user could say, "I want to sell 300 shares of Ultrasoft." If the system cannot determine the exact context, another option is to list the functions that are available at the current branch and have the user point the system in the right direction. Here's an example:

Computer:	Let's start over. You can forward, reply, delete, or create a new message. What do you want to do?
User:	I want to forward a message.
Computer:	To forward a message, say the first and last name of another subscriber. For example, say, "Forward this message to John Smith."
User:	Forward this message to Alishia Grant.

If the system returns a recognition error now, it is easy to establish that the name is at fault because we already know the function. This dialogue might continue:

Computer:	Oh dear. I didn't catch that name. Just say the first name.
User:	Alishia.
Computer:	I think you said, "Alice." Is that correct?
User:	No, that is not correct.
Computer:	Thanks for your patience with this. Please spell the first name.
User:	A-L-I-S-H-I-A.

Now the system can repeat a similar process for the last name, or can look up all subscribers with a first name of Alishia and offer the user a set of possible last names from which to choose.

Handling Errors

How a system handles recognition errors can dramatically affect the quality of a user's experience. If either the application or the user detects an error, an effective speech user interface should provide one or more mechanisms for correcting the error. Although this seems obvious, correcting a speech input error is not always easy! If the user speaks a word or phrase again, the same error could recur, depending on the cause of the error (Yankelovich, Levow, & Marx, 395).

Recognition errors can be divided into three categories: (a) rejection, (b) substitution, and (c) insertion (Schmandt, 394; Ballentine & Morgan, 399). A rejection error occurs when the recognizer has no hypothesis about what the user said. A substitution error involves the recognizer mistaking the user's utterance for a different valid utterance, as when "send a message" is interpreted as "seventh message." With an insertion error, the recognizer either interprets noise as a valid utterance, or decodes multiple words when only one was spoken. This can be caused by other people talking nearby or by the user inadvertently tapping the telephone or microphone.

Rejection errors.   In handling rejection errors, designers want to avoid the "brick wall" effect, when every rejection is met with the same "I didn't understand" response. Users get frustrated very quickly when faced with repetitive error messages. Instead, give *progressive assistance*: a short error message the first couple of times, and if errors persist, offer more detailed assistance. For example, here is one progression of error messages that a user might encounter: "Sorry?", "What did you say?", "Sorry. Please rephrase.", "I didn't understand. Speak clearly, but do not overemphasize.", "Still no luck. Wait for the prompt tone before speaking." Progressive assistance does more than bring the error to the user's attention; the user is guided toward speaking a valid utterance with successively more informative error messages that consider possible causes of the error.

Substitution errors.   Although rejection errors are frustrating, substitution errors can be damaging. If the user asks a weather application for "Kauai," but the recognizer hears "Goodbye" and hangs up, the interaction could be completely terminated. In situations like this, the system should explicitly verify that the user's utterance was correctly understood. Verification should be commensurate with the cost of the action that would be effected by the recognized utterance. Reading the wrong stock quote or calendar entry will make the user wait a few seconds, but hanging up or sending a confidential message to the wrong person by mistake could have serious consequences.

Insertion errors.   Spurious recognition typically occurs because of background noise. The illusory utterance will either be rejected or mistaken for an actual command; in either case, the previous methods can be applied. The real challenge is to prevent insertion errors. One option is to provide users with a keypad command to turn off the speech recognizer to talk to someone, sneeze, or simply gather their thoughts. Pressing the keypad command again can restart the recognizer with a simple prompt, such as "What now?" to indicate that the recognizer is listening again.

Whatever the type of error, a general technique for avoiding errors in the first place is to filter recognition results for unlikely user input. For example, a scheduling application might assume that an error has occurred if the user appears to want to schedule a meeting for three o'clock in the morning.

Correction Strategies

If errors do occur, it is important to provide a means for the user to correct the error (assuming they notice it). Flexible correction mechanisms that allow a user to correct a portion of the input are helpful. For example, if the user asks for a weather forecast for Boston for Tuesday, the system might respond, "Thursday's weather for Boston is . . . ." A flexible correction mechanism would allow the user to just correct the day: "No, I said Tuesday."

When possible, using an alternate form of input, such as the telephone keypad, can alleviate the user's frustration. If the user is at a prompt where three choices are available and the user has encountered several rejection errors, the user could be instructed: "Press any key when you hear the option you want." The telephone keypad also works well when the requested input is numeric (e.g., telephone numbers, account or social security numbers). Getting users to type alphabetic text using a telephone keypad is not a good idea; speech systems are usually recommended to avoid this type of input in the first place.

Another strategy is to have the system take its best guess at the requested function. This is a good tactic to take when the number of functions enabled at a particular branch in the dialogue is too large to list for the user. A reasonable prompt is, "I think you are trying to create a message, is that correct?" If the user answers in the affirmative, the conversation moves forward and the system can present the user with a sample valid utterance for that function. If the best guess is wrong, however, it is not a good idea to keep iterating through the N-best choices because this only leads to user frustration. If the response to the best guess is negative, a better solution is to reprompt with a restricted set of choices. Be sure to eliminate the choice that is definitely wrong (e.g., creating a message in the previous example). The goal is to move away from a very general prompt such as, "I'm sorry, I do not understand. Please try again" towards a directed prompt that will increase the likelihood of success. A series of errors in a row is a clear indication that simply having the user repeat the utterance, or rephrase it, is not working.

The best-guess tactic can be combined with another correction strategy, switching to more constrained grammar, to increase its likelihood of success. For example, in the prior prompt, a directive of what utterances are available to the user can be added: "I think you are trying to create a message. Is that correct? Please say, 'Yes' or 'No.'"

TESTING AND ITERATING

Once the preliminary application design is complete, a Wizard of Oz study can help test and refine the interface. The speech data collected from this study can be used to help refine the grammar (Rudnicky, 395) or, in the case of an NLU system, to statistically train the engine on potential utterances. In these studies, a human wizard—often using software tools—simulates the speech interface. Major usability problems are often uncovered with these types of simulations.

A Wizard of Oz study usually involves bringing participants into a lab and telling them they will be interacting with a computer. If it is a telephony application, they can be asked to call a telephone number, at which point a human "wizard" picks up the phone and manipulates software so that recordings of synthesized speech are played to the participant. As the participant makes requests to the computer, the wizard carries out the operations and has the computer speak the responses. Often, none of the participants suspect that they are not interacting with a real speech system (Dahlback, 393). Because computer tools are usually necessary to carry out a convincing simulation, Wizard of Oz studies are more time-consuming and complicated to run than natural dialogue studies. If a prototype of the final application can be built quickly, it may be more cost-effective to move directly to a usability study.

With speech applications, usability studies are particularly important for uncovering problems because of recognition errors, which are difficult to simulate effectively in a Wizard of Oz study, but are a leading cause of usability problems. The effectiveness of an application's error recovery functionality must be tested in the environments in which real users will use the application. Conducting usability tests of speech applications can be a bit tricky. Two standard techniques used in tests of graphical applications—(a) facilitated discussions and (b) speak-aloud protocols—cannot be used effectively with speech applications. It is best to have study participants work in isolation, speaking only into a telephone or microphone. A tester should not intervene unless the participant becomes completely stuck. A follow-up interview can be used to collect the participant's comments and reactions. Ultimately, the system needs to be tested with real users, accomplishing real tasks in a realistic environment. This type of testing will provide data that will allow the designer to modify the grammar and prompts as necessary.

CONCLUSION

The design and development life cycle for an effective speech application differs from a traditional GUI application. An effective speech application is one that uses speech to enhance a user's performance of a task or enable an activity that cannot be done without it; however, the design process does share common elements with the design of a successful GUI application, such as the need to first understand the task that the users are trying to accomplish. In the case of speech, modeling the task can usually be accomplished by listening to how humans accomplish the task today. Sometimes this involves listening to the interactions between a call center employee and a caller; sometimes it involves constructing a situation between two humans and asking them to converse to accomplish the task. Observing users during the task-modeling phase helps a designer to understand who the users are and what their goals are. Listening carefully to users while conducting natural dialogue studies shows how they speak in the context of the task. A natural dialogue study also ensures that prompts and feedback follow the conversational conventions that users expect in a cooperative interaction.

Once the task is modeled and well understood, and the business goals have been defined, the designer/developer must select the appropriate technology for the task, users and context of use. Conversational interfaces that use NLU technologies capitalize on human expertise in interacting with each other through language. Not all tasks will be able to use NLU, given the resource requirements and the need for a constrained domain; however, even speech applications that are built with grammars can carry many of the usability advantages of NLU and have a conversational feel to them, if the prompts are carefully crafted. Successful dialogs will move between user-initiated, system-initiated, and mixed-initiative styles depending on the state of the task and the history of the interaction.

Another key to the success of the design of an effective speech user interface is to follow an iterative process. Not even the most experienced designer will craft a perfect dialogue in the first iteration. Once an application is designed, Wizard of Oz and usability studies provide opportunities to test interaction techniques and refine application behavior based on feedback from prototypical users. To ensure that the system will be used over time, the designer must (a) focus on users and (b) modify the design as new data is collected.

Finally, testing the design with target users ensures that the prompts are clear, that feedback is appropriate, and that errors are caught and corrected. In addition, as part of the testing, verifying that the design accomplishes the business goals that were set out to be achieved helps the application gain approval from the sponsors. If problems are uncovered during testing, the design should be revised and tested again. By focusing on users and iterating on the design, one can produce an effective, polished speech interface design.

References

Ballentine, B., & Morgan, D. (399). *How to build a speech recognition application: A style guide for telephony dialogues.* San Ramon, CA: Enterprise Integration Group, Inc.

Chai, J. (2001). *Natural language sales assistant—a Web-based dialog system for online sales.* To be presented at the 13th Conference on Innovative Applications of Artificial Intelligence.

Clark, H. (393). *Arenas of language use.* Chicago: University of Chicago Press.

Cohen, M., Giangola, J., & Balogh, J. (2004). *Voice user interface design.* Boston: Addison-Wesley.

Dahlback, N., Jonsson, A., & Ahrenberg, L. (393). Wizard of Oz studies—why and how. *Proceedings of the Third Conference on Applied Natural Language Processing,* Trento, Italy, April 393.

Feng, J., Sears, A., & Karat, C. (2006). A longitudinal evaluation of hands-free speech-based navigation during dictation. *International Journal of Human Computer Studies, 64,* 6, 553–569.

Francis, A., & Nusbaum, H. (399). Evaluating the quality of synthetic speech. In D. Gardner-Bonneau (Ed.), *Human factors and voice interactive systems* (pp. 25–47). Boston: Kluwer Academic Publishers.

Halverson, C. A., Horn, D. A., Karat, C., & Karat, J. (399). The beauty of errors: Patterns of error correction in desktop speech systems. In M. A. Sasse & C. Johnson (Eds.), *Human-computer interaction—INTERACT '99* (pp. 133–140). Edinburgh: IOS Press.

Kamm, C. (394). User interfaces for voice applications. In D. B. Roe & J. G. Wilpon (Eds.), *Voice communication between humans and machines* (pp. 422–442). Washington, DC: National Academy Press.

Karat, C., Horn, D., Halverson, C., & Karat, J. (399). Patterns of entry and correction in large vocabulary continuous speech recognition systems. In M. Altom & M. Williams (Eds.), *Human Factors in Computing Systems—CHI 99 Conference Proceedings* (pp. 568–576). New York: ACM.

Karat, J., Horn, D., Halverson, C., & Karat, C. (2000). Overcoming unusability: Developing efficient strategies in speech recognition systems. In Karat, C.-M., Lund, A., Coutaz, J., & Karat, J. (Eds), *Human Factors in Computing Systems—CHI 2000 Conference Proceedings* New York: ACM.

Karat, J., Lai, J., Danis, C., & Wolf, C. (399). Speech user interface evolution. In D. Gardner-Bonneau (Ed.), *Human factors and voice interactive systems* (pp. 1–35). Boston: Kluwer.

Lai, J., Wood, D., & Considine, M. (2000). The effect of task conditions on the comprehensibility of speech. *Human Factors in Computing Systems—CHI 2000 Conference Proceedings.* The Hague, Netherlands.

Price, K., & Sears, A. (2005). Speech-based text entry for mobile handheld devices: An analysis of efficacy and error correction techniques for server-based solutions. *International Journal of Human-Computer Interaction, 3*(3), 279–304.

Price, K. J., Lin, M., Feng, J., Goldman, R., Sears, A., & Jacko, J. A. (in press). Motion does matter: An examination of speech-based text entry on the move. *Universal Access in the Information Society.*

Rabiner, L. R. (389). A tutorial on hidden Markov models and selected applications in speech recognition. *Proceedings of IEEE, 77,* 257–286.

Roe, D. B., & Wilpon, J. G. (393). Wither speech recognition: The next 25 years. *IEEE Communications Magazine, 11,* 54–62.

Rudnicky, A. (395). The design of spoken language interfaces. In A. Syrdal, R. Bennett, & S. Greenspan (Eds.), *Applied speech technology* (pp. 18–39). CRC Press.

Schmandt, C. (394). *Voice communication with computers: Conversational systems.* New York: Van Nostrand Reinhold.

Sears, A., Karat, C., Oseitutu, K., Kaimullah, A., & Feng, J. (2000). Productivity, satisfaction, and interaction strategies of individuals with spinal cord injuries and traditional users interacting with speech recognition software. *Universal Access in the Information Society, 1,* 5–25.

Weinshenk, S., & Barker, D. T. (2000). *Designing effective speech user interfaces.* New York: Wiley.

Yankelovich, N., Levow, G. A., & Marx, M. (395, May 7–11). Designing SpeechActs: Issues in speech user interfaces. *Proceedings of the '95 Conference on Human Factors in Computing Systems,* Denver, CO.

Yankelovich, N. (396, November/December). How do users know what to say? *ACM Interactions, 3* (pp. 32–43).

Yankelovich, N. (in press). Using natural dialogs as the basis for speech interface design. In D. Gardner-Bonneau & H. E. Blanchard (Eds.), *Human factors and voice interactive systems* (2nd ed.). Springer.

•4•

MULTIMEDIA USER INTERFACE DESIGN

Alistair Sutcliffe
Manchester Business School
University of Manchester

Definitions and Terminology 66
Cognitive Background 67
   Perception and Comprehension 67
   Selective Attention 67
   Emotion and Arousal 68
   Learning and Memorization 68
Design Process 69
   Users, Requirements, and Domains 70
   Information Architecture 70
   Media Selection and Combination 72
      Media selection 73
      Engagement and attractiveness 75

Aesthetic design 75
Image and identity 76
Affective effects 76
Interaction and Navigation 77
   Metaphors and interaction design 77
   Navigation 78
Design for Attention 79
   Still image media 81
   Moving image media 81
   Linguistic media (text and speech) 81
Conclusions **82**
References **82**

Design of multimedia interfaces currently leaves a lot to be desired. As with many emerging technologies, it is the fascination with new devices, functions, and forms of interaction that has motivated design rather than ease of use, or even utility of practical applications. Poor usability limits the effectiveness of multimedia products that might look good, but do not deliver effective use (Scaife, Rogers, Aldrich, & Davies, 1997). The multimedia market has progressed beyond the initial hype, and customers are looking for well-designed, effective, and mature products.

The distinguishing characteristics of multimedia are information-intensive applications that have complex design spaces for presenting information to people. Designers, therefore, must start by modeling information requirements. This chapter describes a design process that starts with an information analysis and then progresses to deal with issues of media selection and integration. The background to the method and its evolution with experience can be found in several publications (Sutcliffe & de Angeli, 405; Faraday & Sutcliffe, 1996, 1997b, 1998b; Sutcliffe & Faraday, 1994). A more detailed description is given in Sutcliffe (403). The time-to-market pressure gives little incentive for design, so at first reading, a systematic approach may seem to be counter to the commercial drivers of development; however, I would argue that if multimedia design does not adopt a usability engineering approach, it will fail to deliver effective and usable products.

Multimedia applications have significant markets in education and training, although dialogue in many systems is restricted to drill and quiz interaction and simple navigation. This approach, however, is oversimplified: for training and education, interactive simulations and microworlds are more effective (Rogers & Scaife, 1998). Multimedia has been used extensively in task-based applications in process control and safety critical systems (Alty, 1991; Hollan, Hutchins, & Weitzman, 1984); however, most transaction processing applications are currently treated as standard interfaces rather than multimedia-based designs. With the advent of the web and e-commerce, this view may change.

Design issues for multimedia user interfaces (UIs) expand conventional definitions of usability (e.g., International Standards Organization, 1997) into five components:

1. *Operational usability* is the conventional sense of usability that concerns design of graphical user interface features such as menus, icons, metaphors, and navigation in hypermedia.
2. *Information delivery* is a prime concern for multimedia or any information-intensive application, and raises issues of media selection, integration, and design for attention.
3. *Learning:* Training and education are both important markets for multimedia; hence, learnability of the product and its content are key quality attributes. Design of educational technology, however, is a complex subject in its own right, and multimedia is only one part of the design problem.
4. *Utility:* In some applications, utility will be the functionality that supports the user's task; in others, information delivery and learning will represent the value perceived by the user.
5. *Engagement and attractiveness:* The attractiveness of multimedia is now a key factor, especially for websites. Multi-

media interfaces have to attract users and deliver a stimulating user experience, as well as being easy to use and learn.

Multimedia design involves several specialisms that are technical subjects in their own right. For instance, design of text is the science (or art) of calligraphy that has developed new fonts over many years; visualization design encompasses the creation of images, either drawn or captured as photographs. Design of moving images, cartoons, video, and film are further specializations, as are musical composition and design of sound effects. Multimedia design lies on an interesting cultural boundary between the creative artistic community and science-based engineering. One implication of this cultural collision is that space precludes "within media" design, or guidelines for design of one particular medium, which is dealt with in depth in this chapter. Successful multimedia design often requires teams of specialists who contribute from their own skill sets (Kristof & Satran, 1995; Mullet & Sano, 1995).

DEFINITIONS AND TERMINOLOGY

Multimedia essentially extends the graphical user interface (GUI) paradigm by providing a richer means of representing information for the user by use of image, video, sound, and speech. Some views of what constitutes multimedia can be found in Bernsen (1994), who proposed a taxonomy of analog versus discrete media, which he called modalities, as well as visual, audio, and tactile dimensions. Heller and Martin (1995) took a more conventional view of classifying image, text, video, and graphics for educational purposes. The following definitions broadly follow those in the ISO standard 14915 on Multimedia User Interface Design (International Standards Organization, 1998). The starting point is to ask about the difference between what is perceived by someone and what is stored on a machine.

Communication concepts in multimedia can be separated into:

- *Message:* The content of communication between a sender and receiver.
- *Medium* (plural, *media*): The means by which that content is delivered. Note that this is how the message is represented rather than the technology for storing or delivering a message. There is a distinction between perceived media and physical media such as CD-ROM, hard disk, and so forth.
- *Modality:* The sense by which a message is sent or received by people or machines. This refers to the senses of vision, hearing, touch, smell, and taste.

A message is conveyed by a medium and received through a modality. A modality is the sensory channel that we use to send and receive messages to and from the world—essentially, our senses. Two principal modalities are used in human-computer communication:

1. *Vision:* All information received through our eyes, including text and image-based media.
2. *Hearing:* All information received through our ears, as sound, music, and speech.

In the future, as multimedia converges with Virtual Reality (VR), we will use other modalities more frequently: *haptic* (sense of touch), *kinesthetic* (sense of body posture and balance), *gustation* (taste) and *olfaction* (smell). These issues are dealt with in chapters 5, Multimodal interfaces (Oviatt, Darves, & Coulston, 404), and 15, Virtual environments (Stanney).

Defining a medium is not simple because it depends on how it was captured in the first place, how it was designed, and how it has been stored. For example, a photograph can be taken on film, developed, and then scanned into a computer as a digitized image. The same image may have been captured directly by a digital camera and sent to a computer as an e-mail file. At the physical level, media may be stored by different techniques.

Physical media storage has usability implications for the quality of image and response time in networked multimedia. A screen image with 640×480 VGA resolution using 24 bits per pixel for good color coding gives 921,600 bytes, so at 30 frames per second, one second needs around 25 megabytes of memory or disk space. Compression algorithms, for instance, MPEG (Moving Pictures Expert Group), reduce this by a factor of 10. Even so, storing more than a few minutes of moving image consumes megabytes. The usability trade-off is between the size of the display footprint (e.g., window size), the resolution measured in dots per inch, and the frame rate. The ideal might be full-screen high resolution (600 dpi) at 30 frames per second; with current technology, a 10-cm window at 300 dpi and 15 frames per second is more realistic. Physical image media constraints become more important on networks, when bandwidth will limit the desired display quality. Sound, in comparison, is less of a problem. Storage demands depend on the fidelity required for replay. Full stereo with a complete range of harmonic frequencies consumes only 100 kilobytes for five minutes, so there are few technology constraints on delivery of high-quality audio.

COGNITIVE BACKGROUND

The purpose of this section is to give a brief overview of cognitive psychology as it affects multimedia design. More details can be found in section II, Humans in HCI.

Perception and Comprehension

Generally, our eyes are drawn to moving shapes, then complex, different, and colorful objects. Visual comprehension can be summarized as "what you see depends on what you look at and what you know." Multimedia designers can influence what users look at by controlling attention with display techniques such as use of movement, highlighting and salient icons; however, designers should be aware that the information people assimilate from an image also depends on their internal motivations, what they want to find and how well they know the domain (Treisman, 1988). A novice will not see interesting plant species in a tropical jungle, whereas a trained botanist will. Selection of visual content therefore has to take the user's knowledge and task into account.

Because the visual sense receives information continuously, it is overwritten in working memory (Baddeley, 1986). This means that memorization of visually transmitted information is not always effective unless users are given time to view and comprehend images. Furthermore, users only extract very high-level or "gist" (general sense) information from moving images. Interpreting visual information relies on memory. In realistic images, this process is automatic; however, with nonrealistic images, we have to think carefully about the meaning, for example, to interpret a diagram. While extraction of information from images is rapid, it does vary according to the complexity of the image and how much we know about the domain. Sound is a transient medium, so unless it is processed quickly, the message can be lost. Even though people are remarkably effective at comprehending spoken language and can interpret other sounds quickly, the audio medium is prone to interference because other sounds can compete with the principal message. Because sound is transient, information in speech will not be assimilated in detail, and so only the gist will be memorized (Gardiner & Christie, 1987).

Selective Attention

We can only attend to a limited number of inputs at once. While people are remarkably good at integrating information received by different senses (e.g., watching a film and listening to the sound track), limits are determined by the psychology of human information processing (Wickens, Sandry, & Vidulich, 1983). Our attention is selective and closely related to perception; for instance, we can overhear a conversation in a room with many people speaking (the cocktail party effect). Furthermore, selective attention differs between individuals and can be improved by learning. For example, a conductor can distinguish the different instruments in an orchestra, whereas a typical listener cannot. However, all users have cognitive resource limitations, which means that information delivered on different modalities (e.g., by vision and sound) have to compete for the same resource. For instance, speech and printed text both require a language-understanding resource, while video and a still image use image-interpretation resources. Cognitive models of information processing architectures (e.g., Interacting Cognitive Subsystems; Barnard, 1985) can show that certain media combinations will not result in effective comprehension because they compete for the same cognitive resources, thus creating a processing bottleneck. We have two main perceptual channels for receiving information: (a) vision and (b) hearing. Information going into these channels has to be comprehended before it can be used. Figure 4.1, based on Card, Moran, and Newell's (1983) Model Human Processor, shows the cognitive architecture of human information processing and resource limitations that lead to multimedia usability problems.

Capacity overflow may happen when too much information is presented in a short period, swamping the user's limited working memory and cognitive processor's capability to comprehend, chunk, and then memorize or use the information. The connotation is to give users control over the pace of information delivery. Integration problems arise when the message on two media is different, making integration into working

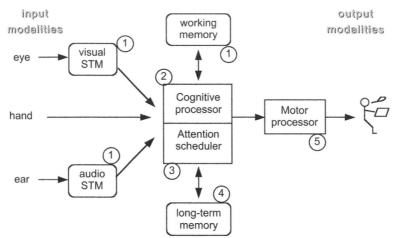

Bottlenecks
1. Capacity overflow: information overload
2. Integration: common message?
3. Contention: conflicting channels
4. Comprehension
5. Multi-tasking input/output

STM = short term memory
Similar to working memory
limited capacity,
continuously overwritten
with new input

FIGURE 4.1. Approximate model of human information processing using a "human as computer system" analogy, based on the Model Human Processor (Card, Moran, & Newell, 1983).

memory difficult; this leads to the thematic congruence principle. Contention problems are caused by conflicting attention between dynamic media, and when two inputs compete for the same cognitive resources, e.g., speech and text both requiring language understanding at once. Comprehension is related to congruence; we understand the world by making sense of it with our existing long-term memory. Consequently, if multimedia content is unfamiliar, we cannot make sense of it. Finally, multitasking makes further demands on our cognitive processing, so we will experience difficulty in attending to multimedia input while performing output tasks.

Making clear a theme in a multimedia presentation involves directing the user's reading and viewing sequence across different media segments. Video and speech are processed in sequence, while text enforces a serial reading order by the syntactic convention of language. In contrast, viewing image media is less predictable since it depends on the size and complexity of the image, the user's knowledge of the contents, task, and motivation (Norman & Shallice, 1986), and the designed effects for salience. Attention-directing effects can increase the probability that the user will attend to an image component, although no guarantee can be given that a component will be perceived or understood.

Emotion and Arousal

The content of image media in particular can evoke an emotional response, which can be used to promote a more exciting and engaging user experience. These issues are dealt with more extensively in other chapters. People treat human photographs, video and even animated characters with similar social responses as they give to real people, so human image content can be used to increase interest and draw attention. Emotional responses can be invoked not only by content, but also by surprise interactive effects (e.g., when a character suddenly appears to challenge the users). Surprise effects, moving image, and unexpected stimulating images all affect the arousal system, which broadly equates with our feeling of excitement. Designs that stimulate our arousal are more likely to be memorable and engaging.

Learning and Memorization

Learning is the prime objective in tutorial multimedia. In these applications, the objective is to create a rich memory schema, which can be accessed easily in the future. We learn more effectively by active problem solving, or learning by doing. This approach is at the heart of constructivist learning theory (Papert, 1980), which has connotations for tutorial multimedia. Interactive microworlds where users learn by interacting with simulations, or constructing and testing the simulation, give a more vivid experience that forms better memories (Rogers & Scaife, 1998). Multiple viewpoints help to develop rich schemata by presenting different aspects of the same problem, so the whole concept can be integrated from its parts. An example might be to explain the structure of an engine, then how it operates, and finally, display a causal model of why it works. Schema integration during memorization fits the separate viewpoints together.

The implications from psychology are summarized in the form of multimedia design principles that amplify and extend those proposed for general user interface design (e.g., ISO 9241 part 10; International Standards Organization, 1997). The principles are high-level concepts that are useful for general guidance, but they have to be interpreted in a context to give more specific advice.

- *Thematic congruence:* Messages presented in different media should be linked together to form a coherent whole. This helps comprehension as the different parts of the message make sense when fitted together. Congruence is partly a matter of designing the content so it follows a logical theme—for example, the script or story line makes sense and does not assume too much about the user's domain knowledge—and partly a matter of attentional design to help the user follow the message thread across different media (Ayers & Sweller, 405).

- *Manageable information loading:* Messages presented in multimedia should be delivered either at a pace that is under the user's control, or at a rate that allows for effective assimilation of information without causing fatigue. The rate of information delivery depends on the quantity and complexity of information in the message, the effectiveness of the design in helping the user extract the message from the media, as well as the user's domain knowledge and motivation. Two ways to reduce information overload are to (a) avoid excessive use of concurrent dynamic media and (b) give the user time to assimilate complex messages.

- *Ensure compatibility with the user's understanding:* Media should be selected that convey the content in a manner compatible with the user's existing knowledge (e.g., the radiation symbol and road sign icons are used to convey hazards and dangers to users who have the appropriate knowledge and cultural background). The user's ability to understand the message is important for designed image media (e.g., diagrams, graphs) when interpretation is dependent on the user's knowledge and background.

- *Complementary viewpoints:* Similar aspects of the same subject matter should be presented on different media to create an integrated whole. Showing different aspects of the same object—for example, picture and design diagram of a ship—can help memorization by developing richer schema and better memory cues.

- *Consistency:* Consistency helps users learn an interface by making the controls, command names, and layout follow a familiar pattern. People recognize patterns automatically, so operating the interface becomes an automatic skill. Consistent use of media to deliver messages of a specific type can help by cueing users with what to expect.

- *Interaction and engagement:* Interaction and engagement help understanding and learning by encouraging the user to problem solve. Memory is an active process. Interaction increases arousal and this makes the user's experience more vivid, exciting, and memorable.

- *Reinforce messages:* Redundant communication of the same message on different media can help learning. Presentation of the same or similar aspects of a message helps memorization by the frequency effect. Exposing users to the same thing in a different modality also promotes rich memory cues.

DESIGN PROCESS

Multimedia design has to address the problems inherent in the design of any user interface, viz. defining user requirements, tasks, dialogue design; however, three issues specifically concern multimedia:

1. *Matching the media to the message* by selecting and integrating media so the user comprehends the information content effectively.
2. *Managing users' attention* so key items in the content are noticed and understood, and the user follows the message thread across several media.
3. *Interaction and navigation* so the user can access, play, and interact with media in an engaging and predictable manner.

Figure 4.2 gives an overview of the design process that addresses these issues.

The method shown in the figure starts by requirements and information analysis to establish the necessary content and communication goals of the application. It then progresses to domain and user characteristic analysis to establish a profile of the user and the system environment. The output from these stages feeds into media selection and integration, which match the logical specification of the content to available media resources. This is interleaved with interaction design, unless the application is restricted to information presentation. Design then progresses to thematic integration of the user's reading/viewing sequence and design to direct the users' attention. The method can be tailored to fit within different development approaches. For

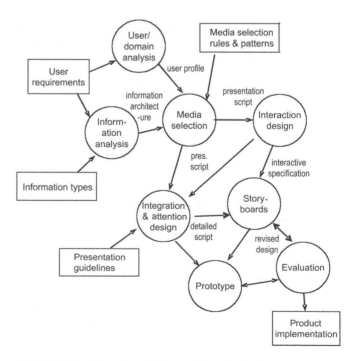

FIGURE 4.2. Overview of the multimedia design process expressed as a data flow diagram.

instance, in rapid application development, storyboards, prototypes, and iterative build-and-evaluate cycles would be used. On the other hand, in a more systematic, software engineering approach, more detailed specifications and scripts will be produced before design commences. Even though the process is described as a sequence, in practice, the stages are interleaved and iterated; however, requirements, information modeling, and media selection should be carried out, even if they are not complete, before subsequent design stages commence.

Design approaches in multimedia tend to be iterative and user-centered. *Storyboards* are a well-known means of informal modeling in multimedia design (Nielsen, 1995; Sutcliffe, 1999). Originating from animation and cartoon design, storyboards are a set of images that represent key steps in a design. Translated to software, storyboards depict key stages in interaction and are used for conducting walkthroughs to explain what happens at each stage. Allowing the users to edit storyboards and giving them a construction kit to build their own encourages active participation. Storyboards are followed by building concept demonstrators using multimedia authoring tools (e.g., Macromedia Director, Toolbook) to rapidly develop early prototypes. *Concept demonstrators* are active simulations that follow a scenario script of interaction; departure from the preset sequence is not allowed. Several variations can be run to support comparison; however, the user experience is passive. In contrast, users can test *interactive prototypes* by running different commands or functions. The degree of interactivity depends on the implementation cost, which increases as prototypes converge with a fully functional product.

Users, Requirements, and Domains

The starting point for multimedia, as in all applications, is requirements analysis. The difference in multimedia lies in the greater emphasis on information requirements. A variety of analytic approaches can be adopted, such as task-analysis, contextual inquiry, or scenario analysis. Requirements are listed and categorized into information, task-related, and nonfunctional classes. These will be expanded in subsequent analyses.

It is important to get a profile of the target user population to guide media selection. There are three motivations for user analysis:

1. *Choice of modalities,* which is important for people with disabilities, but also for user preferences. Some people prefer verbal-linguistic material over image.
2. *Tuning the content* presented to the level of users' existing knowledge, which is particularly important for training and educational applications.
3. *Capturing users' expectations* so the experience can be geared toward their backgrounds (e.g., different styles for younger, older people, cultures, and socioeconomic audiences).

Acquiring information about the level of experience possessed by the potential user population is important for customization. User profiles are used to design training applications to ensure

that the right level of tutorial support is provided, and to assess the users' domain knowledge so that appropriate media can be selected. This is particularly important when symbols, designed images, and diagrammatic notations may be involved. The roles and backgrounds of users will have an important bearing on design. For example, marketing applications will need simple, focused content and more aesthetic design, whereas tutorial systems need to deliver detailed content. Information kiosk applications need to provide information, as do task-based applications, but decision support and persuasive systems (Fogg, 1998) also need to ensure users comprehend and are convinced by messages. Domain knowledge, including use of conventions, symbols, and terminology in the domain, is important because less experienced users will require that more complete information is presented.

The context and environment of a system will also have an important bearing on design. For example, tourist information systems in outdoor public areas will experience a wide range of lighting conditions, which can make image and text hard to read. High levels of ambient noise in public places or factory floors can make audio and speech useless. Hence, it is important to gather information on (a) the location of use (e.g., office, factory floor, public/private space, and hazardous locations), (b) pertinent environmental variables (e.g., ambient light, noise levels, and temperature), (c) usage conditions (e.g., single user, shared use, and broadcast) and (d) expected range of locations (e.g., countries, languages, and cultures). Choice of language, icon conventions, interpretation of diagrams, and choice of content all have a bearing on design of international user interfaces.

As well as gathering general information about the system's context of use, domain modeling can prove useful for creating the system metaphor. Domain models are recorded as sketches of the work environment, showing the layout and location of significant objects and artifacts, accompanied by lists of environmental factors. Structural metaphors for organizing information and operational metaphors for controls and devices have their origins in domain analysis.

Information Architecture

This activity consists of several activities that will differ according to the type of application. Some applications might have a strong task model (e.g., a multimedia process control application where the tasks are monitoring a chemical plant, diagnosing problems, and supporting the operator in controlling plant operation). In task-driven applications, information requirements are derived from the task model. In information provision applications, such as websites with informative roles, information analysis involves categorization, and the architecture generally follows a hierarchical model. In the third class of explanatory or thematic applications, analysis is concerned with the story or argument (e.g., how the information should be explained or delivered). Educational multimedia and websites with persuasive missions fall into the last category.

In task-driven applications, information needs are annotated on the task model following a walkthrough that asks users what information they need to (a) complete the task subgoal, (b) take a decision at this step, or (c) provide as input (see Sutcliffe,

1997). In information provision applications, classification of the content according to one or more user views defines the information architecture; for example, most university departments have an information structure with upper-level categories for research, undergraduate courses, postgraduate courses, staff interests, departmental organization, mission and objectives, and so forth. A theme or story line should be developed for explanatory applications. This will depend on the application's objectives and the message the owner wishes to deliver. An example thematic map from a hypothetical health awareness application is illustrated in Fig. 4.3.

The requirement is to convince people of the dangers of heart disease. The theme is a persuasive argument that first tries to convince people of the dangers (e.g, smoking, poor diet, stressed lifestyles, etc.), and then convince them of ways to improve their lifestyles to prevent heart disease, followed by reinforcing the message with the benefits of a healthy lifestyle (e.g., lower health insurance, saving money, longer life, etc.). Subthemes are embedded at different points so users can explore the facts behind heart disease, the statistics and their exposure, how to get help, and so forth. Information is then gathered for each node in the thematic map. How this architecture will be delivered depends on interaction design decisions: it could become an interactive story to explore different lifestyle choices, combined with a quiz. The outcome of information architecture analysis will be an information-enhanced task model, a thematic map, or a hierarchy/network to show the structure and relationships of information categories. The next step is to analyze the information content by classifying it by types.

Information types are amodal, conceptual descriptions of information components that elaborate the content definition. Information types specify the message to be delivered in a multimedia application and are operated on by mapping rules to select appropriate media resources. The following definitions are based on the Task-Based Information Analysis Method (Sutcliffe, 1997) and ISO 14915, part 3 (International Standards Organization, 400).

The information types are used in "walkthroughs", in which the analyst progresses through the task/scenario/use case asking questions about information needs. This can be integrated with data modeling (or object/class modeling), so that the information in objects and their attributes can be categorized by the following types, using the decision tree shown in Fig. 4.4. The first question is whether information represents concrete facts about the real world or more abstract, conceptual information; this is followed by questions about the information that relates to change in the world or describes permanent states. Finally, the decision tree gives a set of ontological categories to classify information that expand on type definitions commonly found in software engineering specifications. More complex ontologies are available (Arens, Hovy, & Van Mulken, 1993; Mann & Thompson, 1988), so the classification presented in Fig. 4.4 is a compromise between complexity and ease of use. A finer-grained classification enables more finely tuned media selection decisions, but at the cost of more analysis effort.

Components are classified by "walking through" the decision tree using the definitions and the following questions:

- Is the information contained in the component physical or conceptual?
- Is the information static or dynamic—for example, does it relate to change or not?
- Which type in the terminal branch of the tree does the information component belong to?

It is important to note that one component may be classified with more than one type; for instance, instructions on how to get to the railway station may contain procedural information (the instructions "turn left," "straight ahead," etc.), and spatial or descriptive information ("the station is in the corner of the square, painted blue"). The information types are "tools for thought" and can be used either to classify specifications of content or to consider what content may be necessary. To illustrate, for the task "navigate to the railway station," the content may be minimally specified as "instructions how to get there," in which case the information types prompt questions in the form "what sort of information does the user need to fulfil the task/user goal?" Alternatively, the content may be specified as a scenario narrative of directions, waymarks to recognize and description of the target. In this case, the types classify components in the

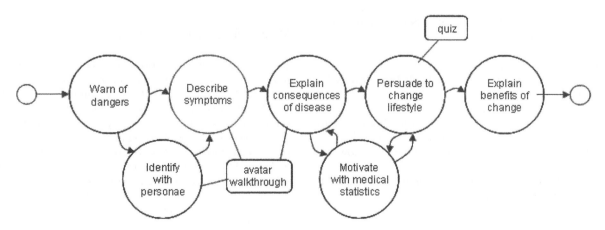

FIGURE 4.3. Thematic map for a healthcare promotion application.

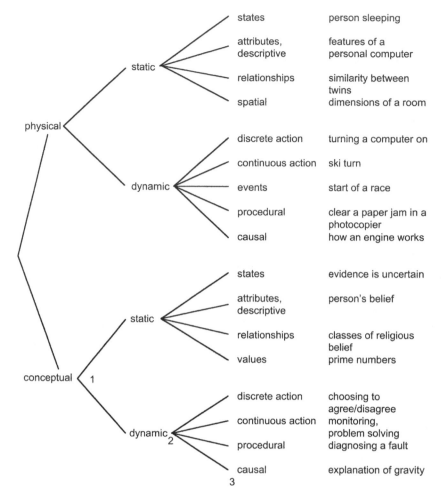

FIGURE 4.4.   Decision tree for classifying information types. The first decision point reflects abstraction from the real world, the second points to change in time and the third categorizes content.

narrative to elucidate the deeper structure of the content. The granularity of components is a matter of the designer's choice and will depend on the level of detail demanded by the application. To illustrate the analysis:

Communication goal: Explain how to assemble a bookshelf from ready-made parts.

Information component 1:
    Parts of the bookshelf: sides, back, shelves, connecting screws
    Mapping to information types:
        Physical-Static-Descriptive; parts of the bookshelf are tangible, don't change, and need to be described
        Physical-Static-Spatial; dimensions of the parts, how they are organized
        Physical-Static-Relationship type could also be added to describe which parts fit together

Information component 2:
    How to assemble parts instructions

Mapping to information types:
    Physical-Dynamic-Discrete action
    Physical-Dynamic-Procedure
    Physical-Static-State; to show final assembled bookshelf

Media Selection and Combination

The information types are used to select the appropriate media resource(s). Media classifications have had many interpretations (Alty, 1997; Bernsen, 1994; Heller & Martin, 1995). The following classification focuses on the psychological properties of the representations, rather than the physical nature of the medium (e.g., digital or analog encoding in video). Note that these definitions are combined to describe any specific medium, so speech is classified as an audio, linguistic medium, while a cartoon is classified as a nonrealistic (designed) moving image.

The definitions may be usefully considered in two dimensions of abstraction: (a) the designer's involvement in creating the medium, and (b) the rate of change. Media resources are classified using the decision tree illustrated in Fig. 4.5. More

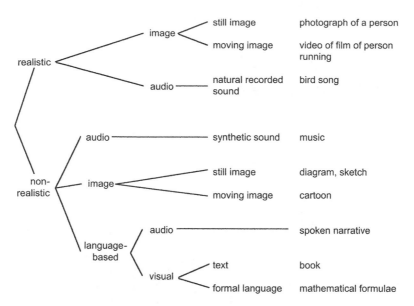

FIGURE 4.5.   Decision tree for classifying media resources.

fine-grained, taxonomic distinctions can be made, for instance, between different signs and symbolic languages (see Bernsen, 1994), but as with information types, richer taxonomies increase specification effort.

The approach to classifying media uses a walkthrough of the decision tree with the following questions that reflect the facets of the classification:

- Is the medium perceived to be realistic or not? Media resources captured directly from the real world will usually be realistic (e.g., photographs of landscapes, birdsong sound recordings). In contrast, nonrealistic media are created by human action; however, the boundary case category that illustrates the dimension is a realistic painting of a landscape.
- Does the medium change over time or not? The boundary case here is the rate of change, particularly in animations where some people might judge ten frames/sec to be a video, but a PowerPoint presentation of five slides in one minute to be a sequence of static images.
- Which modality does the resource belong to? In this case, the categories are orthogonal, although one resource may exhibit two modalities (e.g., a film with a sound track communicates in both visual and audio modalities).

Classification of media resources facilitates mapping of information types to media resources; however, the process of selection may also guide the acquisition or creation of appropriate media resources. So if the selection process indicates the need for a resource that is not in the designer's media resource library, the classification guides the necessary acquisition/creation. Cost trade-offs will naturally be considered in this process. Finally, the classification provides a mechanism for indexing media resource libraries.

Media selection.   Recommendations for selecting and influencing the user's attention have to be interpreted according to the user's task and design goal. If information provision is the main design goal (e.g., a tourist kiosk information system), then persistence of information and drawing attention to specific items is not necessarily as critical as in tutorial applications. Task and user characteristics influence media choice; for instance, verbal media are more appropriate to language-based and logical reasoning tasks, while visual media are suitable for spatial tasks involving moving, positioning, and orienting objects. Some users may prefer visual media, while image is of little use for blind users. Media resources may be available for selection, or have to be purchased from elsewhere. If existing media can be edited and reused, this is usually preferable to creating new media from scratch. Graphical images can be particularly expensive to draw, whereas capture of images by scanning is usually quick and cheap.

The suggested mappings from information types to media resources are given in Table 4.1. Information types are mapped to a simple categorization of media as (a) static, nonstreaming media; (b) still image, text, diagrams, graphs and dynamic media; (c) audio: natural sounds, speech, music, and designed sounds; and (d) video and animation. Several media map to each type, although the preferred choice is shown in italics. Since most components in the information architecture will have multiple information types and each information type maps to several media, the selection process encourages multimedia integration. For example, when a procedure for explaining a physical task is required, first a series of realistic images will be selected, followed by video and speech to integrate the steps, then text to summarize the key points. The mappings are supplemented by the following heuristics:

- To convey detail, use static media (e.g., text for language-based content, diagrams for models, or still image for physical detail of objects; Booher, 1975; Faraday & Sutcliffe, 1998b).
- To engage the user and draw attention, use dynamic media (e.g., video for physical information, animation, or speech).

TABLE 4.1. Media Selection Example

Media Type \ Information Type	Causation	Conceptual	Continuous Action	Descriptive	Discrete Action	Event	Physical	Procedure	Relationship	Spatial Information	State	Value
Realistic audio	Sound of rain and storms		Sound of skiing		Click of ON switch	Sound of the starting gun	*Noise of a tornado*			Echoes in a cave	Sound of snoring	Musical note encodes a value
Nonrealistic audio		Rising tone illustrates increasing magnetic force	Continuous tone signals progress of action	Morse code describes a ship	Tones signal open/close door	*Alarm siren*			Tones associate two objects	Sonar and Doppler effect	Continuous sound in a heart beat monitor	
Speech	*Tell someone why El Nino happens*	Tell someone about your religious beliefs	Tell someone what a ski turn looks like	Verbal description of a person	Tell someone how to turn computer on	*Tell someone race has started*	Tell someone how it feels to be in a storm	Speak instructions on engine assembly	Tell someone Jack and Jill are related	Tell someone pathway to and location of railway station	Tell someone "Jane's asleep"	Verbal report of numbers, figures
Realistic still image	*Photograph of El Nino storms and ocean currents*	Statue of Liberty photograph represents "freedom"	Set of photographs showing snap shots of action	*Overview and detail photographs of a car*	*Photograph of computer ON switch*	Photograph of the start of a race	*Photograph of a person's face*	*Photographs showing engine assembly*		*Photograph of a landscape*	Photograph of a person sleeping	
Nonrealistic still image	Diagrams of ocean currents and sea temp. to explain El Nino	*Hierarchy diagram of plant taxonomy*	Diagram with arrow depicting ski turn motion	Histogram of ageing population	Diagram showing where and how to press ON switch	Event symbol in a race sequence diagram		Explode parts diagram of engine with assembly numbers	*Graphs, histograms, ER diagrams*	*Map of the landscape*	Waiting state symbol in race sequence diagram	*Charts, graphs, scatter plots*
Text	*Describe reasons for El Nino storms*	*Explain taxonomy of animals*	Describe ski turn action	*Describe a person's appearance*	Describe how to turn computer on	Report that the race has started	Report of the storm's properties	Bullet point steps in assembling engine	*Describe brother and sister relationship*	Describe dimensions of a room	*Report that the person is asleep*	*Written number one, two*
Realistic moving image	Video of El Nino storms and ocean currents		*Movie of person turning while skiing*	Aircraft flying		*Movie of the start of a race*	*Movie of a storm*	*Video of engine assembly sequence*		Fly through landscape	Video of a person sleeping	
Nonrealistic moving image	Animation of ocean temperature change and current reversal	Animated diagram of force of gravity	Animated mannequin doing ski turn		Animation showing operation of ON switch	Animation of start event symbol in diagram		*Animation of parts diagram in assembly sequence*	Animation of links on ER diagram			
Language-based formal, numeric	Equations, functions formalizing cause and effect	*Symbols denoting concepts, e.g., pi*			Finite state automata	Event based notations		Procedural logics, process algebras	Functions, equations, grammars		State based languages, e.g., Z	Numeric symbols

Note: The italics denote the preferred mappings for media and information types, while ordinary text shows other potential media uses for the information type.

- For spatial information, use diagrams, maps with photographic images to illustrate detail, animations to indicate pathways (Bieger & Glock, 1984; May & Barnard, 1995).
- For values and quantitative information, use charts and graphs for overviews and trends, supplemented by tables for detail (Bertin, 1983; Tufte, 1997).
- Abstract concepts, relationships, and models should be illustrated with diagrams explained by text captions and speech to give supplementary information.
- Complex actions and procedures should be illustrated as a slideshow of images for each step, followed by a video of the whole sequence to integrate the steps. Text captions on the still images and speech commentary provide supplementary information (Hegarty & Just, 1993). Text and bullet points summarize steps at the end, so choice trade-offs may be constrained by cost and quality considerations.
- To explain causality, still and moving image media need to be combined with text (e.g., the cause of a flood is explained by text describing excessive rainfall with an animation of the river level rising and overflowing its banks; Narayanan & Hegarty, 1998, 402). Causal explanations of physical phenomena may be given by introducing the topic using linguistic media, showing the cause and effect by a combination of still image and text with speech captions for commentary; integrating the message by moving image with voice commentary and providing a bullet point text summary.

The end point of media selection is media integration: one or more media will be selected for each information group to present complementary aspects of the topic. Some examples of media combination that amplify the basic selection guidelines are given in Table 4.1. The table summarizes the media selection and combinations for each information type; the first preference media choice for each is shown in italics.

Engagement and attractiveness. The previous guidelines were oriented to a task-driven view of media. Media selection, however, can also be motivated by aesthetic choice. These considerations may contradict some of the earlier guidelines because the design objective is to please the user and capture the user's attention, rather than deliver information effectively. First, a health warning should be noted: the old saying "beauty is in the eye of the beholder" has good foundation. Judgments of aesthetic quality suffer from considerable individual differences. A person's reaction to a design is a function of their motivation, individual preferences, knowledge of the domain, and exposure to similar examples, to say nothing of peer opinion and "fashion." Furthermore, attractiveness is often influenced more by content than the choice of media or presentation format. The following heuristics should therefore be interpreted with care and their design manifestations tested with users:

- *Dynamic media*, especially video, have an arousing effect and attract attention; hence, video and animation are useful in improving the attractiveness of presentations. Animation must be used with care, however, as gratuitous video that cannot be turned off quickly offends (Spool, Scanlon, Snyder, Schroeder, & de Angelo, 1999).

- *Speech* engages attention because we naturally listen to conversation. Choice of voice depends on the application: female voices for more restful and information effects, male voices to suggest authority and respect (Reeves & Nass, 1996).
- *Image* must be selected with careful consideration of content. Images may be selected for aesthetic motivations, to provide a restful setting for more important foreground information (Mullet & Sano, 1995). Backgrounds in half shades and low saturation color provide more depth and interest in an image.
- *Music* has an important emotive appeal, but it must be used with care. Classical music may be counterproductive for a younger audience, while older listeners will not find heavy metal pop attractive.
- *Natural sounds* such as running water, wind in trees, birdsongs, and waves on a seashore have restful properties and hence decrease arousal.

Media integration rules may also be broken for aesthetic reasons. If information transfer is not at a premium, use of two concurrent video streams might be arousing for a younger audience, as MTV (music television) and pop videos indicate. Multiple audio and speech tracks can give the impression of complex, busy, and interesting environments.

Aesthetic design. If the user profile and domain analysis indicate that having a pleasing and attractive design is important for the user's perception, then aesthetics need to be considered in depth; however, aesthetics should be considered as a design criterion for all applications, since poor appearance and interaction design may provoke adverse reaction (Norman, 404). Some studies suggest that aesthetic design is an important component of usability and overall preference (Tractinsky, 1997; Tractinsky, Shoval-Katz, & Ikar, 400); however, others have shown that aesthetic preferences are open to contextual effects on users' judgment (Sutcliffe & de Angeli, 405). Judging when aesthetics may be important is not easy. For example, in e-commerce applications with high-value, designer-label products, aesthetic presentation is advisable, as is similar when selling to a design-oriented audience; however, the decision is not clear-cut in many applications.

Aesthetic design primarily concerns graphics and visual media. Evaluation questionnaires assess design on classic aesthetics, which broadly equate with conventional usability guidelines on structured and consistent layout, and expressive aesthetics that capture the more creative aspects of visual design (Lavie & Tractinsky, 404); however, these measure user reaction to general design aspects such as "original, fascinating, clear, and pleasant." The following heuristics provide more design-directed guidance, but they may also be employed for evaluation (Sutcliffe, 402; Sutcliffe & de Angeli, 405):

- *Judicious use of color:* Color use should be balanced, and low saturation pastel colors should be used for backgrounds. Designs should not use more than two to three fully saturated intense colors.
- *Symmetry:* Symmetrical layouts (e.g., bilateral, radial organization) that can be folded over to show the symmetrical match.

- *Simplicity and space:* Uncluttered, simple layouts that use space to separate and emphasize key components.
- *Shape:* Curved shapes convey an attractive visual style when contrasted with rectangles.
- *Structured and consistent layout:* Use of grids to structure image components and promote consistency between pages.
- *Depth of field:* Use of layers in an image and washed out background images stimulates curiosity and can be attractive by promoting a peaceful effect.
- *Design of unusual or challenging images* that stimulate the users' imagination and increase attraction: Unusual images often disobey normal laws of form and perspective.

Several sources provide more detailed advice on aesthetics and visual design (Kristof & Satran, 1995; Mullet & Sano, 1995); however, advice is usually given as examples of good design rather than specific guidelines.

Image and identity. Design of media for motivation is a complex area in its own right, and this topic is dealt with in more depth in Fogg (chap. 7, HCI Handbook); the treatment here will focus on media selection issues.

Simple photographs or more complex interactive animations (talking heads or full body mannequins) have an attractive effect. We appear to ascribe human properties to computers when interfaces give humanlike visual cues (Reeves & Nass, 1996); however, the effectiveness of media representing people depends on the characters' appearance and voice (see Fig. 4.6, courtesy of Sapient). In human-human conversation, we modify our reactions according to our knowledge, or assumptions about, the other person's role, group identification, culture, and intention (Clark, 1996). For example, reactions to a military mannequin will be very different from those to the representation of a parson. Male voices tend to be treated as more authoritative than female voices.

Use of humanlike forms is feasible with prerecorded video and photographs; however, the need depends on the application. Video representation of the lecturer can augment presentations, and video communication helps interactive dialogue. A good speaker holds our attention by a variety of tricks, such as maintaining eye contact, varying the voice tone, and using simple and concise language, as well as delivering an interesting message. These general effects can be reinforced by projected personality. Friendly people are preferred over colder, more hostile individuals. TV announcers, who tend to be middle-aged and confident, but avuncular, characters, have attention drawing power of a dominant yet friendly personality. Both sexes pay attention to extrovert, young personalities, while the male preference for beautiful young women is a particularly strong effect. These traits have been exploited by advertisers for a long time. There are lessons here for multimedia designers as the Web and interactive TV converge. Multimedia will increasingly convey persuasive messages (Reeves & Nass, 1996).

Affective effects. Media design for affect (emotional response and arousal) involves both choice of context and interaction. Arousal is increased by (a) interactive applications, (b) surprising events during interaction, (c) use of dynamic media, and (d) challenging images. In contrast, if the objective is to calm the users, arousal can be decreased by choice of natural

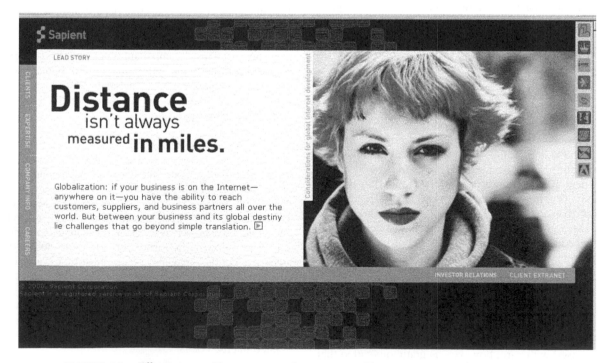

FIGURE 4.6. Effective use of human image for attraction. The picture attracts by the direction of gaze to the user as well as by the appearance of the individual. Courtesy of Sapient.

images and sounds, and soothing music. The most common emotional responses that designers may want to invoke are pleasure, anxiety, fear, and surprise. Pleasure, anxiety, and fear usually depend on our memory of agents, objects, and events (Ortony, Clore, & Collins, 1988), which will be presented, so content selection is the important determinant; however, anxiety can be evoked by uncertainty in interaction and cues to hidden effects. The emotional response of fear or pleasure will depend on the user's previous experience.

Although the bounds of media selection are only set by the creative imagination of the designer, some fundamentals don't change: design still needs to be motivated by users' goals, which may be related to work, learning, or having fun; nevertheless, testing with users is still essential.

The outcome of media selection is a first-cut script. The presentation sequence can be planned using bar charts to illustrate the presentation order, organized according to the task, or application requirements (e.g., lesson plan); see Fig. 4.7. The script gives an outline plan of the presentation, which can be annotated with navigational and dialogue controls. The information model defines the high-level presentation order. Key information items are annotated onto presentation bar charts when planning the sequence and duration of media delivery. The information types are ordered in a first-cut sequence. The selected media are added to the bar chart to show which media stream will be played over time. Decisions on timing depend on the content of the media resource (e.g., length of a video clip, frame display rate). Timeline diagrams can be augmented by hierarchy diagrams to show the classification of information groups that can be mapped to menu style access paths, or network diagrams to help planning hypermedia links.

Interaction and Navigation

Although discussion of interactive multimedia has been delayed until now, in practice, dialogue and presentation design proceed hand in hand. Task analysis provides the basis for dialogue design, as well as specification of navigation controls. Dialogue controls will be included in early storyboards. Navigational and control dialogues allow flexible access to the multimedia content, and give users ability to control how media are played. Dialogue design may also involve specifying how users interact with tools, agents, and objects in interactive microworlds.

Metaphors and interaction design. While task and domain analysis can provide ideas for interaction design, this is also a creative process. Interaction design is essentially a set of choices along a dimension, from simple controls such as menus and buttons, where the user is aware of the interface, to embodiment, in which the user becomes involved as part of the action by controlling an avatar or other representation of their presence. At this end of the dimension multimedia interaction converges with virtual reality (see chap. 15). Interactive metaphors occupy the middle ground.

Some interactive metaphors are generally applicable, such as (a) timelines to move through historical information, (b) the use of a compass to control direction of movement in

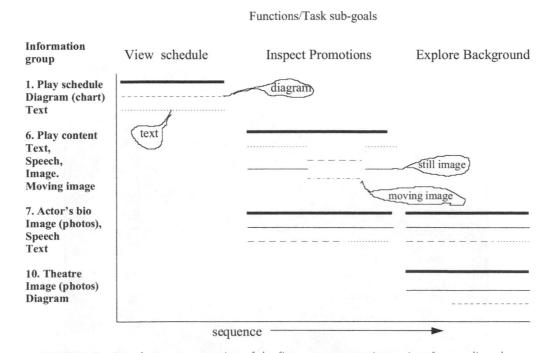

FIGURE 4.7. Bar chart representation of the first-cut presentation script after media selection. Information types are placed on the *y* axis with the task sub-goals that determine the script order on the *x* axis with media resources represented by bars. For this browsing task, three separate sequences are planned for the view schedule, promotions, and so forth.

an interactive space, and (c) controls based on automobiles (steering wheels) or ships (rudders). Others will be more specific, such as selecting and interacting with different characters (young, old, male, female, overweight, fit, etc.) in a health promotion application. Design of interaction also involves creating the microworld within which the user moves, as well as interactive objects that can be selected and manipulated.

Interaction via characters and avatars can increase the user's sense of engagement first by selecting or even constructing the character, although some users may not have the patience to build their own avatar using a graphical paint program. In character-based interaction, the user can either see the world from an egocentric viewpoint (e.g., from their character's position) or exocentric viewpoint (e.g., when they see their character in the graphical world). The sophistication in control of movement and interaction will depend on the hardware available (e.g., joystick, wand, or standard mouse and keyboard). While mimicking physical interaction via datagloves and tracking requires virtual reality technology, relatively complex interaction (e.g., actions in a football game: pass, head ball in direction north/south/east/west) can be programmed into button and function keys. Engagement is also promoted by surprise and unexpected effects, so as the user moves into a particular area a new subworld opens up, or system-controlled avatars appear. These techniques are well known to games programmers; however, they are also applicable to other genres of multimedia applications. The design concepts for engagement can be summarized as follows:

- Character-driven interaction: Provides the user with a choice of avatars or personae they can adopt as representations of themselves within the interactive virtual world.

- Tool-based interaction: Places tools in the world that users can pick up, and the tool becomes the interface (e.g., a virtual mirror magnifies, and a virtual helicopter flies; Tan, Robertson, & Czerwinski, 401).

- Collaborative characters: In computer mediated communication, these characters may represent other users; in other applications, system-controlled avatars appear to explain, guide, or warn the user.

- Surprise effects: Although conventional HCI guidelines should encourage making the affordances and presence of interactive objects explicit, when designing for engagement, hiding and surprise are important.

Interaction design for an explanatory/tutorial application is illustrated in Fig. 4.8 (www.bbc.co.uk/sn/). This is an interactive microworld in which the user plays the role of a dinosaur character, illustrating use of the engagement concepts. A compass navigation metaphor allows the user to act as the dinosaur, moving around the landscape illustrated in the photographs. The user is given feedback on the characteristics of other predators and prey in the vicinity and has to decide whether to attack or avoid them. Other controls that might be added to such interactive microworlds could be settings to change the environment (e.g., add more predators, change the weather, etc.). Engagement can be taken even further by giving the user facilities to actually design the MacWorld so the application becomes a domain-oriented design environment (Fischer, Giaccardi, Ye, Sutcliffe, & Mehandjiev, 404).

Navigation. In information-intensive multimedia where access to content is the main design goal, hypermedia dialogues

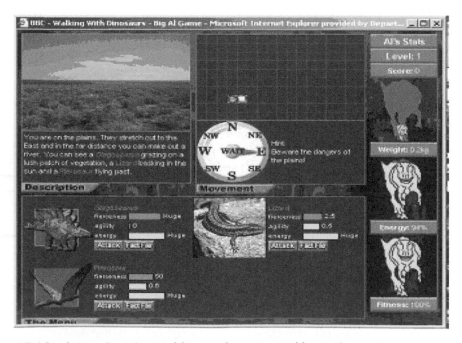

FIGURE 4.8. Interactive microworld—Big Al game (www.bbc.co.uk/sn/). The user plays the dinosaur role by navigating with the compass metaphor. The photograph updates with each move and the user is given a choice of attacking or avoiding other dinosaurs in the virtual world.

that link content segments will be appropriate. Good hypertext design is based on a sound information analysis that specifies the pathways between related items, and the use of cues to show the structure of the information space to the user. In document-based hypermedia (e.g., HTML and the Web), links can only access the whole media resource rather than point to components within it. The design issues are to (a) plan the overall structure, (b) segment complex structures into a hierarchy of subnetworks, and then (c) plan the implementation of each subnetwork. The access structure of most hypermedia will be hierarchical, organized according to the information model and categorization of content (e.g., information grouped by function, organization, task usage, or user preference). Navigation design transforms the user's conceptual model of an information space into a hypermedia structure. Unfortunately, individual users have different models, so this may not be an easy task. Implementing too many links in order to satisfy each user's view will make the system too complex and increase the chance of the user getting lost. Too few links will frustrate users who cannot find the associations they want. Unfortunately, hypermedia systems assume a fixed link structure so the user is limited to the pathways provided by the designer. More open-ended hypermedia environments (e.g., Microcosm: Lowe & Hall, 1998) provide more flexibility via links with query facilities that allow access to databases. Dynamic links attached to hotspots in images or nodes in text documents provide access paths to a wider variety of data.

One problem with large hypermedia systems is that users get lost in them. Navigation cues, waymarks, and mini-map overviews can help to counter the effects of disorientation. *Mini-maps* give an overview of the hypertext area and a reference context for where users are in large networks. *Filters* help to reduce complexity by showing only a subset of nodes and links that the user is interested in. Having typed links helps filtering views because the user can guess the information content from the link type (e.g., reference, example, source, related work, etc.). Other navigation facilities are *visit lists* containing a history of nodes traversed in a session, and *bookmarks,* so users can tailor a hypermedia application with their own navigation aide-memoires (Nielsen, 1995). Once the structure has been designed, link cues need to be located within media resources, so the appropriate cues need to be considered for each medium. For example:

- *Text media:* The Web convention is to underline and highlight text in a consistent color, such as blue or purple.
- *Images:* Link cues can be set as stand-alone icons or as active components in images. Icons need to be tested with users because the designer's assumed meaning can be ambiguous. Active components should signal the link's presence by captions or pop-up hover text so the user can inspect a link before deciding whether to follow it.
- *Moving images:* Links from animation and film are difficult to design because the medium is dynamic; however, link buttons can be placed below the video window. Active components, such as overlaid buttons within a moving image, are technically more challenging programs. Buttons may also be timed to pop up at appropriate times during the video.

- *Audio links:* Sound and speech links present the same difficulties as moving images. One solution is to use visual cues, possibly synchronized with the sound or speech track. If speech recognition is available, then voice commands can act as links, but these commands need to be explained to the user.

Navigation controls provide access to the logical content of multimedia resources; however, access to logical components may be constrained by limitations of physical media resources. For example, a movie may be logically composed of several scenes, but it can only be accessed by an approximate timer set to the beginning of the whole film. Worse still, a video clip may only be playable as a single segment, making implementation of navigation requirements impossible. It is advisable to give the user control whenever possible.

In many cases, the media-rendering device, such as a video player for .avi files or QuickTime movies, will provide controls. If controls have to be implemented from scratch, the following list should be considered for each media type:

- *Static media:* Size and scale controls to zoom and pan; page access, if the medium has page segmentation, as in text and diagrams; the ability to change attributes such a color and display resolution; and font type and size in text.
- *Dynamic media:* The familiar video controls of stop, start, play, pause, fast forward and rewind, as well as the ability to address a particular point or event in the media stream by a time marker or an index (e.g., "go to" component/marker, etc.).

Navigation controls employ standard user interface components (e.g., buttons, dialogue boxes, menus, icons, and sliders) and techniques (e.g., form filling, dialogue boxes, and selection menus). For more guidance, see ISO 9241, parts 12, 14 & 17 (International Standards Organization, 1997) and ISO 14915, part 2 (International Standards Organization, 1998).

Design for Attention

Having selected the media resources, the designer must now ensure that the user will extract the appropriate information. An important consideration of multimedia design is to link the thread of a message across several different media. This section gives recommendations on planning the user's reading/viewing sequence, and guidelines for realizing these recommendations in presentation sequences, hypermedia dialogues, and navigation controls. The essential differences are timing and user control. In a presentation design, the reading/viewing sequence and timing are set by the designer, whereas the reading/viewing sequence in hypertext implementation and interactive dialogues is under user control.

Presentation techniques help to direct the user's attention to important information and specify the desired order of reading or viewing. The need for thematic links between information components is specified, and attention-directing techniques are selected to implement the desired effect.

The design issues are:

- To plan the overall thematic thread of the message.
- To draw the user's attention to important information.
- To establish a clear reading/viewing sequence.

- To provide clear links when the theme crosses from one medium to another.

Design for attention is particularly important for images. User attention to time-varying media is determined by the medium itself (e.g., we have little choice but to listen to speech or to view animations in the order in which they are presented). The reading sequence is directed by the layout of text, although this is culturally dependent (e.g., western languages read left to right, Arabic languages in the opposite direction). Viewing order in images, however, is unpredictable unless the design specifically selects the user's attention.

In some cases, a common topic may be sufficient for directing the user's reading/viewing sequence; however, when the thread is important or hard to follow, designed effects for attention, or contact points, are necessary to aid the user's perception. The term *contact point* refers to a reference from one medium to another and comes from the research of Hegarty and Just (1993) and Narayanan and Hegarty (1998), which demonstrated that comprehension is improved by reinforcing the links between information in different media. Two types of contact points are distinguished:

1. *Direct contact points:* Attention-directing effects are implemented in both the source and destination medium (e.g., in the text, an instruction such as, "Look at the oblong component in Figure 1" is given while the component is highlighted). Direct contact points create a strong cross-reference between two media, but can become intrusive if overused. An example of a direct contact point is shown in Fig. 4.9 (Faraday & Sutcliffe, 1997a).

2. *Indirect contact points* implement an attention-directing effect only in the source, or less frequently the destination, medium (e.g., "In Figure 1, the assembly is shown" is spoken, with no highlighting being used in the image.). Indirect contact points have less attention-directing force and work by temporal sequencing or spatial juxtapositoning. Indirect contact points are less intrusive, so they may be used more frequently without becoming disruptive.

In most cases, the attentional effect in a direct contact point will be actuated in sequence, although occasionally, both effects may be presented concurrently if the order of the association is not important. In hypermedia implementations, direct contact points become a link cue in the source medium and a highlight anchor in the destination medium.

Multiple contact points may be organized in a logical order to follow the theme and connect a thread of topics (Faraday & Sutcliffe, 1998a, 1999). For instance, in a biology tutorial, the explanation of parts of a cell is organized with interleaved

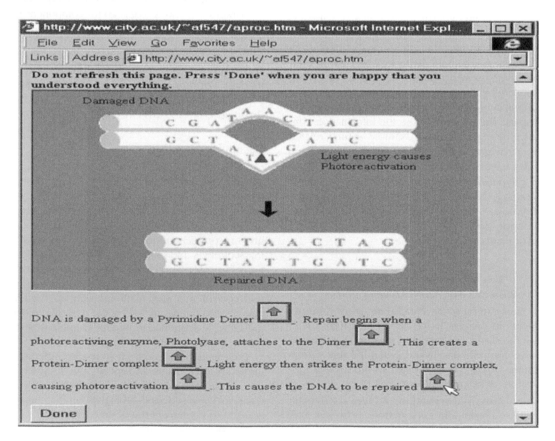

FIGURE 4.9. Direct contact point between text and video segments, reinforced by speech. The arrows play video clips linked to the text description. The user can either play the whole explanation or step through it using the contact point links (Faraday & Sutcliffe, 1997a).

speech segments and a diagram describing the cell's components top to bottom, left to right. Highlighting techniques locate each component in turn, following the order of the spoken explanation. The attention-directing techniques described in the following section are used to implement contact points.

The design problem is how to direct the user's attention to the appropriate information at the correct level of detail. Initially, users will tend to extract information from images at the scene level, (e.g, major objects will be identified but with very little descriptive detail; Treisman, 1988). Regular layout grids help design composite images (Mullet & Sano, 1995) and encourage viewing sequences in image sets. Alternatively, the window frame can be set to control which parts of an image are viewed. Larger window frames will be attended to before smaller areas. A list of the key components that the user needs to focus on and the facts that should be extracted is checked against the initial presentation design to see if the key components will attract sufficient attention, or whether the user is likely to be confused by extraneous detail.

Still image media. Highlighting techniques for designed and natural images, organized in approximate power of their effect, are summarized in Table 4.2. A common highlighting technique will pick out spatially distributed objects, such as when all related objects are changed to the same color; co-located objects can be grouped by using a common color or texture for their background, or by drawing a box around them. The highlighted area will set the granularity of the user's attention. Captions linked to objects in an image are another useful means of drawing attention and providing supplementary information (e.g., identity). Dynamic revealing of captions is particularly effective for directing the user's viewing sequence. Sequential highlighting is also useful for showing the pathways or navigational instructions.

Moving image media. Directing attention to components within moving images is difficult because of the dynamic nature of the medium. Design of film and video is an extensive subject in its own right, so treatment here will necessarily be brief. The following design advice is based on Hochberg (1986). The design objectives, as for other media, are to draw the user's attention to key components within the video or animation.

First, the content should be structured into scenes that correspond to the information script. To structure animation sequences and make scene boundaries obvious, use a cut, wipe, or dissolve to emphasize that a change in the content structure has taken place; however, cuts should be used with care and continuity maintained between the two sequences if they are to be integrated. *Continuity* is manifest as the same viewpoint and subject matter in two contiguous shots. Change in background or action (e.g., an individual walking left in one clip and walking right in the next) is quickly noticed as a change. An establishing shot that shows the whole scene helps to introduce a new sequence and provide context. To provide detail of a newly introduced object or context, the object is shown filling the frame with a small amount of surrounding scene, while to imply a relationship or compare two objects, a tight two-shot with both objects together in the same frame is advisable.

Linguistic media (text and speech). As with moving image, the literature is extensive, so the following heuristics are a brief summary; see Levie and Lentz (1982) for more detail. Text may be structured to indicate subsections by indentation, formatting into paragraphs or columns, or segmented by background color. Bullet points or numbered sections indicate order more formally, such as for procedures. Different voices help to structure speech while also attracting attention. If language is being used to set the context for accompanying media, it is important that the correct level of identification is set. For instance, a higher level concept, or the whole scene in an accompanying image, is described at the beginning of a script, and then lower level topics reset the user's focus. Discourse markers can make phrases and sentences more salient.

Adding contact points and attention-directing effects completes the design process; however, as with all user interfaces, there is no substitute for usability testing. Designs are constructed

TABLE 4.2. Attention-Directing Techniques for Different Media

	Attention Orienting Techniques in Approximate Order of Power	Notes
Still image: designed and natural	Movement of or change in the shape/size/color of an object. Use of bold outline. Object marked with a symbol (e.g., arrow) or icon. Draw boundary, use color, shape, size, or texture to distinguish important objects.	Some effects may compromise natural images because they overlay the background image with new components (e.g., arrows, arcs, icons). Group objects by a common technique.
Moving image	Freeze frame followed by applying a still image highlight. Zoom, close-up shot of the object. Cuts, wipes, and dissolve effects.	Change in topographic motion, in which an object moves across the ground of an image, is more effective than internal movement of an object's components. Size and shape may be less effective for highlighting a moving object.
Text	Bold, font size, type, color, or underlining. To direct attention to larger segments of text, use formatting, bullet points, subsections, and/or indentation.	Formatting techniques are paragraphs; headings/titles as entry points; and indents to show hierarchical nesting, with bullet points and lists.
Speech/sound	Familiar voice. Silence followed by onset of sound. Different voices, or a change in voice prosody (tonality), amplitude (loudness), change and variations in pitch (frequency), voice rate, change source direction, and alarm sounds (police sirens).	Voices familiar to the user (e.g., close relatives) attract attention over nonfamiliar speech. Discourse markers such as "Next," "Because," "So," and so forth draw attention to subsequent phrases.

incrementally by iterations of design and evaluation that checks for usability using standard methods, with additional memory and comprehension tests for multimedia. Therefore, when testing a design, ask the user to tell you what they understood the message to be. This can be done during the presentation with a think-aloud protocol to check that users did attend to key items, and afterwards, by a memory test. If a user cannot remember key components in the message, then the design may need to be improved.

CONCLUSIONS

Multimedia still poses many issues for further research. The design method described in this chapter, coupled with user-centred design, can improve quality; however, there is still a need for experts to create specific media resources (e.g., film/video and audio experts). Furthermore, considerable research is still necessary before we fully understand the psychology of multimedia interaction. Design for motivation and attractiveness is still poorly understood, and personality effects in media may not be robust when usability errors impede communication. The process by which people extract information from complex images still requires extensive research, although the increasing number of eyetracking studies is beginning to throw some light on this topic. In the future, language and multimodal communication will change our conception of multimedia from its current CD-ROM or Web-based form into conversational and multisensory interfaces. Multimedia will become part of wearable and ubiquitous user interfaces, where the media are part of our everyday environment. Design for multisensory communication will treat media and artifacts (e.g., our desks, clothing, walls in our homes) as a continuum, while managing the diverse inputs to multimedia from creative design, technology, and usability engineering will be one of the many interesting future challenges.

References

Alty, J. L. (1991). Multimedia: what is it and how do we exploit it? In D. Diaper & N. V. Hammond (Eds.), *Proceedings of HCI '91: People and Computers VI* (pp. 31–41). Cambridge, U.K.: Cambridge University Press.

Alty, J. L. (1997). Multimedia. In A. B. Tucker (Ed.), *Computer science and engineering handbook* (pp. 1551–1570). New York: CRC Press.

Arens, Y., Hovy, E., & Van Mulken, S. (1993). Structure and rules in automated multimedia presentation planning. *Proceedings: IJCAI-93: Thirteenth International Joint Conference on Artificial Intelligence.* Chambéry, France (pp. 231–235). San Mateo, CA: Morgan-Kaufmann.

Ayers P., & Sweller, J. (405). The split attention principle for multimedia learning. In R. E. Mayer (Ed.), *The Cambridge Handbook for Multimedia Learning* (pp 135–146). Cambridge, U.K.: Cambridge University Press.

Baddeley, A. D. (1986). *Working memory.* Oxford, U.K.: Oxford University Press.

Barnard, P. (1985). Interacting cognitive subsystems: A psycholinguistic approach to short term memory. In A. Ellis (Ed.), *Progress in psychology of language: Vol. 2.* (pp. 197–258). London: Lawrence Erlbaum Associates.

Bernsen, N. O. (1994). Foundations of multimodal representations: a taxonomy of representational modalities. *Interacting with Computers, 6*(4), 347–371.

Bertin, J. (1983). *Semiology of graphics.* Madison, WI: University of Wisconsin Press.

Bieger, G. R., & Glock, M. D. (1984). The information content of picture-text instructions. *Journal of Experimental Education, 53*, 68–76.

Booher, H. R. (1975). Relative comprehensibility of pictorial information and printed word in proceduralized instructions. *Human Factors, 17*(3), 266–277.

Card, S. K., Moran, T. P., & Newell, A. (1983). *The psychology of human computer interaction.* Hillsdale, NJ: Lawrence Erlbaum Associates.

Clark, H. H. (1996). *Using language.* Cambridge, U.K.: Cambridge University Press.

Faraday, P., & Sutcliffe, A. G. (1996). An empirical study of attending and comprehending multimedia presentations. *Proceedings ACM Multimedia 96: 4th Multimedia Conference,* Boston (pp. 265–275). New York: ACM Press.

Faraday, P., & Sutcliffe, A. G. (1997a). Designing effective multimedia presentations. *Proceedings: Human Factors in Computing Systems CHI'97,* Atlanta, GA (pp. 272–279). New York: ACM Press.

Faraday, P., & Sutcliffe, A. G. (1997b). Multimedia: design for the moment. *Proceedings: Fifth ACM International Multimedia Conference,* Seattle, WA (pp. 183–192). New York: ACM Press.

Faraday, P., & Sutcliffe, A. G. (1998a). Providing advice for multimedia designers. In C. M. Karat, A. Lund, J. Coutaz, & J. Karat, (Eds.), *CHI 98 Conference Proceedings: Human Factors in Computing Systems,* Los Angeles (pp. 124–131). New York: ACM Press.

Faraday, P., & Sutcliffe, A. G. (1998b). Using contact points for Web page design. *People and Computers XIII, Proceedings: BCS-HCI Conference,* Sheffield (pp. 20–22). Berlin: Springer-Verlag.

Faraday, P., & Sutcliffe, A. G. (1999). Authoring animated Web pages using contact points. *Proceedings of CHI'99: Human Factors in Computing Systems,* Pittsburgh, PA (pp. 458–465). New York: ACM Press.

Fischer G., Giaccardi E., Ye, Y.,Sutclifffe A.G., & Mehandjiev, N. (2004), Meta-Design: A manifesto for End User development. *Communications of the ACM, 47*(9), pp. 33–37.

Fogg, B. J. (1998). Persuasive computer: perspectives and research directions. *Proceedings: Human Factors in Computing Systems: CHI '98,* Los Angeles (pp. 225–232). New York: ACM Press.

Gardiner, M., & Christie, B. (1987). *Applying cognitive psychology to user interface design.* Chichester, NY: Wiley.

Hegarty, M., & Just, M. A. (1993). Constructing mental models of text and diagrams. *Journal of Memory and Language, 32,* 717–742.

Heller, R. S., & Martin, C. (1995). A media taxonomy. *IEEE Multimedia* (Winter), 36–45.

Hochberg, J. (1986). Presentation of motion and space in video and cinematic displays. In K. R. Boff, L. Kaufman, & J. P. Thomas (Eds.), *Handbook of perception and human performance, 1: Sensory processes and perception.* New York: Wiley.

Hollan, J. D., Hutchins, E. L., & Weitzman, L. (1984). Steamer: an interactive inspectable simulation-based training system. *AI Magazine, 5*(2), 15–27.

International Standards Organization. (1997). *ISO 9241: Ergonomic requirements for office systems with visual display terminals (VDTs).* Geneva: ISO.

International Standards Organization. (1998). *ISO 14915 Multimedia user interface design software ergonomic requirements, Part 1: Introduction and framework; Part 3: Media combination and selection.* Geneva: ISO.

International Standards Organization. (400). *ISO 14915-3: Software ergonomics for multimedia user interfaces. Part 3: Media selection and combination. Draft international standard.* Geneva: ISO.

Kristof, R., & Satran, A. (1995). *Interactivity by design: creating and communicating with new media.* Mountain View, CA: Adobe Press.

Lavie, T., & Tractinsky, N. (404). Assessing dimensions of perceived visual aesthetics of websites. *International Journal of Human-Computer Studies, 60*(3), 269–298.

Levie, W. H., & Lentz, R. (1982). Effects of text illustrations: a review of research. *Educational Computing and Technology Journal, 30*(4), 159–232.

Lowe, D., & Hall, W. (1998). *Hypermedia and the Web.* Chichester, NY: Wiley.

Mann, W. C., & Thompson, S. A. (1988). Rhetorical Structure Theory: toward a functional theory of text organisation. *Text, 8*(3), 243–281.

May, J., & Barnard, P. (1995). Cinematography and interface design. In K. Nordbyn, P. H. Helmersen, D. J. Gilmore, & S. A. Arnesen, (Eds.), *Proceedings: Fifth IFIP TC 13 International Conference on Human-Computer Interaction,* Lillehammer, Norway (pp. 26–31). London: Chapman & Hall.

Mullet, K., & Sano, D. (1995). *Designing visual interfaces: communication oriented techniques.* Englewood Cliffs, NJ: SunSoft Press.

Narayanan, N. H., & Hegarty, M. (1998). On designing comprehensible interactive hypermedia manuals. *International Journal of Human-Computer Studies, 48,* 267–301.

Narayanan, N. H., & Hegarty, M. (402). Multimedia design for communication of dynamic information. *International Journal of Human-Computer Studies, 57*(4), 279–316.

Nielsen, J. (1995). *Multimedia and hypertext: the internet and beyond.* Boston: AP Professional.

Norman, D. A. (404). *Emotional design: Why we love (or hate) everyday things.* New York: Basic Books.

Norman, D. A., & Shallice, T. (1986). Attention to action: willed and automatic control of behaviour. In G. E. Davidson, & G. E. Schwartz (Eds.), *Consciousness and self-regulation, Vol. 4* (pp. 1–18). New York: Plenum.

Ortony, A., Clore, G. L., & Collins, A. (1988). *The cognitive structure of emotions.* Cambridge, U.K.: Cambridge University Press.

Oviatt S., Darves C., & Coulston R. (404). Towards adaptive conversational interfaces: Modelling speech convergence with animated personas. *ACM Transactions of Computer Human Interaction, 11*(3), 300–328.

Papert, S. (1980). *Mindstorms: children, computers, and powerful ideas.* New York: Basic Books.

Reeves, B., & Nass, C. (1996). *The media equation: how people treat computers, television and new media like real people and places.* Stanford, CA/Cambridge: CLSI/Cambridge University Press.

Rogers, Y., & Scaife, M. (1998). How can interactive multimedia facilitate learning? In J. Lee (Ed.), *Intelligence and multimodality in multimedia interfaces: research and applications.* Menlo Park, CA: AAAI Press.

Scaife, M., Rogers, Y., Aldrich, F., & Davies, M. (1997). Designing for or designing with? Informant design for interactive learning environments. In Pemberton S. (Ed.), *Proceedings: Human Factors in Computing Systems CHI'97,* Atlanta, GA (pp. 343–350). New York: ACM Press.

Spool, J. M., Scanlon, T., Snyder, C., Schroeder, W., & de Angelo, T. (1999). *Web site usability: a designer's guide.* San Francisco: Morgan Kaufmann.

Sutcliffe, A. G. (1997). Task-related information analysis. *International Journal of Human-Computer Studies, 47*(2), 223–257.

Sutcliffe, A. G. (1999). User-centred design for multimedia applications. *Proceedings Vol. 1: IEEE Conference on Multimedia Computing and Systems,* Florence, Italy (pp. 116–123). Los Alamitos, CA: IEEE Computer Society Press.

Sutcliffe, A. G. (402). Assessing the reliability of heuristic evaluation for website attractiveness and usability. *Proceedings HICSS-35: Hawaii International Conference on System Sciences,* Honolulu (pp. 1838–1847). Los Alamitos, CA: IEEE Computer Society Press.

Sutcliffe, A. G. (403). *Multimedia and virtual reality: designing multisensory user interfaces.* Mahwah, NJ: Lawrence Erlbaum Associates.

Sutcliffe, A. G., & Faraday, P. (1994). Designing presentation in multimedia interfaces. In B. Adelson, S. Dumais, & J. Olson, (Eds.), *CHI '94 Conference Proceedings: Human Factors in Computing Systems 'Celebrating Interdependence',* Boston (pp. 92–98). New York: ACM Press.

Sutcliffe, A. G., & de Angeli, A. (405). Assessing interaction styles in web user interfaces. In M. F. Costabile & F. Paterno (Eds.), *Proceedings of Human computer Interaction- INTERACT 05,* Rome (pp. 405–417) Berlin: Springer-Verlag.

Sutcliffe, A. G., Kurniawan, S., & Shin, J-E. (406). A Method and advisor tool for multimedia user interface design. *International Journal of Human Computer Studies, 64*(4), 375–312.

Tan, D. S., Robertson, G. R., & Czerwinski, M. (401). Exploring 3D navigation: Combining speed coupled flying with orbiting. In J. A. Jacko, A. Sears, M. Beaudouin-Lafon, & R. J. K. Jacob, (Eds.), *CHI 401 Conference Proceedings: Conference on Human Factors in Computing Systems,* Seattle, WA, (pp. 418–425). New York: ACM Press.

Tractinsky, N. (1997). Aesthetics and apparent usability: Empirically assessing cultural and methodological issues. In Pemberton S. (Ed.), *Human Factors in Computing Systems: CHI 97 Conference Proceedings,* Atlanta, GA (pp. 115–122). New York: ACM Press.

Tractinsky, N., Shoval-Katz, A., & Ikar, D. (400). What is beautiful is usable. *Interacting with Computers, 13*(2), 127–145.

Treisman, A. (1988). Features and objects: fourteenth Bartlett memorial lecture. *Quarterly Journal of Experimental Psychology, 40A*(2), 41–237.

Tufte, E. R. (1997). *Visual explanations: images and quantities, evidence and narrative.* Cheshire, CT: Graphics Press.

Wickens, C. D., Sandry, D., & Vidulich, M. (1983). Compatibility and resource competition between modalities of input, output and central processing. *Human Factors, 25,* 227–248.

·5·

MULTIMODAL INTERFACES

Sharon Oviatt
Oregon Health & Science University

What are Multimodal Systems, and Why
Are We Building Them? 86

What Types of Multimodal Interfaces Exist, and
What Is Their History and Current Status? 86

What Are the Goals and Advantages
of Multimodal Interface Design? 89

What Methods and Information Have Been
Used to Design Novel Multimodal Interfaces? 92

What Are the Cognitive Science Underpinnings
of Multimodal Interface Design? 93

    When Do Users Interact Multimodally? 93

    What Are the Integration and Synchronization
    Characteristics of Users' Multimodal Input? 94

    What Individual Differences Exist in Multimodal Interaction,
    and What Are the Implications for
    Designing Systems for Universal Access? 95

Is Complementarity or Redundancy
the Main Organizational Theme That
Guides Multimodal Integration? 96

What Are the Primary Features of Multimodal
Language? 96

What Are the Basic Ways in Which
Multimodal Interfaces Differ From
Graphical User Interfaces? 97

What Basic Architectures and Processing
Techniques Have Been Used to Design
Multimodal Systems? 97

What Are the Main Future Directions
for Multimodal Interface Design? 99

Acknowledgments 100

References 100

WHAT ARE MULTIMODAL SYSTEMS, AND WHY ARE WE BUILDING THEM?

Multimodal systems process two or more combined user input modes—such as speech, pen, touch, manual gestures, gaze, and head and body movements—in a coordinated manner with multimedia system output. This class of systems represents a new direction for computing, and a paradigm shift away from conventional WIMP interfaces. Since the appearance of Bolt's (1980) "Put That There" demonstration system, which processed speech in parallel with touch-pad pointing, a variety of new multimodal systems has emerged. This new class of interfaces aims to recognize naturally occurring forms of human language and behavior, which incorporate at least one recognition-based technology (e.g., speech, pen, vision). The development of novel multimodal systems has been enabled by the myriad input and output technologies currently becoming available, including new devices and improvements in recognition-based technologies. This chapter will review the main types of multimodal interfaces, their advantages and cognitive science underpinnings, primary features and architectural characteristics, and general research in the field of multimodal interaction and interface design.

The growing interest in multimodal interface design is inspired largely by the goal of supporting more transparent, flexible, efficient, and powerfully expressive means of human-computer interaction. Multimodal interfaces are expected to be easier to learn and use, and are preferred by users for many applications. They have the potential to expand computing to more challenging applications, to be used by a broader spectrum of everyday people, and to accommodate more adverse usage conditions than in the past. Such systems also have the potential to function in a more robust and stable manner than unimodal recognition systems involving a single recognition-based technology, such as speech, pen, or vision.

The advent of multimodal interfaces based on recognition of human speech, gazes, gestures, and other natural behaviors represents only the beginning of a progression toward computational interfaces capable of relatively humanlike sensory perception. Such interfaces eventually will interpret continuous input from a large number of different visual, auditory, and tactile input modes, which will be recognized as users engage in everyday activities. The same system will track and incorporate information from multiple sensors on the user's interface and surrounding physical environment in order to support intelligent adaptation to the user, task, and usage environment. Future adaptive multimodal-multisensor interfaces have the potential to support new functionality, to achieve unparalleled robustness, and to perform flexibly as a multifunctional and personalized mobile system.

WHAT TYPES OF MULTIMODAL INTERFACES EXIST, AND WHAT IS THEIR HISTORY AND CURRENT STATUS?

Multimodal systems have developed rapidly during the past decade, with steady progress toward building more general and robust systems, as well as more transparent human interfaces than ever before (Benoit, Martin, Pelachaud, Schomaker, & Suhm, 2000; Oviatt et al., 2000). Major developments have occurred in the hardware and software needed to support key component technologies incorporated within multimodal systems, as well as in techniques for integrating parallel input streams. Multimodal systems also have diversified to include new modality combinations, including speech and pen input, speech and lip movements, speech and manual gesturing, and gaze tracking and manual input (Benoit & Le Goff, 1998; Cohen et al., 1997; Stork & Hennecke, 1995; Turk & Robertson, 2000; Zhai, Morimoto, & Ihde, 1999). In addition, the array of multimodal applications has expanded extremely rapidly in recent years. Among other areas, it presently includes multimodal map-based systems for mobile and in-vehicle use; multimodal browsers; multimodal interfaces to virtual reality systems for simulation and training; multimodal person-identification/verification systems for security purposes; multimodal medical, educational, military, and web-based transaction systems; and multimodal access and management of personal information on handhelds and cell phones (Cohen & McGee, 2004; Iyengar, Nock, & Neti, 2003; McGee, 2003; Neti, Iyengar, Potamianos, & Senior, 2000; Oviatt, 2003; Oviatt, Flickner, & Darrell, 2004; Oviatt & Lunsford, 2005; Oviatt et al., 2000; Pankanti, Bolle, & Jain, 2000; Reithinger et al., 2003).

In one of the earliest multimodal concept demonstrations, Bolt had users sit in front of a projection of "Dataland" in "the Media Room" (Negroponte, 1978). Using the "Put That There" interface (Bolt, 1980), they could use speech and pointing on an armrest-mounted touchpad to create and move objects on a 2D large-screen display. For example, the user could issue a command, such as "Create a blue square there," with the intended location of "there" indicated by a 2D cursor mark on the screen. Semantic processing was based on the user's spoken input, and the meaning of the deictic "there" was resolved by processing the x/y coordinate indicated by the cursor at the time "there" was uttered. Since Bolt's early prototype, considerable strides have been made in developing a wide variety of different types of multimodal systems.

Among the earliest and most rudimentary multimodal systems were ones that supported speech input along with a standard keyboard and mouse interface. Conceptually, these multimodal interfaces represented the least departure from traditional graphical user interfaces (GUIs). Their initial focus was on providing richer natural language processing to support greater expressive power for the user when manipulating complex visuals and engaging in information extraction. As speech recognition technology matured during the late 1980s and 1990s, these systems added spoken input as an alternative to text entry via the keyboard. As such, they represent early involvement of the natural language and speech communities in developing the technologies needed to support new multimodal interfaces. Among the many examples of this type of multimodal interface are CUBRICON, Georal, Galaxy, XTRA, Shoptalk, and Miltalk (Cohen et al., 1989; Kobsa et al., 1986; Neal & Shapiro, 1991; Seneff, Goddeau, Pao, & Polifroni, 1996; Siroux, Guyomard, Multon, & Remondeau, 1995; Wahlster, 1991).

Several of these early systems were multimodal-multimedia map systems to which a user could speak or type and point with a mouse to extract tourist information or engage in military situation assessment (Cohen et al., 1989; Neal & Shapiro, 1991; Seneff et al., 1996; Siroux et al., 1995). For example, using the CUBRICON system a user could point to an object on a map and ask, "Is this <point> an air base?" CUBRICON was an expert system with extensive domain knowledge, as well as natural language processing capabilities that included referent identification and dialogue tracking (Neal & Shapiro, 1991). With the Georal system, a user could query a tourist-information system to plan travel routes using spoken input and pointing via a touch-sensitive screen (Siroux et al., 1995). In contrast, the Shoptalk system permitted users to interact with complex graphics representing factory production flow for chip manufacturing (Cohen et al., 1989). Using Shoptalk, a user could point to a specific machine in the production layout and issue the command: "Show me all the times when this machine was down." After the system delivered its answer as a list of time ranges, the user could click on one to ask the follow-up question: "What chips were waiting in its queue then, and were any of them hot lots?" Multimedia system feedback was available in the form of a text answer, or the user could click on the machine in question to view an exploded diagram of the machine queue's contents during that time interval.

More recent multimodal systems have moved away from processing simple mouse or touchpad pointing, and have begun designing systems based on two parallel input streams that each are capable of conveying rich semantic information. These multimodal systems recognize two natural forms of human language and behavior, for which two recognition-based technologies are incorporated within a more powerful bimodal user interface. To date, systems that combine either speech and pen input (Oviatt & Cohen, 2000) or speech and lip movements (Benoit et al., 2000; Stork & Hennecke, 1995; Rubin, Vatikiotis-Bateson, & Benoit, 1998; Potamianos, Neti, Gravier, & Garg, 2003) constitute the two most mature areas within the field of multimodal research. In these cases, the keyboard and mouse have been abandoned. For speech and pen systems, spoken language sometimes is processed along with complex pen-based gestural input involving hundreds of different symbolic interpretations beyond pointing[1] (Oviatt et al., 2000). For speech and lip movement systems, spoken language is processed along with corresponding human lip movements during the natural audio-visual experience of spoken interaction. In both of these sub-literatures, considerable work has been directed toward quantitative modeling of the integration and synchronization characteristics of the two rich input modes being processed, and innovative time-sensitive architectures have been developed to process these patterns in a robust manner.

Multimodal systems that recognize speech and pen-based gestures first were designed and studied in the early 1990s (Oviatt, Cohen, Fong, & Frank, 1992), with the original QuickSet system prototype built in 1994. The QuickSet system is an agent-based collaborative multimodal system that runs on a hand-held PC (Cohen et al., 1997). With QuickSet, for example, a user can issue a multimodal command such as "Airstrips . . . facing this way <draws arrow>, and facing this way <draws arrow>,"

using combined speech and pen input to place the correct number, length, and orientation (e.g., SW, NE) of aircraft landing strips on a map. Other research-level systems of this type were built in the late 1990s. Examples include the Human-centric Word Processor, Portable Voice Assistant, QuickDoc and MVIEWS (Bers, Miller, & Makhoul, 1998; Cheyer, 1998; Oviatt et al., 2000; Waibel, Suhm, Vo, & Yang, 1997). These systems represent a variety of different system features, applications, information fusion, and linguistic processing techniques. For illustration purposes, a comparison of five different speech and gesture systems is summarized in Fig. 5.1. In most cases, these multimodal systems jointly interpreted speech and pen input based on a frame-based method of information fusion and a late semantic fusion approach, although QuickSet used a statistically-ranked unification process and a hybrid symbolic/statistical architecture (Wu, Oviatt, & Cohen, 1999). Other recent systems also have adopted unification-based multimodal fusion and a hybrid architectural approach for processing multimodal input (Bangalore & Johnston, 2000; Denecke & Yang, 2000; Pfleger, 2004; Wahlster, 2001) and even multimodal output (Kopp, Tepper, & Cassell, 2004).

Multimodal systems that process speech and continuous 3D manual gesturing are emerging rapidly, although these systems remain less mature than ones that process 2D pen input (Encarnacao & Hettinger, 2003; Flanagan & Huang, 2003; Sharma, Pavlovic, & Huang, 1998; Pavlovic, Sharma, & Huang, 1997). This primarily is because of the significant challenges associated with segmenting and interpreting continuous manual movements, compared with a stream of x/y ink coordinates. Because of this difference, multimodal speech and pen systems have advanced more rapidly in their architectures, and have progressed further toward commercialization of applications. However, a significant cognitive science literature is available for guiding the design of emerging speech and 3D-gesture prototypes (Condon, 1988; Kendon, 1980; McNeill, 1992), which will be discussed further later in this chapter. Among the earlier systems to begin processing manual pointing or 3D gestures combined with speech were developed by Koons and colleagues (Koons, Sparrell, & Thorisson, 1993), Sharma and colleagues (Sharma et al., 1996), Poddar and colleagues (Poddar, Sethi, Ozyildiz, & Sharma, 1998), and by Duncan and colleagues (Duncan, Brown, Esposito, Holmback, & Xue, 1999).

Historically, multimodal speech and lip movement research has been driven by cognitive science interest in intersensory audio-visual perception and the coordination of speech output with lip and facial movements (Benoit & Le Goff, 1998; Bernstein & Benoit, 1996; Cohen & Massaro, 1993; Massaro & Stork, 1998; McGrath & Summerfield, 1985; McGurk & MacDonald, 1976; McLeod & Summerfield, 1987; Robert-Ribes, Schwartz, Lallouache, & Escudier, 1998; Sumby & Pollack, 1954; Summerfield, 1992; Vatikiotis-Bateson, Munhall, Hirayama, Lee, & Terzopoulos, 1996). Among the many contributions of this literature has been a detailed classification of human lip movements (visemes) and the viseme-phoneme mappings that occur during articulated speech. Existing systems that have processed combined

---

[1]However, other recent pen/voice multimodal systems that emphasize mobile processing, such as MiPad and the Field Medic Information System (Holzman, 1999; Huang et al., 2000), still limit pen input to pointing.

Multimodal System Characteristics:	QuickSet	Human-Centric Word Processor	VR Aircraft Maintenance Training	Field Medic Information	Portable Voice Assistant
Recognition of simultaneous or alternative individual modes	Simultaneous & individual modes	Simultaneous & individual modes	Simultaneous & individual modes	Alternative individual modes[1]	Simultaneous & individual modes
Type & size of gesture vocabulary	Pen input, Multiple gestures, Large vocabulary	Pen input, Deictic selection	3D manual input, Multiple gestures, Small vocabulary	Pen input, Deictic selection	Pen input, Deictic selection[2]
Size of speech vocabulary[3] & type of linguistic processing	Moderate vocabulary, Grammar-based	Large vocabulary, Statistical language processing	Small vocabulary, Grammar-based	Moderate vocabulary, Grammar-based	Small vocabulary, Grammar-based
Type of signal fusion	Late semantic fusion, Unification, Hybrid symbolic/statistical MTC framework	Late semantic fusion, Frame-based	Late semantic fusion, Frame-based	No mode fusion	Late semantic fusion, Frame-based
Type of platform & applications	Wireless handheld, Varied map & VR applications, digital paper	Desktop computer, Word processing	Virtual reality system, Aircraft maintenance training	Wireless handheld, Medical field emergencies	Wireless handheld, Catalogue ordering
Evaluation status	Proactive user-centered design & iterative system evaluations	Proactive user-centered design	Planned for future	Proactive user-centered design & iterative system evaluations	Planned for future

[1]The FMA component recognizes speech only, and the FMC component recognizes gestural selections or speech. The FMC also can transmit digital speech and ink data, and can read data from smart cards and physiological monitors.

[2]The PVA also performs handwriting recognition.

[3]A small speech vocabulary is up to 200 words, moderate 300–1,000 words, and large in excess of 1,000 words. For pen-based gestures, deictic selection is an individual gesture, a small vocabulary is 2–20 gestures, moderate 20–100, and large in excess of 100 gestures.

FIGURE 5.1. Examples of functionality, architectural features, and general classification of different speech and gesture multimodal applications.

speech and lip movements include the classic work by Petajan (1984), Brooke and Petajan (1986), and others (Adjoudani & Benoit, 1995; Bregler & Konig, 1994; Silsbee & Su, 1996; Tomlinson, Russell, & Brooke, 1996). Additional examples of speech and lip movement systems, applications, and relevant cognitive science research have been detailed elsewhere (Benoit et al., 2000). Researchers in this area have been actively exploring adaptive techniques for improving system robustness, especially in noisy environmental contexts (Dupont & Luettin, 2000; Meier, Hürst, & Duchnowski, 1996; Potamianos, Neti, Gravier, & Garg, 2003; Rogozan & Deglise, 1998), which is an important future research direction. Although this literature has not emphasized the development of applications, nonetheless its quantitative modeling of synchronized phoneme/viseme patterns has been used to build animated characters that generate text-to-speech output and coordinated lip movements. These new animated characters are being used as an interface design vehicle for facilitating users' multimodal interaction with next-generation conversational interfaces (Cassell, Sullivan, Prevost, & Churchill, 2000; Cohen & Massaro, 1993).

While the main multimodal literatures to date have focused on either speech and pen input or speech and lip movements, recognition of other modes also is maturing and beginning to be integrated into new kinds of multimodal systems. In particular, there is growing interest in designing multimodal interfaces that incorporate vision-based technologies, such as interpretation of gaze, facial expressions, head nodding, gesturing, and large body movements (Flanagan & Huang, 2003; Morency, Sidner, Lee, & Darrell, 2005; Morimoto, Koons, Amir, Flickner, & Zhai, 1999; Pavlovic, Berry, & Huang, 1997; Turk & Robertson, 2000; Zhai et al., 1999). These technologies unobtrusively or *passively* monitor user behavior and need not require explicit user commands to a "computer." That contrasts with *active input modes*, such as speech or pen, which the user deploys intentionally as a command issued to the system (see Fig. 5.2). While passive modes may be "attentive" and less obtrusive, active modes generally are more reliable indicators of user intent.

As vision-based technologies mature, one important future direction will be the development of *blended* multimodal interfaces that combine both passive and active modes. These interfaces typically will be *temporally cascaded*, so one goal in designing new prototypes will be to determine optimal processing strategies for using advance information from the first mode (e.g., gaze) to constrain accurate interpretation of the following modes (e.g., gesture, speech). This kind of blended multimodal interface potentially can provide users with greater transparency and control, while also supporting improved robustness and broader application functionality (Oviatt & Cohen, 2000; Zhai et al., 1999). As this collection of technologies matures, there also is strong interest in designing new types of pervasive and mobile interfaces, including ones capable of adaptive processing to the user and environmental context.

As multimodal interfaces gradually evolve toward supporting more advanced recognition of users' natural activities in context, they will expand beyond rudimentary bimodal systems to ones that incorporate three or more input modes, qualitatively different modes, and more sophisticated models of multimodal interaction. This trend already has been initiated within biometrics research, which has combined recognition of multiple behavioral input modes (e.g., voice, handwriting) with physiological ones (e.g., retinal scans, fingerprints) to achieve reliable person identification and verification in challenging field conditions (Choudhury, Clarkson, Jebara, & Pentland, 1999; Jain et al., 1999; Jain & Ross, 2002; Pankanti et al., 2000).

Apart from these developments within research-level systems, multimodal interfaces also are being commercialized as products, especially in areas like personal information access and management on handhelds and cell phones. Microsoft's handheld Mipad for personal information management, and Kirusa's cell phone interface for directory assistance, messaging, and so on, are just two examples of the many mobile commercial products that are being developed with multimodal interfaces. Both include spoken language processing and a stylus for tapping on fields to constrain and guide the natural language processing. In some cases, keyboard input is supported as a third option, as well as multimedia output in the form of visualizations and text-to-speech. Another visible growth area for multimodal interfaces involves in-vehicle control of navigation, communication, and entertainment systems, which has emerged in both domestic and import cars. Mobile map-based systems and systems for safety-critical medical and military applications also are being commercialized by companies like Natural Interaction Systems (e.g., Rasa, NISMap, and NISChart applications), which places an emphasis on developing tangible multimodal interfaces that preserve users' existing work practice, minimize cognitive load, and provide backups in case of system failure (Cohen & McGee, 2004; McGee, 2003).

WHAT ARE THE GOALS AND ADVANTAGES OF MULTIMODAL INTERFACE DESIGN?

Over the past decade, numerous advantages of multimodal interface design have been documented. Unlike a traditional keyboard-and-mouse interface or a unimodal recognition-based interface, multimodal interfaces permit flexible use of input modes. This includes the choice of which modality to use for conveying different types of information, to use combined input modes, or to alternate between modes at any time. Since individual input modalities are well suited in some situations, and less ideal or even inappropriate in others, modality choice is an important design issue in a multimodal system. As systems become more complex and multifunctional, a single modality simply does not permit all users to interact effectively across all tasks and environments.

Since there are large individual differences in ability and preference to use different modes of communication, a multimodal interface permits diverse user groups to exercise selection and control over how they interact with the computer (Fell et al., 1994; Karshmer & Blattner, 1998). In this respect, multimodal interfaces have the potential to accommodate a broader range of users than traditional interfaces, including users of different ages, skill levels, native language status, cognitive styles, sensory impairments, and other temporary illnesses or permanent handicaps. For example, a visually impaired user or one with repetitive stress injury may prefer speech input and text-to-speech output. In contrast, a user with a hearing impairment

Multimodal interfaces process two or more combined user input modes— such as speech, pen, touch, manual gestures, gaze, and head and body movements— in a coordinated manner with multimedia system output. They are a new class of interfaces that aim to recognize naturally occurring forms of human language and behavior, and which incorporate one or more recognition-based technologies (e.g., speech, pen, vision).

Active input modes are ones that are deployed by the user intentionally as an explicit command to a computer system (e.g., speech).

Passive input modes refer to naturally occurring user behavior or actions that are recognized by a computer (e.g., facial expressions, manual gestures). They involve user input that is unobtrusively and passively monitored, without requiring any explicit command to a computer.

Blended multimodal interfaces are ones that incorporate system recognition of at least one passive and one active input mode. (e.g., speech and lip movement systems).

Temporally-cascaded multimodal interfaces are ones that process two or more user modalities that tend to be sequenced in a particular temporal order (e.g., gaze, gesture, speech), such that partial information supplied by recognition of an earlier mode (e.g., gaze) is available to constrain interpretation of a later mode (e.g., speech). Such interfaces may combine only active input modes, only passive ones, or they may be blended.

Mutual disambiguation involves disambiguation of signal or semantic-level information in one error-prone input mode from partial information supplied by another. Mutual disambiguation can occur in a multimodal architecture with two or more semantically rich recognition-based input modes. It leads to recovery from unimodal recognition errors within a multimodal architecture, with the net effect of suppressing errors experienced by the user.

Simultaneous integrator refers to a user who habitually presents two input signals (e.g., speech, pen) in a temporally overlapped manner when communicating multimodal commands to a system.

Sequential integrator refers to a user who habitually separates their multimodal signals, presenting one before the other with a brief pause intervening.

Multimodal hypertiming refers to the fact that both sequential and simultaneous integrators will further accentuate their basic multimodal integration pattern when under duress (e.g., as task difficulty or system recognition errors increase).

Visemes refers to the detailed classification of visible lip movements that correspond with consonants and vowels during articulated speech. A *viseme-phoneme mapping* refers to the correspondence between visible lip movements and audible phonemes during continuous speech.

Feature-level fusion is a method for fusing low-level feature information from parallel input signals within a multimodal architecture, which has been applied to processing closely synchronized input such as speech and lip movements.

Semantic-level fusion is a method for integrating semantic information derived from parallel input modes in a multimodal architecture, which has been used for processing speech and gesture input.

FIGURE 5.2. Multimodal interface terminology.

or accented speech may prefer touch, gesture, or pen input. The natural alternation between modes that is permitted by a multimodal interface also can be effective in preventing overuse and physical damage to any single modality, especially during extended periods of computer use (Markinson[2], personal communication, 1993).

Multimodal interfaces also provide the adaptability that is needed to accommodate the continuously changing conditions of mobile use. In particular, systems involving speech, pen, or touch input are suitable for mobile tasks and, when combined, users can shift among these modalities from moment to mo-ment as environmental conditions change (Holzman, 1999; Oviatt, 2000b, 2000c). There is a sense in which mobility can induce a state of temporary disability, such that a person is unable to use a particular input mode for some period. For example, the user of an in-vehicle application may frequently be unable to use manual or gaze input, although speech is relatively more available. In this respect, a multimodal interface permits the modality choice and switching that is needed during the changing environmental circumstances of actual field and mobile use.

A large body of data documents that multimodal interfaces satisfy higher levels of user preference when interacting with

---

[2]R. Markinson, University of California at San Francisco Medical School, 1993.

simulated or real computer systems. Users have a strong preference toward interacting multimodally, rather than unimodally, across a wide variety of different application domains, although this preference is most pronounced in spatial domains (Hauptmann, 1989; Oviatt, 1997). For example, 95% to 100% of users preferred to interact multimodally when they were free to use either speech or pen input in a map-based spatial domain (Oviatt, 1997). During pen/voice multimodal interaction, users preferred speech input for describing objects and events, sets and subsets of objects, out-of-view objects, conjoined information, and past and future temporal states, and for issuing commands for actions or iterative actions (Cohen & Oviatt, 1995; Oviatt & Cohen, 1991). However, their preference for pen input increased when conveying digits, symbols, graphic content, and especially when conveying the location and form of spatially oriented information on a dense graphic display such as a map (Oviatt & Olsen, 1994; Oviatt, 1997; Suhm, 1998). Likewise, 71% of users combined speech and manual gestures multimodally, rather than using one input mode, when manipulating graphic objects on a CRT screen (Hauptmann, 1989).

During the early design of multimodal systems, it was assumed that efficiency gains would be the main advantage of designing an interface multimodally, and that this advantage would derive from the ability to process input modes in parallel. It is true that multimodal interfaces sometimes support improved efficiency, especially when manipulating graphical information. In simulation research comparing speech-only with multimodal pen/voice interaction, empirical work demonstrated that multimodal interaction yielded 10% faster task-completion time during visual-spatial tasks, but no significant efficiency advantage in verbal or quantitative task domains (Oviatt, 1997; Oviatt, Cohen, & Wang, 1994). Likewise, users' efficiency improved when they combined speech and gestures multimodally to manipulate 3D objects, compared with unimodal input (Hauptmann, 1989). In another early study, multimodal speech-and-mouse input improved efficiency in a line-art drawing task (Leatherby & Pausch, 1992). Finally, in a study that compared task-completion times for a graphical interface versus a multimodal pen/voice interface, military domain experts averaged four times faster at setting up complex simulation scenarios on a map when they were able to interact multimodally (Cohen, McGee, & Clow, 2000). This latter study was based on testing of a fully functional multimodal system, and it included time required to correct recognition errors.

One particularly advantageous feature of multimodal interface design is its superior error handling, both in terms of error avoidance and graceful recovery from errors (Oviatt & van Gent, 1996; Oviatt, Bernard, & Levow, 1999; Oviatt, 1999a; Rudnicky & Hauptmann, 1992; Suhm, 1998; Tomlinson et al., 1996). There are user-centered and system-centered reasons why multimodal systems facilitate error recovery, when compared with unimodal recognition-based interfaces. For example, in a multimodal speech and pen-based gesture interface users will select the input mode that they judge to be less error prone for particular lexical content, which tends to lead to error avoidance (Oviatt & van Gent, 1996). They may prefer speedy speech input, but will switch to pen input to communicate a foreign surname. Secondly, users' language often is simplified when interacting multimodally, which can substantially reduce the complexity of

natural language processing and thereby reduce recognition errors, as described later in this chapter (Oviatt & Kuhn, 1998). In one study, users' multimodal utterances were documented to be briefer, to contain fewer complex locative descriptions, and 50% fewer spoken disfluencies, when compared with a speech-only interface. Thirdly, users have a strong tendency to switch modes after system recognition errors, which facilitates error recovery. This error resolution occurs because the confusion matrices differ for any given lexical content for the different recognition technologies involved in processing (Oviatt et al., 1999; Oviatt, 2002).

In addition to these user-centered reasons for better error avoidance and resolution, there also are system-centered reasons for superior error handling. A well-designed multimodal architecture with two semantically rich input modes can support *mutual disambiguation* of input signals. For example, if a user says "ditches" but the speech recognizer confirms the singular "ditch" as its best guess, then parallel recognition of several graphic marks can result in recovery of the correct plural interpretation. This recovery can occur in a multimodal architecture even though the speech recognizer initially ranks the plural interpretation "ditches" as a less preferred choice on its n-best list. Mutual disambiguation involves recovery from unimodal recognition errors within a multimodal architecture, because semantic information from each input mode supplies partial disambiguation of the other mode, thereby leading to more stable and robust overall system performance (Oviatt, 1999a, 2000a, 2002). Another example of mutual disambiguation is shown in Fig. 5.3. To achieve optimal error handling, a multimodal interface ideally should be designed to include complementary input modes, and

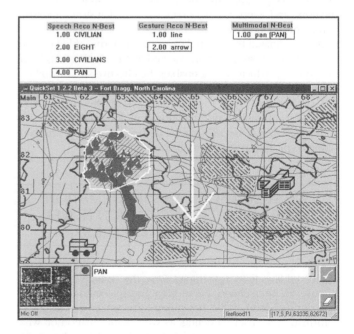

FIGURE 5.3. Multimodal command to "pan" the map, which illustrates mutual disambiguation occurring between incoming speech and gesture information, such that lexical hypotheses were pulled up on both *n*-best lists to produce a correct final multimodal interpretation.

the alternative input modes should provide duplicate functionality such that users can accomplish their goals using either mode.

In two recent studies involving over 4,600 multimodal commands, a multimodal architecture was found to support mutual disambiguation and error suppression ranging between 19% and 41% (Oviatt, 1999a, 2000a, 2002). Improved robustness also was greater for "challenging" user groups (accented vs. native speakers) and usage contexts (mobile vs. stationary use). These results indicate that a well-designed multimodal system not only can perform more robustly than a unimodal system, but also in a more stable way across varied real-world users and usage contexts. Finally, during audio-visual perception of speech and lip movements, improved speech recognition also has been demonstrated for both human listeners (McLeod & Summerfield, 1987) and multimodal systems (Adjoudani & Benoit, 1995; Tomlinson et al., 1996).

Another recent focus has been on the advantages of multimodal interface design for minimizing users' cognitive load. As task complexity increases, there is evidence that users self-manage their working memory limits by distributing information across multiple modalities, which in turn enhances their task performance during both perception and production (Calvert, Spence, & Stein, 2004; Mousavi, Low, & Sweller, 1995; Oviatt, 1997; Oviatt, Coulston, & Lunsford, 2004; Tang, McLachlan, Lowe, Saka, & MacLean, 2005). These predictions and findings are based on Wickens and colleagues' cognitive resource theory and Baddeley's theory of working memory (Baddeley, 1992; Wickens, Sandry, & Vidulich, 1983). The latter maintains that short-term or working memory consists of multiple independent processors associated with different modes. This includes a visual-spatial "sketch pad" that maintains visual materials such as pictures and diagrams in one area of working memory, and a separate phonological loop that stores auditory-verbal information. Although these two processors are believed to be coordinated by a central executive, in terms of lower-level modality processing they are viewed as functioning largely independently, which is what enables the effective size of working memory to expand when people use multiple modalities during tasks (Baddeley, 1992). So with respect to management of cognitive load, the inherent flexibility of multimodal interfaces is well suited to accommodating the high and changing load conditions typical of realistic mobile use.

WHAT METHODS AND INFORMATION HAVE BEEN USED TO DESIGN NOVEL MULTIMODAL INTERFACES?

The design of new multimodal systems has been inspired and organized largely by two things. First, the cognitive science literature on intersensory perception and intermodal coordination during production is beginning to provide a foundation of information for user modeling, as well as information on what systems must recognize and how multimodal architectures should be organized. For example, the cognitive science literature has provided knowledge of the natural integration patterns that typify people's lip and facial movements with speech output (Benoit, Guiard-Marigny, Le Goff, & Adjoudani, 1996; Ekman, 1992; Ekman & Friesen, 1978; Fridlund, 1994; Hadar, Steiner, Grant, & Rose, 1983; Massaro & Cohen, 1990; Stork & Hennecke, 1995; Vatikiotis-Bateson et al., 1996), and their coordinated use of manual or pen-based gestures with speech (Kendon, 1980; McNeill, 1992; Oviatt, DeAngeli, & Kuhn, 1997). Given the complex nature of users' multimodal interaction, cognitive science has played and will continue to play an essential role in guiding the design of robust multimodal systems. In this respect, a multidisciplinary perspective will be more central to successful multimodal system design than it has been for traditional GUI design. The cognitive science underpinnings of multimodal system design are described later in this chapter.

Secondly, high-fidelity automatic simulations also have played a critical role in prototyping new types of multimodal systems (Dahlbäck, Jëonsson, & Ahrenberg, 1992; Oviatt et al., 1992). When a new multimodal system is in the planning stages, design sketches and low-fidelity mock-ups may initially be used to visualize the new system and plan the sequential flow of human-computer interaction. These tentative design plans then are rapidly transitioned into a higher-fidelity simulation of the multimodal system, which is available for proactive and situated data collection with the intended user population. High-fidelity simulations have been the preferred method for designing and evaluating new multimodal systems, and extensive data collection with such tools preferably is completed before a fully functional system ever is built.

During high-fidelity simulation testing, a user interacts with what she believes is a fully functional multimodal system although the interface is actually a simulated front-end designed to appear and respond as the fully functional system would. During the interaction, a programmer assistant at a remote location provides the simulated system responses. As the user interacts with the front end, the programmer tracks her multimodal input and provides system responses as quickly and accurately as possible. To support this role, the programmer makes use of automated simulation software that is designed to support interactive speed, realism with respect to the targeted system, and other important characteristics. For example, with these automated tools, the programmer may be able to make a single selection on a workstation field to rapidly send simulated system responses to the user during a data-collection session.

High-fidelity simulations have been the preferred method for prototyping multimodal systems for several reasons. Simulations are relatively easy and inexpensive to adapt, compared with building and iterating a complete system. They also permit researchers to alter a planned system's characteristics in major ways (e.g., input and output modes available), and to study the impact of different interface features in a systematic and scientific manner (e.g., type and base-rate of system errors). In comparison, a particular system with its fixed characteristics is a less flexible and suitable research tool, and the assessment of any single system basically amounts to an individual case study. Using simulation techniques, rapid adaptation and investigation of planned system features permit researchers to gain a broader and more principled perspective on the potential of newly emerging technologies. In a practical sense, simulation research can assist in the evaluation of critical performance trade-offs and in making decisions about alternative system designs, which designers must do as they strive to create more usable multimodal systems.

To support the further development and commercialization of multimodal systems, additional infrastructure that will be needed in the future includes (a) simulation tools for rapidly building and reconfiguring multimodal interfaces, (b) automated tools for collecting and analyzing multimodal corpora, and (c) automated tools for iterating new multimodal systems to improve their performance (see Oviatt et al., 2000, for further discussion).

WHAT ARE THE COGNITIVE SCIENCE UNDERPINNINGS OF MULTIMODAL INTERFACE DESIGN?

This section discusses the growing cognitive science literature that provides the empirical underpinnings needed to design next-generation multimodal interfaces. The ability to develop multimodal systems depends on knowledge of the natural integration patterns that typify people's combined use of different input modes. In particular, the design of new multimodal systems depends on intimate knowledge of the properties of different modes and the information content they carry, the unique characteristics of multimodal language and its processability, and the integration and synchronization characteristics of users' multimodal interaction. It also relies on accurate prediction of when users are likely to interact multimodally, and how alike different users are in their specific integration patterns. The relevant cognitive science literature on these topics is very extensive, especially when consideration is given to all of the underlying sensory perception and production capabilities involved in different input modes currently being incorporated in new multimodal interfaces. As a result, this section will be limited to introducing the main cognitive science themes and findings that are relevant to the more common types of multimodal system.

This cognitive science foundation also has played a key role in identifying computational "myths" about multimodal interaction, and replacing these misconceptions with contrary empirical evidence. Figure 5.4 summarizes 10 common myths

---

Ten Myths of Multimodal Interaction

Myth #1: *If you build a multimodal system, users will interact multimodally*

Myth #2: *Speech & pointing is the dominant multimodal integration pattern*

Myth #3: *Multimodal input involves simultaneous signals*

Myth #4: *Speech is the primary input mode in any multimodal system that includes it*

Myth #5: *Multimodal language does not differ linguistically from unimodal language*

Myth #6: *Multimodal integration involves redundancy of content between modes*

Myth #7: *Individual error-prone recognition technologies combine multimodally to produce even greater unreliability*

Myth #8: *All users' multimodal commands are integrated in a uniform way*

Myth #9: *Different input modes are capable of transmitting comparable content*

Myth #10: *Enhanced efficiency is the main advantage of multimodal systems*

(taken from Oviatt, 1999b)

FIGURE 5.4. Ten myths of multimodal interaction: Separating myth from empirical reality.

about multimodal interaction, which are addressed and discussed in more detail elsewhere (Oviatt, 1999b). As such, the literature summarized in this section aims to provide a more accurate foundation for guiding the design of next-generation multimodal systems.

When Do Users Interact Multimodally?

During natural interpersonal communication, people are always interacting multimodally. Of course, in this case the number of information sources or modalities that an interlocutor has available to monitor is essentially unlimited. However, all multimodal systems are constrained in the number and type of input modes they can recognize. Also, a user can compose active input during human-computer interaction that either is delivered multimodally or that is delivered entirely using just one mode. That is, although users in general may have a strong preference to interact multimodally rather than unimodally, this is no guarantee that they will issue every command to a system multimodally, given the particular type of multimodal interface available. Therefore, the first nontrivial question that arises during system processing is whether a user is communicating unimodally or multimodally.

In the case of speech- and pen-based multimodal systems, users typically mix unimodal and multimodal expressions. In one study involving a visual-spatial domain, users' commands were expressed multimodally 20% of the time, with others just spoken or written (Oviatt et al., 1997). In contrast, in other spatial domains, the ratio of users' multimodal interaction often is 65% to 70% (Oviatt, 1999b; Oviatt et al., 2004). Predicting whether a user will express a command multimodally also depends on the type of action she is performing. In particular, users usually express commands multimodally when describing spatial information about the location, number, size, orientation, or shape of an object. In one study, users issued multimodal commands 86% of the time when they had to add, move, modify, or calculate the distance between objects on a map in a way that required specifying spatial locations (Oviatt et al., 1997). They also were moderately likely to interact multimodally when selecting an object from a larger array, for example, when deleting a particular object from the map. However, when performing general actions without any spatial component, such as printing a map, users expressed themselves multimodally less than 1% of the time. These data emphasize that future multimodal systems will need to distinguish between instances in which users are and are not communicating multimodally, so that accurate decisions can be made about when parallel input streams should be interpreted jointly versus individually. They also suggest that knowledge of the type of actions to be included in an application, such as whether the application entails manipulating spatial information, should influence the basic decision of whether to build a multimodal interface at all.

Findings from a more recent study reveal that multimodal interface users spontaneously respond to dynamic changes in their own cognitive load by shifting to multimodal communication as load increases with task difficulty and communicative complexity (Oviatt, Coulston, & Lunsford, 2004). Given a flexible multimodal interface, users' ratio of multimodal (versus unimodal) interaction increased substantially from 18.6% when referring to established dialogue context to 77.1% when required to establish a new context, a +315% relative increase. Likewise, the ratio of users' multimodal interaction increased significantly as the tasks became more difficult, from 59.2% during low difficulty tasks, to 65.5% at moderate difficulty, 68.2% at high, and 75.0% at very high difficulty, an overall relative increase of +27%. These adaptations in multimodal interaction levels reflect users' efforts to self-manage limitations in their working memory as discourse-level demands and task complexity increased. As discussed earlier, users accomplished this by distributing communicative information across multiple modalities in a manner compatible with a cognitive load theory of multimodal interaction. This interpretation is consistent with Baddeley's theory of working memory (Baddeley, 1992), as well as the growing literatures within education (Mousavi, Low, & Sweller, 1995; Sweller, 1988), linguistics (Almor, 1999), and multisensory perception (Calvert, Spence, & Stein, 2004; Ernst & Bulthoff, 2004). Recent work on visual and haptic processing under workload also indicates that presentation of haptic feedback during a complex task can augment users' ability to handle visual information overload (Tang, McLachlan, Lowe, Saka, & MacLean, 2005).

In a multimodal interface that processes passive or blended input modes, there always is at least one passively tracked input source providing continuous information (e.g., gaze tracking, head position). In these cases, all user input would by definition be classified as multimodal, and the primary problem would become segmentation and interpretation of each continuous input stream into meaningful actions of significance to the application. In the case of blended multimodal interfaces (e.g., gaze tracking and mouse input), it still may be opportune to distinguish active forms of user input that might be more accurately or expeditiously handled as unimodal events.

What Are the Integration and Synchronization Characteristics of Users' Multimodal Input?

The past literature on multimodal systems has focused largely on simple selection of objects or locations in a display, rather than considering the broader range of multimodal integration patterns. Since the development of Bolt's (1980) "Put That There" system, speak-and-point has been viewed as the prototypical form of multimodal integration. In Bolt's system, semantic processing was based on spoken input, but the meaning of a deictic term such as "that" was resolved by processing the x/y coordinate indicated by pointing at an object. Since that time, other multimodal systems also have attempted to resolve deictic expressions using a similar approach—for example, using gaze location instead of manual pointing (Koons et al., 1993).

Unfortunately, this concept of multimodal interaction as point-and-speak makes only limited use of new input modes for selection of objects—just as the mouse does. In this respect, it represents the persistence of an old mouse-oriented metaphor. In contrast, modes that transmit written input, manual gesturing, and facial expressions are capable of generating symbolic information that is much more richly expressive than simple pointing or selection. In fact, studies of users' integrated

pen/voice input indicate that a speak-and-point pattern only comprises 14% of all spontaneous multimodal utterances (Oviatt et al., 1997). Instead, pen input more often is used to create graphics, symbols and signs, gestural marks, digits, and lexical content. During interpersonal multimodal communication, linguistic analysis of spontaneous manual gesturing also indicates that simple pointing accounts for less than 20% of all gestures (McNeill, 1992). Together, these cognitive science and user-modeling data highlight the fact that any multimodal system designed exclusively to process speak-and-point will fail to provide users with much useful functionality. For this reason, specialized algorithms for processing deictic-point relations will have only limited practical use in the design of future multimodal systems. It is clear that a broader set of multimodal integration issues needs to be addressed in future work. Future research also should explore typical integration patterns between other promising modality combinations, such as speech and gaze.

It also is commonly assumed that any signals involved in a multimodal construction will co-occur temporally. The presumption is that this temporal overlap then determines which signals to combine during system processing. In the case of speech and manual gestures, successful processing of the deictic term "that square" in Bolt's original system relied on interpretation of pointing when the word "that" was spoken in order to extract the intended referent. However, one empirical study indicated that users often do not speak deictic terms at all, and when they do, the deictic frequently is not overlapped in time with their pointing. In fact, it has been estimated that as few as 25% of users' commands actually contain a spoken deictic that overlaps with the pointing needed to disambiguate its meaning (Oviatt et al., 1997).

Beyond the issue of deixis, a series of studies has shown that users' input frequently does not overlap at all during multimodal commands to a computer (Oviatt, 1999b; Oviatt et al., 2003; Oviatt et al., 2005; Xiao, Girand, & Oviatt, 2002; Xiao, Lunsford, Coulston, Wesson, & Oviatt, 2003). In fact, there are two distinct types of user with respect to integration patterns: *simultaneous* integrators and *sequential* ones. A user who habitually integrates her speech and pen input in a *simultaneous* manner overlaps them temporally, whereas a sequential integrator finishes one mode before beginning the second, as summarized in Fig. 5.2. These two types of user integration patterns occur across the lifespan from children through the elderly (Oviatt et al., 2005; Xiao et al., 2002; Xiao et al., 2003). They also can be detected almost immediately during multimodal interaction, usually on the very first input. Users' habitual integration pattern remains strikingly highly consistent during a session, as well as resistant to change following explicit instructions or attempts at training (Oviatt et al., 2003; Oviatt et al., 2005). This bimodal distribution of user integration patterns has been observed in different task domains (e.g., map-based real estate selection, crisis management, educational applications with animated characters), and also when using different types of interface (e.g., conversational, command style) (Oviatt, 1999b; Xiao et al., 2002; Xiao et al., 2003). In short, empirical studies have demonstrated that this bimodal distinction between users in their fundamental integration pattern generalizes widely across different age groups, task domains, and types of interface.

One interesting discovery in recent work is the phenomenon of *multimodal hypertiming*, which refers to the fact that both sequential and simultaneous integrators will entrench further or accentuate their habitual multimodal integration pattern (*e.g.*, increasing their intermodal *lag* during sequential integrations, or *overlap* during simultaneous integrations, as summarized in Fig. 5.2) during system error handling or when completing increasingly difficult tasks. In fact, users will progressively increase their degree of entrenchment by 18% as system errors increase, and by 59% as task difficulty increases (Oviatt et al., 2003). As such, changes in the degree of users' multimodal hypertiming provide a potentially sensitive means of evaluating their cognitive load during real-time interactive exchanges. In the context of system error handling, the phenomenon of multimodal hypertiming basically replaces the hyperarticulation that is typically observed in users during error-prone speech-only interactions.

Given the bimodal distribution of user integration patterns, *adaptive temporal thresholds* potentially could support more tailored and flexible approaches to fusion. Ideally, an adaptive multimodal system would detect, automatically learn, and adapt to a user's dominant multimodal integration pattern, which could result in substantial improvements in system processing speed, accuracy of interpretation, and synchronous interchange with the user. For example, it has been estimated that system delays could be reduced to approximately 40% to 50% of what they currently are by adopting user-defined thresholds (Oviatt et al., 2005). Recent research has begun comparing different learning-based models for adapting a multimodal system's temporal thresholds to an individual user in real time (Huang & Oviatt, in press).

Unfortunately, users' multimodal integration patterns have not been studied as extensively or systematically for other input modes, such as speech and manual gesturing. Linguistics research on interpersonal communication patterns has revealed that both spontaneous gesturing and signed language often precede their spoken lexical analogues during human communication (Kendon, 1980; Naughton, 1996), when considering word-level integration pattern. In fact, the degree to which gesturing precedes speech is greater in topic-prominent languages such as Chinese than it is in subject-prominent ones like Spanish or English (McNeill, 1992). Even in the speech and lip-movement literature, close but not perfect temporal synchrony is typical, with lip movements occurring a fraction of a second before the corresponding auditory signal (Abry, Lallouache, & Cathiard, 1996; Benoit, 2000). However, when considering the whole user utterance as the unit of analysis, some other studies of speech and manual gesturing have found a higher rate of simultaneity for these modes (Epps, Oviatt, & Chen, 2004). Learning-based approaches that are capable of accurately identifying and adapting to different multimodal integration patterns, whether due to differences among users, modality combinations, or applications and usage contexts, will be required in order to generalize and speed up multimodal system development in the future.

In short, although two input modes may be highly interdependent and synchronized during multimodal interaction, synchrony does not imply simultaneity. The empirical evidence reveals that multimodal signals often do not co-occur temporally at all during human-computer or natural human communication. Therefore, multimodal system designers cannot necessarily count on conveniently overlapped signals in order to achieve successful processing in the multimodal architectures they build. Future research needs to explore the integration patterns and temporal cascading that can occur among three or more input modes—such as gaze, gesture, and speech—so that more advanced multimodal systems can be designed and prototyped.

In the design of new multimodal architectures, it is important to note that data on the order of input modes and average time lags between input modes has been used to determine the likelihood that an utterance is multimodal versus unimodal, and to establish temporal thresholds for fusion of input. In the future, weighted likelihoods associated with different utterance segmentations, for example, that an input stream containing speech, writing, speech should be segmented into [S/W S] rather than [S W/S], and with intermodal time lag distributions, will be used to optimize correct recognition of multimodal user input (Oviatt, 1999b). In the design of future time-critical multimodal architectures, data on users' integration and synchronization patterns will need to be collected for other mode combinations during realistic interactive tasks, so that temporal thresholds can be established for performing multimodal fusion.

What Individual Differences Exist in Multimodal Interaction, and What Are the Implications for Designing Systems for Universal Access?

There are large individual differences in users' multimodal interaction patterns, beginning with their overall preference to interact unimodally versus multimodally, and which mode they generally prefer (e.g., speaking versus writing) (Oviatt et al., 2004). As outlined above, there likewise are striking differences among users in adopting either a sequential or simultaneous multimodal integration pattern. Recent research has revealed that these two patterns are associated with behavioral and linguistic differences between the groups (Oviatt et al., 2005). Whereas in an interactive task context their performance speed was comparable, sequential integrators were far less error-prone and excelled during new or complex tasks. Although their speech rate was no slower, sequential integrators also had more precise articulation (e.g., less disfluency). Finally, sequential integrators were more likely to adopt terse and direct command-style language, with a smaller and less varied vocabulary, which appeared focused on achieving error-free communication. These user differences in interaction patterns have been interpreted as deriving from fundamental differences among users in their reflective-impulsive cognitive style (Oviatt et al., 2005). Based on this work, one goal of future multimodal interface design will be to support the poorer attention span and higher error rate of impulsive users—especially for mobile in-vehicle, military, and similar application contexts in which the cost of committing errors is unacceptably high.

Apart from these individual differences, cultural differences also have been documented between users in modality integration patterns. For example, substantial individual differences have been reported in the temporal synchrony between speech and lip movements (Kricos, 1996) and, in addition, lip movements during speech production are known to be less exaggerated among Japanese speakers than Americans (Sekiyama & Tohkura, 1991). In fact, extensive inter-language differences have been observed in the information available from lip movements during audio-visual speech (Fuster-Duran, 1996). These findings have implications for the degree to which disambiguation of speech can be achieved through lip movement information in noisy environments or for different user populations. Finally, nonnative speakers, the hearing impaired, and elderly listeners all are more influenced by visual lip movement than auditory cues when processing speech (Fuster-Duran, 1996; Massaro, 1996). These results have implications for the design and expected value of audio-visual multimedia output for different user groups in animated character interfaces. With respect to support for universal access, recent work also has shown the advantage of combined audio-visual processing for recognition of impaired speech (Potamianos & Neti, 2001).

Finally, gender, age, and other individual differences are common in gaze patterns, as well as speech and gaze integration (Argyle, 1972). As multimodal interfaces incorporating gaze become more mature, further research will need to explore these gender and age-specific patterns, and to build appropriately adapted processing strategies. In summary, considerably more research is needed on multimodal integration and synchronization patterns for new mode combinations, as well as for diverse and disabled users for whom multimodal interfaces may be especially suitable for ensuring universal access.

Is Complementarity or Redundancy the Main Organizational Theme That Guides Multimodal Integration?

It frequently is claimed that the propositional content conveyed by different modes during multimodal communication contains a high degree of redundancy. However, the dominant theme in users' natural organization of multimodal input actually is complementarity of content, not redundancy. For example, speech and pen input consistently contribute different and complementary semantic information, with the subject, verb, and object of a sentence typically spoken, and locative information written (Oviatt et al., 1997). In fact, a major complementarity between speech and manually oriented pen input involves visual-spatial semantic content, which is one reason these modes are an opportune combination for visual-spatial applications. Whereas spatial information is uniquely and clearly indicated via pen input, the strong descriptive capabilities of speech are better suited for specifying temporal and other nonspatial information. Even during multimodal correction of system errors, when users are highly motivated to clarify and reinforce their information delivery, speech and pen input express redundant information less than 1% of the time. Finally, during interpersonal communication linguists also have documented that spontaneous speech and manual gesturing involve complementary rather than duplicate information between modes (McNeill, 1992).

Other examples of primary multimodal complementarities during interpersonal and human–computer communication have been described in past research (McGurk & MacDonald,

1976; Oviatt & Olsen, 1994; Wickens et al., 1983). For example, in the literature on multimodal speech and lip movements, natural feature-level complementarities have been identified between visemes and phonemes for vowel articulation, with vowel rounding better conveyed visually, and vowel height and backness better revealed auditorally (Massaro & Stork, 1998; Robert-Ribes et al., 1998).

In short, actual data highlight the importance of complementarity as a major organizational theme during multimodal communication. The designers of next-generation multimodal systems therefore should not expect to rely on duplicated information when processing multimodal language, although in certain contexts (such as teaching) a greater percentage of duplicate content than usual may be expected to exist. In multimodal systems involving both speech and pen-based gestures and speech and lip movements, one explicit goal has been to integrate complementary modalities in a manner that yields a synergistic blend, such that each mode can be capitalized upon and used to overcome weaknesses in the other mode (Cohen et al., 1989). This approach to system design has promoted the philosophy of using modes and component technologies to their natural advantage, and of combining them in a manner that permits mutual disambiguation. One advantage of achieving such a blend is that the resulting multimodal architecture can function more robustly than an individual recognition-based technology or a multimodal system based on input modes lacking natural complementarities.

What Are the Primary Features of Multimodal Language?

Communication channels can be tremendously influential in shaping the language transmitted within them. From past research, there now is cumulative evidence that many linguistic features of multimodal language are qualitatively very different from that of spoken or formal textual language. In fact, it can differ in features as basic as brevity, semantic content, syntactic complexity, word order, disfluency rate, degree of ambiguity, referring expressions, specification of determiners, anaphora, deixis, and linguistic indirectness. In many respects, multimodal language is simpler linguistically than spoken language. In particular, comparisons have revealed that the same user completing the same map-based task communicates significantly fewer words, briefer sentences, and fewer complex spatial descriptions and disfluencies when interacting multimodally, compared with using speech alone (Oviatt, 1997). One implication of these findings is that multimodal interface design has the potential to support more robust future systems than a unimodal design approach. The following is an example of a typical user's spoken input while attempting to designate an open space using a map system: "Add an open space on the north lake to b—include the north lake part of the road and north." In contrast, the same user accomplished the same task multimodally by encircling a specific area and saying, "Open space."

In previous research, hard-to-process, disfluent language has been observed to decrease by 50% during multimodal interaction with a map, compared with a more restricted speech-only interaction (Oviatt, 1997). This drop occurs mainly because people have difficulty speaking spatial information, which precipitates disfluencies. In a flexible multimodal interface, they instead use pen input to convey spatial information, thereby avoiding the need to speak it. Further research is needed to establish whether other forms of flexible multimodal communication also generally ease users' cognitive load, which may be reflected in a reduced rate of disfluencies.

During multimodal pen/voice communication, the linguistic indirection that is typical of spoken language frequently is replaced with more direct commands (Oviatt & Kuhn, 1998). In the following example, a study participant made a disfluent indirect request using speech input while requesting a map-based distance calculation: "What is the distance between the Victorian Museum and the, uh, the house on the east side of Woodpecker Lane?" When requesting distance information multimodally, the same user encircled the house and museum while speaking the following brief direct command: "Show distance between here and here." In this research, the briefer and more direct multimodal pen/voice language also contained substantially fewer referring expressions, with a selective reduction in co-referring expressions that instead were transformed into deictic expressions. This latter reduction in coreference would simplify natural language processing by easing the need for anaphoric tracking and resolution in a multimodal interface. Also consistent with fewer referring expressions, explicit specification of definite and indefinite reference is less common in multimodal language (Oviatt & Kuhn, 1998). Current natural language processing algorithms typically rely heavily on the specification of determiners in definite and indefinite references in order to represent and resolve noun-phrase reference. One unfortunate by-product of the lack of such specifications is that current language processing algorithms are unprepared for the frequent occurrence of elision and deixis in multimodal human-computer interaction.

In other respects, multimodal language clearly is different from spoken language, although not necessarily simpler. For example, users' multimodal pen/voice language departs from the canonical English word order of S-V-O-LOC (e.g., Subject-Verb-Object-Locative constituent), which is observed in spoken language and formal textual language. Instead, users' multimodal constituents shift to a LOC-S-V-O word order. A recent study reported that 95% of locative constituents were in sentence-initial position during multimodal interaction. However, for the same users completing the same tasks while speaking, 96% of locatives were in sentence-final position (Oviatt et al., 1997). It is likely that broader analysis of multimodal communication patterns, which could involve gaze and manual gesturing to indicate location rather than pen-based pointing, would reveal a similar reversal in word order.

One implication of these many differences is that new multimodal corpora, statistical language models, and natural language-processing algorithms will need to be established before multimodal language can be processed optimally. Future research and corpus-collection efforts also will be needed on different types of multimodal communication, and in other application domains, so that the generality of previously identified multimodal language differences can be explored.

WHAT ARE THE BASIC WAYS IN WHICH MULTIMODAL INTERFACES DIFFER FROM GRAPHICAL USER INTERFACES?

Multimodal research groups currently are rethinking and redesigning basic user interface architectures because a whole new range of architectural requirements has been posed. First, graphical user interfaces typically assume that a single event stream controls the underlying event loop, with any processing sequential in nature. For example, most GUIs ignore typed input when a mouse button is depressed. In contrast, multimodal interfaces typically can process continuous and simultaneous input from parallel incoming streams. Secondly, GUIs assume that the basic interface actions, such as selection of an item, are atomic and unambiguous events. In contrast, multimodal systems process input modes using recognition-based technologies, which are designed to handle uncertainty and entail probabilistic methods of processing. Thirdly, GUIs often are built to be separable from the application software that they control, although the interface components usually reside centrally on one machine. In contrast, recognition-based user interfaces typically have larger computational and memory requirements, which often makes it desirable to distribute the interface over a network so that separate machines can handle different recognizers or databases. For example, cell phones and networked PDAs may extract features from speech input, but transmit them to a recognizer that resides on a server. Finally, multimodal interfaces that process two or more recognition-based input streams require time stamping of input, and the development of temporal constraints on mode fusion operations. In this regard, they involve uniquely time-sensitive architectures.

WHAT BASIC ARCHITECTURES AND PROCESSING TECHNIQUES HAVE BEEN USED TO DESIGN MULTIMODAL SYSTEMS?

Many early multimodal interfaces that handled combined speech and gesture, such as Bolt's "Put That There" system (Bolt, 1980), have been based on a control structure in which multimodal integration occurs during the process of parsing spoken language. As discussed earlier, when the user speaks a deictic expression such as "here" or "this," the system searches for a synchronized gestural act that designates the spoken referent. While such an approach is viable for processing a point-and-speak multimodal integration pattern, as discussed earlier, multimodal systems must be able to process richer input than just pointing, including gestures, symbols, graphic marks, lip movements, meaningful facial expressions, and so forth. To support more broadly functional multimodal systems, general processing architectures have been developed since Bolt's time. Some of these recent architectures handle a variety of multimodal integration patterns, as well as the interpretation of both unimodal and combined multimodal input. This kind of architecture can support the development of multimodal systems in which modalities are processed individually as input alternatives to one another, or those in which two or more modes are processed as combined multimodal input.

For multimodal systems designed to handle joint processing of input signals, there are two main subtypes of multimodal architecture. One subtype integrates signals at the *feature level* (e.g., "early fusion"). The other integrates information at a *semantic level* (e.g., "late fusion"). Examples of systems based on an early feature-fusion processing approach include those developed by Bregler and colleagues (Bregler, Manke, Hild, & Waibel, 1993), Vo and colleagues (Vo et al., 1995), and Pavlovic and colleagues (Pavlovic, Sharma, & Huang, 1997; Pavlovic & Huang, 1998). In feature-fusion architecture, the signal-level recognition process in one mode influences the course of recognition in the other. Feature fusion is considered more appropriate for closely temporally synchronized input modalities, such as speech and lip movements (Stork & Hennecke, 1995; Rubin et al., 1998).

In contrast, multimodal systems using the late semantic fusion approach have been applied to processing multimodal speech and pen input or manual gesturing, for which the input modes are less coupled temporally. These input modes provide different but complementary information that typically is integrated at the utterance level. Late semantic integration systems use individual recognizers that can be trained using unimodal data, which are easier to collect and are already publicly available for speech and handwriting. In this respect, systems based on semantic fusion can be scaled up more easily in number of input modes or vocabulary size. Examples of systems based on semantic fusion include Put That There (Bolt, 1980), ShopTalk (Cohen et al., 1989), QuickSet (Cohen et al., 1997), CUBRICON (Neal & Shapiro, 1991), Virtual World (Codella et al., 1992), Finger-Pointer (Fukumoto, Suenaga, & Mase, 1994), VisualMan (Wang, 1995), Human-Centric Word Processor, Portable Voice Assistant (Bers et al., 1998), the VR Aircraft Maintenance Training System (Duncan et al., 1999) and Jeanie (Vo & Wood, 1996).

As an example of multimodal information processing flow in a late-stage semantic architecture, Fig. 5.5 illustrates two input modes (e.g., speech and manual or pen-based gestures) recognized in parallel and processed by an understanding component. The results involve partial meaning representations that are fused by the multimodal integration component, which also is influenced by the system's dialogue management and interpretation of current context. During the integration process, alternative lexical candidates for the final multimodal interpretation

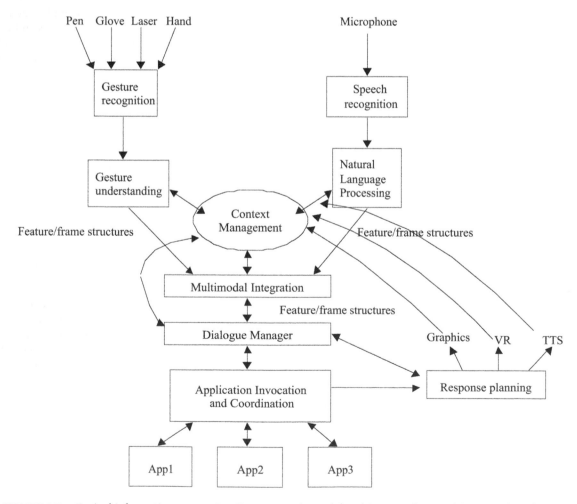

FIGURE 5.5. Typical information processing flow in a multimodal architecture designed for speech and gesture.

are ranked according to their probability estimates on an *n*-best list. The best-ranked multimodal interpretation then is sent to the application invocation and control component, which transforms this information into a series of commands to one or more back-end application systems. System feedback typically includes multimedia output, which may incorporate text-to-speech and non-speech audio; graphics and animation; and so forth. For examples of feature-based multimodal processing flow and architectures, especially as applied to multimodal speech and lip movement systems, see Benoit et al. (2000).

There are many ways to realize this information processing flow as architecture. One common infrastructure that has been adopted by the multimodal research community involves *multi-agent architectures*, such as the Open Agent Architecture (Cohen, Cheyer, Wang, & Baeg, 1994; Martin, Cheyer, & Moran, 1999) and Adaptive Agent Architecture (Kumar & Cohen, 2000). In a multi-agent architecture, the many components needed to support the multimodal system (e.g., speech recognition, gesture recognition, natural language processing, multimodal integration) may be written in different programming languages, on different machines, and with different operating systems. Agent communication languages are being developed that can handle asynchronous delivery, triggered responses, multi-casting, and other concepts from distributed systems, and that are fault-tolerant (Kumar & Cohen, 2000). Using a multi-agent architecture, for example, speech and gestures can arrive in parallel or asynchronously via individual modality agents, with the results recognized and passed to a facilitator. These results, typically an *n*-best list of conjectured lexical items and related time-stamp information, then are routed to appropriate agents for further language processing. Next, sets of meaning fragments derived from the speech and pen signals arrive at the multimodal integrator. This agent decides whether and how long to wait for recognition results from other modalities, based on the system's temporal thresholds. It fuses the meaning fragments into a semantically- and temporally-compatible whole interpretation before passing the results back to the facilitator. At this point, the system's final multimodal interpretation is confirmed by the interface, delivered as multimedia feedback to the user, and executed by any relevant applications. In summary, multi-agent architectures provide essential infrastructure for coordinating the many complex modules needed to implement multimodal system processing, and they permit doing so in a distributed manner that is compatible with the trend toward mobile computing.

The core of multimodal systems based on semantic fusion involves algorithms that integrate common meaning representations derived from speech, gesture, and other modalities into a combined final interpretation. The semantic fusion operation requires a common meaning-representation framework for all modalities, and a well-defined operation for combining partial meanings that arrive from different signals. To fuse information from different modalities, various research groups have independently converged on a strategy of recursively matching and merging attribute/value data structures, although using a variety of different algorithms (Vo & Wood, 1996; Cheyer & Julia, 1995; Pavlovic & Huang, 1998; Shaikh et al., 1997). This approach is considered a *frame-based integration* technique. An alternative logic-based approach derived from computational linguistics (Carpenter, 1990, 1992; Calder, 1987) involves the use of *typed*

feature structures and *unification-based integration*, which is a more general and well-understood approach. Unification-based integration techniques also have been applied to multimodal system design (Cohen et al., 1997; Johnston et al., 1997; Wu et al., 1999). Feature-structure unification is considered well suited to multimodal integration, because unification can combine complementary or redundant input from both modes, but it rules out contradictory input. Given this foundation for multimodal integration, more research still is needed on the development of canonical meaning representations that are common among different input modes that will need to be represented in new types of multimodal systems.

When statistical processing techniques are combined with a symbolic unification-based approach that merges feature structures, then the multimodal architecture that results is a *hybrid symbolic/statistical* one. Hybrid architectures represent one major new direction for multimodal system development. Multimodal architectures also can be hybrids in the sense of combining Hidden Markov Models (HMMs) and Neural Networks (NNs). New hybrid architectures potentially are capable of achieving very robust functioning, compared with either an early- or late-fusion approach alone. For example, the Members-Teams-Committee (MTC) hierarchical recognition technique, which is a hybrid symbolic/statistical multimodal integration framework trained over a labeled multimodal corpus, recently achieved 95.26% correct recognition performance, or within 1.4% of the theoretical system upper bound (Wu et al., 1999). Other architectural approaches and contributions to processing multimodal information have been summarized elsewhere (Oliver & Horvitz, 2005; Potamianos et al., 2003).

WHAT ARE THE MAIN FUTURE DIRECTIONS FOR MULTIMODAL INTERFACE DESIGN?

The computer science community is just beginning to understand how to design innovative, well integrated, and robust multimodal systems. To date, most multimodal systems remain bimodal, and recognition technologies related to several human senses (e.g., haptics, smell, taste) have yet to be well represented within multimodal interfaces. The design and development of new types of systems that include such modes will not be achievable through intuition. Rather, it will depend on knowledge of the natural integration patterns that typify people's combined use of various input modes. This means that the successful design of multimodal systems will continue to require guidance from cognitive science on the coordinated human perception and production of natural modalities. In this respect, multimodal systems only can flourish through multidisciplinary cooperation, as well as teamwork among those representing expertise in the component technologies.

Most of the systems outlined in this chapter have been built during the past 15 years, and they are research-level systems. However, in some cases they have developed well beyond the prototype stage, and are being integrated with other software at academic and federal sites, or appearing as newly shipped products. To achieve wider commercialization of multimodal interfaces, such systems will need to develop more powerful and general methods of natural language and dialogue processing, and temporal modeling and processing of incoming signals. In addition, multimodal datasets and tools are very much needed to build applications more rapidly in a wide range of domains, including for newly emerging collaborative multimodal applications such as meeting support and education (Barthelmess, Kaiser, Huang, & Demirdjian, 2005; Cohen & McGee, 2004; Danninger et al., 2005; Gatica-Perez, Lathoud, Odobez, & McCowan, 2005; McGee, 2003; Pentland, 2005). The many mobile multimodal interfaces currently being built also will require active adaptation to the user, task, ongoing dialogue, and environmental context, which is another very active area of recent work (Gorniak & Roy, 2005; Gupta, 2004; Huang & Oviatt, 2005; Jain & Ross, 2002; Potamianos et al., 2003; Xiao et al., 2003). To facilitate the speed and generality of multimodal interface adaptation to these important variables, future work will need to integrate new machine learning techniques that are now being developed to handle asynchronous and heterogeneous data (Bengio, 2004; McCowan et al., 2005). Finally, in the future a coherent theoretical framework needs to be developed to account for multimodal interaction patterns. This will be invaluable for proactively guiding the design of new multimodal interfaces to be compatible with human capabilities and limitations. Current work in cognitive neuroscience and multisensory perception is beginning to provide an empirical and theoretical basis for this future interface design (Calvert et al., 2004; Ernst & Bulthoff, 2004).

In conclusion, multimodal interfaces are just beginning to model human-like sensory perception. They are recognizing and identifying actions, language, and people that have been seen, heard, or in other ways experienced in the past. They literally reflect and acknowledge the existence of human users, empower them in new ways, and create for them a "voice." They also can be playful and self-reflective interfaces that suggest new forms of human identity as we interact face to face with animated personas representing our own kind. In all of these ways novel multimodal interfaces, as primitive as their early bimodal instantiations may be, represent a new multidisciplinary science, a new art form, and a sociopolitical statement about our collective desire to humanize the technology we create.

ACKNOWLEDGMENTS

I'd like to thank the National Science Foundation for their support over the past decade, which has enabled me to pursue basic exploratory research on many aspects of multimodal interaction, interface design, and system development. The preparation of this chapter has been supported by NSF Grants IRI-9530666, IIS-0117868, and NSF Special Extension for Creativity (SEC) Grant IIS-9530666. This work also has been supported by contracts DABT63-95-C-007, N66001-99-D-8503, and NBCHD030010 from DARPA's Information Technology and Information Systems Office, and Grant No. N00014-99-1-0377 from ONR. Any opinions, findings or conclusions expressed in this chapter are those of the author and do not necessarily reflect the views of the federal agencies sponsoring this work.

I'd also like to thank Phil Cohen and others in the Center for Human-Computer Communication for many insightful discussions, and Dana Director and Rachel Coulston for expert assistance with manuscript preparation. Finally, I wish to acknowledge LEA, Inc. for giving permission to reprint Figs. 5.1 and 5.5, and to acknowledge ACM for allowing the reprint of Figs. 5.3 and 5.4.

References

Abry, C., Lallouache, M.-T., & Cathiard, M.-A. (1996). How can coarticulation models account for speech sensitivity to audio-visual desynchronization? In D. G. Stork & M. E. Hennecke (Eds.), *Speechreading by Humans and Machines: Models, Systems and Applications* (pp. 247–255). New York: Springer Verlag.

Adjoudani, A., & Benoit, C. (1995). Audio-visual speech recognition compared across two architectures. *Proc. of the Eurospeech Conference Vol. 2* (pp. 1563–1566). Madrid, Spain.

Almor, A. (1999). Noun-phrase anaphora and focus: The informational load hypothesis. *Psychological Review, 106,* 748–765.

Argyle, M. (1972). nonverbal communication in human social interaction. In R. Hinde (Ed.), *Nonverbal Communication* (pp. 243–267). Cambridge, UK: Cambridge University Press.

Baddeley, A. (1992). Working memory. *Science, 255,* 556–559.

Bangalore, S., & Johnston, M. (2000). Integrating multimodal language processing with speech recognition. In B. Yuan, T. Huang & X. Tang (Eds.), *Proceedings of the International Conference on Spoken Language Processing (ICSLP'2000) Vol. 2* (pp. 126–129). Beijing: Chinese Friendship Publishers.

Barthelmess, P., Kaiser, E., Huang, X., & Demirdjian, D. (2005). Distributed pointing for multimodal collaboration over sketched diagrams, *Proceedings of the Seventh International Conference on Multimodal Interfaces* (pp. 10–17) New York, N.Y.: ACM Press.

Bengio, S. (2004). Multimodal speech processing using asynchronous Hidden Markov Models, *Information Fusion, 5*(2), 81–89.

Benoit, C., Guiard-Marigny, T., Le Goff, B., & Adjoudani, A. (1996). Which components of the face do humans and machines best speechread? In D. G. Stork & M. E. Hennecke, (Eds.), *Speechreading by Humans and Machines: Models, Systems, and Applications: Vol. 150 of NATO ASI Series. Series F: Computer and Systems Sciences* (pp. 315–325). Berlin, Germany: Springler-Verlag.

Benoit, C., & Le Goff, B. (1998). Audio-visual speech synthesis from French text: Eight years of models, designs and evaluation at the ICP. *Speech Communication, 26,* 117–129.

Benoit, C., Martin, J.-C., Pelachaud, C., Schomaker, L., & Suhm, B. (2000). Audio-visual and multimodal speech-based systems. In D. Gibbon, I. Mertins, & R. Moore (Eds.), *Handbook of Multimodal and Spoken Dialogue Systems: Resources, Terminology and Product Evaluation* (pp. 102–203). Amsterdam, the Netherlands: Kluwer.

Bernstein, L., & Benoit, C. (1996). For speech perception by humans or machines, three senses are better than one. *Proceedings of the International Conference on Spoken Language Processing, (ICSLP 96) Vol. 3* (pp. 1477–1480). New York: IEEE press.

Bers, J., Miller, S., & Makhoul, J. (1998). Designing conversational interfaces with multimodal interaction. *DARPA Workshop on Broadcast News Understanding Systems,* 319–321.

Bolt, R. A. (1980). Put-that-there: Voice and gesture at the graphics interface. *Computer Graphics, 14*(3), 262–270.

Bregler, C., & Konig, Y. (1994). Eigenlips for robust speech recognition. *Proceedings of the International Conference on Acoustics Speech and Signal Processing (IEEE-ICASSP) Vol. 2*, 669–672.

Bregler, C., Manke, S., Hild, H., & Waibel, A. (1993). Improving connected letter recognition by lipreading. *Proceedings of the International Conference on Acoustics, Speech and Signal Processing (IEEE-ICASSP) Vol. 1*, 557–560.

Brooke, N. M., & Petajan, E. D. (1986). Seeing speech: Investigations into the synthesis and recognition of visible speech movements using automatic image processing and computer graphics. *Proceedings International Conference Speech Input and Output: Techniques and Applications, 258*, 104–109.

Calder, J. (1987). Typed unification for natural language processing. In E. Klein & J. van Benthem (Eds.), *Categories, Polymorphisms, and Unification* (pp. 65–72). Edinburgh, Scotland: Center for Cognitive Science, University of Edinburgh.

Calvert, G., Spence, C., & Stein, B. E. (Eds.). (2004). *The Handbook of Multisensory Processing*. Cambridge, MA: MIT Press.

Carpenter, R. (1990). Typed feature structures: Inheritance, (in)equality, and extensionality. *Proceedings of the ITK Workshop: Inheritance in Natural Language Processing*, 9–18. Tilburg the Netherlands: Institute for Language Technology and Artificial Intelligence, Tilburg University.

Carpenter, R. (1992). *The logic of typed feature structures*. Cambridge, England: Cambridge University Press.

Cassell, J., Sullivan, J., Prevost, S., & Churchill, E. (Eds.). (2000). *Embodied conversational agents*. Cambridge, MA: MIT Press.

Cheyer, A. (1998, January). MVIEWS: Multimodal tools for the video analyst. *International Conference on Intelligent User Interfaces (IUI'98)*, (pp. 55–62). New York: ACM Press.

Cheyer, A., & Julia, L. (1995, May). Multimodal maps: An agent-based approach. *International Conference on Cooperative Multimodal Communication, (CMC'95)*, (pp. 103–113). Eindhoven, The Netherlands.

Choudhury, T., Clarkson, B., Jebara, T., & Pentland, S. (1999). Multimodal person recognition using unconstrained audio and video. *Proceedings of the 2nd International Conference on Audio-and-Video-based Biometric Person Authentication* (pp. 176–181). Washington, DC:.

Codella, C., Jalili, R., Koved, L., Lewis, J., Ling, D., Lipscomb, J., et al. (1992). Interactive simulation in a multi-person virtual world. *Proceedings of the Conference on Human Factors in Computing Systems (CHI'92)* (pp. 329–334). New York: ACM Press.

Cohen, M. M., & Massaro, D. W. (1993). Modeling coarticulation in synthetic visual speech. In M. Magnenat-Thalmann & D. Thalmann (Eds.), *Models and Techniques in Computer Animation*. Tokyo: Springer-Verlag.

Cohen, P. R., Cheyer, A., Wang, M., & Baeg, S. C. (1994). An open agent architecture. *AAAI '94 Spring Symposium Series on Software Agents*, 1–8. (Reprinted in Huhns and Singh (Eds.). (1997). *Readings in Agents* (pp. 197–204). San Francisco: Morgan Kaufmann.)

Cohen, P. R., Dalrymple, M., Moran, D. B., Pereira, F. C. N., Sullivan, J. W., Gargan, R. A., et al. (1989). Synergistic use of direct manipulation and natural language. *Proceedings of the Conference on Human Factors in Computing Systems (CHI'89)*, (pp. 227–234). New York: ACM Press. Reprinted in Maybury & Wahlster (Eds.), (1998). *Readings in Intelligent User Interfaces* (pp. 29–37). San Francisco: Morgan Kaufmann.)

Cohen, P. R., Johnston, M., McGee, D., Oviatt, S., Pittman, J., Smith, I., et al. (1997). Quickset: Multimodal interaction for distributed applications. *Proceedings of the Fifth ACM International Multimedia Conference* (pp. 31–40). New York: ACM Press.

Cohen, P. R., & McGee, D. (2004, January). Tangible multimodal interfaces for safety-critical applications, *Communications of the ACM, Vol. 47*. (pp. 41–46). New York, NY: ACM Press.

Cohen, P. R., McGee, D. R., & Clow, J. (2000). The efficiency of multimodal interaction for a map-based task. *Proceedings of the Language Technology Joint Conference (ANLP-NAACL 2000)* (pp. 331–338). Seattle, WA: Association for Computational Linguistics Press.

Cohen, P. R., & Oviatt, S. L. (1995). The role of voice input for human-machine communication. *Proceedings of the National Academy of Sciences, 92*(22), (pp. 995–9927). Washington, DC: National Academy of Sciences Press.

Condon, W. S. (1988). An analysis of behavioral organization. *Sign Language Studies, 58*, 55–88.

Dahlbäck, N., Jëonsson, A., & Ahrenberg, L. (1992, January). Wizard of Oz studies—why and how. In W. D. Gray, W. E. Hefley, & D. Murray (Eds.), *Proceedings of the International Workshop on Intelligent User Interfaces* (pp. 193–200). New York: ACM Press.

Danninger, M., Flaherty, G., Bernardin, K., Ekenel, H., Kohler, T., Malkin, R., et al. (2005). The connector- Facilitaing context-aware communication, *Proceedings of the International Conference on Multimodal Interfaces* (pp. 69–75) New York: ACM.

Denecke, M., & Yang, J. (2000). Partial Information in Multimodal Dialogue. *Proceedings of the International Conference on Multimodal Interaction*, (pp. 624–633). Beijing, China:Friendship Press.

Duncan, L., Brown, W., Esposito, C., Holmback, H., & Xue, P. (1999). Enhancing virtual maintenance environments with speech understanding. Boeing M&CT TechNet.

Dupont, S., & Luettin, J. (2000, September). Audio-visual speech modeling for continuous speech recognition. *IEEE Transactions on Multimedia, 2*(3), 141–151. Piscataway, NJ: Institute of Electrical and Electronics Engineers.

Ekman, P. (1992, January). Facial expressions of emotion: New findings, new questions. *American Psychological Society, 3*(1), 34–38.

Ekman, P., & Friesen, W. (1978). *Facial Action Coding System*. Mountain View, CA: Consulting Psychologists Press.

Encarnacao, L. M., & Hettinger, L. (Eds.) (2003, September/October) Perceptual Multimodal Interfaces (special issue), *IEEE Computer Graphics and Applications*. Mountain View, CA.

Epps, J., Oviatt, S., & Chen, F. (2004). Integration of speech and gesture input during multimodal interaction. *Proceedings of the Australian International Conference on Computer-Human Interaction (OzCHI)*.

Ernst, M., & Bulthoff, H. (2004, April). Merging the sense into a robust whole percept. *Trends in Cognitive Science, 8*(4), 162–169.

Fell, H., Delta, H., Peterson, R., Ferrier, L., Mooraj, Z., & Valleau, M. (1994). Using the baby-babble-blanket for infants with motor problems. *Proceedings of the Conference on Assistive Technologies (ASSETS'94)*, 77–84.

Flanagan, J., & Huang, T. (Eds.). (2003, September). Multimodal Human Computer Interfaces (special issue), *Proceedings of IEEE, Vol. 91* (9).

Fridlund, A. (1994). *Human facial expression: An evolutionary view*. New York: Academic Press.

Fukumoto, M., Suenaga, Y., & Mase, K. (1994). Finger-pointer: Pointing interface by image processing. *Computer Graphics, 18*(5), 633–642.

Fuster-Duran, A. (1996). Perception of conflicting audio-visual speech: an examination across Spanish and German, In D. G. Stork & M. E. Hennecke (Eds.), *Speechreading by Humans and Machines: Models, Systems and Applications* (pp. 135–143). New York: Springer Verlag.

Gatica-Perez, D., Lathoud, G., Odobez, J.-M., & McCowan, I. (2005). Multimodal multispeaker probabilistic tracking in meetings. *Proceedings of the Seventh International Conference on Multimodal Interfaces*, (pp. 183–190). New York: ACM.

Gorniak, P., & Roy, D. (2005). Probabilistic grounding of situated speech using plan recognition and reference resolution. *Proceedings of the Seventh International Conference on Multimodal Interfaces* (pp. 138–143). New York: ACM.

Gupta, A. (2004). Dynamic time windows for multimodal input fusion, *Proceedings of the International Conference on Spoken Language Processing (ICSLP'04)*. 1009–1012.

Hadar, U., Steiner, T. J., Grant, E. C., & Rose, C. F. (1983). Kinematics of head movements accompanying speech during conversation. *Human Movement Science, 2*, 35–46.

Hauptmann, A. G. (1989). Speech and gestures for graphic image manipulation. *Proceedings of the Conference on Human Factors in Computing Systems (CHI'89), Vol. 1* (pp. 241–245). New York: ACM Press.

Holzman, T. G. (1999). Computer-human interface solutions for emergency medical care. *Interactions, 6*(3), 13–24.

Huang, X., Acero, A., Chelba, C., Deng, L., Duchene, D., Goodman, J., et al. (2000). MiPad: A next-generation PDA prototype. *Proceedings of the International Conference on Spoken Language Processing (ICSLP 2000) Vol. 3* (pp. 33–36). Beijing, China: Chinese Military Friendship Publishers.

Huang, X. & Oviatt, S. Toward adaptive information fusion in multimodal systems, *Second Joint Workshop on Multimodal Interaction and Related Machine Learning Algorithms (MIML'05)*, Springer Lecture Notes in Computer Science, ed. by Steve Renals & Samy Bengio, Springer-Verlag GmbH, 2006, vol. 3869, 15–27.

Iyengar, G., Nock, H., & Neti, C. (2003). Audio-visual synchrony for detection of monologues in video archives. *Proceedings of ICASSP*.

Jain, A., Hong, L., & Kulkarni, Y. (1999). A multimodal biometric system using fingerprint, face and speech, *2nd International Conference on Audio- and Video-based Biometric Person Authentication* (pp. 182–187). Washington DC.

Jain, A., & Ross, A. (2002). Learning user-specific parameters in a multibiometric system. *Proceedings of the International Conference on Image Processing (ICIP)*, September 22–25, 2002. (pp. 57–60) Rochester, NY.

Johnston, M., Cohen, P. R., McGee, D., Oviatt, S. L., Pittman, J. A., & Smith, I. (1997). Unification-based multimodal integration. *Proceedings of the 35th Annual Meeting of the Association for Computational Linguistics* (pp. 281–288). San Francisco: Morgan Kaufmann. Karshmer, A. I., & Blattner, M. (editors). (1998). *Proceedings of the 3rd International ACM Proceedings of the Conference on Assistive Technologies (ASSETS'98)*. (URL http://www.acm.org/sigcaph/assets/assets98/assets98index.html).

Kendon, A. (1980). Gesticulation and speech: Two aspects of the process of utterance. In M. Key (Ed.), *The Relationship of Verbal and Nonverbal Communication* (pp. 207–227). The Hague, Netherlands: Mouton.

Kobsa, A., Allgayer, J., Reddig, C., Reithinger, N., Schmauks, D., Harbusch, K., et al. (1986). Combining Deictic Gestures and Natural Language for Referent Identification. *Proceedings of the 11th International Conference on Computational Linguistics* (pp. 356–361). Bonn, Germany Proceedings of COLING '86, August 25–29, 1986, University of Bonn, Germany. Published by Institut für angewandte Kommunikations-und Sprachforschung e.V. (IKS).

Kopp, S., Tepper, P., & Cassell, J. (2004). Towards integrated microplanning of language and iconic gesture for multimodal output. *Proceedings of the 6th International Conference on Multimodal Interfaces*, (pp. 97–104). New York: ACM.

Koons, D., Sparrell, C., & Thorisson, K. (1993). Integrating simultaneous input from speech, gaze, and hand gestures. In M. Maybury (Ed.), *Intelligent Multimedia Interfaces* (pp. 257–276). Cambridge, MA: MIT Press.

Kricos, P. B. (1996). Differences in visual intelligibility across talkers. In D. G. Stork & M. E. Hennecke (Eds.), *Speechreading by Humans and Machines: Models, Systems and Applications* (pp. 43–53). New York: Springer Verlag.

Kumar, S., & Cohen, P. R. (2000). Towards a fault-tolerant multi-agent system architecture. *Fourth International Conference on Autonomous Agents 2000* (pp. 459–466). Barcelona, Spain: ACM Press.

Leatherby, J. H., & Pausch, R. (1992, July). Voice input as a replacement for keyboard accelerators in a mouse-based graphical editor: An empirical study. *Journal of the American Voice Input/Output Society, 11*(2).

Martin, D. L., Cheyer, A. J., & Moran, D. B. (1999). The open agent architecture: A framework for building distributed software systems. *Applied Artificial Intelligence, 13*, 91–128.

Massaro, D. W. (1996). Bimodal speech perception: A progress report. In D. G. Stork & M. E. Hennecke (Eds.), *Speechreading by Humans and Machines: Models, Systems and Applications* (pp. 79–101). New York: Springer Verlag.

Massaro, D. W., & Cohen, M. M. (1990, January). Perception of synthesized audible and visible speech. *Psychological Science, 1*(1), 55–63.

Massaro, D. W., & Stork, D. G. (1998). Sensory integration and speechreading by humans and machines. *American Scientist, 86*, 236–244.

McCowan, I., Gatica-Perez, D., Bengio, S., Lathoud, G., Barnard, M., & Zhang, D. (2005). Automatic analysis of multimodal group actions in meetings, *IEEE Transactions on Pattern Analysis and Machine Intelligence (PAMI), 27*(3), 305–317.

McGee, D. (2003, June). *Augmenting environments with multimodal interaction*. Unpublished manuscript, Oregon Health & Science University.

McGrath, M., & Summerfield, Q. (1985). Intermodal timing relations and audio-visual speech recognition by normal-hearing adults. *Journal of the Acoustical Society of America, 77*(2), 678–685.

McGurk, H., & MacDonald, J. (1976). Hearing lips and seeing voices, *Nature, 264*, 746–748.

McLeod, A., & Summerfield, Q. (1987). Quantifying the contribution of vision to speech perception in noise. *British Journal of Audiology, 5*, 131–141.

McNeill, D. (1992). *Hand and mind: What gestures reveal about thought*. Chicago: University of Chicago Press.

Meier, U., Hürst, W., & Duchnowski, P. (1996). Adaptive bimodal sensor fusion for automatic speechreading. *Proceedings of the International Conference on Acoustics, Speech and Signal Processing (IEEE-ICASSP)* (pp. 833–836). Los Alamitos, CA: IEEE Press.

Morency, L.-P., Sidner, C., Lee, C., & Darrell, T. (2005). Contextual recognition of head gestures. *Proceedings of the Seventh International Conference on Multimodal Interfaces* (pp. 18–24). New York: ACM.

Morimoto, C., Koons, D., Amir, A., Flickner, M., & Zhai, S. (1999). Keeping an Eye for HCI. *Proceedings of SIBGRAPI'99, XII Brazilian Symposium on Computer Graphics and Image Processing*, 171–176.

Mousavi, S. Y., Low, R., & Sweller, J. (1995). Reducing cognitive load by mixing auditory and visual presentation modes. *Journal of Educational Psychology, 87*(2), 319–334.

Naughton, K. (1996). Spontaneous gesture and sign: A study of ASL signs co-occurring with speech. In L. Messing (Ed.), *Proceedings of the Workshop on the Integration of Gesture in Language & Speech* (pp. 125–134). Newark, DE:University of Delaware

Neal, J. G., & Shapiro, S. C. (1991). Intelligent multimedia interface technology. In J. Sullivan & S. Tyler (Eds.), *Intelligent User Interfaces* (pp.11–43). New York: ACM Press.

Negroponte, N. (1978, December). *The Media Room*. Cambridge, MA: MIT Press.

Neti, C., Iyengar, G., Potamianos, G., & Senior, A. (2000). Perceptual interfaces for information interaction: Joint processing of audio and visual information for human-computer interaction. In B. Yuan, T. Huang & X. Tang (Eds.), *Proceedings of the International Conference on Spoken Language Processing (ICSLP'2000), Vol. 3* (pp. 11–14). Beijing, China: Chinese Friendship Publishers.

Oliver, N., & Horvitz, E. (2005). S-SEER: Selective perception in a multimodal office activity system. *International Journal of Computer Vision and Image Understanding*.

Oviatt, S. L. (1997). Multimodal interactive maps: Designing for human performance. *Human-Computer Interaction [Special issue on Multimodal Interfaces], 12,* 93–129.

Oviatt, S. L. (1999a). Mutual disambiguation of recognition errors in a multimodal architecture. *Proceedings of the Conference on Human Factors in Computing Systems (CHI'99)* (pp. 576–583). New York: ACM Press.

Oviatt, S. L. (1999b). Ten myths of multimodal interaction. *Communications of the ACM, 42*(11), 74–81.

Oviatt, S. L. (2000a). Multimodal system processing in mobile environments. *Proceedings of the Thirteenth Annual ACM Symposium on User Interface Software Technology (UIST'2000)* (pp. 5–30). New York: ACM Press.

Oviatt, S. L. (2000b). Taming recognition errors with a multimodal architecture. *Communications of the ACM, 43*(9), 45–51.

Oviatt, S. L. (2000c). Multimodal signal processing in naturalistic noisy environments. In B. Yuan, T. Huang, & X. Tang (Eds.), *Proceedings of the International Conference on Spoken Language Processing (ICSLP'2000), Vol. 2* (pp. 696–699). Beijing, China: Chinese Friendship Publishers.

Oviatt, S. L. (2002). Breaking the robustness barrier: Recent progress in the design of robust multimodal systems. *Advances in Computers 56,* 305–341.

Oviatt, S. L. (2003, September/October). Advances in robust multimodal interfaces, *IEEE Computer Graphics and Applications,* 62–68.

Oviatt, S. L., Bernard, J., & Levow, G. (1999). Linguistic adaptation during error resolution with spoken and multimodal systems. *Language and Speech, 41*(3–4), 415–438.

Oviatt, S. L., & Cohen, P. R. (1991). Discourse structure and performance efficiency in interactive and noninteractive spoken modalities. *Computer Speech and Language, 5*(4), 297–326.

Oviatt, S. L., & Cohen, P. R. (2000, March). Multimodal systems that process what comes naturally. *Communications of the ACM, 43*(3), 45–53.

Oviatt, S. L., Cohen, P. R., Fong, M. W., & Frank, M. P. (1992). A rapid semiautomatic simulation technique for investigating interactive speech and handwriting. *Proceedings of the International Conference on Spoken Language Processing, 2,* 1351–1354.

Oviatt, S. L., Cohen, P. R., & Wang, M. Q. (1994). Toward interface design for human language technology: Modality and structure as determinants of linguistic complexity. *Speech Communication, 15,* 283–300.

Oviatt, S. L., Cohen, P. R., Wu, L., Vergo, J., Duncan, L., Suhm, B., et al. (2000). Designing the user interface for multimodal speech and gesture applications: State-of-the-art systems and research directions. *Human Computer Interaction, 15*(4), 263–322.

Oviatt, S. L., Coulston, R., & Lunsford, R. (2004). When do we interact multimodally? Cognitive load and multimodal communication patterns, *Proceedings of the Sixth International Conference on Multimodal Interfaces (ICMI'04)* 129–136.

Oviatt, S. L., Coulston, R., Shriver, S., Xiao, B., Wesson, R., Lunsford, R., et al. (2003). Toward a theory of organized multimodal integration patterns during human-computer interaction, *Proceedings of the International Conference on Multimodal Interfaces (ICMI'03)* (pp. 44–51). New York: ACM Press.

Oviatt, S. L., DeAngeli, A., & Kuhn, K. (1997). Integration and synchronization of input modes during multimodal human-computer interaction. *Proceedings of Conference on Human Factors in Computing Systems (CHI'97)* (pp. 415–422). New York: ACM Press.

Oviatt, S. L, Flickner, M., & Darrell, T. (Eds.) (2004, January). Multimodal interfaces that flex, adapt and persist, *Communications of the ACM 47*(1), 30–33.

Oviatt, S. L., & Kuhn, K. (1998). Referential features and linguistic indirection in multimodal language. *Proceedings of the International Conference on Spoken Language Processing, 6,* (pp. 2339–2342). Syndey, Australia: ASSTA, Inc.

Oviatt, S. L. & Lunsford, R. (2005). Multimodal interfaces for cell phones and mobile technology, *International Journal of Speech Technology, 8*(2), Springer & Springer-link on-line, 127–132

Oviatt, S. L., Lunsford, R., & Coulston, R. (2005). Individual differences in multimodal integration patterns: What are they and why do they exist? *Proceedings of the Conference on Human Factors in Computing Systems (CHI'05), CHI Letters* (pp. 241–249). New York: ACM Press.

Oviatt, S. L., & Olsen, E. (1994). Integration themes in multimodal human-computer interaction. In Shirai, Furui, & Kakehi (Eds.), *Proceedings of the International Conference on Spoken Language Processing, 2,* 551–554.

Oviatt, S. L., & van Gent, R. (1996). Error resolution during multimodal human-computer interaction. *Proceedings of the International Conference on Spoken Language Processing, 2,* 204–207.

Pankanti, S., Bolle, R.M., & Jain, A. (Eds.). (2000). Biometrics: The future of identification. *Computer, 33*(2), 46–80.

Pavlovic, V., Berry, G., & Huang, T. S. (1997). Integration of audio/visual information for use in human-computer intelligent interaction. *Proceedings of IEEE International Conference on Image Processing,* 15–124.

Pavlovic, V., & Huang, T. S. (1998). Multimodal prediction and classification on audio-visual features. *AAAI'98 Workshop on Representations for Multi-modal Human-Computer Interaction* (pp. 55–59). Menlo Park, CA: AAAI Press.

Pavlovic, V., Sharma, R., & Huang, T. (1997). Visual interpretation of hand gestures for human-computer interaction: A review. *IEEE Transactions on Pattern Analysis and Machine Intelligence, 19*(7) 677–695.

Pentland, S. (2005, March). Socially aware computation and communication, *IEEE Computer,* 63–70.

Potamianos, G., & Neti, C. (2001). Automatic speechreading of impaired speech, *Proceedings of the International Conference on Auditory-Visual Speech Processing,* 177–182.

Potamianos, G., Neti, C., Gravier, G., & Garg, A. (2003, September). Automatic recognition of audio-visual speech: Recent progress and challenges, *Proceedings of the IEEE, Vol. 91* (9). September, 2003.

Petajan, E. D. (1984). *Automatic Lipreading to Enhance Speech Recognition.* Unpublished doctoral dissertation, University of Illinois at Urbana-Champaign.

Pfleger, N. (2004). Context-based multimodal fusion. *Proceedings of the 6th International Conference on Multimodal Interfaces* (pp. 265–272). New York: ACM.

Poddar, I., Sethi, Y., Ozyildiz, E., & Sharma, R. (1998, November). Toward natural gesture/speech HCI: A case study of weather narration. In M. Turk, (Ed.), *Proceedings 1998 Workshop on Perceptual User Interfaces (PUI'98)* (pp. 1–6). San Francisco, CA.

Reithinger, N., Alexandersson, J., Becker, T., Blocher, A., Engel, R., Lockelt, M., et al. (2003). Multimodal architectures and frameworks: SmartKom: adaptive and flexible multimodal access to multiple applications, *Proceedings of the 5th International Conference on Multimodal Interfaces* (pp. 101–108). New York: ACM.

Robert-Ribes, J., Schwartz, J.-L., Lallouache, T., & Escudier, P. (1998). Complementarity and synergy in bimodal speech: Auditory, visual, and auditory-visual identification of French oral vowels in noise. *Journal of the Acoustical Society of America, 103*(6) 3677–3689.

Rogozan, A., & Deglise, P. (1998). Adaptive fusion of acoustic and visual sources for automatic speech recognition. *Speech Communication, 26*(1–2), 149–161.

Rubin, P., Vatikiotis-Bateson, E., & Benoit, C. (Eds.). (1998). Audio-visual speech processing *Speech Communication, 26,* 1–2.

Rudnicky, A., & Hauptman, A. (1992). Multimodal interactions in speech systems. In M. Blattner & R. Dannenberg (Eds.), *Multimedia Interface Design* (pp. 147–172). New York: ACM Press.

Sekiyama, K., & Tohkura, Y. (1991). McGurk effect in non-English listeners: Few visual effects for Japanese subjects hearing Japanese syllables of high auditory intelligibility. *Journal of the Acoustical Society of America, 90*, 1797–1805.

Seneff, S., Goddeau, D., Pao, C., & Polifroni, J. (1996). Multimodal discourse modelling in a multi-user multi-domain environment. In T. Bunnell & W. Idsardi (Eds.), *Proceedings of the International Conference on Spoken Language Processing, Vol. 1* (pp. 192–195).

Shaikh, A., Juth, S., Medl, A., Marsic, I., Kulikowski, C., & Flanagan, J. (1997). An architecture for multimodal information fusion. *Proceedings of the Workshop on Perceptual User Interfaces (PUI'97)*, 91–93.

Sharma, R., Huang, T. S., Pavlovic, V. I., Schulten, K., Dalke, A., Phillips, J., et al. (1996, August). Speech/gesture interface to a visual computing environment for molecular biologists. *Proceedings of 13th International Conference on Pattern Recognition (ICPR 96), Vol. 3*, 964–968.

Sharma, R., Pavlovic, V. I., & Huang, T. S. (1998). Toward multimodal human-computer interface. *Proceedings IEEE, 86*(5), 853–860.

Silsbee, P. L., & Su, Q. (1996). Audiovisual sensory intergration using Hidden Markov Models. In D. G. Stork & M. E. Hennecke (Eds.), *Speechreading by Humans and Machines: Models, Systems and Applications* (pp. 489–504). New York: Springer Verlag.

Siroux, J., Guyomard, M., Multon, F., & Remondeau, C. (1995, May). Modeling and processing of the oral and tactile activities in the Georal tactile system. *International Conference on Cooperative Multimodal Communication, Theory & Applications*. Eindhoven, Netherlands.

Stork, D. G., & Hennecke, M. E. (Eds.). (1995). *Speechreading by Humans and Machines*. New York: Springer Verlag.

Suhm, B. (1998). *Multimodal interactive error recovery for non-conversational speech user interfaces*. Doctoral thesis. Fredericiana University, Germany: Shaker Verlag.

Sumby, W. H., & Pollack, I. (1954). Visual contribution to speech intelligibility in noise. *Journal of the Acoustical Society of America, 26*, 52–55.

Summerfield, A. Q. (1992). Lipreading and audio-visual speech perception, *Philosophical Transactions of the Royal Society of London, Series B, 335*, 71–78.

Sweller, J. (1988). Cognitive load during problem solving: Effects on learning. *Cognitive Science, 12*, 257–285.

Tang, A., McLachlan, P., Lowe, K., Saka, C., & MacLean, K. (2005). Perceiving ordinal data haptically under workload, *Proceedings of the Seventh International Conference on Multimodal Interfaces*, (pp. 317–324). New York: ACM.

Tomlinson, M. J., Russell, M. J., & Brooke, N. M. (1996). Integrating audio and visual information to provide highly robust speech recognition. *Proceedings of the International Conference on Acoustics, Speech and Signal Processing (IEEE-ICASSP) Vol. 2*, 85–824.

Turk, M., & Robertson, G. (Eds.). (2000). Perceptual user interfaces. *Communications of the ACM, 43*(3), 32–70.

Vatikiotis-Bateson, E., Munhall, K. G., Hirayama, M., Lee, Y. V., & Terzopoulos, D. (1996). The dynamics of audiovisual behavior of speech. In D. G. Stork and M. E. Hennecke, (Eds.), *Speechreading by Humans and Machines: Models, Systems, and Applications, Vol. 150 of NATO ASI Series. Series F: Computer and Systems Sciences* (pp. 25–232). Berlin, Germany: Springler-Verlag.

Vo, M. T., & Wood, C. (1996). Building an application framework for speech and pen input integration in multimodal learning interfaces. *Proceedings of the International Conference on Acoustics Speech and Signal Processing (IEEE-ICASSP) Vol. 6*, 3545–3548.

Vo, M. T., Houghton, R., Yang, J., Bub, U., Meier, U., Waibel, A., et al. (1995). Multimodal learning interfaces. *Proceedings of the DARPA Spoken Language Technology Workshop.*

Wahlster, W. (1991). User and discourse models for multimodal communciation. In J. W. Sullivan & S. W. Tyler (Eds.), *Intelligent User Interfaces* (pp. 45–67). New York: ACM Press.

Wahlster, W. (2001, March). SmartKom: multimodal dialogs with mobile web users. *Proceedings of the Cyber Assist International Symposium*, 33–34.

Waibel, A., Suhm, B., Vo, M. T., & Yang, J. (1997, April). Multimodal interfaces for multimedia information agents. *Proceedings of the International Conference on Acoustics Speech and Signal Processing (IEEE-ICASSP) Vol. 1*, 167–170.

Wang, J. (1995). Integration of eye-gaze, voice and manual response in multimodal user interfaces. *Proceedings of IEEE International Conference on Systems, Man and Cybernetics*, 3938–3942.

Wickens, C. D., Sandry, D. L., & Vidulich, M. (1983). Compatibility and resource competition between modalities of input, central processing, and output. *Human Factors, 25*, 227–248.

Wu, L., Oviatt, S., &. Cohen, P. (1999). Multimodal integration—A statistical view. *IEEE Transactions on Multimedia, 1*(4), 334–341.

Xiao, B., Lunsford, R., Coulston, R., Wesson, R., & Oviatt, S. L. (2003). Modeling multimodal integration patterns and performance in seniors: Toward adaptive processing of individual differences, *Proceedings of the International Conference on Multimodal Interfaces (ICMI'03)* (pp. 265–272). New York: ACM Press.

Xiao, B., Girand, C., & Oviatt, S. (2002, September). Multimodal integration patterns in children. *Proceedings of the International Conference on Spoken Language Processing (ICSLP'02)*, 629–632.

Zhai, S., Morimoto, C., & Ihde, S. (1999). Manual and gaze input cascaded (MAGIC) pointing. *Proceedings of the Conference on Human Factors in Computing Systems (CHI'99)* (pp. 246–253). New York: ACM Press.

·6·

ADAPTIVE INTERFACES AND AGENTS

Anthony Jameson
German Research Center for Artificial Intelligence (DFKI)
and International University in Germany

Introduction 106
   Concepts 106
Functions: Supporting System Use 107
   Taking Over Parts of Routine Tasks 107
   Adapting the Interface 108
   Helping With System Use 109
   Mediating Interaction with the Real World 110
   Controlling a Dialogue 111
Functions: Supporting Information
Acquisition 112
   Helping Users to Find Information 112
      Support for browsing 113
      Support for query-based search or filtering 114
   Spontaneous Provision of Information 114
   Recommending Products 114
   Tailoring Information Presentation 115
   Supporting Collaboration 116
   Supporting Learning 117
Usability Challenges 118
   Threats to Predictability and Comprehensibility 119
   Threats to Controllability 119
   Obtrusiveness 120
   Threats to Privacy 120
   Breadth of Experience 120
   Dealing With Trade-offs 120
Obtaining Information About Users 121
   Explicit Self-Reports and -Assessments 121
      Self-reports about objective personal
      characteristics 121
      Self-assessments of interests and knowledge 121

      Self-reports on specific evaluations 122
      Responses to test items 122
   Nonexplicit Input 122
   Naturally Occurring Actions 122
   Previously Stored Information 122
   Low-Level Indices of Psychological States 123
   Signals Concerning the Current Surroundings 123
Special Considerations Concerning
Empirical Methods 123
   Use of Data Collected With a Nonadaptive System 123
   Early Studies of Usage Scenarios and
   User Requirements 124
   Wizard of Oz Studies 124
   Comparisons With the Work of Human Designers 125
   Experimental Comparisons of Adaptive
   and Nonadaptive Systems 125
   Taking Into Account Individual Differences 126
   Checking Usability Under Realistic Conditions 126
The Future of User-Adaptive Systems 126
   Growing Need for User-Adaptivity 126
      Diversity of users and contexts of use 126
      Number and complexity of interactive systems 126
      Scope of information to be dealt with 126
   Increasing Feasibility of Successful Adaptation 127
      Ways of acquiring information about users 127
      Advances in techniques for learning,
      inference, and decision 127
      Attention to empirical methods 127
Acknowledgments 127
References 127

INTRODUCTION

This chapter covers a broad range of interactive systems. They all have one idea in common: It can be worthwhile for a system to learn something about each individual user and adapt its behavior to them in some nontrivial way.

An example familiar to most readers is shown in Fig. 6.1. A visitor to Amazon.com has just explicitly requested recommendations, without having specified a particular type of product. During the user's past visits, Amazon.com has learned something about his or her interests, on the basis of items he or she has purchased and ratings he or she has made. Therefore, the system can make recommendations that are especially likely to appeal to this particular user.

Concepts

The key idea embodied in Amazon.com's recommendations and the other systems discussed in this chapter is that of adaptation to the individual user. Depending on their functions and forms, systems that adapt to their users have been given labels ranging from *adaptive interfaces* through *user modeling systems* to *software agents* or *intelligent agents*. Starting in the late 1990s, the broader term *personalization* became popular, especially in connection with commercially deployed systems. In order to discuss the common issues that all of these systems raise, we will refer to them with the term *user-adaptive systems*, which describes their common properties explicitly. Fig. 6.2 introduces some concepts that can be applied to any user-adaptive system; Fig. 6.3 shows the form that they take in Amazon.com's recommendations.

A user-adaptive system makes use of some type of information about the current individual user, such as the products that the user has bought. In the process of *user model acquisition*, the system performs some type of learning and inference on the basis of the information about the user in order to arrive at some sort of user model, which in general concerns only limited aspects of the user (such as his or her interest in particular types of product). In the process of *user model application*, the system applies the user model to the relevant features of the current situation in order to determine how to adapt its behavior to the user.

A user-adaptive system can be defined as an interactive system that adapts its behavior to individual users on the basis of processes of user model acquisition and application that involve some form of learning, inference, or decision making.

This definition distinguishes user-adaptive systems from *adaptable* systems: ones that the individual user can explicitly tailor to his or her own preferences (for example, by choosing options that determine the appearance of the user interface). The relationship between adaptivity and adaptability will be discussed at several places in this chapter.

FIGURE 6.1. Part of a screen showing a list of recommendations generated on request by Amazon.com. (Screen shot made from http://amazon.com/ in December 2005.)

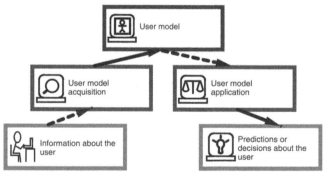

FIGURE 6.2. General schema for the processing in a user-adaptive system. (Dotted arrows: use of information; solid arrows: production of results.)

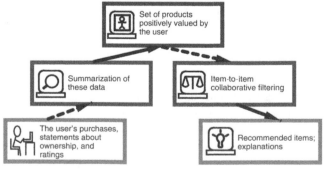

FIGURE 6.3. Overview of adaptation in Amazon.com.

The next two sections of this chapter address the following question: What can user-adaptivity be good for? They examine in turn 10 different functions that can be served by user-adaptivity, giving examples ranging from familiar commercially deployed systems to research prototypes. The following section discusses some usability challenges that are especially important in connection with user-adaptive systems and that have stimulated most of the controversy that has surrounded these systems. The next section considers a key design decision: What types of information about each user should be collected? The final major section looks at several distinctive aspects of the empirical study of user-adaptive systems. The chapter concludes with comments on the reasons why their importance is likely to continue to grow.[1]

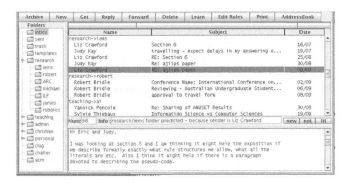

FIGURE 6.4. Partial screen shot from the intelligent e-mail sorting system i-ems. (Screen shot courtesy of Eric McCreath.)

FUNCTIONS: SUPPORTING SYSTEM USE

Some of the ways in which user-adaptivity can be helpful involve support for a user's efforts to operate a system successfully and effectively. This section considers five types of support.

Taking Over Parts of Routine Tasks

The first function of adaptation involves taking over some of the work that the user would normally have to perform himself or herself—routine tasks that may place heavy demands on a user's time, though typically not on his or her intelligence and knowledge. Maybe the most obvious task of this sort is organizing e-mail, which takes up a significant proportion of the time of many office workers. The classic early work of Maes's (1994) group on "agents that reduce work and information overload" addressed this task.

A more recent effort (Fig. 6.4) is found in the prototype "intelligent electronic mail sorter" (i-ems; McCreath, Kay, & Crawford, 2006; see also Crawford, Kay, & McCreath, 2002; McCreath & Kay, 2003), which is designed to expedite the tedious task of filing incoming e-mail messages into folders. By observing and analyzing the way an individual user files messages, the system learns to predict the most likely folder for any new message. In the overview of messages in the user's inbox (shown in the top part of the screen shot), i-ems tentatively sorts the new messages into categories that correspond to the most likely folders. When an individual message is being displayed, the one-line field in the middle of the screen shows an explanation of the folder prediction. If the user agrees with the prediction, he or she can click on the "Archive" button at the top of the screen to have the message moved into the predicted folder. To file it away in another folder, he or she drags it to the icon for the folder in the left-hand panel, just as he or she would with a system that did not make any predictions.

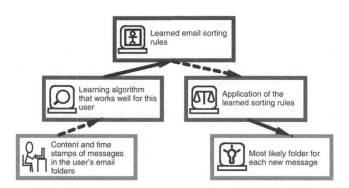

FIGURE 6.5. Overview of adaptation in i-ems.

One reason why research on systems like i-ems has continued for so long is that the problem raises a number of challenges. For example, since different users apply radically different principles for creating categories of e-mail, i-ems supplies several different methods for learning a user's implicit rules, each of which may show a different degree of success with different users. It also allows the learned rules to operate alongside any handcrafted rules that the user may have defined so that the strengths of both types of rules can be exploited (an evaluation is discussed below in the empirical methods section). Since even the best set of learned rules will sometimes incorrectly predict how the user would classify a message, the interface must be designed in such a way that incorrect predictions will have minimal consequences. New approaches to the general problem continue to appear (e.g., Surendran, Platt, & Renshaw, 2005). Two systems that have been fairly widely deployed have been SwiftFile (Segal & Kephart, 1999, 2000), which was incorporated into Lotus Notes, and POPFile (available in early 2006 from http://popfile.sourceforge.net/cgi-bin/wiki.pl), a public domain program that is used mainly for spam filtering but that can also learn to sort messages into a limited number of user-specific folders.[2]

---

[1]The version of this chapter in the first edition of the *HCI Handbook* included a section about some of the machine learning and artificial intelligence techniques that are most commonly used for user model acquisition and application. There is no such section in this book, because (a) it seemed more important to expand the material in the other sections, and (b) the range of techniques used has grown to the point where a brief summary would have limited value. Discussions of the relevant methods will be found in many of the works cited in the chapter.

[2]For examples of approaches to support for e-mail management that do not involve adaptation to individual users, see Gruen et al. (2004) and Bälter and Sidner (2002).

Another traditional task in this category is the scheduling of meetings and appointments (Mitchell, Caruana, Freitag, McDermott, & Zabowski, 1994; Maes, 1994; Horvitz, 1999; Gervasio, Moffitt, Pollack, Taylor, & Uribe, 2005). By learning the user's preferences for particular meeting types, locations, and times of day, a system can tentatively perform part of the task of entering appointments in the user's calendar.

The primary benefits of this form of adaptation are saving time and effort for the user. The potential benefits are greatest where the system can perform the entire task without input from the user. In most cases, however, the user is kept in the loop (as with i-ems), because the system's ability to predict what the user would want done is limited (cf. the section on usability challenges below).

Adapting the Interface

A different way of helping a person to use a system more effectively is to adapt the user interface so that it fits better with the user's way of working with the system. Interface elements that have been adapted in this way include menus, icons, and the system's processing of signals from input devices such as keyboards.

An example familiar to most readers is provided by the Smart Menus feature that has been found in Microsoft operating systems since Windows 2000. Fig. 6.6 illustrates the basic mechanism: An infrequently used menu option is initially hidden from view; it appears in the main part of a menu only after the user has selected it for the first time. (It will be removed later if the user does not select it often enough.) The idea is that, in the long run, the menus should contain just the items that the user accesses frequently (at least recently), so that the user needs to spend less time searching within menus.

Some informative studies related to Smart Menus have been conducted by McGrenere, Baecker, and Booth (2002). In a field

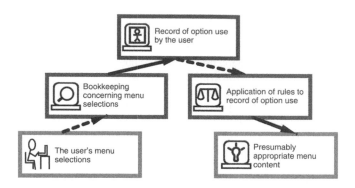

FIGURE 6.7. Overview of adaptation in Smart Menus.

study with experienced users of Microsoft Word 2000, McGrenere et al. compared the Smart Menus of Word 2000 with (a) traditional static menus and (b) an alternative approach to reducing the number of functions that confront users. Their variant MSWord Personal is an adaptable system: It provides a reasonably intuitive and convenient way for users to add and remove menu functions. After working with MSWord Personal for several weeks, most of the users in the study preferred this adaptable system to the normal Word 2000 with Smart Menus. The users who had been classified as feature-shy appeared to benefit most, but as is typical in studies like this (as will be discussed), quite a variety of attitudes about the relative merits of the three approaches to adapting menu content were shown by the subjects. As the authors pointed out, it seems worthwhile to consider design solutions that combine some degree of adaptivity and adaptability. For example, instead of automatically adapting the menus, the system might recommend possible adaptations on the basis of its analysis of the user's menu selections (cf. Bunt, Conati, & McGrenere, 2004).

A more direct experimental comparison by Findlater and McGrenere (2004) involving adaptive menus like Smart Menus is discussed in the section on empirical methods below.

One promising application of both adaptable and adaptive methods involves taking into account special perceptual or physical impairments of individual users to allow them to use a system more efficiently, with minimal errors and frustration. A system in which the two approaches are combined is the Web Adaptation Technology of IBM Research (Hanson & Crayne, 2005), which aims to facilitate web browsing by older adults. With regard to most of the adaptations, such as the reformatting of multicolumn text in a single column, the system is adaptable: It provides convenient ways for the user to request the changes. (It would in fact be difficult for a system to determine automatically whether a given user would benefit from one-column formatting.) Several changes in the keyboard settings are achieved via automatic adaptation (for a more detailed discussion, see Trewin, 2004). For example, the *key repeat delay* interval is a parameter that determines how long a key (e.g., the left-arrow key) has to be held down before the system starts repeating the associated action (e.g., moving the cursor to the left). Some users require a relatively long key repeat delay because of a tendency to hold keys down relatively long even when they do not want repetition. But asking the user to specify the key repeat delay is not an attractive option: It can be time-consuming to

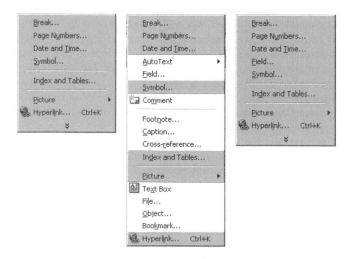

FIGURE 6.6. Example of adaptation in Smart Menus. (The user accesses the "Insert" menu. Not finding the desired option, the user clicks on the extension arrows and selects the "Field" option. When the user later accesses the same menu, "Field" now appears in the main section.)

explain what the parameter means; the user him- or herself may have no idea what the best setting is for him or her; trial and error with different settings can be time-consuming and frustrating; and for some users, the optimal setting can change from day to day. The Dynamic Keyboard component of the Web Adaptation Technology therefore includes an algorithm that analyzes a user's typing behavior to determine an optimal key repeat delay (as well as other parameters); the system then adjusts the parameter in a relatively conservative fashion. Although, under some circumstances, automatic adjustment of keyboard parameters could make the keyboard unpredictable and hard to use, results obtained in the context of Web Adaptation Technology (Trewin, 2004) revealed no problems of this sort.

Helping With System Use

Instead of suggesting (or executing) changes to the interface of a given application, a user-adaptive system can adaptively offer information and advice about how to use that application, and perhaps also perform some of the necessary actions itself. Various tendencies make it increasingly difficult for users to attain the desired degree of mastery of the applications they use. A good deal of research into the development of systems that can take the role of a knowledgeable helper was conducted in the 1980s, especially in connection with the complex operating system Unix.[3] During the 1990s, such work became less frequent, perhaps partly because of a recognition of the fundamental difficulties involved. In particular, it is often difficult to recognize a user's goal when the user is not performing actions that tend to lead toward that goal. The Office Assistant, an ambitious attempt at adaptive help introduced in Microsoft Office 97, was given a mixed reception, partly because of the inherent difficulty of its task but especially because of its widely perceived obtrusiveness (cf. the section on usability challenges below).

Most adaptive help systems to date have been based on the paradigm called *keyhole recognition*: (passively) observing the user and attempting to make useful inferences about her goals and tasks. By contrast, Fig. 6.8 shows an example of an alternative approach to intelligent help that has been developed by researchers at Mitsubishi Electric Research Laboratory, which is based on a collaborative dialogue paradigm (Rich et al., 2005; for a presentation of the theoretical and technical background, see Rich & Sidner, 1998; Rich, Sidner, & Lesh, 2001). In this demonstration scenario, DiamondHelp is collaborating with the user of a feature-rich programmable washer-dryer.[4] Instead of working independently on the problem, the user conducts a dialogue with the help system, the goal of the dialogue being the execution of the user's task. The dialogue contributions of the help system and the user are shown in the chat balloons at the left- and right-hand sides, respectively, of the screen. The user's possible dialogue contributions are automatically gener-

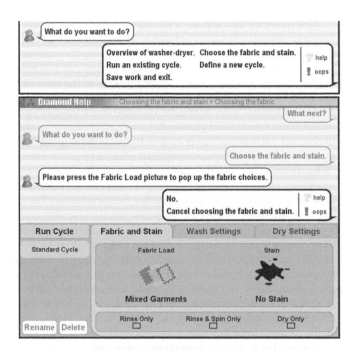

FIGURE 6.8. Example of collaborative assistance offered by DiamondHelp. (Explanation in text. Screen shots courtesy of Charles Rich.)

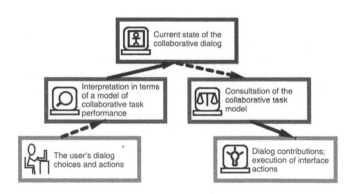

FIGURE 6.9. Overview of adaptation in DiamondHelp.

ated from the current collaborative dialogue state and offered in a menu inside of his or her balloon. The user can choose what he or she wants to say by either touching the appropriate phrase or saying it using speech recognition.[5] For example, in the top part of the figure, the user is offered a choice among three possible top-level tasks, the most complex of which is defining a new cycle. In a typical exchange, the user specifies a goal or subgoal that he or she would like to achieve, and the system responds by giving instructions and perhaps offering further possible utterances for the user.

---

[3]A collection of papers from this period appeared in a volume edited by Hegner, McKevitt, Norvig, and Wilensky (2001).

[4]The interface shown in the figures may be displayed on the washer-dryer itself or remotely accessed via a home network.

[5]DiamondHelp does not support unrestricted natural language or speech understanding. In a Wizard of Oz study (a type of study that will be discussed in the section on empirical methods) involving a prototype help system of this general sort, DeKoven (2004) found that users who were able to employ unrestricted speech would have preferred to have more guidance about what they could say to the system.

This dialogue in some ways resembles the interaction with the more familiar type of wizard that is often employed for complex, unfamiliar tasks such as software installation. The main difference is that the dialogues with DiamondHelp can be more flexible, because the system has explicit models of the tasks that the user can perform and is capable of making use of these models in various ways during the dialogue. For example, after pressing the Fabric Load picture, the user can continue manipulating the GUI in the lower half of the screen by himself or herself until he or she requests guidance again (e.g., by asking "What next?"). Because the user's actions with the interface are reported to the help system, the help system can keep track of how far the user has progressed in the performance of his or her task. In other words, the help system incorporates a restricted form of the sort of goal and plan recognition that featured prominently in earlier intelligent help systems. In DiamondHelp, recognition of the user's actions is relatively likely to be accurate, because of the information that the user has supplied about his or her goals (e.g., Lesh, Rich, & Sidner, 1999). Depending on the experience and the preferences of the user, therefore, the user can rely on the help system to various degrees, ranging from ignoring it, occasionally asking for a hint, or being led step by step through the entire task.

Mediating Interaction With the Real World

Whereas an intelligent help system aids the user as he or she uses a complex interactive system, some recently developed systems help the user to cope with the world itself. They do so by acquiring and processing evidence concerning the user's cognitive and emotional state and taking actions designed to mitigate any conflict between this state and the demands of the environment.

One common function of systems in this category is to protect people from the flood of incoming messages (via cell phone, instant messaging, e-mail, and other channels) whose number and diversity are increasing with advances in communication technology. When a potential recipient is focusing on some particular task or activity, an adaptive assistant causes messages to be discouraged, delayed, or otherwise buffered until some more appropriate time. One strategy is to provide to the potential initiators of communication information about the state of the recipient. The experimental prototype Lilsys (cf. Fig. 6.10) illustrates this strategy. The system continuously updates a user model that contains assessments of its user's availability for communication. The assessments are based on a number of cues that have been found in previous research to be useful predictors of a person's physical presence and availability: whether the user (or someone else in the room) is moving or speaking; whether the door is open; whether the user is using the phone or the computer keyboard and mouse; and what events are scheduled in the user's calendar. A handcrafted model uses this information to arrive at a global assessment of the user's avail-

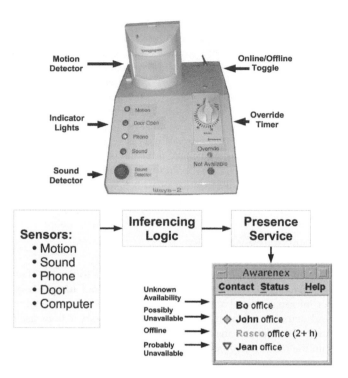

FIGURE 6.10. Above: Lilsys's sensor and data acquisition module; below: Lilsys's data flow and a screen shot of the user interface. (Adapted from Figures 1 and 2 of: "Lilsys: Sensing unavailability," by J. Begole, N. E., Matsakis, & J. C. Tang, 2004, In J. Herbsleb & G. Olson (Eds.), *Proceedings of the 2004 Conference on Computer-Supported Cooperative Work*, pp. 511–514, New York: ACM. Copyright 2004 by ACM. Adapted with permission.)

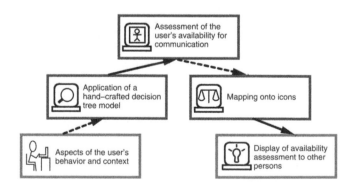

FIGURE 6.11. Overview of adaptation in Lilsys.

ability; this assessment is in turn displayed to potential communicators.[6] A field study with a small number of users indicated that other persons do in fact adapt their behaviors to take into account a Lilsys user's availability, often by changing the nature of their communications rather than by postponing it. Lilsys users appreciated the possibility of having their availabilities

---

[6]In many systems in this category, such as the ones mentioned in the following paragraph, the model for the interpretation of evidence is acquired via machine learning methods on the basis of relevant training data.

sensed automatically, as opposed to having explicitly to specify availability themselves. For example, they virtually never used the timer switch (visible in Fig. 6.10) that allowed them to specify that they were going to be unavailable for a particular period. More generally, the modeling of a user's changing cognitive or emotional state appears to be a task for which automatic adaptation is an especially promising approach, simply because people are typically not willing or able to update an explicit self-assessment continually.

In some other availability management systems, decisions about when and how to present messages are made by the system itself on the basis of the user model (e.g., Horvitz, Koch, Sarin, Apacible, & Subramani, 2005). A good deal of research has examined effective cues for the recognition of availability and interruptibility (e.g., Fogarty et al., 2005; Ho & Intille, 2005; Iqbal, Adamczyk, Zheng, & Bailey, 2005).

A related line of research has focused on the recognition of the mental states of drivers, which is especially important because of safety issues. The modeling of interruptibility is important here as well (e.g., Schneider & Kiesler, 2005). But even when no other persons are involved, there are reasons to try to recognize safety-relevant states, like drowsiness and stress, so that the system can intervene, for example, by waking the driver up or by playing soothing music. Stress and emotion are manifested in physiological indicators (e.g., Healey & Picard, 2000; Lisetti & Nasoz, 2004) and in speech (e.g., Fernandez & Picard, 2000; Jones & Jonsson, 2005). Products along these lines have begun to appear in cars, beginning with relatively simple detection methods such as the recognition of long or frequent eye closures. An example of a more complex and comprehensive approach to the modeling of drivers' affective state is found in the work of Li and Ji (2005).

A general problem with adaptation for the purpose of safety is that the user may come to rely on the adaptation, reducing his or her own attention to safety.[7] For example, a driver may make less effort to avoid distraction or to remain alert, expecting that the assistant will recognize any dangerous situation and warn him or her in time. Especially since the recognition of a person's mental states is almost always error-prone, this tendency can eliminate the potential safety benefits of monitoring systems unless appropriate measures are taken (e.g., making warning sounds so unpleasant that the driver will want to avoid relying on them more than is necessary).

Controlling a Dialogue

Much of the early research on user-adaptive systems concerned systems that conducted natural language dialogues with their users (see, e.g., Kobsa & Wahlster, 1989). During the 1990s, attention shifted to interaction modalities that were more widely available and that made it possible in many cases to implement adaptation straightforwardly. Toward the year 2000, advances in

the technology of natural language and speech processing led to a recent reawakening of interest in user-adaptive dialogue systems (see, e.g., Haller, McRoy, & Kobsa, 1999; Zukerman & Litman, 2001; Litman & Pan, 2002).

Natural language dialogue has served as an interaction modality in connection with most of the functions of user-adaptivity discussed in this and the following sections, such as the provision of help and the recommendation of products. One type of adaptivity is largely characteristic of natural language dialogue: adaptation of the system's dialogue strategy, which is a policy for determining when and how the system should provide information, acquire information from the user, and perform other dialogue acts.

Adaptation is especially important in spoken dialogue systems, such as those that offer information about train departures or flight arrivals via the phone. Novice users may require extensive explanations and frequent confirmations, but these elements can be unnecessarily time consuming and frustrating to experienced users. Many deployed systems apply simple adaptation principles that distinguish between new and more experienced users. For example, if a phone-based mail-order system knows that the current caller has previously ordered a product, it may adopt a dialogue style that presupposes familiarity with the system. Since it is not always this easy to classify the user, and since asking the user for a self-assessment can be awkward and time consuming, researchers have looked for ways of adapting to the user on the basis of his or her behavior during the current dialogue (e.g., Litman & Pan, 2002). A recent example is shown in Fig. 6.12, which shows a translation of an example dialogue conducted with the Kyoto City Bus Information System (Komatani, Ueno, Kawahara, & Okuno, 2005). The system adjusts its assessments of three properties of the caller on the basis of each utterance of the caller: level of skill at conducting dialogues with this system; level of knowledge concerning the domain (e.g., Kyoto and its buses); and desire to complete the dialogue quickly ("urgency"). The evidence used is a set of properties such as the amount of information specified in each utterance; whether the user refers to bus stops by their exact names; and whether the user barges in, interrupting the system before the system has completed an utterance. The dialogue excerpt includes several examples that illustrate how the system's dialogue behavior is adapted on the basis of the user model. In a controlled experiment, each of 20 subjects conducted dialogues with the system either with the adaptive version or with a version that assumed that all callers had the same properties. Dialogues were completed more quickly with the adaptive variant, and subjects expressed greater satisfaction.[8]

Another popular idea about how to adapt dialogues to the user concerns the recognition of negative emotions like anger and frustration, the goal being to transfer callers who express these emotions to human call agents before they are lost as customers. Although there has been a lot of research on the recognition of mental states on the basis of speech in dialogues

---

[7]A brief overview of the theory of *risk homeostasis* is given by Wilde (1998), while a recent alternative perspective is offered by Fuller (2005).
[8]See the section on empirical methods for a discussion of the difficulties involved in the interpretation of the results of experiments like this one. Although the Kyoto City Bus Information System is accessible to the public, the adaptive features described here are normally turned off, because they lead to slower processing given the currently available infrastructure.

```
case:   [skill level: low] [knowledge level: low] [urgency: low]
Sys:    What is your current bus stop, your destination, or
        specific bus route number?
User:   Maruyama Park.
Sys:    Let me confirm these in turn.
        (← Explanation is added by [skill level: low].)
        Will you get on at Maruyama Park?
        (← Content is confirmed by [urgency: low].)
        Reply with yes or no.
        (← Instruction is added by [skill level: low].)
User:   Yes.
Sys:    Where will you get off the bus?
        Specify by bus stop name or a well-known landmark.
        (← Explanation is added by [skill level: low].)
User:   Tofukuji Temple.
Sys:    Will you get off the bus at Tofukuji Temple?
        Reply with yes or no.
User:   Yes.
Sys:    Following buses are located.
        (← Explanation is added by [skill level: low].)
        Nearest bus stop to Maruyama Park is Gion.
        (← Domain information is added by [knowledge level: low].)
        Bus number 202 bound for Nishioji-Kujo has departed
        Higashiyama-Sanjo, which is two stops away.
        ...
```

FIGURE 6.12.   Excerpt from a dialog with the Kyoto City Bus Information System. (Figure 2 of "User modeling in spoken dialogue systems to generate flexible guidance," by K. Komatani, S. Ueno, T. Kawahara, & H. G. Okuno, 2005, *User Modeling and User-Adapted Interaction*, 15, pp. 169–183. Copyright 2005 by Springer Verlag. Reproduced with permission.)

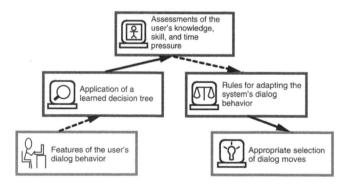

FIGURE 6.13.   Overview of adaptation in the Kyoto City Bus Information System.

(e.g., Yacoub, Simske, Lin, & Burns, 2003; Liscombe, Riccardi, & Hakkani-Tür, 2005), it remains to be seen how widespread this particular application will become. One possible drawback is that with some systems callers might find it worthwhile to adapt to the adaptation (as with safety-relevant adaptations), feigning emotion in order to get quicker attention.

Other dialogue adaptations that are being explored concern stable personal characteristics like gender and age. Since it is possible to recognize these characteristics reasonably well on the basis of speech, a system might adopt a voice or dialogue style that designers thought appropriate for the age and gender in question (e.g., Müller, Wittig, & Baus, 2003).

FUNCTIONS: SUPPORTING INFORMATION ACQUISITION

We are constantly hearing that information overload is a typical problem of our age, especially because of the explosive growth of the Internet and in particular the World Wide Web. In addition to the vast number of electronic documents of various sorts, users now have access to a vast number of products available for sale, people that they can get in touch with, and systems that can teach them about some topic. The second major type of function of user-adaptive systems is to help people to find what they need in a form that they can deal with.

Helping Users to Find Information

We will first look at the broad class of systems that help the user to find relevant electronic documents, which may range from brief news stories to complex multimedia objects.

As an especially clear example, consider the situation of a user who, in the year 2006, heard that a lot of interesting facts and opinions can be found in weblogs (blogs), of which millions are accessible. The user would like to be able to read, each day, the articles in blogs that are of particular interest. But how is he or she to find these? The user does not know in advance which blogs are especially likely to produce material of interest (as, for example, he or she could specify a well-known newspaper as a promising source of online stories if he or she were interested in news). The user could submit queries to a search engine that indexes blogs, but in general, he or she cannot know in advance what topics of interest will be covered by the latest blog articles; and given the low quality and lack of authority of many blogs, the user will not be confident of receiving good results on any given topic.

An approach to this problem that relies heavily on adaptation to the individual user (called "personalization" in this context) is realized at the time of this writing in the Website Findory (http://findory.com), which offers access to both blogs and news articles. To the new user visiting the blog section of the website, Findory offers a page that shows the first few lines of a number of blog articles on different topics (cf. Fig. 6.14). The user can then click to read the articles that interest him or her most. Each selection causes the system to update its model of the user's interests and adapt the selection of blog articles accordingly. For example, if the user chooses an article discussing a copyright infringement suit against a search engine company, further articles concerning copyright issues and search engines are likely to appear, marked with the sunburst icon that is visible in Fig. 6.14. If the user clicks on the icon for a recommended article, a page is displayed that explains the recommendation in a style similar to that of Amazon.com (cf. Fig. 6.1 and the next subsection) with a list of articles that the user has read in the past that are similar to the recommended item in terms of their content (which the system can characterize on the

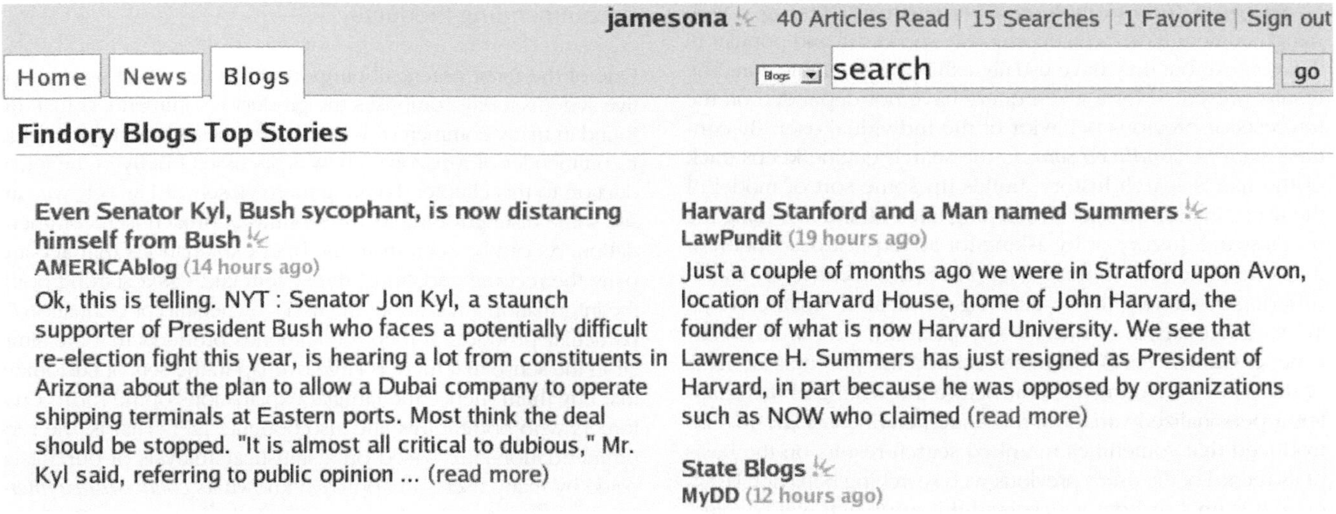

Home | News | Blogs

Blogs ▼ **search** go

Findory Blogs Top Stories

Even Senator Kyl, Bush sycophant, is now distancing himself from Bush ⟟
AMERICAblog (14 hours ago)
Ok, this is telling. NYT : Senator Jon Kyl, a staunch supporter of President Bush who faces a potentially difficult re-election fight this year, is hearing a lot from constituents in Arizona about the plan to allow a Dubai company to operate shipping terminals at Eastern ports. Most think the deal should be stopped. "It is almost all critical to dubious," Mr. Kyl said, referring to public opinion ... (read more)

Harvard Stanford and a Man named Summers ⟟
LawPundit (19 hours ago)
Just a couple of months ago we were in Stratford upon Avon, location of Harvard House, home of John Harvard, the founder of what is now Harvard University. We see that Lawrence H. Summers has just resigned as President of Harvard, in part because he was opposed by organizations such as NOW who claimed (read more)

State Blogs ⟟
MyDD (12 hours ago)

FIGURE 6.14.   A small part of a personalized display of Findory (http://findory.com, March 2006). (The icon that appears after a title indicates that the entry has been recommended on the basis of the articles that the user has read previously.)

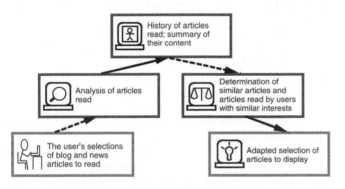

FIGURE 6.15.   Overview of adaptation in Findory.

evant key words; and (b) since interests change over time and as a function of current developments, the user would have to keep updating the descriptions.

More generally speaking, user-adaptive systems that help users find information[9] typically draw from the vast repertoire of techniques for analyzing textual information (and to a lesser extent, information presented in other media) that have been developed in the field of information retrieval. The forms of adaptive support are in part different in three different situations, the first two of which can arise with Findory:

Support for browsing.   In the World Wide Web and other hypermedia systems, users often actively search for desired information by examining information items and pursuing cross-references among them. A user-adaptive hypermedia system can help focus the user's browsing activity by recommending or selecting promising items or directions of search on the basis of what the system has been able to infer about the user's information needs. An especially attractive application scenario is that of mobile information access, where browsing through irrelevant pages can be especially time-consuming and expensive. In this context, the best approach may be for the system to omit entirely links that it expects to be less interesting to the individual user. Billsus and Pazzani (2007) describe a case study of an adaptive news server that operates in this way. Stationary systems with greater communication bandwidth tend to include all of the same links that would be presented by a nonadaptive system, highlighting the ones that are most likely to be of interest or presenting separate lists of recommended links. As is argued and illustrated by Tsandilas and Schraefel (2004), this approach makes it easier for the user to remedy incorrect assessments of the user's interests on the part of the system.

basis of the words in the text) or in terms of the users who have previously read them (G. Linden, personal communication, February, 2006). If the user sees in this way an article that he or she does not want to be used for recommendations in the future, he or she can delete it from the reading history.

Findory's approach relies on the user being able to identify, early in his or her use of the system, some articles that interest him or her and that can therefore serve as examples for the system's learning. This process is facilitated by the opportunities that the user has to issue explicit queries with keywords and to consult trusted sources in the News section of the website. The main advantage of this approach is that the user need not make any effort to specify explicitly in what types of content he or she is interested. Any such effort would be problematic anyway in that (a) it can be difficult and tedious to describe a large number of general interests accurately, for example by specifying rel-

---

[9]Surveys of parts of this large area are provided by, among others, Kelly and Teevan (2003) and several chapters in the collection edited by Brusilovsky, Kobsa, and Nejdl (2007).

Support for query-based search or filtering. Web search engines have been enormously successful and popular in this context, but they have usually exhibited one limitation: The results presented for a given query have not depended on the interests or previous behavior of the individual user. By contrast, with *personalized search*, the search engine keeps track of the user's search history, builds up some sort of model of the user's interests (either by keeping track of and analyzing the user's search history or by asking for an explicit description of interests), and biases the results presented accordingly by reordering or filtering the results. A good deal of research (e.g., Teevan, Dumais, & Horvitz, 2005; Micarelli, Gasparetti, Sciarrone, & Gauch, 2007) has demonstrated the potential benefits of this strategy. During the year before the writing of this chapter, a personalized variant of the search engine Google was introduced that sometimes reranked search results on the basis of its record of the user's previous web searching behavior. However, it is unclear how widespread this approach will become. The added value of personalization is less obvious when the user has given an explicit query than when the user is simply looking for something interesting, as is often the case with Findory. With an explicit query, it is feasible and worthwhile for the user to think about an informative description of his or her interests and to modify his or her query (perhaps repeatedly) on the basis of the results obtained.

An interesting variant on personalized search is found in the system I-Spy (e.g., Smyth et al., 2005): This community-oriented search engine tailors the results of web search queries to an entire community of users, such as the employees of a particular company. It moves upward in the search results list results that have been clicked on by previous users in the community who had issued the same or similar queries.

Spontaneous Provision of Information

A number of systems present information that may be useful to the user even while the user is simply working on some task and is making no effort to find information. A recent prototype that has been deployed at a research laboratory is the FXPAL Bar (Billsus, Hilbert, & Maynes-Aminzade, 2005). While an employee visits websites in the course of normal work, the system searches in the background for potentially relevant information (e.g., about company visitors and internal publications). A central design issue for this and similar systems concerns the methods for making the retrieved information available to the user. Presentation of results via means like pop-up windows risks being obtrusive (cf. the section on usability challenges below), but if the presentation is too subtle, users will often ignore the recommendations and derive little or no benefit from the system. Moreover, the optimal solution in general differs from one user to the next. Billsus et al. (2005) reported studies with a variety of interface solutions for the FXPAL Bar, some of which are adaptable by the user (e.g., the size of a translucent pop-up window that describes a potentially relevant document).[10]

Recommending Products

One of the most practically important categories of user-adaptive systems today comprises the product recommenders that are found in many commercial websites. The best-known system, the recommender of Amazon.com, was discussed briefly in the introduction to this chapter. Looking more closely at Fig. 6.1, we can see some distinguishing features of this approach to recommendation. As can be seen from the brief explanations that accompany the recommendations, the system takes as a starting point the information it has about the user's ownership or evaluation of particular products. It then recommends products that are similar in the sense that there is large overlap in the sets of customers that buy them (hence the familiar explanations of the form "Customers who bought this title also bought . . ."). That is, the recommendations are based on a statistical analysis of purchases made by many users, an approach known as *collaborative filtering* (see, e.g., Schafer, Frankowski, Herlocker, & Sen, 2007, for an overview). The products recommended in this way may also be similar in the sense of having the same author or a similar title (as in the examples in the figure), but similarities of this sort can also be conspicuously absent: In the category "Coming soon," the user of Fig. 6.1 was recommended the DVD *The Island* because of having positively rated a Sony VAIO notebook PC. As is explained by Linden, Smith, and York (2003), the details of this particular variant of collaborative filtering are due largely to the constraint that it has to be able to cope with Amazon.com's millions of products and customers. Although it is generally acknowledged that the recommendations are not always accurate, they can yield notable benefits simply by being better than the generic recommendations (e.g., of top-selling items) that would be presented without personalization.

Some product recommenders allow and require the user to specify his or her evaluation criteria explicitly, instead of simply rating or purchasing individual items. For example, with the Active Buyers Guide in Fig. 6.16, the user has specified how he or she intends to use the digital camera that he or she would like to buy, and the system has recommended three cameras, explaining why each one is suitable in terms of the user's requirements. This needs-based approach to recommendation offers a natural alternative to the purely statistical approach of systems like Amazon.com when relatively complex and important decisions are involved for which it is worthwhile for the user to think carefully about the attributes of the products in question. However, it does require that a good deal of knowledge about the features of products and their relationships to user requirements be incorporated in the system.

An intermediate approach between these two extremes is the critiquing paradigm (for an early exposition, see Burke, Hammond, & Young, 1997; for an evaluation of some recent advances, see McCarthy, Reilly, McGinty, & Smyth, 2005). The distinguishing feature is an iterative cycle in which the system proposes a product (e.g., a restaurant in a given city), the user criticizes the proposal (e.g., asking for a more casual restaurant), and the system proceeds to propose a similar product that takes the critique into account.

---

[10]Influential earlier systems in this category include those of Rhodes (2000) and Budzik, Hammond, and Birnbaum (2001).

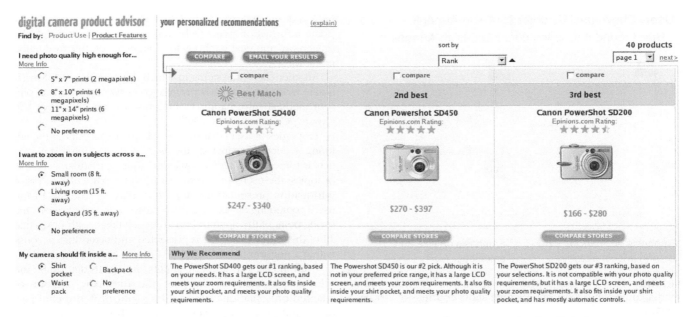

FIGURE 6.16. Partial screen shot from the Active Buyers Guide recommender for digital cameras. (Screen shot made from http://www.activebuyersguide.com in December 2005.)

Since some products are often used by groups of users (e.g., movies, vacations), a number of systems have been developed that explicitly address groups (for an overview, see Jameson & Smyth, 2007). The need to address a group rather than an individual has an impact on several aspects of the recommendation process. Users may want to specify their preferences in a collaborative way; there must be some appropriate and fair way of combining the information about the various users' preferences; and the explanations of the recommendations may have to refer to the preferences of the several individual group members.

Product recommenders of these various types address several problems that computer users typically experience when they search for products:

1. The user may not know what aspects of the products to attend to or what criteria should determine his or her decision. Some recommenders either (a) make it less necessary for the user to be explicitly aware of his or her evaluation criteria (as when collaborative filtering is used), or (b) help the user to learn about his or her own criteria during the course of the interaction with the system.
2. If the user is unfamiliar with the concepts used to characterize the products, he or she may be unable to make effective use of any search or selection mechanisms that may be provided. Product recommenders generally reduce this communication gap by allowing the user to specify his or her criteria (if this is necessary at all) in terms that are more natural to him or her. For example, in Fig. 6.16, the user does not need to know in advance how many megapixels his or her camera requires, since the question about photo quality is formulated in everyday terms.
3. The user may have to read numerous product descriptions in various parts of the website, integrating the information

found in order to arrive at a decision. Once a product recommender has acquired an adequate user model, the system can take over a large part of this work, often examining the internal descriptions of a much larger number of products than the user could deal with.

From the point of view of the vendors of the products concerned, the most obvious potential benefit is that users will find one or more products that they consider worth buying, instead of joining the notoriously large percentage of browsers who never become buyers. A related benefit is the prospect of cross-selling: the system's model of the user can be employed for the recommendation of further products that the user might not have considered him- or herself. Finally, some vendors aim to build up customer loyalty with recommenders that acquire long-term models of individual customers: If the user believes that the system has acquired an adequate model of him or her, the user will tend to prefer to use the system again rather than starting from scratch with some other system.

Tailoring Information Presentation

Even after a system has decided which documents or products to present to a user, the question may remain of exactly how to present them. In some cases, this question should be answered differently for different users. In Fig. 6.16, the verbal descriptions of the recommended products refer explicitly to the preferences that the user has expressed; if they did not, the user would have to invest more effort to judge how well each product met his or her requirements.

An example from a research prototype is found in the system RIA (Fig. 6.17; Zhou & Aggarwal, 2004), a multimodal system that helps users search for real estate, often presenting

User: Show ranches unter $800K in Armonk.
Ria: I found 4 ranches under $800K in Armonk.

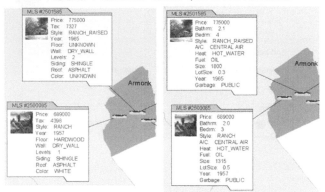

Emphasis on financial, exterior, and interior aspects

Emphasis on size and amenities

FIGURE 6.17. Two cropped screen shots from the RIA multimedia conversation system. (Screen shots courtesy of Vikram Aggarwal.)

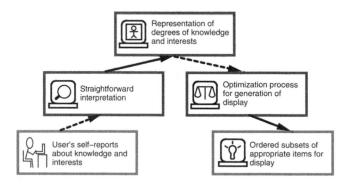

FIGURE 6.18. Overview of adaptation in RIA.

information about houses on a map. As Fig. 6.17 shows, the amount of space available for describing a house is limited, so it is important to select the information that is most likely to help the user to decide how to proceed; otherwise, the user may have to request additional information explicitly, which would slow down the interaction. The process of user model acquisition (the three left-most nodes in Fig. 6.18) is quite straightforward in RIA: Before the interaction begins, the user is asked a small number of questions about his or her interest and knowledge concerning houses. The sophisticated part of the system is the way in which it uses this information—along with information about the houses that satisfy the query—to decide which information to select for presentation and how it is to be ordered. The problem is one of optimizing the presentation with respect to a large set of constraints.[11] In an evaluation, the displays generated by this method were similar to those generated by a

human designers, who found the task of selecting the appropriate information items to be quite time-consuming (cf. the section on empirical methods for comments on this evaluation method).

Another class of systems in which the tailoring of information to individual users has promise comprises systems that present medical information to patients (for an overview, see Cawsey, Grasso, & Paris, 2007).

Properties of users that may be taken into account in the tailoring of documents include the user's degree of interest in particular topics; the user's knowledge about particular concepts or topics; the user's preference or need for particular forms of information presentation; and the display capabilities of the user's computing device (e.g., web browser vs. cell phone). One strong point of the optimization approach taken with RIA is that all of these factors can be represented and taken into account within a uniform framework.

Even in cases where it is straightforward to determine the relevant properties of the user, the automatic creation of adapted presentations can require sophisticated techniques of natural language generation (e.g., Bontcheva & Wilks, 2005) and multimedia presentation generation. Various less complex ways of adapting hypermedia documents to individual users have also been developed (see Bunt, Carenini, & Conati, 2007).

Supporting Collaboration

The increasing tendency for computer users to be linked via networks has made it increasingly feasible for users to collaborate, even in a spontaneous way and without prior acquaintance. A system that has models of a large number of users can facilitate such collaboration by taking into account the ways in which users match or complement each other.

A striking, though not typical, example is the system AgentSalon (Sumi & Mase, 2001, 2002), shown in Fig. 6.19. The system is used at conferences, in conjunction with a handheld guide (PalmGuide) that collects information about exhibits that the user has visited and ratings that he or she has given of them (the purpose within PalmGuide being to make recommendations to the user about other exhibits). When two visitors agree to work with AgentSalon, the information about them is transferred from their handhelds to AgentSalon. Like a host at a party, AgentSalon then looks for topics about which the two visitors might hold an interesting conversation, such as an exhibit about which they gave different ratings. The system tries to get a conversation going by having two animated agents simulate a conversation between these two visitors.

User modeling has been applied in connection with several (partially overlapping) types of collaboration:

• In computer-supported learning environments, in which the idea of *collaborative learning* has gained popularity in recent years (e.g., Soller, 2007)

---

[11]Decisions about what modalities to use for presentation—for example, text or speech output—are made in a similar way (cf. Zhou, Wen, & Aggarwal, 2005).

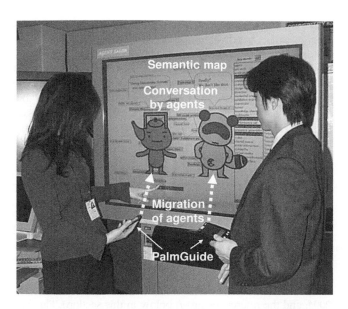

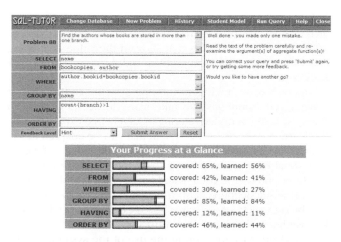

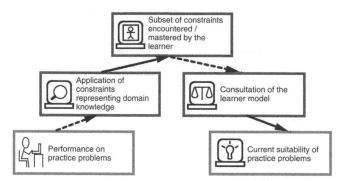

FIGURE 6.21. Screen shots from the SQL-Tutor. (Above: the main interface; below: display of the learner model. Screen shots courtesy of Antonija Mitrovic.)

FIGURE 6.19. Attempt by AgentSalon to stimulate discussion between two conference visitors. (The system has identified an interesting topic by comparing the records of their conference experiences that have been stored on their PDAs. Figure 1 of "AgentSalon: Facilitating face-to-face knowledge exchange through conversations among personal agents," by Y. Sumi & K. Mase, 2001, in *Proceedings of the Fifth International Conference on Autonomous Agents*, pp. 393–400, New York: ACM. Research conducted at ATR Media Information Science Laboratories, Kyoto. Copyright 2001 by the Association for Computing Machinery, Inc. Reproduced with permission.)

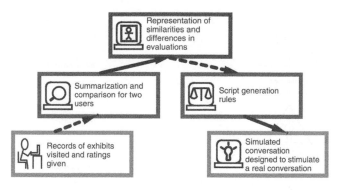

FIGURE 6.22. Overview of adaptation in the SQL Tutor.

Supporting Learning

Research on *student modeling*—or *learner modeling*, as it has been called more often in recent years—aims to add user-adaptivity to computer-based tutoring systems and learning environments (cf. Corbett, A. T., Koedinger, & Anderson, 1997).[12]

Increasingly, learning environments are being made available on the World Wide Web. An example is the SQL-Tutor (e.g., Mitrovic, Suraweera, Martin, & Weerasinghe, 2004), which teaches the database query language SQL.[13] The top part of Fig. 6.21 illustrates how the tutor presents a database querying problem and gives feedback on the learner's solution. The lower part of the figure shows a simple visualization of the learner model, which indicates the learner's degree of mastery of each of the six clauses of the SQL SELECT statement. It also shows a suggestion by the system about what type of problem the learner should attempt next—one of several ways in which the system helps the learner to pick a problem to work on next.

FIGURE 6.20. Overview of adaptation in AgentSalon.

- As a way of providing intelligent help for complex tasks (e.g., Vivacqua & Lieberman, 2000; Aberg & Shahmehri, 2001); putting a human expert into the loop is a way of avoiding some of the difficulties associated with fully automatic adaptive help systems that were discussed above

- In environments for computer-supported cooperative work within organizations (e.g., Terveen & McDonald, 2005)

---

[12]Good sources of literature include the *International Journal of Artificial Intelligence in Education* and the proceedings of the biennial Conferences on Artificial Intelligence in Education (e.g., Looi, McCalla, Bredeweg, & Breuker, 2005).

[13]The tutor is available to registered students via Addison-Wesley's Website Database Place (http://www.aw-bc.com/databaseplace/).

A number of different aspects of the SQL-Tutor were evaluated in 10 studies (e.g., Mitrovic et al., 2004; Mitrovic & Ohlsson, 1999), including 2 studies that showed the value of helping learners to choose the next problem and showing learners their learner model.

Interaction in intelligent tutoring systems and intelligent learning environments can take many forms, ranging from tightly system-controlled tutoring to largely free exploration by the learner. Aspects of the system that can be adapted to the individual user include (a) the selection and the form of the instructional information presented, (b) the content of problems and tests, and (c) the content and timing of hints and feedback.

Learner-modeling systems may adapt their behaviors to any of a broad variety of aspects of the user, such as (a) the user's knowledge of the domain of instruction, including knowledge acquired prior to and during the use of the system, (b) the user's learning style, motivation, and general way of looking at the domain in question, and (c) the details of the user's current processing of a problem.

The underlying assumption is that the adaptation of the system's behavior to some of these properties of the learner can lead to more effective and more enjoyable learning. One series of studies that directly demonstrated the added value of learner-adaptive tutoring is described by A. Corbett (2001). Many evaluations, however, do not focus on measuring the benefits of adaptivity but rather on comparing alternative variants of the same adaptive system. In some cases, it has been found that the modeling of the learner, however well realized, is not necessary for the effective functioning of the learning environment (see, e.g., VanLehn et al., 2005).

USABILITY CHALLENGES

Some of the typical properties of user-adaptive systems can lead to usability problems that may outweigh the benefits of adaptation to the individual user. Discussions of these problems have been presented by a number of authors (e.g., Norman, 1994; Wexelblat & Maes, 1997; Höök, 2000; Tsandilas & Schraefel, 2004; and the references given below in this section). Fig. 6.23 gives a high-level summary of many of the relevant ideas, using

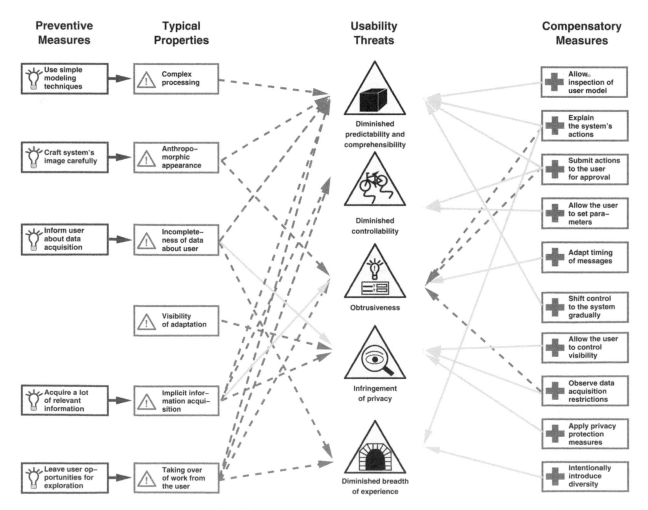

FIGURE 6.23. Overview of usability challenges for user-adaptive systems and of ways of dealing with them. (Dashed arrows denote threats and solid arrows mitigation of threats, respectively; further explanation is given in the text.)

the metaphor of signs that give warnings and advice to persons who enter a dangerous area.

The Usability Threats, shown in the third column, concern several generally desirable properties of interactive systems. Those referred to by the top three signs (concerning predictability and comprehensibility, controllability, and unobtrusiveness) correspond to general usability principles. The remaining two threats, to privacy and to breadth of experience, are especially relevant to user-adaptive systems.

The Typical Properties column lists some frequently encountered (though not always necessary) properties of user-adaptive systems, each of which has the potential of creating particular usability threats.

Each of the remaining two columns shows a different strategy for avoiding or mitigating one or more usability threats: Each of the Preventive Measures aims to ensure that one of the Typical Properties is not present in such a way that it would cause problems. Each of the Compensatory Measures aims to ward off one or more threats once it has arisen.

A discussion of all of the relationships indicated in Fig. 6.23 would exceed the scope of this chapter, but some remarks will clarify the main ideas.

Threats to Predictability and Comprehensibility

The concept of *predictability* refers to the extent to which a user can predict the effects of his or her actions. *Comprehensibility* is the extent to which the user can understand system actions and/or has a clear picture of how the system works.[14] These goals are grouped together here because they are associated with largely the same set of other variables.

Users can try to predict and understand a system on several different levels of detail.

1. *Exact layout and responses.* Especially detailed predictability is important when interface elements are involved that are accessed frequently by skilled users—for example, icons in control panels or options in menus (cf. the discussion of interface adaptations above). If the layout and behavior of the system is highly predictable—in fact, essentially identical—over time, skilled users can engage in *automatic processing* (see, e.g., Hammond, 1987): They can use the parts of the interface quickly, accurately, and with little or no attention. In this situation, even minor deviations from complete predictability on a fine-grained level can have the serious consequence of making automatic processing impossible or error-prone.
2. *Success at specific subtasks.* Users may desire only more global predictability and comprehensibility when the system is performing some more or less complex task on the user's behalf (e.g., searching for suitable products on the web): In the extreme case, the user may want only to predict (or evaluate) the quality of the result of a complex system action.
3. *Overall competence.* The most global form of predictability and comprehensibility concerns the user's ability to assess the system's overall level of competence: the degree to

which the system tends in general to perform its tasks successfully. With many types of systems, high overall competence can be taken for granted; but as we have seen, the processes of acquiring and applying user models do not in general ensure a high degree of accuracy. If the user seriously overestimates the system's competence, he or she may rely on the system excessively; if the user underestimates the system, he or she will not derive the potential benefits that the system can provide. A factor that is especially important with regard to this global level is the way in which the adaptive part of the system is presented to the user. Some user-adaptive systems, such as AgentSalon (which was discussed above) and the well-known Microsoft Office Assistant, have employed lifelike characters, for various reasons. As has often been pointed out, such anthropomorphic representations can invoke unrealistically high expectations concerning system competence—not only with regard to capabilities like natural language understanding but also with regard to the system's ability to understand and adapt to the user.

In general, the levels and degrees of predictability and comprehensibility that are necessary or desirable in a given case can depend on many factors, including the function that is being served by the adaptation and the user's level of skill and experience. The same is true of the choice of the strategies that are most appropriate for the achievement of predictability and comprehensibility.

Threats to Controllability

Controllability refers to the extent to which the user can bring about or prevent particular actions or states of the system if he or she has the goal of doing so. Although controllability tends to be enhanced by comprehensibility and predictability, these properties are not perfectly correlated. For example, when the user clicks on a previously unused option in Smart Menus, he or she can predict with certainty that it will be moved to the main part of its menu; but the user has no control over whether this change will be made.

A typical measure for ensuring some degree of control is to have the system submit any action with significant consequences to the user for approval. This measure causes a threat of obtrusiveness (see below); so it is an important interface design challenge to find ways of making recommendations in an unobtrusive fashion that still makes it easy for the user to notice and follow up on them (cf. the earlier discussion of FXPAL Bar).

Like predictability and comprehensibility, controllability can be achieved on various levels of granularity. Especially since the enhancement of controllability can come at a price, it is important to consider what kinds of control will really be desired. For example, there may be little point in submitting individual actions to the user for approval if the user lacks the knowledge or interest required to make the decisions. Wexelblat and Maes (1997) recommended making available several alternative types of control from which users can choose.

---

[14]The term *transparency* is sometimes used for this concept, but it can be confusing, because it also has different, incompatible meanings.

Obtrusiveness

We will use the term *obtrusiveness* to refer to the extent to which the system places demands on the user's attention that reduce the user's ability to concentrate on her primary tasks. This term—and the related words *distracting* and *irritating*—are often heard in connection with user-adaptive systems. Fig. 6.23 shows that (a) there are several different reasons why user-adaptive systems can easily turn out to be obtrusive and (b) there are equally many corresponding strategies for minimizing obtrusiveness. Some of these measures can lead straightforwardly to significant improvements—for example, when it is recognized that distracting lifelike behaviors of an animated character are not really a necessary part of the system.

Threats to Privacy

User-adaptive systems typically (a) gather data about individual users and (b) use these data to make decisions that may have more or less serious consequences. Users may accordingly become concerned about the possibility that their data will be put to inappropriate use. Privacy concerns tend to be especially acute in e-commerce contexts (e.g., Cranor, 2004) and with some forms of support for collaboration (e.g., Terveen & McDonald, 2004), because in these cases (a) data about the user are typically stored on computers other than the user's own, (b) the data often include personally identifying information, and (c) there may be strong incentives to use the data in ways that are not dictated by the user's own interests. As will be discussed in the next section, different means of acquiring information about users can have different consequences with regard to privacy. On the other hand, many of the measures that can be taken to protect privacy—for example, a policy of storing as little personally identifying data as possible—are not specific to user-adaptive systems.

Breadth of Experience

When a user-adaptive system helps the user with some form of information acquisition (cf. the second major section of this chapter) much of the work of examining the individual documents, products, and people involved is typically taken over by the system. A consequence can be that the user ends up learning less about the domain in question than she would with a nonadaptive system (cf. Lanier, 1995). For example, if the Amazon.com visitor for whom recommendations are shown in Fig. 6.1 relies heavily on such recommendations (as opposed to browsing freely), the user is likely to learn a lot about the books of Frederick Forsyth and about closely related products but little about the full range of books and other media that are available. One point of view here (see, e.g., the remarks of Maes in Shneiderman & Maes, 1997, p. 53) is that it should be up to the user to decide whether he or she prefers to learn about a given domain or to save time by delegating work to a system. It may be worthwhile to give the user a continuous spectrum of possibilities between complete control over a task and complete delegation of it. For example, many product recommen-

dation systems, such as Amazon.com's, allow users to alternate freely between pursuing the system's recommendations and browsing through product descriptions in the normal way.

Reduction of breadth of experience is especially likely if the system relies more heavily than is necessary on an incomplete user model. The user of Fig. 6.1 had (understandably) informed the system about only a tiny proportion of the books that he had ever read and liked; and since the system tends to recommend—and offer a chance to rate—similar items, the system may never obtain much evidence about all of the user's other literary interests. Some systems mitigate this problem by systematically proposing solutions that are not dictated by the current user model (for a method that is directly applicable to recommendation lists such as Amazon.com's, see Ziegler, McNee, Konstan, & Lausen, 2005; for methods realized in different types of recommenders, see Linden, Hanks, & Lesh, 1997; Shearin & Lieberman, 2001).

Dealing With Tradeoffs

As can be seen in Fig. 6.23, the designer who attempts to combat a particular usability threat will often have to accept a greater threat to some other usability goal. The most obvious tradeoffs involve unobstrusiveness. In particular, steps taken to enhance control or to protect privacy often require the user to perform additional actions, input additional information, and pay attention to additional system messages. Dealing with tradeoffs of this sort is complicated by the fact that users often differ markedly in the relative priority that they assign to each of the conflicting goals.

Fig. 6.24 illustrates some general points, referring for concreteness to a recently developed prototype Office Control

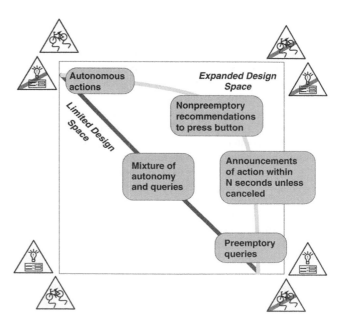

FIGURE 6.24. Illustration of strategies for dealing with tradeoffs among usability goals in user-adaptive systems. (Explanation in text.)

System (Cheverst et al., 2005). This system first observes how the occupant of an office tends to operate various devices such as the fan and the window shades; it then tries to help the user by performing some actions autonomously (e.g., opening the window when certain weather conditions prevail and there is no visitor in the office). An early version offered two ways of dividing the work between the user and the system for each type of action: The system could either perform actions of that type autonomously or request the user's permission with a preemptory dialogue box on the user's normal computer screen. Users chose the latter option for different proportions of the available actions, reflecting different priorities for the goals of unobtrusiveness and controllability respectively. In terms of the tradeoff graph shown in Fig. 6.24, users chose different points on the straight diagonal line. Despite this freedom of choice, users were often not satisfied with the overall usability of the prototype. A significant improvement in acceptance was achieved when the designers expanded the design space: They introduced a separate small screen for the Office Control System, in which information and requests for confirmation could be offered in several ways that are not very familiar in everyday graphical user interfaces (though they are familiar in industrial and traffic contexts; see, e.g., Wickens & Hollands, 2000). As Fig. 6.24 indicates, these additional forms of interaction represented a more favorable combination of degrees of unobtrusiveness and controllability at least for some of the users some of the time.

Consistent with more complex examples (see Jameson & Schwarzkopf, 2002; Billsus & Pazzani, 2007), this small case study illustrates two general points: (a) When dealing with tradeoffs among the usability goals discussed here, it is important to offer alternative solutions for users with different priorities, and (b) it is equally important to consider relatively novel interface design solutions that may spare users the need to choose among unsatisfactory alternatives.

OBTAINING INFORMATION ABOUT USERS

Some of the usability challenges discussed in the previous section are closely connected with the ways in which information about individual users is acquired—a consideration that also largely determines the success of a system's adaptation. The next two subsections will look, respectively, at (a) information that the user supplies to the system explicitly for the purpose of allowing the system to adapt, and (b) information that the system obtains in some other way.

Explicit Self-Reports and -Assessments

Self-reports about objective personal characteristics. Information about objective properties of the user (such as age, profession, and place of residence) sometimes has implications that are relevant for system adaptation—for example, concerning the topics in which the user is likely to be knowledgeable or interested. This type of information also has the advantage of changing relatively infrequently. Some user-

adaptive systems request information of this type from users, but the following caveats apply:

1. Specifying information such as profession and place of residence may require a fair amount of tedious menu selection and/or typing.
2. Since information of this sort can often be used to determine the user's identity, the user may justifiably be concerned about privacy issues. Even in cases where such concerns are unfounded, they may discourage the user from entering the requested information.

A general approach is (a) to restrict requests for personal data to the few pieces of information (if any) that the system really requires, and (b) to explain the uses to which the data will be put. Cranor (2004) gave a number of suggestions about how the use of personally identifying data can be minimized. An especially creative approach was tried in the web-based Lifestyle Finder prototype (Fig. 6.25; Krulwich, 1997), which was characterized by a playful style and an absence of requests for personally identifying information. Of the users surveyed, 93% agreed that the Lifestyle Finder's questions did not invade their privacy.

It is sometimes possible to avoid requests for explicit input about personal characteristics by accessing sources where similar information has already been stored (a strategy that will be discussed later in this section).

Self-assessments of interests and knowledge. It is sometimes helpful for a user-adaptive system to have an assessment of a property of the user that can be expressed naturally as a position on a particular general dimension: the level of the user's interest in a particular topic, the level of his or her knowledge about it, or the importance that the user attaches to a particular evaluation criterion. Often an assessment is arrived at through inference on the basis of indirect evidence, as with the assessments of a learner's knowledge in the SQL-Tutor (Fig. 6.21). But it may be necessary or more efficient to ask the user

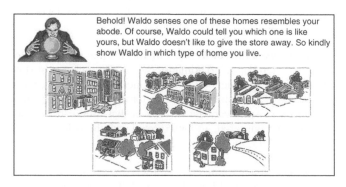

FIGURE 6.25. Example of a screen with which the LifeStyle Finder elicits demographic information. (Figure 3 of "Lifestyle Finder: Intelligent user profiling using large-scale demographic data," by B. Krulwich, 1997, AI *Magazine*, 18(2), pp. 37–45. Research conducted at the Center for Strategic Technology Research of Andersen Consulting (now Accenture Technology Labs). Copyright 1997 by the American Association for Artificial Intelligence. Adapted with permission.)

for an explicit assessment. For example, it would be difficult for the Active Buyers Guide recommender shown in Fig. 6.16 to estimate the importance that the user attaches to photo quality without asking the user directly. The scales in this figure illustrate good practice in that they make clear the meaning of the various possible answers, instead of asking, "How important is photo quality to you (on a scale from 1 to 5)?"

Because of the effort involved in this type of self-assessment, it is generally worthwhile to consider ways of minimizing such requests, making responses optional, ensuring that the purpose is clear, and integrating the self-assessment process into the user's main task (for some innovative ideas about how to achieve these goals, see Tsandilas & Schraefel, 2004).

Self-reports on specific evaluations. Instead of asking the user to describe his or her interests explicitly, some systems try to infer the user's position on the basis of his or her explicitly evaluative responses to specific items. Amazon.com's rating scales (Fig. 6.1) illustrate one form that this type of input can take; other forms include checkboxes and icons (e.g., "thumbs-up" and "thumbs-down"). The items that the user evaluates can be (a) items that the user is currently experiencing directly (e.g., the current website); (b) actions that the system has just performed, which the user may want to encourage or discourage (see, e.g., Wolfman, Lau, Domingos, & Weld, 2001); (c) items that the user must judge on the basis of a description (e.g., the abstract of a talk; a table listing the attributes of a physical product); or (d) the mere name of an item (e.g., a movie) that the user may have had some experience with in the past (see, e.g., Fig. 6.1). The cognitive effort required depends in part on how directly available the item is: In the third and fourth cases just listed, the user may need to perform memory retrieval and inference in order to arrive at an evaluation.

Even when the effort is minimal, users often do not like to bother with explicit evaluations that do not constitute a necessary part of the task they are performing. For this reason, many designers try to get by with the type of nonexplicit input discussed in the following section. For example, Findory (Fig. 6.14) could allow the user to rate the various news stories and blogs presented, but instead it just interprets the user's behavior in selecting items to read them.

Responses to test items. In systems that support learning, it is often natural to administer tests of knowledge or skill. In addition to serving their normal educational functions, these tests can yield valuable information for the system's adaptation to the user. An advantage of tests is that they can be constructed, administered, and interpreted with the help of a large body of theory, methodology, and practical experience (e.g., Wainer, 2000).

Outside of a learning context, users are likely to hesitate to invest time in tests of knowledge or skill, unless these can be presented in an enjoyable form (e.g., the color discrimination test used by Gutkauf, Thies, & Domik, 1997, to identify perceptual limitations relevant to the automatic generation of graphs). Trewin (2004, p. 76) reported experience with a brief typing test that was designed to identify helpful keyboard adaptations: Some users who turned out to require no adaptations were disappointed that their investment in the test had yielded no ben-

efit. As a result, Trewin decided that adaptations should be based on the users' naturally occurring typing behavior.

Nonexplicit Input

The previous subsection has given some examples of why designers often look for ways of obtaining information about the user that does not require any explicit input by the user.

Naturally Occurring Actions

The broadest and most important category of information of this type includes all of the actions that the user performs with the system that do not have the express purpose of revealing information about the user to the system. These actions may range from major actions like purchasing an expensive product to minor ones like scrolling down a web page. The more significant actions tend to be specific to the particular type of system that is involved (e.g., e-commerce websites vs. learning environments). Within some domains, there has been considerable research on ways of interpreting particular types of naturally occurring user actions. For example, researchers interested in adaptive hypertext navigation support have developed a variety of ways of analyzing the user's navigation actions to infer the user's interests and to propose navigation shortcuts (e.g., Mobasher, 2007).

In their purest forms, naturally occurring actions require no additional investment by the user, because they are actions that the user would perform anyway. The main limitation is that they are hard to interpret; for example, the fact that a given web page has been displayed in the user's browser for four minutes does not reveal with certainty which (if any) of the text displayed on that page the user has actually read. Some designers have tried to deal with this trade-off by designing the user interface in such a way that the naturally occurring actions are especially easy to interpret. For example, a web-based system might display just one news story on each page, even if displaying several stories on each page would normally be more desirable.

The interpretation of naturally occurring actions by the system can raise privacy and comprehensibility issues (cf. Fig. 6.23) that do not arise in the same way with explicit self-reports and self-assessments of the types discussed earlier in this section: Whereas the latter way of obtaining information about the user can be compared with interviewing, the former way is more like eavesdropping—unless the user is informed about the nature of the data that are being collected and the ways in which they will be used (cf. Cranor, 2004).

Previously Stored Information

Sometimes a system can access relevant information about the user that has been acquired and stored independently of the system's interaction with the user:

1. If the user has some relationship (e.g., patient, customer) with the organization that operates the system, this organization

may have information about the user that it has stored for reasons unrelated to any adaptation, such as the user's medical record (see Cawsey et al., 2007, for examples) or address.

2. Relevant information about the user may be stored in publicly available sources such as electronic directories or web home pages. For example, Pazzani (1999) explored the idea of using a user's web home page as a source of information for a restaurant recommending system.

3. If there is some other system that has already built up a model of the user, the system may be able to access the results of that modeling effort and try to apply them to its own modeling task. There is a line of research that deals with *user modeling servers* (e.g., Kobsa, 2007): systems that store information about users centrally and supply such information to a number of different applications. Some of the major commercial personalization software is based on this conception.

Relative to all of the other types of information about users, previously stored information has the advantage that it can be applied in principle right from the start of the first interaction of a given user with a given system. To be sure, the interpretability and usefulness of the information in the context of the current application may be limited. Moreover, questions concerning privacy and comprehensibility may be even more important than with the interpretation of naturally occurring actions.

Low-Level Indices of Psychological States

The next two categories of information about the user have become practically feasible only in recent years, with advances in the miniaturization of sensing devices.

The first category of sensor-based information (discussed at length in the classic book of Picard, 1997) comprises data that reflect aspects of a user's psychological state. Some of the application scenarios in which this type of information can be useful were discussed in the section on systems that mediate interaction with the real world.

Two categories of sensing devices have been employed: (a) devices attached to the user's body (or to the computing device itself) that transmit physiological data, such as electromyogram signals, the galvanic skin response, blood volume pressure, and the pattern of respiration (for an overview, see Lisetti & Nasoz, 2004); and (b) video cameras and microphones that transmit psychologically relevant information about the user, such as features of his or her facial expressions (e.g., Bartlett, Littlewort, Fasel, & Movellan, 2003) or his or her speech (e.g., Liscombe et al., 2005).

With both categories of sensors, the extraction of meaningful features from the low-level data stream requires the application of pattern recognition techniques. These typically make use of the results of machine learning studies in which the relationships between low-level data and meaningful features have been learned.

One advantage of sensors is that they supply a continuous stream of data, the cost to the user being limited to the physical and social discomfort that may be associated with the carrying or wearing of the devices. These factors are still significant now, but further advances in miniaturization—and perhaps changing attitudes as well—seem likely to reduce their importance.

Signals Concerning the Current Surroundings

As computing devices become more portable, it is becoming increasingly important for a user-adaptive system to have information about the user's current surroundings. Here again, two broad categories of input devices can be distinguished (for a discussion of a number of specific types of devices, see Krüger, Baus, Heckmann, Kruppa, & Wasinger, 2007).

1. Devices that receive explicit signals about the user's surroundings from specialized transmitters. Some mobile systems that are used outdoors employ GPS (Global Positioning System) technology. More specialized transmitters and receivers are required, for example, if a portable museum guide system is to be able to determine which exhibit the user is looking at.

2. More general sensing or input devices. For example, Schiele, Starner, Rhodes, Clarkson, and Pentland (2000) described the use of a miniature video camera and microphone (each roughly the size of a coin) that enable a wearable computer to discriminate among different types of surroundings (e.g., a supermarket vs. a street). The use of general-purpose sensors eliminates the dependence on specialized transmitters. On the other hand, the interpretation of the signals requires the use of sophisticated machine learning and pattern recognition techniques.

SPECIAL CONSIDERATIONS CONCERNING EMPIRICAL METHODS

The full repertoire of empirical methods in HCI is in principle applicable to user-adaptive systems. This section will focus on some methods that are more important for user-adaptive systems than for other types and on some typical problems that need to be dealt with. This focused discussion should not obscure the fact that a lot of empirical work with user-adaptive systems looks the same as with other systems.[15]

Use of Data Collected With a Nonadaptive System

The key difference between user-adaptive systems and other interactive systems is the inclusion of some method for acquiring and exploiting a user model. This feature gives rise to a type of empirical study that is largely unique to user-adaptive

---

[15]More extended discussions of empirical methods for user-adaptive systems are provided by Gena and Weibelzahl (2007), Höök (2000), and Langley and Fehling (1998).

systems: studies in which the accuracy of the modeling methods is evaluated.

This type of evaluation can often be performed even if no user-adaptive systems that employ the user modeling method in question exist. What is needed are (a) some implementation of the adaptation algorithm, not necessarily embedded in any interactive system, and (b) a database of behavioral data on how a number of users have used a relevant nonadaptive system. The researcher can then apply the modeling method to the data in order to determine how well the system would adapt to the users in question.

A number of studies of this type were conducted with the i-ems system (see McCreath et al., 2006, and the discussion earlier in this chapter). In one case, the researchers wanted to find out whether a user who had defined a number of handcrafted e-mail sorting rules could benefit from having automatically learned rules applied to messages that were not covered by the handcrafted rules. One simulation was performed on 5,100 e-mail messages that had been sorted by a single user within a nonadaptive e-mail client over three months. In the order in which the messages had been received, they were presented to one of the system's learning algorithms in batches of 100, along with information about how the user had sorted them, so that the system could continually refine its set of learned rules. After each batch of 100 messages, the accuracy of the updated set of rules was evaluated: The system was asked to make predictions about the next batch of messages before being told how the user had in fact sorted them, and these predictions were compared with the user's actual behavior. Several indices of the system's performance were computed, one of which is shown in Fig. 6.26: the percentage of messages for which the learned rules made no prediction (e.g., where the appropriate folder was unknown to the system). The middle curve in the graph shows how this percentage gradually decreased over time. The uppermost curve shows the corresponding results for the case where only the user's handcrafted rules were used for prediction; the lowest curve shows the results where both types of rules were applied, with the handcrafted rules taking prece-

dence in the case of disagreement. It can be seen that the joint use of both sets of rules gave the best results in terms of avoiding "unknown" predictions. Since according to other indices the accuracy of the combined set of rules was as good as that of the handcrafted rules alone, the system's performance was best overall with the combined set.

Note that it was not necessary, for this evaluation, to create three different versions of i-ems and have them used for months by different users. In addition to being extremely time-consuming, this procedure would allow less direct accuracy comparisons. By contrast, using existing e-mail corpora, McCreath et al. (2006) performed simulations that shed light on many properties of the algorithms used.

The appeal of this type of evaluation, in terms of being able to yield numerous interpretable results with minimal involvement of actual users, is so great that researchers often seem to lose sight of the fact that studies with real users are likewise essential. No simulation study, for example, can reveal how well the design of the i-ems interface shown in Fig. 6.4 corresponds to the way users like to deal with incoming e-mail, or how helpful users find the explanations that the system offers for its predictions.

Early Studies of Usage Scenarios and User Requirements

In the field of HCI as a whole, it is expected that user-centered design should begin with a study of contexts of use, usage scenarios, properties of users, and user requirements. To date, this strategy has been applied less frequently in the design of user-adaptive systems—perhaps because the designers less frequently come from an HCI background, often specializing instead in the development of the necessary adaptive algorithms. Early user studies are actually at least as important with novel user-adaptive systems as with other types of systems: It is often not clear in advance whether adaptation will yield added value and achieve acceptance in a particular context. Careful attention to the requirements and contexts of users may greatly increase the likelihood of success—or at least warn the designers at an early stage that a particular usage scenario is not promising for the sort of adaptive interaction that they envision.

A positive example of early attention to user requirements is found in the development of the museum guide HyperAudio, which was developed as a prototype in the 1990s (for a retrospective discussion, see Petrelli & Not, 2005). Studying the attitudes and behavior of museum visitors at an early stage in the system's design, the researchers found that many visitors enjoy guided tours but that few visitors want to spend time interacting with technical devices. These two findings, along with others, led to a modification of the original conception of HyperAudio, in that they suggested the appropriateness of a museum guide that selects information for presentation on the basis of the user's behavior and location, requiring little or no explicit input.

Wizard of Oz Studies

Systems that adapt to their users are in one methodological respect similar to systems that make use of speech: They attempt to realize a capability that is so far, at least in many contexts,

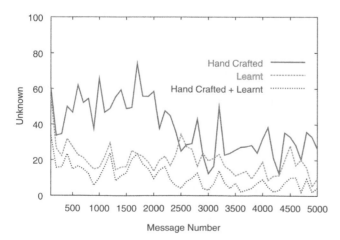

FIGURE 6.26. Results of one of the simulation experiments performed with i-ems's algorithms. (Explanation in text. Figure courtesy of Eric McCreath.)

possessed to the highest degree by humans. Consequently, as with speech interfaces, valuable information can sometimes be obtained from a Wizard of Oz study: In a specially created setting, a human takes over a part of the processing of the to-be-developed system for which humans are especially well suited.

One example is the Wizard of Oz study that was conducted early in the development of the Lumière intelligent help system (Horvitz, Breese, Heckerman, Hovel, & Rommelse, 1998), which formed the basis for the Office Assistant, which was introduced in Microsoft Office 97. In this study, subjects working with a spreadsheet were told that an experimental help system would track their activities and make guesses about how to help them. They received the advice via computer monitors. The advice was actually provided by usability experts who, working in a separate room, viewed the subjects' activity via monitors and conveyed their advice by typing.

This type of study can yield an upper-bound estimate of the highest level of modeling accuracy that might be attainable given the available information—as long as one can assume that the human wizards are more competent at the type of assessment in question than a fully automatic system is likely to be in the foreseeable future. In this example study, the expert advisers showed some ability to identify the users' goals and needs; if they had not done so, perhaps the entire project would have been reconsidered.

Aside from accuracy, a Wizard of Oz study can also shed light on problems with the acceptability and usability of the new system, as long as they concern the content and basic nature of the adaptations performed, as opposed to interface details that are not faithfully reproduced in the study. In our example study, it turned out that even incorrect advice was often taken seriously by the users, who wasted time following up on irrelevant suggestions. One design implication is that users should be made aware of the fact that the system's advice is not necessarily relevant.

Comparisons With the Work of Human Designers

Just as humans can sometimes be employed as a surrogate for a user-adaptive system in the early stages of design, humans can sometimes also serve as a standard of comparison for the evaluation of an implemented system. This method makes most sense when the system is performing a task at which human authors or designers are likely to be experienced and skilled, such as the tailoring of the content of a presentation to the individual user. An example of such a study was mentioned earlier: a comparison of the displays generated by the real estate recommender system RIA with those generated by two experienced designers. Instead of performing the same task as the system, the human designers may simply act as judges of the appropriateness of the system's output. In either case, the comments of the designers can yield valuable qualitative information to complement the objective results.

Experimental Comparisons of Adaptive and Nonadaptive Systems

Many experimental studies involving user-adaptive systems compare an adaptive variant with some nonadaptive one. This

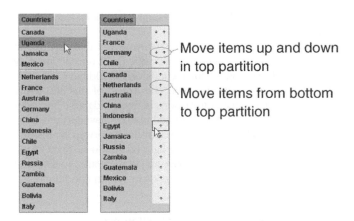

FIGURE 6.27. The adaptive (left) and adaptable (right) split menus used in an experimental comparison. (Adapted from Figure 1 of "A comparison of static, adaptive, and adaptable menus," by L. Findlater & J. McGrenere, 2004, in E. Dykstra-Erickson & M. Tscheligi (Eds.), *Human factors in computing systems*: CHI 2004 *conference proceedings*, pp. 89–96, New York: ACM, Copyright 2004 by the Association for Computing Machinery, Inc. Adapted with permission.)

strategy is understandable given the doubt that often exists as to whether the additional overhead required for user-adaptivity is justified by any sort of improvement in the interaction. However, studies like this are trickier to conduct and interpret than they may seem at first glance.

As an example, consider the experiment by Findlater and McGrenere (2004), which examined whether subjects could work faster using (a) an adaptive menu system somewhat like Smart Menus (cf. the discussion of adaptive menus above); (b) an adaptable system in which the users explicitly determined the content of the menus themselves; or (c) a conventional static menu system. For each of these three types of menu, a realization was chosen that seemed optimal in the context of the experiment. All three menus were a special type of split menu (Sears & Shneiderman, 1994): The four most frequent items in each menu were placed in a special section at the top of the menu for quick access (instead of being temporarily hidden, as in Smart Menus). The items selected for the upper part of the static menu were optimally chosen in that they reflected the actual frequency of the items in the experimental tasks. The adaptive menu was initially identical to the static menu, but the arrangement of the items changed as a function of the user's behavior, favoring the most frequently and recently used menu items. For the adaptable menu, the upper part was initially empty, so that users would be encouraged to perform some adaptations. Though the overall pattern of results is complex, it tends to speak in favor of the adaptable menu. Note that the conditions did not give the adaptive variant much of a chance to provide any benefit. Since the initial menu was already the best possible single menu for the experimental tasks, adaptation could improve performance only by taking advantage of any local concentrations of commands within a particular period of time (e.g., the need to execute the same command several times in succession). By contrast, in normal usage situations, an adaptive menu can also improve performance by reflecting increasingly the user's longer-term patterns of use.

The difficulties in interpreting the results of this experiment could not easily have been avoided with a different design: Any other way of realizing the three conditions would have left some different set of questions open. The lesson of this and many other examples is that comparisons between adaptive and nonadaptive variants of a system should not be viewed as empirical tests whose results can be interpreted straightforwardly. Instead, they should be seen as shedding light on various aspects of the ways in which people use adaptive and nonadaptive systems and on the effectiveness of these methods in certain conditions.

Taking Into Account Individual Differences

Individual differences among users show up in just about every user study in the field of HCI; but with user-adaptive systems, they are especially important because of the wide range of subjective reactions that user-adaptivity tends to invoke (illustrated in the experiment discussed in the previous subsection). As a result, asking whether people like a particular type of user-adaptive system is in many cases like asking whether the voters in a given country prefer progressive or conservative policies. Even if, in a given sample, a statistically significant tendency in one direction or the other can be found, important minority points of view should be understood and reported. As in politics, the goal should be to take into account the range of different preferences in a way that is satisfactory to at least a large proportion of the potential user group.

When individual differences are present, it may be tempting to try to find correlations with demographic characteristics or with general personality variables. Although some relationships of this sort can be found (see, for example, Graziola, Pianesi, Zancanaro, & Goren-Bar, 2005, with regard to personality variables), it is not always worthwhile to look for them. The relationships tend to be weak, since different responses can also be due to more specific causes, such as degree of familiarity with the particular type of system and particular conditions under which a user performs a task.

Checking Usability Under Realistic Conditions

With just about any type of interactive system, new lessons are likely to be learned when a working prototype (or the finished system itself) is tested in realistic situations, even if the system has been studied thoroughly in earlier stages. With user-adaptive systems, realistic testing is especially advisable because of the issues discussed in the section on usability challenges, whose importance for a given system can often be assessed only in real use. For example, an obtrusive proactive recommender might be considered quite acceptable, or even amusing, when the user is performing some artificial assigned task in a laboratory; in the real world, when he or she is under pressure to complete an important task quickly, the interruptions may be evaluated quite differently. Similarly, privacy issues are serious mainly when real data about the user are involved and when misuse could have real consequences.

THE FUTURE OF USER-ADAPTIVE SYSTEMS

This chapter has shown that adaptive interfaces, agents, and other user-adaptive systems do not represent a smooth and easy shortcut to more successful HCI: They present a complex set of usability challenges, and they require carefully designed methods of acquiring information about users, as well as relatively sophisticated computational techniques that are not needed in other types of interactive systems. Even when all of these requirements have been dealt with, it is often tricky to prove empirically that user-adaptivity has actually added any value. It is no wonder that some experts believe that the interests of computer users are better served by continued progress within more familiar paradigms of user-centered system design.

On the other hand, our understanding of the complex challenges raised by user-adaptive systems has been growing steadily, and they are now familiar and valued elements in a number of types of systems as the survey in the first two major sections of this chapter showed.

Growing Need for User-Adaptivity

Increases in the following variables suggest that the functions served by user-adaptivity will continue to grow in importance:

Diversity of users and contexts of use. Computing devices are being used by an ever-increasing variety of users in an increasing variety of contexts. It is therefore becoming harder to design a system that will be suitable for all users and contexts without some sort of user-adaptivity or user-controlled adaptability; and as has been discussed at several points in this chapter, user-controlled adaptability has its limitations.

Number and complexity of interactive systems. The functions of user-adaptivity discussed in the first major section of this chapter involve helping users to deal effectively with interactive systems and tasks even when they are not able or willing to gain complete understanding and control in each individual case. This goal becomes increasingly important as the number—and in some cases the complexity—of the systems that people must deal with continues to increase—because of factors ranging from the growth of the World Wide Web to the proliferation of miniature interactive computing devices.

Scope of information to be dealt with. Even when using a single, relatively simple system, users today can often access a much larger and more diverse set of objects of interest than they could a few years ago—be they documents, products, or potential collaborators. It is therefore becoming relatively more attractive to delegate some of the work of dealing with these objects—even to a system that has an imperfect model of the user's requirements. In the early 1990s, the idea that an e-mail sorting agent, such as the one described by Maes (1994), might delete an incoming message without consulting the user seemed preposterous to many people. After the huge increase in the amount of (largely unwanted) e-mail that has occurred since then, many people now regularly allow dozens of their incoming messages to be deleted unseen.

Increasing Feasibility of Successful Adaptation

As the need for user-adaptivity increases, fortunately so does its feasibility, largely because of advances in the following areas:

Ways of acquiring information about users. Most of the methods discussed in the section about acquiring information about users are becoming more powerful with advances in technology and research. They therefore offer the prospect of substantial increases in the quality of adaptation—although methods for ensuring users' privacy call for equal attention.

Advances in techniques for learning, inference, and decision. In addition to the more general progress in the fields of machine learning and artificial intelligence, communities of researchers have been focusing on the specific requirements of computational techniques that support user-adaptivity. Consequently, noticeable progress is being made every year.

Attention to empirical methods. The special empirical issues and methods that are involved in the design and evaluation of user-adaptive systems have been receiving increasing attention from researchers, as emphasis has shifted from high technical sophistication to ensuring that the systems enhance the users' experience.

Despite these tendencies, it is actually unlikely that the number of deployed systems associated with labels like "user-adaptive" will increase. Once an adaptation technique has left the research laboratory and started playing some genuinely useful role in people's lives, it tends to be described in terms of the function that it serves rather than in terms of the techniques that it uses. Awareness of the commonalities discussed in this chapter should help both to increase the number of systems that succeed in this way and to recognize them despite the new labels that are placed on them.

ACKNOWLEDGMENTS

Preparation of this chapter was supported by the German Ministry of Education and Research (projects Specter and Collate). The following colleagues provided valuable assistance in the form of comments on an earlier draft, supplementary information about their research, and/or graphics depicting their systems: Vikram Aggarwal and Michelle Zhou; James 'Bo' Begole and Nicholas Matsakis; Vania Dimitrova; Tatsuya Kawahara; Bruce Krulwich; Greg Linden; Eric McCreath and Judy Kay; Joanna McGrenere and Leah Findlater; Antonija Mitrovic; Charles Rich and Elyon DeKoven; Yasuyuki Sumi and Kenji Mase; Karin Vogel.

References

Aberg, J., & Shahmehri, N. (2001). An empirical study of human web assistants: Implications for user support in web information systems. In J. A. Jacko, A. Sears, M. Beaudouin-Lafon, & R. J. Jacob (Eds.), *Human factors in computing systems: CHI 2001 conference proceedings* (pp. 400–411). New York: ACM.

Bälter, O., & Sidner, C. L. (2002). Bifrost Inbox Organizer: Giving users control over the inbox. *Proceedings of NORDICHI 2002*, 111–118.

Bartlett, M. S., Littlewort, G., Fasel, I., & Movellan, J. R. (2003). Real time face detection and facial expression recognition: Development and applications to human computer interaction. *Proceedings of the Workshop on Computer Vision and Pattern Recognition for Human-Computer Interaction at the 2003 Conference on Computer Vision and Pattern Recognition*, 53.

Billsus, D., & Pazzani, M. (2007). Adaptive news access. In P. Brusilovsky, A. Kobsa, & W. Nejdl (Eds.), *The adaptive web: Methods and strategies of web personalization* (pp. 550–572).

Billsus, D., Hilbert, D., & Maynes-Aminzade, D. (2005). Improving proactive information systems. In J. Riedl, A. Jameson, D. Billsus, & T. Lau (Eds.), *IUI 2005: International Conference on Intelligent User Interfaces* (pp. 159–166). New York: ACM.

Bontcheva, K., & Wilks, Y. (2005). Tailoring automatically generated hypertext. *User Modeling and User-Adapted Interaction, 15*, 135–168.

Brusilovsky, P., Kobsa, A., & Nejdl, W. (Eds.). (2007). *The adaptive web: Methods and strategies of web personalization*. Berlin: Springer.

Budzik, J., Hammond, K., & Birnbaum, L. (2001). Information access in context. *Knowledge-Based Systems, 14*, 37–53.

Bunt, A., Carenini, G., & Conati, C. (2007). Adaptive presentation for the web. In P. Brusilovsky, A. Kobsa, & W. Nejdl (Eds.), *The adaptive web: Methods and strategies of web personalization* (pp. 409–432). Berlin: Springer.

Bunt, A., Conati, C., & McGrenere, J. (2004). What role can adaptive support play in an adaptable system? In C. Rich & N. J. Nunes (Eds.), *IUI 2004: International Conference on Intelligent User Interfaces* (pp. 117–124). New York: ACM.

Burke, R. D., Hammond, K. J., & Young, B. C. (1997). The FindMe approach to assisted browsing. *IEEE Expert, 12*(4), 32–40.

Cawsey, A., Grasso, F., & Paris, C. (2007). Adaptive information for consumers of healthcare. In P. Brusilovsky, A. Kobsa, & W. Nejdl (Eds.), *The adaptive web: Methods and strategies of web personalization* (pp. 465–484). Berlin: Springer.

Cheverst, K., Byun, H. E., Fitton, D., Sas, C., Kray, C., & Villar, N. (2005). Exploring issues of user model transparency and proactive behaviour in an office environment control system. *User Modeling and User-Adapted Interaction, 15*(3/4), 235–273.

Corbett, A. (2001). Cognitive computer tutors: Solving the two-sigma problem. In M. Bauer, P. Gmytrasiewicz, & J. Vassileva (Eds.), *UM2001, User Modeling: Proceedings of the Eighth International Conference* (pp. 137–147). Berlin: Springer.

Corbett, A. T., Koedinger, K. R., & Anderson, J. R. (1997). Intelligent tutoring systems. In M. Helander, T. K. Landauer, & P. V. Prabhu (Eds.), *Handbook of human-computer interaction* (pp. 849–874). Amsterdam: North-Holland.

Cranor, L. F. (2004). 'I didn't buy it for myself': Privacy and e-commerce personalization. In C. Karat, J. Blom, & J. Karat (Eds.), *Designing personalized user experiences for eCommerce* (pp. 57–74). Dordrecht, The Netherlands: Kluwer.

Crawford, E., Kay, J., & McCreath, E. (2002). An intelligent interface for sorting electronic mail. In Y. Gil & D. B. Leake (Eds.), *IUI 2002: International Conference on Intelligent User Interfaces* (pp. 182–183). New York: ACM.

DeKoven, E. A. (2004). *Help me help you: Designing support for person-product collaboration.* Unpublished doctoral dissertation, Delft University of Technology.

Fernandez, R., & Picard, R. W. (2000). Modeling drivers' speech under stress. *Proceedings of the ISCA Workshop on Speech and Emotions,* Belfast.

Findlater, L., & McGrenere, J. (2004). A comparison of static, adaptive, and adaptable menus. In E. Dykstra-Erickson & M. Tscheligi (Eds.), *Human factors in computing systems: CHI 2004 conference proceedings* (pp. 89–96). New York: ACM.

Fogarty, J., Ko, A. J., Aung, H. H., Golden, E., Tang, K. P., & Hudson, S. E. (2005). Examining task engagement in sensor-based statistical models of human interruptibility. In W. Kellogg, S. Zhai, G. van der Veer, & C. Gale (Eds.), *Human factors in computing systems: CHI 2005 conference proceedings* (pp. 331–340). New York: ACM.

Fuller, R. (2005). Towards a general theory of driver behaviour. *Accident Analysis & Prevention, 37*(3), 461–472.

Gena, C., & Weibelzahl, S. (2007). Usability engineering for the adaptive web. (2006). In P. Brusilovsky, A. Kobsa, & W. Nejdl (Eds.), *The adaptive web: Methods and strategies of web personalization* (pp. 720–766). Berlin: Springer.

Gervasio, M. T., Moffitt, M. D., Pollack, M. E., Taylor, J. M., & Uribe, T. E. (2005). Active preference learning for personalized calendar scheduling assistance. In J. Riedl, A. Jameson, D. Billsus, & T. Lau (Eds.), *IUI 2005: International Conference on Intelligent User Interfaces* (pp. 90–97). New York: ACM.

Graziola, I., Pianesi, F., Zancanaro, M., & Goren-Bar, D. (2005). Dimensions of adaptivity in mobile systems: Personality and people's attitudes. In J. Riedl, A. Jameson, D. Billsus, & T. Lau (Eds.), *IUI 2005: International Conference on Intelligent User Interfaces* (pp. 223–230). New York: ACM.

Gruen, D., Rohall, S. L., Minassian, S., Kerr, B., Moody, P., Stachel, B. B., et al. (2004). Lessons from the ReMail prototypes. In J. Herbsleb & G. Olson (Eds.), *Proceedings of the 2004 Conference on Computer-Supported Cooperative Work* (pp. 152–161). New York: ACM.

Gutkauf, B., Thies, S., & Domik, G. (1997). A user-adaptive chart editing system based on user modeling and critiquing. In A. Jameson, C. Paris, & C. Tasso (Eds.), *UM97 User modeling: Proceedings of the Sixth International Conference* (pp. 159–170). Vienna: Springer Wien New York.

Höök, K. (2000). Steps to take before IUIs become real. *Interacting with Computers, 12*(4), 409–426.

Haller, S., McRoy, S., & Kobsa, A. (Eds.). (1999). *Computational models of mixed-initiative interaction.* Dordrecht, The Netherlands: Kluwer.

Hammond, N. (1987). Principles from the psychology of skill acquisition. In M. M. Gardiner & B. Christie (Eds.), *Applying cognitive psychology to user interface design* (pp. 163–188). Chichester, England: Wiley.

Hanson, V. L., & Crayne, S. (2005). Personalization of web browsing: Adaptations to meet the needs of older adults. *Universal Access in the Information Society, 4,* 46–58.

Healey, J., & Picard, R. (2000). SmartCar: Detecting driver stress. *Proceedings of the Fifteenth International Conference on Pattern Recognition* (pp. 4218–461), Barcelona.

Hegner, S. J., McKevitt, P., Norvig, P., & Wilensky, R. L. (Eds.). (2001). *Intelligent help systems for UNIX.* Dordrecht, The Netherlands: Kluwer.

Ho, J., & Intille, S. S. (2005). Using context-aware computing to reduce the perceived burden of interruptions from mobile devices. In W. Kellogg, S. Zhai, G. van der Veer, & C. Gale (Eds.), *Human factors in computing systems: CHI 2005 conference proceedings* (pp. 909–918). New York: ACM.

Horvitz, E. (1999). Principles of mixed-initiative user interfaces. In M. G. Williams, M. W. Altom, K. Ehrlich, & W. Newman (Eds.), *Human factors in computing systems: CHI '99 conference proceedings* (pp. 159–166). New York: ACM.

Horvitz, E., Breese, J., Heckerman, D., Hovel, D., & Rommelse, K. (1998). The Lumière project: Bayesian user modeling for inferring the goals and needs of software users. In G. F. Cooper & S. Moral (Eds.), *Uncertainty in Artificial Intelligence: Proceedings of the Fourteenth Conference* (pp. 256–265). San Francisco: Morgan Kaufmann.

Horvitz, E., Koch, P., Sarin, R., Apacible, J., & Subramani, M. (2005). Bayesphone: Precomputation of context-sensitive policies for inquiry and action in mobile devices. In L. Ardissono, P. Brna, & A. Mitrovic (Eds.), *UM2005, User Modeling: Proceedings of the Tenth International Conference* (pp. 251–260). Berlin: Springer.

Iqbal, S. T., Adamczyk, P. D., Zheng, X. S., & Bailey, B. P. (2005). Towards an index of opportunity: Understanding changes in mental workload during task execution. In W. Kellogg, S. Zhai, G. van der Veer, & C. Gale (Eds.), *Human factors in computing systems: CHI 2005 conference proceedings* (pp. 311–318). New York: ACM.

Jameson, A., & Schwarzkopf, E. (2002). Pros and cons of controllability: An empirical study. In P. De Bra, P. Brusilovsky, & R. Conejo (Eds.), *Adaptive hypermedia and adaptive web-based systems: Proceedings of AH 2002* (pp. 193–202). Berlin: Springer.

Jameson, A., & Smyth, B. (2007). Recommendation for groups. In P. Brusilovsky, A. Kobsa, & W. Nejdl (Eds.), *The adaptive web: Methods and strategies of web personalization* (pp. 596–627). Berlin: Springer.

Jones, C. M., & Jonsson, I. (2005). Automatic recognition of affective cues in the speech of car drivers to allow appropriate responses. *Proceedings of OZCHI 2005.*

Kelly, D., & Teevan, J. (2003). Implicit feedback for inferring user preference: A bibliography. *ACM SIGIR Forum, 37*(2), 18–28.

Kobsa, A. (2007). Generic user modeling systems. In P. Brusilovsky, A. Kobsa, & W. Nejdl (Eds.), *The adaptive web: Methods and strategies of web personalization* (pp. 291–324). Berlin: Springer.

Kobsa, A., & Wahlster, W. (Eds.). (1989). *User models in dialogue systems.* Berlin: Springer.

Komatani, K., Ueno, S., Kawahara, T., & Okuno, H. G. (2005). User modeling in spoken dialogue systems to generate flexible guidance. *User Modeling and User-Adapted Interaction, 15,* 169–183.

Krüger, A., Baus, J., Heckmann, D., Kruppa, M., & Wasinger, R. (2007). Adaptive mobile guides. In P. Brusilovsky, A. Kobsa, & W. Nejdl (Eds.), *The adaptive web: Methods and strategies of web personalization* (pp. 521–549). Berlin: Springer.

Krulwich, B. (1997). Lifestyle Finder: Intelligent user profiling using large-scale demographic data. *AI Magazine, 18*(2), 37–45.

Langley, P., & Fehling, M. (1998). *The experimental study of adaptive user interfaces* (Tech. Rep. 98-3). Palo Alto, CA: Institute for the Study of Learning and Expertise.

Lanier, J. (1995). Agents of alienation. *interactions, 2*(3), 66–72.

Lesh, N., Rich, C., & Sidner, C. L. (1999). Using plan recognition in human-computer collaboration. In J. Kay (Ed.), *UM99, User modeling: Proceedings of the Seventh International Conference* (pp. 23–32). Vienna: Springer Wien New York.

Li, X., & Ji, Q. (2005). Active affective state detection and user assistance with dynamic Bayesian networks. *IEEE Transactions on Systems, Man, and Cybernetics—Part A: Systems and Humans, 35*(1), 93–105.

Linden, G., Hanks, S., & Lesh, N. (1997). Interactive assessment of user preference models: The automated travel assistant. In A. Jameson, C. Paris, & C. Tasso (Eds.), *UM97 User modeling: Proceedings of the Sixth International Conference* (pp. 67–78). Vienna: Springer Wien New York.

Linden, G., Smith, B., & York, J. (2003, February). Amazon.com recommendations: Item-to-item collaborative filtering. *IEEE Internet Computing, 7*(1), 76–80.

Liscombe, J., Riccardi, G., & Hakkani-Tür, D. (2005). Using context to improve emotion detection in spoken dialog systems. *Proceedings of Eurospeech 2005, the Ninth European Conference on Speech Communication and Technology,* Barcelona, 1845–1848.

Lisetti, C. L., & Nasoz, F. (2004). Using noninvasive wearable computers to recognize human emotions from physiological signals. *EURASIP Journal on Applied Signal Processing, 11,* 1672–1687.

Litman, D. J., & Pan, S. (2002). Designing and evaluating an adaptive spoken dialogue system. *User Modeling and User-Adapted Interaction, 12*(2–3), 111–137.

Looi, C., McCalla, G., Bredeweg, B., & Breuker, J. (Eds.). (2005). *Artificial intelligence in education: Supporting learning through intelligent and socially informed technology.* Amsterdam: IOI Press.

Maes, P. (1994). Agents that reduce work and information overload. *Communications of the ACM, 37*(7), 30–40.

McCarthy, K., Reilly, J., McGinty, L., & Smyth, B. (2005). Experiments in dynamic critiquing. In J. Riedl, A. Jameson, D. Billsus, & T. Lau (Eds.), *IUI 2005: International Conference on Intelligent User Interfaces* (pp. 175–182). New York: ACM.

McCreath, E., & Kay, J. (2003). I-ems: Helping users manage e-mail. In P. Brusilovsky, A. Corbett, & F. de Rosis (Eds.), *UM2003, User Modeling: Proceedings of the Ninth International Conference* (pp. 263–272). Berlin: Springer.

McCreath, E., Kay, J., & Crawford, E. (2006). I-ems—an approach that combines hand-crafted rules with learnt instance-based rules. *Australian Journal of Intelligent Information Processing Systems, 9*(1), 49–63.

McGrenere, J., Baecker, R. M., & Booth, K. S. (2002). An evaluation of a multiple interface design solution for bloated software. In L. Terveen, D. Wixon, E. Comstock, & A. Sasse (Eds.), *Human factors in computing systems: CHI 2002 conference proceedings* (pp. 164–170). New York: ACM.

Micarelli, A., Gasparetti, F., Sciarrone, F., & Gauch, S. (2007). Personalized search on the World Wide Web. In P. Brusilovsky, A. Kobsa, & W. Nejdl (Eds.), *The adaptive web: Methods and strategies of web personalization* (pp. 155–194). Berlin: Springer.

Mitchell, T., Caruana, R., Freitag, D., McDermott, J., & Zabowski, D. (1994). Experience with a learning personal assistant. *Communications of the ACM, 37*(7), 81–91.

Mitrovic, A., & Ohlsson, S. (1999). Evaluation of a constraint-based tutor for a database language. *International Journal of Artificial Intelligence in Education, 10,* 238–256.

Mitrovic, A., Suraweera, P., Martin, B., & Weerasinghe, A. (2004). DB-Suite: Experiences with three intelligent, web-based database tutors. *Journal of Interactive Learning Research, 15*(4), 409–432.

Mobasher, B. (2007). Data mining for web personalization. In P. Brusilovsky, A. Kobsa, & W. Nejdl (Eds.), *The adaptive web: Methods and strategies of web personalization* (pp. 90–135). Berlin: Springer.

Müller, C., Wittig, F., & Baus, J. (2003). Exploiting speech for recognizing elderly users to respond to their special needs. *Proceedings of the Eighth European Conference on Speech Communication and Technology,* Geneva, pp. 1305–1308.

Norman, D. A. (1994). How might people interact with agents? *Communications of the ACM, 37*(7), 68–71.

Pazzani, M. J. (1999). A framework for collaborative, content-based and demographic filtering. *Artificial Intelligence Review, 13,* 393–408.

Petrelli, D., & Not, E. (2005). User-centred design of flexible hypermedia for a mobile guide: Reflections on the HyperAudio experience. *User Modeling and User-Adapted Interaction, 15*(3/4), 303–338.

Picard, R. W. (1997). *Affective computing.* Cambridge, MA: MIT Press.

Rhodes, B. J. (2000). *Just-in-time information retrieval.* Unpublished doctoral dissertation, School of Architecture and Planning, Massachusetts Institute of Technology.

Rich, C., & Sidner, C. L. (1998). COLLAGEN: A collaboration manager for software interface agents. *User Modeling and User-Adapted Interaction, 8,* 315–350.

Rich, C., Sidner, C. L., & Lesh, N. (2001). COLLAGEN: Applying collaborative discourse theory to human-computer interaction. *AI Magazine, 6*(4), 15–25.

Rich, C., Sidner, C., Lesh, N., Garland, A., Booth, S., & Chimani, M. (2005). DiamondHelp: A collaborative interface framework for networked home appliances. *5th International Workshop on Smart Appliances and Wearable Computing, IEEE International Conference on Distributed Computing Systems Workshops,* 514–519.

Schafer, J. B., Frankowski, D., Herlocker, J., & Sen, S. (2007). Collaborative filtering recommender systems. In P. Brusilovsky, A. Kobsa, & W. Nejdl (Eds.), The adaptive web: Methods and strategies of web Personalization (pp. 291–234). Berlin: Springer.

Schiele, B., Starner, T., Rhodes, B., Clarkson, B., & Pentland, A. (2001). Situation aware computing with wearable computers. In W. Barfield & T. Caudell (Eds.), *Fundamentals of wearable computers and augmented reality* (pp. 511–538). Mahwah, NJ: Erlbaum.

Schneider, M., & Kiesler, S. (2005). Calling while driving: Effects of providing remote traffic context. In W. Kellogg, S. Zhai, G. van der Veer, & C. Gale (Eds.), *Human factors in computing systems: CHI 2005 conference proceedings* (pp. 561–570). New York: ACM.

Sears, A., & Shneiderman, B. (1994). Split menus: Effectively using selection frequency to organize menus. *ACM Transactions on Computer-Human Interaction, 1,* 27–51.

Segal, R. B., & Kephart, J. O. (1999). MailCat: An intelligent assistant for organizing e-mail. *Proceedings of the Third International Conference on Autonomous Agents,* Seattle, 276–282.

Segal, R. B., & Kephart, J. O. (2000). Incremental learning in SwiftFile. In P. Langley (Ed.), *Machine Learning: Proceedings of the 2000 International Conference* (pp. 863–870). San Francisco: Morgan Kaufmann.

Shearin, S., & Lieberman, H. (2001). Intelligent profiling by example. In J. Lester (Ed.), *IUI 2001: International Conference on Intelligent User Interfaces* (pp. 145–151). New York: ACM.

Shneiderman, B., & Maes, P. (1997). Direct manipulation vs. interface agents. *Interactions, 4*(6), 42–61.

Smyth, B., Balfe, E., Freyne, J., Briggs, P., Coyle, M., & Boydell, O. (2005). Exploiting query repetition and regularity in an adaptive community-based web search engine. *User Modeling and User-Adapted Interaction, 14*(5), 383–423.

Soller, A. (2007). Adaptive support for distributed collaboration. In P. Brusilovsky, A. Kobsa, & W. Nejdl (Eds.), *The adaptive web: Methods and strategies of web personalization* (pp. 573–595). Berlin: Springer.

Sumi, Y., & Mase, K. (2001). AgentSalon: Facilitating face-to-face knowledge exchange through conversations among personal agents. *Proceedings of the Fifth International Conference on Autonomous Agents,* Montreal, 393–400.

Sumi, Y., & Mase, K. (2002). Supporting the awareness of shared interests and experiences in communities. *International Journal of Human-Computer Studies, 56*(1), 127–146.

Surendran, A. C., Platt, J. C., & Renshaw, E. (2005). Automatic discovery of personal topics to organize e-mail. *Proceedings of the Second Conference on E-mail and Anti-Spam,* Stanford, CA.

Teevan, J., Dumais, S., & Horvitz, E. (2005). Personalizing search via automated analysis of interests and activities. *Proceedings of 28th Annual International ACM SIGIR Conference on Research and Development in Information Retrieval,* 449–456.

Terveen, L., & McDonald, D. W. (2005). Social matching: A framework and research agenda. *ACM Transactions on Computer-Human Interaction, 12*(3), 401–434.

Trewin, S. (2004). Automating accessibility: The dynamic keyboard. *Proceedings of ASSETS 2004,* 71–78.

Tsandilas, T., & schraefel, m. (2004). Usable adaptive hypermedia. *New Review of Hypermedia and Multimedia, 10*(1), 5–29.

VanLehn, K., Lynch, C., Schulze, K., Shapiro, J. A., Shelby, R., Taylor, L., et al. (2005). The Andes physics tutoring system: Lessons learned. *International Journal of Artificial Intelligence in Education, 15*(3), 147–204.

Vivacqua, A., & Lieberman, H. (2000). Agents to assist in finding help. In T. Turner, G. Szwillus, M. Czerwinski, & F. Paternò (Eds.), *Human factors in computing systems: CHI 2000 conference proceedings* (pp. 65–72). New York: ACM.

Wainer, H. (Ed.). (2000). *Computerized adaptive testing: A primer.* Hillsdale, NJ: Erlbaum.

Wexelblat, A., & Maes, P. (1997). Issues for software agent UI. Unpublished manuscript, available from http://citeseer.ist.psu.edu/128549.html on 18 March 2007. Cited with permission.

Wickens, C. D., & Hollands, J. G. (2000). *Engineering psychology and human performance.* Upper Saddle River, NJ: Prentice Hall.

Wilde, G. J. (1998). Risk homeostasis theory: An overview. *Injury Prevention, 4*(2), 89–91.

Wolfman, S. A., Lau, T., Domingos, P., & Weld, D. S. (2001). Mixed initiative interfaces for learning tasks: SMARTedit talks back. In J. Lester (Ed.), *IUI 2001: International Conference on Intelligent User Interfaces* (pp. 167–174). New York: ACM.

Yacoub, S., Simske, S., Lin, X., & Burns, J. (2003). Recognition of emotions in interactive voice response systems. *Proceedings of Eurospeech 2003, the Eighth International Conference on Speech Communication and Technology,* Geneva, 729–732.

Zhou, M. X., & Aggarwal, V. (2004). An optimization-based approach to dynamic data content selection in intelligent multimedia interfaces. *Proceedings of UIST 2004,* 67–236.

Zhou, M. X., Wen, Z., & Aggarwal, V. (2005). A graph-matching approach to dynamic media allocation in intelligent multimedia interfaces. In J. Riedl, A. Jameson, D. Billsus, & T. Lau (Eds.), *IUI 2005: International Conference on Intelligent User Interfaces* (pp. 114–121). New York: ACM.

Ziegler, C., McNee, S. M., Konstan, J. A., & Lausen, G. (2005). Improving recommendation lists through topic diversification. *Proceedings of the 2003 International World Wide Web Conference,* 6–32.

Zukerman, I., & Litman, D. (2001). Natural language processing and user modeling: Synergies and limitations. *User Modeling and User-Adapted Interaction, 11,* 129–158.

·7·

MOBILE INTERACTION DESIGN IN THE AGE OF EXPERIENCE ECOSYSTEMS

Marco Susani
Advanced Concepts Group, Motorola Experience Design

The Mobile Experience **132**
Interacting With Networked Ecosystems 132
Content, Communication, Control 132
From Interaction to Experience 132
Spheres of Interaction **133**
Sphere 1: Of Humans and Devices 133
Cars and Motorcycles 133
The Ritual Power of Hands 134
Form Follows Gesture 134
The Sense of Touch 134
Product Species 135
Sphere 2: Mental Models
(and Their Representation) 136
Mental Models 136

Sphere 3: Context 136
Physical Context 136
Smart Objects and Active Ambient 137
Sphere 4: Communities 137
Sphere 5: Content 138
Deconstructed Content 138
Narrowcast 138
Seamless Experience Ecosystems 139
Adapting experiences to evolving behaviors 139
Shaping an experience without controlling
the full ecosystem 139
Embracing the idea of human-to-human
(and human-to-content) mediation rather
than human-machine interaction 139

THE MOBILE EXPERIENCE

Interacting With Networked Ecosystems

The parallel evolution of network technologies and social behaviors is transforming the character of interactive systems.

The conventional model of the individual working in front of a personal computer is challenged by innovative ways of accessing content and communication; networked mobile wireless devices grow their functionality, handheld interfaces become richer, mobile distributed delivery of content will be fully integrated with conventional media, and new social forms of communication are emerging.

The mobile component of this system of interactions, centered on the device formerly known as the cell phone has been among the most impressive techno-cultural phenomena of the past 10 years, and its evolution is still changing drastically the way human interaction is conceived. Mobility is only one aspect of this revolution; mobile devices also happen to be much more related to the individuals that use them, and follow constantly both their private and work life. Thus, the interaction with these tools is characterized by a combination of functional and emotional aspects.

Present interaction models, even the ones developed respecting the rules of usability, often lack fundamental emotional qualities that are needed to support and increase existing social behaviors and rituals.

The paradigm of pervasive and ubiquitous networking, instead of gravitating around the personal computer, is taking the shape of an ecosystem of mobile and static platforms, for individual and collective use, all connected to the network, and their coordination is taking the shape of a choreography of interactions to provide access to integrated content and communication.

"Convergence" seems finally to take shape, centered on three platforms that dominate this experience ecosystem:

- Virtually all personal computers are on the net, and most of them can access the net via mobile wireless technologies.
- Two billion cell phones, most of them capable to network both voice and net, are in the hands of people of all cultures and statuses.
- The predictable transition of cable TV to IP networked TV is about to transform the paradigm of interacting with video, movies, and TV content. Around the TV, other distributed devices (the evolution of the "device formerly known as the remote control") will increase the mobility within the home.
- Embedded networking capabilities are growing in other "appliances," such as cars, or in any "thing" (via the RFID technology), and in active environments.
- All of the above is going to be connected with a multiplicity of technologies; the integration of wired and wireless, and the integration of different technologies within each one of these domains (e.g., cellular networks talking to local wireless networks at different scales, from Bluetooth to wifi to RFID), transforms the picture of a network into a flexible, capable, seamless, "opportunistic" infrastructure that can take advantage of any available connectivity.

Similar social and cultural disruptions are happening in parallel with this fully developed technology disruption:

- Format and genres of communication integrate and hybridize;
- Content becomes deconstructed and pervasive;
- Based on the intersection of the two aspects above, new types of communities are emerging, and new social models growing around the generation and distribution of content explore the space between one-to-one communication (the evolution of the telephone) and broadcast (the evolution of mass media).

Devices of different types become multiple points of access to the network, and the Net is much more embedded in the everyday reality, both because it is more integrated with spaces and artifacts of everyday life, and because it influences more and more human activities and social systems.

Content, Communication, Control

The original interaction with personal computers focused on doing (writing a document, make calculations on a spread sheet, and so on), and its physical context was a single individual user sitting in front of his computer, isolated from the surrounding physical space. Today, the merging of computer technologies with telecommunication technologies and wireless network access calls for a completely different paradigm: computers, mobile devices, and networks are not only a complex set of tools to "do" things, to perform tasks. Instead, they are above all gates to communication media that connect people, hybrid physical and digital collaborative spaces to meet others, and tools to access published, dynamic, and ubiquitous content.

The interfaces of these devices aren't anymore exclusively mediators between a machine and an individual, but much more often are the interfaces that structure the user's communication mental model, or interfaces that organize and categorize in an accessible way the vast space of networked content.

Simplicity, the key attribute that allowed in the past easy access to the functionality of a "machine," becomes today a much more sophisticated concept, because a "simple" access to the user's sphere of communication and content cannot come only from a functional, efficient organization of the interaction, but is much more rooted in subjective and cultural aspects, such as the mental construct of the user's social universe.

In this chapter we'll investigate the specifics of interaction with a multiplicity of mobile networked devices, and we'll see how the whole paradigm of networked reality changes the way we think about HCI. We'll look at any networked device, whether mono-functional or multi-functional, such as cell phones, networked photo-cameras, networked digital-music players, digital TV, and so on, as a point of access to this vast ecosystem of networked content, communication, and control.

From Interaction to Experience

In a networked universe that has no center, we'll use the human user as a subjective viewer of the whole experience ecosystem.

Around this hypothetical user, we'll describe different concentric "spheres" of interaction. The first "sphere," closest to the user, corresponds to what usually is taken into consideration when designing a computer system: the physical interaction with a device. The other "spheres," growing progressively to include multiple elements of this experience ecosystem, will include topics related to human-to-human communication or access to content, or interacting with objects in a space.

When mobile devices are enriched by multiple functions and applications, this whole ecosystem is what actually influences the overall experience of the user. This is the new territory for designers to act and define the character of a product, the features of a service, or the identity of a medium.

SPHERES OF INTERACTION

Sphere 1: Of Humans and Devices

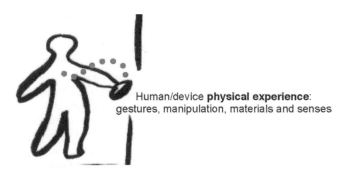

Human/device **physical experience**: gestures, manipulation, materials and senses

The first sphere of interaction, the most intimate connection between a human and a device, is the physical manipulation. The very nature of handheld devices makes this manipulation a key component of the relationship between user and system. Although handheld devices like mobile phones have conventional keypads and a screen that sometimes mimics the navigation in the GUI (Graphical User Interface), the overall experience is extremely different from using a PC. The manipulation and relation with the hand, and in some respect with the rest of the body (such as through placing the device in a bag or in the pocket), roots these objects in the history of tools that are real extensions of the hand, such as pens. Also, the handheld nature of mobile devices highlights social aspects, such as the proxemic role of the object, like handing over a phone to show a picture to a friend, for example. In this context, rich and sophisticated sociocultural aspects complement the basic ergonomic components of the interaction.

Cars and Motorcycles

If you talk to a passionate biker, you'll most often hear a comparison between the emotional aspects of driving a motorcycle compared to the "dullness" of driving a car; riding a bike is a richer sensorial experience that involves the whole body. While one drives a car by "codified" controls, proper interfaces that act on the machine, riding a bike involves an extension of the body.

Actually, riding a bike is a hybrid between some coded controls (such as clutch and brakes) and "natural" proprioceptive controls (such as moving the body while making a turn).

Consequently, a car may be perceived as "passive," while riding a bike is perceived as an active and engaging experience. Furthermore, riding a bike puts you in direct, immediate connection with the surrounding environment. From a certain point of view, the same can be said of personal computers and cell phones. Computers' only relationship with the body is the flat space of the desk, the movements of fingers on the keyboard and the hand on the mouse. We use computers while sitting alone at a desk, our bodies still, and our minds isolated from the surrounding environment. Mobile devices such as cell phones are handheld, and the relationship with the hands is a fundamental character of the interaction. A PC provides an experience that atrophies our body to minuscule movements of the fingers and the wrist, while handling a cell phone is an experience that may involve larger gestures and has a richer relationship with the body. We use mobile phones while walking, sitting, standing, and driving. We use them in isolation or in a crowd. The experience of communicating via a mobile phone is only partially immersive; the environmental noise and the distractions of the environment balance this immersion, and provide meaningful context to our activity. Cell phones have transformed the way we approach others and the power of the context enriches the communication; rather than asking, "How are you?" when we start a conversation, we ask, "*Where* are you?" as if the environment was an integral part of the conversation. The limited size of the device, which is in some respects a limit to the interaction because the screen is small and crowded, actually helps provide a further element of context; with such a small screen, we're forced to see what's around it, and the interaction is always peripheral, never immersive, and always hybridized by the "real" experience of the world around.

The Ritual Power of Hands

Consider the way people take pictures with a cell phone. Sometimes the camera phone is pointed toward the subject of the photo like a camera, and sometimes the phone is used in a very different way to take pictures that are much more spontaneous and transient. To be fair, the way camera phones are used had been anticipated by a phenomenon grown around very conventional film-based cameras. In the early 1990s, users of a camera called Lomo started to take very spontaneous pictures by not aiming in the viewfinder. The Lomo phenomenon transformed the genre of photography with a disruption that was purely "behavioral": a new gesture, such as handling the camera without bringing it near the eye, revolutionized photography and made the images much more "instantaneous" and genuine. This gave birth to a new photographic genre that was called Lomography. In Lomography, the gesture *is* the photo. The way people manipulate camera phones is very similar. In this case, the instantaneous photo genre is reinforced by the fact that photos are taken to be shared, not to be kept. While in origin photography was conceived as a way of "conserving" reality by documenting it, networked photos are transient, not perennial. And the informality of the gesture of "catching" them is a fundamental part of both the interaction and the narratives of this genre of photography. Designing a form that facilitated or denied this spontaneity would be for a designer a way of confirming, or denying, this behavior, and as a consequence a way of influencing the genre of photographs taken with such a device. In this case, the whole form and shape of the device, rather than an interface or a control, is the fundamental component of the interaction. Better, the shape of the device *is* the interface. While this is relatively new in the history of human computer interaction, it is not new in the history of objects.

Form Follows Gesture

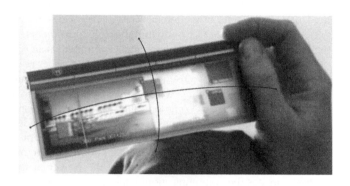

In the history of material culture, a history that includes, for example, ancient cooking utensils and working tools, the relationship between objects and hands is fundamental, and ends up defining the ritual relationship between humans and their tools. We are facing a kind of "calligraphic" use of an object. The idea of calligraphy, etymologically a "beautiful way of drawing or writing," is very much related to the idea of dignity of the gesture.

In addition to being a fundamental component of the use of the device, the overall shape, proportions, and mechanics of the object are also establishing the "character" of the object, and with that they influence the social dynamics around the usage of the object. The way John Travolta pulls the antenna of his flip phone with his teeth in *Pulp Fiction* (in the topical moment when his friend just shot a guy by error in a car) is an integral part of his character, and of the violence in that scene. How different is this from the way Anna Magnani "hugs" the handset in a cinema rendition of Jean Cocteau's theatrical piece *La Voix Humaine* a love monologue "designed" around the relationship between the woman—a desperate lover who has been abandoned by her partner—and the handset of her phone, the only thread left between her and her lover. Or consider how different is a large phone hanging from the holster of a worker from a tiny phone held constantly in the hands of a minute Japanese teenager.

The manipulation of handheld devices also has a social role: sharing a photo sometimes means also grabbing the digital camera or the phone and handing it to somebody to watch at the screen. Even if the screen is small, sharing it with a friend is a rich social act. Handheld devices are part of the proxemic relationship between humans, other humans, and the social space that surrounds them.

When devices such as cell phones are becoming so miniaturized that their shape is no more influenced by their technical components, their form would rather be defined by the way people manipulate them. And, as we described previously, the manipulation is only in part delimited by the mere rules of usability.

The Sense of Touch

The diffusion of keyboards and keypads has led to a kind of "buttonification" of the human being. This is definitely a reduction of the power of the sense of touch. Our senses are way more sophisticated than the basic on/off controls that they operate. In the case of handheld devices, this sensorial atrophy is more evident for at least two reasons. The first one is that while we hold the device with the full hand, five fingers plus the palm, we only use one finger to actually operate the keypad and receive tactile feedback. The second one is that, for the "peripheral" nature of the interaction with handhelds that we mentioned before, our interaction with a keypad ignores other fundamental spatial and proprioceptual clues, such as the horizon, our wrist position, and gravity. Some recent advances in sensing and feedback technologies have finally fought back this kind of sensorial deprivation that keypads have provoked. With the introduction of haptics, the primitive vibrators used originally only to replace the ringer tone of phones are used to provide a broader tactile feedback. Haptic actuators amplified by the resonance of the whole mass of the handheld can stimulate the whole palm of the hand, or they could be used in a denser "resolution" to provide localized feedback and a more meaningful haptic "message" to parts of the hand. Similarly, the introduction of accelerometers and gravity sensors allows the device to be "conscious" of its absolute position in space and its relative position to the hand of the user. This allows users to take full advantage of the proprioceptual character of handheld manipulation, and provides a fuller tactile stimulation as a feedback to this manipulation. When combined with a proper reflection of the proprioceptual character of the GUI, the overall combination of enriched tactile and visual contextualized experience finally takes full advantage of the handheld nature of mobile devices. Localized buttons will remain for a long time, but combined with these additional sensory messages they will provide a much richer sensorial experience.

Product Species

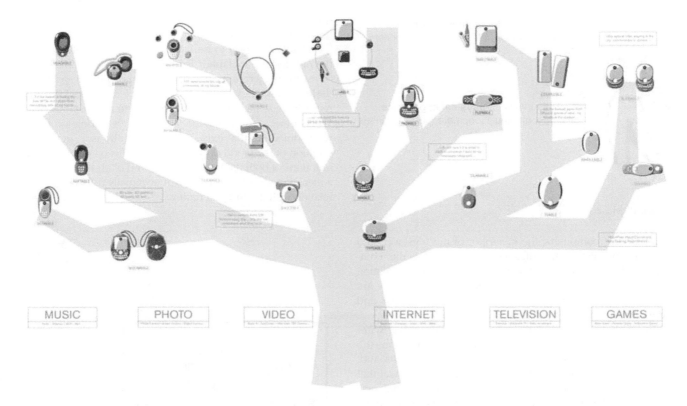

MUSIC   PHOTO   VIDEO   INTERNET   TELEVISION   GAMES

Even when the interaction with devices is properly designed and based on the principles of usability, sometimes there seems to be an additional difficulty in understanding the functions of an object, a subtler barrier to its adoption. An object may fail to deliver a clear message of its functionality to the user. The issue in this case is not "how easy is it to perform a task with this device?" but "what does this device do?" or, even more, "what is this?" In the past, a good rule to introduce a new object has been to leverage an analogy with a known, familiar object.

The reference to familiar devices and interactions could still apply today, but can also become too big of a limitation, a restriction of new functionalities or richer paradigms that simply do not have any parallel with anything familiar. Too many references are challenged: networked TV has no channels; mobile phones and house phones hybridize; e-mail and IM integrate; camera, telephone, and video functionalities converge in the same device. Communication paradigms such as blogging, or content distribution models such as podcasting, simply do not have any reference to past, established, familiar models.

Inventing new object archetypes, as much as it is an oxymoron, may be the only way out of this impasse—developing new product types and new interaction paradigms when playing analogy to existing archetypes is just not applicable. The same applies to mono-functional devices in comparison to multifunctional ones; the history of interaction design is full of successes of simple, mono-functional devices that leveraged this analogy.

From the Palm PDA to the iPod, all fall in the category of devices that "do one thing and do it right." In the future, however, reducing the functionality in order to achieve simplicity of use may be no longer possible. Multifunctionality may offer too much of an advantage to be compromised, and new generations of users, younger hyper-taskers, wouldn't necessarily appreciate oversimplification.

So, the challenge for a designer is to ensure that an object would be "understood" while preserving multifunctionality and richness of experience. Designers would on the one side develop a richer perspective on visual affordances, one that could accommodate multiple and partially ambiguous affordances related to different usages. On the other side, beyond affordances, designers need to develop a better understanding of what constitutes a recognizable, understandable "species" of objects. That is to say, rather than seeking the familiarity with a known object or function "as-is," developing a more sophisticated understanding of how familiarity(ies), or just components of this, can be carried over along the evolution of an object, or carried across a discontinuity in the evolution of such an object or functionality. The result will be the transformation of the simple concept of affordance into a richer notion of "objecthood," defined as the ability of humans to recognize and perceive correctly the world of objects, and the mutual ability of objects to be understandable and to overtly disclose their functionality, that is, for both humans and objects to develop their relationship around a sixth sense, the "sense of meaning."

Sphere 2: Mental Models (and Their Representation)

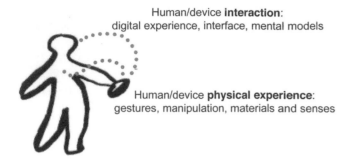

Human/device interaction:
digital experience, interface, mental models

Human/device physical experience:
gestures, manipulation, materials and senses

Mental Models

In personal computers, the metaphor of the "desktop" and the "point and click" in that digital space starts to feel the "competition" of other models of interaction, namely the information architectures organized around e-mail and IM that today are often dominant compared to the desktop access to files and folders.

The mental construct of the interaction with mobile wireless media is intrinsically centered on communication; the phonebook, and its extension to include presence, status, and, in perspective, location, is often the master mental model that represents our communication universe. The phonebook acts as a point of departure for a broad series of actions: starting a call, generating a message, initiating a textual chat, checking the online status of our "buddies," or sending a picture to somebody. However, with the diffusion of other ways of contextualizing the interaction, such as location detection, a user may decide to switch to a different model and, for example, "view by location" and reorganize the communication, mapping it over the actual physical surroundings. In this model, a user may reorganize the phonebook to check first "who's around me" in that moment. Even if the idea of communicating in physical proximity may seem paradoxical, cell phones are often used to create an additional communication channel (texting or sharing images, for example) with people who are not remote, or to manage meetings with people in physical proximity (e.g., a group of friends in the same city arranging a dinner together).

Other mental models use incoming events as an organizational context; an incoming call may trigger a reorganization of the interface around the person who calls (bringing in the foreground previous messages or pictures exchanged with that person); or walking in front of a place may trigger some location-based content and contextual actions around that content; or an approaching event, such as a meeting, can trigger other information such as the names of the participants, their status, and their location.

While the taxonomies arranged around the sphere of communication may work better for phone-first devices, the proliferation of functions (imaging, video, music management, e-mail, etc.) could suggest taxonomies based around a clear distinction of applications, or around categories of content (my photos, my music, my phonebook, etc.).

The dominance of the former model (centered on communication) or the latter (centered on content or applications) is not yet clear. For a long time we'll probably continue to have hybrids with the co-presence of multiple categorizations.

What is common to both these models and different from the old desktop metaphor is that they have a less direct analogy with an interaction space—such as the physical space of the desktop—and they have more reference to mental constructs. As such, these models could be referred to as "cosmographies." While a metaphor like the desktop implies having a reference with something that exists, a cosmography is a "general description of the world or universe," and this could embrace descriptions of both the material world and of abstract concepts—a mental model of something that eventually does not have any relation with the material reality.

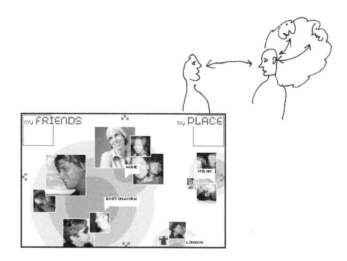

Sphere 3: Context

Interacting with/in the physical space

Physical Context

Mobile interaction cannot be isolated from its context. While the interaction between a user and a personal computer tends to be immersive and central, and is agnostic of the physical context of the user, the interaction with mobile devices is typically contextual and peripheral.

Primitive graphic menus and icons arranged in the small screen, a kind of translation of the conventional PC, are the dominant GUI paradigm in wireless mobile media, but they are only the most superficial aspect of the interaction with devices like cell phones. Multiple other aspects of interaction with cell phones are less tangible, yet more important, than the keypad-and-screen UI. The whole experience ecosystem of the interaction with wireless media includes the combination of the interaction with the physical device, the interaction with its services, and with the physical environment.

Context is what makes mobile interaction much richer than computer interaction. Mobility means that to be *more connected* to the physical place, not *more disconnected* from it; communication relates to *where* we are and *who* is around us. Location-based services, mobile imaging, and mobile blogging extend this concept to content itself: content accessed and generated from a mobile platform has the potential to be heavily contextualized over the physical environment.

Smart Objects and Active Ambient

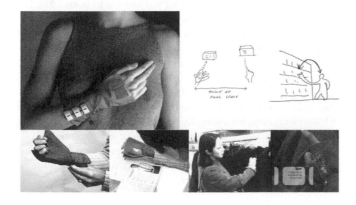

Smart tags are very tiny, low-cost, wireless microcomputers. They only cost a few cents and they can be attached to anything—items of clothing, food cans, spare parts for cars, electric irons, and so on. The tags can be programmed to provide information about the items to which they are attached. They are able to identify a product and network this information, and can

also take note of their environment, recording changes in temperature and so on. Their broad adoption is changing the way logistics are organized around raw materials and products traveling to transformation and to markets.

Smart tags also enable a minute level of interaction with the physical space. Cell phones equipped with RFID tags are already enabling financial transactions such as paying a ticket on the underground or replacing a credit card. The extension of this level of interaction may support the paradigm of "point and click at real environments." Handheld devices could be used to trigger contextual information when pointed at other objects or parts of the environment.

Finally, the connection of these tags to an integrated information system able to tell, at any moment, about of the life of any item of produce, for example, may allow the quality, the origin, or the value, to be tracked in real-time. Produce that is fresh or older, in-season or off-season, near or far from its origin, with or without a track record of its growth, will have a different value. In this scenario, smart tags may enable the creation of *digital word-of-mouth*, similar to the spreading of information and knowledge by human word-of-mouth, connected to the system of goods. Attached to items will be information and knowledge. It is possible to describe this as a *knowledge aura*, and this is the most interesting aspect of future material-knowledge systems. Experiments on connecting *collective wisdom* with digital systems *on the field* have been undertaken in India since the 1990s. The fully fledged scenario will allow replication in a digital environment of the wisdom of word-of-mouth know-how.

The whole idea of using advertisement before the purchase, and separate from the purchase experience, could be revolutionized by the presence of an aura of information that follows the product until the point of purchase. Through this aura of digital information, the product would "advertise itself" from the shelves of a shop. Even more intriguing is the idea that deeper information about a product, connected to similar experiences of purchase and use—what is known as a recommendation on a website—will be readily available at any moment that a product is encountered. The lure and temptations that attract a buyer will come directly from the product, from the background voices of people that bought and used it, and by the life history of the product—a kind of very detailed, dynamic, digital certificate of authenticity.

Sphere 4: Communities

Networked media are characterized by the hybridization of a multiplicity of communication genres. While 20 years ago the evolution of telephone technologies seemed to go toward hyper-realism—technologies focused on replicating, at best, face-to-face communication—the actual evolution of telecommunication has denied this trend and pointed toward mediating communication in a form *different* from face-to-face. Text messaging via SMS created a communication channel that was simply impossible to achieve with face-to-face—often more intimate, more enigmatic, than face-to-face, and complementary to that. Instant messaging has added the management of multiple one-to-one communications, also not possible in real face-to-face, and, later, blogging created a hybrid between one-to-one expression and broadcast. Finally, the multimedia capabilities of devices like camera phones allow the integration between communication and content sharing, making it possible to share experiences in an almost permanent way.

Mobile wireless media also are the ideal platform to support tribal social structures; mobile phones have favored the birth of tribal forms of few-to-few and few-to-many communication, and they revitalize the scale of the small circle of friends, of groups of people who are in constant touch, exchanging information and passions, and sharing images and music.

Different from both an individual who communicates one-to-one with his interlocutor and from a broadcaster with a passive mass audience, tribal models of communication define a dense flow of exchanges within a restricted circle of friends. The space for tribal relations, superimposition of the physical space and of the "digital territory of belonging," also amplifies information and messages while building a shared memory. The permanence of content in a shared space creates not a simple exchange of messages but the sedimentation, the memory, and the *construction* of a shared permanent experience.

Sphere 5: Content

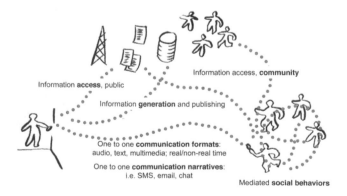

"Convergence" between media, for long just a buzzword, is happening for real, and this means a much richer interaction for the user. Richer communication—messaging, voice, and so on—is more and more integrated with content access, such as sharing music, videos, and images.

Deconstructed Content

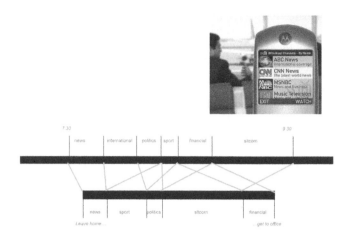

Mobile imaging is an already-established form of hybridization between communication and content capture. Camera phones as an imaging platform are better integrated with both the PC and the TV. Video content will soon follow, and this, combined with the integration of web content with TV productions, will create yet another "triple point" (mobile, PC, and TV and home systems) interaction with content. As soon as digital music players become networked themselves and mobile phones download music over the air, the music-content ecosystem will have continuity between environments and platforms.

The first consequence this will have from an interaction point of view will be the separation of content creation from content delivery and the consequent "deconstruction" of content. The possibility of capturing images with different devices (camera phones or digital cameras) and accessing them in different modalities (in the device itself, shared with others or published on the web, or printed or accessed from a PC or a TV) are already evident signs of separation between creation and access, and examples of truly networked access to content. Mobile TV will not be an exact replica of TV delivered to home. The latter has already been transformed by digital video recorders, and broadcasts have been separated from real time. Mobile TV will be even more disconnected from the real time of broadcast—often buffered and stored locally—but also re-associated with the "relative" time of the user; the possibility of caching content on a mobile device, combined with the need for "snacking" shorter versions of the same content during leftover time while commuting, for example, will make the idea of a personal, adapted palimpsest more realistic. Content will thus be deconstructed in different multiple formats (filling up many lengths, between the short trailer and the full-length movie) and consumed from different platforms.

Narrowcast

Podcasting and blogging are the signal of a larger trend that opens the possibility for everybody to become a publisher. In both cases the broadcast paradigm evolves from one-to-many (where the one is an author, an institutional publisher such as a TV station) to any-to-many (where anybody, with a limited

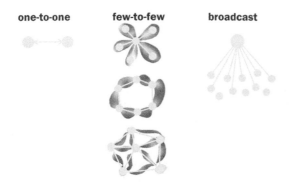

one-to-one   few-to-few   broadcast

3. A spatial ecosystem, because location-based services and communication strongly "ground" the interaction via mobile devices in the physical space;
4. A business ecosystem, because no single company or designer can control the overall experience, but can only shape the interaction from a partial point of control; and last, but not least;
5. A platform ecosystem, since mobile interaction is integrated with interaction with other platforms, such as personal computers and digital TV.

The complexity of these ecosystems puts additional challenges to designers who want to shape the interaction.

Among these challenges are the following:

Adapting experiences to evolving behaviors. Convergence will change drastically the way users perceive their sphere of content and communication, and sometimes will give up familiar mental models to embrace new paradigms may actually facilitate the ease of use. This is in apparent contrast with one of the rules of user-centered design, which is to refer to familiar models and avoid disruptions. For example, the potential elimination of "channels" when TV content moves over IP may help embrace a new paradigm that has no reference to the old "zapping." In the same way, embracing the paradigm of one-phone-number-per-person may confuse users at first, but will actually facilitate the transition to a communication paradigm where the user, not the device, is the one to be reached. A well-crafted, user-oriented approach may distinguish between the respect of familiarities that are structurally embedded in human perception and pushing back resistance of old mental habits that may actually hinder innovation and cause disruptions.

Shaping an experience without controlling the full ecosystem. The design culture has long insisted on the protection of the brand and on establishing a "signature" experience to a device or a service. In the future it will be very rare, however, that a single brand, thus a single designer, would control the full experience ecosystem. Two approaches to this are possible: one is a resistance to sharing components of the experience with other brands; another is embracing the complexity of the system and finding lean, smart ways to shape the experience without actually owning and controlling all the elements. The latter seems to be a more open-minded, relaxed way to accept changes in the environment a designer operates, while the former may end up in a stubborn, desperate attempt to protect a territory that simply cannot be protected anymore. In other words, it is better to adapt and shape an experience by seeding signature elements in few key areas rather than working on a more rigid model and risking losing all the control.

Embracing the idea of human-to-human (and human-to-content) mediation rather than human-machine interaction. The HCI discipline is rooted in the interaction between a human and a machine (or a system) and focused on the shift from a mechanical interaction to a logical one. While this competence remains true for the broad majority of devices we are designing today, it is also true that another aspect, the one of designing interfaces to access other humans and content

investment and a much leaner technology, can become a publisher). But there is more. Both blogging and podcasting are creating communities of adepts that do not behave like a conventional passive audience of broadcast mass media. They feed back and create a closer circle of communication that constitutes an actual "narrowcast" environment, a scale of publishing unheard of before. There is no solution of continuity between the one-to-one paradigm of telephony that extends to the larger scale of a tribe of friends and the blogging and podcasting narrowcast that takes broadcast to a smaller scale. Also, communication and content are so hybridized that they become indistinguishable.

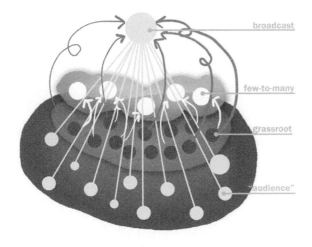

broadcast

few-to-many

grassroot

"audience"

Seamless Experience Ecosystems

Today no interaction can be isolated in the confined space of human computer interaction; any consideration on the interaction between a user and a mobile device cannot ignore the complexity of the relational "ecosystems" that exist around this interaction.

The components of this complex networked experience are:

1. A social ecosystem, as mobile devices go beyond the one-to-one communication to mediate a very rich and complex system of social relationships;
2. A content ecosystem, since mobile devices allow access to images, music, and videos; and peer-to-peer content generation and sharing introduce a completely new perspective of content authoring;

via a mediating interface, is becoming more and more dominant. A deeper understanding of human patterns of communication, and of the interactions between interfaces of devices and genres of communication, is needed. In the same way, the understanding of how innovative ways of interacting with content may shape new types and formats of content is also needed. Media studies were born in a condition in which the media genre (e.g., the TV programs), the transport technology (via aerial or cable TV), the device of access (the TV set), the method of interaction (zapping through channels), the method of delivery (broadcast) and the interaction environment (a couch potato in a living room) of a given medium were linked and homogeneous. Today, a certain type of content (e.g., a TV program) may be delivered in a multiplicity of formats (real time or cached, in snippets or in a linear fashion) to extremely different platforms and environments (at home or while mobile, to a passive individual or annotated and redistributed within a community). This poses new challenges to designers who need to shape at the same moment an innovative interaction and an innovative medium genre.

Experience ecosystems are reshaping the way human knowledge is grown and diffused. Interaction designers have the responsibility of both shaping a pleasurable, meaningful experience and of facilitating the chain of innovations that flows through new behaviors, new business models, and new technologies. Designers need to be the first to drive innovation in their own disciplines.

Images from interaction design projects developed by the Advanced Concepts Group in Motorola between 2000 and 2005.

·8·

TANGIBLE USER INTERFACES

Hiroshi Ishii
MIT Media Laboratory

Introduction 142
From GUI to TUI 142
Urp: An Example of TUI 142
Basic Model of Tangible User Interface 144
GUI 144
TUI 144
Tangible Representation as Control 144
Intangible Representation 145
Key Properties of TUI 145
Computational coupling of tangible
representations to underlying digital
information and computation 145
Embodiment of mechanisms for interactive
control with tangible representations 145
Perceptual coupling of tangible
representations to dynamic intangible
representations 146
Genres of TUI Applications 146
Tangible Telepresence 146
Tangibles with Kinetic Memory 146
Constructive Assembly 146
Tokens and Constraints 146
Interactive Surfaces—Tabletop TUI 147
Continuous Plastic TUI 147
Augmented Everyday Objects 147
Ambient Media 147
TUI Instances 147
InTouch: Tangible TelePresence Through
Distributed Synchronized Physical Objects 148

Curlybot: A Toy to Record and Play 148
Topobo: 3D Constructive Assembly
With Kinetic Memory 149
mediaBlocks: Token and Constraint Approach 149
Digital Desk: Pioneer of Tabletop TUI 150
Sensetable and AudioPad: Tabletop TUI
for Real-Time Music Performance 150
IP Network Design Workbench:
Event Driven Simulation on Sensetable 151
Actuated Workbench: Closing a Loop
of Computational Actuation and Sensing 152
SandScape: Continuous TUI for
Landscape Design 152
musicBottles: Transparent Interface Based
on Augmented Glass Bottles 153
Pinwheels: Ambient Interface to Information 154
Contributions of TUIs 155
Double Interactions Loop: Immediate
Tactile Feedback 155
Persistency of Tangibles 155
Coincidence of Input and Output Spaces 155
Special Purpose vs. General Purpose 156
Space-Multiplexed Input 156
Conclusion 156
Acknowledgments 157
References 157

INTRODUCTION

Where the sea meets the land, life has blossomed into a myriad of unique forms in the turbulence of water, sand, and wind. At another seashore, between the land of atoms and the sea of bits, we are now facing the challenge of reconciling our dual citizenships in the physical and digital worlds. Our visual and auditory sense organs are steeped in the sea of digital information, but our bodies remain imprisoned in the physical world. Windows to the digital world are confined to flat square screens and pixels, or "painted bits." Unfortunately, one cannot feel and confirm the virtual existence of this digital information through one's hands and body.

Imagine an iceberg, a floating mass of ice in the ocean. That is the metaphor of Tangible User Interfaces. A Tangible User Interface gives physical form to digital information and computation, salvaging the bits from the bottom of the water, setting them afloat, and making them directly manipulatable by human hands.

FROM GUI TO TUI

People have developed sophisticated skills for sensing and manipulating their physical environments. However, most of these skills are not employed in interaction with the digital world today. A Tangible User Interface (TUI) is built upon those skills and situates the physically embodied digital information in a physical space. Its design challenge is a seamless extension of the physical affordance of the objects into digital domain (Ishii & Ullmer, 1997, 2000).

Interactions with digital information are now largely confined to Graphical User Interfaces (GUIs). We are surrounded by a variety of ubiquitous GUI devices such as personal computers, handheld computers, and cellular phones. The Graphical User Interface (GUI) has been in existence since the 1970s and first appeared commercially in the Xerox 8010 Star System in 1981 (Smith, 1982). With the commercial success of the Apple Macintosh and Microsoft Windows, the GUI has become the standard paradigm for Human Computer Interaction (HCI) today.

GUIs represent information (bits) with pixels on a bitmapped display. Those graphical representations can be manipulated with generic remote controllers such as mice and keyboards. By decoupling representation (pixels) from control (input devices) in this way, GUIs provide the malleability to emulate a variety of media graphically. By utilizing graphical representation and "see, point, and click" interaction, the GUI made a significant improvement over its predecessor, the CUI (Command User Interface), which required the user to "remember and type" characters.

However, interactions with pixels on these GUI screens are inconsistent with our interactions with the rest of the physical environment within which we live. The GUI, tied down as it is to the screen, windows, mouse, and keyboard, is utterly divorced from the way interaction takes place in the physical world. When we interact with the GUI world, we cannot take advantage of our dexterity or utilize our skills for manipulating various physical objects, such as building blocks, or our ability to shape models out of clay.

Tangible User Interfaces (TUIs) aim to take advantage of these haptic interaction skills, which is a significantly different approach from GUI. The key idea of TUIs is to give physical forms to digital information. The physical forms serve as both representations and controls for their digital counterparts. TUI makes digital information directly manipulable with our hands, and perceptible through our peripheral senses, by physically embodying it.

Tangible User Interface serves as a special-purpose interface for a specific application using explicit physical forms, while GUI serves as a general-purpose interface by emulating various tools using pixels on a screen.

TUI is an alternative to the current GUI paradigm, demonstrating a new way to materialize Mark Weiser's (1991) vision of Ubiquitous Computing of weaving digital technology into the fabric of a physical environment and making it invisible. Instead of making pixels melt into an assortment of different interfaces, TUI uses tangible physical forms that can fit seamlessly into a user's physical environment.

This chapter introduces the basic concept of TUI in comparison with GUI, early prototypes of TUI that highlight the basic design principles and design challenges that TUI needs to overcome.

URP: AN EXAMPLE OF TUI

To illustrate basic TUI concepts, we introduce "Urp" (Urban Planning Workbench) as an example of TUI (Underkoffler & Ishii, 1999). Urp uses scaled physical models of architectural buildings to configure and control an underlying urban simulation of shadow, light reflection, wind flow, and so on. (Photo 8.1). In addition to a set of building models, Urp also provides a variety of interactive tools for querying and controlling the parameters of the urban simulation. These tools include a clock tool to change the position of sun, a material wand to change the building surface between bricks and glass (with light reflection), a wind tool to change the wind direction, and an anemometer to measure wind speed.

The physical building models in Urp cast digital shadows onto the workbench surface (via video projection), corresponding to solar shadows at a particular time of day. The time of day, representing the position of the sun, can be controlled by turning the physical hands of a "clock tool" (Photo 8.2). The building models can be moved and rotated, with the angle of their corresponding shadows transforming according to their position and the time of day.

Correspondingly, moving the hands of the clock tool can cause Urp to simulate a day of shadow movement between the situated buildings. Urban planners can identify and isolate intershadowing problems (shadows cast on adjacent buildings), and reposition buildings to avoid areas that are needlessly dark, or maximize light between buildings.

A "material wand" alters the material surface properties of a building model. By touching the material wand to a building

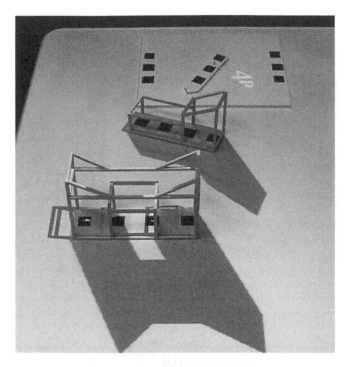

PHOTO 8.1.   Urp and shadow stimulation.

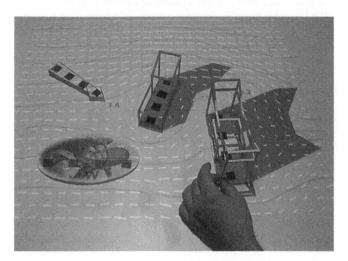

PHOTO 8.2.   Urp and wind stimulation.

model, the building surface material is switched from bricks to glass, and a projected reflection of sunlight appears to bounce off the walls of the building. Moving the building allows urban designers to be aware of the relationship between the building reflection and other infrastructure. For example, the reflection off the building at sundown might result in distraction to drivers on a nearby highway. The designer can then experiment with altering the angles of the building to oncoming traffic or moving the building farther away from the roadway. Tapping again with the material wand changes the material back to brick, and the sunlight reflection disappears, leaving only the projected shadow.

PHOTO 8.3.   inTouch.

By placing the "wind tool" on the workbench surface, a wind flow simulation is activated based on a computational fluid dynamics simulation, with field lines graphically flowing around the buildings. Changing the wind tool's physical orientation correspondingly alters the orientation of the computationally simulated wind. Urban planners can identify any potential wind problems, such as areas of high pressure that may result in hard-to-open doors or unpleasant walking environments. An "anemometer" object allows point monitoring of the wind speed (Photo 8.3). By placing the anemometer onto the workspace, the wind speed of that point is shown. After a few seconds, the point moves along the flow lines, to show the wind speed along that particular flow line. The interaction between the buildings and their environment allows urban planners to visualize and discuss inter-shadowing, wind, and placement problems.

In Urp, physical models of buildings are used as tangible representations of digital models of the buildings. To change the location and orientation of buildings, users simply grab and move the physical model as opposed to pointing and dragging a graphical representation on a screen with a mouse. The physical forms of Urp's building models, and the information associated with their position and orientation upon the workbench, represent and control the state of the urban simulation.

Although standard interface devices for GUIs, such as keyboards, mice, and screens, are also physical in form, the role of the physical representation in TUI provides an important distinction. The physical embodiment of the buildings to represent the computation involving building dimensions and location allows a tight coupling of control of the object and manipulation of its parameters in the underlying digital simulation.

In Urp, the building models and interactive tools are both physical representations of digital information (shadow dimensions and wind speed) and computational functions (shadow interplay). The physical artifacts also serve as controls of the underlying computational simulation (specifying the locations of objects). The specific physical embodiment allows a dual use in representing the digital model and allowing control of the digital representation. In the next section, the model of TUI is introduced in comparison with GUI to illustrate this mechanism.

BASIC MODEL OF TANGIBLE USER INTERFACE

The interface between people and digital information requires two key components: input and output, or control and representation. *Controls* enable users to manipulate the information, while *external representations* are perceived with the human senses. Figure 8.1 illustrates this simple model of a user interface consisting of control, representation, and information.

In the Smalltalk-80 programming language (Burbeck, 1992; Goldberg, 1984), the relationship between these components is illustrated by the "model-view-controller" or "MVC" archetype, which has become a basic interaction model for GUIs.

Drawing from the MVC approach, we have developed an interaction model for both GUI and TUI. We carry over the "control" element from MVC, while dividing the "view" element into two subcomponents: tangible and intangible representations, and renaming "model" as "digital information" to generalize this framework to illustrate the difference between GUI and TUI.

In computer science, the term *representation* often relates to the programs and data structures serving as the computer's internal representation (or model) of information. In this article, the meaning of "representation" centers upon external representations—the external manifestations of information in fashions directly perceivable by the human senses that include visual, hearing, and tactile senses.

GUI

In 1981, the Xerox Star workstation set the stage for the first generation of GUI (Johnson et al., 1989; Smith, 1982), establishing the "desktop metaphor," which simulates a desktop on a bitmapped screen. The Star workstation was the first commercial system that demonstrated the power of a mouse, windows, icons, property sheets, and modeless interaction. The Star also set several important HCI design principles, such as "seeing and pointing vs. remembering and typing," and "what

you see is what you get (WYSIWYG)." The Apple Macintosh brought this new style of HCI to the public's attention in 1984, creating a new trend in the personal computer industry. Now, the GUI is widespread, largely through the pervasiveness of Microsoft Windows, PDAs, and cellular phones.

GUI uses windows, icons, and menus made of pixels on bitmapped displays to visualize information. This is an intangible representation. GUI pixels are made interactive through general "remote controllers" such as mice, tablets, or keyboards. In the pursuit of generality, GUI introduced a deep separation between the digital (intangible) representation provided by the bitmapped display, and the controls provided by the mouse and keyboard.

Figure 8.2 illustrates the current GUI paradigm in which generic input devices allow users to remotely interact with digital information. Using the metaphor of seashore that separates a sea of bits from the land of atoms, the digital information is illustrated at the bottom of the water, and mouse and screen are above sea level in the physical domain. Users interact with the remote control, and ultimately experience an intangible external representation of digital information (display pixels and sound).

TUI

Tangible User Interface (TUI) aims at a different direction from GUI by using tangible representations of information that also serve as the direct control mechanism of the digital information. By representing information in both tangible and intangible forms, users can more directly control the underlying digital representation using their hands.

Tangible Representation as Control

Figure 8.3 illustrates this key idea of TUI to give tangible (physical and graspable) external representation to the digital information. The tangible representation helps bridge the boundary between the physical and digital worlds. Also notice that the tangible representation is computationally coupled to the control to the underlying digital information and computational models. Urp illustrates examples of such couplings, including

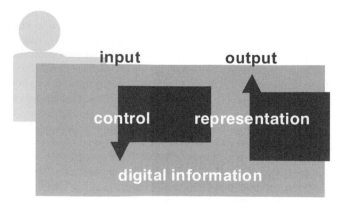

FIGURE 8.1. User interface model.

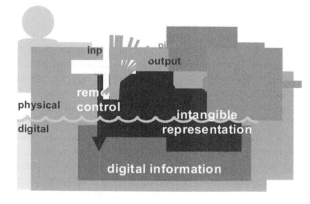

FIGURE 8.2. GUI model.

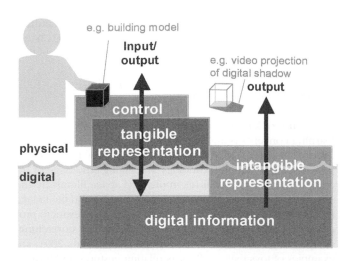

FIGURE 8.3. TUI model.

the binding of graphical geometries (digital data) to the physical building models, and computational simulations (operations) to the physical wind tool. Instead of using a GUI mouse to change the location and angle graphical representation of a building model by pointing, selecting handles, and keying in control parameters, an Urp user can grab and move the building model to change both location and angle.

The tangible representation functions as an interactive physical control. TUI attempts to embody the digital information in physical form, maximizing the directness of information by coupling manipulation to the underlying computation. Through physically manipulating the tangible representations, the digital representation is altered. In Urp, changing the position and orientation of the building models influences the shadow simulation, and the orientation of the "wind tool" adjusts the simulated wind direction.

Intangible Representation

Although the tangible representation allows the physical embodiment to be directly coupled to digital information, it has limited ability to represent or change many material or physical properties. Unlike malleable pixels on the computer screen, it is very hard to change a physical object in its form, position, or properties (e.g., color or size) in real time. In comparison with malleable "bits," "atoms" are extremely rigid, taking up mass and space.

To complement this limitation of rigid "atoms," TUI also utilizes malleable representations, such as video projections and sounds, to accompany the tangible representations in the same space to give dynamic expression of the underlying digital information and computation. In the Urp, the digital shadow that accompanies the physical building models is such an example.

The success of a TUI often relies on a balance and strong perceptual coupling between the tangible and intangible representations. It is critical that both tangible and intangible representations be perceptually coupled to achieve a seamless interface that actively mediates interaction with the underlying digital information, and appropriately blurs the boundary between phys-

ical and digital. Coincidence of input and output spaces and real-time response are important requirements to accomplish this goal.

Note: There exist certain types of TUIs that have actuation of the tangible representation (physical objects) as the central means of feedback. Examples are inTouch (Brave, Ishii, & Dahley, 1998), curlybot (Frei, Su, Mikhak, & Ishii, 2000), and topobo (Raffle, Parkes, & Ishii, 2004). This type of force-feedback-TUI does not depend on "intangible" representation since active feedback through the tangible representation serves as the main display channel.

Key Properties of TUI

While Fig. 8.2 illustrates the GUI's clear distinction between graphical representation and remote controls, the model of TUI illustrated in Fig. 8.3 highlights TUI's integration of physical representation and control. This model provides a tool for examining the following important properties and design requirements of tangible interfaces (Ullmer & Ishii, 2000).

Computational coupling of tangible representations to underlying digital information and computation. The central characteristic of tangible interfaces is the coupling of tangible representations to underlying digital information and computational models. One of the challenges of TUI design is how to map physical objects and their manipulation to digital computation and feedback in a meaningful and comprehensive manner.

As illustrated by the Urp example, a range of digital couplings and interpretations is possible, such as the coupling of data to the building models, operations to the wind tool, and property modifiers to the material wand.

Deciding the embodiment and mapping of the controller is dictated by the type of application envisioned. We give example cases in which a range of specificity of embodiment is used. In some applications, more abstract forms of physical objects (such as round pucks) are used as generic controllers that are reusable to control a variety of parameters by rotating and pushing a button (Patten, Ishii, Hines, & Pangaro, 2001). When a puck is used as a dial to control a simulation parameter, graphical feedback is given to complement the information, such as scale of the dial.

Embodiment of mechanisms for interactive control with tangible representations. The tangible representations of TUIs serve simultaneously as interactive physical controls. Tangibles may be physically inert, moving only as directly manipulated by a user's hands. Tangibles may also be physically actuated, whether through motor-driven force-feedback approaches (e.g., inTouch, Curlybot) or magnet-driven approaches such as Actuated Workbench (Pangaro, Maynes-Aminzade, & Ishii, 2002).

Tangibles may be unconstrained and manipulated in free space with six degrees of freedom. They may also be weakly constrained through manipulation on a planar surface, or tightly constrained, as in the movement of the abacus beads with one degree of freedom.

In order to make interaction simple and easy to learn, TUI designers need to utilize the physical constraints of the chosen physical embodiment. Because the physical embodiment, to some extent, limits the interaction choices, a designer must design the interaction so that the actions supported by the object are based on well-understood actions related to the physical object. For example, if a bottle shape is chosen, then opening the bottle by pulling out a cork is a well-understood mechanism (Ishii, Mazalek, & Lee, 2001). This understanding of the culturally common manipulation techniques helps disambiguate the users' interpretation of how to interact with the object.

Perceptual coupling of tangible representations to dynamic intangible representations. Tangible interfaces rely on a balance between tangible and intangible representations. Although embodied tangible elements play a central, defining role in the representation and control of a TUI, there is a supporting role for the TUI's intangible representation. A TUI's intangible representation—usually graphics and audio—often mediates much of the dynamic information provided by the underlying computation.

The real-time feedback of the intangible representation corresponding to the manipulation of the tangible representation is critical to insure perceptual coupling. The coincidence of inputs and output spaces (spatial continuity of tangible and intangible representations) is also an essential requirement to enhance perceptual coupling. For example, in Urp, the building models (tangible representation) are always accompanied by a "digital shadow" (intangible representation) without noticeable temporal or spatial gaps. That convinces users of an illusion that the shadows are cast by the building models (rather than by the video projector).

Genres of TUI Applications

By giving physical form to digital information to enhance an experience, TUIs have a wide variety of application domains. This section gives an overview of seven genres for promising TUI applications. For a more exhaustive survey of TUIs in a historical context, I would encourage the readers to refer to Ullmer and Ishii (2000), Holmquist, Redström, and Ljungstrand (1999), and Fishkin (2004). Zuckerman, Arida, and Resnick (2005) also provided a useful taxonomy and frameworks to analyze the design space of TUIs.

Tangible Telepresence

One such genre is an interpersonal communication taking advantage of haptic interactions using mediated tangible representation and control. This genre relies on mapping haptic input to haptic representations over a distance. Also called "tangible telepresence," the underlying mechanism is the synchronization of distributed objects and the gestural simulation of "presence" artifacts, such as movement or vibration, which allow remote participants to convey their haptic manipulations of distributed physical objects. The effect is to give a remote user the sense of ghostly presence, as if an invisible person was manipulating a shared object. InTouch (Brave & Dahley, 1997), HandJive (Fogg, Cutler, Arnold, & Eisbach, 1998), and ComTouch (Chang, O'Modhrain, Jacob, Gunther, & Ishii, 2002) are examples of this.

Tangibles with Kinetic Memory

The use of kinesthetic gestures and movement to promote learning concepts is another promising domain. Educational toys to materialize, record, and play concepts have been also explored using actuation technology and taking advantage of i/o coincidence of TUI. Gestures in physical space illuminate the symmetric mathematical relationships in nature, and the kinetic motions can be used to teach children concepts relevant to programming and differential geometry as well as storytelling. Curlybot (Frei et al., 2000) and topobo (Raffle et al., 2004) are examples of toys that distill ideas relating gestures and form to dynamic movement, physics, and storytelling.

Constructive Assembly

Another domain is a constructive assembly approach that draws inspiration from LEGO™ and building blocks, building upon the interconnection of modular physical elements. This domain is mainly concerned with the physical fit between objects, and the kinetic relationships between these pieces that enable larger constructions and varieties of movement.

Constructive assembly was pioneered by Aish and Frazer in the late 1970s. Aish developed BBS (Aish, 1979; Aish & Noakes, 1984) for thermal performance analysis, and Frazer developed a series of intelligent modeling kits such as "Universal Constructor" (Frazer, 1994; Frazer, Frazer, & Frazer, 1980) for modeling and simulation. Recent examples include GDP (Anagnostou, Dewey, & Patera, 1989), AlgoBlock (Suzuki & Kato, 1993), Triangles (Gorbet, Orth, & Ishii, 1998), Blocks (Anderson et al., 2000), ActiveCube (Kitamura, Itoh, & Kishino, 2001), and System Blocks (Zuckerman & Resnick, 2004). Topobo (Raffle et al., 2004) is a unique instance that inherits the properties of both "constructive assembly" and "tangibles with kinetic memory."

Tokens and Constraints

"Tokens and constraints" is another TUI approach to operate abstract digital information using mechanical constraints (Ullmer, Ishii, & Jacob, 2005). Tokens are discrete, spatially reconfigurable physical objects that represent digital information or operations. Constraints are confining regions within which tokens can be placed. Constraints are mapped to digital operations or properties that are applied to tokens placed within their confines. Constraints are often embodied as physical structures that mechanically channel how tokens can be manipulated, often limiting their movement to a single physical dimension.

The Marble Answering Machine (Crampton Smith, 1995) is a classic example which influenced many following research. MediaBlocks (Ullmer, Ishii, & Glas, 1998), LogJam (Cohen, Withgott, & Piernot, 1999), DataTile (Rekimoto, Ulmer, & Oba, 2001), and Tangible Query Interface (Ullmer, Ishii, & Jacob, 2003) are other recent examples of this genre of development.

Interactive Surfaces—Tabletop TUI

Interactive surfaces are another promising approach to support collaborative design and simulation that has been explored by many researchers in the past years to support a variety of spatial applications (e.g., Urp). On an augmented workbench, discrete tangible objects are manipulated and their movements are sensed by the workbench. The visual feedback is provided on the surface of the workbench, keeping input/output space coincidence. This genre of TUI is also called "tabletop TUI" or "tangible workbench."

Digital Desk (Wellner, 1993) is the pioneering work in this genre, and a variety of tabletop TUIs were developed using multiple tangible artifacts within common frames of horizontal work surface. Examples are metaDesk (Ullmer & Ishii, 1997), InterSim (Arias, Eden, & Fisher, 1997), Illuminating Light (Underkoffler & Ishii, 1998), Urp (Underkoffler & Ishii, 1999), Build-It (Rauterberg et al., 1998), Sensetable (Patten et al., 2001), AudioPad (Patten, Recht, & Ishii, 2002), and IP Network Design Workbench (Kobayashi, Hirano, Narita, & Ishii, 2003).

One limitation of the above systems is the computer's inability to move objects on the interactive surfaces. To address this problem, the Actuated Workbench was designed to provide a hardware and software infrastructure for a computer to smoothly move objects on a table surface in two dimensions (Pangaro et al., 2002), providing an additional feedback loop for computer output, and helping to resolve inconsistencies that otherwise arise from the computer's inability to move objects on the table.

Continuous Plastic TUI

Fundamental limitation of previous TUIs was the lack of capability to change the forms of tangible representations during the interactions. Users had to use predefined finite sets of fixed-form objects, changing only the spatial relationship among them but not the form of individual objects themselves.

Instead of using predefined discrete objects with fixed forms, the new type of TUI systems utilizing continuous tangible material such as clay and sand were developed for rapid form-giving and sculpting for the landscape design. Examples are Illuminating Clay (Piper, Patti, & Ishii, 2002), and SandScape (Ishii, Ratti, Piper, Wang, Biderman, & Ben-Joseph, 2004). Later this interface was applied to the browsing of 3D volume metric data in Phoxel-Space project (Ratti, Wang, Piper, Ishii, & Biderman, 2004).

Augmented Everyday Objects

Augmentation of familiar everyday objects is an important design approach of TUI to lower the floor and to make it easy to understand the basic concepts. Examples are the Audio Notebook (Stifelman, 1996), musicBottles (Ishii et al., 1999), HandScape (Lee, Su, Ren, & Ishii, 2000), LumiTouch (Chang, Resner, Koerner, Wang, & Ishii, 2001), Designers' Outpost (Klemmer,

Thomsen, Phelps-Goodman, Lee, & Landay, 2002), and I/O Brush (Ryokai, Marti, & Ishii, 2004). It is a challenge for industrial designers to improve upon a product by adding some digital augmentation to an existing digital object. This genre is open to much eager interpretation by artists and designers, to have our everyday physical artifacts evolve with technology.

Ambient Media

In the early stages of TUI research, we were exploring ways of improving the quality of interaction between people and digital information. We employed two approaches to extending interaction techniques to the physical world:

- Allowing users to "grasp and manipulate" foreground information by coupling bits with physical objects;
- Enabling users to be aware of background information at the periphery using ambient media in an augmented space.

At that time, HCI research had been focusing primarily on foreground activity on the screen and neglecting the rest of the user's computing environment (Buxton, 1995). However, in most situations, people are subconsciously receiving ambient information from their peripheral senses without attending to it explicitly. If anything unusual is noticed, it immediately comes to their attention, and they could decide to bring it to the foreground. For example, people subconsciously are aware of the weather outside their window. If they hear thunder, or a sudden rush of wind, the user can sense that a storm is on its way, out of his or her peripheral attention. If it was convenient, the user could then look outside, or continue working without distraction.

Ambient media describes the class of interfaces that is designed to smooth the transition of the users' focus of attention between background and foreground. Natalie Jeremijenko's Live Wire in 1995, at Xerox Parc, was a spinning wire that moved to indicate network traffic. Designing simple and adequate representations for ambient media using tangible objects is a key part of the challenge of Tangible Bits (Ishii & Ullmer, 1997).

The ambientROOM is a project that explores the ideas of ambient media constructing a special room equipped with embedded sensors and ambient displays (Ishii et al., 1998). This work was a preliminary investigation into background/peripheral interfaces, and led to the design of stand-alone ambient fixtures such as Pinwheels and Walter Lamps that make users aware of "digital wind" and "bits of rain" at their peripheral senses (Dahley, Wisneski, & Ishii, 1998).

Strictly speaking, ambient media is not a kind of TUI, since in many cases there are no direct interactions. Rather, ambient media serve as background information displays that complement tangible/graspable media that users manipulate in the foreground. TUI's approach to ambient media is concerned with the design of simple mappings that give easy-to-understand form to cyberspace information and represent change in a subtle manner. We started experimenting with a variety of ambient media such as sound, light, airflow, and water movement for background interfaces for awareness of cyberspace at the periphery of human perception.

This concept of "ambient media" is now widely studied in the HCI community as a way to turn the architectural/physical spaces into an ambient and calm information environment. Another design space is low-attention interfaces for interpersonal communication through ambient media (Chang et al., 2001). Ambient devices further commercialized the domain of low-attention ambient media interfaces by developing the Ambient Orb and Weather Beacon, exploring the new genre of "glanceable interfaces" (http://www.ambientdevices.com/).

TUI Instances

In this section, 10 TUI examples are presented to illustrate the potential application domains described in a previous section, and to highlight unique features of TUIs. However, given the limited space and rapid growth of TUI research in HCI community in recent years, the collection of examples introduced here can only cover a relatively small portion of the representative works of TUIs.

InTouch: Tangible TelePresence Through Distributed Synchronized Physical Objects

InTouch is a project to explore new forms of interpersonal communication over distance through touch by preserving the physical analog movement of synchronized distributed rollers (Brave & Dahley, 1997; Brave et al., 1998). Force-feedback is employed to create the illusion that people, separated by distance, are interacting with a shared physical object. The "shared" object provides a haptic link between geographically distributed users, opening up a channel for physical expression over distance.

Two identical mechanisms were built with three freely rotating rollers (Photo 8.3). Each roller is synchronized to the corresponding roller on the distant mechanism using force-feedback, so that when one roller is moved the other corresponding roller also moves. If the movement of one roller is held, then the roller transmits that resistance to the other roller. They are in a sense connected by a stiff computational spring. Two users separated by distance can then play, moving or tapping the rollers or more passively feeling the other person's manipulation of the object. The presence of the other person is represented tangibly through physical interaction with the inTouch device.

Force-feedback is conventionally used to allow a user to "touch" virtual objects in the computer screen through a single point. InTouch applies this technology to realize a link for interpersonal haptic communication, instead of just touching virtual objects. InTouch allows people to feel as if they are connected, through touching the rollers, to another person. Instead of touching inanimate objects, each person is touching a dynamic, moving object that is shared.

Important features of inTouch from HCI points of view can be summarized as follows:

1. No boundary between "input" and "output" (i/o coincidence: the wooden rollers are force displays as well as input devices);
2. Principal human input/output organs are hands, not eyes or ears (with the sense of touch being the primary mode);

3. Information can be sent and received simultaneously through one's hand.

Past communication media such as video-telephony set themselves the ultimate goal of reproducing the voice or the image of the human face and body as realistically as possible in order to create the illusion of "being there" for each interlocutor. InTouch takes the opposite approach by making users aware of the other person without ever rendering him or her in bodily terms, and creating what we call a "tangible presence" or "ghostly presence." By seeing and feeling an object being moved in a human fashion on its own, we imagine a ghostly body. The concept of the ghostly presence provides us with a different approach to the conventional notion of telepresence.

Curlybot: A Toy to Record and Play

Curlybot is a toy that can record and play back physical motion (Photo 8.4). As one plays with it, it remembers how it has been moved and can replay that movement with all the intricacies of the original gesture; every pause, acceleration, and even the shaking in the user's hand, is recorded. Curlybot then repeats that gesture indefinitely, creating beautiful and expressive patterns. Children can use curlybot to gain strong intuition for advanced mathematical and computational concepts, like differential geometry, through play outside of traditional computers (Frei et al., 2000).

The forced-feedback technology used for real-time simultaneous communication in inTouch was employed in curlybot for the recording and playback of gestures. Two motors equipped with an optical encoder enable free rotation in addition to forward and backward movement.

When the user presses the button a red LED is illuminated to indicate the recording mode. The user then moves the curlybot around; meanwhile an encoder is recording this gesture information. Pushing the button a second time terminates recording and a green LED alights to indicate the playback mode. The microprocessor compares the current position with the stored positions and instructs the motors to retrace the steps recorded in the curlybot's memory.

PHOTO 8.4. Curlybot.

This project contributes to both interface design and education. As a tangible interface it blurs the boundary between input and output, as inTouch does. Curlybot itself is both an input device to record gestures and a physical display device to reenact them. By allowing the user to teach it gestures with his or her hand and body, and then reenacting those gestures in a physical space around the body, curlybot enables a strong connection between body and mind not obtainable from anything expressed on a computer screen.

From an educational standpoint, curlybot allows very young children to explore "advanced" mathematical and computational concepts. Curlybot supports new ways of thinking about geometric shapes and patterns. Children can also use curlybot to explore some of the basic ideas behind computational procedures, like how complexity can be built from simple parts. This is similar to what is possible with the Logo programming language, but does not require children to read or write and thus makes advanced ideas accessible to younger children. Curlybot also draws strongly on children's intuition about their own physical actions in the world to learn—what Seymour Papert called "body syntonic learning" (Papert, 1980). In addition, the direct input and beautifully expressive patterns that result through curlybot's repetition of the gestures keep children playing and engaged.

Topobo: 3D Constructive Assembly With Kinetic Memory

Topobo, a combination of "topology" and "robotics," is a 3D constructive assembly system with kinetic memory, and the ability to record and play back physical motion (Raffle et al., 2004). By snapping together a combination of passive (static) and active (motorized) components, people can quickly assemble dynamic biomorphic forms like animals and skeletons with Topobo. Topobo allows users to animate those forms by recording the movement of pushing, pulling, and twisting them, and later observe the system play back those motions repeatedly. This "record and play" function was inherited from the prior curlybot project, and the constructive assembly function was inherited from the commercial toy, Zoob™.

For example, a dog can be constructed and then taught to gesture and walk by twisting its body and legs. The dog will then repeat those movements and walk repeatedly. The same way people can learn about static structures playing with regular building blocks, they can learn about dynamic structures by playing with Topobo. Topobo works like an extension of the body, giving one gestural fluency. Topobo embeds computation within a dynamic building system so that gestural manipulation of the material becomes a programming language (Photo 8.5).

Topobo was inspired by current trends in computational media design and by artists and empiricists using visual explorations and models of natural phenomena to more deeply appreciate patterns found in the natural world. In this spirit, Topobo is designed to allow people to use experimentation, play, and self-expression to discover and explore common natural relationships between natural forms and dynamic motion. Building toys and educational manipulatives have been used for years by children to learn about the world though model making.

PHOTO 8.5. Topobo.

Unique among modeling systems is Topobo's coincident physical input and output behaviors (which is also common to inTouch and curlybot). The system is comprised of 10 different primitives that can be snapped together in a variety of ways. Nine of these primitives are called "passive" because they form static connections. These static connections constrain the form and the range of motion available to the structure. One "active" primitive is built with an embedded encoder and motor that are programmed by demonstration. These motorized components are the only ones that move, so the system is able to faithfully record and replay every dynamic manipulation to the structure.

mediaBlocks: Token and Constraint Approach

The mediaBlocks system is a tangible interface for manipulating lists of online digital media such as video clips and images (Ullmer et al., 1998). Whereas Urp provides a spatial interface for leveraging object arrangements consistent with real-world building configurations, the mediaBlocks system provides a relational interface for manipulating more abstract digital information.

The mediaBlocks are small, digitally tagged blocks, dynamically bound to lists of online media elements. The mediaBlocks support two major modes of use. First, they function as capture, transport, and playback mechanisms for moving online media between different media devices. In this mode, conference room cameras, digital whiteboards, wall displays, printers, and other devices are outfitted with mediaBlock slots. Inserting one of the mediaBlocks into the slot of a recording device (e.g., a camera) activates the recording of media into online space, and the dynamic binding of the media to the physical block.

Similarly, inserting one of the bound mediaBlocks into a playback device (e.g., video display) activates playback of the associated online media. Inserting mediaBlocks into slots mounted on computer monitors provides an intermediate case, allowing mediaBlock contents to be exchanged bi-directionally with traditional computer applications using the GUI drag-and-drop operation.

The second functionality of mediaBlocks uses the blocks as physical controls on a media-sequencing device (Photo 8.6). A mediaBlock "sequence rack" (partially modeled after the tile racks of the Scrabble game) allows the media contents of multiple adjacent mediaBlocks to be dynamically bound to a new mediaBlock carrier. Similarly, a second "position rack" maps the physical position of a block to an indexing operation upon its contents. When mediaBlocks are positioned on the left edge of the position rack, the first media element of the block is selected. Intermediate physical positions on the rack provide access to later elements in the associated media list of the block.

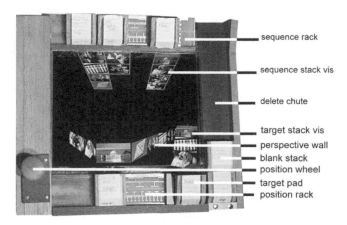

sequence rack

sequence stack vis

delete chute

target stack vis
perspective wall
blank stack
position wheel
target pad
position rack

PHOTO 8.6. mediaBlocks: media sequencing device.

PHOTO 8.7. AudioPad running on Sensetable platform.

Digital Desk: Pioneer of Tabletop TUI

Digital Desk (Wellner, 1993) is a pioneering work to demonstrate a way to integrate physical and digital document processing on a table. Wellner brought some of the functionality we typically associate with GUIs onto the physical desktop. This table used a camera and a microphone to detect finger presses on a graphical interface displayed on a desk with a video projector. Wellner used this desk for tasks such as graphic design and spreadsheet computations on physical paper. This system also employed some physical props, such as a scanner that would scan items and place them directly on the tabletop interaction surface.

Wellner's research pointed the way toward enabling the computer to perform some of the operations we traditionally associate with GUIs in a tabletop environment. The Digital Desk also illustrates some of the compelling reasons for considering computer interfaces based on horizontal interactive surfaces. Because many work surfaces in our environment are already planar, horizontal or nearly horizontal surfaces, integrating computer interfaces into these surfaces may provide an opportunity for new types of relationships between computation and physical objects, and may help create computer systems that are more relevant to problem domains with established work practices based on tabletops.

The Digital Desk inspired many tabletop tangible interfaces including the Luminous Room project (Underkoffler, Ullmer, & Ishii, 1999) from which Urp (Underkoffler & Ishii, 1999) was created. Sensetable (Patten et al., 2001) is another example.

Sensetable and AudioPad: Tabletop TUI for Real-Time Music Performance

Sensetable (Patten et al., 2001) is a system that wirelessly tracks the positions of multiple objects on a flat display surface. The Sensetable serves as a common platform for a variety of tabletop TUI applications such as Audio Pad and IP Network Design Workbench.

Audiopad (Patten et al., 2002) is a composition and performance instrument for electronic music that tracks the positions of objects on a tabletop surface and converts their motion into music. One can pull sounds from a giant set of samples, juxtapose archived recordings against warm synthetic melodies, cut between drum loops to create new beats, and apply digital processing all at the same time on the same table. Audiopad not only allows for spontaneous reinterpretation of musical compositions, but also creates a visual and tactile dialogue between itself, the performer, and the audience.

Audiopad is based on the Sensetable platform that has a matrix of antenna elements that track the positions of electronically tagged objects on a tabletop surface. Software translates the position information into music and graphical feedback on the tabletop. Each object represents either a musical track or a microphone (Photo 8.8).

Experience of Audiopad with tangible user interface through a series of live performances suggests that interacting with electromagnetically tracked objects on a tabletop surface with graphical feedback can be a powerful and satisfying tool for musical expression. The integration of input and output spaces gives the

PHOTO 8.8.   IP network design workbench running on Sensetable platform.

performer a great deal of flexibility in terms of the music that can be produced. At the same time, this seamless integration allows the performer to focus on making music, rather than using the interface. Spatial multiplexed inputs of TUI also supported two performers playing music simultaneously and collaboratively (Photo 8.8).

IP Network Design Workbench: Event Driven Simulation on Sensetable

The IP Network Design Workbench (IPNWDWB) is the collaborative project between NTT Comware and the Tangible Media Group. The IP Network Design Workbench supports collaborative network design and simulation by a group of experts and customers (Kobayashi et al., 2003). This system is also based on the Sensetable platform, which can wirelessly detect the location and orientation of physical pucks. Simulation engine is based on the event-driven simulation model. Using the Sensetable system, users can directly manipulate network topologies for modeling, control simulation parameters of nodes and links using physical pucks on the sensing table, and simultaneously see the simulation results projected onto the table in real time (Photo 8.9).

The goal of IPNWDWB is to make simulation tools more accessible for non-experts, so that they can join the network design process and interact with experts more easily than using traditional GUI computer. This system was commercialized and has been used for collaborative network design with customers to ensure their understanding of the performance and cost of network enhancements dealing with the increases of network traffic caused by Voice over IP and/or streaming video, for example. Because of the large tiling horizontal work surface and TUI interaction that invites all the participants to touch and manipulate pucks simultaneously, the process of decision making becomes much more democratic and more convincing than ordinary PowerPoint presentations through conventional GUI.

If we compare IPNWDWB with Urp, we notice a big difference in the nature of applications. In Urp, we used physical scale models of buildings, which humans have used for thousands of years to design cities, as tangible representations of urban models. Therefore, it is very natural to apply TUIs to such domains (e.g., urban planning, landscape design) in which physical models have been used long before the birth of digital computers.

In contrast, IP Network Design is based on event-driven simulation models, which are quite abstract and new. This modeling technique requires digital computers. IPNWDWB is important

PHOTO 8.9.   Actuated workbench used for distributed collaboration.

since it demonstrated that TUI can empower the design process even in abstract and computational application domain that does not have straightforward mappings from abstract concepts to physical objects. There is a wide range of modeling and simulation techniques such as System Dynamics and Event-Driven Simulation that uses 2D graph representation. We learned that many of these abstract computational applications can be supported by Sensetable-like TUI platforms in the collaborative design sessions. For example, simultaneously changing parameters, transferring control between different people or different hands, and distributing the adjustment of simulations dynamically are interactions enabled by TUI.

Actuated Workbench: Closing a Loop of Computational Actuation and Sensing

The aforementioned tabletop TUI systems share a common weakness. While input occurs through the physical manipulation of tangible objects, output is displayed only through sound or graphical projection on and around the objects. As a result, the objects can feel like loosely coupled handles to digital information rather than physical manifestations of the information itself.

In addition, the user must sometimes compensate for inconsistencies when links between the digital data and the physical objects are broken. Such broken links can arise when a change occurs in the computer model that is not reflected in a physical change of its associated object. With the computer system unable to move the objects on the table surface, it cannot undo physical input, correct physical inconsistencies in the layouts of the objects, or guide the user in the physical manipulation of the objects. As long as this is so, the physical interaction between human and computer remains one-sided.

To address this problem, the Actuated Workbench was designed to provide a hardware and software infrastructure for a computer to smoothly move objects on a table surface in two dimensions (Pangaro et al., 2002).

The Actuated Workbench is a new technology that uses magnetic forces to move objects on a table in two dimensions. It is intended for use with existing tabletop tangible interfaces, providing an additional feedback loop for computer output, and helping to resolve inconsistencies that otherwise arise from the computer's inability to move objects on the table.

Actuation enables a variety of new functions and applications. For example, a search and retrieve function could respond to a user query by finding matching data items and either moving them to another place on the tabletop or wiggling them to get the user's attention. A more powerful function would be one in which the computer could physically sort and arrange pucks on the table according to user-specified parameters. This could help the user organize a large number of data items before manually interacting with them. As a user makes changes to data through physical input, he or she may wish to undo some changes. A physical undo in this system could move the pucks back to their positions before the last change. It could also show the user the exact sequence of movements she had performed. In this sense, both "undo" and "rewind" commands are possible.

One advantage that tabletop tangible user interfaces offer is the ease with which multiple users can make simultaneous changes to the system. Users can observe each other's changes, and any user can reach out and physically change the shared layout without having to grab a mouse or other pointing device. This is not the case, however, when users are collaborating remotely. In this scenario, a mechanism for physical actuation of the pucks becomes valuable for synchronizing multiple, physically separated workbench stations (Photo 8.9). Without such a mechanism, real-time physical synchronization of the two tables would not be possible, and inconsistencies could arise between the graphical projection and the physical state of the pucks on the table.

In addition to facilitating the simple synchronization of these models, the Actuated Workbench can recreate remote users' actual gestures with objects on the table, adding greatly to the "Ghostly Presence" (Brave et al., 1998) sought in remote-collaboration interfaces.

Actuated Workbench is helpful in teaching students about physics by demonstrating the attraction and repulsion of charged particles represented by pucks on the table. As a student moves the pucks around on the table, the system could make them rush together or fly apart to illustrate forces between the objects.

SandScape: Continuous TUI for Landscape Design

SandScape (Ishii et al., 2004) is a tangible interface for designing and understanding landscapes through a variety of computational simulations using sand. Users view these simulations as they are projected on the surface of a sand model that represents the terrain. The users can choose from a variety of different simulations that highlight the height, slope, contours, shadows, drainage, or aspect of the landscape model (Photo 8.10).

The users can alter the form of the landscape model by manipulating sand while seeing the resultant effects of computational analysis generated and projected on the surface of sand in real time. The project demonstrates how TUI takes advantage of our natural ability to understand and manipulate physical forms while still harnessing the power of computational simulation to help in our understanding of a model representation.

The SandScape configuration is based on a box containing 1 m-diameter glass beads lit from beneath with an array of 600 high-power infrared LEDs. Four IR mirrors are placed around the LED array to compensate for the uneven radiance distribution on the boundary. A monochrome infrared camera is mounted 2 m above the surface of the beads and captures the intensity of light passing through the volume. The intensity of transmitted light is a function of the depth of the beads, and a look-up table can be used to convert surface radiance values into surface elevation values. The system has been calibrated to work with a specific bead size and the optical properties of the material used (absorption and scattering coefficients) are critical to its successful functioning. Owing to the exponential decay of the IR light passing through the glass beads (or any other material) the intensity at the top surface can vary greatly and sometimes exceed the dynamic range of the video camera. This problem can be solved by taking several images with different exposure times

PHOTO 8.10.   SandScape.

and combining them to recover the effective radiance of the scene. SandScape is less accurate than its predecessor Illuminating Clay, which used laser rangefinders to capture the geometry of a clay model (Piper et al., 2002).

SandScape and Illuminating Clay show the potential advantages of combining physical and digital representations for landscape modeling and analysis. The physical clay and sand models convey spatial relationships that can be intuitively and directly manipulated by hand. Users can also insert any found physical objects directly under the camera. This approach allows users to quickly create and understand highly complex topographies that would be difficult and time-consuming to produce with conventional CAD tools. We believe that this "Continuous TUI" approach makes better use of our natural abilities to discover solutions through the manipulation of physical objects and materials.

At the same time, the projected graphics give the user real-time feedback. While tracked physical models interfaced with a computer are not a novelty, we believe that SandScape and Illuminating Clay offer a new contribution, by using the continuous surface geometry of the model itself to act as the input/output mechanism. In so doing we hope to give the projected information the same tangible immediacy as the clay/sand material itself and allow quantitative data to support the intuitive understanding of the landscape.

Landscape architecture, as well as urban and architectural design, requires the collaboration of a number of specialists. These include earth engineers, water engineers, agrarian managers, land economists, and transport engineers—to name just a few. In the current process of design, the collaboration happens at different stages, and sometimes without much direct and synchronous interaction. SandScape and Illuminating Clay provide a common platform for collaboration, centered on the table workspace. Numerous representations and analyses can be combined in a single design environment, potentially offering a greater cohesion between different specialists and streamlining the process of design.

musicBottles: Transparent Interface Based on Augmented Glass Bottles

musicBottles introduces a tangible interface that deploys bottles as containers and controls for digital information (Photo 8. 11). The system consists of a specially designed table and three corked bottles that "contain" the sounds of the violin, the cello, and the piano in Edouard Lalo's Piano Trio in C Minor op. 7. Custom-designed electromagnetic tags embedded in the bottles enable each one to be wirelessly identified.

When a bottle is placed onto the stage area of the table, the system identifies each bottle, and lights up the stage to show that the bottles have been recognized. The opening and closing of a bottle is also detected, and as the cork is removed, the corresponding instrument becomes audible. A pattern of colored light is rear-projected onto the table's translucent surface to reflect changes in pitch and volume for each instrument. The interface allows users to structure the experience of the musical composition by physically manipulating the different soundtracks.

Humans have used glass bottles for thousands of years. Through the seamless extension of physical affordances and metaphors of the bottles into the digital world, the bottles project explores the transparency of the interface (Ishii, 2004).

A wide variety of contents, including music, weather reports, poems, and stories have been designed to test the concept (Ishii et al., 1999). The bottles lined up on a specially designed table, the feel of the glass as we open them, and the music and light from the LED lamps that come out of them together create a unique aesthetic experience. This is a pleasure not to be had from the mere click of a mouse.

Potential applications are not limited to music alone. One might imagine perfume bottles filled with poetry or wine bottles that decant stories (Mazalek, Wood, & Ishii, 2001). More practical applications might include a medicine chest full of bottles that tell the user how and when to take them and let the hospital

PHOTO 8.11. musicBottles.

know when they do. As an intimate part of our daily lives, glass bottle interfaces offer a simple and transparent interface.

Pinwheels: Ambient Interface to Information

Pinwheels are an example of ambient media that demonstrate a new approach to interfacing people with online digital information through subtle changes in sound and movement, which can be processed in the background of awareness. Pinwheels spin in a "bit wind" and represent an invisible flow of digital information such as network traffic as physical movement within an architectural spaces (Photo 8.12).

Nature is filled with subtle, beautiful, and expressive ambient media that engage each of our senses. The sounds of rain and the feeling of warm wind on our cheeks help us understand and enjoy the weather even as we engage in other activities. Similarly, we are aware of the activity of neighbors through passing sounds and shadows at the periphery of our attention. Cues like an open door or lights in an office help us subconsciously understand the activities of other people, and communicate our own activity and availability.

Current personal computing interfaces, however, largely ignore these rich ambient spaces, and squeeze vast amounts of digital information into small rectangular screens. Information is presented as "painted bits" (pixels) on flat screens that must be in the center (foreground) of a user's focus to be processed. In order to broaden the concept of "display" to make use of the entire physical environment as an interface, using ambient media, information can be manifested as subtle changes in form, movement, sound, color, smell, temperature, or light. We call them "ambient displays."

The Pinwheels evolved from the idea of using airflow as ambient media. However, we found that the flow of air itself was difficult to control and to convey information. As an alternative, we envisioned that a visual/physical representation of airflow based on the "spinning pinwheels" could be legible and poetic.

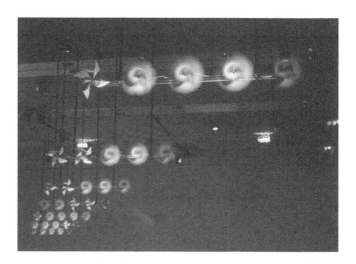

PHOTO 8.12. Pinwheels.

The Pinwheels spin in the "bit wind" at different speeds based upon their input information source.

Ambient displays are envisioned as being all around and suited to the display of:

1. People's presence (awareness of remote people's status/ activities);
2. Atmospheric and astronomical phenomena;
3. General states of large and complex systems (e.g., an atomic power plant).

For instance, an atmospheric scientist might map patterns of solar wind into patterns of Pinwheel spins in a room.

There are many design challenges surrounding ambient displays. One of them is the mapping of information to the physical motion and other ambient media. A designer of ambient displays must transform the digital data into a meaningful pattern of physical motion that successfully communicates the information. The threshold between foreground and background is another key issue. Ambient displays are expected to go largely unnoticed until some change in the display or user's state of attention makes it come into the foreground of attention. How to keep the level of display at the threshold of a user's attention is an open design issue.

Contributions of TUIs

TUI is generally built from systems of physical artifacts with digital coupling with computation. Taken together as ensembles, TUI has several important advantages over traditional GUI as well as limitations. This section summarizes those contributions of TUIs and required design considerations.

Double Interactions Loop: Immediate Tactile Feedback

One important advantage of TUI is that users receive passive haptic feedback from the physical objects as they grasp and manipulate them. Without waiting for the digital feedback (mainly visual), users can complete their input actions (e.g., moving a building model to see the interrelation of shadows).

Typically there are two feedback loops in TUI, as shown in Fig. 8.4:

1. The passive haptic feedback loop provides the user with an immediate confirmation that he or she has grasped and moved the object. This loop exists within a physical domain, and it does not require any sensing and processing by a computer. Thus, there is no computational delay. The user can begin manipulating the object as desired without having to wait for the second feedback loop, the visual confirmation from the interface. In contrast, when the user uses a mouse with a GUI computer, he or she has to wait for the visual feedback (second loop) to complete an action.
2. The second loop is a digital feedback loop that requires sensing of physical objects moved by users, computation based

on the sensed data, and displaying the results as visual (and auditory) feedback. Therefore, this second loop takes longer than the first loop.

Many of the frustrations of using current computers come from the noticeable delay of digital feedback as well as a lack of tactile confirmation of actions taken by computers. We believe the double loops of TUI give users a way to ease those frustrations.

Note: Actuation technology introduced in Actuated Workbench will contribute to add another loop, that of physical actuation. Figure 8.5 illustrates the third loop introduced into the TUI model by computer-controlled actuation and sensing. The third loop allows the computer to give feedback on the status of the digital information as the model changes or responds to internal computation.

Persistency of Tangibles

As physical artifacts, TUI tangibles are persistent. Tangibles also carry physical state, with their physical configurations tightly coupled to the digital state of the systems they represent. The physical state of tangibles embodies key aspects of the digital state of an underlying computation.

For example, the physical forms of the Urp building models, as well as their position and orientation on the workbench of the system, serve central roles in representing and controlling the state of the underlying digital simulation system. Even if the mediating computers, cameras, and projectors of Urp are turned off, many aspects of the state of the system are still concretely expressed by the configuration of its physical elements.

In contrast, the physical form of the mouse holds little representational significance because GUIs represent information almost entirely in visual form.

Coincidence of Input and Output Spaces

Another important feature (and design principle) of TUI is coincidence of input and output spaces to provide seamless information representation that spans both tangible (physical) and intangible (digital) domains.

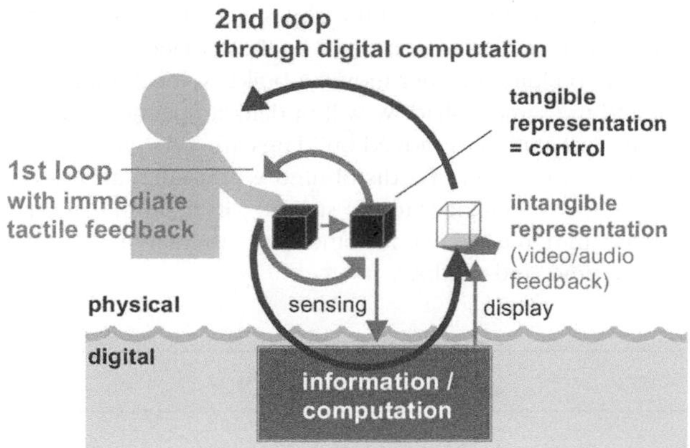

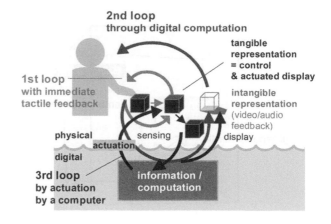

FIGURE 8.4. TUI's double feedback loops.

FIGURE 8.5. TUI with actuation (Actuated workbench).

GUI utilizes the mouse and keyboard as generic "remote" controllers (input), and the screen serves as the main output medium. Thus, there is spatial discontinuity between those two spaces. There is also multimodal inconsistency, as touch is the main input while vision is the only output.

TUI tries to coincide input space and output space as much as possible to realize seamless coupling of physical and digital worlds (Ishii & Ullmer, 1997). An example of this seamless coupling is Underkoffler's Urp (Underkoffler & Ishii, 1999). A series of architectural models serve as the input devices, and output in the form of a wind-and-shadow simulation is projected down onto the same tabletop surface, on top of and around the building models. Illuminating Clay (Piper et al., 2002) and SandScape (Ishii et al., 2004) demonstrate another example of i/o coincidence using continuous flexible material: sand. Curlybot and topobo demonstrate the same concept using the contact surface of the tangibles as input and output to digitize the person's physical motion.

Special Purpose vs. General Purpose

GUIs are fundamentally general-purpose interfaces that are supposed to emulate a variety of applications visually using dynamic pixels on a screen and generic remote controllers such as the mouse and keyboard. On the other hand, TUIs are relatively specific interfaces tailored to certain types of applications in order to increase the directness and intuitiveness of interactions.

The selection of the correct and specific application domain is critical to apply TUI successfully to take advantage of existing skills and work practices (e.g., use of physical models in urban planning).

One notable aspect of Urp is its use of objects with very application-specific physical forms (scaled building models) as a fundamental part of the interface. Physical building models represent the buildings themselves in the interactive simulation. Thus they give the user important visual and tactile information about the computational object they represent. Indicators such as a clock and weather vane work in reverse in the Urp system. Instead of the clock hands moving to indicate the passage of time, the user can move the clock hands to change the time of day for the shadow study (Photo 8.1). Likewise, he or she can change the orientation of the weather vane to control the direction of the wind (Photo 8.2).

In the design of TUI, it is important to give an appropriate form to each tangible tool and object so that the form will give an indication of the function available to the users. For example, the clock hands allow people to automatically make the assumption that they are controlling time.

Of course, this special-purposeness of TUIs can be a big disadvantage if users would like to apply it to a wide variety of applications, since customized physical objects tailored to certain applications cannot be reused for most other applications. By making the form of objects more abstract (e.g., a round puck), you lose the legibility of tangible representation and the object will become a generic handle rather than the representation of underlying digital information. It is important to attain a balance between specific/concrete vs. generic/abstract to give a form to digital information and computational function.

Space-Multiplexed Input

Another distinct feature of TUI is space-multiplexed input (Fitzmaurice, Ishii, & Buxton, 1995a). Each tangible representation serves as a dedicated controller occupying its own space, and encourages two-handed and multi-user simultaneous interaction with an underlying computational model. Thus, TUI is suitable for collocated collaboration allowing concurrent manipulation of information by multiple users.

GUI, in contrast, provides time-multiplexed input that allows users to use one generic device to control different computational functions at different points in time. For instance, the mouse is used for menu selection, scrolling windows, pointing, and clicking buttons in a time-sequential manner.

TUI can support not only collocated collaboration, but also remote collaboration using an actuation mechanism to synchronize the physical states of tangibles over distance. Actuated Workbench is an example of such a technology that extends TUI for remote collaboration (Pangaro et al., 2002).

In the Urp scenario, applying the Actuated Workbench technology, it is possible to have two distributed Urp tables in different locations, connected and synchronized over the Internet. One Urp can be in Tokyo, while the other Urp can be in Boston, and the shadows are synchronized as the urban planning team moves the buildings around the Urp space. The movement of buildings can be also synchronized by the actuation mechanism. When the building planner moves a building location, both the local and the remote shadow will update simultaneously; position and orientation of moved buildings are also synchronized. This synchronization of a distributed workbench allows both teams to discuss changes to the situation in real time, and provides a common reference for otherwise ethereal qualities such as wind, time, and shadow.

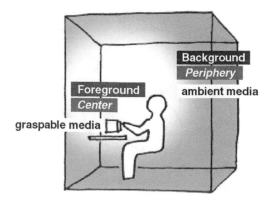

FIGURE 8.6. Center and periphery of user's attention within physical space.

CONCLUSION

The author met a highly successful computational device called the "abacus" when he was two years old (Photo 8.13). He could enjoy the touch and feel of the "digits" physically represented as

PHOTO 8.13.   Abacus.

mediately understand what they can do with this artifact without reading a manual.

TUI pursues these features further into the digital domain by giving physical form to digital information and computation, employing physical artifacts both as representations and controls for computational media. Its design challenge is a seamless extension of the physical affordances of the objects into the digital domain.

This chapter introduced the basic concept of TUI and a variety of examples of TUI applications to address the key properties of TUI and its design challenges. TUI is still in its infancy, and extensive research is required to identify the killer applications, scalable TUI toolkits, and a set of strong design principles.

The research of TUI, which gives physical forms to digital information/computation, naturally crosses with the paths of industrial/product design as well as environmental/architectural design. It has also made an impact on the media-arts/interactive-arts community. The author hopes that TUI design will contribute to promote those interdisciplinary design research initiatives in the HCI community to bring strong design culture as well as media-arts perspective to the scientific/academic world.

Mark Weiser's (1991) seminal paper on ubiquitous computing started with the following paragraph: *"The most profound technologies are those that disappear. They weave themselves into the fabric of everyday life until they are indistinguishable from it."*

I do believe that TUI is one of the promising paths to his vision of invisible interface.

arrays of beads. This simple abacus was not merely a digital computational device. Because of its physical affordance, the abacus also became a musical instrument, imaginary toy train, and a backscratcher. He was captivated by the sound and tactile interaction with this simple artifact.

His childhood abacus became a medium of awareness too. When his mother kept household accounts, he was aware of her activities by the sound of her abacus, knowing he could not ask her to play with him while her abacus made its music.

This abacus suggests to us a new direction of Human-Computer Interaction (HCI) that we call Tangible User Interfaces (TUI). First, it is important to note that the abacus makes no distinction between "input" and "output." Instead, the beads, rods, and frame serve as physical representations of numerical information and computational mechanism. They also serve as directly manipulatable physical controls to compute on numbers.

Second, the simple and transparent mechanical structure of the abacus (without any digital black boxes) provides rich physical affordances (Norman, 1999) so that even children can im-

ACKNOWLEDGMENTS

The author would like to thank many of the former and current TMG students and colleagues in the MIT Media Lab for their contributions to the TUI research in the past 10 years at the MIT Media Lab. Especially thanks are due to Dr. John Underkoffler, Dr. Brygg Ullmer, Dr. Kimiko Ryokai, Dr. James Patten, Angela Chang, Hayes Raffle, Amanda Perks, and Oren Zuckerman. Thanks are also due to Things That Think and Digital Life Consortia at the Media Lab for their support of our research. The author also would like to thank Prof. Bill Buxton and Dr. George Fitzmaurice for our collaboration on the Bricks project in 1994 at the University of Toronto (Fitzmaurice, Ishii, & Buxton, 1995b) which laid the foundation of TUI, and Prof. Mitchel Resnick for his vision of Digital Manipulatives (Resnick et al., 1998), which influenced our design of educational TUI applications.

References

Aish, R. (1979). 3D input for CAAD systems. *Computer-Aided Design, 11*(2), 66–70.

Aish, R., & Noakes, P. (1984). Architecture without numbers—CAAD based on a 3D modelling system. *Computer-Aided Design, 16*(6), 321–328.

Anagnostou, G., Dewey, D., & Patera., A. (1989). Geometry-defining processors for engineering design and analysis. *The Visual Computer, 5*, 304–315.

Anderson, D., Frankel, J. L., Marks, J., Agarwala, A., Beardsley, P., Hodgins, J., Leigh, D., Ryall, K., Sullivan, E., & Yedidia, J. S. (2000). Tangible interaction + graphical interpretation: a new approach to 3D modeling. *Proceedings of the 27th Annual Conference on Computer Graphics and Interactive Techniques* (pp. 393–402). ACM Press/Addison-Wesley Publishing Co.

Arias, E., Eden, H., & Fisher, G. (1997). Enhancing communication, facilitating shared understanding, and creating better artifacts by integrating physical and computational media for design. *Proceedings of the Conference on Designing Interactive Systems: Processes, Practices, Methods, and Techniques* (pp. 1–12). ACM Press.

Brave, S., & Dahley, A. (1997). inTouch: A medium for haptic interpersonal communication. *Conference on Human Factors in Computing Systems (CHI '97)* Atlanta, GA (pp. 363–364). ACM.

Brave, S., Ishii, H., & Dahley, A. (1998). Tangible interfaces for remote collaboration and communication. *Proceedings of the ACM Conference on Computer Supported Cooperative Work* (pp. 169–178). ACM Press.

Burbeck, S. (1992). Applications Programming in Smalltalk-80™: How to use Model-View-Controller (MVC). http://st-www.cs.uiuc.edu/users/smarch/st-docs/mvc.html

Buxton, W. (1995). Integrating the Periphery and Context: A New Model of Telematics. *Proceedings of Graphics Interface '95* (pp. 239–246).

Chang, A., O'Modhrain, S., Jacob, R., Gunther, E., & Ishii, H. (2002). ComTouch: design of a vibrotactile communication device. *Proceedings of the Conference on Designing Interactive Systems: Processes, Practices, Methods, and Techniques* (pp. 312–320). ACM Press.

Chang, A., Resner, B., Koerner, B., Wang, X., & Ishii, H. (2001). LumiTouch: an emotional communication device. *CHI '01 Extended Abstracts on Human Factors in Computing Systems* (pp. 313–314). ACM Press.

Cohen, J., Withgott, M., & Piernot, P. (1999). LogJam: a tangible multi-person interface for video logging. *Proceedings of the SIGCHI Conference on Human Factors in Computing Systems: the CHI is the limit*, Pittsburgh, PA (pp. 128–135). ACM Press.

Crampton Smith, G. (1995). The Hand That Rocks the Cradle. *I.D.*, 60–65.

Dahley, A., Wisneski, C., & Ishii, H. (1998). Water Lamp and Pinwheels: Ambient Projection of Digital Information into Architectural Space. *Conference on Human Factors in Computing Systems*, Los Angeles (pp. 269–270). ACM.

Fishkin, K. P. (2004). A taxonomy for and analysis of tangible interfaces. *Personal Ubiquitous Comput, 8*, 347–358.

Fitzmaurice, G. W., Ishii, H., & Buxton, W. A. S. (1995a). Bricks: Laying the Foundations for Graspable User Interfaces. *Conference on Human Factors in Computing Systems*, Denver, Colorado (pp. 442–449). ACM.

Fitzmaurice, G. W., Ishii, H., & Buxton, W. A. S. (1995b). Bricks: laying the foundations for graspable user interfaces, *Proceedings of the SIGCHI Conference on Human factors in computing systems* (pp. 442–449). ACM Press/Addison-Wesley Publishing Co.

Fogg, B., Cutler, L. D., Arnold, P., & Eisbach, C. (1998). HandJive: a device for interpersonal haptic entertainment. *Proceedings of the SIGCHI Conference on Human Factors in Computing Systems* (pp. 57–64). ACM Press/Addison-Wesley Publishing Co.

Frazer, J. (1994). *An Evolutionary Architecture Architectural Association*, London.

Frazer, J., Frazer, J., & Frazer, P. (1980). Intelligent physical threedimensional modelling system. *Computer Graphics 80*, North Holland (pp. 359–370).

Frei, P., Su, V., Mikhak, B., & Ishii, H. (2000). curlybot: designing a new class of computational toys. *Proceedings of the SIGCHI Conference on Human factors in Computing Systems*, the Netherlands (pp. 129–136). ACM Press.

Goldberg, A. (1984). *Smalltalk-80: The Interactive Programming Environment*. Addison-Wesley.

Gorbet, M., Orth, M., & Ishii, H. (1998). Triangles: Tangible Interface for Manipulation and Exploration of Digital Information Topography. *Conference on Human Factors in Computing Systems (CHI '98)* (pp. 49–56). ACM.

Holmquist, L. E., Redström, J., & Ljungstrand, P. (1999). Token-Based Access to Digital Information. *Proceedings of the 1st international symposium on Handheld and Ubiquitous Computing* (pp. 234–245). Springer-Verlag.

Ishii, H. (2004). Bottles: A Transparent Interface as a Tribute to Mark Weiser. *IEICE Transactions on Information and Systems E87-D, 6*, 1299–1311.

Ishii, H., Fletcher, H. R., Lee, J., Choo, S., Berzowska, J., Wisneski, C., Cano, C., Hernandez, A., & Bulthaup, C. (1999). musicBottles. *ACM SIGGRAPH 99 Conference abstracts and applications* Los Angeles (p. 174). ACM Press.

Ishii, H., Mazalek, A., & Lee, J. (2001). Bottles as a minimal interface to access digital information. *CHI '01 Extended Abstracts on Human Factors in Computing Systems* (pp. 187–188). ACM Press.

Ishii, H., Ratti, C., Piper, B., Wang, Y., Biderman, A., & Ben-Joseph, E. (2004). Bringing clay and sand into digital design—continuous tangible user interfaces. *BT Technology Journal, 22*(4), 287–299.

Ishii, H., & Ullmer, B. (1997). Tangible Bits: Towards Seamless Interfaces between People, Bits and Atoms. *Conference on Human Factors in Computing Systems (CHI '97)*, Atlanta, GA (pp. 234–241). ACM.

Ishii, H., Wisneski, C., Brave, S., Dahley, A., Gorbet, M., Ullmer, B., & Yarin, P. (1998). ambientROOM: Integrating Ambient Media with Architectural Space (video). *Conference on Human Factors in Computing Systems (CHI '98)* (pp. 173–174). ACM.

Johnson, J., Roberts, T. L., Verplank, W., Smith, D. C., Irby, C. H., Beard, M., & Mackey, K. (1989). The Xerox Star: a retrospective. *IEEE Computer, 22*(9), 11–26, 28–29.

Kitamura, Y., Itoh, Y., & Kishino, F. (2001). Real-time 3D interaction with ActiveCube. *CHI '01 Extended Abstracts on Human Factors in Computing Systems* (pp. 355–356). ACM Press.

Klemmer, S. R., Thomsen, M., Phelps-Goodman, E., Lee, R., & Landay, J. A. (2002). Where do websites come from?: capturing and interacting with design history. *Proceedings of the SIGCHI Conference on Human factors in Computing Systems: Changing our world, changing ourselves* (pp. 1–8). ACM Press.

Kobayashi, K., Hirano, M., Narita, A., & Ishii, H. (2003). A tangible interface for IP network simulation. *CHI '03 extended abstracts on Human Factors in Computing Systems* (pp. 800–801). ACM Press.

Lee, J., Su, V., Ren, S., & Ishii, H. (2000). HandSCAPE: a vectorizing tape measure for on-site measuring applications. *Proceedings of the SIGCHI Conference on Human Factors in Computing Systems* (pp. 137–144). ACM Press.

Mazalek, A., Wood, A., & Ishii, H. (2001). genieBottles: An Interactive Narrative in Bottles. *Conference Abstracts and Applications of SIGGRAPH '01* (pp. 189). ACM Press.

Norman, D. A. (1999). Affordance, conventions, and design. *Interactions*, pp. 38–43.

Pangaro, G., Maynes-Aminzade, D., & Ishii, H. (2002). The actuated workbench: computer-controlled actuation in tabletop tangible interfaces. *Proceedings of the 15th annual ACM symposium on User Interface Software and Technology* (pp. 181–190). ACM Press.

Papert, S. (1980). *Mindstorms: Children, Computers, and Powerful Ideas*. New York: Basic Books.

Patten, J., Ishii, H., Hines, J., & Pangaro, G. (2001). Sensetable: a wireless object tracking platform for tangible user interfaces. *Proceedings of the SIGCHI Conference on Human Factors in Computing Systems* (pp. 253–260). ACM Press.

Patten, J., Recht, B., & Ishii, H. (2002). Audiopad: A tag-based interface for musical performance. In *Proceedings of the 2002 Conference on New Interfaces for Musical Expression* (Dublin, Ireland,

May 24–26, 2002). E. Brazil, Ed. New Interfaces for Musical Expression. National University of Singapore, Singapore, 1–6.

Piper, B., Ratti, C., & Ishii, H. (2002). Illuminating clay: a 3-D tangible interface for landscape analysis, *Proceedings of the SIGCHI Conference on Human Factors in Computing Systems: Changing our world, changing ourselves* (pp. 355–362). ACM Press.

Raffle, H. S., Parkes, A. J., & Ishii, H. (2004). Topobo: a constructive assembly system with kinetic memory. *Proceedings of the SIGCHI Conference on Human Factors in Computing Systems* (pp. 647–654). ACM Press.

Ratti, C., Wang, Y., Piper, B., Ishii, H., & Biderman, A. (2004). PHOXEL-SPACE: an interface for exploring volumetric data with physical voxels. *Proceedings of the Conference on Designing Interactive Systems: Processes, Practices, Methods, and Techniques* (pp. 289–296). ACM Press.

Rauterberg, M., Fjeld, M., Krueger, H., Bichsel, M., Leonhardt, U., & Meier, M. (1998). BUILD-IT: a planning tool for construction and design. *CHI '98 Conference Summary on Human Factors in Computing Systems* (pp. 177–178). ACM Press.

Rekimoto, J., Ullmer, B., & Oba, H. (2001). DataTiles: a modular platform for mixed physical and graphical interactions. *Proceedings of the SIGCHI Conference on Human Factors in Computing Systems* (pp. 269–276). ACM Press.

Resnick, M., Martin, F., Berg, R., Borovoy, R., Colella, V., Kramer, K., & Silverman, B. (1998). Digital manipulatives: new toys to think with. *Proceedings of the SIGCHI Conference on Human Factors in Computing Systems* (pp. 281–287). ACM Press/Addison-Wesley Publishing Co.

Ryokai, K., Marti, S., & Ishii, H. (2004). I/O brush: drawing with everyday objects as ink. *Proceedings of the SIGCHI Conference on Human Factors in Computing Systems* (pp. 303–310). ACM Press.

Smith, D. (1982). Designing the Star User Interface. *Byte*, 242–282.

Stifelman, L. J. (1996). Augmenting real-world objects: a paper-based audio notebook. Conference companion on Human Factors in Computing Systems: Common ground (pp. 199–200). ACM Press.

Suzuki, H., & Kato, H. (1993). AlgoBlock: A tangible programming language—a tool for collaborative learning. *The 4th European Logo Conference* (pp. 297–303).

Ullmer, B., & Ishii, H. (1997). The metaDESK: Models and Prototypes for Tangible User Interfaces. *Symposium on User Interface Software and Technology (UIST '97)* (pp. 223-232). ACM Press.

Ullmer, B., & Ishii, H. (2000). Emerging frameworks for tangible user interfaces. *IBM Systems Journal, 39*(3–4), 915–931.

Ullmer, B., Ishii, H., & Glas, D. (1998). mediaBlocks: physical containers, transports, and controls for online media. *Proceedings of the 25th annual conference on Computer graphics and interactive techniques* (pp. 379–386). ACM Press.

Ullmer, B., Ishii, H., & Jacob, R. J. K. (2003). Tangible Query Interfaces: Physically Constrained Tokens for Manipulating Database Queries. INTERACT 2003 Conference, IFIP.

Ullmer, B., Ishii, H., & Jacob, R. J. K. (2005). *Token+constraint systems for tangible interaction with digital information, 12*, 81–118.

Underkoffler, J., & Ishii, H. (1998). Illuminating Light: An Optical Design Tool with a Luminous-Tangible Interface. *Conference on Human Factors in Computing Systems (CHI '98)* (pp. 542–549). ACM Press/Addison-Wesley Publishing Co.

Underkoffler, J., & Ishii, H. (1999). Urp: a luminous-tangible workbench for urban planning and design. *Proceedings of the SIGCHI Conference on Human Factors in Computing Systems: the CHI is the limit* (pp. 386–393). ACM Press.

Underkoffler, J., Ullmer, B., & Ishii, H. (1999). Emancipated pixels: real-world graphics in the luminous room. *Proceedings of the 26th annual Conference on Computer Graphics and Interactive Techniques* (pp. 385–392). ACM Press/Addison-Wesley Publishing Co.

Weiser, M. (1991). The computer for the 21st Century. *Scientific American, 265*(3), 94–104.

Wellner, P. (1993). Interacting with Paper on the DigitalDesk. *Communications of the ACM, 36*(7), 87–96.

Zuckerman, O., Arida, S., & Resnick, M. (2005). Extending tangible interfaces for education: digital montessori-inspired manipulatives. *Proceedings of the SIGCHI Conference on Human Factors in Computing Systems* (pp. 859–868). ACM Press.

Zuckerman, O., & Resnick, M. (2004). Hands-on modeling and simulation of systems. *Proceedings of the 2004 conference on Interaction Design and Children: Building a community* (pp. 157–158). ACM Press.

·9·

ACHIEVING PSYCHOLOGICAL SIMPLICITY: MEASURES AND METHODS TO REDUCE COGNITIVE COMPLEXITY

John C. Thomas and John T. Richards
IBM T. J. Watson Research Center

Scope and Structure of this Chapter **162**
   Why Study Psychological Complexity? 162
   The Nature of Psychological Complexity
   as a Variable 162
Complexity and Related Concepts **165**
   Relationship of Complexity and Ease of Use 165
   Relationship of Simplicity and Complexity 165
   Relationship of Complexity and Uncertainty 166
   Relationship of Complexity to Number 166
   Relationship of Complexity to Nonlinearity 167
   Relationship of Complexity to Distribution 167
   Relationship of Complexity to Nature of Elements 167
   Relationship of Complexity to Naming Scheme 167
   Relationship of Complexity to Obscurity 168
   Relationship of Complexity to Structural Framework 168
   Complexity of System vs. Task Complexity 168
   Complexity of System vs. Contextual Complexity 169
   Complexity, Feedback, and Interactivity 169
Four Sources of Difficulty for HCI **170**
   Understanding the Syntax and Semantics
   of Communicating with the Computer 170
   Making Tacit Knowledge Explicit 170
   Making the Computation Efficient and Effective 170
   Making the System Understandable
   and Maintainable 170

**Sources of Complexity in the
Development Process** **171**
   Radical Iteration in the Field 171
   Problem Finding 171
   Problem Formulation and Requirements 171
   Design 172
   Development 172
   Testing 173
   Deployment 173
   Service 173
   Maintenance 174
Ways to Measure Complexity **174**
   A Priori Mathematical Models 174
   Linear Regression 174
   Subjective Measures 174
   Textual Analysis of Documentation 175
   Iterative Design and Testing 175
Possible Future Approaches **175**
**Complexity Reduction in Practice:
Case Studies** **175**
   Case Study 1: Query By Example 175
   Case Study 2: Web Accessibility Technology 176
Conclusions **177**
References **177**

SCOPE AND STRUCTURE OF THIS CHAPTER

In this chapter, we explain why psychological complexity is (or should be) of interest to the designers of human computer systems. We then distinguish between intrinsic complexity and undue complexity. We presume that undue complexity is generally (but not universally) counterproductive in that it leads to more errors, frustration, and greater task completion times. Generally, these are all things to be avoided in a work-oriented context. However, in a more aesthetic, recreational, or pleasure-oriented context, increased psychological complexity can often be desirable. We examine the sources of undue complexity. We hypothesize that undue complexity can arise intentionally, through incompetence, or (most commonly) as an unintended side effect of normal socialization processes.

We differentiate complexity from many related concepts such as uncertainty, obscurity, and difficulty. Then, we review various approaches to measuring complexity. We suggest where and how complexity may be introduced during the overall development of human-computer systems and several approaches that may help minimize undue complexity. Finally, we speculate on some possible future developments in the field of psychological complexity and briefly discuss two case studies.

Why Study Psychological Complexity?

Although mathematicians have treated complexity as a topic for a long time, more recently this interest has spread into many fields (e.g., Holland, 1995; Bar-Yam, 1997, 2000). The attempts to understand fields as diverse as economics, ecology, biology, and machine learning, among others, relate to similar mathematical treatments, often under the general rubric of Complex Adaptive Systems. Although there are still many unsolved problems in this field, one might raise the issue of why there needs to be a separate inquiry into the nature of psychological complexity. There are two basic reasons why psychological complexity deserves a separate treatment. First, what may be thought of as objectively complex may or may not be psychologically complex. Indeed, an exploration of these differences forms a major part of this chapter. Second, when we consider complexity in the more specific context of human-computer interaction (HCI), it is useful to differentiate intrinsic complexity from undue or gratuitous complexity. Some tasks are inherently complex. We may help people perform these tasks via work aids, education, documentation, rule-based systems, or the clever design of work groups, but some considerable intrinsic complexity may remain. In contrast, although regrettable, it still seems to be a fact that many systems, applications, and artifacts are unnecessarily complex. Poor design, for instance, burdens users with complexity beyond what is required by the nature of the task. In this chapter, we will explore sources of this undue complexity, as well as ways to prevent or mitigate it. While this chapter will reference some of the relevant literature on the more general topics of complexity and psychological complexity, the focus of the chapter is on how we can use these concepts to improve HCIs, in most cases by reducing undue complexity.

In terms of the underlying cause, there seem to be three main reasons for undue complexity. First, undue complexity sneaks into systems even though people consciously try to prevent it. This can happen for numerous reasons, which we will explore in detail below (e.g., lack of appropriate methodology). Second, seemingly undue complexity can be injected into systems intentionally. For instance, systems commonly enforce rules governing both password structure and password lifetime. These are often perceived as an annoyance by end users but are intentional policy decisions aimed at improving overall system security. A third and more subtle reason for undue complexity relates to ordinary social processes. As a group of people work together or live together, they develop and evolve ways of referring to things that become convenient shorthands for the in-group but become increasingly obscure or difficult for the out-group. In the extreme, this results in different accents and dialects and eventually distinctly different natural languages, customs, and assumptions. Disparate groups of people end up not simply using different terms, but thinking about the world differently (Abley, 2005). This process is typically, though not exclusively, unconscious. This type of socialization begins early; experiments indicate that children as young as several weeks old are already less capable than they were at birth of making auditory distinctions within the phonemic categories deemed equivalent by their linguistic community. In one experiment, as children living in an English-speaking environment aged from 6 months to 12 months, their percentage of discriminations of a non-English distinction decreased from 80% correct to 10% (Werker & Tees, 1984). Categorical speech perception is only one dramatic form of a process that is happening all the time and at many levels in normal social interaction.

The Nature of Psychological Complexity as a Variable

Why are people interested in psychological complexity? It is an important intervening variable that simplifies analysis and prediction. On one hand, a number of factors, explored in detail in this chapter, impact psychological complexity, and on the other hand, a number of dependent variables depend on psychological complexity.

One might yet question the utility of complexity as a unifying variable. Although Card, Moran, and Newell (1983) talked about complex behavior (and the principle of rationality to help explain how to build models of complex behaviors from simple ones), they do not treat complexity as a variable, per se. Despite this, under a wide variety of conditions, accurate predictions of complex human behavior are possible (see also, John, 1990; Gray, John, Stuart, Lawrence, & Atwood, 1990). However, modeling at this level of detail requires a substantial amount of work. The hope for complexity is that one might develop predictive measures of complexity that could be applied easily without special expertise that would still be highly predictive of human behavior. In the ideal case, for instance, a development team might calculate the complexity of several alternative design ideas for a human-computer system and make at least some of the many necessary design decisions without the need for assessing each one with costly behavioral observations.

We expect intuitively that increased psychological complexity is positively correlated with increased errors, increased time to complete a task, and decreased productivity, as well as increased frustration in a results-oriented work environment. (Even in such work contexts, there may be exceptions; e.g., when lengthy vigilance at an overly simple task may actually produce more errors than a somewhat more complex one.) Nonetheless, the overall presumption, both in the public mind and in the scientific literature, is that unnecessary complexity is to be generally avoided on the grounds of reducing errors and improving productivity. The cost of complexity can be considerable. For example, half of the consumer products returned as malfunctioning are actually functioning as designed; they are just too complex for the customer to use (den Ouden, 2006). A well-meaning attempt to improve a homeland security system resulted in more errors, longer decision times, and a decreased probability that people would even use the system (Coskun & Grabowski, 2005). Although details vary, Fig. 9.1 shows the general relationship of complexity to performance and frustration.

However, the relationship of psychological complexity to more ludic (pleasure-oriented) and aesthetic variables is more complex. In general, the prevailing wisdom is that a moderate amount of complexity is aesthetically pleasing. Stimuli, responses, mappings between stimuli and responses or contexts that are either too complex or too simple are not as interesting, pleasurable, or engaging as those that are moderately so. This seems to be a robust phenomenon applying to infants as well as adult humans and to various species. Of course, it is still the case that what constitutes moderate complexity for an individual depends upon previous knowledge as well as personality predilections for more or less complexity. In any case, psychological complexity is potentially a powerful intervening variable that, on the one hand, promises to collapse the impact of numerous separate independent variables into one (e.g., complexity) and, on the other hand, to expand the implications of complexity onto a number of dependent variables. In the case of performance-related variables, the relationship is thought to be monotonic; in the case of subjective variables, the relationship is thought generally to be an inverted U-function, although it may vary according to the sense and the individual.

In the last few decades, most HCI work has examined work contexts in which productivity, in the broadest sense, has been

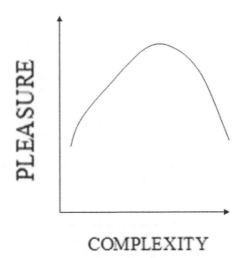

FIGURE 9.2. The typical relationship between complexity of a stimulus and the associated aesthetic pleasure in ludic situations.

a primary focus. As computing technology has become more intertwined with home life and entertainment, the importance of also using psychological complexity as a predictor of pleasure and preference is increasing.

In the domain of things that are meant to taste good, for instance, increasing complexity often seems to be presumed a good thing. For example, Vintage Cocktails (2007) created a subjective five-point scale for cocktails: simple, delicious but not daunting, sophisticated and smart, provocative and profound, and unusual and unforgettable. A blog of commentary on Macallan Elegancia scotch also indicates that complexity is presumed to be a good thing. For instance:

The nose is very good balances, with pleasant sweet nutty and vanilla hints. The elegant balance does not mean any lack at complexity, with discrete wood and cut flower fragrances. In the palate, a clear nut taste with some slightly acid hints (citrus fruits) on a background of slightly bitter malt. A spicy and relatively long finish. (The Macallan: Speyside distillery (2007)

On the other hand, it is generally thought that visual and sonic patterns of moderate complexity are more often preferred. In the domains of art and architecture, some attempts have been made to quantify what makes for interesting or aesthetically pleasing patterns. For example, Christopher Alexander's (2002) work included an examination of and ordering of visual binary patterns, and Klinger and Salingaros (2000) suggested a metric for measuring the visual complexity of simple arrays. Other recent approaches attempt to relate preferences to fractal numbers (Taylor, Spehar, Wise, Clifford, Newell, Hagerhall, et al., 2005). Other investigators have tried to predict musical complexity based on objective factors (e.g., Shmulevich & Povel, 2000; Shmulevich, Yli-Harja, Coyle, Povel, & Lemstrom, 2001).

In more work-oriented contexts, psychological complexity is clearly related to other intervening variables such as stress, difficulty, and workload. However, there are distinctions among these. Stress depends not just on the complexity of the problem

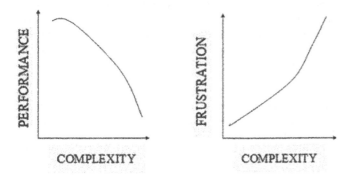

FIGURE 9.1. The general relationship between complexity and performance, and complexity and frustration.

situation itself but also on external factors such as the pay-off matrix and internal factors like neuroticism. For instance, a specific puzzle might be thought to have a certain level of complexity. The amount of stress that one experiences in attempting to solve it, however, would depend upon the perceived rewards, punishments, and time pressure as well as the individual's internalized habits for viewing how various outcomes relate to overall life goals; for example, proclivities for awfulizing (Ellis, 2001) tend to increase stress far more than what might be objectively justified.

The more complex the task, the more difficult we would expect it to be; other things being equal. However, difficulty can also accrue to a task for a number of other reasons including environmental stressors such as extreme heat or cold, vibration, noise, or the necessity of applying large forces.

The term *workload* is more commonly applied to an individual or team across tasks. One can imagine an individual with a large number of tasks, each of which is simple, but whose job includes many interruptions and context switches from one simple task to another. In such a case, the workload may be high even though any particular task may be simple.

There is another sense in which the term *psychological complexity* is sometimes used. Social critics may use the term to refer not to a single task or system but to the totality of life experiences. Modern life in the information age is sometimes deemed too complex in its totality. This increased complexity may apply to politics, personal relationships, child rearing, transportation, healthcare, food choices, and so on, extending potentially to every aspect of modern life. Even providing for our entertainment needs has become a complex endeavor (Grinter & Edwards, 2005). In addition to the notion that each separate aspect of life is becoming more complex, people often find themselves multitasking and task switching (e.g., Gonzalez & Mark, 2005). While this is a potentially interesting avenue to explore in its own right, it is beyond the scope of this chapter, which instead will focus on psychological complexity as it applies to individual systems or tasks that have information technology aspects.

Early attempts to measure task difficulty with objective metrics include Attneave's (1957) figure complexity. There have

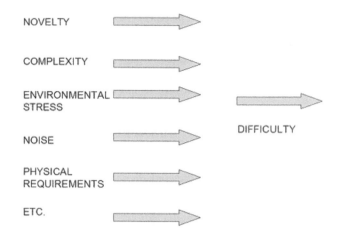

FIGURE 9.4. A number of variables, including complexity, can add to task difficulty.

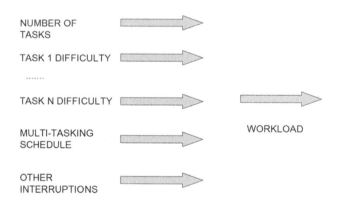

FIGURE 9.5. Overall workload is influenced by the difficulty of various tasks, the number of tasks, the scheduling of the tasks and other factors.

been attempts to provide measures for the meaningfulness of trigram nonsense stimuli, responses, and the mapping between the stimulus and response in the verbal learning literature. Others have attempted to extend this work to real words; English words have been rated for frequency of occurrence (Kucera & Francis, 1967; Brown, G. D. A., 1984), concreteness (Paivio, Yuille, & Madigan, 1968), and so on. In one case, (e.g., Rubin, 1980), 51 dimensions of words were measured. Such measures do provide predictability in a wide range of laboratory tasks. For instance, the log of word frequency is closely related to decreased latency in a simple naming task where subjects are shown pictures and asked to name the pictured object as quickly as possible (Thomas, Fozard, & Waugh, 1977). Such empirical relationships as these, however, are problematic in terms of applicability to HCI. For one thing, when measures are taken on real words in English, many dimensions tend to be correlated. Carroll and White (1973) used a linear regression model and showed that age of acquisition is actually a more powerful predictor than frequency of occurrence (though the two are highly correlated). Neural net models of learning also demonstrate this effect (Smith, Cottrell, & Anderson, 2001). How might one apply such findings to HCI? For example, given a choice

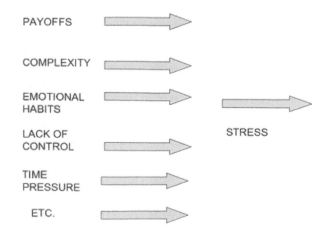

FIGURE 9.3. A number of variables, including complexity, can increase stress.

between two otherwise equally appropriate words, it is generally better to use the word that is more frequent in the language, or even better, one that is learned early in life. However, the relevance of this heuristic to the design of instructions, web pages, error messages, and so on, is very limited because in real design situations, we seldom have a choice between two otherwise equally appropriate words. More frequent words also often have the property of being more ambiguous, both semantically and syntactically (e.g., can, run), than longer, less frequent words. Words learned early in life often have little relevance to most tasks that adult users engage in. Knowledge of the particular users, their vocabularies, the task demands, conventions, contexts, and so on, dominates more general properties of the language in suggesting the best wording.

COMPLEXITY AND RELATED CONCEPTS

In this section, we will attempt to further define psychological complexity by distinguishing it from a number of similar concepts.

Relationship of Complexity and Ease of Use

It would seem that ease of use and simplicity constitute closely related concepts. However, they do differ in several ways. First, ease of use implies an empirical orientation. There are a number of interesting issues involved in measuring ease of use, such as choosing the sets of reference users, choosing representative tasks, and choosing the dependent variables to be measured (time to complete, errors, quality of result, etc.). In principle, ease of use can be measured objectively in terms of human behavior. Complexity, on the other hand, does not have the same degree of conceptual consensus. There are approaches that focus on the formal, intrinsic properties of a system or stimulus (e.g., Halstead, 1977; Alexander, 2002). Other approaches focus on the reactions of a human being to the system or stimulus. These can include passive measures, such as pupil dilation, staring time, or heart rate deceleration. They can also include more subjective judgments or attempts to measure various aspects of task-oriented behavior; for example, the number of steps in a process or the number of variables that must be remembered (e.g., Brown, A. B. & Hellerstein, 2004; Brown, A. B., Keller, & Hellerstein, 2005).

Nonetheless, other things being equal, one would expect something that is less complex to be easier to use. However, other things are typically not equal. Ease of use often depends heavily on people's expectations that may result from cultural conventions; often varying widely with time and place. Today, a new application that uses the typical GUI widgets, such as pull-down menus, may be easy to use for experienced users, primarily because people have a fair amount of positive transfer from earlier, similar experiences. A unique and unfamiliar but less complex interface may not prove as easy to use.

In addition, a design that maximizes visual simplicity through symmetry and minimalism may actually increase the difficulty of use. For example, a shower arrangement with one, clearly marked handle for hot water and another for cold water, combined with a two-position lever for shower and bath is more complex visually than a single, unlabeled knob with several degrees of freedom. At least initially, however, the former will be easier to use. Similarly, a normal bottle opener is not particularly elegant or simple in shape. However, its asymmetries, as well as its commonality, make it easy for people (in our culture) to use. An alternative bottle opener (See Fig. 9.6) consists of a metal cone with a thin protruding knob. This device is physically simpler, but harder to use, at least initially.

Relationship of Simplicity and Complexity

Simplicity and complexity seem to be opposites. However, there is an interesting asymmetry in their opposition. It is as though absolute simplicity is the center point of an n-dimensional sphere, while absolute complexity is the surface. In other words, there are many ways of moving away from simplicity toward complexity. Figure 9.7 illustrates just a few of the many ways that a system can become more complicated. For simplicity, we illustrate a two-dimensional rather than n-dimensional space.

FIGURE 9.6. A "simple" but highly non-intuitive bottle opener.

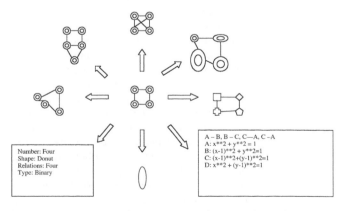

FIGURE 9.7. A "simple" system can be complicated in many different ways. (Here are just a few.)

This is more than an idle philosophical observation. In the real-world iterative cycle of design and behavioral observation, it is relatively difficult to move inward toward greater simplicity while improving ease of use. It is relatively easy to move along the surface of a sphere, resolving some types of complexity while simultaneously introducing others and therefore not increasing either overall simplicity or ease of use. For example, in designing a word-processing system, you might find that people are having difficulty finding desired items in long pull-down menus. You decide to make each menu simpler by changing the pull-down options depending on whether one clicks or clicks with the shift key depressed. This makes each menu simpler, but introduces another kind of complexity. Much of the art and the importance of experience in HCI can be conceived of as being able to move in from the surface rather than around the surface. (For a related discussion with respect to simplicity and accessability, see Lewis, in press.)

Relationship of Complexity and Uncertainty

Increased complexity is associated with increased uncertainty, *ceteris paribus*. The term *complexity* can be reasonably applied to stimuli, responses, or to the mapping between them. The concept of uncertainty, however, seems to apply only to mapping. One can have uncertainty about what a stimulus is, or uncertainty about what response to make, but fundamentally, it has to do with what action is appropriate, given the current circumstances.

Complexity, as mentioned earlier, can be thought of either as something objectively measurable or as something subjective. The term *uncertainty* also has an objective meaning. If one turns over a randomly chosen card from an ordinary deck, there is more objective uncertainty about the outcome than if one flips a coin. The term *uncertainty* may also be applied in a purely subjective way. In this subjective sense, a given person may be completely certain that a randomly chosen card will be the ace of spades (because of nothing more than a compelling intuition) and completely uncertain about the outcome of a coin toss. The concept of subjective uncertainty has even further shades of application. One may watch the movie *Apollo 13* on numerous occasions. At one level, the outcome is already predefined and completely known. There is neither objective nor subjective uncertainty. Yet, within the inner context of watching the story play out, the viewers allow themselves (each time) to feel the subjective uncertainty of outcome.

Subjective uncertainty in HCI can apply at numerous levels. For example, in order to make touch typing a felicitous interaction, one subjectively presumes that each keystroke will be correctly transmitted to the computer and that what appears on the screen is an accurate mapping of what is actually being stored inside the computer. Objective certainty about the states of electromechanical systems probably never reaches 100%. However, as a strategy for partitioning effort, it is often useful to ignore very small error probabilities in subjective judgments. The first time the word-processing software changed Thomas's (this chapter) *the* to *The*, certainty changed. Thomas began to wonder what else the application might be doing. Later, a much worse problem surfaced in the form of an automatic update to

styles. In response to something typed on page 1, something might change appearance on page 15. Subjective uncertainty also bears some resemblance to the concept of trust. You might have 1,000 interactions with someone, each of which provides evidence of trustworthiness, yet even a single interaction that shows a person to be untrustworthy may color a relationship for years to come. Similarly, if an interaction between a user and the computer system violates an assumed behavioral norm by doing something completely unexpected, the level of subjective uncertainty created may be much greater than warranted in any objective sense. Moreover, if the user's experience includes such a violation, the scope of subjective uncertainty may be difficult to predict. One user may simply generalize uncertainty to a specific function; another may come to distrust a specific application (have a high degree of uncertainty) while a third may come to distrust computers in general.

Relationship of Complexity to Number

One would expect that if systems *A* and *B* are equivalent in other respects, but system *A* is more extensive than system *B* in terms of the number of items, that system *A* is more complex than system *B*. The degree to which an increase in number produces an increase in complexity, however, depends greatly on the structure of the systems. Suppose that systems *A* and *B* consist of 5 and 10 buttons respectively, combined with an OK light. The user has to use trial and error to discover the correct button to push to cause the OK light to light. On average, it will take 2.5 trials to find the right button for system *A* and 5 trials to find the right button for system *B*. Here, doubling the number of choices doubles the average number of trials. On the other hand, imagine that the user must find a sequence of 5 button pushes for system *A* and 10 button pushes for system *B*. When the user hits each correct button, the OK light comes on. On average, it will take the user $2.5 + 2 + 1.5 + 1 = 7$ attempts to find the right sequence for system *A* and $4.5 + 4 + 3.5 + 3 + 2.5 + 2 + 1.5 + 1 = 21$ attempts to find the right sequence for system *B*. Here a doubling in the number of buttons results in three times the mean number of trials. For a final case, imagine that the OK light comes on only after all buttons are pushed in the correct sequence. There are 5! possible sequences for system *A* and the user will find the correct sequence on average in 5!/2 trials or 60 trials. There are 10! possible sequences for system *B* and the user will find the correct sequence, on average, in 1,814,400 trials. Here a doubling in number has led to an increase in trials by a factor of 30,240. In practical terms, system *A*, though cumbersome, could be solved by hand in a few minutes. System *B* would probably extend beyond anyone's patience. A quantitative difference in complexity actually would result in the qualitative difference between something doable and something not doable. These examples illustrate that the impact of number on complexity greatly depends on the nature of the system. In particular, hidden states are usually a bad thing in terms of complexity, but just how bad they become rises quickly with increasing number.

Another example has to do with working memory and relates to George Miller's (1956) famous paper "The magic number seven plus or minus two." Though the size of working memory

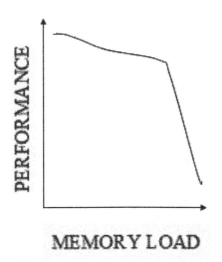

FIGURE 9.8. How memory load can have a minor effect on performance, up to a point, after which increasing memory load still more yields a sharp drop in performance.

varies somewhat depending on the number of dimensions and the type of material, it is not large. A system that allows a person to operate within that span can be nearly error free, while a system that forces a person to operate beyond that span will be extremely error prone. A modest increase in memory load can thus result in a very large increase in time and errors.

Relationship of Complexity to Nonlinearity

One of the defining characteristics of complex adaptive systems is that they exhibit nonlinearity. From the perspective of survival, this nonlinearity is a good thing. From the perspective of someone trying to interact with a computer system, nonlinearity can be a major source of difficulty in understanding, predicting, and controlling the behavior of the system. For example, Thomas (this chapter) once had to debug a PDP-8 program that was intended to collect reaction time data. There was only a single bit wrong in the entire program (which probably had roughly 5,000 bits total). In a linear system, you would expect that a .02% error would result in a small behavioral problem; for example, the reaction time would be off by a constant .02%. Of course, nothing of the kind happened. The program did not collect reaction times at all. Instead, it eventually caused the interpreter to crash. The bit that was wrongly set was the indirect bit so that instead of storing the reaction time, the program used the reaction time as an address in which to store the contents of a register. Furthermore, since every time the author tried to test the program, the reaction time varied slightly, the precise behavioral path of the program was different every single time. This example is not unusual, but it illustrates the property that computer systems are typically highly nonlinear in their behaviors. Generally, higher layers of software are designed to make the system, as seen by the user, as being more nearly linear (and therefore, more predictable). However, even in commonplace high-level applications, many nonlinear aspects typically remain.

Relationship of Complexity to Distribution

Halstead (1977) claimed that complexity C is equal to a constant k times the sum of the number of unique operators U times the natural log of the total number of operators u plus the number of unique operands O multiplied by the natural log of the total number of operands o.

$$C = k * (U + \ln u + O + \ln o)$$

This is not an unreasonable approximation. We have already seen how number can increase complexity. We could also consider, however, how these operators (or operands) are distributed throughout a program. Unlike Halstead (1977), we speculate that (a) AAAAAAAABBBBBBBBCCCCCCCC is easier to comprehend, use, edit, and so on, than (b) AABBCAABCBBCCAABCAACCCB.

It has been known for some time that repeated operations are typically faster than alternating ones (e.g., Fozard, Thomas, & Waugh, 1976). There can be exceptions to this general rule based on rhythm, fatigue, or satiation, especially for extremely simple tasks. However, for complex tasks, we generally expect a distribution such as (a) above to be much easier than (b). For example, in the Stroop task, one can be asked either to name the colors of ink (while ignoring the word) or to read the names of colors (while ignoring the color of ink). Alternating between the two every line is much slower than doing either task *en masse* (e.g., Philips, Bull, Adams, & Fraser, 2002). In the Brown et al. (2005) model, this source of complexity is modeled as context shifts.

Relationship of Complexity to Nature of Elements

Given the same number and structure of elements, two systems that are formally isomorphic may be quite different in terms of psychological complexity based on differences in human biology, learning, or both. For example, it seems clear that the human visual system is especially well tuned to handle the complexity of human faces. For example, Johnson and Morton (1991) presented evidence that babies as young as three hours after birth can recognize human faces. Chernoff (1973) suggested that this ability might be used to represent underlying complexity in a way that is easier for humans to deal with. In the auditory domain, it is clear that the difficulty of dealing with (recalling, rearranging, etc.) a string of phonemes that follow the phonological rules of one's native language is much less than dealing with an equal length string that violates those rules.

Relationship of Complexity to Naming Scheme

Consider two programs that manifest identical structures. The program asks the user to input an integer from 1 to 100. For each integer, the program then displays a corresponding unique horoscope. Structurally, this is a very simple program. However, imagine that in one version, the user's input number is called, "User's Input Number" and in another it is called "NXBNM." Further, suppose the horoscopes in the first program are labeled "horoscope1," "horoscope2," "horoscope3," and so on for each corresponding integer. In the second program, they are each labeled with a random two-letter string such as "HA," "IB," "JK," and so on. It is hopefully clear that these two programs, though formally identical, would be vastly different in how easy they would be to understand, modify, debug, and so on.

While the above example is somewhat contrived, developing a consistent and intuitive naming scheme is a real-world problem. Often, when a programmer (or any other kind of designer) begins to name things, they do not have a complete understanding of the space of things they will have to name. For instance, when Thomas (this chapter) began working on a recent project, he named a file folder "ELFL" for "Electronic Learning Flow Language." As the project grew, one folder was much too generic for all the nuances that emerged. Furthermore, the focus of the project changed, and no one even talked about Electronic Learning Flow Language any more. Furnas, Gomez, Landauer, and Dumain (1982) found that not only are two different people likely to come up with different names for things, but that even the same person over time is unlikely to spontaneously come up with the same name. They suggest tables of synonyms to help alleviate this problem. Real computer systems are often rife with names that are difficult to comprehend or to recall precisely.

Relationship of Complexity to Obscurity

Consider the following two sequences of binary digits.

A: 0000 0001 0010 0011 0100 0101 0110 0111 1000
B: 1000 0101 0100 0001 0111 0110 0011 0010 0000

These two sequences each contain the same number of binary digits and the same number of groupings. The first sequence is essentially a counting sequence from zero to eight. This is an obvious sequence if one knows how to count in binary. The second sequence would presumably be much harder to memorize for most people. However, it also represents the numbers zero to eight; however, they are arranged in alphabetical order according to the English names. Once one realizes this rule, memorizing or reconstructing the second sequence becomes much easier. There is a sense in which the first sequence strikes us as a natural ordering because it is based on what we can see right before us. Numbers are likely to be highly associated with counting, after all. The rule that orders the second sequence strikes us as more obscure, however, since it depends on some other representation (the English names of the numbers) that is not naturally and strongly associated with the numbers *qua* numbers. This notion of obscurity is closely associated with the contrary concept of affordance (Norman, 1988). If the perceptible properties of an object or system immediately make it clear what it can do and how to do it we can say the object or system has good affordances. Of course, whether something appears obscure or not depends heavily on the user's previous experience, both cultural and personal, and may include inborn factors as well. In Fig. 9.2, the equations version of the simple diagram in the middle illustrates obscurity.

Relationship of Complexity to Structural Framework

One of the interesting aspects of perceived complexity is that when it comes to human beings, sometimes more is less. For example, most English readers are capable of distinguishing *WORD* from *CORD* at a briefer presentation time than they can distinguish *W* from *C* (Reicher, 1969; Hildebrandt, Caplan, Sokol, & Torreano, 1995). Learning a set of paired associates, presented along with mnemonic visual images, is much easier than learning the set of paired associates alone (Thomas & Ruben, 1973). Subjects presented with a set of sentences and a context-setting theme or picture have a much easier time learning and remembering a story than those who do not have the picture (Bransford & Johnson, 1972). Recalling a story that fits with our cultural patterns is much easier than recalling one that does not (Bartlett, 1932). In all of the above cases, the advantage of the structural framework is that it relates new items to something that already exists in memory.

In other cases, however, the advantage of structure seems to have more to do with the underlying nature of the nervous system. For example, in attempts to develop useful and novel representations of speech signals in order to improve synthesis, one technique developed by Pickover (1985) mapped an autocorrelation function into polar coordinates scaled to fit within a 30 degree angle. The resulting dot patterns (see Figure 9.9) proved singularly difficult to interpret in any useful way. However, when the pattern was reflected to include a mirror image and the resulting 60 degree angle repeated six times around a central point to produce a snowflake-like pattern; the representation easily differentiated various vowel sounds as well as some subtleties, like anticipatory nasalization, that were virtually invisible in the traditional sonogram. Of interest in this context is the fact that the useful representation contained no more information than did the useless one. Arguably, one could say the speech-flake pattern was more complex. However, the complexity apparently allowed visual symmetry detectors to come into play and produced far more distinguishable and memorable patterns. These Symmetricized Dot Patterns have since been applied to numerous other domains as diverse as cardiac signals, mechanical stress, and even the crunchiness of cereals.

Complexity of System vs. Task Complexity

It is easy to fall into the trap of conflating the complexity of an interactive computer system with the complexity of the task that one is attempting to perform while using that system. In fact, many studies in the HCI literature compare two or more versions of systems by having subjects perform a small number of tasks in the two systems. Often these tasks are treated as fixed factors when they are actually a small sample (probably not randomly chosen) of all possible tasks. In principle, it is very difficult to generalize about the superiority of one system over another based on task performance. One cannot really know that a different set of tasks may not have completely obliterated or even reversed any observed trend.

As a general heuristic, one might suppose that the complexity of the system should reflect the complexity of the necessary tasks to be supported. That is, if the user only needs to

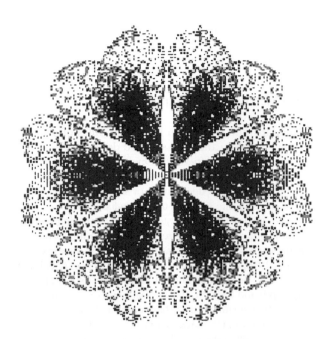

FIGURE 9.9. A symmetrized dot pattern.

perform very simple tasks, it may be enough to provide a simple system. However, if the user needs to perform very complex tasks, then a more complex tool may be required. In general, there may be some truth to this heuristic, but it must be applied with care. One caveat is that the function associated with this increased system complexity must actually be useful and allow the user to focus attention on the task. Otherwise, additional complexity in a system may actually be more disruptive when the task is complex than when it is simple. For example, consider the tasks of writing a two-page trip report and writing a full-length novel. Writing the novel (we assume) is a much more complex task. It may be that the various functions, fonts, formatting, and options in a complex word processor are unnecessary and even more distracting for someone trying to keep in mind a complex set of subplots and characters than for someone writing a short trip report.

Another important distinction is between the underlying complexity of a system and the complexity that is surfaced to the user. A huge amount of research and technological sophistication may be involved in the development of an automatic transmission but from the driver's perspective, the transmission is simple. From the perspective of the user, modern search engines provide a simple, commonplace interface. Most have no notion of the complexity of the underlying processes. Of course, attempts to do the user favors by providing hidden intelligence may sometimes backfire.

The ideal case is probably to provide a layered interface in which the system reveals only the complexity needed for the task. Early examples of this approach include the concept of "training wheels" (Carroll & Carrithers, 1984). The interfaces for another related set of projects, the Speech Filing System, the Audio Distribution System, and the Olympic Message System (Gould & Boies, 1984), also provided layered interfaces. However, in many real systems, it often happens that the user inadvertently falls through the user interface into a deeper layer or

must do so to accomplish their real task. In addition, undue complexity in the underlying system can indirectly hurt the user experience by making code harder to test, debug, document, and maintain. Therefore, we consider undue complexity as something to be generally avoided both internally and in terms of what is to be made visible to the end user.

Complexity of System vs. Contextual Complexity

A person may be using a system that in and of itself is simple, such as a bicycle, and their task may be simple, such as riding in a straight line without falling or crashing. However, riding that bicycle in a straight line on an empty, dedicated bike path through a park might be quite different from riding on a busy city street. It is somewhat of a judgment call to draw a line between task and context. In this example, one might just as well say that the two tasks are different. However, collapsing all such variations into the task complexity is probably counterproductive because the types of actions that can be taken and the power to take those actions are typically quite different in the case of simplifying the task versus simplifying the context.

Contextual complexity could increase overall complexity by providing secondary tasks or by bombarding the person with extraneous stimuli. Contextual complexity might also produce internal distracting material. Imagine two students using a word processor to complete an essay take-home exam. One of them is also worried about the outcome of some diagnostic medical tests. Consider two executives who are using presentation software to construct a sales pitch. One of them is operating in an organizational context that makes it necessary to pass the presentation through four layers of management, each with very different ideas about what makes a good sales presentation. It seems much more natural to consider such differences to be differences in contextual complexity rather than task complexity. We might expect the design of the presentation software to help the user deal with the task complexity in this example but not the contextual complexity.

Complexity, Feedback, and Interactivity

Finally, the extent to which we can easily understand, control, and predict the behavior of a system has much to do not only with how complex it is but also how readily we can interact with it and receive timely, unambiguous feedback. For example, the "dynamic query" system (Ahlberg, Williamson, & Shneiderman, 1992) does not decrease the complexity of the underlying data that one is attempting to understand, nor does it decrease the complexity of the interface by which one browses that data. In fact, the interface is actually more complex than a static browsing interface might be. However, it does allow fast, continuous, unambiguous feedback. Conversely, it is well known that delayed auditory feedback (or delayed visual feedback possible with television circuits) is highly disruptive to performance. In some settings, feedback on performance can still present complexity in the form of credit assignment. If you lose a game of chess, it is unambiguous that you lost, but determining what caused you to lose is difficult. This is not just a human problem; the "credit assignment" issue is crucial to any complex adaptive system (Holland, 1995). In golf, putting can be a very difficult skill to improve. The main reason is that it is difficult to determine which of many possible factors is responsible for success or failure in putting in a real game situation. For example, if you miss a putt to the left it might mean that you misread the slope of the green; misread the grain of the green; hit the ball off center of the putter; hit the ball with a curved arc swing; hit the ball with a blade not normal to the path; and all combinations of these factors. In the domain of HCI (and elsewhere), this lack of unambiguous feedback about which subset of actions is causing which difficulties has sometimes been referred to as "tangled problems" (e.g., Carroll & Mack, 1984). In the domain of golf, Dave Pelz (2000), an MIT physicist and ex-astronaut, has invented a series of devices to give the learner unambiguous and differentiated feedback about these various possible sources of error. Studying these devices might be a valuable exercise for system designers hoping to provide a similar untangling for users.

FOUR SOURCES OF DIFFICULTY FOR HCI

One goal related to psychological complexity is to provide a system that will allow non-programmers to do what is essentially programming; that is, allow them to get a computer system to do what they want it to, not just in terms of choosing from pre-existing options but to allow them to have more open-ended productive control. A whole succession of projects at Carnegie-Mellon University (e.g., Myers, McDaniel, Kosbie, & Marquise, 1993), the LOGO and related projects at MIT (e.g., Papert, 1993), and the long series of projects by Alan Kay and others, leading most recently to Squeak and Etoys (Kay, 2007), have attempted to reach this goal. Much of the focus of these projects was based on trying either to simplify the syntax and semantics of communicating with the computer or allowing it to take place in a manner more nearly like the way people communicate with each other. It is certainly true that the detailed and often obscure syntax of ordinary computer languages can provide a barrier to use. Weinberg (1971) pointed out, for example, that FORTRAN then had one set of rules for what constituted a valid arithmetic expression when used in an assignment statement and another, more restrictive set of rules for what constituted a valid arithmetic expression in an array index. Typically, programming languages are full of these kinds of details and, no doubt, they provide unnecessary complexity for someone who does not spend a large amount of time using such a language. However, it is important to note that such details are only one of four potential sources of difficulty people may have in communicating their desires to a computing system. For concreteness, consider a scenario in which a chess grandmaster without programming experience desires to write a computer program that plays excellent chess.

Understanding the Syntax and Semantics of Communicating with the Computer

As stated, the chess grandmaster will probably have to learn the somewhat arbitrary syntax and semantics of a programming language. Although there is still promise in techniques aimed at programming by example and natural language communication, at this point, such systems are insufficient for writing something as complex as a chess program.

Making Tacit Knowledge Explicit

Assuming that the expert can master the rules of some programming language and its associated development environment, there are other difficulties. One of these is that the chess expert, like experts in many fields, is unlikely to be consciously aware of all the knowledge that he or she brings to bear on the game of chess. While some of this knowledge is somewhat explicit, such as develop the center, protect your king, and look for double attacks, much more of it exists in the form of patterns that have developed over the course of many experiences (deGroot, 1978; Simon & Barenfeld, 1969; Simon & Gilmartin, 1973). Making this tacit or implicit knowledge explicit, for fields of any depth and breadth, will probably prove a far more difficult and time-consuming task than learning the details of a specific programming language. The methods and techniques of knowledge programming and expert systems may be of some help here along with storytelling. Typically, one may elicit additional tacit knowledge (that is, transform it from tacit to explicit) by encouraging the expert to recall specific instances, asking them to generalize and then asking for counter-examples. It can also be useful to involve pairs or small groups in exchanging stories about their experiences.

Making the Computation Efficient and Effective

Another set of issues revolves around the efficiency of what is going on under the covers. Although computers are becoming ever more powerful and cheaper, it is still easy for nonexperts to write programs that are so inefficient as to be unworkable. A chess program that theoretically makes good moves but takes years for each move is unworkable and unusable. In fact, a naïve approach to writing a chess program is simply to do an exhaustive search to all finishes and work backwards. It may or may not be obvious to someone untrained in mathematics or computer science that such an approach would be completely unworkable.

While the chess case may be extreme, far less extreme cases can still be quite problematic. If one is writing an interactive program, timing issues are important and how to achieve good timing may require a great deal of expertise and knowledge. If systems involve multiple people and/or computers, it is quite easy to introduce inconsistencies, deadlocks, thrashing, and so on, even if a nonprogrammer can master syntax and make implicit knowledge explicit. For a deeper examination of the relationship of usability and choices in the underlying architecture, see Bass and John (2003).

Making the System Understandable and Maintainable

In the ongoing stream of behavior, we make what seem to be obvious choices, such as where to put the car keys. If, a day later, we are prone to forget this obvious choice, imagine how much more difficult it can be to understand and recall the numerous decisions that must be made in designing and implementing a computer system. A nonprofessional programmer may have very little knowledge of documentation, help systems, updates, security, compatibility issues, the process for reporting and fixing bugs, and so on. Consequently, even if the first three hurdles are overcome and a workable program is produced, its half-life may be very short indeed.

To summarize, the goal of having nonprogrammers directly instruct computers in a generative way has often focused on designing a very simple and consistent syntax for a programming language. In the above section, we indicated that such a focus only addresses the first of four major obstacles end-users face: to wit, learning a complex syntax.

While the possibility of programming by nonprogrammers is intriguing and probably quite a bit more difficult than it at first appears, there is a slight variant that is becoming more commonplace. There are instances where a community of practice includes some individuals who, although perhaps not professional programmers, are quite proficient at finding and modifying code in order to fulfill certain functions. In some cases, they may be using programming-like functions in an application program such as a spreadsheet. In other cases, they may actually be using full-fledged programming languages. By modifying existing code incrementally, these semi-expert users are often able to address all four of the issues mentioned above. An interesting early example of such a community was "Moose Crossing" (Bruckman, 1997), which provided a shared environment for kids to teach each other object-oriented programming in a special simplified language. Nonetheless, there remain many applications and systems that users interact with that are developed by professional programmers working in software companies. An examination of the way in which products are developed, in turn, can provide some insight into various places where undue complexity may be injected into systems (and therefore, how it might be reduced).

SOURCES OF COMPLEXITY IN THE DEVELOPMENT PROCESS

Radical Iteration in the Field

While this topic is dealt with in more detail elsewhere in this handbook (see also Greene, Jones, Matchen, & Thomas, 2003), it is worth at least a brief mention here that radical iteration in the field can greatly help avoid, prevent, and address all of the sources of complexity that we are about to enumerate. We will discuss two examples of this approach at the end of this chapter. By working as closely as possible with the potential users of a system doing their real work in their real context, one can avoid unnecessary complexity that might otherwise be injected during problem finding, problem formulation, design, development, deployment, and maintenance.

Problem Finding

Most of our typical education focuses on solving problems that other people have already found and formulated. Great leverage

can arise from finding and formulating problems. Conversely, unnecessary complexity in a system can be injected from the very beginning by finding a problem that is not really a problem (for the users). A computer scientist, for instance, might discover that users intuitively choose a sequence of tasks to complete that is not quite as efficient as the theoretically optimal sequence. The users choose routes that are "good enough" and "satisficing" (Simon, 1962). However, the computer scientist (or accountant) might see the inefficiency as a problem that needs to be solved. The result may be a system that requires users to key in a long list of tasks to be done, resources to be used, the data required by the tasks and so on. The system that solves the (imaginary) problem could easily take far longer than the labor supposedly saved. Of course, there is the even more insidious problem that in order to model the efficiency of a process in the first place, certain simplifying assumptions must be made and these simplifications may well lead to a solution that is less nearly optimal in the real world than the original behavior. For example, in a large telecommunications company, a route optimization program was developed to dictate schedules to repair people. The program failed because it did not take into account many specific details that the repair people knew about the times and schedules and constraints of others with whom they had to coordinate.

Inadequate observation or starting with untested assumptions about what problems exist may also prevent people from even noticing easily solvable problems. On one factory tour in the early 1980s, Thomas (this chapter) was shown someone whose job was to precisely align two silver needles. The person was sitting in a very awkward position and behind the silver needles was a silver background. It was explained to me later that since the error rate was so high, they were working on a project to use machine vision to replace this position. The author pointed out that they might first try providing the person with ergonomic seating and a different background against which to align the needles. Automation proved unnecessary.

Problem Formulation and Requirements

One popular anecdote that illustrates the importance of proper problem formulation recounts a modern high-rise office building in which the office workers kept complaining about the slowness of the elevators. Computer programmers were called in and they reprogrammed the algorithms. The complaints increased. Engineers put in heavier duty cables and motors so that the elevators could move faster. Complaints remained as strong as ever, and several important multifloor tenants threatened to move out. In desperation, the building owner was considering sacrificing some of the floor space and adding additional elevator shafts at what would obviously be a high price. Someone suggested putting mirrors on all the floors near the elevators and the complaints ceased. In this case, people initially assumed that the problem to be solved was that the elevators ran too slowly and focused on various ways to increase the speed. Eventually, someone came along who realized that the real problem was that people were unhappy about the elevator speed. Mirrors gave them something to do and the time waiting did not seem so onerous.

In many real-world cases, this step is made more complex because typically there are a number of stakeholders, each of which may have very different perceptions of the problems to be solved. A sociotechnical pattern, "Who Speaks for Wolf?" (Thomas, Danis, & Lee, 2002), based on a Native American story, transcribed by Paula Underwood (1991), suggested both the importance of finding all the relevant perspectives and stakeholders early in development and various techniques to try to accomplish that. Briefly, one of the members of a tribe was called "Wolf" because he made a life study of wolves. Once, while Wolf and a few other braves were on an extended hunting expedition, the tribe held council and decided they needed to move. A location was chosen, and the tribe moved; however, a few months later, it became obvious that the tribe had moved into the midst of the spring breeding ground of the wolves. They had to decide whether to move again, post guards, or destroy the wolves. They finally decided to move again, but asked themselves, "What did we learn from this, and how can we avoid this kind of error in the future?" Someone pointed out that if Wolf had been present at the first council, he would have advised against the location. From then on, they decided that whenever they made a major decision, they would ask themselves, "Who speaks for Wolf?" to see whether there were missing stakeholders or perspectives that needed to be taken into account. Many projects could profit from such a process.

It is typically well understood that the earlier in the process of development an error is caught, the less expensive it is. An error in problem formulation can be extremely long-lived and expensive because it sets the context for measuring success. A product or system may be developed that appears to be successful at every step because the wrong thing is being measured so that ultimate failure goes unnoticed until too late. A classic example is the replacement of Coca-Cola with New Coke. According to Gladwell (2005), executives at the Coca-Cola Company were worried because they were losing some market share to Pepsi and blind taste tests indicated that Pepsi was more often preferred. As a result, a sweeter version of Coke that tasted more like Pepsi was developed and put on the market as a replacement for Coke. The reaction was surprising, immediate, passionate, and nearly disastrous for Coke. People wanted the old Coke back. The entire story probably involves brand loyalty, memory, and cognitive dissonance, but one fundamental problem was that the Coca-Cola executives were assuming that the goal was to develop a new product that was sweeter so that it would be preferred in taste tests over Pepsi. And, they succeeded—at solving the wrong problem. What tastes best when you take a decontextualized sip is not necessarily what you prefer (for a variety of reasons, not all directly related to taste) day after day, month after month. In contrast to the taste test, the case test shows what people actually buy over a long period of time, and of course, it is the latter measure that is actually important to profitability. In this case, the use of too simple a measure resulted in the wrong problem being solved. This is probably a common situation. Another example of this type occurred when the manufacturer for a new terminal for telephone operators did usability and productivity tests on operators using the terminal but failed to take into account the phone company customer who was also on the line and whose behavior actually turned out to be on the critical path most of the time. The entire design, development, testing, and so on was predicated on optimizing a system that consisted of the computer and the operator when the system that really mattered was the operator, the computer, and the customer. Again, an initial oversimplification resulted in a much more complex total solution than a more inclusive (and somewhat more complex) initial formulation would have produced.

However, undue complexity can also be introduced by formulating the problem in overly complex terms. The idea of (and failure of) detailed centralized economic planning may be the quintessential example.

Finally, undue complexity can be introduced by beginning with the wrong problem formulation. Even if a better formulation is discovered later, unless the development team is willing to throw out everything and start over, remnants of the original formulation will tend to persist into design, development, deployment, testing, and so on, making both the resulting system and its associated elements (e.g., sales and marketing materials, documentation, education packages, problem determination aids, etc.) more complex than they need have been.

Design

In an attempt to break down a complex problem into manageable subproblems, development teams, quite reasonably, divide into subteams to deal with various subsystems. Unfortunately, this can result in inconsistencies in basic functionality as well as in the user interface. For example, in one word processor, under certain conditions, the user would be faced with a message that said, in essence, "You cannot delete that file because it does not exist." An attempt to create another file by the same name however, resulted in a message that said, in essence, "You cannot create that file because it already exists." It seems clear that no commonly agreed upon definition of what it meant for a file to exist held sway through the whole of the development team.

At a more superficial level, one approach to reduce unnecessary discrepancies in the way that the user interface functions is to provide a style guide so that diverse developers or business partners working on various aspects or functions of an application suite will tend to provide a similar look and feel. There is much to be said for this approach if it is applied with perspective and intelligence. In the worst case, development teams may blindly follow what was meant as a guideline and interpret it as an ironclad rule. For instance, Thomas (this chapter) became involved in the development of an application for the service representatives for a large telecommunications company. The corporate development team insisted that we must follow some guidelines that claimed users should choose an object before choosing an action. For users who must move between multiple applications, there is a relative advantage of having consistency among applications. In this specific context, however, the users did not use multiple applications and it was clear that the way that they naturally thought of and interacted with the task, choosing an action first, was far more intuitive, quicker, and less error prone than choosing an object first. Nonetheless, the management of the development team believed that the provided guidelines were received truth, and therefore they must be followed.

Development

Perhaps the greatest contributor to undue complexity is that the development team (and to a lesser extent, management, associated marketing and sales, documentation specialists, etc.) becomes so familiar with a system that almost everything about it becomes obvious and easy, regardless of how it might be perceived by an end user. Once one sees a hidden figure (such as a pig in the clouds), it is nearly impossible not to see it. Over the lifecycle of a product, tens, hundreds, or even thousands of little conventions and assumptions become second nature to those associated with the development.

The cure for this malady is to continually test the scenarios of use, the designs, paper mock-ups, screen shots, prototypes, and beta versions with naïve users; that is, folks who are representative of potential users but not part of the design team. If, for any reason, this is not feasible (e.g., security, the users do not yet exist) other techniques can provide some amelioration. For instance, heuristic evaluation (Nielsen & Landauer, 1993) is likely to catch a fair proportion of actual problems. This is best done with HCI experts who are already familiar with both the technology and the application area. Typically, five to eight experts provide the most value. A variation on this technique involves having people successively take on different personae while interacting with a system and this may increase the number of errors found in the same period of time (DeSurvire & Thomas, 1993).

Testing

All too often, testing schedules become compressed. Partly as a result of time pressure and partly as a result of the fact that the development team has become accustomed to the high level and low level design, the error codes (or error messages) are often cryptic and designed for the productivity and convenience of the testing team. Unfortunately, in many cases, these error messages persist into the product that the ultimate end user experiences. Testers are often testing functions in a very well-specified context so that an error message is quite interpretable to them. However, to an end user, these same error messages are completely incomprehensible and often the output comes from dropping down several layers in the software stack. Thus, an end user attempting to hit a button in a high-level application may see an error message that not only does not specify what corrective action to take, but it mentions software elements that the end user is not even aware existed.

Another issue with testing, of course, is that testers who work in the development organization may only test reasonable combinations of function. Thomas (this chapter) designed the user experience for a "Dynamic Learning Environment" (Farrell, Thomas, Rubin, Gordin, Katriel, O'Donnell, et al., 2004). In testing the code, he tried altering the URL returned by the system in order to find what he was looking for. The developers all knew that you could not do this and it caused the system to crash—but it is an often-used strategy. For instance, if you know that http://www.umich.edu/~person1 is the URL of person 1, and you are trying to find the website of their colleague at Michigan, person 2, you might reasonably suspect that their URL might be http://www.umich.edu/~person2, and indeed this often works. Happily, he was able to convince the developers to prevent this ploy from crashing the system. More generally, it is important to have potential users test the system as well as those steeped in the cultural assumptions of IT generally and a particular company or product specifically.

Deployment

Today, many applications include "Wizards" to help unzip, install, and even use a product. Often, the distribution of applications is via websites and, therefore, finding the right application, finding the right version of the application, finding out whether one has all the prerequisites and necessary patches, and deciding which features and options to turn on or off can prove as complex, or even more complex, than actually using the application. Often, the instructions and interfaces associated with deployment suffer obscurities from the same root cause as those introduced by the development team—being overly familiar with the application or system. For example, in recently attempting to download an upgrade to a system, the instructions cautioned the user to be sure to use one specific URL and not another. This URL, however, did not refer to the actual URL for finding the desired download. The required URL laid three layers down a menu structure in a different application. To the folks who deployed this update, the context in which this specific URL was to be used was so obvious as to not bother making that context explicit.

Service

With the spread of an ever cheaper, higher bandwidth telecommunications infrastructure, service has experienced the dual trends of centralization of function and service being geographically distant from the users they are trying to support. While centralization offers some benefits in economy of scale and knowledge sharing, having service personnel geographically distant can mean a decrease in shared context between service people and the users they support. For example, local telephone operators used to be able to answer problems such as, "I need the number for that gas station across from the theater downtown." Such problems are currently unanswerable from distant, centralized locations. In some cases, service personnel and users or customers may even have different native languages and come from different cultures. There is a possibility that some of this shared context may soon be reinstated, such as from geographical information systems and from systems that allow service people to view screens and actions remotely. It may be that a shared view of the real visual world is unnecessary; a shared representation of the salient features, assuming they can be identified, may be enough (O'Neill, Castellani, Grasso, Roulland, & Tolmie, 2005).

From a user's perspective, undue complexity is often injected into the process of joint problem solving with service personnel because the service personnel are often organized according to the underlying system that they are trying to support rather than the symptom experienced by a user. For instance, an end user may attempt to print something on a new printer and get an error message indicating that the print job failed. The user calls the help desk, and the top-level menu asks whether they are having a problem with connectivity, the operating system, or an application program, any of which may be the appropriate answer.

An apparently successful innovation in online help is the "Answer Garden" (Ackerman & Malone, 1990), which allows expert users over time with minimal disruption to grow a more detailed tree of FAQs. Other trends making the end user's job of getting help simpler include the use of search engines as well as Wikis and weblogs to support communities of practice and communities of interest.

Maintenance

Consider the maintenance of an automobile circa 1950. While such machines were complex, the parts were all large enough to be visible, and when a part broke, it could be replaced by a similar part. Such repairs might be simple or they might be complex in that replacing a part might require moving other parts in order to reach the part to be replaced. Modern software systems (and such systems are becoming ubiquitously embedded in all other technologies) introduce a set of new and often complex maintenance issues. Before installing a new upgraded version of one piece of software, for instance, it is often necessary to check for several prerequisite pieces of software. Each of these may in turn require still other prerequisites. Even if you already have one of the prerequisites installed, say version X of program Y, you may also need to install a fair number of patches to X. In some cases, you need to uninstall software, taking care throughout to reboot as needed in order to clear out persistent memory. Indeed, if this entire tree is followed, the case still may be that installing some new piece of software causes something else not to work any longer. While the use of wizards has made maintenance much simpler in many cases, if something does fail, diagnosing and fixing the failure may require delving into several layers of software beneath the GUI that is supposed to provide the simple interface for the end user.

WAYS TO MEASURE COMPLEXITY

A Priori Mathematical Models

We have already mentioned an attempt to measure the complexity of programs by Halstead (1977). Other measures have been proposed; probably the best known are those by McCabe and Butler (1986; McCabe, 1976) and Jones (1996). These metrics aim to capture the inherent psychological complexity of the program and not just the behavior that the program surfaces to the end user. We might think to find useful analogues here to measures for the complexity of the user's interaction with a program as well. Unfortunately, the measures do not currently deal with several major factors of psychological complexity. First, the models do not deal with the complexity of the visual (or other sensory) stimuli presented to the user. Second, the models do not deal with the complexity of the response required of the user. Third, the models do not take into account the many complicating factors mentioned above that impact the correspondence between what is required of the user's behavior and what the user already knows.

One of the first attempts to quantitatively predict ease of use *a priori* tried to measure the complexity of the internal structure necessary to generate behavior (Reisner, 1984). A similar approach has been used to try to predict learning time (Bovair, Kieras, & Polson, 1990). Basically, both approaches consider the complexity of the underlying rule systems. Reisner's claim is that, other things being equal, it takes longer to execute a more

complex rule set. Bovair, Kieras, and Polson claimed that it takes longer to learn a more complex rule set as well as to execute it once learned. Both of these claims seem justified, though only with the caution of *ceteris paribus* (see Rautenberg, 1996 for a meta-analysis of quantification attempts of user interfaces; see Ivory & Hearst, 2001 for a broader review of attempts to automate all or part of the process of usability evaluation).

Linear Regression

Psychological tests have often been constructed by taking a very large number of diverse items and then seeing which ones correlate with a desired criterion. Of course, one needs to revalidate these items with a new sample, but it is interesting that somewhat predictive tests have been constructed with this method for many domains. In fact, linear predictive models have been successfully applied to a large number of domains and are generally better than human experts (see Dawes, 1982; Dawes, Faust, & Meehl, 1992). Walston & Felix (1977) used such a technique to predict the effort needed to complete a wide selection of software development projects. Interestingly, the most important factors had to do with the sociopolitical aspects of projects rather than the technical ones. A widely used, flexible method to predict various aspects of software development efforts is COCOMO (Boehm, 1981; 2000). Though not strictly linear, the calculations are based on an empirical analysis of actual software development projects.

Perhaps such approaches can be profitably extended to psychological complexity as well. In fact, at least one attempt to measure the "goodness" (related at least in part to complexity) of websites used a similar approach (Ivory & Hearst, 2001) and showed some evidence of success. However, in their study, somewhat different factors were predictive for different topic areas. Further, because conventions and underlying technology keep changing, it is likely that linear predictive equations might have to be updated on a frequent basis. In general, this is a potential limitation of linear predictive models. They can be quite accurate and quite robust, but only provided the underlying conditions remain relatively constant. The absence of a theoretical underpinning can lead to predictions that can become inaccurate without warning.

Subjective Measures

An entirely different approach to measuring cognitive complexity is to use subjective measures; in effect, to ask people to rate or rank items in terms of cognitive complexity. In judgments of simple stimulus sequences, it appears that complexity and randomness are highly related (Falk & Konold, 1997). An interesting and sophisticated attempt to provide a more differentiated subjective view of complexity applicable to real-world systems is that of Cognitive Dimensions of Complexity (Green, 1989; Green & Petre, 1996). This approach identifies 14 different dimensions of complexity. It can be used as a practical tool to focus the attention of developers successively on various areas of design that can lead to undue complexity. More recently, related work has focused on mismatches between the concepts inherent in the design of an artifact and the way users conceptualize (Connell, Blandford, & Green, 2004).

Textual Analysis of Documentation

In general, we expect the complexity of a description of something to correlate highly with the complexity of that thing. Of course, it is possible to describe a simple thing with undue complexity, but in most circumstances, both extrinsic reward structures and the rules of conversation tend to push description to be sufficiently complex to describe something accurately but sufficiently simple to be understood. Therefore, one promising ersatz measure of the complexity of something, such as a computer system or a procedure using that system, is to measure the complexity of the description. One commonly used measure is the Flesch Formula (Flesch, 1948), which basically uses the average length of words in letters and the length of sentences in words in a document to provide as a score the educational reading level required to understand that document. Another metric is the Gunning Fog index (Gunning, 1952), which seeks to measure the clarity of writing. A still more sophisticated automated approach is possible with the DICTION program (Hart, 2001) which gives 31 primary and several derived dimensions to describe the rhetorical style of a document. One of the primary dimensions is labeled complexity and several of the others might relate. Although this approach offers promise as a metric to be applied after the fact, caution must be applied to using it as an in-process metric. Otherwise, writers could use various tricks to make the description of an artifact appear simple while in reality making it incomprehensible. (In the absurd limiting case, one could provide a manual for a complex product that simply said, "Use it.") Clearly, documents must also be constrained to be complete in order to apply a fair and useful application of complexity metrics to descriptions.

Iterative Design and Testing

Clearly, much of the rest of this handbook expands on this topic. While considerations of complexity may prove useful, at this point, there is no substitute for interacting with real users doing real work in their real context as a general method for designing, developing, and deploying truly useful and usable systems. Thinking about and measuring complexity may well help in this process, but it will probably never substitute for it.

POSSIBLE FUTURE APPROACHES

One can imagine that ever more sophisticated techniques for brain activity imaging and other physiological measures may someday render a reliable and easy-to-use objective index of psychological complexity. Somewhat more likely is the continued evolution of modeling approaches, such as SOAR (Laird, Newell, & Rosenbloom, 1987) and EPIC (Kieras & Meyer, 1997). The follow-on programs might someday conceivably and automatically calculate a set of complexity numbers for combinations of user, task, system, and context. In the nearer term, it seems likely that continued improvement of heuristic approaches that do require some human intervention will prove useful (e.g., Dimensions of Cognitive Complexity; Green, 1989) and the Brown, Keller, and Hellerstein (2005) model in helping to design better user experiences. The latter model assumes that an expert path is known and complexity depends upon the number of actions, the memory load associated with retaining needed parameters, and the number of context shifts.

In our own work, we are attempting to extend this model in order to account more completely for points of uncertainty and the difficulties of decision making. In addition, we are building tools that enable complexity metrics to be produced as a side effect of normal development processes. In this way, developers may gain feedback that is both timelier and more differentiated with respect to their design decisions. In the short term, such metrics should provide useful feedback. As the tools come to be used over time, we expect the development teams to internalize the implications so that better initial design decisions are made.

COMPLEXITY REDUCTION IN PRACTICE: CASE STUDIES

Clearly, complexity is itself a complex topic. However, it is not intractable. Numerous examples of good design and development leading to suitable levels of complexity may be found. We conclude this chapter with two case studies from our own laboratory that illustrate how complexity may be considerably reduced in a real world setting.

Case Study 1: Query By Example

Query By Example, invented by Moshe Zloof (1975) in the early 1970s, was an attempt to provide a query language easy enough for computer-naïve novices to learn to use. Early studies indicated that nonprogramming high school and college students were fairly successful in translating English questions into the syntax of this formal query language (Thomas & Gould, 1975). One probable reason for the relatively fast times and low error rates was that, in many cases, users only had to select the right place to write information rather than having to write all information from scratch (see Fig. 9.10). Another likely reason was that, in Halstead's (1977) terms, Query By Example was quite "dense." That is, almost all of the symbols required for a given query were symbols that would be required in any conceivable query language. There was little overhead. In other words, the intrinsic complexity of any given query remained, but there was no gratuitous or undue complexity introduced by the system itself.

Employees			
Name	Age	Salary	Year of Hire
p.	>65		

"Print the names of everyone who is more than 65 years of age." Only the "p." and the ">65" is entered by the user.

FIGURE 9.10. A simple query in "Query By Example."

In some cases, students continued to have some difficulties, but further studies showed that these difficulties did not spring from the query language per se, but from general difficulties with logic (Thomas, 1976). For example, students had some difficulties with quantifiers (*all* and *some*), logical connectives (*and, or,* and *not*) and with operational distinctions (*sum, count,* and *count unique*). These same difficulties occurred in the absence of any requirements of the query language. Additional difficulties were encountered when the structure of the database did not match well with the structure of the question. For instance, if the database contained a column labeled "Year of Hire," a question asking for the names of people "hired after 1970" was easy. A question asking for the names of people "who have worked for more than 35 years," however, was much more problematic.

In a follow-on study (Thomas, 1983), students were not given English questions to translate but instead were shown a data base structure and given problem statements. They were then to generate their own English questions relevant to the problem and then translate their own questions into Query By Example. These results were much less encouraging. These students had very little concept of what types of questions were reasonable to put to the computer. One problem given was, "Some of the younger faculty members feel that they are not paid enough relative to the older faculty. Write a question that you think might shed light on this issue." A typical question was, "Are the younger faculty paid enough?" In other words, many of the students expected the computer to simply "formulate and solve the problem" for them.

In the 1970s, novice computer users not only might have trouble writing a correct query; they might also write an incorrect query that would consume a huge amount of resource. For this reason, it was considered worthwhile to try to predict the chances that a query was correct before submitting it to the search engine. Based on the formal properties of the query alone (number of operators, etc.), we were able to predict 54% of the variance in query accuracy. If we added the time taken to write the query and the student's own confidence rating, we achieved an accuracy of 75%. Finally, if we added a term for whether the mapping between the English and query was straightforward, more than 90% of the variance was predictable. All these measures may be thought of as factors related to psychological complexity. While the interface and language were developed and improved iteratively, our studies also showed that remaining user difficulties were due to intrinsic task complexity.

Case Study 2: Web Accessibility Technology

With the web growing in importance to both recreation and work, it became clear to us some time ago that barriers to web use needed to be lowered. It was no longer acceptable for people with visual and motor disabilities to be denied effective web access.

We began our explorations in this area by working directly with older adults, a population needing both web content modifications (text and image enlargement, navigational simplification, visual contrast modifications, etc.) and input adaptations (to filter out hand tremors, remove extraneous key presses and

mouse clicks, etc.). Of course, modern browsers and operating systems provide a number of these modifications and adaptations, but access to them is distributed throughout the system, inconsistently surfaced, and often inaccessible to those most in need. Our challenge lay in finding a way to reduce the complexity of so much functionality simultaneously while adding still more accessibility features not otherwise available.

Our approach was iterative and user centered. Early on, it became clear that our users did not want a simplified browser. Nor did they want to be restricted to a subset of the web. After multiple attempts (Richards and Hanson, 2004; Hanson and Richards, 2005), we converged on a design that added a set of very simple control panels to the bottom of an otherwise normal browser, each control panel being dedicated to one class of adaptation or transformation. Figure 9.11 shows one such panel in the context of the full browser.

The only change to the browser's normal interface is the addition of a single "Settings" button. Clicking on this button brings up the first of the series of control panels. Clicking on one of the buttons in a control panel applies an immediate change to the web content. If the change is desired it is kept by doing nothing more with the panel. Otherwise, it is cancelled by clicking on a "None" or "Standard" button depending on the panel. The panels are organized in a conceptual ring. Clicking on the ">" button brings up the next panel. Clicking on the "<" button brings up the previous panel. Clicking on the "?" button brings up an interactive help page keyed to the current panel, which allows the user to easily explore the full range of modification the panel exposes.

A number of mechanisms lie beneath this simple surface. Some modifications are applied by altering browser settings stored in the system's global registry. Others are applied by automatically creating and installing a user style sheet. Others are applied by adjusting parameters in the underlying accessibility settings of the system. Others are applied by modifying the web page's in-memory Document Object Model. Still others are applied by means of always-running agents examining input event streams. All of this complexity is hidden from the user.

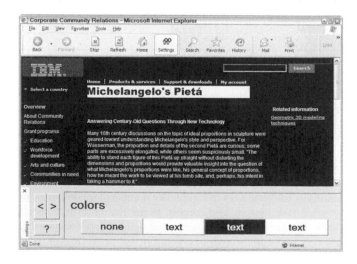

FIGURE 9.11. The "colors" panel for web accessibility software.

For a concrete example of the resulting simplification, consider the case of changing the colors of web pages. This is a modification often found useful for people with visual disabilities and for people with various forms of dyslexia. To change just the text foreground and background colors on a popular browser requires 16 separate steps. To ensure that link, visited link, and hover colors are suitably contrasting requires 24 additional steps. In contrast, in our design, setting two of the most popular color combinations—black text on white backgrounds or white text on black backgrounds—requires only a single click on a button showing the desired color combination. Optimally contrasting link, visited link, and hover colors are automatically set as a side effect of this choice. For those needing a broader range of choice, another simple panel allows the direct setting of the foreground and background RGB values. Again, the link, visited link, and hover colors are set automatically.

This case study illustrates a number of approaches to reducing software complexity. First, iterative user-centered design allowed for the rapid exploration of a range of alternative interfaces. Second, the addition of a new interface layer allowed for the unification of a number of disparate mechanisms. Third, careful analysis of the needs of users allowed for some choices

to be surfaced prominently and others to be removed. Fourth, the provision of immediate feedback allowed users to directly experience the effects of any choice, trying them alone or in combination until their overall web experience was optimal.

CONCLUSIONS

Not surprisingly, psychological complexity is itself a complex topic. It should be of interest to designers of HCI because they typically need to reduce undue complexity. There are occasional exceptions, for instance, in entertainment and learning applications where the designer may want to increase psychological complexity. In this chapter, we have distinguished complexity from many related but different concepts such as uncertainty. Also, we have explored how undue complexity may be introduced into the development process and have suggested approaches to measure and reduce undue complexity. As the intrinsic complexity of the world seems to be on the increase, the importance of reducing undue complexity will continue to increase.

References

Abley, M. (2005), *Spoken here: Travels among threatened languages*. London: Arrow Books.

Ackerman, M., & Malone, T. (1990, April). Answer garden: A tool for growing organizational memory. *Proceedings of the ACM Conference on Office Information Systems*, Cambridge, MA, 31–39.

Ahlberg, C., Williamson, C., & Shneiderman, B. (1992). Dynamic queries for information exploration: An implementation and evaluation, *Proceedings of the ACM CHI'92: Human Factors in Computing Systems* (pp. 619–626). New York: ACM.

Alexander, C. A. (2002). *The nature of order: An essay on the art of building and the nature of the universe. Book One: The phenomenon of life*. Berkeley, CA: Center for Environmental Structure.

Attneave, F. (1957). Physical determinants of the judged complexity of shape. *Journal of Experimental Psychology, 53*, 221–227.

Barlett, F. C. (1932). *Remembering: An experimental and social study*. Cambridge: Cambridge University Press.

Bar-Yam, Y. (1997). *Dynamics of complex systems (Studies in nonlinearity)*. Boulder, CO: Westview Press.

Bar-Yam, Y. (2000). Unifying themes in complex systems. *Proceedings of the international conference on complex systems*. New York, NY: Westview Press.

Bass, L., & John, B. E. (2003). Linking usability to software architecture patterns through general scenarios. *The Journal of Systems and Software, 66*, 187–197.

Boehm, B. W. (1981). *Software engineering economics*. Upper Saddle River, New Jersey: Prentice Hall.

Boehm, B. W. (2000). *Software cost estimation with COCOMO 2000*. Upper Saddle River, New Jersey: Prentice Hall.

Bovair, S., Kieras, D. E., & Polson, P. G. (1990). The acquisition and performance of text-editing skill: A cognitive complexity analysis. *Human-Computer Interaction, 5*(1), 1–48.

Bransford, J., & Franks, J. (1972). Contextual prerequisites for understanding: Some investigations of comprehension and recall. *Journal of Verbal Learning and Verbal Behavior, 11*, 717–726.

Brown, A. B., & Hellerstein, J. L. (2004, September). An approach to benchmarking configuration complexity. *Proceedings of the 11th ACM SIGOPS European Workshop*.

Brown, A. B., Keller, A., & Hellerstein, J. L. (2005, May). A model of configuration complexity and its application to a change management system. *Proceedings of the Ninth IFIP/IEE International Symposium on Integrated Network Management (IM 2005)* (pp. 631–644). Nice, FR.

Brown, G. D. A. (1984). A frequency count of 190,000 words in the London-Lund Corpus of English Conversation. *Behavioral Research Methods Instrumentation and Computers, 16*, 502–532.

Bruckman, A. S. (1997). Moose crossing: Construction, community, and learning in a networked virtual world for kids. *Dissertation Abstracts International*. (UMI No. AAI0598541).

Carroll, J., & Carrithers, C. (1984). Training wheels in a user interface. *Communications of the ACM, 27*(8), 800–806.

Carroll, J., & Mack, R. L. (1984). Learning to use a word processor by doing, by thinking, and by knowing. In J. Thomas & M. Schneider (Eds.), *Human Factors in Computer Systems* (pp. 13–51). Norwood, New Jersey: Ablex.

Carroll, J., & White, M. (1973). Age-of-acquisition norms for 220 picturable nouns. *Journal of Verbal Learning and Verbal Behavior, 12*(5), 563–576.

Card, S. K., Moran, T. P., & Newell, A. (1983). *The psychology of human-computer interaction*. Hillsdale, New Jersey: Lawrence Erlbaum Associates.

Chernoff, H. (1973). The use of faces to represent points in k-dimensional space graphically. *Journal of the American Statistical Society, 68*, 361–368.

Connell, I., Blandford, A., & Green, T. G. R. (2004). CASSM and cognitive walkthrough: Usability issues with ticket vending machines. *Behavior and Information Technology, 23*(5), 307–320.

Coskun, E., & Grabowski, M. (2005). Impacts of user interface complexity on user acceptance and performance in safety-critical systems. *Emergency Management, 2*(1), 1–29.

Dawes, R. (1982). The robust beauty of improper linear models in decision-making. In D. Kahneman, P. Slovic, & A. Tversky (Eds.), *Judgment under uncertainty: Heuristics and biases* (pp. 391–407). Cambridge: Cambridge University Press.

Dawes, R., Faust, D., & Meehl, P. (1989). Clinical versus actuarial judgment. *Science, 241*, 1668–1674.

DeGroot, A. D. (1978). *Thought and choice in chess (2nd edition)*. The Hague, Netherlands: Mouton.

DeSurvire, H., & Thomas, J. C. (1993). Enhancing the performance of interface evaluation using non-empirical usability methods. *Proceedings of the 37th Annual Meeting of the Human Factors and Ergonomics Society* (pp. 1132–1136). Santa Monica, CA.

Ellis, A. (2001). *Overcoming destructive beliefs, feelings and behaviors: New directions in rational emotive behavior therapy*. New York: Albert Ellis Institute.

Falk, R., & Konold, C. (1997). Making sense of randomness: Implicit encoding as a basis of judgment. *Psychological Review, 104*, 301–318.

Farrell, R., Thomas, J., Rubin, B., Gordin, D., Katriel, A., O'Donnell, et al. (2004, November). Personalized just-in-time dynamic assembly of learning objects. *E-learning 2004* (pp. 607–614).

Flesch, R. F. (1948). A new readability yardstick. *Journal of Applied Psychology,32*, 221–233.

Fozard, J. L., Thomas, J. C., & Waugh, N. C. (1976). Effects of age and frequency of stimulus repetitions on two-choice reaction time. *Journal of Gerontology, 31*(5), 556–563.

Furnas, G. W., Gomez, L. M., Landauer, T. K., & Dumais, S. (1982, March). Statistical semantics: How can a computer use what people name things to guess what things people mean when they name things? In J. Nichols & M. Schneider (Ed.), *Proceedings of the SIGCHI conference on Human factors in computing systems* (pp. 251–253). Gaithersburg, MD: ACM.

Gladwell, M. (2005). *Blink: The power of thinking without thinking*. New York: Little Brown.

Gonzalez, V. M., & Mark, G. (2005). Managing currents of work: Multitasking among multiple collaborations. *ECSCW 2005: Proceedings of the Ninth European Conference on Computer-Supported Cooperative Work* (pp. 143–162). Netherlands: Springer.

Gould, J. D., & Boies, S. J. (1984). Speech filing: An office system for principals. *IBM Systems Journal, 23*(1), 65–81.

Gray, W. D., John, B. E., Stuart, R., Lawrence, D., & Atwood, M. E. (1990). GOMS meets the phone company: Analytic modeling applied to real-world problems. *Proceedings of IFIP Interact '90: Human Computer Interaction* (pp. 29–34). Cambridge, UK.

Greene, S. L., Jones, L., Matchen, P., & Thomas, J. C. (2003). Iterative development in the field. *IBM Systems Journal, 42*(4), 594–612.

Green, T. G. R. (1989). Cognitive dimensions of notations. In A. Sutcliffe & L. Macaulay (Eds.), *People and computers V* (pp. 443–460). Cambridge: Cambridge University Press.

Green, T. G. R., & Petre, M. (1996). Usability analysis of visual programming environments: A "cognitive dimensions" approach. *Journal of visual languages and computing, 7*, 131–174.

Grinter, R. E., & Edwards, W. K. (2005). The work to make a home network work. *ECSCW 2005: Proceedings of the Ninth European Conference on Computer-Supported Cooperative Work* (pp. 469–488). Netherlands: Springer.

Gunning, R. (1952). *The Technique of Clear Writing*. New York: McGraw-Hill International Book Co.

Halstead, M. H. (1977). *Elements of software science* (Operating and programming systems series). New York: Elsevier.

Hanson, V. L., & Richards, J. T. (2005, May–June). Achieving a more usable World Wide web. *Behaviour & Information Technology, 24*(3), 231–246.

Hart, R. P. (2001). Redeveloping diction: Theoretical considerations. In M. West (Ed.), *Theory, method and practice in computer content analysis* (pp. 43–60). Westport, CT: Ablex.

Hildebrandt, N., Caplan, D., Sokol, S., & Torreano, L. (1995, January). Lexical factors in the word-superiority effect. *Memory and Cognition, 23*(1), 23–33.

Holland, J. J. (1995). *Hidden order: How adaptation builds complexity*. New York: Basic Books.

Ivory, M. Y., & Hearst, M. A. (2001). The state of the art in automating usability evaluation of user interfaces. *ACM Computing Surveys, 33*(4), 470–516.

John, B. E. (1990). Extensions of GOMS analyses to expert performance requiring perception of dynamic visual and auditory information. *Proceedings of CHI'90* (pp. 107–115). New York.

Johnson, M. H., & Morton, J. (1991). *Biology and cognitive development: The case of face recognition*. Cambridge, MA: Blackwell.

Jones, C. (1996). *Applied software measurement*. New York: McGraw-Hill.

Kay, A. Squeak Etoys Authoring & Media. Retrieved March 2007 from www. squeakland.org

Kieras, D., & Meyer, D. E. (1997). An overview of the EPIC architecture for cognition and performance with application to human-computer interaction. *Human-Computer Interaction, 12*, 391–438.

Klinger, A., & Salingaros, N. A. (2000). A pattern measure. *Environment and Planning: Planning and Design, 27*, 537–547.

Kucera, H., & Francis, W. N. (1967). *Computational analysis of present-day American English*. Providence, RI: Brown University Press.

Laird, J. E., Newell, A., & Rosenbloom, P. S. (1987). SOAR: An architecture for general intelligence. *Artificial Intelligence, 33*, 1–64.

Lewis, C. (in press). Simplicity in cognitive assistive technology: A framework and agenda for research. *International Journal on Universal Access*.

McCabe, T. J. (1976). A Complexity Measure. *IEEE Transactions on Software Engineering* , SE-2 4, 308–320.

McCabe, T. J., & Butler, C. W. (1989). Design complexity measurement and testing. *Communications of the ACM, 32*(12).

Miller, G. A. (1956). The magic number seven, plus or minus two: Some limits on our capacity for processing information. *Psychological Review, 63*, 81–97.

Myers, B. A., McDaniel, R. G., & Kosbie, D. S. Marquise (1993). Creating complete user interfaces by demonstration. *Proceedings of INTERCHI'93: Human Factors in Computing Systems* (pp. 293–300). Amsterdam.

Nielsen, J., & Landauer, T. K. (1993). A mathematical model of the finding of usability problems. *Proceedings of INTERCHI'93 Human Factors in Computing Systems* (pp. 206–213). Amsterdam.

Norman, D. A. (1988). *The psychology of everyday things*. New York: Basic Books.

O'Neill, J., Castellani, S., Grasso, A., Roulland, F., & Tolmie, P. (2005). Representations can be good enough. *ECSCW 2005: Proceedings of Ninth European Conference on Computer-Supported Cooperative Work* (pp. 267–286). Netherlands: Springer.

Ouden, E., den, Lu, Y., Sonnemanns, P., & Brombacher, A. (2006). Quality and reliability problems from a consumer's perspective: An increasing problem overlooked by businesses? *Quality and Reliability Engineering International, 22*(7), 821–838.

Paivio, A., Yuille, J. C., & Madigan, S. A. (1968). Concreteness, imagery and meaningfulness values for 925 words. *Journal of Experimental Psychology Monograph Supplement, 76*(3)2, 1–25.

Papert, S. (1993). *The children's machine: Rethinking school in the age of the computer*. New York: Basic Books.

Pelz, D. (2000). *Dave Pelz's putting Bible*. New York: Doubleday.

Phillips, L. H., Bull, R., Adams, E., & Fraser, L. (2002). Positive mood and executive function: Evidence from Stroop and fluency tasks. *Emotion, 2*(1), 12–22.

Pickover, C. (1985). Tusk: A versatile graphics workstation for speech research. *IBM Research Report: RC 11497.* Armonk, NY: IBM.

Rauterberg, M. (1996). A concept to quantify different measures of user interface attributes: A meta-analysis of empirical studies. *Proceedings of the 1996 IEEE International Conference on Systems, Man and Cybernetics* (pp. 2799–2804). Beijing, China.

Reicher, G. M. (1969). Perceptual recognition as a function of meaningfulness of stimulus material. *Journal of Experimental Psychology, 81,* 275–280.

Reisner, P. (1984). Formal grammar as a tool for analyzing ease of use: Some fundamental concepts. In J. Thomas & M. Schneider (Eds.), *Human factors in computer systems* (pp. 53–78). Norwood, NJ: Ablex.

Richards, J. T., & Hanson, V. L. (2004). Web accessibility: A broader view. *Proceedings of the Thirteenth International ACM World Wide Web Conference, WWW2004,* New York: ACM, 2004.

Rubin, D. C. (1980). 51 Properties of 125 words: A unit analysis of verbal behavior. *J. Verbal Learn & Verbal Behavior, 19,* 736–755.

Shmulevich, I., & Povel, D. J. L. (2000). Complexity measures of musical rhythms. In P. W. M. Desain & W. L. Windsor (Eds.), *Rhythm perception and production* (pp. 239–244). Lisse: Swets & Zeitlinger.

Shmulevich, I., Yli-Harja, O., Coyle, E., Povel, D., & Lemstrom, K. (2001). Perceptual issues in music pattern recognition: Complexity of rhythm and key finding. *Computers and Humanities, 35,* 23–35.

Simon, H. A. (1962). The architecture of complexity. *Proceedings of the American Philosophical Society. 106,* 467–482. Philadelphia: American Philosophical Society.

Simon, H. A., & Barenfeld, M. (1969). Information processing analysis of perceptual processes in problem solving. *Psychological Review, 76,* 473–483.

Simon, H. A., & Gilmartin, K. J. (1973). A simulation of memory for chess positions. *Cognitive Psychology, 5,* 29–46.

Smith, M. A., Cottrell, G. W., & Anderson, K. L. (2001). The early word catches the weights. *Advances in Neural Information Processing Systems, 13,* 52–58.

Taylor, R. P., Spehar, B., Wise, J. A., Clifford, C. W. G., Newell, B. R., Hagerhall, C. M., et al. (2005). Perceptual and physiological responses to the visual complexity of Pollock's dripped fractal patterns. *Nonlinear Dynamics, Psychology & Life Sciences, 9,* 89–114.

The Macallan: Speyside distillery—Scotch Whiskey. Retrieved March 5, 2007 from http://www.whiskey-distilleries.info/Macallan_EN.shtml

Thomas, J. C., Danis, C. M., & Lee, A. (2002). Who speaks for wolf? *IBM Research Report: RC22644.* Armonk, NY: IBM.

Thomas, J. C., & Gould, J. D. (1975). A psychological study of Query By Example. *National Computer Conference Proceedings,* New York, *44,* 439–445.

Thomas, J. C. (1976). Quantifiers and question-asking. *IBM Research Report, RC-5886.* Armonk, NY: IBM.

Thomas, J. C. (1983). Psychological issues in the design of database query languages. In M. E. Sime & M. J. Coombs (Eds.), *Designing for human-computer communication* (pp. 173–206). London: Academic Press.

Thomas, J. C., & Ruben, H. (1973, November 8). *Age and mnemonic techniques in paired-associate learning.* Paper presented at the Gerontological Society Meeting, Miami Beach, Florida.

Thomas, J. C., Fozard, J. L., & Waugh, N. C. (1977). Age-related differences in naming latency. *American Journal of Psychology, 90*(30), 499–509.

Underwood, P. (1991). Who speaks for wolf: A Native American learning story. San Anselmo, CA: Tribe of Two Press.

Vintage Cocktails. (2007). Retrieved March 5, 2007, from http://www.vintagecocktails.com/taste.html.

Walston, C. E., & Felix, C. P. (1977). A method of programming measurement and estimation. *IBM Systems Journal, 16,* 54–73.

Weinberg, G. M. (1971). *The psychology of computer programming.* New York: Van Nostrand Reinhold.

Werker, J. F., & Tees, R. C. (1984). Cross-language speech perception: evidence of perceptual reorganization during the first year of life. *Infant Behavior and Development, 7,* 49–63.

Zloof, M. (1975). Query by example. *Proceedings of the AFIPS National Computer Conference, 44,* 431–438. Montvale, NJ: AFIPS Press.

·10·

INFORMATION VISUALIZATION

Stuart Card
Xerox PARC

Introduction **182**

Example 1: Finding Videos with the FilmFinder 182

Example 2: Monitoring Stocks with TreeMaps 184

Example 3: Sensemaking with Permutation

Matrices 185

What Is Information Visualization? 187

Why Does Visualization Work? 187

Historical Origins 188

The Visualization Reference Model **191**

Mapping Data to Visual Form 191

Data Structures 191

Visual Structures 192

Spatial substrate 192

Marks 193

Connection and enclosure 194

Retinal properties 194

Temporal encoding 194

Expressiveness and Effectiveness 195

Taxonomy of Information Visualizations 195

Simple Visual Structures **196**

1-Variable 197

2-Variables 198

3-Variables and Information Landscapes 198

n-Variables 199

Trees 199

Connection 199

Enclosure 202

Networks 202

Composed Visual Structures **203**

Single-axis composition 203

Double-axis composition .204

Mark composition and case composition 205

Recursive composition .205

Interactive Visual Structures **206**

Dynamic queries 206

Magic lens (movable filter) 206

Overview + detail 206

Linking and brushing .206

Extraction and comparison 208

Attribute explorer 208

Focus + Context Attention-Reactive

Abstractions **208**

Data-Based Methods 208

Filtering 208

Selective aggregation 209

View-Based Methods 209

Micro-macro readings 209

Highlighting 209

Visual transfer functions 209

Perspective distortion 210

Alternate geometries 211

Sensemaking With Visualization **211**

Knowledge Crystallization 211

Acquire information .212

Make sense of it 212

Create something new 212

Act on it 212

Levels for Applying Information Visualization 212

Acknowledgment **214**

References **214**

INTRODUCTION

The working mind is greatly leveraged by interaction with the world outside it. A conversation to share information, a grocery list to aid memory, a pocket calculator to compute square roots—all effectively augment a cognitive ability otherwise severely constrained by what is in its limited knowledge, by limited attention, and by limitations on reasoning. But the most profound leverage on cognitive ability is the ability to invent new representations, procedures, or devices that augment cognition far beyond its unaided biological endowment—and bootstrap these into even more potent inventions.

This chapter is about one class of inventions for augmenting cognition, collectively called "information visualization." Other senses could be employed in this pursuit—audition, for example, or a multi-modal combination of senses—the broader topic is really *information perceptualization*; however, in this chapter, we restrict ourselves to visualization. Visualization employs the sense with the most information capacity; recent advances in graphically agile computers have opened opportunities to exploit this capacity, and many visualization techniques have now been developed. A few examples suggest the possibilities.

Example 1: Finding Videos with the FilmFinder

The use of information visualization for finding things is illustrated by the FilmFinder (Ahlberg & Shneiderman, 1994a,

1994b). Unlike typical movie-finder systems, the FilmFinder is organized not around searching with keywords, but rather around rapid browsing and reacting to collections of films in the database. Figure 10.1 shows a scattergraph of 2000 movies, plotting rated quality of the movie as a function of year when it was released. Color differentiates type of movies—comedy from drama and the like. The display provides an overview, the entire universe of all the movies, and some general features of the collection. It is visually apparent, for example, that a good share of the movies in the collection were released after 1965, but also that there are movies going back as far as the 1920s. Now the viewer "drills down" into the collection by using the sliders in the interface to show only movies with Sean Connery that are between 1 and 4½ hours in length (Fig. 10.2). As the sliders are moved, the display zooms in to show about 20 movies. It can be seen that these movies were made between 1960 and 1995, and all have a quality rating higher than 4. Since there is now room on the display, titles of the movies appear. Experimentation with the slider shows that restricting maximum length to 2 hours cuts out few interesting movies. The viewer chooses the highly rated movie, "Murder on the Orient Express" by double-clicking on its marker. Up pop details in a box (Fig. 10.3) giving names of other actors in the movie and more information. The viewer is interested in whether two of these actors, Anthony Perkins and Ingrid Bergman, have appeared together in any other movies. The viewer selects their names in the box, and then requests another search (Fig. 10.4). The result is a new display of two movies. In addition to the movie the viewer knew about, there is one other movie, a drama entitled "Goodbye,

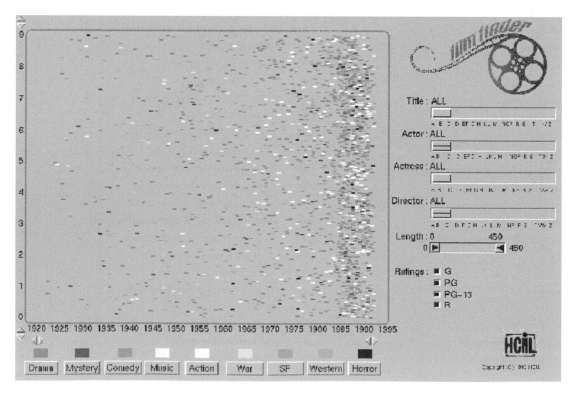

FIGURE 10.1. FilmFinder overview scattergraph. Courtesy University of Maryland.

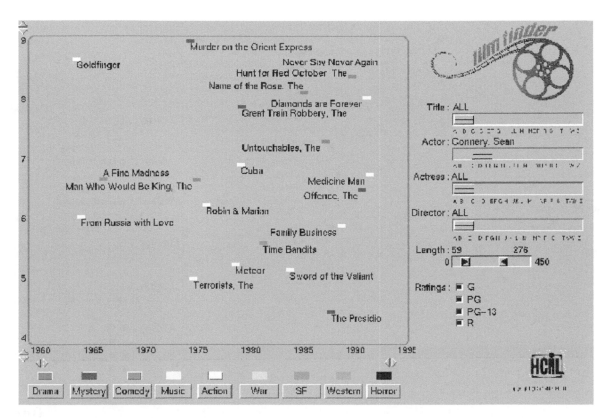

FIGURE 10.2.   FilmFinder scattergraph zoom-in. Courtesy University of Maryland.

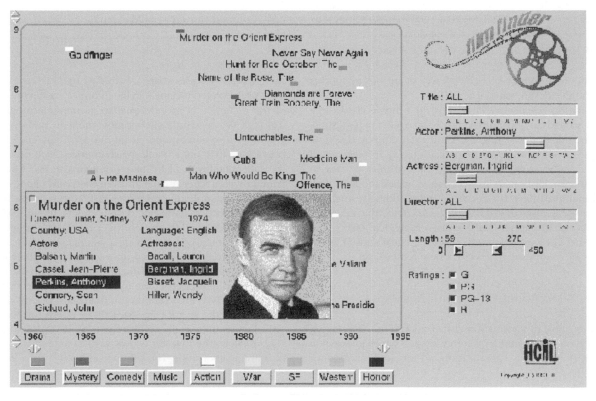

FIGURE 10.3.   FilmFinder details on demand. Courtesy University of Maryland.

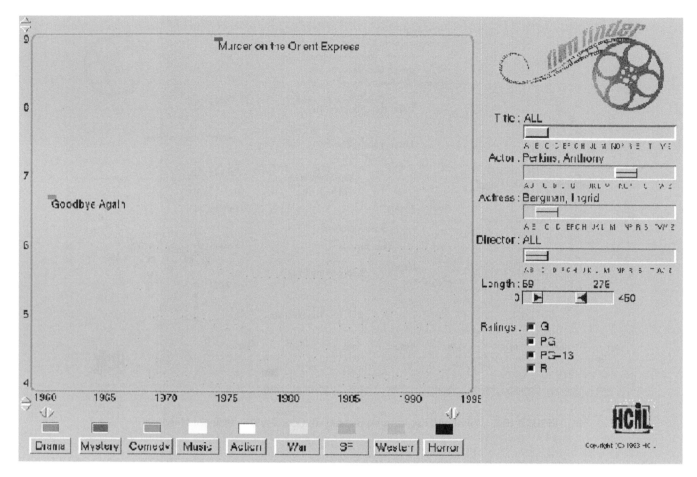

FIGURE 10.4. FilmFinder retrieval by example. Courtesy University of Maryland.

Again," made around 1960. The viewer is curious about this movie and decides to watch it.

Information visualization has allowed a movie viewer in a matter of seconds to find a movie he or she could not have specified at the outset. To do this, the FilmFinder employed several techniques from information visualization: (a) an *overview* of the collection showing its structure; (b) *dynamic queries*, in which the visualization seems to change instantaneously with control manipulations; (c) *zooming in* by adding restrictions to the set of interest; (d) *details on demand*, in which the user can display temporarily detail about an individual object, and (e) *retrieval by example*, in which selected attributes of an individual object are used to specify a new retrieval set.

Example 2: Monitoring Stocks with TreeMaps

Another example of information visualization is the TreeMap visualization on the SmartMoney.com website,[1] which is shown in Fig. 10.5(a). Using this visualization, an investor can monitor

more than 500 stocks at once, with data updated every 15 minutes. Each colored rectangle in the figure is a company. The size of the rectangle is proportional to its market capitalization. Color of the rectangle shows movement in the stock price. Bright yellow corresponds to about a 6% increase in price, bright blue to about a 6% decrease in price. Each business sector is identified with a label like "Communications." Those items marked with a letter *N* have an associated news item.

In this example, the investor's task is to monitor the day's market and notice interesting developments. In Fig. 10.5(a), the investor has moved the mouse over one of the bright yellow rectangles, and a box identifying it as Erickson, with a +9.28% gain for the day, has popped up together with other information. Clicking on a box gives the investor a pop-up menu for selecting even more detail. The investor can either click to go to World Wide Web links on news or financials, or drill down, for example, to the sector (Fig. 10.5[b]), or down further to individual companies in the software part of the technology sector (Fig. 10.5[c]). The investor is now able to immediately note interesting relationships. The software industry is now larger than

---

[1]www.smartmoney.com

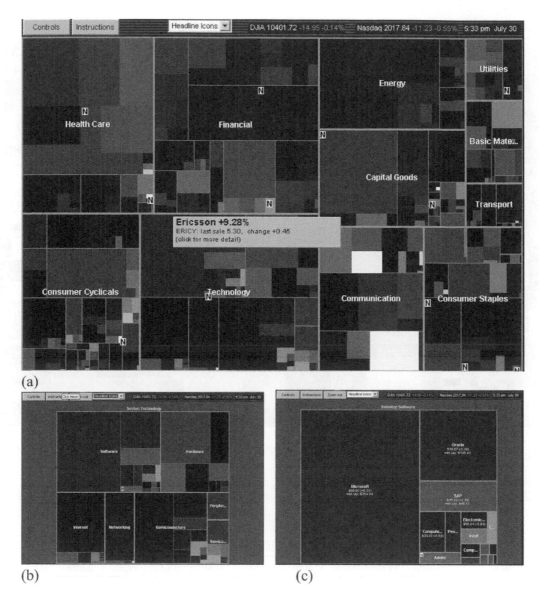

FIGURE 10.5. TreeMap of daily stock prices. Courtesy SmartMoney.com.

the hardware industry, for example, and despite a recent battering at the time of this figure, the Internet industry is also relatively large. Microsoft is larger than all the other companies in its industry combined. Selecting a menu item to look at year-to-date gains (Fig. 10.6), the investor immediately notes interesting patterns: Microsoft stock shows substantial gains, whereas Oracle is down; Dell is up, but Compaq is down; Tiny Advanced Micro is up, whereas giant Intel is neutral. Having noticed these relationships, the investor drills down to put up charts or analysts' positions for companies whose gains in themselves, *or in relation to a competitor*, are interesting. For example, the investor is preparing a report on the computer industry for colleagues and notices how AMD is making gains against Intel, or how competition for the Internet is turning into a battle between Microsoft and AOL/Time Warner.

Example 3: Sensemaking with Permutation Matrices

As a final information visualization example, consider the case proposed by Bertin (1977/1981) of a hotel manager who wants to analyze hotel occupancy data (Table 10.1) to increase her return. In order to search for meaningful patterns in her data, she represents it as a permutation matrix (Fig. 10.7[a]). A permutation matrix is a graphic rendition of a cases x variables display. In Fig. 10.7(a), each cell of Table 10.1 is a small bar of a bar chart. The bars for cells below the mean are white; those above the bar are black. By permuting rows and columns, patterns emerge that lead to making sense of the data.

In Fig. 10.7(a), the set of months, which form the cases, are repeated to reveal periodic patterns across the end of the cycle. By visually comparing the pairs of rows, one can find rows

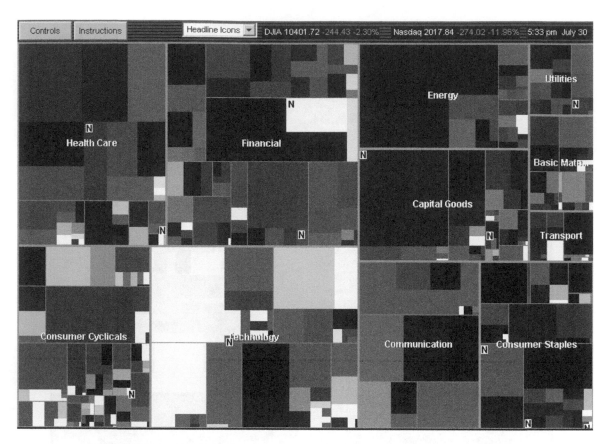

FIGURE 10.6. TreeMap of year-to-date stock prices. Courtesy SmartMoney.com.

TABLE 10.1. Data for Hotel Occupancy (Based on Bertin (1977/1981))

ID	VARIABLE	JAN	FEB	MAR	APR	MAY	JUNE	JULY	AUG	SEPT	OCT	NOV	DEC
1	% Female	26	21	26	28	20	20	20	20	20	40	15	40
2	% Local	69	70	77	71	37	36	39	39	55	60	68	72
3	% USA	7	6	3	6	23	14	19	14	9	6	8	8
4	% South America	0	0	0	0	8	6	6	4	2	12	0	0
5	% Europe	20	15	14	15	23	27	22	30	27	19	19	17
6	% M.East/Africa	1	0	0	8	6	4	6	4	2	1	0	1
7	% Asia	3	10	6	0	3	13	8	9	5	2	5	2
8	% Businessmen	78	80	85	86	85	87	70	76	87	85	87	80
9	% Tourists	22	20	15	14	15	13	30	24	13	15	13	20
10	% Direct Reservations	70	70	75	74	69	68	74	75	68	68	64	75
11	% Agency Reservations	20	18	19	17	27	27	19	19	26	27	21	15
12	% Air Crews	10	12	6	9	4	5	7	6	6	5	15	10
13	% Under 20	2	2	4	2	2	1	1	2	2	4	2	5
14	% 20–35	25	27	37	35	25	25	27	28	24	30	24	30
15	% 35–55	48	49	42	48	54	55	53	51	55	46	55	43
16	% Over 55	25	22	17	15	19	19	19	19	19	20	19	22
17	Price of rooms	163	167	166	174	152	155	145	170	157	174	165	156
18	Length of stay	1.7	1.7	1.7	1.91	1.9	2	1.54	1.6	1.73	1.82	1.66	1.44
19	% Occupancy	67	82	70	83	74	77	56	62	90	92	78	55
20	Conventions	0	0	0	1	1	1	0	0	1	1	1	1

that are similar. These are reordered and grouped (Fig. 10.7[b]). By this means, it is discovered that there seem to be two patterns of yearly variation. One pattern in Fig. 10.7(b) is semiannual, dividing the year into the cold months of October through April and the warm months of May through September. The other pattern breaks the year into four distinct regions. We have thus found the beginnings of a *schema*—that is, a framework in terms of which we can encode the raw data and describe it in a more compact language. Instead of talking about the events of the year in terms of individual months, we can now talk in terms of two series of periods, the semiannual one, and the four distinct periods. As we do so, there is a *residue* of information not included as part of our descriptive language. Sensemaking proceeds by the *omission and recoding of information into more compact form* (see Resnikoff, 1989). This residue of information may be reduced by finding a better or more articulated schema, or it may be left as noise. Beyond finding the basic patterns in the data, the hotel manager wants to make sense of the data relative to a purpose: she wants to increase the occupancy of the hotel. Therefore, she has also permuted general indicators of activity in Fig. 10.7(b), such as % Occupancy and Length of Stay, to the top of the diagram and put the rows that correlate with these below them. This reveals that Conventions, Businessmen, and Agency Reservations, all of which generally have to do with convention business, are associated with higher occupancy. This insight comes from the match in patterns *internal* to the visualization; it also comes from noting why these variables might correlate as a consequence of factors *external* to the visualization. She also discovers that marked differences exist between the winter and summer guests during the slow periods. In winter, there are more local guests, women, and age differences. In summer, there are more foreign tourists and less variation in age.

This visualization was useful for sensemaking on hotel occupancy data, but it is too complicated to communicate the high points. The hotel manager therefore creates a simplified diagram, Fig. 10.7(c). By graying some of the bars, the main points are more readily graspable, while still preserving the data relations. A December convention, for example, does not seem to have the effect of the other conventions in bringing in guests. It is shown in gray as residue in the pattern. The hotel manager suggests moving the convention to another month, where it might have more effect on increasing the occupancy of the hotel.

What Is Information Visualization?

The FilmFinder, the TreeMap, and the permutation matrix hotel analysis are all examples of the use of information visualization. We can define information visualization as "the use of computer-supported, interactive, visual representations of abstract data in order to amplify cognition" (Card, Mackinlay, & Shneiderman, 1999).

Information visualization needs to be distinguished from related areas: *scientific visualization* is like information visualization, but it is applied to scientific data and typically is physically based. The starting point of a natural geometrical substrate for the data, whether the human body or earth geography, tends to emphasize finding a way to make visible the invisible (say, velocity of air flow) within an existing spatial framework. The chief problem for information visualization, in contrast, is often finding an effective mapping between abstract entities and a spatial representation. Both information visualization and scientific visualization belong to the broader field of *data graphics*, which is the use of abstract, nonrepresentational visual representations to amplify cognition. Data graphics, in turn, is part of *information design*, which concerns itself with external representations for amplifying cognition. At the highest level, we could consider information design a part of *external cognition*, the uses of the external world to accomplish some cognitive process. Characterizing the purpose of information visualization as *amplifying cognition* is purposely broad. Cognition can be the process of writing a scientific paper or shopping on the Internet for a cell phone. Generally, it refers to the intellectual processes in which information is obtained, transformed, stored, retrieved, and used. All of these can be advanced generally by means of external cognition, and in particular by means of information visualization.

Why Does Visualization Work?

Visualization aids cognition not because of some mystical superiority of pictures over other forms of thought and communication, but rather because visualization helps the user by making the world outside the mind a resource for thought in specific ways. We list six groups of these in Table 10.2 (Card et al., 1999): Visualization amplifies cognition by (a) increasing the memory and processing resources available to the users, (b) reducing search for information, (c) using visual representations to enhance the detection of patterns, (d) enabling perceptual inference operations, (e) using perceptual attention mechanisms for monitoring, and (f) by encoding information in a manipulable medium. The FilmFinder, for example, allows the representation of a large amount of data in a small space in a way that allows patterns to be perceived visually in the data. Most important, the method of instantly responding in the display to the dynamic movement of the sliders allowed users to rapidly explore the multidimensional space of films. The TreeMap of the stock market allows monitoring and exploration of many equities. Again, much data is represented in little space. In this case, the display manages the user's attention, drawing it to those equities with unusually large changes, and supplying the means to drill down into the data to understand why these movements may be happening. In the hotel management case, the visual representation makes it easier to notice similarities of behavior in a multidimensional attribute space, then to cluster and rerepresent these. The final product is a compact (and simplified) representation of the original data that supports a set of forward decisions. In all of these cases, visualization allows the user to (a) examine a large amount of information, (b) keep an overview of the whole while pursuing details, (c) keep track of (by using the display as an external working memory) many things, and (d) produce an abstract representation of a situation through the omission and recoding of information.

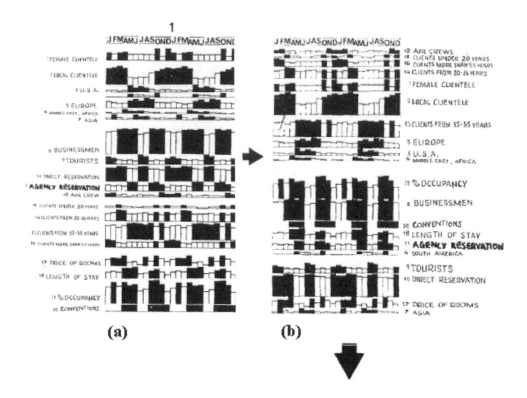

(a) **(b)**

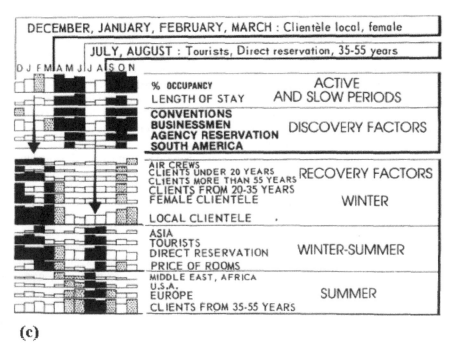

(c)

FIGURE 10.7. Permutation matrix representation of hotel data (Berlin, 1977/1981). (a) Initial matrix of variables. (b) Permuted matrix to group like patterns together.

Historical Origins

Drawn visual representations have a long history. Maps go back millennia. Diagrams were an important part of Euclid's books on geometry. Science, from earliest times, used diagrams to (a) record observations, (b) induct relationships, (c) explicate methodology of experiments, and (d) classify and conceptualize phenomena (for a discussion, see Robin, 1992). For example, Fig. 10.8 is a hand-drawn illustration, in Newton's first scientific publication, illustrating how white light is really composed of

TABLE 10.2. How Information Visualization Amplifies Cognition

1. Increased Resources	
High-bandwidth hierarchical interaction	Human moving gaze system partitions limited channel capacity so that it combines high spatial resolution and wide aperture in sensing the visual environments (Larkin & Simon, 1987).
Parallel perceptual processing	Some attributes of visualizations can be processed in parallel compared to text, which is serial.
Offload work from cognitive to perceptual system	Some cognitive inferences done symbolically can be recoded into inferences done with simple perceptual operations (Larkin & Simon, 1987).
Expanded working memory	Visualizations can expand the working memory available for solving a problem (Norman, 1993).
Expanded storage of information	Visualizations can be used to store massive amounts of information in a quickly accessible form (e.g., maps).
2. Reduced Search	
Locality of processing	Visualizations group information used together reducing search (Larkin & Simon, 1987).
High data density	Visualizations can often represent a large amount of data in a small space (Tufte, 1983).
Spatially-indexed addressing	By grouping data about an object, visualizations can avoid symbolic labels (Larkin & Simon, 1987).
3. Enhanced Recognition of Patterns	
Recognition instead of recall	Recognizing information generated by a visualization is easier than recalling that information by the user.
Abstraction and aggregation	Visualizations simplify and organize information, supplying higher centers with aggregated forms of information through abstraction and selective omission (Card, Robertson, & Mackinlay, 1991); (Resnikoff, 1989).
Visual schemata for organization	Visually organizing data by structural relationships (e.g., by time) enhances patterns.
Value, relationship, trend	Visualizations can be constructed to enhance patterns at all three levels (Bertin, 1967/1983).
4. Perceptual Inference	
Visual representations make some problems obvious	Visualizations can support a large number of perceptual inferences that are very easy for humans (Larkin & Simon, 1987).
Graphical computations	Visualizations can enable complex specialized graphical computations (Hutchins, 1996).
5. Perceptual Monitoring	Visualizations can allow for the monitoring of a large number of potential events if the display is organized so that these stand out by appearance or motion.
6. Manipulable medium	Unlike static diagrams, visualizations can allow exploration of a space of parameter values and can amplify user operations.

Source: Card, Mackinlay, & Shneiderman, 1999.

many colors. Sunlight enters from the window at right and is refracted into many colors by a prism. One of these colors can be selected (by an aperture in a screen) and further refracted by another prism, but the light stays the same color, showing that it has already been reduced to its elementary components. As in Newton's illustration, early scientific and mathematical diagrams generally had a spatial, physical basis and were used to reveal the hidden, underlying order in that world.

Surprisingly, diagrams of abstract, nonphysical information are apparently rather recent. Tufte (1983) dates abstract diagrams to Playfair (1786) in the 18th century. Figure 10.9 is one of Playfair's earliest diagrams. The purpose was to convince readers that English imports were catching up with exports. Starting with Playfair, the classical methods of plotting data were developed—graphs, bar charts, and the rest.

Recent advances in the visual representation of abstract information derive from several strands that became intertwined. In 1967, Bertin (1967/1983, 1977/1981), a French cartographer, published his theory of *The Semiology of Graphics*. This theory identified the basic elements of diagrams and their combination. Tufte (1983, 1990, 1997), from the fields of visual design and data graphics, published a series of seminal books that set forth principles for the design of data graphics and emphasized maximizing the density of useful information. Both Bertin's and Tufte's theories became well known and influential. Meanwhile, within statistics, Tukey (1977) began a movement on exploratory data analysis. His emphasis was not on the quality of graphical

presentation, but on the use of pictures to give rapid, statistical insight into data relations. For example, "box and whisker plots" allowed an analyst to get a rapid characterization of data distributions. Cleveland and McGill (1988) wrote an influential book, *Dynamic Graphics for Statistics*, explicating new visualizations of data with particular emphasis on the visualization of multidimensional data.

In 1985, NSF launched an initiative on *scientific visualization* (McCormick & DeFanti, 1987). The purpose of this initiative was to use advances in computer graphics to create a new class of analytical instruments for scientific analysis, especially as a tool for comprehending large, newly produced datasets in the geophysical and biological sciences. Meanwhile, the computer graphics and artificial intelligence communities were interested in the automatic design of visual presentations of data. Mackinlay's (1986a, 1986b) thesis APT formalized Bertin's design theory, added psychophysical data, and used these to build a system for automatically generating diagrams of data, tailored for some purpose. Roth and Mattis (1990) built a system to do more complex visualizations, such as some of those from Tufte. Casner (1991) added a representation of tasks. This community was interested not so much in the quality of the graphics as in the automation of the match between data characteristics, presentational purpose, and graphical presentation. Finally, the user interface community saw advances in graphics hardware opening the possibility of a new generation of user interfaces. The first use of the term "information visualization" was probably in

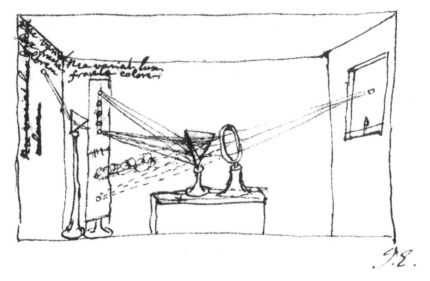

FIGURE 10.8.   Newton's optics illustration (from Robin, 1992).

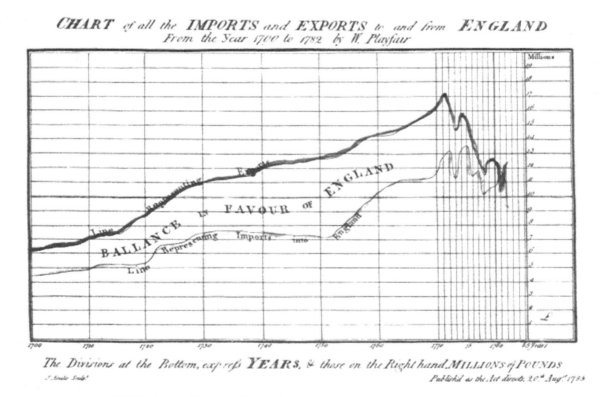

FIGURE 10.9.   Playfair's charts of English imports and exports (Tufte, 1983).

Robertson, Card, and Mackinlay (1989). Early studies in this community focused on user interaction with large amounts of information: Feiner and Beshers (1990) presented a method, worlds within worlds, for showing six-dimensional financial data in an immersive virtual reality. Shneiderman (1992) developed a technique called "dynamic queries" for interactively selecting subsets of data items and TreeMaps, a space-filling representa-

tion for trees. Robertson, Card, and Mackinlay (1993) presented ways of using animation and distortion to interact with large data sets in a system called the Information Visualizer, which used *focus + context* displays to nonuniformly present large amounts of information. The emphasis for these studies was on the means for cognitive amplification, rather than on the quality of the graphics presentations.

The remainder of this chapter will concentrate on the techniques that have been developed for mapping abstract information to interactive visual form to aid some intellectual task. The perceptual foundations of this effort are beyond the scope of this chapter, but are covered in Ware (2000). Further details on information visualization techniques are addressed in a text by Spence (2000). The classic papers in information visualization are collected in Card et al. (1999).

THE VISUALIZATION REFERENCE MODEL

Mapping Data to Visual Form

Despite their seeming variability, information visualizations can be systematically analyzed. Visualizations can be thought of as adjustable mappings from data to visual form to the human perceiver. In fact, we can draw a simple Visualization Reference Model of these mappings (Fig. 10.10). Arrows follow from *Raw Data* (data in some idiosyncratic format) on the left, though a set of *Data Transformations* into *Data Tables* (canonical descriptions of data in a variables x cases format extended to include metadata). The most important mapping is the arrow from Data Tables to *Visual Structures* (structures that combine values an available vocabulary of visual elements—spatial substrates, marks, and graphical properties). Visual Structures can be further transformed by *View Transformations*, such as visual distortion or 3D viewing angle, until it finally forms a *View* that can be perceived by human users. Thus, Raw Data might start out as text represented as indexed strings or arrays. These might be transformed into *document vectors*, normalized vectors in a space with dimensionality as large as the number of words. Document vectors, in turn, might be reduced by multidimensional scaling to create the analytic abstraction to be visualized, expressed as a Data Table of x, y, z coordinates that could be displayed. These coordinates might be transformed into a Visual Structure—that is, a surface on an information landscape—which is then viewed at a certain angle.

Similar final effects can be achieved by transformations at different places in the model: When a point is deleted from the visualization, has the point been deleted from the dataset? Or is it still in the data merely not displayed? Chi and Riedl (1998) called this the *view-value distinction*, and it is an example of just one issue where identifying the locus of a transformation using the Visualization Reference Model helps to avoid confusion.

Information visualization is not just about the creation of visual images, but also the interaction with those images in the service of some problem. In the Visualization Reference Model, another set of arrows flow back from the human at the right into the transformations themselves, indicating the adjustment of these transformations by user-operated controls. It is the rapid reciprocal reaction between the generation of images by machine and the selection and parametric adjustment of those images, giving rise to new images that gives rise to the attractive power of interactive information visualization.

Data Structures

It is convenient to express Data Tables as tables of objects and their attributes, as in Table 10.3. For example, in the FilmFinder, the basic objects (or "cases") are films. Each film is associated with a number of attributes or variables, such as title, stars, year of release, genre type, and so forth. The vertical double black line in the table separates data in the table to the left of the line

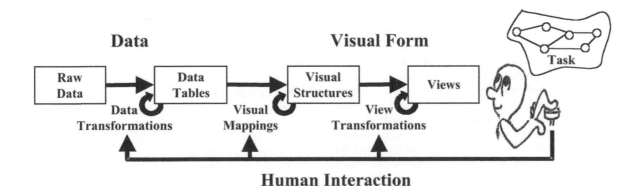

FIGURE 10.10. Reference model for visualization (Card et al., 1999). Visualization can be described as the mapping of data to visual form that supports human interaction in a workplace for visual sense making.

TABLE 10.3. A Data Table About Films

FilmID	230	105	540	...
Title	Goldfinger	Ben Hur	Ben Hur	...
Director	Hamilton	Wyler	Niblo	...
Actor	Connery	Heston	Novarro	...
Actress	Blackman	Harareet	McAvoy	...
Year	1964	1959	1926	...
Length	112	212	133	...
Popularity	7.7	8.2	7.4	
Rating	PG	G	G	...
FilmType	Action	Action	Drama	...

Source: (Card et al., 1999).

from the metadata, expressed as variable names, to the left of the line. The horizontal black line across the table separates input variables from output variables—that is, the table can be thought of as a function.

$$f(\text{input variables}) = \text{output variables}$$

So,

$$Year\ (FilmID = 105) = 1959$$

Variables imply a scale of measurement, and it is important to keep these straight. The most important to distinguish are:

N = *Nominal (are only = or ≠ to other values)*
O = *Ordinal (obeys a < relation)*
Q = *Quantitative (can do arithmetic on them)*

A nominal variable N is an unordered set, such as film titles {Goldfinger, Ben Hur, Star Wars}. An ordinal variable O is a tuple (ordered set), such as film ratings ⟨G, PG, PG-13, R⟩. A quantitative variable Q is a numeric range, such as film length [0, 360].

In addition to the three basic types of variables, subtypes represent important properties of the world associated with specialized visual conventions. We sometimes distinguish the subtype *Quantitative Spatial* (Q_s) for intrinsically spatial variables common in scientific visualization and the subtype *Quantitative Geographical* (Q_g) for spatial variables that are specifically geophysical coordinates. Other important subtypes are similarity metrics *Quantitative Similarity* (Q_m), and the temporal variables *Quantitative Time* (Q_t) and *Ordinal Time* (O_t). We can also distinguish Interval Scales (I) (like Quantitative Scales, but since there is not a natural zero point, it is not meaningful to take ratios). An example would be dates. It is meaningful to subtract two dates (June 5, 2002 – June 3, 2002 = 2 days), but it does not make sense to divide them (June 5, 2002 ÷ June 23, 2002 = Undefined). Finally, we can define an *Unstructured* Scale (U), whose only value is present or absent (e.g., an error flag). The scales are summarized in Table 10.4.

Scale types can be altered by transformations, and this practice is sometimes convenient. For example, quantitative variables can be mapped by data transformations into ordinal variables

$$Q \rightarrow O$$

by dividing them into ranges. For example, film lengths [0, 360] minutes (type Q) can be broken into the ranges (type O),

$$[0, 360]\ \text{minutes} \rightarrow \langle \text{Short, Medium, Long} \rangle.$$

This common transformation is called "classing," because it maps values onto classes of values. It creates an accessible summary of the data, although it loses information. In the other direction, nominal variables can be transformed to ordinal values

$$N \rightarrow O$$

based on their name. For example, film titles {Goldfinger, Ben Hur, Star Wars} can be sorted lexicographically

$$\{\text{Goldfinger, Ben Hur, Star Wars}\} \rightarrow$$
$$\langle \text{Ben Hur, Goldfinger, Star Wars} \rangle$$

Strictly speaking, we have not transformed their values, but in many uses (e.g., building alphabetically arranged dictionaries of words or sliders in the FilmFinder), we can act as if we had.

Variable scale types form an important class of metadata that, as we shall see, is important for proper information visualization. We can add scale type to our Data Table in Table 10.3 together with cardinality or range of the data to give us essentially a codebook of variables as in Table 10.5.

Visual Structures

Information visualization maps data relations into visual form. At first, it might seem that a hopelessly open set of visual forms can result. Careful reflection, however, reveals what every artist knows: that visual form is subject to strong constraints. Visual form that reflects the systematic mapping of data relations onto visual form, as in information visualization or data graphics, is subject to even more constraints. It is a genuinely surprising fact, therefore, that most information visualization involves the mapping data relations onto only a half dozen components of visual encoding:

1. *Spatial substrate*
2. *Marks*
3. *Connection*
4. *Enclosure*
5. *Retinal properties, or*
6. *Temporal encoding*

Of these mappings, the most powerful is how data are mapped onto the spatial substrate—that is, how data are mapped into spatial position. In fact, one might say that the design of an information visualization consists first of deciding which variables are going to get the spatial mappings, and then how the rest of the variables are going to make do with the coding mappings that are left.

Spatial substrate. As we have just said, the most important choice in designing an information visualization is which variables are going to map onto spatial position. This decision

TABLE 10.4. Classes of Data and Visual Elements

	Data Classes			Visual Classes	
Class	Description	Example		Description	Example
U	Unstructured (can only distinguish presence or absence)	ErrorFlag		Unstructured (no axis, indicated merely whether something is present or absent)	Dot
N	Nominal (can only distinguish whether two values are equal)	{Goldfinger, Ben Hur, Star Wars}		Nominal Grid (a region is divided into subregions, in which something can be present or absent)	Colored circle
O	Ordinal (can distinguish whether one value is less or greater but not difference or ratio)	{Small, Medium, Large}		Ordinal Grid (order of the subregions is meaningful)	Alpha slider
I	Interval (can do subtraction on values, but no natural zero and can't compute ratios)	|10 Dec. 1978– 4 Jun. 1982|		Interval Grid (region has a metric but no distinguished origin)	Year axis
Q	Quantitative (can do arithmetic on values)	|0–100| kg		Quantitative Grid (a region has a metric)	Time slider
Q_s	—Spatial variables	|0–20| m		—Spatial grid	
Q_m	—Similarity	|0–1|		—Similarity space	
Q_g	—Geographical coord.	|30°N–50°N|Lat.		—Geographical coord.	
Q_t	—Time variable	|10–20| μsec		—Time grid	

TABLE 10.5. Data Table with Meta-Data Describing the Types of the Variables

FilmID	N	230	105	
Title	N	Goldfinger	Ben Hur	. . .
Director	N	Hamilton	Wyler	. . .
Actor	N	Connery	Heston	. . .
Actress	N	Blackman	Harareet	. . .
Year	Q_t	1964	1959	. . .
Length	Q	112	212	. . .
Popularity	Q	7.7	8.2	. . .
Rating	O	PG	G	. . .
FilmType	N	Action	Action	. . .

Source: (Card et al., 1999).

gives importance to spatially encoded variables at the expense of variables encoded using other mappings. Space is perceptually dominant (MacEachren, 1995); it is good for discriminating values and picking out patterns. It is easier, for example, to identify the difference between a sine and a tangent curve when encoded as a sequence of spatial positions than as a sequence of color hues.

Empty space itself, as a container, can be treated as if it had metric structure. Just as we classified variables according to their scale type, we can think of the properties of space in terms of the scale type of an axis of space (cf. Engelhardt, Bruin, Janssen, & Scha, 1996). Axis scale types correspond to the variable scale types (see Table 10.4). The most important axes are:

U = Unstructured (no axis, indicated merely whether something is present or absent)
N = Nominal Grid (a region is divided into subregions, in which something can be present or absent)
O = Ordinal Grid (the ordering of these subregions is meaningful), and
Q = Quantitative Grid (a region has a metric)

Besides these, it is convenient to make additional distinctions for frequently used subtypes, such as Spatial axes (Qs).

Axes can be linear or radial; essentially, they can involve any of the various coordinate systems for describing space. Axes are an important building block for developing Visual Structures. Based on the Data Table for the FilmFinder in Table 10.5, we represent the scatterplot of as composed of two orthogonal quantitative axes:

$$Year \rightarrow Q_x,$$
$$Popularity \rightarrow Q_y.$$

The notation states that the Year variable is mapped to a quantitative X-axis and the Popularity variable is mapped to a quantitative Y-axis. Other axes are used for the FilmFinder query widgets. For example, an ordinal axis is used in the radio buttons for film ratings,

$$Ratings \rightarrow O_y.$$

and a nominal axis is used in the radio buttons for film type,

$$FilmType \rightarrow N_x.$$

Marks. Marks are the visible things that occur in space. There are four elementary types of marks (Fig. 10.11):

1. P = Points (0D),
2. L = Lines (1D),
3. A = Areas (2D), and
4. V = Volumes (3D).

Area marks include surfaces in three dimensions, as well as 2D-bounded regions.

Unlike their mathematical counterpart, point and line marks actually take up space (otherwise, they would be invisible) and may have properties such as shape.

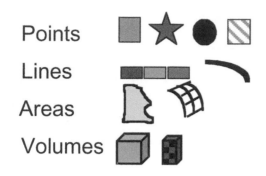

Points Lines Areas Volumes

FIGURE 10.11. Types of marks.

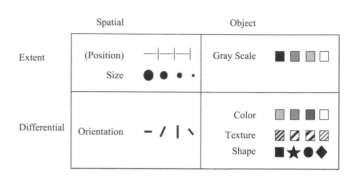

FIGURE 10.12. Retinal properties (Card et al., 1999). The six retinal properties can be grouped by whether they form a scale with a natural zero point (extend) and whether they deal with spatial distance or orientation (spatial).

Connection and enclosure. Point marks and line marks can be used to signify other sorts of topological structure: graphs and trees. These allow showing relations among objects without the geometrical constraints implicit in mapping variables onto spatial axes. Instead, we draw explicit lines. Hierarchies and other relationships can also be encoded using enclosure. Enclosing lines can be drawn around subsets of items. Enclosure can be used for trees, contour maps, and Venn diagrams.

Retinal properties. Other graphical properties were called retinal properties by Bertin (1967/1983), because the retina of the eye is sensitive to them independent of position. For example, the FilmFinder in Fig. 10.1 uses color to encode information in the scatterplot:

$$FilmID(FilmType) \rightarrow P(Color)$$

This notation says that the FilmType attribute for any FilmID case is visually mapped onto the color of a point.

Figure 10.12 shows Bertin's six "retinal variables" separated into spatial properties and object properties according to which area of the brain they are believed to be processed (Kosslyn, 1994). They are sorted according to whether the property is good for expressing the extent of a scale (has a natural zero point), or whether its principal use is for differentiating marks (Bertin, 1977/1981). Spatial position, discussed earlier as basic visual substrate, is shown in the position it would occupy in this classification.

Other graphical properties have also been proposed for encoding information. MacEachren (1995) has proposed (a) crispness (the inverse of the amount of distance used to blend two areas or a line into an area), (b) resolution (grain with raster or vector data will be displayed), (c) transparency, and (d) arrangement (e.g., different ways of configuring dots). He further proposed dividing color into (a) value (essentially, the gray level of Fig. 10.12), (b) hue, and (c) saturation. Graphical properties from the perception literature that can support preattentive processing have been suggested candidates for coding variables such as curvature, lighting direction, or direction of motion (see Healey, Booth, and Enns, 1995). All of these suggestions require further research.

Temporal encoding. Visual Structures can also temporally encode information; human perception is very sensitive to changes in mark position and the mark's retinal properties. We need to distinguish between temporal data variables to be visualized:

$$Q_t \rightarrow some\ visual\ representation$$

and animation, that is, mapping a variable into time,

$$some\ variable \rightarrow Time.$$

Time as animation could encode any type of data (whether it would be an effective encoding is another matter). Time as animation, of course, can be used to visualize time as data.

$$Q_t \rightarrow Time$$

This is natural, but not always the most effective encoding. Mapping time data into space allows comparisons between two points in time. For example, if we map time and a function of time into space (e.g., time and accumulated rainfall),

$$Q_t \rightarrow Qx\ [make\ time\ be\ the\ X\text{-}axis]$$
$$f(Q_t) \rightarrow Qy,\ [make\ accumulated\ rainfall\ be\ the\ Y\text{-}axis],$$

then we can directly experience rates as visual linear slope, and we can experience changes in rates as curves. This encoding of time into space for display allows us to make much more precise judgments about rates than would be possible from encoding time as time. Another use of time as animation is similar to the unstructured axes of space. Animation can be used to enhance the ability of the user to keep track changes of view or visualization. If the user clicks on some structure, causing it to enlarge and other structures to become smaller, animation can effectively convey the change and the identity of objects across the change, whereas simply viewing the two end states is confusing. Another use is to enhance a visual effect. Rotating a complicated object, for example, will induce 3D effects (hence, allow better reading of some visual mappings).

Expressiveness and Effectiveness

Visual mappings transform Data Tables into Visual Structure and then into a visual image. This image is not just an arbitrary image. It is an image that has a particular meaning it must express. That meaning is the data relation of which it is the visual transformation. We can think of the image as a sentence in a visual language (Mackinlay, 1986b) that expresses the relations in the Data Table. To be a good information visualization, the mappings must satisfy some constraints. The first constraint is that the mapping must be expressive. A visualization is said to be *expressive* if and only if it encodes all the data relations intended and no other data relations. The first part of expressiveness turns out to be easier than the second. Suppose we plot Film-Type against Year using the data-to-visual mapping in Fig. 10.13. The problem of this mapping is that the nominal movie rating data are expressed by a quantitative axis. That is, we have tried to map:

$$FilmType(N) \rightarrow Position(Q).$$

In so doing, we have visually expressed all the data relation, but the visualization also implies relationships that do not exist. For example, the 1959 version of Ben Hur does not have a film type that is five times greater than the 1926 version of Ben Hur, as implied in the figure. Wisely, the authors of the FilmFinder chose the mapping:

$$FilmType(N) \rightarrow Color(N).$$

Of course, there are circumstances in which color could be read as ordinal, or even possibly quantitative, but the miscellaneous order of the buttons in Fig. 10.1 discourages such an interpretation and the relatively low effectiveness of color for this purpose in Table 10.7 also discourages this interpretation.

Table 10.6 shows the mappings chosen by authors of the FilmFinder. The figure shows the Data Table's metadata and data

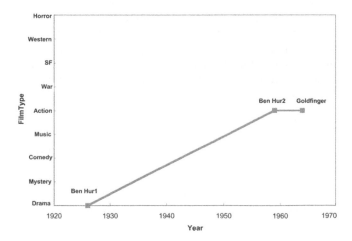

FIGURE 10.13. Mapping from data to visual form that violates expressiveness criterion.

and how they are mapped onto the Visual Structure. Note that the nominal data of the PG ratings is mapped onto a nominal visualization technique (colors). Note also, that names of directors and stars (nominal variables) are raised to ordinal variables (through alphabetization), and then mapped onto an ordinal axis. This is, of course, a common way to handle searching among a large number of nominal items.

Some properties are more effective than others for encoding information. Position is by far the most effective all-around representation. Many properties are more effective for some types of data than for others. Table 10.7 gives an approximate evaluation for the relative effectiveness of some encoding techniques based on (MacEachren, 1995). We note that spatial position is effective for all scale types of data. Shape, on the other hand, is only effective for nominal data. Gray scale is most effective for ordinal data. Such a chart can suggest representations to a visualization designer.

Taxonomy of Information Visualizations

We have shown that the properties of data and visual representation generally constrain the set of mappings that form the basis for information visualizations. Taken together, these constraints form the basis of a taxonomy of information visualizations. Such a taxonomy is given in Table 10.8. Visualizations are grouped into four categories. First are *Simple Visual Structures*, the static mapping of data onto multiple spatial dimensions, trees, or networks plus retinal variables, depicted in Fig. 10.10. Here it is worth distinguishing two cases. There is a perceptual barrier at three (or, in special cases, four) variables, a limit of the amount of data that can be perceived as an immediate whole. Bertin (1977, 1981) called this elementary unit of visual data perception the "image". Although this limit has not been definitively established in information visualization by empirical research, there must be a limit somewhere or else people could simultaneously comprehend a thousand variables. We therefore divide visualizations into those that can be comprehended in an elementary perceptual grasp (three, or in special cases, four variables)—let us call these *direct reading visualizations*—and those more complex than that barrier—which we call *articulated reading visualizations*, in which multiple actions are required.

Beyond the perceptual barrier, direct composition of data relationships in terms of 1, 2, or 3 spatial dimensions plus remaining retinal variables is still possible, but rapidly diminishes in effectiveness. In fact, the main problem of information visualization as a discipline can be seen as devising techniques for accelerating the comprehension of these more complex *n*-variable data relations. Several classes of techniques for *n*-variable visualization, which we call *Composed Visual Structures*, are based on composing Simple Visual Structures together by reusing their spatial axes. A third class of Visual Structures—*Interactive Visual Structures*—comes from using the rapid interaction capabilities of the computer. These visualizations invoke the parameter-controlling arrows of Fig. 10.10. Finally, a fourth class of visualizations—*Attention-Reactive Visual Structures*—comes from interactive displays where the system reacts to user actions

TABLE 10.6. Meta-Data and Mappings of Data onto Visual Structure in the FilmFinder

		Data							Visual Form		
Variable	Type	Range	Case$_i$	Case$_j$	Case$_k$	...		Type	Visual Structure	Control	Transformation Affected
FilmID	N	All-IDs	230	105	540	...	→	N	Points	Button	All (details)
Title	N	All-titles	Goldfinger	Ben Hur	Ben Hur	...	→sort	O		Alphaslider	Select cases
Director	N	All-directors	Hamilton	Wyler	Niblo	...	→sort	O		Alphaslider	Select cases
Actor	N	All-actors	Connery	Heston	Novarro	...	→sort	O		Alphaslider	Select cases
Actress	N	All-actresses	Blackman	Haraneet	McAvoy	...	→sort	O		Alphaslider	Select cases
Year	Q	[1926, 1989]	1964	1959	1926	...	→	Q	X-axis	Axis	Clip range
Length	Q	[0, 450]	112	212	133	...	→	Q		Two-sided slider	Clip range
Popularity	Q	[1, 9]	7.7	8.2	7.4	...	→	Q	Y-axis	Axis	Clip range
Rating	O	{G, PG, PG-13, R}	PG	G	G	...	→	O		Radio buttons	Select cases
Film Type	N	{Drama, Mystery, Comedy, Music, Action, War, SF, Western, Horror}	Action	Action	Drama		→	N	Color	Radio buttons	Select cases

Source: (Card et al., 1999).

TABLE 10.7. Relative Effectiveness of Position and Retinal Encodings

	Spatial	Q	O	N	Object	Q	O	N
Extent	(Position)	●	●	●	Gray Scale	◐	●	○
	Size	●	●	●	Color	◐	◐	●
Differential	Orientation	◐	◐	●	Texture	◐	◐	●
					Shape	○	○	●

Source: (Card et al., 1999).

by changing the display, even anticipating new displays, to lower the cost of information access and sensemaking to the user. To summarize,

I. *Simple Visual Structures*
   Direct Reading
   Articulated Reading
II. *Composed Visual Structures*
   Single-Axis Composition
   Double-Axis Composition
   Recursive Composition
III. *Interactive Visual Structure*
IV. *Attention-Reactive Visual Structure*

These classes of techniques may be combined to produce visualizations that are more complex. To help us keep track of the variable mapping into visual structure, we will use a simple shorthand notation for listing the element of the Visual Structure that the Data Table has mapped into. We will write, for example, [XYR^2] to note that variables map onto the X-axis, the Y-axis, and two retinal encodings. [OX] will indicate that the variables map onto one spatial axis used to arrange the objects (that is, the

cases), while another was used to encode the objects' values. Examples of this notation appear in Table 10.8 and Fig. 10.21.

SIMPLE VISUAL STRUCTURES

The design of information visualizations begins with mappings from variables of the Data Table into the Visual Structure. The basic strategy for the visualization designer could be described as follows:

1. *Determine which variables of the Analytic Abstraction to map into spatial position in the Visual Structure.*
2. *Combine these mappings to increase dimensionality (e.g., by folding).*
3. *Use retinal variables as an overlay to add more dimensions.*
4. *Add controls for interaction.*
5. *Consider attention-reactive features to expand space and manage attention.*

We start by considering some of the ways in which variables can be mapped into space.

TABLE 10.8.   Taxonomy of Information Visualization Techniques

I. SIMPLE VISUAL STRUCTURES	Trees
	  Node and link trees
	  Enclosure trees
	  TreeMaps
	  Cone trees
	Networks
	Time
		III. INTERACTIVE VISUAL STRUCTURES

Direct Reading
1-Variable [X]
  Lists
  1D object charts
  1D scatterplots
  Pie charts
  Folded dimensions
  Distributions
  Box plots
2-Variable [XY]
  2D object charts
  2D scatterplots
3-Variable
  [XYR]
  Retinal scatterplot
  Kahonen diagrams
  Retinal topographies
  [(XY)Z]
  Information landscapes
  Information surfaces
  [XYZ]
  3D scatterplots
4-Variable
  [XYZR]
  3D retinal scatterplots
  3D topographies
 	
—*Barrier of Perception*—
Articulated Reading
n-Variable
  [XYR^{n-2}]
  2D Retinal scatterplots
  [XYZR^{n-3}]
  2D Retinal scatterplots
	II. COMPOSED VISUAL STRUCTURES
	
	Single-axis
	Composition [XY^n]
	  Permutation matrices
	  Parallel coordinates
	Double-axis
	Composition [XY]
	  Graphs
	Recursive Composition
	2D in 2D [(XY)^{XY}]
	  Scatterplot matrices
	  Prosection matrices
	  Hierarchical axes
	Marks in 2D [(XY)^R]
	  Stick figures
	  Color icons
	  Shape coding
	  Keim spirals
	3D in 3D [(XYZ)^{XYZ}]
	  Worlds within worlds
		Dynamic queries
		Magic lens
		Overview + detail
		Linking and brushing
		Extraction & comparison
		Attribute explorer
		
		IV. FOCUS + CONTEXT ATTENTION-REACTIVE VISUAL ABSTRACTION
		
		Data-based Methods
		  Filtering
		  Selective aggregation
		View-based Methods
		  Micro-macro readings
		  Highlighting
		  Visual transfer functions
		  Perspective distortion
		  Alternate geometries

1-Variable

One-variable visual displays may actually use more than one visual dimension. This is because the data variable or attribute is displayed against some set of objects using some mark and because the mark itself takes space. Or, more subtly, it may be because one of the dimensions is used for arranging the objects and another for encoding via position of the variable. A simple example would be when the data are just visually mapped into a simple text list as in Fig. 10.14(a). The objects form a sequence on the Y-dimension, and the width of the marks (the text descriptor) takes space in the X-dimension. By contrast, a one-dimensional scattergraph (Fig. 10.14[b]) does not use a dimension for the objects. Here, the Y-axis is used to display the attribute variable (suppose these are distances from home of gas stations); the objects are encoded in the mark (which takes a little bit of the X-dimension).

More generally, many single-variable visualizations are in the form $v = f(o)$, where v is a variable attribute and o is the object. Figure 10.14(c) is of this form and uses the Y-axis to encode the variable and the X-axis for the objects. Note that if the objects are, as usual, nominal, then they are reorderable: sorting the objects on the variable produces easily perceivable visual patterns. For convenience, we have used rectangular coordinates, but any other orthogonal coordinates could be used as the basis of decomposing space. Figure 10.14(d) uses θ from polar coordinates to encode, say, percentage voting for different presidential candidates. In Fig. 10.14(e), a transformation on the data side has transformed variable o into a variable representing the distribution, then mapped that onto points on the Y-axis. In Fig. 10.14(f), another transformation on the data side has mapped this distribution into 2nd quartiles, 3rd quartiles, and outlier points, which is then mapped on the visual side into a box plot on the Y-axis. Simple as they are, these techniques can be very useful, especially in combination with other techniques.

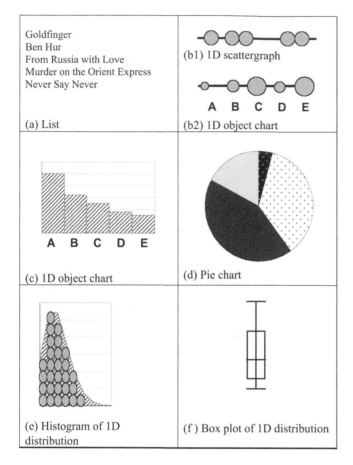

Goldfinger
Ben Hur
From Russia with Love
Murder on the Orient Express
Never Say Never

(a) List

(b1) 1D scattergraph

A B C D E

(b2) 1D object chart

A B C D E

(c) 1D object chart

(d) Pie chart

(e) Histogram of 1D distribution

(f) Box plot of 1D distribution

FIGURE 10.14.   1-variable visual abstractions.

One special, but common, problem is how to visualize very large dimensions. This problem occurs for single-variable visualizations, but may also occur for one dimension of a multi-variable visualization. Figure 10.15 shows several techniques for handing the problem. In Fig. 10.15(a) (Freeman & Fertig, 1995), the visual dimension is laid out in perspective. Even though each object may take only one or a few pixels on the axis, the objects are actually fairly large and selectable in the diagram. In Fig. 10.15(b) (Eick, Steffen, & Sumner, 1992), the objects (representing lines of code) are laid out on a *folded* Y-axis. When the Y-axis reaches the bottom of the page, it continues offset at the top. In Fig. 10.15(c) (Keim & Kriegel, 1994), the axis is wrapped in a square spiral. Each object is a single pixel, and its value is coded as the retinal variable color hue. The objects have been sorted on another variable; hence, the rings show the correlation of this attribute with that of the sorting attribute.

One-variable visualizations are also good parts of controls. Controls, in the form of slides, also consume considerable space on the display (for example, the controls in Fig. 10.1) that could be used for additional information communication. Figure 10.15(d) shows a slider on whose surface is a distribution representation of the number of objects for each value of the input variable, thereby communicating information about the slider's sensitivity in different data ranges. The slider on the left

of Fig. 10.15(b) has a one-variable visualization that serves as a legend for the main visualization: it associates color hues with dates and allows the selection of date ranges.

2-Variables

As we increase the number of variables, it is apparent that their mappings form a combinatorial design space. Figure 10.16 schematically plots the structure of this space, leaving out the use of multiple lower variable diagrams to plot higher variable combinations. Two-variable visualizations can be thought of as a composition of two elementary axes (Bertin, 1977, 1981; Mackinlay, 1986b), which use a single mark to encode the position on both those axes. Mackinlay called this *mark composition*, and it results in a 2D scattergraph (Fig. 10.16[g]). Note that instead of mapping onto two positional visual encodings, one positional axis could be used for the objects, and the data variables could be mapped onto a position encoding and a retinal encoding (size), as in Fig. 10.16(f).

3-Variables and Information Landscapes

By the time we get to three data variables, a visualization can be produced in several ways. We can use three separate visual dimensions to encode the three data variables in a *3D scattergraph* (Fig. 10.16[j]). We could also use two spatial dimensions and one retinal variable in a *2D retinal scattergraph* (Fig. 10.16[k]). Or we could use one spatial dimension as an object dimension, one as a data attribute dimension, and two retinal encodings for the other variables, as in an *object chart* such as in Fig 10.16(i). Because Fig. 10.16(i) uses multiple retinal encodings, however, it may not be as effective as other techniques. Notice that because they all encode three data variables, we have classified 2D and 3D displays together. In fact, one popular 3-variable information visualization that lies between 2D and 3D is the *information landscape* (Fig. 10.16[m]). This is essentially a 2D scattergraph with one data variable extruded into the third spatial dimension. Its essence is that two of the spatial dimensions are more tightly coupled and often relate to a 2D visualization. For example, the two dimensions might form a map with the bars showing the GDP of each region.

Another special type of 3-variable information visualization is a 2D *information topography*. In an information typography, space is partly defined by reference to external structure. For example, the topography of Fig. 10.17(a) is a map of San Francisco, requiring two spatial variables. The size of blue dots indexes the number of domain names registered to San Francisco street addresses. Looking at the patterns in the visualization shows that Internet addresses have especially concentrated in the Mission and South of Mission districts. Figure 10.17(a) uses a topography derived from real geographical space. Various techniques, such as multidimensional scaling, factor analysis, or connectionist self-organizing algorithms, can create abstract spaces based on the similarities among collections of documents or other objects. These abstract similarity spaces can function like a topography. An example can be seen in Fig. 10.17(b), where the pages in a website are depicted as regions in a similarity

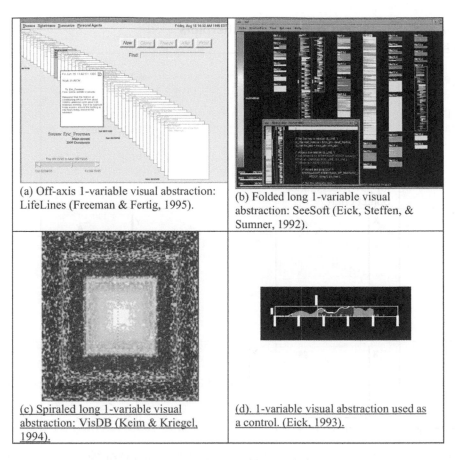

(a) Off-axis 1-variable visual abstraction: LifeLines (Freeman & Fertig, 1995).

(b) Folded long 1-variable visual abstraction: SeeSoft (Eick, Steffen, & Sumner, 1992).

(c) Spiraled long 1-variable visual abstraction: VisDB (Keim & Kriegel, 1994).

(d). 1-variable visual abstraction used as a control. (Eick, 1993).

FIGURE 10.15.   Uses of 1-variable visual abstractions.

space. To create this diagram[2], a web crawler crawls the site and indexes all the words and pages on the site. Each page is then turned into a document vector to represent the semantic content of that page. The regions are created using a neural network learning algorithm (see Lin, Soergel, & Marchionini (1991)). This algorithm organizes the set of web pages into regions. A visualization algorithm then draws boundaries around the regions, colors them, and names them. The result, called a *Kahonen diagram* after its original inventor, is a type of *retinal similarity topography*.

Information landscapes can also use marks that are surfaces. In Fig. 10.18(a), topics are clustered on a similarity surface, and the strength of each topic is indicated by a 3D contour. A more extreme case is Fig. 10.18(b), where an information landscape is established in spherical coordinates, and the amount of ozone is plotted as a semitransparent overlay on the ρ-axis.

n-Variables

Beyond three variables, direct extensions of the methods we have discussed become less effective. It is possible, of course to make plots using two spatial variables and *n*–2 retinal variables, and the possibilities for four variables are shown in Fig. 10.16. These diagrams can be understood, but at the cost of progressively more effort as the number of variables increases. It would be very difficult to understand an [XYR[20]] retinal scattergraph, for example.

Trees

An interesting alternative to showing variable values by spatial positioning is to use explicitly drawn linkages of some kind. Trees are the simplest form of these. Trees map cases into subcases. One of the data variables in a Data Table (for example, the variable Reports To in an organization chart) is used to define the tree. There are two basic methods for visualizing a tree: (a) Connection and (b) Enclosures.

Connection.   Connection uses lines to connect marks signifying the nodes of the tree. Logically, a tree could be drawn merely by drawing lines between objects randomly positioned on the plane, but such a tree would be visually unreadable. Positioning in space is important. Figure 10.20(a) is a tree from Charles Darwin's notebook (Robin, 1992) drawn to help

---

[2]This figure is produced by a program called SiteMap by Xia Lin and Associates. See http://faculty.cis.drexel.edu/sitemap/index.html.

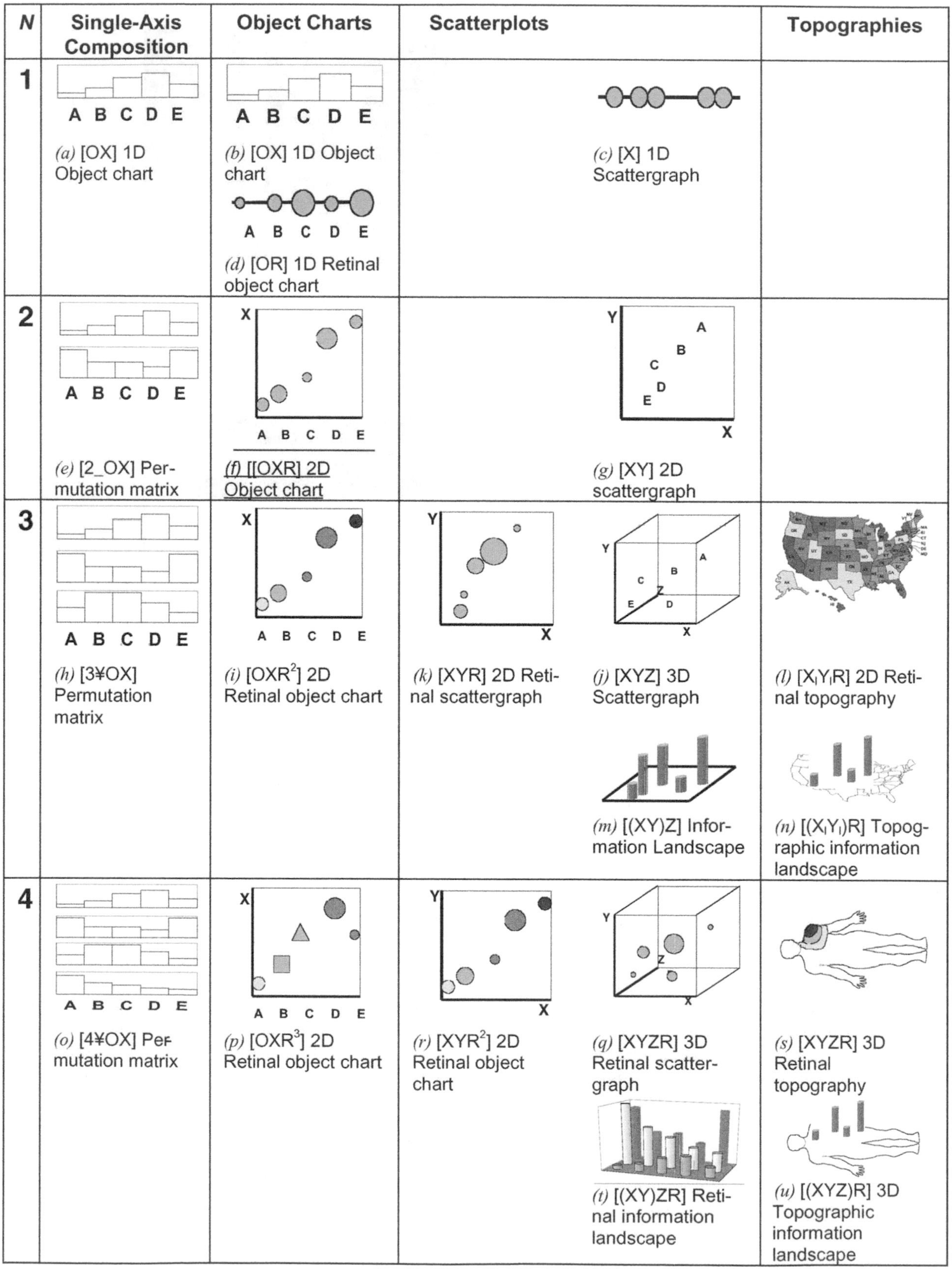

N	Single-Axis Composition	Object Charts	Scatterplots		Topographies
1	 A B C D E *(a)* [OX] 1D Object chart	 A B C D E *(b)* [OX] 1D Object chart A B C D E *(d)* [OR] 1D Retinal object chart	 *(c)* [X] 1D Scattergraph		
2	 A B C D E *(e)* [2_OX] Permutation matrix	 A B C D E <u>*(f)* [[OXR] 2D Object chart</u>	 *(g)* [XY] 2D scattergraph		
3	 A B C D E *(h)* [3¥OX] Permutation matrix	 A B C D E *(i)* [OXR^2] 2D Retinal object chart	 *(k)* [XYR] 2D Retinal scattergraph	 *(j)* [XYZ] 3D Scattergraph *(m)* [(XY)Z] Information Landscape	 *(l)* [X$_l$Y$_l$R] 2D Retinal topography *(n)* [(X$_l$Y$_l$)R] Topographic information landscape
4	 A B C D E *(o)* [4¥OX] Permutation matrix	 A B C D E *(p)* [OXR^3] 2D Retinal object chart	 *(r)* [XYR^2] 2D Retinal object chart	 *(q)* [XYZR] 3D Retinal scattergraph *(t)* [(XY)ZR] Retinal information landscape	 *(s)* [XYZR] 3D Retinal topography *(u)* [(XYZ)R] 3D Topographic information landscape

FIGURE 10.16. Simple Visual Structures.

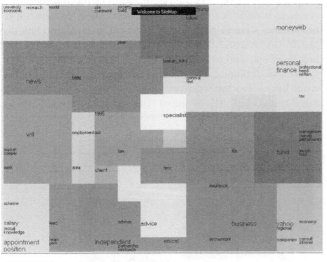

(a) X_lY_lR Retinal topography

(b) X_sY_sR Retinal similarity topography

FIGURE 10.17.   Retinal information topographies.

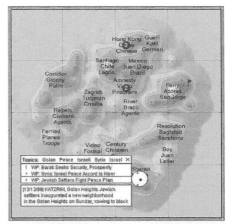

(a) News stories based on ThemeScapes (Wise et al., 1995). Courtesy NewsMaps.com.

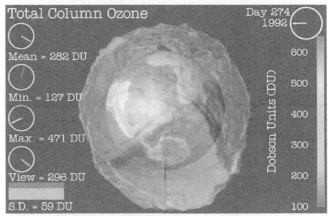

(b) Ozone layer surrounding earth. L. Treinish. Courtesy IBM.

FIGURE 10.18.   3D information surface topographies.

him work out the theory of evolution. Lines proceed from ancestor species to new species. Note that even in this informal setting intended for personal use that the tree uses space systematically (and opportunistically). There are no crossed lines. A common way of laying out trees is to have the depth in the tree map onto one ordinal access as in Fig. 10.20(b), while the other axis is nominal and used to separate nodes. Of course, trees could also be mapped into other coordinate systems: for example, there can be circular trees in which the r-axis represents depth and the θ-axis is used to separate nodes as in the representation of the evolution species in Fig. 10.20(c).[3] It is because trees have no cycles that one of the spatial dimensions can be used to encode tree depth. This partial correlation of

tree structure and space makes trees relatively easy to lay out and interpret, compared to generalized networks. Hierarchical displays are important not only because many interesting collections of information, such as organization charts or taxonomies, are hierarchical data, but also because important collections of information, such as websites, are *approximately* hierarchical. Whereas practical methods exist for displaying trees up to several thousand nodes, no good methods exist for displaying general graphs of this size. If a visualization problem involves the displaying of network data, a practical design heuristic is to see whether the data might not be forced into a display as a modified tree, such as a tree with a few non-tree links. A significant disadvantage of trees is that as they get large, they acquire an extreme aspect ratio, because the nodes expand exponentially as a function with depth. Consequently, any sufficiently large tree (say, >1000 nodes) resembles a straight line. Circular trees such as Fig. 10.20(c) are one way of trying to buy more space to mitigate this problem. Another disadvantage of trees is the significant empty space between nodes to make their organization easily readable. Various tricks can be used to

---

[3]This figure is from David Hillis, University of Texas.

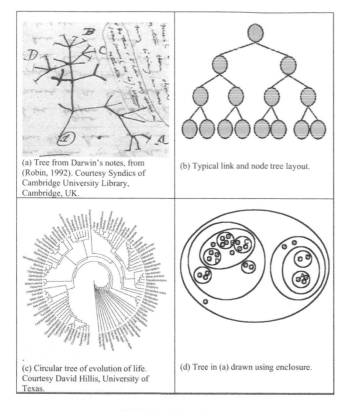

(a) Tree from Darwin's notes, from (Robin, 1992). Courtesy Syndics of Cambridge University Library, Cambridge, UK.

(b) Typical link and node tree layout.

(c) Circular tree of evolution of life. Courtesy David Hillis, University of Texas.

(d) Tree in (a) drawn using enclosure.

FIGURE 10.19. Trees.

wrap parts of the tree into this empty space, but at the expense of the tree's virtues of readability.

Enclosure. Enclosure uses lines to hierarchically enclose nested subsets of the tree. Figure 10.20(d) is an enclosure tree encoding of Darwin's tree in Fig. 10.20(a). We have already seen one attempt to use tree enclosure, TreeMaps (Fig. 10.5). TreeMaps make use of all the space and stay within prescribed space boundaries, but they do not represent the nonterminal nodes of the tree very well and similar leaves can have wildly different aspect ratios. Recent variations on TreeMaps found ways to "squarify" nodes (Shneiderman & Wattenberg, 2001), mitigating this problem.

Networks

Networks are more general than trees and may contain cycles. Networks may have directional links. They are useful for describing communication relationships among people, traffic in a telephone network, and the organization of the Internet. Containment is difficult to use as a visual encoding for network relationships, so most networks are laid out as node and link diagrams. Unfortunately, straightforward layouts of large node and link diagrams tend to resemble a large wad of tangled string.

We can distinguish the same types of nodes and links in network Visual Structures that we did for spatial axes: (a) Unstructured (unlabeled), (b) Nominal (labeled), (c) Ordinal (labeled

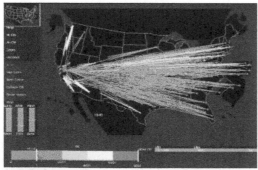

(a) Telephone traffic after California earthquake. (Becker, Eick, & Wilks, 1995).

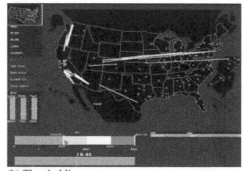

(b) Thresholding.

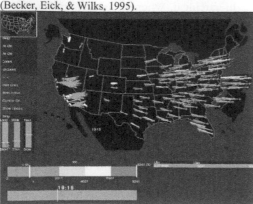

(c) Line shortening

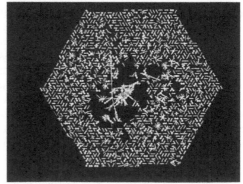

(d) Visualization to detect telephone fraud. (Cox, Eick, & Wills, 1997).

FIGURE 10.20. Network methods.

with an ordinal quantity), or (d) Quantitative (weighted links). Retinal properties, such as size or color, can be used to encode information about links and nodes. As in the case of trees, spatial positioning of the nodes is extremely important. Network visualizations escape from the strong spatial constraints of simple Visual Structures only to encounter another set of strong spatial constraints of node links crossing and routing. Networks and trees are not so much an alternative of the direct graphical mappings we have discussed so far as they are another set of techniques that can be overlaid on these mappings. Small node and link diagrams can be laid out opportunistically by hand or by using graph drawing algorithms that have been developed (Battista, Eades, Tamassia, & Tollis, 1994; Cruz & Tamassia, 1998; Tamassia, 1996) to optimize minimal link crossing, symmetry, and other aesthetic principles.

For very large node and link diagrams, additional organizing principles are needed. If there is an external topographic structure, it is sometimes possible to use the spatial variables associated with the nodes. Figure 10.20(a) shows a network based on call traffic between cities in the United States (Becker, Eick, & Wilks, 1995). The geographical location of the cities is used to lay out the nodes of the network. Another way to position nodes is by associating nodes with positions in a similarity space, such that the nodes that have the strongest linkages to each other are closest together. There are several methods for computing node nearness in this way. One is to use multidimensional scaling (MDS) (Fairchild, Poltrock, & Furnas, 1988). Another is to use a "spring" technique, in which each link is associated with a Hooke's Law spring weighted by strength of association and the system of springs is solved to obtain node position. Eick and Willis (1993) have argued that the MDS technique places too much emphasis on smaller links. They have derived an alternative that gives clumpier (and hence, more visually structured) clusters of nodes. If positioning of nodes corresponds perfectly with linkage information, then the links do not add more visual information. If positioning does not correspond at all with linkage information, then the diagram is random and obscure. In large graphs, node positions must have a partially correlated relationship to linkage in order to allow the emergence of visual structure. Note that this is what happens in the telephone traffic diagram, Fig. 10.20(a). Cities are positioned by geographical location. Communication might be expected to be higher among closer cities, so the fact that communications is heavy between coasts stands out.

A major problem in a network such as Fig. 10.20(a) is that links may obscure the structure of the graph. One solution is to route the links so that they do not obscure each other. The links could even be drawn outside the plane in the third dimension; however, there are limits to the effectiveness of this technique. Another solution is to use *thresholding*, as in Fig. 10.20(b). Only those links representing traffic greater than a certain threshold are included; the others are elided, allowing us to see the most important structure. Another technique is *line shortening*, as in Fig. 10.20(c). Only the portion of the line near the nodes is drawn. At the cost of giving up the precise linkage, it is possible to read the density of linkages for the different nodes. Figure 10.20(d) is a technique used to find patterns in an extremely large network. Telephone subscribers are represented as nodes on a hexagonal array. Frequent pairs are located near each other on the array. Suspicious patterns are visible because of the sparseness of the network.

The insightful display of large networks is difficult enough that many information visualization techniques depend on interactivity. One important technique, for example, is node aggregation. Nodes can be aggregated to reduce the number of links that have to be drawn on the screen. Which nodes are aggregated can depend on the portion of the network on which the user is drilling down. Similarly, the sets of nodes can be interactively restricted (e.g., telephone calls greater than a certain volume) to reduce the visualization problem to one within the capability of current techniques.

COMPOSED VISUAL STRUCTURES

So far, we have discussed simple mappings from data into spatial position axes, connections and enclosures, and retinal variables. These methods begin to run into a barrier around three variables as the spatial dimensions are used up and as multiples of the less efficient retinal variables needed. Most interesting problems involve many variables. We shall therefore look at a class of methods that reuse precious spatial axes to encode variables. This is done by composing a compound Visual Structure out of several simple Visual Structures. We will consider five subclasses of such composition: (a) single-axis composition, (b) double-axis composition, (c) mark composition, (d) case composition, and (e) recursive composition. Schematically, we illustrate these possibilities in Fig. 10.21.

Single-axis composition. In single-axis composition, multiple variables that share a single axis are aligned using that axis, as illustrated in Fig. 10.21(a). An example of single-axis composition is a method due to Bertin called *permutation matrices* (Bertin, 1977/1981). In a permutation matrix (Fig. 10.16[o], for example), one of the spatial axes is used to represent the cases and the other a series of bar charts (or rows of circles of different size or some other depiction of the value of each variable) to represent the values. In addition, bars for values below average may be given a different color, as in Fig. 10.7, in order to enhance the visual patterns. The order of the objects and the order of the variables may both be permuted until patterns come into play. Permutation matrices were used in our hotel analysis example. They give up direct reading of the data space in order to handle a larger number of variables. Of course, as the number of variables (or objects) increases, manipulation of the matrices becomes more time-consuming and visual interpretation more complex. Still, permutation matrices or their variants are one of the most practical ways of representing multi-variable data.

If we superimpose the bar charts of the permutation matrix atop one another, and then replace the bar chart with a line linking together the tops of the bars, we get another method for handling multiple variables by single-axis composition—*parallel coordinates* (Inselberg, 1997; Inselberg & Dimsdale, 1990), as shown in Fig. 10.22. A problem is analyzed in parallel coordinates by interactively restricting the objects displayed (the lines) in order to look at cases with common characteristics.

(a) Single-axis composition		
(b) Double-axis composition		
(c) Mark composition		
(d) Case composition		
(e) Recursive composition		

FIGURE 10.21. Composition types.

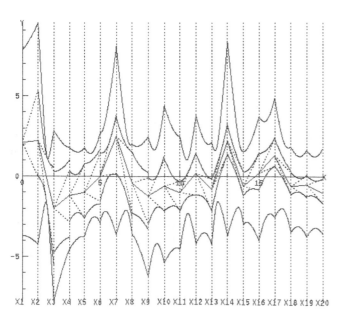

FIGURE 10.22. Single-axis composition: parallel coordinates.

In Fig. 10.22, parallel coordinates are used to analyze the problem of yield from a certain processor chip. X1 is chip yield, X2 is quality, X3 through X12 are defects, and the rest of the variables are physical parameters. The analysis, looking at those subsets of data with high yield and noticing the distribution of lines on the other parameters, was able to solve a significant problem in chip processing.

Both permutation matrices and parallel coordinates allow analyses in multi-dimensional space, because they are efficient in the use (and reuse) of spatial position and the plane. Actually, they also derive part of their power from being interactive. In the case of permutation matrices, interactivity comes in reordering the matrices. In the case of parallel coordinates, interactivity comes in selecting subsets of cases to display.

Double-axis composition. In double-axis composition, two visual axes must be in correspondence, in which case the cases are plotted on the same axes as a multivariable graph (Fig. 10.21[b]). Care must be taken that the variables are plotted on a comparable scale. For this reason, the separate scales of the variables are often transformed to a common proportion change scale. An example would be change in price for various stocks.

The cases would be the years, and the variables would be the different stocks.

Mark composition and case composition. Composition can also fuse diagrams. We discussed that each dimension of visual space can be said to have properties as summarized in Table 10.4. The visual space of a diagram is composed from the properties of its axis. In *mark composition* (Fig. 10.21[c]), the mark on one axis can fuse with the corresponding mark on another axis to form a single mark in the space formed by the two axes. Similarly, two object charts can be fused into a single diagram by having a single mark for each case. We call this latter form *case composition* Fig. 10.21(d).

Recursive composition. Recursive composition divides the plane (or 3D space) into regions, placing a subvisualization

in each region (Fig. 10.21[e]). We use the term somewhat loosely, since regions have different types of subvisualizations. The FilmFinder in Fig. 10.1 is a good example of a recursive visualization. The screen breaks down into a series of simple Visual Structures and controls: (a) a 3-variable retinal scattergraph (Year, Rating, FilmType) + (b) a 1-variable slider (Title) + (c) a 1-variable slider (Actors) + (d) a 1-variable slider (Actresses) + (e) a 1-variable slider (Director) + (f) a 1-variable slider (FilmLength) + (g) a 1-variable radio button control (Rating) + (h) a 1-variable button-set (FilmType).

Three types of recursive composition deserve special mention: (a) 2D-in-2D, (b) marks-in-2D, and (c) 3D-in-3D. An example of *2D-in-2D composition* is the "prosection matrix" (Tweedie, Spence, Dawkes, & Su, 1996) shown in Fig. 10.23(a). Each smaller square in the prosection matrix represents a pair of parameters plotted against each other. The coloring shows

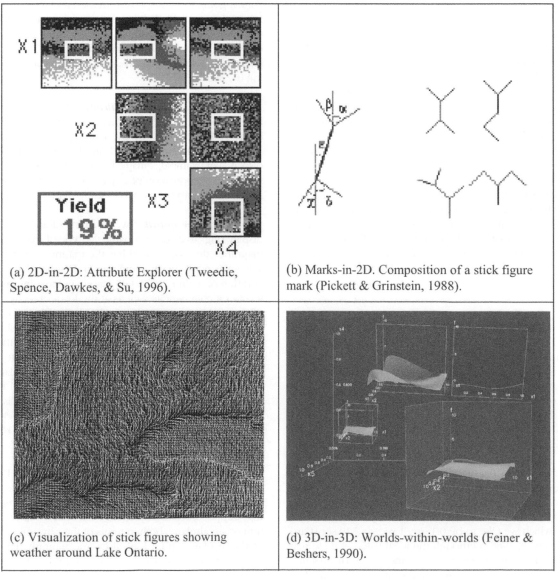

(a) 2D-in-2D: Attribute Explorer (Tweedie, Spence, Dawkes, & Su, 1996).

(b) Marks-in-2D. Composition of a stick figure mark (Pickett & Grinstein, 1988).

(c) Visualization of stick figures showing weather around Lake Ontario.

(d) 3D-in-3D: Worlds-within-worlds (Feiner & Beshers, 1990).

FIGURE 10.23.   Recursive composition.

which values of the plotted pair give excellent (red region) or partly good (gray regions) performance for the design of some device. The arrangement of the individual matrices into a supermatrix redefines the spatial dimensions (that is, associates it with different variables) within each of the cells, and the cells themselves are arranged in an overall scheme that systematically uses space. In this way, the precious spatial dimension is effectively expanded to where all the variables can reuse it. An important property of techniques similar to this one is that space is defined at more than one *grain size*, and these levels of grain become the basis for a *macro-micro reading*.

An example of *marks-in-2D composition* is the use of "stick figure" displays. This is an unusual type of visualization in which the recursion is within the mark instead of within the use of space. Figure 10.23(b) shows a mark that is itself composed of submarks. The mark is a line segment with four smaller line segments protruding from the ends. Four variables are mapped onto angles of these smaller line segments and a fifth onto the angle of the main line segment. Two additional variables are mapped onto the position of this mark in a 2D display. A typical result is the visualization in Fig. 10.23(c), which shows five weather variables around Lake Ontario, the outline of which clearly appears in the figure.

Feiner and Beshers (1990) provided an example of the third recursive composition technique, *3D-in-3D composition*. Suppose a dependent variable is a function of six continuous variables, $y = f(x, y, z, w, r, s)$. Three of these variables are mapped onto a 3D coordinate system. A position is chosen in that space, say, $x1, y1, z1$. At that position, a new 3D coordinate system is presented with a surface defined by the other three variables (Fig. 10.23[d]). The user can thus view $y = f(x1, y1, z1, w, r, s)$. The user can slide the second-order coordinate system to any location in the first, causing the surface to change appropriately. Note that this technique combines a composed visual interaction with interactivity on the composition. Multiple second-order coordinate systems can be displayed at the space simultaneously, as long as they do not overlap by much.

INTERACTIVE VISUAL STRUCTURES

In the examples we have considered so far, we have often seen that information visualization techniques were enhanced by being interactive. Interactivity is what makes visualization a new medium, separating it from generations of excellent work on scientific diagrams and data graphics. Interactivity means controlling the parameters in the visualization reference model (Fig. 10.10). This naturally means that there are different types of interactivity, because the user could control the parameters to data transformations, to visual mappings, or to view transformations. It also means that there are different forms of interactivity based on the response cycle of the interaction. As an approximation, we can think of there being three time constants that govern interactivity, which we take to be 0.1 sec, 1 sec, and 10 sec (Card, Moran, & Newell, 1986) (although the ideal value of these may be somewhat less, say, 0.07 sec, 0.7 sec, and 7 sec). The first time constant is the time in which a system response must be made, if the user is to feel that there is a direct physical manipulation of the visualization. If the user clicks on a button or moves a slider, the system needs to update the display in less than 0.1 sec. Animation frames need to take less than 0.1 sec. The second time constant, 1 sec, is the time to complete an immediate action, for example, an animated sequence such as zooming in to the data or rotating a tree branch. The third time constant 10 sec (meaning somewhere in the 5 to 30 sec interval) is the time for completing some cognitive action, for example deleting an element from the display. Let us consider a few well-known techniques for interactive information visualizations.

Dynamic queries. A general paradigm for visualization interaction is dynamic queries, the interaction technique used by the FilmFinder in Fig. 10.1. The user has a visualization of the data and a set of controls, such as sliders, by which subsets of the Data Table can be selected. For example, Table 10.9 shows the mappings of the Data Table and controls for the FilmFinder. The sliders and other controls will select which subset of the data is going to be displayed. In the FilmFinder, the control for Length is a two-sided slider. Setting one end to 90 minutes and the other end to 120 minutes will select for display only those cases of the Data Table whose year variable lies between these limits. The display needs to change within the 0.1 sec of changing the slider.

Magic lens (movable filter). Dynamic queries is one type of interactive filter. Another type is a movable filter that can be moved across the display, as in Fig. 10.24(a). These *magic lenses* are useful when it is desired to filter only some of the display. For example, a magic lens could be used with a map that showed the population of any city it was moved over. Multiple magic lenses can be used to cascade filters.

Overview + detail. We can think of an overview + detail display (Fig. 10.24[b]) as a particular type of magic lens, one that magnifies the display and has the magnified region off to the side so as not to occlude the region. Displays have information at different grain sizes. A GIS map may have information at the level of a continent as well as at the level of a city. If the shape of the continent can be seen, the display is too coarse to see the roadways of a city. Overview + detail displays show that data at more than one level, but they also show where the finer grain display fits into the larger grain display. In Fig. 10.24(b), from SeeSoft (Eick et al., 1992), a system for visualizing large software systems, the amount of magnification in the detail view is large enough that two concatenated overview + detail displays are required. Overview + detail displays are thus very helpful for data navigation. Their main disadvantage is that they require coordination of two visual domains.

Linking and brushing. Overview + detail is an example of coordinating dual representations of the same data. These can be coordinated interactively with *linking and brushing*. Suppose, for example, we wish to show power consumption on an airplane, both in terms of the physical representation of the airplane and a logical circuit diagram. The two views could be shown and *linked* by using the same color for the same component types. Interactivity itself can be used for a dynamic form of linking called *brushing*. In brushing, running the

TABLE 10.9.   Visual Marks and Controls for FilmFinder

| | | **Data** | | | | | | **Visual Form** | | |
Variable	Type	Range	Case$_i$	Case$_j$	Case$_k$	...	Type	Visual Structure	Control	Transformation Affected
FilmID	N	All-IDs	230	105	540	...	→ N	Points	Button	All (details)
Title	N	All-titles	Goldfinger	Ben Hur	Ben Hur	...	→sort O		Alphaslider	Select cases
Director	N	All-directors	Hamilton	Wyler	Niblo	...	→sort O		Alphaslider	Select cases
Actor	N	All-actors	Connery	Heston	Novarro	...	→sort O		Alphaslider	Select cases
Actress	N	All-actresses	Blackman	Harareet	McAvoy	...	→sort O		Alphaslider	Select cases
Year	Q	[1926, 1989]	1964	1959	1926	...	→ Q	X-axis	Axis	Clip range
Length	Q	[0, 450]	112	212	133		→ Q		Two-sided slider	Clip range
Popularity	Q	[1, 9]	7.7	8.2	7.4	...	→ Q	Y-axis	Axis	Clip range
Rating	O	{G, PG, PG-13, R}	PG	G	G	...	→ O		Radio buttons	Select cases
Film Type	N	{Drama, Mystery, Comedy, Music, Action, War, SF, Western, Horror}	Action	Action	Drama		→ N	Color	Radio buttons	Select cases

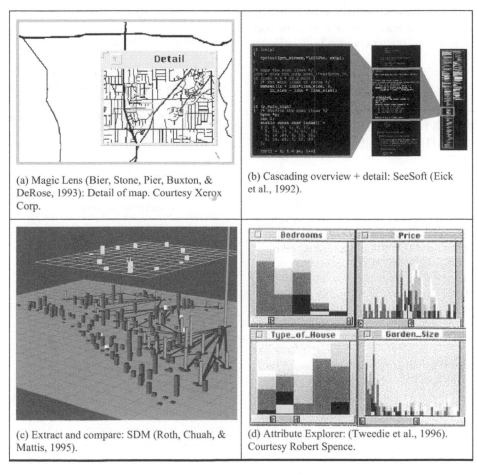

(a) Magic Lens (Bier, Stone, Pier, Buxton, & DeRose, 1993): Detail of map. Courtesy Xerox Corp.

(b) Cascading overview + detail: SeeSoft (Eick et al., 1992).

(c) Extract and compare: SDM (Roth, Chuah, & Mattis, 1995).

(d) Attribute Explorer: (Tweedie et al., 1996). Courtesy Robert Spence.

FIGURE 10.24.   Interaction techniques.

cursor over a part of one of the views causes highlighting both in that view and in the other view.

Extraction and comparison. We can also use interaction to extract a subset of the data to compare with another subset. An example of this is in the SDM system (Chuah, Roth, Mattis, & Kolojejchick, 1995) in Fig. 10.24(c). The data are displayed in a 3D information landscape, but the perspective interferes with the ability to compare it. Information is therefore *extracted* from the display (leaving ghosts behind) and placed in an orthogonal viewing position where it can be *compared* using 2D. It could also be dropped into another display. Interactivity makes possible these manipulations, while keeping them coordinated with the original representations.

Attribute explorer. Several of these interactive techniques are combined in the Attribute Explorer (Tweedie et al., 1996). Figure 10.24(d) shows information on four attributes of houses. Each attribute is displayed by a histogram, where each square making up the histogram represents an individual house. The user selects a range of some attribute, say price. Those pixels making up the histogram on price have their corresponding pixels linked representing houses highlighted on the other attributes. Those houses meeting all the criteria are highlighted in one color; those houses meeting, say, all but one are highlighted in another color. In this way, the user can tell about the "near misses." If the users were to relax one of the criteria only a little (say, reducing price by $100), then the user might be able to gain more on another criterion (say, reducing a commute by 20 miles).

FOCUS + CONTEXT ATTENTION-REACTIVE ABSTRACTIONS

So far, we have considered visualizations that are static mappings from Data Table to Visual Structure and those where the mappings Data Table to Visual Structure are interactively controlled by the user. We now consider visualizations in which the machine is no longer passive, but its mappings from Visual Structure to View are altered by the computer according to its model of the user's *degree of interest*. We can, in principle, associate a cost of access with every element in the Data Table. Take the FilmFinder in Figure 10.3. Details about the movie "Murder on the Orient Express" are accessible at low cost in terms of time because they are presently visible on the screen. Details of "Goldfinger," a movie with only a mark on the display, take more time to find. Details of "Last Year at Marienbad," a movie with no mark on the display, would take much more time. The idea is that with a model for predicting users' changes in interest, the system can adjust its displays to make costs lower for information access. For example, if the user wants some detail about a movie, such as the director, the system can anticipate that the user is more likely to want other details about the movie as well and therefore display them all at the same time. The user does not have to execute a separate command; the cost is therefore reduced.

Focus + context views are based on several premises: First, the user needs both overview (context) and detail information (focus) during information access, and providing these in separate screens or separate displays is likely to cost more in user time. Second, information needed in the overview may be different from that needed in the detail. The overview needs to provide enough information to the user to decide where to examine next or to give a context to the detailed information rather than the detailed information itself. As Furnas (1981) has argued, the user's interest in detail seems to fall away in a systematic way with distance as information objects become farther from current interest. Third, these two types of information can be combined within a single dynamic display, much as human vision uses a two-level focus and context strategy. Information broken into multiple displays (separate legends for a graph, for example) seems to degrade performance due to reasons of visual search and working memory.

Furnas (1981) was the first to articulate these ideas systematically in his theory of *fish-eye views*. The essence of focus + context displays is that the average cost of accessing information is reduced by placing the most likely needed information for navigation and detail where it is fastest to access. This can be accomplished by working on either the data side or the visual side of the visual reference model, Fig. 10.10. We now consider these techniques in more detail.

Data-Based Methods

Filtering. On the data side, focus + context effects can be achieved by filtering out which items from the Data Table are actually displayed on the screen. Suppose we have a tree of categories taken from *Roget's Thesaurus*, and we are interacting with one of these, "Hardness."

> Matter
> > ORGANIC
> > > *Vitality*
> > > > *Vitality in general*
> > > > *Specific vitality*
> > > *Sensation*
> > > > *Sensation in general*
> > > > *Specific sensation*
> > INORGANIC
> > > *Solid*
> > > > **Hardness**
> > > > *Softness*
> > > *Fluid*
> > > > *Fluids in general*
> > > > *Specific fluids*

Of course, this is a small example for illustration. A tree representing a program listing or a computer directory or a taxonomy could easily have thousands of lines, a number that would vastly exceed what could fit on the display and hence would have a high cost of accessing. We calculate a degree-of-interest (DOI) for each item of the tree, given that the focus is on the node Hardness. To do this, we split the DOI into an intrinsic part and a part that varies with distance from the current center of interest and use a formula from Furnas (1981).

$$DOI = Intrinsic\ DOI + Distance\ DOI$$

Figure 10.25 shows schematically how to perform this computation for our example. We assume that the intrinsic DOI of a node is just its distance of the root (Fig. 10.25[a]). The distance part of the DOI is just the traversal distance to a node from the current focus node (Fig. 10.25[b]; it turns out to be convenient to use negative numbers for this computation, so that the maximum amount of interest is bounded, but not the minimum amount of interest). We add these two numbers together (Fig. 10.25 [c]) to get the DOI of each node in the tree. Then we apply a minimum threshold of interest (−5 in this case) and only show nodes more interesting than that threshold. The result is the reduced tree:

> *Matter*
> > *INORGANIC*
> > *ORGANIC*
> > > *Solid*
> > > > **Hardness**
> > > > *Softness*
> > > *Fluid*

The reduced tree gives local context around the focus node and progressively less detail farther away. But it does seem to give the important context.

Selective aggregation. Another focus+context technique from the data side is selective aggregation. Selective aggregation creates new cases in the Data Table that are aggregates of other cases. For example, in a visualization of voting behavior in a presidential election, voters could be broken down by sex, precinct, income, and party affiliation. As the user drills down on, say, male Democrats earning between $25,000 and $50,000, other categories could be aggregated, providing screen space and contextual reference for the categories of immediate interest.

View-Based Methods

Micro-macro readings. Micro-macro readings are diagrams in which "detail cumulates into larger coherent structures" (Tufte, 1990). The diagram can be graphically read at the level of larger contextual structure or at the detail level. An example is Fig. 10.26. The micro reading of this diagram shows three million observations of the sleep (lines), wake (spaces), and feeding (dots) activity of a newborn infant. Each day's activity is repeated three times on a line to make the cyclical aspect of the activity more clearly visible. The macro reading of the diagram, emphasized by the thick lines, shows the infant transitioning from the natural human 25-hour cycle at birth to the 24-hour solar day. The macro reading serves as context and index into the micro reading.

Highlighting. Highlighting is a special form of micro-macro reading in which focal items are made visually distinctive in some way. The overall set of items provides a context for the changing focal elements.

Visual transfer functions. We can also warp the view with viewing transformations. An example is a visualization called the *bifocal lens* (Spence & Apperley, 1982). Fig. 10.27(a) shows a set of documents the user would like to view, but which is too large to fit on the screen. In a bifocal lens, documents not in a central focal region are compressed down to a smaller size. This could be a strict visual compression. It could also involve a change in representation. We can talk about the visual compression in terms of a visual transfer function Fig. 10.27(b), sometimes conveniently represented in terms of its first derivative in Fig. 10.27(c). This function shows how many units of an

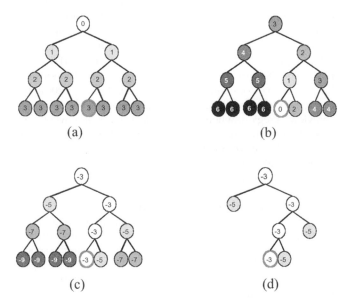

(a) (b)

(c) (d)

FIGURE 10.25. Degree-of-Interest calculation for fish-eye visualization.

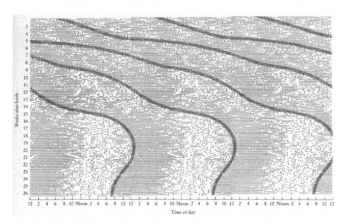

FIGURE 10.26. Micro-macro reading. (Winfree, 1987). Courtesy Scientific American Library.

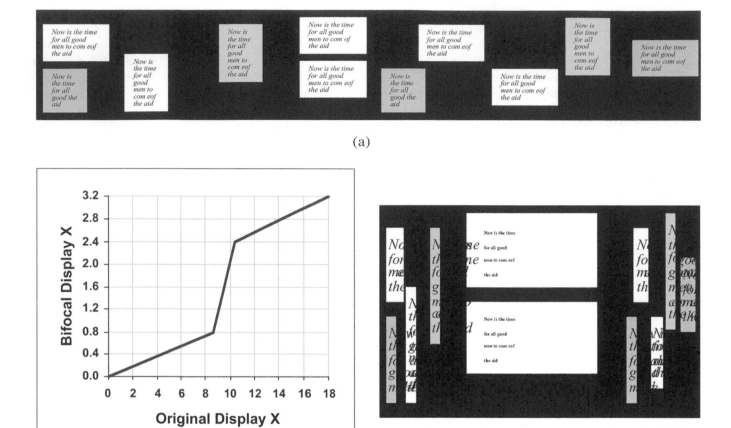

FIGURE 10.27. Bifocal + transfer function.

axis in the original display are mapped into how many units in the resultant display. The result could be compression or enlargement of a section of the display. As a result of applying this visual transfer function to Fig. 10.27(a), the display is compressed to Fig. 10.27(d). Actually, the documents in the compressed region have been further altered by using a semantic zooming function to give them a simplified visual form. The form of Fig. 10.27(c) shows that this is essentially a step function of two different slopes. An example of a two-dimensional step function is the Table Lens (Fig. 10.28[a]). The Table Lens is a spreadsheet in which the columns of selected cells are expanded to full size in X and the rows of selected cells are expanded to full size in Y. All other cells are compressed, and their content represented only by a graphic. As a consequence, spreadsheets up to a couple orders of magnitude larger can be represented.

By varying the visual transfer function (see, for example, the review by Leung and Apperley (1994)), a wide variety of distorted views can be generated. Figure 10.28(b) shows an application in which a visual transfer function is used to expand a bubble around a local region on a map. The expanded space in the region is used to show additional information about that region.

Distorted views must be designed carefully so as not to damage important visual relationships. Bubble distortions of maps may change whether roads appear parallel to each other. However, distorted views can be designed with "flat" and "transition" regions to address this problem. Figure 10.27(a) does not have curvilinear distortions. Focus+context visualizations can be used as part of compact user controls. Keahey (2001) has created an interactive scheme in which the bubble is used to "preview" a region. When the user releases a button over the region, the system zooms in far enough to flatten out the bubble. Bederson has developed a focus+context pull-down menu (Bederson, 2000) that allows the viewing and selection of large lists of typefaces in text editor Fig. 10.27(c).

Perspective distortion. One interesting form of distorting visual transfer functions is 3D perspective. Although it can be described with a 2D distorting visual transfer function, it is usually not experienced as distorting by users due to the special perceptual mechanisms humans have for processing 3D. Figure 10.28(c) shows the Perspective Wall (Mackinlay, Robertson, & Card, 1991). Touching any place on the walls animates its transition into the central focal area. The user perceives the context

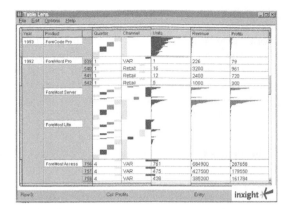

(a) Table Lens. Courtesy of Inxight Software.

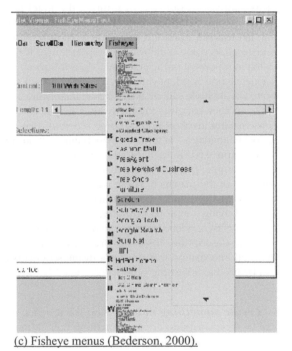

(c) Fisheye menus (Bederson, 2000).

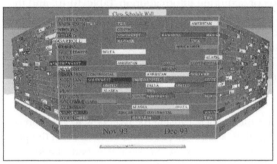

(b) Nonlinear distortion of UK.. Courtesy Alan Keahey

(d) Perspective Wall (Mackinlay, Robertson, & Card, 1991).

FIGURE 10.28. Attention-Reactive Visualizations.

area of the wall as an undistorted 2D image in a 3D space, rather than as a distorted 2D image; however, the same sort of compression is still achieved in the nonfocus area.

Alternate geometries. Instead of altering the size of components, focus+context effects can also be achieved by changing the geometry of the spatial substrate itself. One example is the hyperbolic tree (Lamping & Rao, 1994). A visualization such as a tree is laid out in hyperbolic space (which itself expands exponentially, just like the tree does), and then projected on to the Euclidean plane. The result is that the tree seems to expand around the focal nodes and to be compressed elsewhere. Selecting another node in the tree animates that portion to the

focal area. Munzner (Munzner & Burchard, 1995) has extended this notion to 3D hyperbolic trees and used them to visualize portions of the Internet.

SENSEMAKING WITH VISUALIZATION

Knowledge Crystallization

The purpose of information visualization is to amplify cognitive performance, not just to create interesting pictures. Information visualizations should do for the mind what automobiles do for the feet. So here, we return to the higher level cognitive operations of which information visualization is a means and a component. A recurrent pattern of cognitive activity to which information visualization would be useful (though not the only one!) is "knowledge crystallization." In knowledge crystallization tasks, there is a goal (sometimes ill-structured) that requires the acquisition and making sense of a body of information, as well as the creative formulation of a knowledge product, decision, or action. Examples would be writing a scientific paper, business or military intelligence, weather forecasting, or buying a laptop computer. For these tasks, there is usually a concrete outcome of the task—the submitted manuscript of a paper, a delivered briefing, or a purchase. Knowledge crystallization does have characteristic processes, however, and it is by amplifying these that information visualization seeks to intervene and amplify the user's cognitive powers. Understanding of this process is still tentative, but the basic parts can be outlined:

Acquire information. Make sense of it. Create something new. Act on it.

In Table 10.10, we have listed some of the more detailed activities these entail. We can see examples of these in our initial examples.

Acquire information. The FilmFinder is concentrated largely on acquiring information about films. *Search* is one of the methods of acquiring information in Table 10.10, and the FilmFinder is an instance of the use of information visualization in search. In fact, Shneiderman (Card et al., 1999) has identified a heuristic for designing such systems:

Overview first, zoom and filter, then details-on-demand

The user starts with an overview of the films, and then uses sliders to filter the movies, causing the overview to zoom in on the remaining films. Popping up a box gives details on the particular films. The user could use this system as part of a knowledge crystallization process, but the other activities would take place outside the system. The SmartMoney system also uses the TreeMap visualization for acquiring information, but this time the system is oriented toward *monitoring,* another of the methods in Table 10.10. A glance at the sort of chart in Fig. 10.5 allows an experienced user to notice interesting trends among the hundreds of stocks and industries monitored. Another method

TABLE 10.10.   Knowledge Crystallization Operators

Acquire Information	Monitor
	Search
	Capture (make implicit knowledge explicit)
Make sense of it	Extract information
	Fuse different sources
	Find schema
	Recode information into schema
Create something new	Organize for creation
	Author
Act on it	Distribute
	Apply
	Act

of acquiring information, *capture*, refers to acquiring information that is tacit or implicit. For example, when users browse the World Wide Web, their paths contain information about their goals. This information can be captured in logs, analyzed, and visualized (Chi & Card, 1999). It is worth making the point that acquiring information is not something that the user must necessarily do explicitly. Search, monitoring, and capture can be implicitly triggered by the system.

Make sense of it. The heart of knowledge crystallization is sensemaking. This process is by no means as mysterious as it might appear. Because sensemaking involving large amounts of information must be externalized, the costs of finding, organizing, and moving information around have a major impact on its effectiveness. The actions of sensemaking itself can be analyzed. One process is *extraction*. Information must be got out of its sources. In our hotel example, the hotel manager extracted information from hotel records. A more subtle issue is that information from different sources must be *fused*—that is, registered in some common correspondence. If there are six called-in reports of traffic accidents, does this mean six different accidents, one accident called in six times, or two accidents reported by multiple callers? If one report merely gives the county, while another just gives the highway, it may not be easy to tell. Sensemaking involves finding some *schema*—that is, some descriptive language—in terms of which information can be compactly expressed (Russell, Stefik, Pirolli, & Card, 1993). In our hotel example, permuting the matrices brought patterns to the attention of the manager. These patterns formed a schema she used to organize and represent hotel stays compactly. In the case of buying a laptop computer, the schema may be a table of features by models. Having a common schema then permits compact description. Instances are *recoded* into the schema. Residual information that does not fit the schema is noted and can be used to adjust the schema.

Create something new. Using the schema, information can be reorganized to create something new. It must be *organized* into a form suitable for the output product and that product must be *authored*. In the case of the hotel example, the manager created the presentation of Fig. 10.7(c).

Act on it. Finally, there is some consequential output of the knowledge crystallization task. That action may be to distribute a report or give a briefing, to act directly in some way, such as setting up a new promotion program for the hotel or buying a laptop on the basis of the analysis, or by giving directives to an organization.

Levels for Applying Information Visualization

Information visualization can be applied to facilitate the various subprocesses of knowledge crystallization just described. It can also be applied at different architectural levels in a system. These have been depicted in Fig. 10.29. At one level is the use of visualization to help users access information outside the

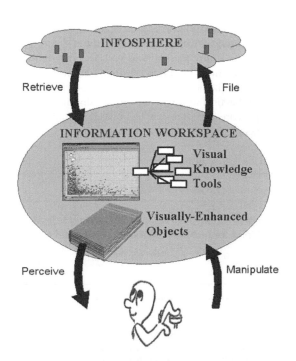

FIGURE 10.29. Levels of use for information visualization.

immediate environment—the *infosphere*—such as information on the Internet or from corporate digital libraries. Figure 10.30(a) shows such a visualization of the Internet (Bray, 1996). Websites are laid out in a space such that sites closer to each other in the visualization tend to have more traffic. The size of the disk represents the number of pages in the site. The globe size represents the number of out-links. The globe height shows the number of in-links.

The second level is the *information workspace*. The information workspace is like a desk or workbench. It is a staging area for the integration of information from different sources. An information workspace might contain several visualizations related to one or several tasks. Part of the purpose of an information workspace is to make the cost of access low for information in active use. Figure 10.30(b) shows a 3D workspace for the Internet, the Web Forager (Card, Robertson, & York, 1996). Pages from the World Wide Web, accessed by users through clicking on URLs or searches, appear in the space. These can be organized into piles or books related to different topics. Figure 10.30 (c) shows another document workspace, STARLIGHT (Risch et al., 1997). Documents are represented as galaxies of points in space such that similar documents are near each other. In the workspace, various tools allow linking the documents to maps and other information and analytical resources.

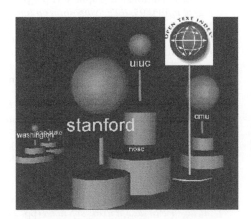

(a) Infosphere: (Bray, 1996).

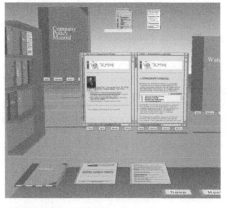

(b) Workspace: Web Forager (Card, Robertson, & York, 1996).

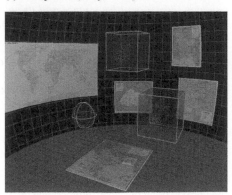

(c) Workspace: STARLIGHT: (Risch et al., 1997).

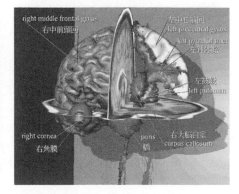

(d) Visually-enhanced object: Voxel-Man. Courtesy of University of Hamburg.

FIGURE 10.30. Information visualization applications.

The third level is *visual knowledge tools*. These are tools that allow schema forming and rerepresentation of information. The permutation matrices in Fig. 10.7, the SeeSoft system for analyzing software in Fig. 10.15(b), and the Table Lens in Fig. 10.27(a) are examples of visual knowledge tools. The focus is on determining and extracting the relationships.

The final level is *visually enhanced objects*, coherent information objects enhanced by the addition of information visualization techniques. An example is Fig. 10.30(d), in which voxel data of the brain have been enhanced through automatic surface rendition, coloring, slicing, and labeling. Abstract data structures representing neural projects and anatomical labels have been integrated into a display of the data. Visually enhanced objects focus on revealing more information from some object of intrinsic visual form.

Information visualization is a set of technologies that use visual computing to amplify human cognition with abstract information. The future of this field will depend on the uses to which it is put and how much advantage it gives to these. Information visualization promises to help us speed our understanding and action in a world of increasing information volumes. It is a core part of a new technology of human interfaces to networks of devices, data, and documents.

ACKNOWLEDGMENT

This chapter benefited from detailed and generous discussion with Jock Mackinlay and tutorial collaborations with Ed Chi of PARC.

References

Ahlberg, C., & Shneiderman, B. (1994a). *Visual information seeking using the filmfinder*. Paper presented at the Conference Companion of CHI'94, ACM Conference on Human Factors in Computing Systems.

Ahlberg, C., & Shneiderman, B. (1994b). *Visual information seeking: Tight coupling of dynamic query filters with starfield displays*. Paper presented at the Proceedings of CHI'94, ACM Conference on Human Factors in Computing Systems, New York.

Battista, G. D., Eades, P., Tamassia, R., & Tollis, I. G. (1994). Annotated bibliography on graph drawing. *Computational Geometry: Theory and Applications, 4*(5), 235–282.

Becker, R. A., Eick, S. G., & Wilks, A. R. (1995, March 1). Visualizing network data. *IEEE Transactions on Visualization and Computer Graphics, 1*, 16–28.

Bederson, B. B. (2000). *Fisheye menus*. Paper presented at the UIST 2000, ACM Symposium on User Interface Software and Technology (CHI Letters, 2(2)).

Bertin, J. (1983). *Semiology of graphics: Diagrams, networks, maps* (W. J. Berg, Trans.). Madison, WI: University of Wisconsin Press. (Original work published 1967)

Bertin, J. (1981). Graphic constructions (P. Scott, Trans.). In *Graphics constructions and graphic information-processing* (pp. 24–31). Berlin, Germany: Walter De Gruyter. (Original work published 1977)

Bray, T. (1996). Measuring the web. *Computer Networks and ISDN Systems, 28*(7–11-May), 992.

Card, S. K., Mackinlay, J. D., & Shneiderman, B. (1999). *Information visualization: Using vision to think*. San Francisco: Morgan Kaufmann Publishers.

Card, S. K., Moran, T. P., & Newell, A. (1986). The model human processor: An engineering model of human performance. In J. Thomas (Ed.), *Handbook of perception and human performance* (pp. 41–35). New York: John Wiley and Sons.

Card, S. K., Robertson, G. G., & York, W. (1996). *The webbook and the web forager: An information workspace for the world-wide web*. Paper presented at the Proceedings of CHI'96, ACM Conference on Human Factors in Computing Systems, New York.

Casner, S. (1991, April 2). Task-analytic approach to the automated design of graphic presentations. *ACM Transactions on Graphics, 10*, 111–151.

Chi, E. H., & Card, S. K. (1999). *Sensemaking of evolving websites using visualization spreadsheets*. Paper presented at the Infovis 1999, IEEE Conference on Information Visualization 1999, San Francisco.

Chi, E. H., & Riedl, J. T. (1998). An operator interaction framework for visualization spreadsheets. Paper presented at the *Proceedings of InfoVis'98, IEEE Symposium on Information Visualization*.

Chuah, M. C., Roth, S. F., Mattis, J., & Kolojejchick, J. A. (1995). Sdm: Malleable information graphics. Paper presented at the *Proceedings of InfoVis'95, IEEE Symposium on Information Visualization*, New York.

Cleveland, W. S., & McGill, M. E. (1988). *Dynamic graphics for statistics*. Pacific Grove, California: Wadsworth and Brooks/Cole.

Cruz, I. F., & Tamassia, R. (1998). *Graph drawing tutorial*. Retrieved from http://www.cs.brown.edu/people/rt/papers/gd-tutorial/gd-constraints.pdf.

Eick, S. G., Steffen, J. L., & Sumner, E. E. (1992, November). Seesoft—a tool for visualizing software. *IEEE Transactions on Software Engineering, 18*(11), 957–968.

Eick, S. G., & Wills, G. J. (1993, October 25–29). Navigating large networks with hierarchies. Paper presented at the *Proceedings of IEEE Visualization'93 Conference*, San Jose, CA.

Engelhardt, Y., Bruin, J. D., Janssen, T., & Scha, R. (1996). The visual grammar of information graphics. In S. University (Ed.), *Artificial intelligence in design workshop notes* (pp. 24–27).

Fairchild, K. M., Poltrock, S. E., & Furnas, G. W. (1988). Semnet: Three-dimensional representations of large knowledge bases. In R. Guindon (Ed.), *Cognitive science and its applications for human-computer interaction* (pp. 201–233). Hillsdale, New Jersey: Lawrence Erlbaum Associates.

Feiner, S., & Beshers, C. (1990). Worlds within worlds: Metaphors for exploring n-dimensional virtual worlds. In *ACM symposium on User Interface Software*.

Freeman, E., & Fertig, S. (1995). Lifestreams: Organizing your electronic life. Paper presented at the *Proceedings of AAAI Fall Symposium on AI Applications in Knowledge Navigation*.

Furnas, G. W. (1981). The fisheye view: A new look at structured files. In B. Shneiderman (Ed.), *Readings in information visualization: Using vision to think* (pp. 312–330). San Francisco: Morgan Kaufmann Publishers, Inc.

Healey, C. G., Booth, K. S., & Enns, J. T. (1995). High-speed visual estimation using preattentive processing. *ACM Transactions on Computer-Human Interaction, 3*(2), 107–135.

Inselberg, A. (1997). Multidimensional detective. Paper presented at the *Proceedings of InfoVis'97, IEEE Symposium on Information Visualization, IEEE Information Visualization*.

Inselberg, A., & Dimsdale, B. (1990). Parallel coordinates : A tool for visualizing multi-dimensional geometry. Paper presented at the *Proceedings of IEEE Visualization '90 Conference*, Los Alamitos, CA.

Keahey, T. A. (2001, October 22–23, 2001). *Getting along: Composition of visualization paradigms*. Paper presented at the Infovis 2001, IEEE Information Visualization 2001, San Diego, California.

Keim, D. A., & Kriegel, H.-P. (1994, September). Visdb: Database exploration using multidimensional visualization. *IEEE Computer Graphics and Applications*, 40–49.

Kosslyn, S. M. (1994). *Image and brain: The resolution of the imagery debate*. Cambridge, MA: The MIT PRess.

Lamping, J., & Rao, R. (1994). *Laying out and visualizing large trees using a hyperbolic space*. Paper presented at the Proceedings of UIST'94, ACM Symposium on User Interface Software and Technology.

Leung, Y. K., & Apperley, M. D. (1994, June). A review and taxonomy of distortion-orientation presentation techniques. *ACM Transactions on Computer-Human Interaction, 1*(2), 126–160.

Lin, X., Soergel, D., & Marchionini, G. (1991). *A self-organizing semantic map for information retrieval*. Paper presented at the Proceedings of SIGIR'91, ACM Conference on Research and Development in Information Retrieval, Chicago, IL.

MacEachren, A. M. (1995). *How maps work*. New York: The Guilford Press.

Mackinlay, J. D. (1986a). *Automatic design of graphical presentations*. Unpublished doctoral dissertation, Stanford University, California.

Mackinlay, J. D. (1986b). Automating the design of graphical presentations of relational information. *ACM Transactions on Graphics, 5*(2), 110–141.

Mackinlay, J. D., Robertson, G. G., & Card, S. K. (1991). *The perspective wall: Detail and context smoothly integrated*. Paper presented at the Proceedings of CHI'91, ACM Conference on Human Factors in Computing Systems, New York.

McCormick, B. H., & DeFanti, T. A. (1987, November 6). Visualization is scientific computing. *Computer Graphics, 21*.

Munzner, T., & Burchard, P. (1995, December 14–15). *Visualizing the structure of the world wide web in 3d hyperbolic space*. Paper presented at the Proceedings of VRML '95.

Playfair, W. (1786). *The commercial and political atlas*. London.

Resnikoff, H. L. (1989). *The illusion of reality*. New York: Springer-Verlag.

Risch, J. S., Rex, D. B., Dowson, S. T., Walters, T. B., May, R. A., & Moon, B. D. (1997). *The starlight information visualization system*. Paper presented at the Proceedings of IEEE International Conference on Information Visualization, London, England.

Robertson, G. G., Card, S. K., & Mackinlay, J. D. (1989). *The cognitive co-processor for interactive user interfaces*. Paper presented at the Proceedings of UIST'89, ACM Symposium on User Interface Software and Technology.

Robertson, G. G., Card, S. K., & Mackinlay, J. D. (1993). Information visualization using 3d interactive animation. *Communications of the ACM, 36*(4), 57–71.

Robin, H. (1992). *The scientific image: From cave to computer*. New York: H. N. Abrams, Inc.

Roth, S. F., & Mattis, J. (1990). *Data characterization for intelligent graphics presentation*. Paper presented at the Proceedings of CHI'90, ACM Conference on Human Factors in Computing Systems, New York.

Russell, D. M., Stefik, M. J., Pirolli, P., & Card, S. K. (1993). *The cost structure of sensemaking*. Paper presented at the Proceedings of INTER-CHI'93, ACM Conference on Human Factors in Computing Systems, Amsterdam.

Shneiderman, B. (1992). Tree visualization with tree-maps: A 2-dimensional space filling approach. *ACM Transactions on Graphics, 11*(1), 92–99.

Shneiderman, B., & Wattenberg, M. (2001). *Ordered tree layouts*. Paper presented at the IEEE Symposium on Information Visualization, San Diego, California.

Spence, R. (2000). *Information visualization*. Harlow, England: Addison-Wesley.

Spence, R., & Apperley, M. (1982). Data base navigation: An office environment for the professional. *Behavior and Information Technology, 1*(1), 43–54.

Tamassia, R. (1996, December 4). Strategic directions in computational geometry working group report. *ACM Computing Surveys, 28*, 591–606.

Tufte, E. R. (1983). *The visual display of quantitative information*. Cheshire, CT: Graphics Press.

Tufte, E. R. (1990). *Envisioning information*. Cheshire, CT: Graphics Press.

Tufte, E. R. (1997). *Visual explanations: Images and quantities, evidence and narrative*. Cheshire, CT: Graphics Press.

Tukey, J. W. (1977). *Exploratory data analysis*. Reading, MA: Addison-Wesley.

Tweedie, L. A., Spence, R., Dawkes, H., & Su, H. (1996). *Externalising abstract mathematical models*. Paper presented at the Proceedings of CHI'96, ACM Conference on Human Factors in Computing Systems.

Ware, C. (2000). *Information visualization: Perception for design*. San Francisco: Morgan Kaufmann Publishers.

·11·

GROUPWARE AND COMPUTER-SUPPORTED COOPERATIVE WORK

Gary M. Olson and Judith S. Olson
University of Michigan

Introduction 218
Adopting Groupware in Context 218
Technical Infrastructure 219
Communication Tools 219
  E-mail 219
  Conferencing Tools: Voice, Video, Text 220
  Blogs 222
Coordination Support 222
  Meeting Support 222
  Workflow 223
  Group Calendars 223
  Awareness 223
Information Repositories 224
  Repositories of Shared Knowledge 224

Wikis 224
  Capture and Replay 224
Social Computing 225
  Social Filtering, Recommender Systems 225
  Trust of People via the Technology 225
Computer-Supported Cooperative Learning 225
Integrated Systems 225
  Media Spaces 225
  Collaborative Virtual Environments 226
  Collaboratories 226
Conclusions 226
Acknowledgments 226
References 226

INTRODUCTION

Computing and communication technologies have provided us with useful and powerful information resources, remote instruments, and tools for interacting with each other. These possibilities have also led to numerous social and organizational effects. These tools are of course just the latest in a long line of modern technologies that have changed human experience. Television and radio long ago broadened our awareness of and interest in activities all over the world. The telegraph and telephone enabled new forms of organization to emerge. The new technologies of Computer-Supported Cooperative Work (CSCW)[1] are giving us greater geographical and temporal flexibility in carrying out our activities. They have also given us new modes of socializing.

Groupware is software designed to run over a network in support of the activities of a group or organization. These activities can occupy any of several combinations of same/different places and same/different times. Groupware has been designed for all four of these combinations. Early groupware applications tended to focus on only one of these cells, but more recently groupware that supports several cells and the transitions among them has emerged. We also do not think of groupware as only dealing with groups. Both the individual members of groups and the organizations in which they are embedded affect and are affected by groupware.

CSCW emerged as a formal field of study in the mid-1980s, with conferences, journals, books, and university courses appearing that used this name. There were a number of important antecedents. The earliest efforts to create groupware used time-shared systems but were closely linked to the development of key ideas that propelled the personal computer revolution. Vannever Bush (1945) described a vision of something similar to today's World Wide Web in an influential essay published shortly after the end of World War II. Doug Engelbart's famous demonstration at the 1968 IFIPS meeting in San Francisco included a number of key groupware components (see Engelbart & English, 1968). These components included support for real-time face-to-face meetings, audio and video conferencing, discussion databases, information repositories, and workflow support. Group decision-support systems and computer-supported meeting rooms were explored in a number of business schools (see McLeod, 1992; Kraemer & Pinsonneault, 1990). Work on office automation included many groupware elements, such as group workflow management, calendaring, e-mail, and document sharing (Ellis & Nutt, 1980). A good summary of early historical trends as well as reprints of key early articles appear in Grief's (1988) important anthology of readings.

Today there are a large number of commercial groupware products. In addition, groupware functions are increasingly appearing as options in operating systems or specific applications (e.g., access to meeting-support tools within Microsoft Office products). Groupware functionality has become widespread and familiar. However, there are still many research issues about how to design such systems and what effects they have on the individuals, groups, and organizations that use them.

ADOPTING GROUPWARE IN CONTEXT

Groupware systems are intended to support groups, which are usually embedded in an organization. As a result, there are a number of issues that bear on groupware success. In a justly famous set of papers, Grudin (1988, 1994) pointed out a number of problems that groupware systems have (see also Markus & Connolly, 1990). In brief, he pointed out that developers of groupware systems need to be concerned about the following issues (Grudin, 1994, p. 97):

1. Disparity in work and benefit. Groupware applications often require additional work from individuals who do not perceive a direct benefit from the use of the application.
2. Critical mass and Prisoner's dilemma problems. Groupware may not enlist the "critical mass" of users required to be useful, or can fail because it is never in any one individual's advantage to use it.
3. Disruption of social processes. Groupware can lead to activity that violates social taboos, threatens existing political structures, or otherwise de-motivates users crucial to its success.
4. Exception handling. Groupware may not accommodate the wide range of exception handling and improvisation that characterizes much group activity.
5. Unobtrusive accessibility. Features that support group processes are used relatively infrequently, requiring unobtrusive accessibility and integration with more heavily used features.
6. Difficulty of evaluation. The almost insurmountable obstacles to meaningful, generalizable analysis and evaluation of groupware prevent us from learning from experience.
7. Failure of intuition. Intuitions in product development environments are especially poor for multi-user applications, resulting in bad management decisions and error-prone design processes.
8. The adoption process. Groupware requires more careful implementation (introduction) in the workplace than product developers have confronted.

However, there are reasons for optimism. In a recent survey of the successful adoption of group calendaring in several organizations, Palen and Grudin (2002) observed that organizational conditions in the 1990s were much more favorable for the adoption of group tools than they were in the 1980s. Further, the tools themselves had improved in reliability, functionality, and usability. There is increased "collaboration readiness" and "collaboration technology readiness." But there are still significant challenges in supporting group work at a distance (Olson & Olson, 2000).

---

[1]This name contains several anachronisms. "Computer" is no longer confined to the familiar desktop device. "Cooperative" does not mean we ignore competitive uses of technology. "Work" is only one of a number of venues that CSCW researchers study.

TECHNICAL INFRASTRUCTURE

Groupware requires networks, and network infrastructure is a key enabler as well as a constraint on groupware. A number of advanced networks are exploring the issues of supporting high-end users (e.g., Abilene, abilene.internet2.edu; National LambdaRail, www.nlr.net; HOPI, networks.internet2.edu/hopi). Wireless networking technology gives users more flexibility. Good access to the Internet is now common in homes, hotels, coffee shops, airports, and many other places. Networking infrastructure is spreading throughout the world. However, heterogeneity in network conditions across both space and time still remains a major technical challenge. For instance, doing web conferencing when some participants are on slow dial-up lines and others are on fast advanced networks requires special coordination. Tanenbaum (2003) is a good resource on the latest developments in networking.

The World Wide Web and its associated tools and standards have had a major impact on the possibilities for groupware (Schatz & Hardin, 1994; Berners-Lee, 1999). Early groupware mostly consisted of stand-alone applications that had to be downloaded and run on each client machine. Increasingly, group tools are written for the web, requiring only a web browser and perhaps some plug-ins. This makes it much easier for the user, and also helps with matters such as version control. It also enables better interoperability across hardware and operating systems.

Security on the Internet is a major challenge for groupware. In some sense the design of Internet protocols is to blame, since the Internet grew up in a culture of openness and sharing (Longstaff et al., 1997; Abbate, 1999; Tanenbaum, 2003). E-commerce and sensitive application domains like medicine have been drivers for advances in security, but there is still much progress to be made (Longstaff et al., 1997; Camp, 2000). Coping with firewalls that block access to certain kinds of organizations can limit the flexibility of web conferencing.

Personal computing was a great enabler of collaborative applications. Today we are liberated from the desktop. Laptops, personal digital assistants, wearables, and cell phones provide access to information and people from almost anywhere. More and more applications are being written to operate across these diverse environments (e.g., Tang et al., 2001; Starner & Rhodes, 2004). These devices vary in computational power, display size and characteristics, network bandwidth, and connection reliability, providing interesting technical challenges to make them all interoperate smoothly. For instance, accessing websites from a cell phone requires special user interface methods to make the tiny displays usable.

Additional flexibility is being provided by the development of infrastructure that lies between the network itself and the applications that run on client workstations, called "middleware." This infrastructure makes it easier to link together diverse resources to accomplish collaborative goals. For instance, the emerging grid technologies allow the marshalling of powerful, scattered computational resources (Foster & Kesselman, 2004). Middleware provides such services as identification, authentication, authorization, directories, and security in uniform ways that facilitate the interoperability of diverse applications. All of these technical elements are components of cyberinfrastructure (Atkins et al., 2003). There is considerable interest in the development of this infrastructure because of its large impact on research, education, and commerce.

COMMUNICATION TOOLS

We now turn to a review of specific kinds of groupware, highlighting their various properties and uses. We have grouped this review under several broad headings. We do not aim to be exhaustive, but rather seek to illustrate the variety of kinds of tools that have emerged to support human collaborative activities over networked systems. We also highlight various research issues pertaining to these tools.

E-mail

E-mail has become a ubiquitous communication tool. The early adoption of standards made it possible for messages to be exchanged across networks and different base machines and software applications. E-mail is now also done from cell phones, PDAs, television sets, and kiosks in public sites. Documents of many types can be easily exchanged. Because of its widespread use, it is often called the first successful groupware application (Sproull & Kiesler, 1991; Satzinger & Olfman, 1992; O'Hara-Devereaux & Johnson, 1994; Anderson, Bikson, Law, & Mitchell, 1995). Indeed, it has become so successful that e-mail overload has become a major problem (Whittaker & Sidner, 1996). And of course it has become a vector for viruses, worms, and other invasive software.

Researchers have shown that this widespread use has had a number of effects on how people behave. It has had large effects on communication in organizations: it changes the social network of who talks to whom (Sproull & Kiesler, 1991; DeSanctis, Jackson, Poole, & Dickson, 1996), the power of people who formerly had little voice in decisions (Finholt, Sproull, & Kiesler, 1990), and the tone of what is said and how it is interpreted (Sproull & Kiesler, 1991). For example, with e-mail, people who were shy found a voice; they could overcome their reluctance to speak to other people by composing text, not speech to another face. This invisibility, however, also has a more general effect—without the social cues in the recipient's face being visible to the sender, people will "flame," send harsh or extremely emotive (usually negative) messages (Arrow et al., 1996; Hollingshead, McGrath, & O'Connor, 1993).

As with a number of other "designed" technologies, people use e-mail for things other than the original intent. People use it for managing time, reminding them of things to do, and keeping track of steps in a workflow (Mackay, 1989; Carley & Wendt, 1991; Whittaker & Sidner, 1996). But because e-mail was not designed to support these tasks, it does not do it very well; people struggle with reading signals about whether they have replied or not (and to whom it was cc'd); they manage folders poorly for reminding them to do things, and so forth.

In addition, because e-mail is so widespread, and it is easy and free to distribute a single message to many people, people experience information overload. Many people get hundreds of e-mail messages each day, many of them mere broadcasts of things for sale or events about to happen, much like "classifieds" in the newspaper. Several early efforts to use artificial intelligence techniques to block and/or sort incoming e-mail were tried, and this has continued to be a very active area of work (Malone, Grant, Lai, Rao, & Rosenblitt, 1987; Winograd, 1988). There are two broad classes of uses of e-mail filters. One use is to automatically sort incoming mail into useful categories. This is relatively easy for mail that has simple properties, such as a person's name. It is more difficult for subtle properties. The other major use is to weed out unwanted mail, such as spam. The state-of-the-art in spam filtering was in the range of 80–90% effectiveness in 2005 (e.g., Federal Trade Commission, 2005). Such filters are so good that many institutions automatically filter mail as it comes in to the organization's gateway, sparing users the need to do it in their own clients. Similarly, many clients now come with built-in spam filters that can be tuned by the user (e.g., Google's Gmail).

These problems have led to the "reinvention" of e-mail (Whittaker, Bellotti, & Moody, 2005). For example, given that e-mail is often used in the context of managing projects, systems have been explored that have a more explicit scheme for task management (Whittaker, 2005; Bellotti, Ducheneaut, Howard, Smith, & Grinter, 2005). To deal with problems of e-mail overload, new schemes for filtering e-mail have been explored, such as routing messages differently to different kinds of clients (e.g., cell phone vs. desktop machine; see Schmandt & Marti, 2005). Another approach has been to explore pricing mechanisms for e-mail that are analogous to pricing for regular mail (Kraut, Sunder, Telang, & Morris, 2005). In such schemes one would pay to send e-mail, with higher prices presumably indicating higher priority, analogous to the difference between first class postage and bulk rates. These schemes are currently exploratory, but are likely to result in new options in future e-mail clients.

Kraut et al. (1998) reported that greater Internet use, which in their sample was mostly e-mail, led to declines in social interactions with family members and an increase in depression and loneliness. Not surprising, these results triggered widespread discussion and debate, both over the substance of the results and the methods used to obtain them. Kraut, Gergle, and Fussell (2002) reported new results that suggest these initial negative effects may not persist. Interpersonal communication is one of the principal uses of the Internet, and the possible implications of this kind of communication for social life are important to understand. Indeed, Putnam (2000) has wondered whether the Internet can be a source of social cohesiveness. These kinds of questions need to be addressed by additional large-scale studies of the kind carried out by Kraut and his colleagues (see Resnick, 2002).

Conferencing Tools: Voice, Video, Text

There are many options available today for on-line conferencing among geographically dispersed members of a group. So-called computer-mediated communication (CMC) has become widespread. There are three principal modes of interaction, but each has numerous subtypes:

Video + Audio
Full-scale video conferencing room; many options for specific design

Individual desktop video; many options for quality, interface

Audio
Phone conference

Voice over IP (see Federal Communications Commission, 2005)

Text
Instant messaging, chat, SMS on mobile phones

There are many studies that compare face-to-face with various forms of CMC. There are some clear generalizations from such work. The main one is that CMC is more difficult to do than FTF, and requires more preparation and care (Hollingshead et al., 1993; McLeod, 1992; Olson, Olson, & Meader, 1995; Siegel, Dubrovsky, Kiesler, & McGuire, 1986; Straus, 1996, 1997; Straus & McGrath, 1994). A variety of things that come for free in FTF are either difficult to support or outright missing in CMC (Kiesler & Cummings, 2002). Backchannel communication, which is important for modulating conversation, is either weak or nonexistent in CMC. Paralinguistic cues that can soften communication are often missing. Participants in CMC tend to have an informational focus, which means there is usually less socializing, less small talk. Over time this can lead to poorer social integration and organizational effectiveness (Nohria & Eccles, 1992).

CMC often introduces delay. This is well-known to be very disruptive to communication (Egido, 1988; Krauss & Bricker, 1966; O'Conaill, Whittaker, & Wilbur, 1993; Ruhleder & Jordan, 2001; Tang & Isaacs, 1993). Participants will communicate less information, be more frustrated with the communication, and actually terminate communication sessions sooner. Delay can be managed, but it takes special care among the participants and turn-taking widgets in the interface of the tools being used. For instance, if there is delay, then full-duplex open communication will not work, since participants will step all over each other's communication. Either the participants must use a social protocol (e.g., like that used in radio communications with spacecraft), or they must employ a mic-passing procedure with interface indications of who wants to talk next.

While it might seem desirable to always have the maximum communication and tool support possible, it is not always possible or even necessary to do so. Research shows that effective real-time collaboration can take place under a number of different arrangements, depending on the task, the characteristics of the participants, the specific geographical dispersion of the participants, and the processes employed to manage the interactions. There are also organizational effects, especially when the real-time collaborations are embedded in ongoing activities, as they almost always are.

For instance, early work (Williams, 1977) showed that, in referential communication tasks, full-duplex audio is just as effective as FTF. Subsequent research comparing audio and video conferencing (see Finn, Sellen, & Wilbur, 1997; Cadiz et al., 2000 recently found similar results for a tutored video-instruction task) showed that for many tasks audio is sufficient, that video adds nothing to task effectiveness though participants usually report they are more satisfied with video. There are important exceptions, however. Negotiation tasks are more effective with video (Short, Williams, & Christie, 1976). This is probably because the more subtle visual cues to the participants' intentions are important in this kind of task. Further, Veinott, Olson, Olson, and Fu (1999) found that when participants have less common ground video helps. In their case, participants were non-native speakers of English who were doing the task in English. For native speakers, video was no better than audio, but non-native speakers did better when they had video. Again, visual cues to comprehension and meaning likely played an important role. Recently, an experimental study by Daly-Jones, Monk, and Watts (1998) showed that high-quality video resulted in greater conversational fluency over just high-quality audio, especially as group size increased. There was also a higher rated sense of presence in the video conditions.

An important lesson to draw from this literature is that there are two broad classes by which we might assess whether video is important in real-time collaboration. On the one hand, except for tasks like negotiation or achieving common ground, groups are able to get their work done effectively with just high-quality audio. However, for things like satisfaction, conversational fluency, and a sense of presence, video adds value. These kinds of factors might be very important for long-term organizational consequences like employee satisfaction. As of yet, no long-term studies have been done to examine this conjecture.

Audio quality is critical. Ever since early literature review (Egido, 1988), it's been reported over and over again that if the audio is of poor quality participants will develop a workaround. For instance, if the audio in a video conferencing system or in a web conferencing system is poor quality, participants will turn to a phone conference.

The social ergonomics of audio and video are also keys to their success. Many of the failures of audio conferencing, especially over the Internet, result from poor-quality microphones, poor microphone placement, poor speakers, and interfering noises like air conditioning. Getting these details right is essential. Similarly, for video, camera placement can matter a lot. For instance, Huang, Olson, and Olson (2002) found that a camera angle that makes a person seem tall (as opposed to actually being tall) affects how influential a person is in a negotiation task. Apparent height matters a lot. Other aspects of camera placement or arrangement of video displays make a big difference as well, but are not well known.

An exception is eye contact, where studies of FTF communication show that eye contact is a key linguistic and social mediator of communication (Argyle & Cook, 1976; Kendon, 1967). It is very difficult to achieve eye contact in CMC systems. Many attempts have been made (Gale & Monk, 2000; Grayson & Monk, 2003; Monk & Gale, 2002; Okada, Maeda, Ichicawaa, & Matsushita, 1994; Vertegaal, 1999; Vertegaal, Slagter, van der Veer, & Nijholt, 2001), and at least the subjective reports are that these

can be effective. But these all require special equipment or setups. And they don't scale very well to multiparty sessions.

While for most situations having at least high-quality audio is essential, there are some special cases where a text-based channel, like chat or instant messaging, can work fine. For instance, in the Upper Atmospheric Research Collaboratory (UARC, later known as the "Space Physics and Aeronomy Research Collaboratory," or SPARC), a chat system worked very well for carrying out geographically distributed observational campaigns, since the flow of events in these campaigns was relatively slow (campaigns went on for several days, key events would take many minutes to unfold). McDaniel, Olson, and Magee (1996) compared chat logs with earlier FTF conversations at a remote site and found many elements of them very similar, including informal socializing. But this kind of ongoing scientific campaign is very unlike the interactions that take place in a typical meeting.

Instant messaging is a new communication modality that is making substantial inroads into organizations. Muller, Raven, Kogan, Millen, and Carey (2003) found in a survey study of three organizations that the introduction of instant messaging led to significantly less use of such communication channels as e-mail, voice-mail, telephone, teleconference, pager, and face-to-face. They also found that instant messaging was used for "substantive business purposes." Furthermore, in one of the organizations where they surveyed users after 24 months of usage they found that the substantive reasons for using IM increased. In a study of IM logs in an organization, Isaacs, Walendowski, Whittaker, Schiano, and Kamm (2002) found that a large proportion of IM conversations involved "complex work discussions." They found that IM users seldom switched to another communication channel once they were engaged in IM. Nardi, Whittaker, and Bradner (2000) observed in a field study that workers used IM for a variety of purposes, not just for information exchange. Such matters as quick questions, scheduling, organizing social interactions, and keeping in touch with others were common uses of IM. Thus, IM has emerged as a significant communication medium in the workplace, and is used even when other, richer communication channels were available.

While IM is a relatively new phenomenon in the workplace, it is clearly established as a useful and widely used tool outside the workplace. This will undoubtedly assist in the development of more sophisticated versions of the tool, as well as its integration into on-line conferencing systems. There is clearly much promise here. We have noticed, for example, that during online conferences IM or chat serves as a backchannel for side conversations or debugging, an extremely useful adjunct to the core audio or video communication taking place in such conferences.

The other key feature of successful remote meetings is the ability to share the objects they are talking about, such as the agenda, the to-do list, the latest draft of a proposal, a view of an object to be repaired, and so on. Many researchers (Fussell, Kraut, & Siegel, 2000; Karsenty, 1999; Kraut, Fussell, & Siegel, 2003; Kraut et al., 2002; Luff, Heath, Kuzuoka, Hindmarsh, & Oyama, 2003; Nardi et al., 1993; Whittaker, 2003; Whittaker, Geelhoed, & Robinson, 1993) have provided experimental evidence of the value of a shared workspace for synchronous audio-supported collaboration. More traditional videoconferencing technologies often offer an "object camera," onto which the participants can put a paper agenda, PowerPoint slides, or a manufactured part. More generally, any form of video can also be used to share work objects (Fussell et al., 2000; Nardi et al., 1993). For digital objects, there are now a number of products that will allow meeting participants to share the screen or, in some cases, the remote operation of an application. Some companies are using electronic whiteboards, both in a collocated meeting and in remote meetings, to mimic the choreography of people using a physical whiteboard. In some "collaboratories," scientists can even operate remote physical instruments from a distance and jointly discuss the results.

Blogs

Weblogs, or more commonly called "blogs," have burst upon the Internet scene in recent years. Blogging software that makes it easy to put up multimedia content has led people to set up sites for all manner of purposes. A site can contain text, pictures, movies, and audio clips. A common social purpose is to keep an on-line diary. Another is to provide commentary on a topic of interest. For instance, blogs played a major role in the 2004 election (Adamic & Glance, 2005). Nardi, Schiano, Gumbrecht, and Swartz (2004) studied why people blog, as it's sometimes puzzling that people would essentially share personal or private information about themselves through the web.

COORDINATION SUPPORT

Meeting Support

An early and popular topic in CSCW was the support of face-to-face meetings. A number of systems were developed and tested. While of late the focus has shifted to the support of geographically distributed meetings, the early work on meeting support led to some important and useful conclusions.

Some meeting-support software imposed structure on the process of the meeting, embodying various brainstorming and voting procedures. Group Decision Support Systems (GDSSs) arose from a number of business schools, focusing on large meetings of stakeholders intent on going through a set series of decisions, such as prioritizing projects for future funding (Nunamaker, Dennis, Valacich, Vogel, & George, 1991). With the help of a facilitator and some technical support, the group was led through a series of stages: brainstorming without evaluating, evaluating alternatives from a variety of positions, prioritizing alternatives, and so on. These meetings were held in specialized rooms in which individual computers were embedded in the tables, networked to central services, and summary displays shown "center stage." A typical scenario involved individuals silently entering ideas into a central repository, and after a certain amount of time, they were shown ideas one at a time from others and asked to respond with a new idea triggered by that one. Later, these same ideas were presented to the individuals who then asked to rank or rate them according to some

fixed criterion, like cost. Aggregates of individuals' opinions were computed, discussed further and presented for vote. The system applied computational power (for voting and rating mechanisms), and networking control (for parallel input) to support typically weak aspects of meetings. These systems were intended to gather more ideas from participants, since one didn't have to wait for another to stop speaking in order to get a turn. And, anonymous voting and rating was intended to insure equal participation, not dominated by those in power.

Evaluations of these GDSSs have been reviewed, producing some generalizations about their value (McLeod, 1992; Kraemer & Pinsoneault, 1990; Hollingshead et al., 1993). The systems indeed fulfill their intentions of producing more ideas in brainstorming and having more evaluative comments because of anonymity. Decisions are rated as higher in quality, but the meetings take longer and the participants are less satisfied than those in traditional meetings.

A second class of technologies to support real-time meetings is less structured, more similar to individual workstation support. In these systems, groups are allowed access to a single document or drawing, and can enter and edit into them simultaneously at will. Different systems enforce different "locking" mechanisms (e.g., paragraph or selection locking) so that one person does not enter while another deletes the same thing (Ellis, Gibbs, & Rein, 1991). Some also allow parallel individual work, where participants view and edit different parts of the same document, but can also view and discuss the same part as well. This kind of unstructured shared editor has been shown to be very effective for certain kinds of free-flowing meetings, like design or requirements meetings (Olson, Olson, Storrosten, & Carter, 1993). The rated quality of the meeting products (e.g., a requirements document or plan) was higher when using these technologies than with traditional whiteboard or paper-and-pencil support, but like working in GDSSs, people were slightly less satisfied. The lower satisfaction here and with GDSSs may reflect the newness of the technologies; people may not have yet learned how to persuade, negotiate, or influence each other in comfortable ways, to harness the powers inherent in the new technologies.

These new technologies did indeed change the way in which people worked. They talked less and wrote more, building on each other's ideas instead of generating far-reaching other ideas. The tool seemed to focus the groups on the core ideas, and keep them from going off on tangents. Many participants reported really liking *doing* work *in* the meetings rather than spending time only *talking about* the work.

A third class of meeting room support appears in electronic whiteboards. For example, the LiveBoard (Elrod et al., 1992), SoftBoard and SmartBoard are approximately 4'×6' rear-projection surfaces that allow pen input, much the way a whiteboard or flipchart does. People at Xerox PARC and Boeing have evaluated the use of these boards in meetings in extended case studies. In both cases, the board was highly valued because of its computational power and the fact that all could see the changes as they were made. At both sites, successful use required a facilitator who was familiar with the applications running to support the meeting. At Xerox, suggestions made in the meeting about additional functionality were built into the system so that it eventually was finely tuned support for their particular needs (Moran et al., 1996). For example, they did a lot of list making of freehand text items. Eventually, the board software recognized the nature of a list and an outline, with simple gestures changing things sensibly. For example, if a freehand text item was moved higher in a list, the other items adjusted their positions to make room for it. The end product was not only a set of useful meeting tools, but also a toolkit to allow people to build new meeting widgets to support their particular tasks.

Meetings are important, though often despised, organizational activities. Laboratory research of the kind just reviewed has shown quite clearly that well-designed tools can improve both work outcomes and participant satisfaction. However, meetings in organizations seldom use such tools. Inexpensive mobile computing and projection equipment combined with many commercial products mean that such tools are within reach of most organizations. But not having these elements readily available in an integrated way probably inhibits their widespread adoption.

While traditional meetings are often viewed as wasteful and frustrating, there can be huge benefits to working together in collocated environments. Kiesler and Cummings (2002) reviewed a number of the characteristics of physical collocation that can benefit performance. In a detailed study of one such situation, Teasley, Covi, Krishnan, and Olson (2002) found that "radical collocation," in which software development teams worked together in a dedicated project room for many weeks, dramatically improved their productivity. Reasons for this included the constant awareness of each other's work status, the associated ability to instantly work on an impasse as a group, and the availability of rich shared artifacts generated by the project.

Workflow

Workflow systems lend technology support to coordinated asynchronous (usually sequential) steps of activities among team members working on a particular task. For example, a workflow system might route a travel reimbursement voucher from the traveler to the approving party to the accounts payable to the bank. The electronic form would be edited and sent to the various parties, their individual to-do lists updated as they received and/or completed the tasks, and permissions and approval granted automatically as appropriate (e.g., allowing small charges to an account if the charges had been budgeted previously or simply if there was enough money in the account). Not only is the transaction flow supported, but also records are often kept about who did what and when they did it. It is this later feature that has potentially large consequences for the people involved, discussed later.

These workflow systems were often the result of work reengineering efforts, focusing on making the task take less time and eliminating the work that could be automated. Not only do workflow systems therefore have a bad reputation in that they often are part of workforce reduction plans, but also for those left, their work is able to be monitored much more closely. The systems are often very rigid, requiring, for example, all of a form to be filled in before it can be handed off to the next in the chain.

They often require a great deal of rework because of this inflexibility. It is because of the inflexibility and the potential monitoring that the systems fall into disuse (e.g., Abbott & Sarin, 1994).

The fact that workflow can be monitored is a major source of user resistance. In Europe, such monitoring is illegal, and powerful groups of organized workers have made sure that such capabilities are not in workflow systems (Prinz & Kovenbach, 1996). In the United States, it is not illegal, but many employees complain about its inappropriate use. For example, in one software engineering team where workflow had just been introduced to track bug reports and fixes, people in the chain were sloppy about noting whom they had handed a piece of work off to. When it was discovered that the manager had been monitoring the timing of the handoffs to assign praise or blame, the team members were justifiably upset (Olson & Teasley, 1996). In general, managerial monitoring is a feature that is not well received by people being monitored (Markus, 1983). If such monitoring is mandated, workers' behavior will conform to the specifics of what is being monitored (e.g., time to pass an item off to the next in the chain) rather than perhaps to what the real goal is (e.g., quality as well as timely completion of the whole process).

Group Calendars

A number of organizations have now adopted online calendars, mainly in order to view people's schedules to arrange meetings. The calendars also allow a form of awareness, allowing people to see if a person who is not present is expected back soon. Individuals benefit only insofar as they offload scheduling meetings to others, like to an administrative assistant, who can write as well as read the calendar. And, in some systems the individual can schedule private time, blocking the time but not revealing to others his or her whereabouts. By this description, on-line calendaring is a classic case of what Grudin (1988) warned against, a misalignment of costs and benefits; the individual puts in the effort to record his/her appointments so that another, in this case a manager or coworker, can benefit from ease of scheduling. However, since the early introduction of electronic calendaring systems, many organizations have found successful adoption (Mosier & Tammaro, 1997; Grudin & Palen, 1995; Palen & Grudin, 2002). Apparently such success requires a culture of sharing and accessibility, something that exists in some organizations and not others (Lange, 1992; Ehrlich, 1987).

Awareness

In normal work, there are numerous occasions in which people find out casually whether others are in and, in some cases, what they are doing. A simple walk down the hall to a printer offers numerous glances into people's offices, noting where their coats are, whether others are talking, whether there is intense work at a computer, and so on. This kind of awareness is unavailable to workers who are remote. Some researchers have offered various technology solutions; some have allowed one to visually walk down the hall at the remote location, taking a five-second glance into each passing office (Bellotti & Dourish,

1997; Fish, Kraut, Root, & Rice, 1993). Another similar system, called "Portholes," provides periodic snapshots instead of full-motion video (Dourish & Bly, 1992). Because of privacy implications, these systems have had mixed success. The places in which this succeeds are those in which the individuals seem to have a reciprocal need to be aware of each other's presence, and a sense of cooperation and coordination. A contrasting case is the instant messaging (IM) system in which the user has control as to what state they wish to advertise to their partners about their availability. The video systems are much more lightweight to the user but more intrusive; the IM ones give the user more control but require intention in action. Another approach investigated by Ackerman, Starr, Hindus, and Mainwaring (1997) looked at shared audio as an awareness tool, though this too has privacy implications.

As mentioned earlier, instant messaging systems provide an awareness capability. Most systems display a list of "buddies" and whether they are currently on-line or not. Nardi et al. (2000) found that people liked this aspect of IM (see also Muller et al., 2003; Isaacs et al., 2002). And, since wireless has allowed constant connectivity of mobile devices like PDAs, this use of tracking others is likely to grow. But again, there are issues of monitoring for useful or insidious purposes, and the issues of trust and privacy loom large (see Godefroid, Herbsleb, Jagadeesan, & Li, 2000).

Another approach to signaling what one is doing occurs at the more micro level. And again, one captures what is easy to capture. When people are closely aligned in their work, there are applications that allow each to see exactly where in the shared document the other is working and what they are doing (Gutwin & Greenberg, 1999). If one is working nearby the other, this signals perhaps a need to converse about the directions each is taking. Empirical evaluations have shown that such workspace awareness can facilitate task performance (Gutwin & Greenberg, 1999).

Studies of attempts to carry out difficult intellectual work within geographically distributed organizations show that one of the larger costs of geographical distribution is the lack of awareness of what others are doing or whether they are even around (Herbsleb, Mockus, Finholt, & Grinter, 2000). Thus, useful and usable awareness tools that mesh well with trust and privacy concerns could be of enormous organizational importance. This is a rich research area for CSCW.

INFORMATION REPOSITORIES

Repositories of Shared Knowledge

In addition to sharing information generally on the web, in both public and intranet settings, there are applications that are explicitly built for knowledge sharing. The goal in most systems is to capture knowledge that can be reused by others, like instruction manuals, office procedures, training, and "boilerplates," or templates of commonly constructed genres, like proposals or bids. Experience shows, however, that these systems are not easy wins. Again, similar to the case of the on-line calendaring systems described above, the person entering information into the system is

not necessarily the one benefiting from it. In a large consulting firm, where consultants were quite competitive in their bid for advancement, there was indeed negative incentive for giving away one's best secrets and insights (Orlikowski & Gash, 1994).

Sometimes subtle design features are at work in the incentive structure. In another adoption of Lotus Notes, in this case to track open issues in software engineering, the engineers slowly lost interest in the system because they assumed that their manager was not paying attention to their contributions and use of the system. The system design, unfortunately, made the manager's actual use invisible to the team. Had they known that he was reading daily what they wrote (though he never wrote anything himself), they would likely have continued to use the system (Olson & Teasley, 1996). A simple design change that would make the manager's reading activity visible to the team would likely have significantly altered their adoption.

The web of course provides marvelous infrastructure for the creation and sharing of information repositories. A variety of tools are appearing to support this. Of particular interest are open source tools that allow for a wider, more flexible infrastructure for supporting information sharing (see www.sakai.org). A major type of collaboratory (see below) are those that provide shared data repositories for a community of scientists. Systematic research on the use of such tools is needed.

Wikis

A wiki is a shared web space that can be edited by anyone who has access to it. They were first introduced by Ward Cunningham in 1995, but have recently become very popular. These can be used in a variety of ways, both for work and for fun. The most famous wiki is Wikipedia (www.wikipedia.org), an online encyclopedia where anyone can generate and edit content. It has grown to have millions of entries, and has versions in at least ten languages. A recent study carried out by *Nature* found that for science articles Wikipedia and the Encyclopedia Britannica were about equally accurate (Giles, 2005). Bryant, Forte, and Bruckman (2005) studied the contributors to Wikipedia, and suggested that a new publishing paradigm was emerging. Viegas, Wattenberg, and Dave (2004) developed imaginative visualizations of Wikipedia authoring and editing behavior over time.

Capture and Replay

Tools that support collaborative activity can create traces of that activity that later can be replayed and reflected upon. The Upper Atmospheric Research Collaboratory (UARC) explored the replay of earlier scientific campaign sessions (Olson et al., 2001), so that scientists could reflect upon their reactions to real-time observations of earlier phenomena. Using a VCR metaphor, they could pause where needed, and fast forward past uninteresting parts. This reflective activity could also engage new players who had not been part of the original session. Abowd (1999) has explored such capture phenomena in an educational experiment called Classroom 2000. Initial experiments focused on reusing educational sessions during the term in college courses. We do not yet fully understand the impact of such promising ideas.

SOCIAL COMPUTING

Social Filtering, Recommender Systems

We often find the information we want by contacting others. Social networks embody rich repositories of useful information on a variety of topics. A number of investigators have looked at whether the process of finding information through others can be automated. The kinds of recommender systems that we find on websites like Amazon.com are examples of the result of such research. The basic principle of such systems is that an individual will tend to like or prefer the kinds of things (e.g., movies, books) that someone who is similar to him/her likes. They find similar people by matching their previous choices. Such systems use a variety of algorithms to match preferences with those of others, and then recommend new items. Resnick and Varian (1997) edited a special issue of the communication of the ACM on recommender systems that included a representative set of examples. Herlocker, Konstan, and Riedl (2001) used empirical methods to explicate the factors that led users to accept the advice of recommender systems. In short, providing access to explanations for why items were recommended seems to be the key. Cosley, Frankowski, Kiesler, Terveen, and Riedl (2005) studied factors that influence people to contribute data to recommender systems. Recommender systems are emerging as a key element of e-commerce (Schafer, Konstan, & Riedl, 2001). Accepting the output of recommender systems is an example of how people come to trust technical systems. This is a complex topic, and relates to issues like security that we briefly described earlier.

Trust of People via the Technology

It has been said that "trust needs touch," and indeed in survey studies, coworkers report that they trust those who are collocated more than those who are remote (Rocco, Finholt, Hofer, & Herbsleb, 2000). Interestingly, those who spend the most time on the phone chatting about non-work related topics with their remote coworkers show higher trust than those they communicate with using only fax and e-mail. But lab studies show that telephone interaction is not as good as face-to-face. People using just the telephone behave in more self-serving, less-trusting ways than they do when they meet face to face (Drolet & Morris, 2000).

What can be done to counteract the mistrust that comes from the impoverished media? Rocco (1998) had people meet and do a team-building exercise the day before they engaged in the social dilemma game with only e-mail to communicate with. These people, happily, showed as much cooperation and trust as those who discussed things face to face during the game. This is important. It suggests that if remote teams can do some face-to-face team-building before launching their project, they will act in a trusting/trustworthy manner.

Since it's not always possible to have everyone on a project meet face to face before they launch into the work, what else will work? Researchers have tried some options, but with mixed success. Zheng, Bos, Olson, Gergle, and Olson (2001) found that using chat for socializing and sharing pictures of each other also led to trustful relations. Merely sharing a resume did not.

When the text is translated into voice, it has no effect on trust, and when it is translated into voice and presented in a moving human-like face, it is even worse than text-chat. (Jensen, Farnham, Drucker, & Kollock, 2000; Kiesler, Sproull, & Waters, 1996). However, Bos, Gergle, Olson, and Olson (2001) found that interactions over video and audio led to trust, albeit of a seemingly more fragile form.

If we can find a way to establish trust without expensive travel, we are likely to see important productivity gains. Clearly the story is not over. However, we must not be too optimistic. In other tasks, video does not produce "being there." There is an overhead to the conversation through video; it requires more effort than working face to face (Olson et al., 1995). And, today's videos over the Internet are both delayed and choppy, producing cues that people often associate with lying. One doesn't trust someone who appears to be lying. Trust is a delicate emotion; today's video might not just do it in a robust enough fashion.

COMPUTER-SUPPORTED COOPERATIVE LEARNING

Obviously the range of CSCW systems that we've been describing can be used for a variety of purposes. One special area that has emerged as a subfield with a distinctive identity is education, using the name Computer-Supported Cooperative Learning (CSCL). This field first emerged in Europe, but has become quite widespread. This area has its own journals and national meetings (see Koschmann, 1996; Koschmann, Hall, & Miyake, 2002; Koschmann, Suthers, & Chan, 2005).

The emergence of this field has coincided with the emergence of a general trend in education toward collaborative learning, of using working with peers in groups as an effective tool for education (Slavin, 1994). The core idea is that by working together with peers the interactions over problem solving and other learning activities will occur at a level that is most easily understood and engaged by the learners. What CSCL adds to this is a focus on software that can facilitate collaboration, particularly across distance.

INTEGRATED SYSTEMS

Media Spaces

As an extension of video conferencing and awareness systems, some people have experimented with open, continuous audio and video connections between remote locations. In a number of cases, these experiments have been called "Media Spaces." For example, at Xerox, two labs were linked with an open video link between two commons areas (Olson & Bly, 1991), the two locations being Palo Alto, California, and Portland, Oregon. Evaluation of these experiments showed that maintaining organizational cohesiveness at a distance was much more difficult than when members are collocated (Finn et al., 1997). However, some connectedness was maintained. Where many of these early systems were plagued with technical difficulties, human factors limitations, or very large communication costs, in today's situation it might actually be possible to overcome these difficulties, making media a possibility for connecting global organizations. A new round of experimental deployments with new tools is needed.

Collaborative Virtual Environments

Collaborative virtual environments are 3D embodiments of multi-user domains (MUDs). The space in which people interact is an analog of physical space, with dimensions, directions, rooms, and objects of various kinds. People are represented as avatars—simplified, geometric, digital representations of people, who move about in the 3D space (Singhal & Zyda, 1999). Similar to MUDs, the users in a meeting situation might interact over some object that is digitally represented, like a mock-up of a real thing (e.g., an automobile engine, an airplane hinge, a piece of industrial equipment) or with visualizations of abstract data (e.g., a 3D visualization of atmospheric data). In these spaces, one can have a sense as to where others are and what they are doing, similar to the simplified awareness systems described above. In use, it is difficult to establish mutual awareness or orientation in such spaces (Hindmarsh, Fraser, Heath, Benford, & Greenhalgh, 1998; Park, Kapoor, & Leigh, 2000; Yang & Olson, 2002). There have even been some attempts to merge collaborative virtual environments with real ones, though with limited success so far (Benford, Greenhalgh, Reynard, Brown, & Koleva, 1998).

What people seem to want is more like the Holodek in "Star Trek." These environments are complicated technically, and perhaps even more complicated socially. In real life, we have developed interesting schemes that trigger behavior and interpretation of others' behavior as a function of real distance, a field called "Proxemics" (Hall, 1982). Only when these subtle behaviors are incorporated into the virtual environment will we have a chance of simulating appropriate interhuman behavior in the virtual 3D world.

Collaboratories

A collaboratory is a laboratory without walls (Finholt & Olson, 1997). From a National Research Council report, a collaboratory is supposed to allow "the nation's researchers [to] perform their research without regard to geographical location—interacting with colleagues, accessing instrumentation, sharing data and computational resources [and] accessing information in digital libraries" (National Research Council, 1993, p. 7). Starting in the early 1990s, these capabilities have been configured into

support packages for a number of specific sciences (see Finholt, 2002). The Science of Collaboratories project (www.science ofcollaboratories.org) has identified more than 200 existing collaboratories and is drawing lessons about why some succeed and others don't (Olson et al., 2004).

A number of companies have also experimented with similar concepts, calling them "virtual collocation." The goal there is to support geographically dispersed teams as they carry out product design, software engineering, financial reporting, and almost any business function. In these cases, suites of off-the-shelf groupware tools have been particularly important and have been used to support a round-the-clock software development among overlapping teams of engineers in time zones around the world. (Carmel, 1999). There have been a number of such efforts, and it is still unclear as to their success or what features make their success more likely. (Olson & Olson, 2000).

CONCLUSIONS

Groupware functionality is steadily becoming more routine in commercial applications. Similarly, suites of groupware functions are being written into operating systems. Many of the functions we have described in this article are becoming ordinary elements of infrastructure in networked computing systems.

Prognosticators looking at the emergence of groupware and the convergence of computing and communication media have forecast that distance will diminish as a factor in human interactions (e.g., Cairncross, 1997). However, to paraphrase Mark Twain, the reports of distance's death are greatly exaggerated. Even with all our emerging information and communications technologies, distance and its associated attributes of culture, time zones, geography, and language will continue to affect how humans interact with each other. Emerging distance technologies will allow greater flexibility for those whose work must be done at a distance, but we believe (see Olson & Olson, 2000) that distance will continue to be a factor in understanding these work relationships.

ACKNOWLEDGMENTS

Preparation of this chapter was facilitated by several grants from the National Science Foundation (research grants IIS-9320543, IIS-9977923, ATM-9873025, IIS-0085951 and cooperative agreement IRI-9216848). We are also grateful to several anonymous reviewers for helpful comments on an earlier draft.

References

Abbate, J. (1999). *Inventing the Internet*. Cambridge, MA: MIT Press.

Abbott, K. R., & Sarin, S. K. (1994). Experiences with workflow management: Issues for the next generation. *Proceedings of the Conference on Computer Supported Cooperative Work* (pp. 113–120). Chapel Hill, NC: ACM Press.

Abowd, G. D. (1999). Classroom 2000: An experiment with the instrumentation of a living educational environment. *IBM Systems Journal, 38*, 508–530.

Ackerman, M. S., Starr, B., Hindus, D., & Mainwaring, S. D. (1997). Hanging on the 'wire': A field study of an audio-only media space. *ACM Transactions on Computer-Human Interaction, 4*(1), 39–66.

Adamic, L. A., & Glance, N. (2005). The political blogosphere and the 2004 U.S. election: Divided they blog. Presented at LinkKDD-2005, Chicago, IL. Retrieved March 13, 2007, from www.blogpulse.com/papers/2005/.

Anderson, R. H., Bikson, T.K., Law, S. A., & Mitchell, B. M. (1995). *Universal access to e-mail: Feasibility and societal implications.* Santa Monica, CA: Rand.

Argyle, M., & Cook, M. (1976). *Gaze and mutual gaze.* New York: Cambridge University Press.

Arrow, H., Berdahl, J. L., Bouas, K. S., Craig, K. M., Cummings, A, Lebei, L., McGrath, J. E., O'Connor, K. M., Rhoades, J. A., & Schlosser, A. (1996). Time, technology, and groups: An integration. *Computer Supported Cooperative Work, 4,* 253–261.

Atkins, D. E., Droegemeier, K. K., Feldman, S. I., Garcia-Molina, H., Klein, M. L., Messerschmitt, D. G., Messina, P., Ostriker, J. P., & Wright, M. H. (2003). *Revolutioning science and engineering through cyberinfrastructure.* Report of the National Science Foundation Blue-Ribbon Advisory Panel on Cyberinfrastructure. National Science Foundation. Arlington, VA.

Bellotti, V., & Dourish, P. (1997). Rant and RAVE: Experimental and experiential accounts of a media space. In K. E. Finn, A. J. Sellen, & S. B. Wilbur (Eds.), *Video-mediated communication* (pp. 245–272). Mahwah, NJ: Lawrence Erlbaum Associates.

Bellotti, V., Ducheneaut, N., Howard, M., Smith, I., & Grinter, R. E. (2005). Quality versus quantity: E-mail centric task management and its relation with overload. *Human-Computer Interaction, 20,* 89–138.

Benford, S., Greenhalgh, C., Reynard, G., Brown, C., & Koleva, B. (1998). Understanding and constructing shared spaces with mixed-reality boundaries. *ACM Transactions on Computer-Human Interaction, 5,* 185–223.

Berners-Lee, T. (1999). *Weaving the web.* New York: Harper Collins.

Bos, N., Gergle, D., Olson, J. S., & Olson, G. M. (2001). *Being there vs. seeing there: trust via video.* Short Paper presented at the Conference on Human Factors in Computing Systems: CHI-2001. Seatle, WA.

Bryant, S. L., Forte, A., & Bruckman, A. (2005). Becoming Wikipedian: Transformation of participation in a collaborative online encyclopedia. *Proceedings of GROUP 2005* (pp. 1–10). New York: ACM.

Bush, V. (1945). As we may think. *Atlantic Monthly, 176*(1), 101–108.

Cadiz, J., Balachandran, A., Sanocki, E., Gupta, A., Grudin, J., & Jancke, G. (2000). *Distance learning through distributed collaborative video viewing.* Paper presented at the CSCW 2000, New York.

Cairncross, F. (1997). *The death of distance: How the communications revolution will change our lives.* Boston: Harvard Business School Press.

Camp, L. J. (2000). *Trust and risk in Internet commerce.* Cambridge, MA: MIT Press.

Carley, K., & Wendt, K. (1991). Electronic mail and scientific communication: A study of the Soar extended research group. *Knowledge: Creation, Diffusion, Utilization, 12,* 406–440.

Carmel, E. (1999). *Global software teams.* Upper Saddle River, NJ: Prentice-Hall.

Cosley, D., Frankowski, D., Kiesler, S., Terveen, L., & Riedl, J. (2005). How oversight improves member-maintained communities. *Proceedings of CHI 2005* (pp. 11–20). New York: ACM.

Daly-Jones, O., Monk, A., & Watts, L. (1998). Some advantages of video conferencing over high-quality audio conferencing: fluency and awareness of attentional focus. *International Journal of Human-Computer Studies, 49*(1), 21–58.

DeSanctis, G., Jackson, B. M., Poole, M. S., & Dickson, G. W. (1996). Infrastructure for telework: Electronic communication at Texaco. *Proceedings of SIGCPR/SIGMIS '96* (pp. 94–102). New York: ACM.

Dourish, P., & Bly, S. (1992). Portholes: Supporting awareness in a distributed work group. *Proceedings of CHI 92* (pp. 541–547). Monterey, CA: ACM Press.

Drolet, A. L. & Morris, M. W. (2000). Rapport in conflict resolution: Accounting for how nonverbal exchange fosters coordination on mutually beneficial settlements to mixed motive conflicts. *Journal of Experimental Social Psychology, 36*(1), 26–50.

Egido, C. (1988). *Video conferencing as a technology to support group work: A review of its failure.* Paper presented at the CSCW '88, New York.

Ehrlich, S. F. (1987). Strategies for encouraging successful adoption of office communication systems. *ACM Transactions on Office Information Systems, 5,* 340–357.

Ellis, C. A., Gibbs, S. J., & Rein, G. L. (1991). Groupware: Some issues and experiences, *CACM, 34*(1), 38–58.

Ellis, C., & Nutt, G. (1980). Office information systems and computer science. *Computing Surveys, 12*(1), 27–60.

Elrod, S., Bruce, R., Gold, R., Goldberg, D., Halasz, F., Janssen, W., Lee, D., McCall, K., Pedersen, E., Pier, K, Tang, J., & Welch, B. (1992). LiveBoard: A large interactive display supporting group meetings, presentations, and remote collaboration. *Proceedings of CHI'92* (pp. 599–607). Monterey, CA: ACM Press.

Engelbart, D., & English, W. (1968). A research center for augmenting human intellect. *Proceedings of FJCC, 33,* 395–410.

Federal Communications Commission (2005). *Voice over internet protocol.* Retrieved January 28, 2006, from at http://www.fcc.gov/voip/

Federal Trade Commission (2005, November). *Email address harvesting and the effectiveness of anti-spam filters.* Report by Federal Trade Commission, Division of Marketing Practices. Washington, DC.

Finholt, T. A. (2002). Collaboratories. In B. Cronin (Ed.), *Annual Review of Information Science and Technology.*

Finholt, T. A., & Olson, G. M. (1997). From laboratories to collaboratories: A new organizational form for scientific collaboration. *Psychological Science, 8,* 28–36.

Finholt, T., Sproull, L., & Kiesler, S. (1990). Communication and performance in ad hoc task groups. In J. Galegher, R. Kraut, & C. Egido (Eds.), *Intellectual teamwork: Social and technological foundations of cooperative work* (pp. 291–325). Hillsdale, NJ: Lawrence Erlbaum Associates.

Finn, K., Sellen, A., & Wilbur, S. (Eds.) (1997). *Video-mediated communication.* Hillsdale, NJ: Lawrence Erlbaum Associates.

Fish, R. S., Kraut, R. E., Root, R. W., & Rice, R. E. (1993). Video as a technology for informal communication. *Communications of the ACM, 36*(1), 48–61.

Foster, I., & Kesselman, C. (2004). *The Grid: Blueprint for a new computing infrastructure* (2nd ed.). San Francisco: Morgan Kaufmann.

Fussell, S. R., Kraut, R. E., & Siegel, J. (2000). *Coordination of communication: Effects of shared visual context on collaborative work.* Paper presented at the CSCW 2000, New York.

Gale, C., & Monk, A. (2000). Where am I looking? The accuracy of video-mediated gaze awareness. *Perception & Psychophysics, 62,* 586–595.

Giles, J. (2005). Internet encyclopaedias go head to head. *Nature, 438,* 900–901.

Godefroid, P., Herbsleb, J. D., Jagadeesan, L. J., & Li, D. (2000). Ensuring privacy in presence awareness systems: An automated verification approach. *Proceedings of CSCW 2000* (pp. 59–68). New York: ACM.

Grayson, D. M., & Monk, A. (2003). Are you looking at me? Eye contact and desktop video conferencing. *ACM Transactions on Computer-Human Interaction, 10*(3), 221–243.

Greif, I. (Ed.) (1988). *Computer-supported cooperative work: A book of readings.* San Mateo, CA: Morgan Kaufmann.

Grudin, J. (1988). Why CSCW applications fail: Problems in the design and evaluation of organizational interfaces. *Proceedings of the*

Conference on Computer Supported Cooperative Work (pp. 85–93). Portland, OR: ACM Press.

Grudin, J. (1994). Groupware and social dynamics: Eight challenges for developers. *Communications of the ACM, 37*(1), 92–105.

Grudin, J., & Palen, L. (1995). Why groupware succeeds: Discretion or mandate? *Proceedings of the European Computer Supported Cooperative Work* (pp. 263–278). Stockholm, Sweden: Springer.

Gutwin, C., & Greenberg, S. (1999). The effects of workspace awareness support on the usability of real-time distributed groupware. *ACM Transactions on Computer-Human Interaction, 6,* 243–281.

Hall, E. T. (1982). *The Hidden Dimension.* New York: Anchor Doubleday Books.

Herbsleb, J. D., Mockus, A., Finholt, T. A., & Grinter, R. E. (2000). Distance, dependencies, and delay in a global collaboration. *Proceedings of CSCW 2000* (pp. 319–328). New York: ACM.

Herlocker, J. L., Konstan, J. A., & Riedl, J. (2000). Explaining collaborative filtering recommendations. *Proceedings of CSCW 2000* (pp. 241–250). New York: ACM.

Hindmarsh, J., Fraser, M., Heath, C., Benford, S., & Greenhalgh, C. (1998). Fragmented Interaction: Establishing Mutual Orientation in Virtual Environments, *Proceedings of Conference on Computer-Supported Cooperative Work* (pp. 217–226). Portland, OR: ACM Press.

Hollingshead, A. B., McGrath, J. E., & O'Connor, K. M. (1993). Group performance and communication technology: A longitudinal study of computer-mediated versus face-to-face work. *Small Group Research, 24,* 307–333.

Huang, W., Olson, J. S., & Olson, G. M. (2002). *Camera angle affects dominance in video-mediated communication.* Paper presented at the CHI 2002, New York.

Isaacs, E., Walendowski, A., Whittaker, S., Schiano, D. J., & Kamm, C. (2002). *The character, functions, and styles of instant messaging in the workplace.* Paper presented at the CSCW 2002, New York.

Jensen, C., Farnham, S. D., Drucker, S. M., & Kollock, P. (2000). The effect of communication modality on cooperation in on-line environments. *Proceedings of CHI '2000.* (pp. 470–477). New York: ACM Press.

Karsenty, L. (1999). Cooperative work and shared visual context: An empirical study of comprehension problems in side-by-side and remote help dialogues. *Human-Computer Interaction, 14,* 283–315.

Kendon, A. (1967). Some functions of gaze direction in social interaction. *Acta Psychologia, 26,* 22–63.

Kiesler, S., & Cummings, J. N. (2002). What do we know about proximity and distance in work groups? A legacy of research. In P. J. Hinds & S. Kiesler (Eds.), *Distributed work* (pp. 57–80). Cambridge, MA: MIT Press.

Kiesler, S., Sproull, L, & Waters, K. (1996). Prisoner's dilemma experiment on cooperation with people and human-like computers. *Journal of Personality and Social Psychology, 70*(1), 47–65.

Koschmann, T. (1996). *CSCL: Theory and practice of an emerging paradigm.* Lawrence Erlbaum Associates. Hillsdale, NJ.

Koschmann, T., Hall, R., & Miyake, N. (2002). *CSCW 2: Carrying forward the conversation.* Lawrence Erlbaum Associates. Hillsdale, NJ.

Koschmann, T., Suthers, D. D., & Chan, T. (2005). *Computer supported collaborative learning 2005: The next 10 years!* Lawrence Erlbaum Associates. Hillsdale, NJ.

Kraemer, K. L., & Pinsonneault, A. (1990). Technology and groups: Assessments of empirical research. In J. Galegher, R Kraut, & C. Egido (Eds.), *Intellectual teamwork: Social and technological foundations of cooperative work* (pp. 373–405). Hillsdale, NJ: Lawrence Erlbaum Associates.

Krauss, R. M., & Bricker, P. D. (1966). Effects of transmission delay and access delay on the efficiency of verbal communication. *Journal of the Acoustical Society, 41,* 286–292.

Kraut, R. E., Fussell, S. R., & Siegel, J. (2003). Visual information as a conversational resource in collaborative physical tasks. *Human-Computer Interaction, 18*(1–2), 13–39.

Kraut, R. E., Gergle, D., & Fussell, S. R. (2002). *The use of visual information in shared visual spaces: Informing the development of virtual co-presence.* Paper presented at the CSCW 2002, New York.

Kraut, R., Kiesler, S., Boneva, B., Cummings, J., Helgeson, V. & Crawford, A. (2002). Internet paradox revisited. *Journal of Social Issues, 58*(1), 49–74.

Kraut, R., Patterson, M., Lundmark, V., Kiesler, S., Mukopadhyay, T., & Scherlis, W. (1998). Internet paradox: A social technology that reduces social involvement and psychological well-being. *American Psychologist, 53,* 1071–1031.

Kraut, R. E., Sunder, S., Telang, R., & Morris, J. (2005). Pricing electronic mail to solve the problem of spam. *Human-Computer Interaction, 20,* 195–223.

Lange, B. M. (1992). Electronic group calendaring: Experiences and expectations. In D. Coleman (Ed.) *Groupware* (pp. 428–432). San Mateo, CA: Morgan Kaufmann.

Longstaff, T. A., Ellis, J. T., Hernan, S. V., Lipson, H. F., McMillan, R. D., Pesanti, L. H., & Simmel, D. (1997). Security on the Internet. In *The Froehlich/Kent Encyclopedia of Telecommunications* (Vol. 15, pp. 231–255). New York: Marcel Dekker.

Luff, P., Heath, C., Kuzuoka, H., Hindmarsh, J., & Oyama, S. (2003). Fractured ecologies: Creating environments for collaboration. *Human-Computer Interaction, 18*(1–2), 51–84.

Mackay, W. E. (1989). Diversity in the use of electronic mail: A preliminary inquiry. *ACM Transactions on Office Information Systems, 6,* 380–397.

Malone, T. W., Grant, K. R., Lai, K. Y., Rao, R., & Rosenblitt, D. A. (1989). The information lens: An intelligent system for information sharing and coordination. In M. H. Olson (Ed.). *Technological support for work group collaboration* (pp. 65–88). Hillsdale, NJ: Lawrence Erlbaum Associates.

Markus, M. L. (1983). *Systems in Organization: Bugs and Features.* San Jose, CA: Pitman.

Markus, M. L., & Connolly, T. (1990). Why CSCW applications fail: Problems in the adoption of interdependent work tools. *Proceedings of the Conference on Computer Supported Cooperative Work* (pp. 371–380). Los Angeles, CA: ACM Press.

McDaniel, S. E., Olson, G. M., & Magee, J. S. (1996). Identifying and analyzing multiple threads in computer-mediated and face-to-face conversations. *Proceeding of the ACM Conference on Computer Supported Cooperative Work* (pp. 39–47). Cambridge, MA: ACM Press

McLeod, P. L. (1992). An assessment of the experimental literature on electronic support of group work: Results of a meta-analysis. *Human-Computer Interaction, 7,* 257–280.

Monk, A., & Gale, C. (2002). A look is worth a thousand words: Full gaze awareness in video-mediated communication. *Discourse Processes, 33*(3), 257–278.

Moran, T. P., Chiu, P., Harrison, S., Kurtenbach, G., Minneman, S., & van Melle, W. (1996). Evolutionary engagement in an ongoing collaborative work process: A case study. *Proceeding of the ACM Conference on Computer Supported Cooperative Work* (pp. 150–159). Cambridge, MA: ACM Press.

Mosier, J. N., & Tammaro, S. G. (1997). When are group scheduling tools useful? *Computer Supported Cooperative Work, 6,* 53–70.

Muller, M. J., Raven, M. E., Kogan, S., Millen, D. R., & Carey, K. (2003). *Introducing chat into business organizations: Toward an instant messaging maturity model.* Paper presented at the GROUP '03, New York.

Nardi, B. A., Schwarz, H., Kuchinsky, A., Leichner, R., Whittaker, S., & Sclabassi, R. (1993). *Turning away from talking heads: the use of video-as-data in neurosurgery.* Paper presented at the CHI 93, New York.

Nardi, B. A., Whittaker, S., & Bradner, E. (2000). Interaction and outeraction: Instant messaging in action. *Proceedings of the ACM Conference on Computer Supported Cooperative Work* (pp. 79–88). Philadelphia, PA: ACM Press.

Nardi, B. A., Schiano, D. J., Gumbrecht, M., & Swartz, L. (2004). Why we blog? *Communications of the ACM, 47*(12), 41–46.

National Research Council. (1993). *National collaboratories: Applying information technology for scientific research*. Washington, D.C.: National Academy Press.

Nohria, N., & Eccles, R. G. (Eds.). (1992). *Networks and organizations: Structure, form, and action*. Boston: Harvard Business School Press.

Nunamaker, J. F., Dennis, A. R., Valacich, J. S., Vogel, D. R., & George, J. F. (1991). Electronic meeting systems to support group work. *Communications of the ACM, 34*(7), 40–61.

O'Conaill, B., Whittaker, S., & Wilbur, S. (1993). Conversations over videoconferences: An evaluation of the spoken aspects of video mediated communication. *Human-Computer Interaction, 8*, 389–428.

O'Hara-Devereaux, M., & Johansen, R. (1994). *Global work: Bridging distance, culture & time*. San Francisco: Jossey-Bass.

Okada, K., Maeda, F., Ichicawaa, Y., & Matsushita, Y. (1994). *Multiparty videoconferencing at virtual social distance: MAJIC design*. Paper presented at the CSCW 94, New York.

Olson, G. M., & Olson, J. S. (2000). Distance matters. *Human-Computer Interaction, 15*, 139–179.

Olson, G. M., Atkins, D., Clauer, R., Weymouth, T., Prakash, A., Finholt, T., Jahanian, F., & Rasmussen, C. (2001). Technology to support distributed team science: The first phase of the Upper Atmospheric Research Collaboratory (UARC) In G. M. Olson, T. Malone, & J. Smith (Eds.), *Coordination theory and collaboration technology* (pp. 761–783). Hillsdale, NJ: Lawrence Erlbaum Associates.

Olson, G. M., Olson, J. S., Bos, N., & the SOC Data Team (2004). International collaborative science on the net. In W. Blanpied (Ed.), *Proceedings of the Trilateral Seminar on Science, Society and the Internet* (pp. 65–77). Arlington, VA: George Mason Univeristy.

Olson, J. S., Olson, G. M., & Meader, D.K. (1995). What mix of video and audio is useful for remote real-time work? *Proceedings of CHI '95* (pp. 362–368). Denver, CO: ACM Press.

Olson, J. S., Olson, G. M., Storrøsten, M., & Carter, M. (1993). Group work close up: A comparison of the group design process with and without a simple group editor. *ACM Transactions on Information Systems, 11*, 321–348.

Olson, J. S., & Teasley, S. (1996). Groupware in the wild: Lessons learned from a year of virtual collocation. *Proceeding of the ACM Conference on Computer Supported Cooperative Work* (pp. 419–427). Cambridge, MA: ACM Press.

Olson, M. H., & Bly, S. A. (1991). The Portland experience: A report on a distributed research group. *International Journal of Man-Machine Studies, 34*, 211–228.

Orlikowski, W. J., & Gash, D. C. (1994). Technological frames: Making sense of information technology in organizations. *ACM Transactions on Information Systems, 12*, 174–207.

Palen, L., & Grudin, J. (2002). Discretionary adoption of group support software: lessons from calendar applications. In B. E. Munkvold (Ed.), *Implementing collaboration technologies in industry* (pp. 159–180). Springer-Verlag. London.

Park, K. S., Kapoor, A., & Leigh, J. (2000). Lessons learned from employing multiple perspective in a collaborative virtual environment for visualizing scientific data. *Proceedings of ACM CVE'2000 Conference on Collaborative Virtual Environments* (pp. 73–82). San Francisco, CA: ACM Press.

Prinz, W., & Kolvenbach, S. (1996). Support for workflows in a ministerial environment. *Proceedings of the Conference on Computer Supported Cooperative Work* (pp. 199–208). Cambridge, MA: ACM Press.

Putnam, R. D. (2000). *Bowling alone: The collapse and revival of American community*. New York: Simon & Schuster.

Resnick, P. (2002). Beyond bowling together: SocioTechnical capital. In J. M. Carroll (Ed.), *Human-computer interaction in the new millennium* (pp. 647–672). New York: ACM Press.

Resnick, P., & Varian, H. R. (Eds.) (1997). Special section: Recommender systems. *Communications of the ACM, 40*(3), 56–89.

Rocco, E. (1998) Trust breaks down in electronic contexts but can be repaired by some initial face-to-face contact. *Proceedings of CHI'98* (pp. 496–502). Los Angeles, CA: ACM Press.

Rocco, E., Finholt, T., Hofer, E. C., & Herbsleb (2000). *Designing as if trust mattered*. (CREW Technical Report). University of Michigan, Ann Arbor.

Ruhleder, K., & Jordan, B. (2001). Co-constructing non-mutual realities: Delay-generated trouble in distributed interaction. *Computer Supported Cooperative Work, 10*(1), 113–138.

Satzinger, J., & Olfman, L. (1992). A research program to assess user perceptions of group work support. *Proceeding of CHI'92* (pp. 99–106). Monterey, CA: ACM Press.

Schafer, J. B., Konstan, J., & Riedl, J., (2001). Electronic commerce recommender applications. *Journal of Data Mining and Knowledge Discovery, 5*(1/2), 115–152.

Schatz, B. R., & Hardin, J. B. (1994). NCSA Mosaic and the World Wide Web: Global hypermedia protocols for the internet. *Science, 265*, 895–901.

Schmandt, C., & Marti, S. (2005). Active Messenger: E-mail filtering and delivery in a heterogeneous network. *Human-Computer Interaction, 20*, 163–194.

Short, J., Williams, E., & Christie, B. (1976). *The social psychology of telecommunications*. New York: Wiley.

Siegel, J., Dubrovsky, V., Kiesler, S., & McGuire, T. W. (1986). Group processes in computer-mediated communication. *Organizational Behavior and Human Decision Processes, 37*(2), 157–187.

Singhal, S., & Zyda, M. (1999). *Networked virtual environments: design and implementation*. New York: Addison-Wesley.

Slavin, R. E. (1994). *Cooperative learning: Theory, research, and practice* (2nd ed.). Boston: Allyn & Bacon.

Sproull, L., & Kiesler, S. (1991). *Connections: New ways of working in the networked organization*. Cambridge, MA: MIT Press.

Starner, T., & Rhodes, B. (2004). Wearable computer. In W. S. Bainbridge (Ed.), *Berkshire encyclopedia of human-computer interaction* (Vol. 2, pp. 797–802). Great Barrington, MA.: Berkshire Publishing Group.

Straus, S. G. (1996). Getting a clue: The effects of communication media and information distribution on participation and performance in computer-mediated and face-to-face groups. *Small Group Research, 1*, 115–142.

Straus, S. G. (1997). Technology, group process, and group outcomes: Testing the connections in computer-mediated and face-to-face groups. *Human-Computer Interaction, 12*(3), 227–266.

Straus, S. G., & McGrath, J. E. (1994). Does the medium matter: The interaction of task and technology on group performance and member reactions. *Journal of Applied Psychology, 79*, 87–97.

Tanenbaum, A. S. (2003). *Computer networks* (4th ed.). Upper Saddle River, NJ: Prentice Hall PTR.

Tang, J. C., & Isaacs, E. (1993). Why do users like video? *Computer Supported Cooperative Work, 1*(3), 163–196.

Tang, J. C., Yankelovich, N., Begole, J., van Kleek, M., Li, F., & Bhalodia, J. (2001). ConNexus to Awarenex: Extending awareness to mobile users. *Proceedings of CHI 2001* (pp. 221–228). New York: ACM.

Teasley, S. D., Covi, L. A., Krishnan, M .S., & Olson, J. S. (2002). Rapid software development through team collocation. *IEEE Transactions on Software Egnineering, 28*, 671–683.

Veinott, E., Olson, J. S., Olson, G. M., & Fu, X. (1999). Video helps remote work: Speakers who need to negotiate common ground benefit from seeing each other. *Proceedings of the Conference on Computer-Human Interaction, CHI'99* (pp. 302–309).Pittsburgh, PA: ACM Press.

Vertegaal, R. (1999). *The GAZE groupware system: Mediating joint attention in multiparty communication and collaboration*. Paper presented at the CHI 99, New York.

Vertegaal, R., Slagter, R., van der Veer, G., & Nijholt, A. (2001). *Eye gaze patterns in conversations: There is more to conversational agents than meets the eye.* Paper presented at the CHI 2001, New York.

Viegas, F. B., Wattenberg, M., & Dave, K. (2004). Studying cooperation and conflict between authors with history flow visualizations. *Proceedings of CHI 2004* (pp. 575–582). New York: ACM.

Wellman, B. (2001). *Design considerations for social networkware: Little boxes, glocalization, and networked individualism.* Draft ms.

Whittaker, S. (2005). Supporting collaborative task management in e-mail. *Human-Computer Interaction, 20,* 49–88.

Whittaker, S., Bellotti, V., & Moody, P. (2005). Introduction to this special issue on revisiting and reinventing e-mail. *Human-Computer Interaction, 10,* 1–9.

Whittaker, S., Geelhoed, E., & Robinson, E. (1993). Shared workspaces: How do they work and when are they useful? *International Journal of Man-Machine Studies, 39*(5), 813–842.

Whittaker, S., & Sidner, C. (1996). Email overload: Exploring personal information management of email. *Proceeding of CHI'96* (pp. 276–283). Vancouver, BC: ACM Press.

Williams, E. (1977). Experimental comparisons of face-to-face and mediated communication: A review. *Psychological Bulletin, 84,* 963–976.

Winograd, T. (1988). A language/action perspective on the design of cooperative work. *Human Computer Interaction, 3,* 3–30.

Yang, H., & Olson, G. M. (2002). Exploring collaborative navigation: the effect of perspectives on group performance. *Proceedings of CVE '02* (pp. 135–142). New York: ACM Press.

Zheng, J., Bos, N., Olson, J. S., Gergle, D. & Olson, G. M. (2001). Trust without touch: Jump-start trust with social chat. *Paper presented at the Conference on Human Factors in Computing Systems CHI-01.* Seatle, WA: ACM Press.

·12·

HCI AND THE WEB

Helen Ashman, Tim Brailsford, Gary Burnett, Jim Goulding,
Adam Moore, and Craig Stewart
University of Nottingham

Mark Truran
University of Teesside

Introduction **232**
What Makes the Web Hard to Use? **232**
   Browsing and Linking: "What's Wrong
   With the World Wide Web?" Revisited 232
   Finding Things (i): Search and Query on the Web 233
   Finding Things (ii): Relevance 234
   User Interface Issues 234
      Context of Use 234
      Navigation Issues 235
   Summary 235
Browsing and Linking **235**
   Broken and Misdirected Links 236
   Personalizing Links 236
Searching and Querying **237**
   Query Formulation 237
      Relevance Feedback 238

The Results List 238
   Community-Based Ranking Algorithms 238
   Improved Visual Interfaces 239
   Document Clustering 239
Summary 239
Personalization, Portals and Communities **239**
   Personalization 240
      Adaptive and Adaptable Systems 240
      Need or Desire? 241
      Methods and Models of Adaptation 241
      User Modeling 241
      Authoring 242
   Portals 242
   Communities 242
Conclusions **243**
References **243**

INTRODUCTION

Pioneering research work in the early 1980s introduced and defined the concept of usability as critical to the success of interactive products and systems (Eason, 1984). From the 1990s there has been a developing relationship between usability and the web, such that for many people the two terms are inextricably linked. Specifically, not only does the web demand good usability, it can be argued that usability necessitates the web.

There are five key reasons why usability and the web have become so closely associated with each other:

1. Web users are ubiquitous. For many countries, the people who access the web are approaching the population as a whole. For instance, within Great Britain 55% of households can now access the Internet from home.[1] For some countries (i.e., United States, Denmark, Australia), it has been estimated that over 70% of the population are regularly using the web.[2] Such diversity raises a large number of issues related to interdependent variables such as age, gender, culture, disability, language abilities, computer skills/knowledge, domain skills/knowledge, and so on. In particular, considerable ongoing research considers the *accessibility* of the web; that is, websites can be designed to account for the needs of as many people as possible (i.e., those with visual impairments) (Abascal & Nicolle, 2001).

2. Web users are largely discretionary users. Apart from the specific case of the work context and company intranets, web users generally do not have to use a particular site (or even the web at all) to achieve their goals. They have alternatives available to them (i.e., visiting another website, making a phone call, visiting a shop), and if they experience usability problems, they do not necessarily have to struggle with or adapt to the poor interface. They are empowered to explore existing options.

3. Web usability problems have a clear relationship with sales. For websites aiming to sell products or services, poor usability directly impacts sales. If a user cannot find the product or relevant information, he or she is unlikely to continue in the transaction. In this case, it has been noted that usability is affecting the experience of a product *prior* to purchase (Nielson, 2000). Consequently, the website information space must consider many of the issues (i.e., navigation, layout of items, labelling) that are relevant to the traditional design of physical space for shops.

4. The web is evolving at a rapid pace. Technical characteristics are continually changing in response to new application/task areas, facilitated by computer processing power and communication speeds. This has an immediate impact on the functionality available to web users (i.e., animations, videos), developments in user interface design (i.e., clickable items, mouse-over navigation), as well as tools (i.e., cookies, plug-ins). In contrast, users' skills, knowledge, and expectations are considerably slower to evolve, leading to an inevitable gulf between users (particularly infrequent users) and the web.

5. Website technical development is easy. It is simple to have a presence on the web when compared with the resources necessary for traditional product-development processes. To generate a website requires few technical skills, and one can do so with no programming experience. This has contributed to the vast number of websites[3] ultimately adding to the complications of navigation, whilst emphasising the importance of usability as a differentiating factor.

In this chapter we first consider what is it that makes the web hard to use. We look specifically at browsing and linking, finding things with search and query, increasing the relevance of things we find with personalization, and user interface issues. The subsequent sections look at some solutions to these problems.

WHAT MAKES THE WEB HARD TO USE?

As successful as the web is for delivering information globally and rapidly, many problems remain that make it an unpleasant or unproductive experience for some users, or even impossible for other users.

Much of the difficulty in using the web lies in the vast quantity of information available—browsing and searching becomes increasingly difficult and imprecise in its results. Then having located what one wants, technical flaws, such as broken links or browser incompatibilities, often render the information unreachable or unreadable. Most importantly, web pages or the browser interface itself can give a poor presentation, and for some classes of users, make the information inaccessible.

Browsing and Linking: "What's Wrong With the World Wide Web?" Revisited

Almost a decade ago, the difficulties in using the web were explored in-depth, and presented a vision of "fourth-generation hypermedia" (Bieber, Vitali, Ashman, Balasubramanian, & Oinas-Kukkonen, 1997). In particular, the importance of hypermedia in the structuring of information was described:

hypermedia provides contextual, navigational access for viewing information and . . . represents knowledge in a form relatively close to the cognitive organizational structures that people use. Thus hypermedia supports understanding.

The judicious use of hypermedia enhances a user's comprehension. When that use is tailored specifically to the user's requirements, i.e., it is *personalized*, the user's comprehension and productivity are maximized.

---

[1]Figures for July 2005 from web.statistics.gov.uk.

[2]Figures for 2005 from www.internetworldstats.com.

[3]Estimated to be over 9 million in 2002—www.oclc.org.

The authors went on to define a table of desired hypermedia features (Bieber et al., 1997). Of particular relevance to HCI and the web are the personalization of links by annotation, computation, and overviews.

Surprisingly, nearly a decade later, many of these desired features are still lacking from the mainstream web applications. In particular, the personalization of links and content is still largely unachievable without specialist tools. In fact, some backsliding on this is evident over the past decade, as the mid-1990s Mosaic browser supported the creation of annotations at both personal and group levels and, more importantly, offered this service by default in a standard browser.[4] Mosaic development, however, was discontinued when Netscape became widespread. W3C's Annotea project[5] offers the same essential functionality but not within a standard browser and not without setting up effort from a technically literate user—annotations are by no means a standard or easily accessible web service.

Creating personalized annotations is challenging enough, but the creation of one's own links over third-party data sources is essentially unknown. This is not due to any shortcomings in the technology; as in fact many solutions have been available within the research community for some time (see "Browsing and Linking"). The earliest plans for web browsers, dating back to 1991 (Cailliau & Ashman, 1999), anticipated a "writeable web" where personalized links and annotations would be easily supported, yet this has not become mainstream web technology.

Another key aspect of personalized links which is inevitably overlooked is that personal links are *private* links. If one creates a web page, complete with links, then those links are necessarily as available as the content. This is not always desirable, especially if links record one's associations and collected thoughts, and the intellectual property represented by those associations has value. The technology for personalized links gives the user private links, over public documents, in much the same way that users' bookmarks are private. For privacy reasons alone, the technology for personalized links should be requisite for all mainstream browsers.

Other desired hypermedia features, such as local overviews, are provided not as a standard web service but as a site-specific courtesy page optionally supplied by site managers. Global overviews seem an increasingly distant feature, as the vast complexity of the web makes the usual, graph-based representations completely unworkable, both computationally and visually. The earlier paper noted that:

Web browsers have no inherent way of presenting the structure and interrelationships of data of any sort. For example, there is no way to visualize even the simple interrelationships of web documents, such as "Where can I go from here?" or "Which documents point to this document?" The reader has no idea of the position of a given document within the corpora unless an author explicitly embeds such details. (Bieber et al., 1997)

So why is it that so little has seemingly been done to address these outstanding problems? It could be because the technology is too inaccessible to the nonspecialist user and that the obvious up-front costs of assimilating the technology outweigh the perceived benefits. It still seems to be true that:

experienced readers learn to get by with the available functionality. However, the use of web technology is in part determined by its capabilities. Readers learn to make do with the available tools instead of demanding better tools, perhaps because they are not aware of how better tools might help them. (Bieber et al., 1997)

Only when we make the web more usable to the nonspecialist user will the use of all but the most basic tools will become commonplace, resulting in higher productivity and better assimilation of information in the user.

So far, we have seen the web does not offer widespread support for personalization of links and other browsing activities. We now look at the other major means of web information access: searching and querying.

Finding Things (i): Search and Query on the Web

One recent estimate has placed the size of the World Wide Web at around 11.5 billion pages (Gulli & Signorini, 2005). This figure, representing nearly 1.8 web pages per person alive today,[6] relates only to static or fixed web pages. It has been suggested that if this survey was extended to include more ephemeral information (i.e., individual auction items or ongoing news stories), then the correct figure is in fact 500 times larger (Bergman, 2001).

With this wealth of information, the web would be untenable without mechanisms to assist navigation and file location. The most common web tool in use today is the search engine. To use a search engine the user must submit a series of terms known as a query. This query in some way formulates and embodies what the user of the search system would like to retrieve information about. The usual output produced by the search engine is a small set of links to a selection of web documents extracted from the billions available. These web pages represent the search engine's response to the user's information need, each document having been determined by the system as probably relevant to the query.

Presented with a set of results, the user selects a web page. If the document fails to meet the user's requirement, they return to the search engine for another result, or modify the query in light of previous results. This continues until the user's need is satisfied or their commitment to the search wanes (Harman, 1992).

From the perspective of a typical user, this has two main problem areas:

1. the confusion and uncertainty surrounding query formulation
2. the impenetrability of seemingly endless results

In "Searching and Querying" we discuss these obstacles and review a range of solutions.

---

[4] In 1994 the Mosaic browser was the "killer application" for the Internet, being the first major web browser, freely available from NCSA. Its developers later spun off the Netscape company.

[5] http://www.w3.org/2001/Annotea/.

[6] http://www.geohive.com approximates the population of the Earth at 6 483 981 671 (5th January 2005).

Finding Things (ii): Relevance

While browsing and search-and-query are excellent tools for finding information on the web, they remain generic functions whose behavior is the same, regardless of the context. What remains to be considered now is how to further filter the mass of information on the web according to its relevance, not to a query, but to the individual user or to a community of users.

Most information systems are designed for a hypothetical "average" user. This "one-size-fits-all" approach ignores diversity in cultural and educational backgrounds, abilities, objectives and aspirations. An information system with a single user interface for all users is conceptually the same as a car manufacturer selling a car in only a single color—"any color so long as it's black".[7]

One solution is to build personalizable information systems, delivering content specific to requirements of different users. Without such systems, effective universal accessibility to information cannot be achieved. With it, every computer user and special interest group will have personally tailored access to information sources.

Personalization can take many forms, and can involve the tailoring of hypertext links (as discussed in "What Makes the Web Hard to Use"), the tailoring of presentation (often an accessibility issue, see "User Interface Issues") and the tailoring of content.

Content personalization has many benefits, for not just users, either as individuals or within a community, but also for the providers of information.

For individuals, the tailoring of content can reflect a wide variety of requirements, but is always aimed at providing information most pertinent to that user in their current context. In e-learning applications for example, more challenging lessons are not served to the user until mastery of prerequisite material is achieved. In e-commerce, a simple form of user consensus (see "Document Clustering") underlies a recommender system that personalizes suggestions for further purchases based on the current users' purchasing history similarity to that of other users.

For groups, personalization of content could help create communities of common interest amongst otherwise disparate users. Personalization can also support existing communities of users, by creating portals for accessing materials which are "personalized" to the interests of the group or community.

Personalization of information can also improve the whole computer use experience for the two major groups of underrepresented users: (a) non-American/English cultural and language groups, and (b) people with special information needs. The web may have been "invented" by Europeans (Cailliau et al., 1999), but its subsequent development from 1993 was U.S.-based and has a distinctly American culture, including a predominance of the English language, linguistic form and alphabets.[8]

For information providers and publishers, personalization could create a delivery of information as suitable as possible to each individual user or special-interest group. In the normal publishing model, it is not economically feasible for publishers or information owners to publish information in a multitude of different forms, but with personalization and adaptation (see "Personalization"), it would be possible to do the personalization "on the fly" at no additional cost to the publishers, bringing enormous economic advantage to publishers using the technology.

This is essentially the same principle encountered in teaching lectures present the information in a fairly general fashion, while tutorials or other small-group teaching give the students a chance to specify exactly what they need to know and to be given examples most helpful to them. In fact, the pressure on teaching resources has partly motivated adaptation and personalization in e-learning, and similarly it is the lack of resources preventing publishers from delivering their materials tailored to special-interest groups and to the individual. With e-learning, adaptation and personalization have arisen because teachers must offer personalized tutoring but are increasingly short of resources. With e-commerce, multicultural groups and special needs groups, adaptation and personalization can also be applied to offer an equally valuable specialisation of information that is just not otherwise available.

User Interface Issues

We have just seen how awareness of context played an important part in the personalization of content and links. However context also needs to be considered in any environment, not just for personalization purposes.

Context of Use

When investigating the usability of web user interfaces, one must first consider the overriding issue of context. Context of use is seen as a critical constituent of usability, defined by ISO (1998) to "consist of the users, tasks, and equipment (hardware, software, and materials) and the physical and social environments in which a product is used." With respect to the web, the user, the task, and the environment, issues are constantly evolving, leading to new challenges for HCI researchers and practitioners. For instance, the traditional environment for web users would have been the workplace, but statistics show that this is no longer the single situation for web use. Eighty-eight percent of UK web users access the Internet primarily from home.[9] Also, an increasing number of people now access the Internet through mobile devices (i.e., laptops, phones, PDAs). In these situations, designers must consider a much wider range of physical environment factors, for instance, varying lighting, noise and thermal conditions, as well as other tasks that users may simultaneously carry out.

---

[7]Henry Ford is credited with saying "You can paint it any color, so long as it's black," but the Model T eventually appeared in many colors. Customer demand can motivate personalization.

[8]The American culture and set of assumptions is reflected in not only issues such as inadvertently rendering Arabic script backwards (left to right) because browsers render left to right by default, but also in American spelling of HTML tags, such as "color" and "gray."

[9]Figures for July 2005 from www.statistics.gov.uk

For example, there are widespread concerns regarding the use of Internet services within road-based vehicles (Lai, Cheng, Green, & Tsimhoni, 2002; Burnett, Summerskill, & Porter, 2004). Access to the web while driving may provide a range of tangible benefits to drivers, some driving-related (i.e., pre-booking a parking space, accessing real-time traffic information), and others oriented toward productivity or entertainment needs (i.e., viewing the latest information on stocks, downloading MP3 files). Traditional access to the web through desktop computers has utilized user interface paradigms that are both highly visual and manual (i.e., scanning a page for a link and then using a mouse to point and then click on the link). Such activities are in clear conflict with the safety/time-critical task of driving, which places heavy demands on the visual modality, while requiring continuous manual responses (i.e., turning the steering wheel). Clearly, designers need to establish fundamentally new interaction styles for use in a driving context of use (i.e., speech recognition, voice output, haptic interfaces; see for example Iwata et al. in this book). As part of this process, designers of interfaces for in-car Internet services must consider whether drivers should be given access to functionality while the vehicle is in motion. Considerable ongoing research is being conducted to determine what constitutes an overly distracting interface (Lai et al., 2002).

Navigation Issues

There is a common assertion that the two biggest difficulties facing web users are download times and navigation (Nielson, 2000; McCracken & Wolfe, 2004). Download times are largely governed by the technical capabilities of Internet connections (which have increased markedly over recent years), together with clear design variables (i.e., the size of individual website pages). Navigation issues are considerably more complex. According to one diary-based study (Lazar, Bessiere, Ceaparu, Robinson, & Shneiderman, 2003), between one-third and one-half of time spent using a computer is unproductive, a situation predominately attributed to problems in web navigation.

In analyzing the navigation problem, we must first consider what is meant by a navigation task. A CHI workshop on this topic from 1997 (Jul & Furnas, 1997) commented that a difficulty with research in the area was that authors tended to define navigation in varying ways. A broad view (adapted from the CHI workshop) is taken here, in which navigation can be said to involve the following:

- Planning "routes." When people navigate through space (whether real or electronic), they must first consider their overall strategy; that is, what methods will be appropriate in the current situation. For the web, a range of methods exist that aim to assist the user when deciding how to navigate across the web (i.e., search engines, directories, URLs) and/or within specific sites (i.e., site maps, navigation menus, links). Many people find it difficult to generate a suitable plan for a

range of reasons, either concerning basic cognitive limitations (such as remembering URLs), a lack of knowledge (choosing appropriate search terms, misunderstanding Boolean logic), or because methods are poorly implemented (i.e., confusing layouts for site maps).

- Following "routes." Once a high-level plan exists, people need to execute the subsequent point-by-point decisions, necessitating a virtual form of locomotion through the information space. In this stage, typical problems facing the web user often relate to the design of linking mechanisms between pages, for example, ambiguous link labelling, unclear graphics or icons, relevant information appearing off-screen, the need to visually scan large numbers of links, and so on.

- Orienting within the "space." The "where am I?" problem of the web is perhaps the one most discussed and researched (Otter & Johnson, 2000). For optimum navigation performance and confidence, people need to have a sense of their current location in relation to their surroundings (i.e., their final destination, their start point, and other key "landmarks," such as a home page). Orientation difficulties are often compounded on the web because users are "dropped" into a specific location when they use search engines and bookmarks.

- Learning the "space." Repeated exposure to any large-scale environment (whether it be real or electronic) will lead to a deepening knowledge of objects within the space (i.e., particular pages), as well as an understanding of the various relationships between the objects (i.e., how pages follow each other, the overall structure of a site). These specific cases of mental models are commonly known as cognitive maps, and facilitate fast and accurate navigation performance. websites with poor differentiation (i.e., all pages appearing to be similar), low visual access (i.e., difficult to see where one can go next) and high path complexity (i.e., many links on a page) will all contribute to a poorly formed cognitive map (Kim & Hurtle, 1995).

Summary

In the subsequent sections, we turn to some solutions for the problems outlined here, looking first at browsing and linking ("Browsing and Linking"), then turning to search ("Searching and Querying") and personalization ("Personalization, Portals, and Communities").

BROWSING AND LINKING

The fundamental mechanism for viewing information on the web is by browsing implemented by hypermedia links.[10] However, this core capability still encounters technical problems that interrupt the user's browsing, such as difficulties in finding relevant material and limitations in the way relevant materials can be presented.

---

[10]The very names of the two most important elements of the web infrastructure, *hypertext* transfer protocol (http) and *HyperText* Markup Language (HTML) indicate this clearly.

In this section, we look at some solutions, including automatic management of broken links, the easy personalization of links, and how links can enable different perspectives on the same data.

Broken and Misdirected Links

Broken links have frustrated users not only of the web, but also of earlier hypertext systems, with some of the first work on link integrity motivating the open hypermedia systems' principle of externalizing links, rather than embedding them into the data (Davis, 1995).

Externalizing links implies that links are stored separately from the data being linked, so that reconciliation of the link to its referent was required, usually immediately before use, as a "late binding" of links (Brailsford, 1999). However, any changes in the data meant that links could be either displaced, pointing to the wrong place in a document (part-of-file error, Davis, 1998), or completely invalidated, with no document now known for the link to apply to (a whole-file error, Davis, 1998).

HTML links suffer from a similar problem, despite being embedded in the data, since it is not only the source of the link (the "from" part of it) that must be reconciled to its referent, but also the destination (the "to" part). It is in the latter that HTML links are frequently wrong, with an estimated 23% of web links being broken within a year (Lawrence, Pennock, Flake, Krovetz, Coetzee, Glover, Nielsen, Kruger, & Giles, 2001).

Broken links can be frustrating for the user, resulting in loss of interest in the website. Fortunately, this is one area that has seen ongoing improvement.

The solutions to broken links can be characterised as being *preventative* (creating infrastructure or procedures that avoid broken links), *corrective* (correcting broken links where they are discovered) or *adaptive* (never storing actual links, only instructions for making them as required) (Ashman, 2000).

From the user's point of view, it might initially seem that preventative solutions are ideal, since the irritation of broken or misdirected links will never happen. However, many of the preventative measures can only guarantee accurate links within a limited scope, and changes outside that scope (such as an entire domain name change) can still result in broken links. Also, they can be functionally limited; for example, it may be impossible to guarantee link integrity in information that is not part of the same preventative scheme.

Corrective solutions tend to be more robust, as they assume breakage will occur and have procedures in place to correct links, where possible, or to otherwise deal with them. These procedures are sometimes computations that aim to discover the new location for the linked document. These often function as mass correction procedures, taking place at intervals, which detect broken links and attempt to correct them, discard them, or notify the user of the problem. From the everyday user's point of view, this is a reasonable form of solution, requiring little or no effort on their part, with breakages often not encountered by the user. However, it is still possible that the user will discover a broken link, especially if it is some time since the most recent correction.

Also, the corrective approach may discard unfixable links that the user has previously required. This leaves the user with the knowledge that a link that was once present is now gone and seemingly unrecoverable. Perhaps a more user-friendly solution to irretrievably broken links is the so-called "soft 404s" (Bar-Yossef, Broder, Kumar, & Tomkins, 2004)—when pages go missing, those pages are replaced by human-readable error messages, which essentially assume the identity of the missing page. They frequently offer the user the option of a search of the site, or perhaps redirect the user to a new location. It is estimated that 25% or more of all dead links are these soft 404s (Bar-Yossef et al., 2004).

The correction of links is very much still a research problem, with much effort expended into double guessing the original link creator's intentions in making the link. However, if the link was originally created by a computation, then the correction often amounts to nothing more than executing the computation again. If, for example, the link computation was to "link every instance of a person's name to his or her home page," and then the name is moved within various pages, or appears or disappears, the links can easily be reinstated to their correct positions.

The adaptive approach assumes that all data is subject to change, as do the corrective solutions. Instead of having procedures for correction in place, it guarantees correct links by never having links in storage. The instructions to create links are stored, and links are created as required from them. This dynamic approach was trialled in experimental systems that created links either on user demand (Davis, Knight, & Hall, 1994) or as a background process (Verbyla & Ashman, 1994).

The adaptive approach is now evident in everyday web usage; for example, scripts are used to create and serve web pages dynamically, and often result from a database query, where links are calculated and inserted at the time of serving. However, it cannot assist with links created and maintained by other users and neither can the preventative or corrective approaches.

In the end, the major obstacle to link integrity in the web is its anarchic nature. There is no central authority that can impose robust linking practice on the mass of users; indeed the failure of the error-tolerant naming solutions such as URNs to capture the imagination shows that it is not even possible to tempt users into good linking practices, let alone force them.[11] Even a direct plea from one of the joint creators of the web has been unable to achieve link robustness (Berners-Lee, 1998). As long as users continue to create links without troubling to maintain them, other users will eventually encounter them as broken links. There are, however, numerous solutions that when combined will help keep the problem at bay, so that users may experience almost 404-free browsing.

Personalizing Links

In the previous section, we saw the utility of computation of links for reducing the number of broken links a user encounters.

---

[11]As evidenced by the lack of recent activity on URNs at the World Wide Web Consortium's *Naming and Addressing* page at http://www.w3.org/Addressing/.

In this section, we will see how computed links, with other technologies, allow the user to personalize the links and types of links, regardless of the source, type, or ownership of the data.

All users at some stage want to create their own links or annotations, generally to record their own associations between data or to ease everyday information access. In fact, easing information access is an obvious but frequently overlooked purpose for links—links function as a form of "user pull" of information, hiding the information while still making its presence known, but providing that information on request with the minimum of user effort.

Users have different needs, and an author of web pages cannot anticipate all such requirements, let alone provide them. Even if all the potentially useful links were provided, not only would users disagree on the facility of the links, but the interface and performance of the web would suffer.

For example, not every user wants a dictionary link, which could give a basic definition of any word selected by the user. However, non-native speakers of a language could find such a link invaluable. Glossary links are essential to a reader not familiar with technical terms, but become intrusive to seasoned readers. In each case, the users want to be able to "switch on or off" links to reflect their own needs.

This is easy with current research technology. Even as far back as 1992 (Davis, Hall, Heath, Hill, & Wilkins, 1992), the technology to provide links over data not owned by the user was available. It was even quite simple to create a form of computed link, for example, to link every occurrence of a name with a bibliography. Within a short time, it even became feasible not just to exploit link computation to automatically create one's own links, but to even create one's own link-computation specifications, so that exactly the right computation, pointing into the right data collections, could be easily linked at the user's wish (Verbyla et al., 1994).

The technology that supports personalization of links has not yet propagated into mainstream web browsers. Yet the different solutions have been trialled in a web context; for example, the Distributed Link Service enabled individuals to make their own private link sets or to contribute to their group's collective link sets (Carr, De Roure, Hall, & Hill, 1995). Even the creation of one's own link-computation specifications was trialled in a web environment (Cawley, Ashman, Chase, Dalwood, Davis, & Verbyla, 1995). Technically, it should be a simple matter to incorporate this feature into all browsers. Whether this happens depends largely on whether users insist on it.

Surprisingly, even the growing awareness of the need for privacy in online actions has not yet motivated users to demand an easy-to-use personalized link facility. Even personalized annotations are a minority facility, only feasible for the reasonably technically literate user. Being able to record one's personalized links enables users to record their private associations and collections, often representing original intellectual ideas or commercially valuable information without advertising it to all. As more of each user's online interactions and transactions are exposed to outside scrutiny, authorized or otherwise, the provision of a feature with the fortunate side effect of offering private means for recording ideas ought to become a priority for future web browser development.

SEARCHING AND QUERYING

Browsing, enabled by hypermedia links, was the first technology for accessing information on the web. However, the enormous success of the web and its rapid uptake by millions of users rapidly rendered browsing alone a manifestly inadequate tool for information access. Searching has now become a major, perhaps *the* major, means of locating information on the web.

However, accessing information via search engines presents difficulties. First, users face the problem of creating an accurate description of their requirement. Second, users must make sense of the results that the search engine produces. Solutions to these difficulties are discussed in this section.

Query Formulation

Experts agree that the typical user will experience difficulties when attempting to formulate an effective query. Three significant factors stand out:

1. *Low user commitment.* Users are reluctant to provide information beyond the bare minimum. In fact, what emerges is that search-engine use is characterized by short queries and limited interaction coupled with unreasonably high expectations (Jansen, Spink, & Saracevic, 2000; Rolker & Kramer, 1999; Kobayashi & Takeda, 2000):

   Real people in the real world, doing real information seeking and in a hurry, use web search engines and give two-word queries to be run against billions of web pages. We expect, and get, sub-second response time and we complain when there are no relevant web pages in the top 10 presented to us (Browne & Smeaton, 2004).

Users, it would appear, are wary of commitment. They expect very high-quality results from every search instance rather than subscribing to the notion of search as an iterative process.

2. *Uncertain information needs.* Users often have an incomplete understanding of their information need, and their initial need will frequently mature during, and in direct response to, the process of searching (Lancaster, 1968):

   Searchers normally start out with an unrefined or vague information need which becomes more sharply focused as their search continues and exposure to information changes their information need. (Browne et al., 2004).

This initial lack of clarity with regard to the object of the search readily translates into imprecise query terms, which in turn begets irrelevant and disappointing results. Some users quit the searching process at a very early stage, confounded by this apparent "failure."

3. *Difficulties in expression.* Users may not know the correct syntax to frame their query or the commands to interact with the search engine. They *may* know in general terms what

web pages they wish to retrieve, but struggle to find the query terms most likely to identify them:

> Except in special circumstances, it is difficult for a user to ask an information retrieval system for what they want, because the user does not, in general, know what is available and does not know from what it has to be differentiated. (Card, Robertson & Mackinlay, 1991)

This is not surprising, as the formulation of a "successful" query requires some awareness of information-retrieval theory, relatively uncommon among typical users. Some of this theory can be guessed by observant users (who might notice, for instance, that query terms that frequently occur in web pages produce poor results), but the remaining users must persevere in an unfamiliar environment, never knowing why their search failed, or indeed why it succeeded.

Relevance Feedback

To counter some of the problems just mentioned, researchers working with search engines have developed strategies designed to improve the quality of a submitted query. One of these strategies is *relevance feedback* (RF):

1. The user submits an initial query and the search engine serves results.
2. The user then identifies *relevant* and *nonrelevant* web pages using associated checkboxes, clickable links, radio buttons etc. This action supplies the search engine with *feedback*.
3. The search engine then automatically *modifies* the original query in response to the feedback. This may involve adding search terms to the query, known as *query expansion*. It may also involve *reweighting* of the query, where information in the relevant and nonrelevant set of documents is used to modify the importance of various query terms.
4. The modified query is run by the search engine and a new set of search results is shown to the user.
5. This process continues until the user's information need is satisfied.

This technique is particularly successful, and it has been repeatedly demonstrated that relevance feedback can improve the performance of a search engine at comparatively little cost to the user or the system (Ruthven & Lalmas, 2003; Harman, 1992b; although see Spink, Jansen, & Ozmultu, 2000, for contrast). For example, Koenemann (1996) noted that the:

> . . . availability and use of relevance feedback increased retrieval effectiveness; and increased opportunity for user interaction with and control of relevance feedback made the interactions more efficient and usable while maintaining or increasing effectiveness.

However, it has also been found that in order for relevance feedback to be effective, the user *must* be offered a meaningful dialogue for participation. Ruthven recently conducted a series of RF experiments examining the performance of user-supplied feedback measured against an automated equivalent. Commenting on the relatively low performance of the users, he identified a failure in the infrastructure supporting the user's search:

> . . . simple term presentation interfaces are not sufficient in providing sufficient support and context to allow good query expansion decisions. Interfaces must support the identification of relationships between relevant material and suggested expansion terms and should support the development of good expansion strategies by the searcher. (Ruthven, 2003)

In summary, many search-engine users find the formulation of an effective query difficult. Relevance feedback offers an effective solution, providing that:

- the users are willing to commit to several search iterations
- a suitable and useable feedback interface is implemented

The Results List

It is well accepted that the results generated by a web search engine will frequently not satisfy the user. Sometimes this negative result will occur because the search engine cannot find *any* documents that match the user's information need. However, it is much more likely to happen because the user has submitted a very broad query, resulting in too many documents. This common difficulty has been identified by one author as the "abundance problem," occurring when "the number of pages that could reasonably be returned as relevant is far too large for a human user to digest" (Kleinberg, 1999).

Many solutions to this challenge have been suggested, but three interesting approaches are (a) community-based ranking algorithms, (b) improved visual interfaces, and (c) document clustering. We now examine each approach in turn.

Community-Based Ranking Algorithms

A search engine normally ranks a set of web pages in order of the likelihood that they will be relevant to the user's information need, the document most likely to be relevant appearing first. This likelihood of relevance is usually calculated using a statistical measure related to the occurrence of the query terms in the documents concerned. Known as "term frequency," this measurement is traditionally normalized with respect to document length and multiplied by a measure reflecting the specificity of each term within the document collection (Sparck Jones, 1972; Aizawa, 2003; Salton, Yang, & Yu, 1975).

Document rankings dependent upon term frequency represent a purely arithmetical evaluation of the web pages concerned. This evaluation can (and does) provide a useful approximation of the likelihood of relevance to the user, but is by no means an authoritative measure. The underlying assumption—that term frequency translates directly into relevance—is exactly that: an assumption. There is no guarantee that a web page containing many query terms will be any more relevant than a second page containing a lower number of the same terms. The latter may have fewer of the important keywords, but might be far superior in other, less-quantifiable respects (i.e., it may be more concise, better illustrated, superior in style, easier to read, more complete in its references, etc.).

One solution to this problem has been to supplement the rudimentary rankings that can be constructed through statistical observations with more sophisticated sources of information. This has led to the development of a general class of ranking algorithms that implement *citation based metrics* for relevance scoring (Garfield, 1972; Pinski & Narin, 1976). In these algorithms, the relative importance of each web page is a function of the number of other web pages that link to it (Brin & Page, 1998). As Kleinberg observed (1999), these "in-links . . . encode a considerable amount of latent human judgment." Accordingly, a web page that is referenced by a good proportion of its peers is a natural candidate for high ranking.

Improved Visual Interfaces

Hearst observed that "long lists littered with unwanted irrelevant material" represent an unwieldy and nonintuitive method for delivering search results to the user (Hearst, 1997). Rather than expending resources developing more sophisticated ranking algorithms, Hearst asserted that the answer to this problem lay in shifting:

. . . the user's mental load from slower, thought intensive processes such as reading to faster, perceptual processes such as pattern recognition. It is easier, for example, to compare bars in a graph than numbers in a list. Color is very useful for helping people quickly select one particular word or object from a sea of others.

The author went on to speculate that in the near future advanced search interfaces may abandon the two-dimensional (2D) page metaphor altogether, adopting "alternatives that allow users to see information on the web from several perspectives simultaneously." A software implementation based on this very idea, known as the Information Visualizer (Card, Robertson, & Mackinlay, 1991) has proved surprisingly powerful; in one experiment an organizational hierarchy requiring 80 printed pages was displayed on just one three-dimensional (3D) screen (Robertson, Mackinlay, & Card, 1991). It seems clear that a visual tool for searching the web that helps the user to "see" a set of search results rather than just "read" them would have considerable utility.

Document Clustering

Another way of simplifying a set of search results is to generalize. The process of generalizing a set of results begins with identifying particular commonalities shared by some members of that set (i.e., use or nonuse of certain terms, subject matter, file type, etc.). Web pages determined to be "similar" in some way are then grouped together under a single category or heading, and the user is subsequently presented with several coherent clusters of web pages rather than the traditional list. Provided the number of clusters is relatively low, this technique quickly reduces the cognitive load of studying the results, allowing a user to "skim" rather than read.

In a good example of this technique, Truran, Goulding, and Ashman (2005) described an attempt to solve the problem of ambiguity in a query-based search. Lexical ambiguity causes considerable problems, complicating the already difficult task of establishing what a user is actually looking for. An analysis of the query log of a very large online encyclopedia recently found that 1 in 20 queries was largely wasted, simply because a user chooses ambiguous query terms (Wen, Nie, & Zhang, 2001).

Discussing this problem, the authors developed a sense categorization service known as SENSAI. This service was remarkable in that it was capable of auto-categorizing a set of search results into believable sense categories without a semantic analysis of the web pages concerned. Neither was an external knowledge source, such as a dictionary or an expert system, consulted. Instead, sense discrimination was the quiet background product of user consensus; the system studied and learned from its clientele, utilizing what Fitzpatrick et al. termed the "information-consuming habits" of its users (Fitzpatrick & Dent, 1997) to create a human-intuitive overview of the distinguishable semantics of the query, like documents grouped with like (Michalski, Stepp, & Diday, 1983).

Summary

Users of a search engine are frequently overawed by the number of results returned when a query is submitted, and rarely proceed beyond the first page of suggestions (Harman, 1992). To ensure that the most relevant documents appear on the first page, more sophisticated ranking algorithms have been developed that exploit the inherently self-referential nature of the web.

However, even with these improvements, search engines are still serving large chunks of unrefined information to their users. This seems likely to change soon. Search engines will almost certainly begin to abstract categories and significant groupings from any given set of results automatically, and visual interfaces that simplify the user's task seem probable.

PERSONALIZATION, PORTALS AND COMMUNITIES

In most circumstances, one of the major impediments to effective use of the web is the disorientation that results from its sheer size. Statistics are commonly quoted as to just how vast the web is with one recent estimate being 11.5 billion pages (Gulli et al., 2005). This, no doubt, will become rapidly out of date, but the actual figures are not important from a user's perspective.

From the perspective of end users (i.e., content consumers), it is increasingly difficult to find what they are looking for amid the sea of inappropriate material. From the perspective of authors (i.e., content providers), it is becoming increasingly difficult to prevent their pages from being lost in the crowd. In this section we will examine approaches of personalization, portals, and community support that are becoming widely used to address these issues.

Although personalization, portals, and communities are quite distinct approaches to web design, they are all aimed at producing an end-user experience that is uniquely appropriate for each individual. This is what might be called "my web." *Personalization* is where the contents of websites are changed (either transparently or under user control), so that the material is appropriate for the needs and requirements of the individual. This involves a wide range of techniques of varying degrees of sophistication, from a user-selectable color scheme at its simplest through to an adaptive content-management system based upon a comprehensive user model. Such systems are commonly used for e-commerce websites, and have been used experimentally for e-learning. *Portals* are gateways to the web that aim to make its vastness more accessible by providing an appropriate set of links and default paths (usually they are continually or at least regularly updated). These may be personalized at the level of individuals, an organization, a special interest, or some combination of these. Where the personalization is operating at the level of organizations, special interests, or other groups, there is an opportunity for the personalized experience to be shared by a *community*. A major change in web design in recent years has been that there are now a variety of tools and techniques to both support community interaction on the web and leverage the power of collaborative effort.

Personalization

Until quite recently, the vast majority of web content has been designed for the needs of content providers rather than end users. In terms of a communication model, a traditional static HTML site has a lot in common with a poster on a billboard. It offers the same message to all passersby, regardless of whether or not it is appropriate to them. The only reason that this model is effective, on either a billboard or the web, is that there are (or at least there can be) many passersby. Modern websites frequently consist of a lot more than static HTML, they are often dynamic front-ends for a wide range of resources and services. However, their communication model is usually declarative—with a "one-size-fits-all" approach to content that treats the user as a hypothetical "average."

This state of affairs has recently begun to change, with the advent of websites that personalize content for each end user. One of the best known examples of this is Amazon.com, which stores information about the customers' interests—gleaned from various sources—to generate a personalized home page and suggest items that are likely to be of interest. Adaptive hypermedia (AH) (Brusilovsky, 2001) is an academic discipline that is dedicated to bringing personalization to the web. Most applications of AH techniques are either simple e-commerce applications, such as that just described, or are in e-learning. Educational adaptive systems (AEH) (Brusilovsky, 2001b) have arisen as a logical progression of using the web for teaching and learning. In traditional education, the experience is very adaptive. A student who doesn't understand a question will ask the teacher for an explanation. If the student doesn't understand the answer, then the teacher explains it in different terms—maybe by simplifying, using an example, or drawing a picture. Exactly how the teacher explains it depends upon the needs of the student, so that the con-

tent is adapted for the individual student. Static web pages generally give the same experience to all students regardless of how effective this may be, good human teachers provide a customised experience for individuals. The ultimate goal of AEH systems is to provide this capability to web-based learning. So far, most such systems are experimental, but they are important because the majority of AH research has been done in this area. See Jameson (in this book) for more detail on adaptive interfaces.

Adaptive and Adaptable Systems

Two quite distinct approaches to the personalization of hypermedia are (Patel & Kinshuk, 1997; Wadge & Schraefel, 2001) the adaptive and the adaptable. Adaptive systems dynamically organize their contents to meet the perceived needs of the user. This process is transparent, and does not require any direct intervention by the user. Adaptable systems make changes to the content (or more commonly the user interface) only as a result of the explicit intervention of the user.

Adaptable systems gather information explicitly from the user, such as iWeaver (Wolf, 2002). This is typically either in the form of some sort of "control-panel" setting, or a questionnaire (commonly completed as part of a user registration process) to elicit a set of preferences from the user. These preferences may be used to determine various aspects of the user interface (i.e., color schemes, screen layout etc), or they may be about the background or perceived needs of the user (i.e., knowledge or areas of interest).

Adaptive systems gather information implicitly by monitoring the behavior of the user (Moore, Stewart, Zakaria, & Brailsford, 2003). The navigation of the user through the system may be monitored (i.e., the links followed) or a quiz may be used to attempt to deduce information about users (i.e., their abilities). The distinction between this and an adaptable system is that the adaptable system is asking users what they want; whereas the adaptive system attempts to deduce what the user needs (not necessarily the same thing) and provide it transparently. Adaptive hypertext is really a type of "attentive system" (Maglio, Barrett, Campbell, & Selker, 2000), which is a system that pays attention to the user's behavior, so that the user can respond appropriately. Because there is always a risk of user behavior being misinterpreted by the system, attentive systems should not be used for critical information (Maglio et al., 2000). The adaptation may well improve the usability and appropriateness of a website, but it is important that any essential information is available independently of any adaptation.

Although these two approaches to personalization are quite distinct, it is possible for a system to be both adaptive and adaptable. Amazon.com is a good example of this because it acquires information from users both implicitly and explicitly. It uses information based upon purchases and follow-ups of search results or suggestions (i.e., click-through), all of which is implicitly derived from user behavior (i.e., from this point of view it is adaptive). However, it also allows users to explicitly rate items, and it uses these ratings as a basis for future suggestions (i.e., from this point of view it is adaptable). AEH systems also can use both methods, for example WHURLE (Brailsford, Ashman, Stewart, Zakaria, & Moore, 2002).

Need or Desire?

Adaptive hypertext systems base their content delivery upon what the system perceives to be a user's need. Adaptable systems, on the other hand, use information obtained directly from the user, which in most case will represent that user's desire. Since need and desire are not necessarily the same thing, there is potential for a conflict of interests. When considering adaptive systems, the underlying assumption is that the system knows better than the user what is most appropriate. Although this sounds disempowering (and if implemented badly it can be), there are many circumstances where this view is justified, particularly if the user doesn't understand the ramifications of a particular choice at a decision point. For example, in educational systems learners may not always have all of the facts or the pedagogic background necessary to take control of their own teaching regime. In "real-life" education, this is why we have teachers rather than just resources. Consider, for instance, a student who has problems with the math in a subject area that makes extensive use of math, although it is not the primary topic of study (engineering or computer science would be good examples here). Given the choice, many such students would prefer fewer math problems; however, what they probably need are *more* (and better explained) math problems. An adaptive system that "knows best" would thus be capable of transparently providing remedial exercises (as would a real-life teacher who spotted the problem). However, an adaptable system runs the risk that the student might choose to avoid the parts that they find the most onerous.

Implementing this "system-knows-best" paradigm without due care creates the potential for serious problems in an adaptive system. It is quite likely to result in frustration or anger if users realize that the system is going against their desires or if the system's assumption of need is incorrect—particularly a problem with a poorly thought-out or incomplete user model (see "User Modeling"). Well-designed adaptive systems do not take control away from the user; rather, they provide suggestions (albeit sometimes strong ones) as to the most appropriate path. This is then a user interface issue, and it may take the form of reordering content or suggesting what content is considered most appropriate (i.e., a common mechanism is to change link color).

Methods and Models of Adaptation

There are two main approaches to implementing adaptation on the web, at the level of either content (i.e., adaptive presentation) or linking (i.e., adaptive navigation support). Within each of these, there are a number of techniques used that are described briefly here but are reviewed in detail elsewhere (Brusilovsky, 1996).

The majority of adaptive presentation operates on text, although in principle any media type may be adapted. Adaptation of modality (Brusilovsky, 1996) is where the adaptation operates at the level of choices between different media types (i.e., text, speech, animation etc.). Although this will likely become important in the future, multimedia adaptation is currently very rarely implemented (one of the few examples is

Fagerjord, 2005). There are a number of ways in which the adaptive presentation of text has been implemented, usually by manipulating fragments of predefined text. Some systems add fragments to a standard minimum body of text; others remove fragments that the system deems inappropriate. When fragments are removed they may be completely hidden, or they may be merely dimmed. Another variant of fragment removal is to use "stretchtext" (a body of text is hidden behind a single word or phrase and expanded upon request). Instead of adding or removing fragments of text, some systems simply reorder the fragments. A major problem with all forms of adaptive-text presentation is the impact that it has upon narrative flow. Different users will necessarily have different narratives, which isn't a problem *per se* other than the fact that it is possible (or even likely) that a web page will no longer form a coherent whole if its text is manipulated. This problem may be ameliorated by writing text in the form of conceptually discrete "atoms" (Brailsford, Stewart, Zakaria, & Moore, 2002), but that causes quite serious problems with authoring (most people find it difficult to write following this sort of constraint) and makes legacy content difficult to adapt.

Adaptive navigation support has fewer problems with narrative flow, because links are (usually) less fundamental to the sense of a page than is text. Adaptive linking has been divided into five main categories (Brusilovsky, 1996): direct guidance, link sorting, link hiding, link annotation, and map adaptation. With *direct guidance*, the system makes a recommendation as to the next node to visit, based upon the adaptation criteria. This is often used for a "guided-tour" type of direction for novice users (i.e., it is tantamount to a linear sequence of "Next" buttons). With *link sorting*, the ordering of links is changed so that the most appropriate one is at the head of the list or in a particularly prominent place. Inappropriate links are not displayed in link hiding. This, of course, can make parts of the system inaccessible to some users—which may or may not be acceptable. A variant of link hiding is *link dimming*, where it is still possible to follow the link, but it is less prominent in the user interface than more relevant links. In a link-annotation system, visual clues are attached to links to give users some idea of their status with respect to the system; for example, font sizes or colors can be used to provide a gradation of relevance. Lastly, in map-adaptation systems a navigational map is generated according to the adaptation rules.

User Modeling

In order to be able to do any adaptation, adaptive systems need to know who is using them and store information about that person. This information is stored in a "user profile" either on the local machine or, far more commonly, in a centralized database. In the user profile, as well as personal information (name, ID, etc.), the criteria that the system will use for adaptation are stored and constitute the user model. Although a detailed discussion of user modelling is beyond the scope of this chapter, the basic principles are very simple. A user model is simply a set of rules that adaptive systems use to make decisions about relevance. One approach to user modelling is that of *stereotype* models (Kaplan, Fenwick, & Chen, 1993). With these, users are grouped together with other like-minded individuals. For example, "second-year chemistry undergraduates" or "people who like cats" are both stereotypes and, as such, it is possible to make reasonable assumptions about members of these groups (i.e., "will know the structure of butane" or "will appreciate photos of cute kittens" respectively). The other main user-modeling methodology is the *overlay* model (Fischer, Mastaglio, Reeves, & Rieman, 1990), where the individual's knowledge, background, and any other relevant data are overlaid onto the sum of all knowledge in the system. Usually, adaptive-hypermedia systems adopt one of these two models. However, they do not have to be mutually exclusive, as it is possible to combine the two approaches so that any given user is initially stereotyped. Then, as the system learns more about them, it uses an overlay model to record their progress (Zakaria & Brailsford, 2002).

User modeling is critical for a usable adaptive system. If the user model is in some way flawed (i.e., if the information stored does not match the reality of users' background and behavior), then content is likely to be maladapted for a user and the system is often extremely difficult to comprehend. This is compounded by the fact that there are no standards for user modelling, and most adaptive systems use their own proprietary user model.

Authoring

To date there are few, if any, mature authoring systems for personalized web systems (AHA! (De Bra, Stash, & Smits, 2004) is one of the more mature authoring systems). This is, no doubt, because most adaptive systems are either experimental or proprietary, and in both cases the authors are either members of, or have direct access to, the development team. However, there are a number of potentially difficult issues that need to be considered when attempting to design authoring interfaces for adaptive web-based systems (Cristea, Carro, & Garzotto, 2005). First, there is a need for a new design paradigm, together with associated metaphors, because personalization breaks the fundamental design paradigm of the web, that of the "page." When authoring a conventional web page, the author concentrates on content, and can reasonably assume that users will see the page as it is created. Hence, most authoring systems for static web pages use a conventional WYSIWYG paradigm, which is quite similar to that of a word processor. In the case of adaptive systems, however, the authoring process requires a separation of content authoring from the authoring of the parameters for adaptation. Depending upon the adaptation model that is used, what the author sees might be somewhat or completely different from what any given end user sees.

Recently, research has started to address these issues, and frameworks (such as LAOS, Cristea, 2005) have been developed for adaptive educational systems that allow the separation of different aspects of the authoring process. The LAOS framework has five layers, which result in different authors being able to create different sections or aspects of adaptive content. For example, there is a "goals & constraints" model where a pedagogic expert can use the content from the "domain model" to structure a lesson that makes use of a "user model," "adaptation model," and a "presentation model." This layered structure not only decreases individual author load, but also encourages the reuse of existing materials (whether they are domain content, adaptation rules, or presentation settings). The great benefit of this approach is that it can combine content created by different authors working at different levels. However, that is also its weakness in that care must be taken to avoid inappropriate use of content originally intended for a different context. Adaptive authoring systems are very much in their infancy, but although there are no mature systems, the current research is at least starting to discover the issues and problems involved with their design.

Portals

A common way to gain a personalized experience of the web is by the use of portals. These are gateways to online resources that are, at their most basic, simply a set of links that provide a starting point for web use. Usually, though, they are highly sophisticated systems that are continually updated and highly personalizable. Portals are hugely popular. At the time of writing, of the top 10 most browsed English language websites, four are portals,[12] and they are often presented as a way for users to create "their space" on the web. They are usually adaptable, allowing for explicit selections of user interface options or content appropriate for specific hardware platforms (i.e., low graphics content for PDAs), although some portals also use techniques of adaptivity (*q.v.*).

Many portals have their roots in either Internet directories or search engines, and therefore a common model of a portal user interface is that of a categorized front page (i.e., a "list of lists"). Some portals are essentially starting points for a collection of resources—typically these are intranet systems where the facilities determined for a particular user are presented on login (i.e., different staff or student resources in a university system). Some portals use semantic web technology to categorize their contents (i.e., Ontoportal[13]).

Communities

The Internet has its roots in communities, in creating links and bridging geographic, cultural, or temporal barriers. Communities on the web are often supported by, or generated through, portals, which provide starting points for people of similar interests. Not only do these provide channels through which users can navigate to find resources, but in doing so they can bring about a shared cultural space. Portals, especially those that are intranet based, are often designed to support communities.

Part of the original vision of computer pioneers, such as Nelson (1982), was that people could easily publish, edit, and comment on as well as read publicly available materials, and this is

---

[12]Alexa Web Search: top sites. http://www.alexa.com/site/ds/top_sites.
[13]Ontoportal—http://www.ontoportal.org.uk/.

now starting to become a reality on the modern web. Weblogs (more commonly referred to as blogs) allow a user to post material, which can then be cited, commented upon, and redistributed throughout the internet. These blogs require no special knowledge of authoring and are usually accessed with a standard web browser.

Another technology that is important in the development of communities on the web are Wikis,[14] which are community websites designed to facilitate collaborative authoring. The most spectacular example of this is Wikipedia, a vast encyclopaedia written collaboratively by volunteers using a Wiki.[15] This project was founded in 2001, and in 2004 it contained over a million articles in more than 100 languages. The system provides facilities to discuss changes made to articles, and authors may vote on controversial issues, the editorial assumption being that the exposure of material to many users will result in the addition of depth and elimination of errors.

CONCLUSIONS

In this chapter we have considered HCI in a specific web context, looking at the issues impeding smooth user interaction with web tools and documents. While some problems remain unsolved, generally the usability of the web has improved greatly since its early days. While many challenges remain, there are at least clear research directions indicating potential solutions.

References

Abascal, J., & Nicolle, C. A. (Eds.). (2001). *Inclusive design guidelines for HCI.* Taylor & Francis.

Aizawa, A. (2003). An information-theoretic perspective of tf-idf measures. *Information Processing and Management, 39*(1), 45–65.

Ashman, H. (2000). Electronic document addressing—dealing with change. *Computing Surveys, 32*(3), 201–212.

Bar-Yossef, Z., Broder, A. Z., Kumar, R., & Tomkins, A. (2004). Sic transit gloria telae: Towards an understanding of the web's decay. In *Proceedings of the 13th International Conference on World Wide Web* (New York, NY, USA, May 17–20, 2004) WWW '04 (pp. 328–337). New York, NY: ACM Press.

Bergman, M. K. (2001). The deep web: Surfacing hidden value. *The Journal of Electronic Publishing.* Retrieved April 23, 2007, from http://www.press.umich.edu/jep/07-01/bergman.html.

Berners-Lee, T. (1998). Cool URIs Don't Change, http://www.w3.org/Provider/Style/URI.

Bieber, M., Vitali, F., Ashman, H., Balasubramanian, V., & Oinas-Kukkonen, H. (1997). Fourth generation hypermedia: Some missing links for the World Wide Web [Special issue on HCI and the web]. *International Journal of Human-Computer Studies, 47*(1), 31–65.

Brailsford. D. (1999). Separable hyperstructure and delayed link binding. *ACM Computing Surveys, 31.*

Brailsford, T. J., Ashman, H. L., Stewart, C. D., Zakaria, M. R., & Moore, A. (2002). *User control of adaptation in an automated web-based learning environment.* First International Conference on Information Technology & Applications (ICITA 2002), Australia.

Brailsford, T., Stewart, C., Zakaria, M., & Moore, A. (2002). Autonavigation, links and narrative in an adaptive web-based integrated learning environment. *Eleventh International World Wide Web Conference.*

Brin, S., & Page, L. (1998). The anatomy of a large-scale hypertextual web search engine. *Proceedings of the Seventh International Conference on World Wide Web (WWW7)*, 107–117.

Browne, P., & Smeaton, A. (2004). Video information retrieval using objects and ostensive relevance feedback. In *Proceedings of the 2004 ACM Symposium on Applied Computing* (pp. 1084–1090). ACM.

Brusilovsky, P. (1996). Methods and techniques of adaptive hypermedia. *User Modelling and User-Adapted Interaction, 6*, 87–129.

Brusilovsky, P. (2001). Adaptive hypermedia. *User Modelling and User Adapted Interaction, 11*(1/2), 87–110.

Brusilovsky, P. (2001). Adaptive educational hypermedia (invited talk). The Tenth International PEG *Conference*, 8–12.

Burnett, G. E., Summerskill, S. J., & Porter, J. M. (2004). "On-the-move" destination entry for vehicle navigation systems—unsafe by any means? *Behaviour and Information Technology, 23*(4), 265–272.

Cailliau, R., & Ashman, H., (1999). A history of hypertext in the web. *Computing Surveys Special Issue on Hypertext and Hypermedia, 31*, ACM.

Card, S. K., Robertson, G. G., & Mackinlay, J. D. (1991). The information visualizer, an information workspace. In *Proceedings of the SIGCHI Conference on Human Factors in Computing Systems (CHI '91)* (pp. 181–186). ACM.

Carr, L., De Roure, D., Hall, W., & Hill, G. (1995). The distributed link service, a tool for publishers, authors and readers. *Paper presented at the 4th International WWW Conference.* Retrieved April 23, 2007 from http://www.w3.org/Conferences/WWW4/Papers/178/.

Cawley, T., Ashman, H., Chase, G., Dalwood, M., Davis, S., & Verbyla, J. (1995). A link server for integrating the web with third-party applications. *Proceedings of the First Australasian World Wide Web Conference*, Australia. Retrieved, from http://ausweb.scu.edu.au/aw95/integrating/cawley/index.html.

Cristea, A. (2005). Authoring of adaptive hypermedia: Adaptive hypermedia and learning environments. In S. Chen & G. Magoulas (Eds.), *Advances in web-based education: Personalized learning environments.* IDEA Publishing Group.

Cristea, A., Carro, R., & Garzotto, F. (2005). A3EH: The 3rd workshop of Authoring of Adaptive and Adaptable Educational Hypermedia, (http://wwwis.win.tue.nl/~acristea/AAAEH05/papers/editorial.pdf).

Davis, H.C. (1995). To embed or not to embed. *Comms of the ACM, 38*(8), 108–109.

Davis, H. C. (1998). Referential integrity of links in open hypermedia systems. In *Proceedings of Hypertext '98* (pp. 207–217). ACM.

Davis, H., Hall, W., Heath, I., Hill, G., & Wilkins, R. (1992). Towards an integrated information environment with open hypermedia

---

[14]Wiki Wiki Web—http://c2.com/cgi/wiki?WikiWikiWeb

[15]Wikipedia—http://www.wikipedia.org/

systems. In *Proceedings of the Second European Conference on Hypertext* (pp. 181–190). ACM.

Davis, H., Knight, S., & Hall, W. (1994). Light hypermedia link services: A study of third party application integration. In *Proceedings of ECHT '94* (pp. 41–50). ACM.

De Bra, P., Stash, N., & Smits, D. (2004, August 23). Creating adaptive applications with AHA!, Tutorial for AHA! version 3.0. Tutorial at the AH2004 Conference, Eindhoven. Computing Science Report 04-20, 2004.

Eason, K. D. (1984). Towards the experimental study of usability: Ergonomics of the user interface. *Behaviour and Information. Technology, 3*(2), 133–143.

Fagerjord, A. (2005). Editing stretchfilm. *Proceedings of the Sixteenth ACM Conference on Hypertext and Hypermedia* (p. 301). ACM.

Fischer, G., Mastaglio, T., Reeves, B., & Rieman, J. (1990). Minimalist explanations in knowledge-based systems. *Proceedings of the 23rd Annual Hawaii International Conference on System Sciences*, 309–317.

Fitzpatrick, L., & Dent, M. (1997) Automatic feedback using past queries: Social searching? In *Proceedngs of the 20th Annual International Conference on Research and Development in Information Retrieval* (pp. 306–313). ACM.

Garfield, E. (1972). Citation analysis as a tool in journal evaluation. *Science, 178*, 471–479.

Gulli, A., & Signorini, A. (2005). The indexable web is more than 11.5 billion pages. Poster at 14th International WWW Conference (WWW2005).

Harman, D. (1992). Relevance feedback revisited. In *Proceedings of the 15th Annual International ACM SIGIR Conference on Research and Development in Information Retrieval* (pp. 1–10). ACM.

Harman, D. (1992). Relevance feedback and other query modification techniques. In W. B. Frakes & R. Baeza-Yates (Eds.), *Information Retrieval. Data Structures & Algorithms* (pp. 241–263). Prentice-Hall.

Hearst, M. (1997). Interfaces for searching the web. *Scientific American, 276*(3), 68–72 (5).

International Organisation for Standardisation (1998). ISO 9241 Ergonomics requirements for work with visual display terminals (VDTs)—Part 11 Guidance on Usability.

Jansen, B. J., Spink, A., & Saracevic, T. (2000). Real life, real users, and real needs: A study and analysis of user queries on the web. *Information Processing and Management, 36*(2), 207–227.

Jul, S., & Furnas, G. W. (1997). Navigation in electronic worlds: A CHI 97 workshop. *SIGCHI Bulletins, 29*(4).

Kaplan, C., Fenwick, J., & Chen, J. (1993). *Adaptive hypertext navigation based on user goals and context.* User Modeling and User-Adapted Interaction. Kluwer Academic.

Kleinberg, J. M. (1999). Authoritative sources in a hyperlinked environment. *Journal of the ACM, 46*(5), 604–632.

Kobayashi, M., & Takeda, K. (2000). Information retrieval on the web. *Computing Surveys, 32*(2), 144–173.

Koenemann, J. (1996). Supporting interactive information retrieval through relevance feedback. In *Conference companion on human factors in computing systems* (pp. 49–50). ACM.

Lai, J., Cheng, K., Green, P., & Tsimhoni, O. (2002). On the road and on the web? Comprehension of synthetic and human speech while driving. In *Proceedings of SIGCHI 2002.* ACM.

Lawrence, S., Pennock, D. M., Flake, G. W., Krovetz, R., Coetzee, F. M., Glover, E., et al. (2001). Persistence of web references in scientific research. *Computer, IEEE, 34*(2), 26–31.

Lazar, J., Bessiere, K., Ceaparu, I., Robinson, J., & Shneiderman, B. (2003). Help! I'm lost: User frustration in web navigation. *IT & Society, 1*(3), 18–26.

Maglio, P., Barrett, R., Campbell, C., & Selker, T. (2000). Suitor: An attentive information system. *Proceedings of the 5th International Conference on Intelligent User Interfaces*, 169–176.

McCracken, D. D., & Wolfe, R. J. (2004). *User-centred website development: A human-computer interaction approach.* NJ: Pearson Prentice Hall.

Michalski, R., Stepp, R., & Diday, E. (1983). Automatic construction of classifications: Conceptual clustering versus numerical taxonomy. *Transactions on Pattern Analysis and Machine Intelligence, IEEE, 5*, 528–552.

Moore, A., Stewart, C. D., Zakaria, M. R., & Brailsford, T. J. (2003). WHURLE—an adaptive remote learning framework. *International Conference on Engineering Education (ICEE-2003)*, Valencia, Spain.

Nelson, T. H. (1982). *Literary machines.* Eastgate Systems Inc.

Nielson, J. (2000). *Designing web usability.* New Riders Publishing.

Otter, M., & Johnson, H. (2000). Lost in hyperspace: metrics and mental models. *Interacting with Computers, 13*, 1–40.

Patel, A., & Kinshuk, (1997). Intelligent tutoring tools in a computer integrated learning environment for introductory numeric disciplines. *Innovations in Education and Training International Journal, 34*(3), 200–207.

Pinski, G., & Narin, F. (1976). Citation influence for journal aggregates of scientific publications: Theory, with application to the literature of physics. *Information Processing and Management, 12*, 297–312.

Robertson, G. G., Mackinlay, J. D., & Card, S. K. (1991). Cone trees: Animated 3D visualizations of hierarchical information. In *Proceedings of the SIGCHI Conference on Human Factors in Computing Systems (CHI '91)* (pp. 189–194). ACM.

Ruthven, I. (2003). Re-examining the potential effectiveness of interactive query expansion. In *SIGIR '03: Proceedings of the 26th Annual International ACM SIGIR Conference on Research and Development in Information Retrieval* (pp. 213–220). ACM.

Ruthven, I., & Lalmas, M. (2003). A survey on the use of relevance feedback for information access systems. *Knowledge Engineering Review, 18*(2), 95–145.

Salton, G., Yang, C. S., & Yu, C. T.(1975). A theory of term importance in automatic text analysis. *Journal of the American Society for Information Science, 26*(1), 33–44.

Sparck Jones, K. (1972). A statistical interpretation of term specificity and its application in retrieval. *Journal of Documentation, 28*(1), 11–20.

Spink, A., Jansen, B., & Ozmultu, H. (2000). Use of query reformulation and relevance feedback by Excite users. *Internet Research: Electronic Networking Applications and Policy, 10*(4), 317–328.

Truran, M., Goulding, J., & Ashman, H. (2005). Co-active intelligence for information retrieval. In *Proceedings of ACM Multimedia '05* (pp. 547–550). ACM.

Verbyla, J. L. M., & Ashman, H. (1994). A user-configurable hypermedia-based interface via the functional model of the link. *Hypermedia, 6*(3), 193–208.

Wadge, W. W., & schraefel, m.c. (2001). A complementary approach for adaptive and adaptable hypermedia: Intensional hypertext. In *Hypermedia: Openness, structural awareness, and adaptivity.* International Workshop OHS-7, SC-3, and AH-3, Aarhus, Denmark.

Wen, J., Nie, J., & Zhang, H. (2001). Clustering user queries of a search engine. In *Proceedings of the 10th International Conference on World Wide Web* (pp. 162–168). ACM.

Wolf, C. (2002). iWeaver: Towards an interactive web-based adaptive learning environment to address individual learning styles. *European Journal of Distance Learning (EuroDL).* Retrieved April 23, 2007, from, http://www.eurodl.org/materials/contrib/2002/2HTML/iWeaver.htm

Zakaria, M. R., & Brailsford, T. J. (2002). User modelling and adaptive educational hypermedia frameworks for education. *New Review of Hypermedia and Multimedia, 8*, 83–97.

HUMAN-CENTERED DESIGN
OF DECISION-SUPPORT SYSTEMS

Philip J. Smith
The Ohio State University

Norman D. Geddes
Applied Systems Intelligence, Incorporated

Roger Beatty
American Airlines

Introduction 246
Approaches to Design 247
  Cognitive Task Analyses and Cognitive
  Walkthroughs 247
  Work-Domain Analyses 249
  Approaches to Design—Summary 249
Human Performance on
Decision-Making Tasks 249
  Errors and Cognitive Biases 250
    Slips 250
    Mistakes 250
    Cognitive Biases 250
    Designer Error 251
    Systems Approaches to Error 251
    Errors and Cognitive Biases—
    Implications for Design 251
  Human Expertise 252
    Descriptive Models 252
    Normative Optimal Models 253
    Human Expertise—Implications for Design 253
  Additional Cognitive Engineering Considerations 254
    The Human Operator as Monitor 254

Complacency and Overreliance 254
Excessive Mental Workload 254
Lack of Awareness or Understanding 254
Lack of Trust and User Acceptance 255
Active Biasing of the User's Cognitive
Processes 255
Distributed Work and Alternative Roles 255
Organizational Failures 255
Case Studies 255
Case Study A—Distributed Work in the National
Airspace System: Ground Delay Programs 256
  The Application Area 256
  The Original Approach to Distributing
  Work in a GDP 256
  An Alternative Approach to the
  Design of GDPs 256
    Strategy for distributing work 256
    Ration-by-schedule (RBS) 257
    Slot swapping 258
    Adapting to deal with uncertainty 259
    Administrative controls 259
  Evaluation of the Implemented Solution 260

Case Study A—Conclusions 260
Case Study B—Human Interactions with
Autonomous Mobile Robotic Systems 261
   The Application Area 261
   The Need for Decision Aiding 262
   The Design Solution 262
   Results 264
      Performance 264
      Acceptance 264
      Communication of beliefs 265
      Communications of intentions in the
      context of a bid 265
      Communication of direct commands 265

Case Study C—Interactive Critiquing
as a Form of Decision Support 265
   The Application Area 265
      Diagnosis as a generic task 265
      Sample problem 266
      The need for decision aiding 267
   The Design Solution 267
      Critiquing—additional design considerations 267
      Complementary strategies to reduce
      susceptibility to brittleness 268
   Evaluation of the Implemented Solution 268
Conclusion **268**
References **269**

INTRODUCTION

Computers can assist decision makers in a variety of different ways. They can, for instance, provide improved access to information or more informative displays of this information, or support more effective forms of communication. They can also use algorithms to actively monitor situations and to generate inferences in order to assist with tasks such as planning, diagnosis, and process control. This chapter focuses on interaction design issues (Preece, Rogers, & Sharp, 2002; Scott, Roth, & Deutsch, 2005) associated with this latter role, in which the software uses numerical computations and/or symbolic reasoning to serve as an active decision-support system or DSS (Turban, Aronson, & Liang, 2004).

The chapter begins with a focus on design methods, emphasizing the value of concrete scenarios and use cases (Bittner, 2002; Cockburn, 2000; Kulak & Guiney, 2003) to define the design problem, and the use of cognitive task analyses and work-domain analyses to guide both the development of the underlying functions in the DSS and the design of the interface that mediates the interactions of the user with these functions.

The chapter then proceeds to discuss aspects of human performance relevant to the design of DSSs. This includes a discussion of factors that limit human performance, as these factors suggest opportunities for improving performance through the introduction of a DSS. This discussion also includes consideration of the influence of a DSS on the user's cognitive processes, emphasizing the need to think in terms of the resultant joint cognitive system that is not a simple combination of the capabilities of the two agents (user and DSS) alone (Hoc, 2000; Hollnagel & Woods, 2005; Jones & Jacobs, 2000).

The discussion further emphasizes the need to consider more than just the surface interactions of the user with the interface to the DSS, and to consider the impact of the broader task context on performance (Amalberti, 1999; C. Miller, Pelican, & Goldman, 2000; Parasuraman, 2000; Rasmussen, Pejtersen, & Goldstein, 1994). This latter emphasis suggests that the design of a DSS needs to be viewed from the perspective of cooperative problem solving, where the computer and the person interact with and influence each other (Beach & Connolly, 2005; Hoc, 2000; Jones & Mitchell, 1995; Larson & C. Hayes, 2005; Mital & Pennathur, 2004; Parasuraman, Sheridan, & Wickens, 2000; P. J. Smith, McCoy, & Layton, 1997). From this viewpoint, the underlying functionality is just as important as the surface-level representation or interface, as the overall interaction determines ultimate performance.

The chapter also frames the design and use of a DSS as a form of cooperative work between several individuals (Bowers, Salas, & Jentsch, 2006; Grabowski & Sanborn, 2003; Hutchins, 1990, 1995; Orasanu & Salas, 1991; G. M. Olson & J. S. Olson, 1997; Rasmussen, Brehner, & Leplat, 1991; P. J. Smith, Billings, et al., 2000), with the computer as the medium through which they cooperate. This teamwork may involve system users who are physically or temporally distributed. In addition, the discussion further suggests that it is useful to think of the design team as working cooperatively with the users, trying to communicate with them and extend or augment their capabilities, and doing so through the software and other artifacts (such as paper documents) that they have developed. This perspective is useful as a reminder that we need to consider the psychology of the designers as well as the psychology of the users of systems, applying that understanding to help us take advantage of the strengths of both groups as they work cooperatively.

Thus, this chapter approaches the design of DSSs from four perspectives. The first emphasizes the need to identify effective approaches to support the design process, including the application of use cases, cognitive task analyses, and work-domain analyses. The second is to provide a discussion of the human factors considerations relevant to the design of a DSS (Helander, Landauer, & Prabhu, 1997; Meister & Enderwick, 2002; Wickens, J. Lee, Liu, Becker, & Gordon, 2004). The third is to provide very practical lists of questions that need to be asked as part of the design process. The fourth is to use case studies to provide concrete, detailed illustrations of how these considerations can be addressed in specific application contexts.

APPROACHES TO DESIGN

A number of complementary and overlapping approaches have been articulated to help ensure adequate consideration of user needs, preferences, and capabilities during the design process, and to gain insights into possible functionality to incorporate in a DSS. These include the use of participatory design (Bodker, Kensing, & Simonsen, 2004; Schuler & Namioka, 1993), scenario-based design (Carroll, 1995), and the completion of various forms of needs assessments (Tobey, 2005; Witkin & Altshuld, 1995), cognitive task analyses (CTAs) and cognitive walkthroughs (Crandall, Klein, & Hoffman, 2006; Jonassen, Tessmer, & Hannum, 1999; Kirwan & Ainsworth, 1992; Rubin, Jones, & Mitchell, 1988; Schragen, Chipman, & Shalin, 2000). It also includes various forms of domain analyses aimed at identifying the critical features and constraints that characterize the domain of interest.

Previous chapters have discussed the use of participatory and scenario-based design as approaches to the design process. The focus of this section will therefore be on various forms of cognitive task and domain analyses. However, it should be noted that the development and results of a cognitive task analysis can in general be very useful in helping to support participatory design, as can the development and use of scenarios. As Mack (1995) noted: "Developing and sharing a set of touchstone stories is an important cohesive element in any social system. . . . Gathering, discussing, and sharing these stories as a group can be an effective means to team building" (p. 368).

Cognitive Task Analyses and Cognitive Walkthroughs

Scenarios and use cases are also important in providing the framing for specific cognitive task analyses, as descriptions or predictions of performance need to be considered in the context of specific tasks and task settings. The starting point for a specific cognitive task analysis is the high-level context provided by a scenario or use case, including the social setting, resources, and goals of users. Thus, to begin a cognitive task analysis, the starting point is specification of the person for whom the product is to be designed (a persona), what goal that person is trying to achieve, and the immediate and broader contexts in which that goal will be pursued.

A number of variations for conducting a cognitive task analysis (Gordon & Gill, 1997; Hollnagel, 2003) have their roots in hierarchical task analysis techniques, decomposing some high-level goal or task into a hierarchy of subgoals (Annett, Duncan, Stammers, & Gray, 1971; Annett & Stanton, 2000; Diaper & Stanton, 2004; G. A. Miller, Galanter, & Pribram, 1960; Preece, Rogers, Sharp, & Benyon, 1994). This approach, originally applied to pure task analyses that were not concerned with representing the underlying cognitive processes, has been extended in a number of ways that maintain a central role for describing goal-subgoal relationships, but that also introduce constructs or methods for describing associated cognitive processes.

One influential example of this evolution is the GOMS (Goals, Operators, Methods and Selection rules) model (Card, Moran, & Newell, 1983; Kieras, 1991, 2004; St. Amant, Freed, Ritter, & Schunn, 2005; Williams, 2005), which makes use of a hierarchical

decomposition, but provides a control structure that introduces operators and rules for selecting which operator to apply in a given goal-driven context. As Preece et al. (1994) noted:

A GOMS analysis of human-system interaction can be applied at various levels of abstraction in much the same way that the hierarchical task analysis splits tasks into subtasks. . . . Three broad levels of granularity determine the GOMS family of models:

- The GOMS model, which describes the general methods for accomplishing a set of tasks.
- The unit task level, which breaks users' tasks into unit tasks, and then estimates the time that it takes for the user to perform these.
- The keystroke level, which describes and predicts the time it takes to perform a task by specifying the keystrokes needed (pp. 419–420).

This move toward explicit computational representation of internal processes has been extended to the point where broad, task-independent cognitive architectures have been developed that can be used to model performance in a given task/system context and generate performance predictions (Byrne & Kirlik, 2005). Three such systems are ACT-R (Anderson, 1993; Anderson et al., 2004), SOAR (Laird, Newell, & Rosenbloom, 1987; Nason, Laird, & Schunn, 2005), and EPIC (Kieras & Meyer, 1997; St. Amant et al., 2005). All three incorporate procedural knowledge that must operate within the constraints of certain constructs, such as working memory (Baddeley, 1998, 1999; Cary & Carlson, 2001; Radvansky, 2005; Seamon, 2005), that are used to model human capabilities and limitations (Meyers, 2004; Wickens et al., 2004).

This work on the development of computational models to describe or predict task performance has been complemented by more qualitative, judgment-based approaches using expert reviews that are similarly rooted in the hierarchical decomposition of a task based on goal-subgoal relationships. These latter techniques are motivated by the assumption that expert judgments about the adequacy of a design will be more accurate and complete if they are made while explicitly considering the context that the user will be dealing with (the user's current goal and subgoal and the display the user is viewing while trying to achieve that goal).

There are a number of variations on how to conduct such qualitative cognitive walkthroughs or predictive cognitive task analyses (Annett & Stanton, 2000; Bisantz, Roth, & Brickman, 2003; Diaper & Stanton, 2004; Gordon & Gill, 1997; Hollnagel, 2003; Jonassen et al., 1999; Kirwan & Ainsworth, 1992; Klein, 2000; Schraagen et al., 2000; Shepherd, 2000; Vicente, 1999). As an example, one such method was outlined by Lewis and Wharton (1997) as follows:

Step 1: Select the context for a use case to be used for an evaluation. Each such use case specifies three primary considerations:

- What are the characteristics of the user of the product that might affect its use? (the persona in the use case)
- What is the broader physical, organizational, social and legal context in which the product will be used?
- What is the high-level task or goal for which the product is being used?

Note that this specification is solution-independent. It essentially defines the design problem for this use case.

Step 2: Specify the normative (correct) paths for completing the goal of this use case (represented as a goal hierarchy) using the current product design, indicating the alternative sequences of steps that the user could take to *successfully* achieve the specified goal. Note that there could be more than one correct path for completing a given task.

Step 3: Identify the state of the product and the associated "world" at each node in the goal hierarchy. In the case of a software product, the state of the product would be the current appearance of the interface and any associated internal states (such as the queue of recently completed actions that would be used should an undo function be applied). The state of the "world" applies if the product or the user actually changes something in the world as part of an action, such as changing the temperature of a glass-manufacturing system through a process-control interface.

Step 4: Generate predictions. For each correct action (node in hierarchy), attempt to:

- Predict all the relevant success stories;
- Predict all the relevant failure stories;
- Record the reasons and assumptions made in generating these stories;
- Identify potential fixes to avoid or assist in recovery from failure stories.

This final step involves walking through the hierarchy along each path that leads to success, and, playing psychologist, generate predicted behaviors at each node (for each associated screen display or product state). This walkthrough could be done by an expert in human-computer interaction or by a domain expert. Because of the differences in their backgrounds relevant to the design and the domain of interest, individuals from each group are likely to generate somewhat different predictions.

Lewis and Wharton (1997) suggested asking four questions to help guide this prediction process:

- "Will the user be trying to achieve the right effect?"
- "Will the user notice that the correct action is available?"
- "Will the user associate the correct action with the desired effect?"
- "If the correct action is performed, will the user see that progress is being made?"

P. J. Smith, Stone, and Spencer (2006) suggested some additional, more detailed questions that can be asked to help the analyst consider how different cognitive processes influence performance (Card et al., 1983; Wickens et al., 2004):

- Selective Attention: What are the determinants of attention? What is most salient in the display? Where will the user's focus of attention be drawn? (See Bennett & Flach, 1992; Eckstein, 2004; Johnston & Dark, 1986; Lowe, 2003; Pashler, Johnston, & Ruthruff, 2001; Yantis, 1998.)

- Automaticity and Controlled Attentional Processes: Does the design support skill-based performance for routine scenarios and controlled, knowledge-based processes for novel situations? (See Logan, 2005; Rasmussen, 1983; Rasmussen et al., 1994.)

- Perception: How will perceptual processes influence the user's interpretation? How, for instance, will the proximity of various items on the screen influence judgments of "relatedness" as predicted by the Gestalt Law of Proximity? (See Bennett & Flach, 1992; Kohler, 1992; Tufte, 1983, 1990, 1997; Vicente, 2002; Watzman, 2003.)

- Learning and Memory: How will the user's prior knowledge influence selective attention and interpretation? Does the knowledge necessary to perform tasks reside in the world or in the user's mind? (See Baddeley, 1998, 1999; Grondin, 2005; Hester & Garavan, 2005; Hutchins, 1995; Norman, 2002; Ormrod, 2003; Radvansky, 2005; Scaife & Rogers, 1996; Seamon, 2005.)

- Information Processing, Mental Models, and Situation Awareness: What inferences/assumptions will the user make? What internal representations will support such processing? (See Endsley, 2003; Endsley & Garland, 2000; Gentner & Stevens, 1983; Johnson et al., 1981; Plous, 1993; Schaeken, 2005.)

- Design-Induced Error: How could the product design and the context of use influence performance and induce errors? (See Bainbridge, 1983; Beach & Connolly, 2005; Johnson et al., 1981; Larson & C. Hayes, 2005; Parasuraman & Riley, 1997; Reason, 1991, 1997; Roth, Bennett, & Woods, 1987; Sheridan, 2002; Skirka, Mosier, & Burdick, 1999; P. J. Smith & Geddes, 2003; P. J. Smith, McCoy, & Layton, 1997.)

- Motor Performance: Can the controls be used efficiently and without error? (See Jagacinski & Flach, 2003.)

- Group Dynamics: How will the system influence patterns of interaction among people? (See Baecker, Grudin, Buxton, & Greenber, 1995; Brehm, Kassin, & Fein, 1999; Forsyth, 1998; Levi, 2001; G. M. Olson & J. S. Olson, 2003.)

Thus, steps 1–3 as outlined above involve describing how a product (such as a DSS) can be used to achieve a specific goal. Step 4 involves predicting how people might perform when using this product, either successfully or unsuccessfully, as they try to achieve this goal. Thus, the "cognitive" part of this analysis is embedded only in step 4. Note also that the goal-subgoal hierarchy does not attempt to represent the further sequences of actions that could arise once the user starts down a failure path. This helps to keep the size of the hierarchy smaller, keeping the analysis task more manageable. (For some applications, it may be desirable to actually expand the goal-subgoal hierarchy to represent the complete path of a failure story.)

Finally, going beyond just the use of hierarchical decomposition techniques, Gordon and Gill (1997) noted that:

CTA [Cognitive Task Analysis] methodologies differ along several dimensions. These include the type of knowledge *representation* or formalism used to encode the information obtained from the analysis, the *methods* used for eliciting or identifying expert (or novice) knowledge, and the *type of task and materials* used in the analysis. Typical representation formats include lists and outlines, matrices and cross-tabulation tables, networks, flowcharts, and problem spaces with alternative correct solution paths (pp. 133–134)

Thus, although the use of hierarchical goal-subgoal representations is one of the more popular approaches, a variety of other complementary approaches are labeled "cognitive task analyses" (Crandall, Klein, & Hoffman, 2006). These alternatives include the use of concept maps and conceptual graph analyses (Gordon & Gill, 1992) and different approaches to extracting and representing the knowledge of experts about critical incidents (Crandall et al., 2006; Klein, Caulderwood, & MacGregor, 1989). The ultimate goal of all of these approaches, however, is to better represent and predict how users' knowledge and control processes or cognitive processes will be applied in a given task environment. This task environment could be the existing world (a descriptive analysis) that is being studied in order to design improved support tools, or an envisioned world that incorporates some specific design proposal intended to improve performance (a predictive analysis).

Work-Domain Analyses

Cognitive task analyses tend to focus on how to support performance for an already identified task or goal. However, in many cases DSSs are designed to support performance in very complex environments (Bar-Yam, 2003, 2005) where it is unlikely that all of the possible scenarios that could arise will be predicted during the design process. Thus, there is also a need for design methodologies that support the development of robust designs that will be effective even when novel situations arise (Vicente, 1999).

To deal with this need, a complementary set of design methodologies have been developed under the label of "work-domain analysis" (Burns, Bizantz, & Roth, 2004; Burns & Hajdukiewicz, 2004; Burns & Vicente, 2001; Reising & Sanderson, 2002; Vicente, 2002; Vicente & Rasmussen, 1990, 1992). In general terms, the emphasis is on understanding and modeling the work domain or application area rather than simply modeling the performance of people on a prespecified set of tasks within this domain. The argument is that the domain analysis can identify the critical parameters or constraints relevant to successful performance in that domain and thus serve as the basis for a design that can handle both routine and novel (unanticipated) situations.

To make this point more concrete, as a very simple but illustrative example, Vicente (2000) contrasted verbal directions on how to travel from one point to another with the use of a map. In this analogy, the directions provide a very efficient but brittle description of how to accomplish the desired goal. The map requires more work (interpretation), but is more flexible in the face of obstacles that might render the directions unusable. Building on this analogy, Vicente noted:

This decrease in efficiency [of maps vs. directions] is compensated for by an increase in both flexibility and generality. Like maps, work domain representations are more flexible because they provide workers with the information they need to generate an appropriate response online in real time to events that have not been anticipated by designers. This is particularly useful when an unforeseen event occurs because task representations, by definition, cannot cope with the unanticipated (p. 115).

The literature on work-domain analyses discusses the use of abstraction (means-end) hierarchies and aggregation (part-whole) hierarchies to complete a domain analysis, and then using the results of this analysis to guide the design of the DSS for supporting operator performance (Burns & Hajdukiewicz, 2004; Vicente, 1999; Vicente & Rasmussen, 1990). The basic logic of this approach is that, in order to design appropriate representations to assist an operator in completing tasks using some system, the complete and correct semantics of the domain must first be appropriately defined. Since such a design is concerned with supporting goal-oriented behaviors, one way to represent the semantics is with an abstraction (means-end) hierarchy that captures the "goal-relevant properties of the work domain" (Vicente & Rasmussen, 1990, p. 210). In addition, because the domains of interest are often quite complex, aggregation hierarchies may be useful to decompose the overall domain into a set of interrelated subparts.

Approaches to Design—Summary

Much of the work on methods for completing cognitive task analyses and work domain analyses has been stimulated by the challenges of developing DSSs to support performance on complex tasks. The description above provides a sense of the range of techniques that have been developed, and emphasizes the need to understand both the constraints of the application context and the nature of the agent(s) (person or people) involved. The insights provided by such analyses can be used to guide the design of the underlying functionality embedded within a DSS, and to guide the design of the interaction between a DSS and its user.

HUMAN PERFORMANCE ON DECISION-MAKING TASKS

There are several reasons why an understanding of human performance is important to the designer of a DSS. The first is that the motivation for developing such systems is to increase efficiency (reduce production costs or time) and/or to improve the quality of performance. Thus, it is important for the design team to be able to efficiently and effectively complete the initial problem definition and knowledge-engineering stages of a project, identifying areas in which improvements are needed. An understanding of human performance makes it possible to, in part, take a top-down approach to this, looking to see whether certain classic human performance phenomena (such as hypothesis fixation) are influencing outcomes in a particular application. A second motivation for understanding human performance is that a human-centered approach to the design of a DSS requires consideration and support of the user's skills (Garb, 2005). To do this effectively, knowledge of human performance (perception, learning and memory, problem solving, decision making, etc.) is essential. A number of these aspects of human performance are covered in previous chapters, so this chapter will just highlight some of the factors most relevant to the design of a DSS, and discuss how such factors are relevant to this design task.

Errors and Cognitive Biases

In very broad terms, human errors (Strauch, 2004; Wiegmann & Shappell, 2003) can be classified as slips and as mistakes (Norman, 1981). Slips arise through a variety of cognitive processes, but are defined as behaviors in which the person's actions do not match his intentions. Generally, this refers to cases where the person has the correct knowledge to achieve some goal but, due to some underlying perceptual, cognitive, or motor process, fails to correctly apply this knowledge. As Norman described it: "Form an appropriate goal but mess up in the performance, and you've made a slip" (1981, p. 106). Mistakes, on the other hand, refer to errors resulting from the accurate application of a person's knowledge to achieve some goal, but where that knowledge is incomplete or wrong.

DSSs are potentially useful for dealing with either of these sources of errors. If slips or mistakes can be predicted by the design team, then tools can be developed to either help prevent them, recover from them, or reduce their impacts.

Slips

Norman (2002) discussed six categories of slips. Knowing something about these different causes can help the designer to look for possible manifestations in a particular application area. The six categories, as defined in Norman, included:

- Capture errors, "in which a frequently done activity suddenly takes charge instead of (captures) the one intended" (p. 107);
- Description errors, where "the intended action has much in common with others that are possible" and "the internal description of the intention was not sufficient . . . often resulting in "performing the correct action on the wrong object" (pp. 107–108);
- Data-driven errors, where an automatic response is triggered by some external stimulus that triggers the behavior at an inappropriate time;
- Associative activation errors where, similar to a data-driven error, something triggers a behavior at an inappropriate time, but in this case the trigger is some internal thought or process;
- Loss-of-activation errors, or "forgetting to do something" (p. 109);
- Mode errors, or performing an action that would have been appropriate for one mode of operation for a system, but is inappropriate for the actual mode or state that the system is in.

Mistakes

As defined above, mistakes are due to incorrect knowledge (the rule, fact, or procedure that the person believes to be true is incorrect, resulting in an error) or incomplete (missing) knowledge.

Cognitive Biases

The literature on human error also provides other useful ways to classify errors in terms of surface level behavior or the underlying cognitive process. Many of these are discussed under the label of "cognitive biases" (Brachman & Levesque, 2004; Bradfield & Wells, 2005; Fraser, P. J. Smith, & J. W. Smith, 1992; Gilovich, Griffin, & Kahneman, 2002; Haselton, Nettle, & Andrews, 2005; Kahneman & Tversky, 2000; Plous, 1993; Poulton, 1989), including:

- Gambler's fallacy (Croson & Sundali, 2005). (Concluding there is a pattern in a series of events that is in reality a random sequence.) Gilovich, Vallone, and Tversky (1985) for instance, found that individuals believe they see streaks in basketball shooting even when the data show that the sequences are essentially random.
- Insensitivity to sample size (Bjork, 1999). (Failing to understand that the law of large numbers implies that the probability of observing an extreme result in an average decreases as the size of the sample increases.) Tversky and Kahneman (1974), for example, found that subjects believed that "the probability of obtaining an average height greater than 6 feet was assigned the same value for samples of 1000, 100 and 10 men." (p. 1127)
- Incorrect revision of probabilities. (Failure to revise probabilities sufficiently when data are processed simultaneously, or, conversely, revising probabilities too much when processing the data sequentially.)
- Ignoring base rates. (Failure to adequately consider prior probabilities when revising beliefs based on new data.)
- Use of the availability heuristic (Blount & Larrick, 2000). (Tversky (1982) suggested that the probability of some type of event is in part judged based on the ability of the person to recall events of that type from memory, thus suggesting that factors like recency may incorrectly influence judgments of probability.)
- Attribution errors. Jones and Nisbett (1971) described this by noting that "there is a pervasive tendency for actors to attribute their actions to situational requirements, whereas observers tend to attribute the same actions to stable personal dispositions." (p. 83)
- Memory distortions due to the reconstructive nature of memory (Blank, 1998; Handberg, 1995). (Loftus (1975) described processes that distort memories based on the activation of a schema as part of the perception of an event, and the use of that schema to reconstruct the memory of the event based on what the schema indicates should have happened rather than what was actually perceived originally. P. J. Smith, Giffin, Rockwell, and Thomas (1986) and Pennington and Hastie (1992) provided descriptions of similar phenomena in decision-making tasks.)

In recent years, there has been considerable controversy regarding the nature and validity of many of these explanations as "cognitive biases" (Fraser et al., 1992; Haselton, 2005; Koehler, 1996), suggesting that it is important to carefully understand the specific context in order to generate predictions as to whether a particular behavior will be exhibited. Using the incorrect revision of probabilities as an illustration, Navon (1978) suggested that in many real-world settings the data aren't independent, and that what appears to be conservatism in revising probabilities in a laboratory setting may be the result of the subjects applying a heuristic or cognitive process that is effective in real-world situations where the data are correlated.

A more detailed example is provided by the popular use of the label "hypothesis fixation." This behavior refers to some process that leads a person to form an incorrect hypothesis and to stick with that hypothesis, failing to collect critical data to assess its validity or to revise it in the face of conflicting data. A variety of cognitive processes have been hypothesized to cause this behavior. One example is called "biased assimilation," in which a person judges a new piece of data as supportive of his hypothesis (increasing the level of confidence in the hypothesis) based simply on the consideration of whether that outcome could have been produced under his hypothesis. This contrasts with inferential processes based on a normative model that suggest that beliefs should be revised based on the relative likelihood of an outcome under the possible competing hypotheses, and that would, in the same circumstance, lead to a reduction in the level of confidence in his hypothesis.

Another example of a cognitive process discussed in the literature as relevant to hypothesis fixation is the so-called "confirmation bias." This phenomenon (described in Wason (1960) and in Mynatt, Doherty, and Tweney (1978)) is concerned with the person's data-collection strategy. For example, in a study asking subjects to discover the rules of particle motion in a simulated world, Mynatt et al. described performance by concluding that there was "almost no indication whatsoever that they intentionally sought disconfirmation." Later studies, however, have suggested that this "bias" is really an adaptive "positive test" strategy that is effective in many real-world settings, and that the unusual nature of the task selected by Mynatt et al. makes it look like an undesirable bias (Klayman & Ha, 1987). P. J. Smith et al. (1986) further suggested that, in real-world task settings where such a strategy might lead to confirmation of the wrong conclusion, experts often have domain-specific knowledge or rules that help them to avoid this strategy altogether, thus ensuring that they collect critical potentially disconfirming evidence.

Designer Error

Many of the error-producing processes discussed above appear to focus on system operators. It is important to recognize, however, that the introduction of a DSS into a system is a form of cooperative work between the users and the design-and-implementation team and that, like the users, the design team is susceptible to errors (Petroski, 1994). These errors may be due to slips or mistakes. In the case of mistakes, it may be due to inadequate knowledge of the application area (the designers may fail to anticipate all of the important scenarios) or incorrect knowledge. It may also be due to a failure to adequately understand or predict how the users will actually apply the DSS, or how its introduction will influence their cognitive processes and

performances. In addition, sources of errors associated with group dynamics may be introduced. Whether such errors are due to a lack of sufficient coordination, where one team member assumes that another is handling a particular issue, or due to the influence of group processes (Brehm et al., 1999; Forsyth, 1998; Janis, 1982; Levi, 2001; McCauley, 1989; Tetlock, 1998; Tetlock, Peterson, McQuire, Chang, & Feld, 1992) on design decisions, the result can be some inadequacy in the design of the DSS, or in the way the user and the DSS work together. (Note that such group processes are also important potential sources of errors when system operators work together as part of teams, with or without technological support.)

Systems Approaches to Error

Reason (1991, 1997) and Wiegmann, Zhang, Von Thaden, Sharma, and Gibbons (2004) cautioned designers against fixating on the immediately preceding "causes" of an accident when trying to prevent future occurrences. They remind us that, in many system failures, a number of conditions must coincide in order for the failure to occur. This presents a variety of leverage points for preventing future occurrences, many of which are preemptive and thus remote from the actual accident or system failure. Many of these changes focus on preventing the conditions that could precipitate the system failure, rather than improving performance once the hazardous situation has been encountered.

Errors and Cognitive Biases—Implications for Design

As described at the beginning of this section, the value of this literature to the designer is the guidance it provides in conducting the initial knowledge-engineering studies that need to be completed in order to identify opportunities for improvement in some existing decision-making process that is to be supported by a new DSS. Familiarity with this literature is also critical to the designer to ensure that the introduction of the DSS doesn't result in new design-induced errors. This literature also serves to emphasize two additional considerations:

- The occurrence of errors is often due to the co-occurrence of a particular problem-solving strategy with a given task environment that is "unfriendly" to that strategy (e.g., situations in which the strategy is not sufficiently robust). Strategies that were adaptive in one setting may no longer be adaptive in the newly designed environment. The potential for such a negative transfer of learning needs to be considered as part of the design.
- For routine situations, people tend to develop expertise that lets them employ knowledge-rich problem-solving strategies (Crandall et al., 2006; Klein, 1993; Kolodner, 1993; Newell, 1990) that avoid the errors that could be introduced by certain general problem-solving strategies. However, in many system designs, people are expected to act as the critical safety net during rare, idiosyncratic events that the design team has failed to anticipate. These are exactly the situations in which people have to fall back on their general problem-solving

strategies, and are thus susceptible to the potential errors associated with these weak problem-solving methods.

In addition, this review emphasizes the need to take a broad systems perspective, recognizing that the design and implementation team are a potential source of error, and understanding that, if the goal is to prevent or reduce the impact of errors, then the solution is not always to change performance at the immediate point where an error was made. It may be more effective to change other aspects of the system so that the potentially hazardous situation never even arises.

Finally, the emphasis on designer error implies that the design process for a DSS needs to be viewed as iterative and evolutionary. Just because a tool has been fielded, it is not safe to assume that no further changes will be needed. As Horton and Lewis (1991) discussed, this has organizational implications (e.g., ensuring adequate communications regarding the actual use of the DSS, budgeting resources to make future revisions), as well as architectural implications (developing a system architecture that enables revisions in a cost-effective manner).

Human Expertise

As discussed in previous chapters, the nature of human information processing imposes a variety of constraints that influence how effectively a person processes certain kinds of information. These include memory, perceptual, and information-processing constraints. As a result, there are certain decision tasks in which the computational complexity or knowledge requirements limit the effectiveness of unaided individual human performance.

From a design perspective, these information-processing constraints offer an opportunity. If important aspects of the application are amenable to computational modeling, then a DSS may provide a significant enhancement of performance, either in terms of efficiency or the quality of the solution. (Note that, even if development of an adequate computational model is not feasible, there may be other technological improvements, such as more effective communications environments, that could enhance performance. However, this chapter is focusing on DSSs that incorporate active information-processing functions by the software.)

A good example of this is flight planning for commercial aircraft. Models of aircraft performance considering payload, winds, distance, and aircraft performance characteristics (for the specific aircraft as well as the general type of aircraft) are sufficiently accurate to merit the application of DSSs that use optimization techniques to generate alternative routes and altitude profiles to meet different goals in terms of time and fuel consumption (P. J. Smith, McCoy, & Layton, 1997). This example also provides a reminder, however, that it is not just human limitations that are relevant to design. It is equally important to consider human strengths, and to design systems that complement and are compatible with users' capabilities. In flight planning, for instance, this includes designing a system that allows the person to incorporate his judgments into the generation and evaluation of alternative flight plans, considering the implications of uncertainty in the weather, air traffic congestion, and other factors that may not be incorporated into the model underlying the DSS.

This cooperative systems perspective has several implications. First, the designer needs to have some understanding of when and how the person should be involved in the alternative generation and selection processes. This requires insights into how the person makes decisions, in terms of problem-solving strategies as well as in terms of access to the relevant knowledge and data, and how the introduction of a DSS can influence these problem-solving processes. It also implies that the strengths underlying human perceptual processes need to be considered through display and representation-aiding strategies (Burns & Hajdukiewicz, 2004; Jones & Schkade, 1995; Kleinmuntz & Schkade, 1993; Larkin, 1989; Tufte, 1997) in order to enhance the person's contributions to the decision-making process by making important relationships more perspicuous or salient.

Descriptive Models

The literature on human problem solving and decision making provides some very useful general considerations for modeling human performance within a specific application. This literature, which is reviewed in more detail in previous chapters, covers a variety of descriptive models of problem solving. This includes the use of heuristic search methods (Clancey, 1985; Dasgupta, Chakrabarti, & Desarkar, 2004; Michalewicz & Fogel, 2004; Michie, 1986; Rayward-Smith, Osman, Reeves, & G. Smith, 1996; Russell & Norvig, 1995). This modeling approach (Newell & Simon, 1972) conceptualizes problem solving as a search through a space of problem states and problem-solving operations to modify and evaluate the new states produced by applying these operations. The crucial insight from modeling problem solving as search is an emphasis on the enormous size of the search space resulting from the application of all possible operations in all possible orders. In complex problems, the size of the space precludes exhaustive searching of the possibilities to select an optimal solution. Thus, some heuristic approach that satisfices, such as elimination by aspects (Tversky, 1972), is needed to guide a selective search of this space, resulting in the identification of an acceptable (if not optimal) solution.

Another aspect of problem solving that Newell and Simon (1972) noted when they formulated the search paradigm was the importance of problem representation. Problem representation issues emphasize that a task environment is not inherently objectively meaningful, but requires interpretation for problem solving to proceed. Some interpretations or representations are more likely to lead to successful solutions than others (for certain agents). In fact, an important component of expertise is the set of features or the representation used to characterize a domain (Lesgold et al., 1988), sometimes referred to as "domain ontology."

Task-specific problem spaces allow problem solvers to incorporate task- or environment-induced constraints, focusing attention on just the operations of relevance to preselected goals (Sewell & Geddes, 1990; Waltz, 1975). For such task-specific problem solving, domain-specific knowledge (represented in computational models as production rules, frames, or some other knowledge representation) may also be incorporated to increase search efficiency or effectiveness (Dechter, 2003;

Ghallib, Nau, & Traverso, 2004; Laird et al., 1987; Shalin, Geddes, Bertram, Szczepkowski, & DuBois, 1997).

These models, based on symbolic reasoning, have been developed for specific generic tasks like diagnosis or planning (Chandrasekaran, 1988; Ghallib et al., 2004; Gordon, 2004; G. A. Miller et al., 1960; Mumford, Schultz, & Van Doorn, 2001; Nareyek, Fourer, & Freuder, 2005). For example, computational models of planning focus on the use of abstraction hierarchies to improve search efficiency (Sacerdoti, 1974), dealing with competing and complementary goals (Wilensky, 1983), and mixed top-down/bottom-up processing (B. Hayes-Roth & F. Hayes-Roth, 1979) to opportunistically take advantage of data as it becomes available. More recent developments have sought to deal with planning in stochastic environments (Madani, Condon, & Hanks, 2003; Majercik & Littman, 2003).

In contrast to such sequential search models, models based on case-based reasoning, recognition-primed decision making, or analogical reasoning (Crandall et al., 2006; Klein, 1993; Kolodner, 1993; Riesbeck & Schank, 1989) focus on prestructured and indexed solutions based on previously experienced situations. These modeling approaches suggest that human experts in complex operational settings rarely describe their cognitions as sequential searches to construct alternative solution states, but rather as recognition processes that match features of the situation to prototypical, preformulated response plans (Zuboff, 1988; Rasmussen, 1983). In complex dynamic domains (Oliver & Roos, 2005) like aviation or firefighting, these preformulated plans incorporate a great deal of implicit knowledge reflecting the constraints of the equipment, the environment, other participants in the work system, and previous experience. Over time, solutions sensitive to these constraints result in familiar, accepted methods that are then triggered by situational cues and modified in small ways as needed to deal with the specific scenario. However, it is important to recognize that such recognition-based processes can cause familiar task features to invoke only a single interpretation, resulting in hypothesis fixation or hindering creative departure from normal solutions. This limitation can be critical when the task environment contains a new, unexpected feature that requires a new approach.

Normative Optimal Models

In addition to these descriptive models, a large amount of literature characterizes human decision making relative to various normative models based on optimal processes, which help to emphasize important factors within the task and task environment that should be considered in making a decision in order to arrive at a better decision. These normative models are based on engineering models such as statistical decision and utility theory, information theory, and control theory (Jagacinski & Flach, 2003; Rouse, 1980; Sheridan & Ferrell, 1974). By contrasting human performance with optimal performance on certain tasks and emphasizing the factors that should influence decision making in order to achieve a high level of performance, this literature helps the designer to look for areas where human performance may benefit from some type of DSS.

Finally, the literature on human expertise (Charness, Feltovich, Hoffman, & Ericsson, 2006) emphasized the ability of

people to learn and adapt to novel situations, and to develop skeletal plans that guide initial decisions or plans, but that are open to adaptation as a situation unfolds (Geddes, 1989; Suchman, 1987; Wilensky, 1983). The literature also emphasizes variability, both in terms of individual differences and in terms of the ability of a single individual to use a variety of decision-making strategies in some hybrid fashion in order to be responsive to the idiosyncratic features of a particular scenario.

Human Expertise—Implications for Design

An understanding of the literature on human expertise is of value to the designer in a number of different ways. First, in terms of the initial problem-definition and knowledge-engineering stages, familiarity with models of human problem solving and decision making can help guide the designer in looking for important features that influence performance within that application. Second, in an effort to provide cognitive compatibility with the users, the design of many of the technologies underlying DSSs is guided by these same computational models of human performance. Third, even if the underlying technology isn't in some sense similar to the methods used by human experts in the application, the designer needs to consider how the functioning of the DSS system should be integrated within the user's decision-making processes.

In terms of some specific emphases, the above discussion leads to some suggestions:

- Don't fixate on active DSS technologies as the only way to improve system performance. Enhancing human performance through changes in procedures, improvements in passive communication tools, better external memory aids, and so on may be more cost-effective in some cases. In addition, such changes may be needed to complement an active DSS in order to make its use more effective.

- Look for ways in which the direct perception of ecological constraints allows people to perform expertly (Flach, Hancock, Caird, & Vicente, 1995). The ability to perceive these critical parameters or constraints directly may make what seems like a very difficult information-processing task much less demanding.

- Related to the consideration of ecological constraints was the earlier suggestion that problem representation has a strong influence on how easily a problem solver can find a good solution. This issue of perspicuity (salience of the problem solution), however, is dependent not only on the characteristics of the task and task environment, but also on the nature of the problem solver. Thus, in order to enhance human performance as part of the decision-making process, it is important to consider alternative ways of representing the situation that will enable human perceptual and cognitive processes to work more effectively. In addition, problem representation can affect the performance of the designer. Consequently, alternative problem representations need to be generated as part of the design process to help the designer think more effectively.

- Consider the applicability of normative optimal models of performance for the task in order to focus attention on factors that should be influencing current performance, and to contrast optimal strategies with the strategies that people are actually using in response to these task-determined factors. Aspects of these normative models may also be appropriate for more direct inclusion in the DSS itself, even though the strategies used by people do not fully reflect the optimal processes highlighted by these normative models. In addition to considering the task structure highlighted by normative optimal models, use knowledge of the variety of different descriptive models of human problem solving and decision making to guide knowledge-engineering efforts, making it easier and more efficient to understand how people are currently performing the tasks. Thus, all of these models of decision making represent conceptual tools that help the designer to more effectively and efficiently understand performance in the existing system, and to develop new tools and procedures to improve performance.

- Whether the DSS is designed to process information "like" the user at some level, or whether it uses some very different processing strategy, the user needs to have an appropriate and effective mental model of what the system is and is not doing (Kotovsky, J. R. Hayes, & Simon, et al., 1985; Lehner & Zirk, 1987; Nickerson, 1988; Zhang, 1997; Zhang & Norman, 1994). To help ensure such cognitive compatibility, the designer needs to understand how people are performing the task.

- Consider design as a prediction task, trying to predict how users will interact with the DSS system, and how it will influence their cognitive processes and performances.

Additional Cognitive Engineering Considerations

The preceding sections focused on our understanding of human performance on decision-making tasks and the implications of that literature for design. Below, some additional considerations based on studies of the use of DSSs are presented.

The Human Operator as Monitor

Numerous studies make it clear that, given the designs of DSSs for complex tasks must be assumed to be brittle, there is a problem with designs that assume that human expertise can be incorporated as a safety net by asking a person to simply monitor the output of the DSS, with the responsibility for overriding the software if a problem is detected. One problem with this role is that, in terms of maintaining a high level of attentiveness, people do not perform well on such sustained attention tasks (Meister & Enderwick, 2002). As Bainbridge (1983) noted:

We know from many 'vigilance' studies (Mackworth, 1950/1961) that it is impossible for even a highly motivated human being to maintain effective visual attention towards a source of information on which very little happens, for more than about half an hour. This means that it is humanly impossible to carry out the basic function of monitoring for unlikely abnormalities. (p. 777)

A related issue is the problem of loss of skill. As Bainbridge (1983) further noted, if the operator has been assigned a passive monitoring role, he "will not be able to take over if he has not

been reviewing his relevant knowledge, or practicing a crucial manual skill" (p. 778). Thus, a major challenge for retaining human expertise within the system is "how to maintain the effectiveness of the human operator by supporting his skills and motivation" (p. 778).

Complacency and Overreliance

Studies reported by Parasuraman and Riley (1997) and Parasuraman (2000) introduced further concerns about assigning the person the role of critiquing the computer's recommendations before acting. These studies discuss how the introduction of a DSS system can lead to overreliance by the human user when the software is generating the initial recommendations (Metzger & Parasuraman, 2005; Skirka et al., 1999).

Excessive Mental Workload

Designs that relegate the person to the role of passive monitor run the risk of a vigilance decrement due to insufficient engagement and mental workload. At the other extreme, designers must recognize that "clumsy automation" that leaves the person with responsibility for difficult parts of the task (such as coping with an emergency), but that adds additional workload (Kushleyeva, F. Salvucci, & Schunn, 2005) due to the awkward interactions now required to access information and functions embedded in the DSS (such as navigating through complex menus to view certain information), can actually impair performance because of the added mental workload of interacting with the DSS (Wiener & Nagel, 1988). One line of research that is now getting increased attention is the potential to design adaptive systems that can monitor the human operator using neurophysiological measures that attempt to detect problems with alertness, mental workload, or attention (Caggiano & Parasuraman, 2004; Freeman, Mikulka, & Scerbo, 2004; Grier, Warm, & Dember, 2003; Luo, Greenwood, & Parasuraman, 2001; Mikulka, Scerbo, & Freeman, 2002). This work seeks to "determine whether a biocybernetic, adaptive system could enhance vigilance performance," with the goal of improving "monitoring performance on critical activities such as air traffic control and radar and sonar operation" (Mikulka et al., 2002, p. 654), or offloading work to software or other people when the computer detects that excessive attentional demands are being placed on an individual.

Lack of Awareness or Understanding

Even if the person is sufficiently engaged with the DSS and the underlying task, designers need to consider how to ensure that the user has an accurate mental model of the situation, and of the functioning and state of the DSS (Billings, 1997; Larkin & Simon, 1987; Mitchell & R. A. Miller, 1986; Roth et al., 1987; Sarter & Woods, 1993). If the person does not have such an understanding, then it may be difficult for him to intervene at appropriate times or to integrate the computer's inputs into his own thinking appropriately (Geddes, 1997b). This concern has implications for selection of the underlying technology and conceptual model for a system, as well as for the design of the visual or verbal displays intended to represent the state of the world and the state of the software for the user, including explanations of how the DSS has arrived at its recommendations (Clancey, 1983; Hasling, Clancey, & Rennels, 1984; Nakatsu & Benbasat, 2003).

Lack of Trust and User Acceptance

As outlined above, overreliance can be a problem with certain assignments of roles to the person and the computer. At the other extreme, lack of trust or acceptance of the technology can eliminate or reduce its value (Clarke, Hardstone, Rouncefield, & Sommerville, 2006; Hollnagel, 1990; Muir, 1987). This lack of acceptance can result in outright rejection of the DSS (in which case it either is not purchased or not used), or in a tendency to underutilize it for certain functions. It is important to note that this lack of acceptance can be due to resistance to change (Cartwright & Zander, 1960; Forsyth, 1998; Levi, 2001) even if there is no intrinsic weakness in the DSS, due to general beliefs held by the operators (rightly or wrongly) about how the software will influence their lives, or due to beliefs about how well such a DSS can be expected to perform (Andes & Rouse, 1992; J. Lee & See, 2004; Marsh & Dibben, 2003; Muir & Moray, 1996; Riegelsberger, Sasse, & McCarthy, 2005).

Active Biasing of the User's Cognitive Processes

Complacency is one way in which a DSS can influence the person's cognitive processing. Studies have also shown, however, that the use of a DSS can also actively alter the user's cognitive processes (Beach & Connolly, 2005; Larson & C. Hayes, 2005; P. J. Smith, McCoy, & Layton, 1997). The displays and recommendations presented by the software have the potential to induce powerful cognitive biases, including biased situation assessment and failures by the user to activate and apply his expertise because normally available cues in the environment are no longer present. The net result is that the person fails to exhibit the expertise that he would normally contribute to the decision-making task if working independently, not because he lacks that expertise but because the computer has influenced his cognitive processes in such a way that this knowledge is never appropriately activated. Studies have shown that these biasing effects can induce practitioners to be 31% more likely to arrive at an incorrect diagnosis on a medical decision-making task (Guerlain et al., 1996) and 31% more likely to select a very poor plan on a flight-planning task (P. J. Smith, McCoy, & Layton, 1997).

Distributed Work and Alternative Roles

Much of the literature focuses on the interactions between a single user and the DSS. Increasingly, however, researchers and system developers are recognizing that one approach to effective performance enhancement is to think in terms of a distributed work paradigm (Schroeder & Axelsson, 2005; P. J. Smith, McCoy, & Orasanu, 2001), in which the software may be one "agent" in this distributed system (Geddes & Lizza, 1999), or the software may be viewed primarily as a mediator to support human-human interactions within virtual or distributed teams (Baecker et al., 1995; Bowers et al., 2006; Caldwell, 2005; Carroll, Neale, Isenhour, & McCrickard, 2003; Handy, 1995; Hertel, Geister, & Konradt, 2005; Hinds & Kiesler, 2005; Hinds & Mortensen, 2005; Hutchins, 1990, 1995; Katz, Lazer, Arrow, & Contractor, 2004; Lurey & Raisinghani, 2001; G. M. Olson & J. S. Olson, 1997, 2003; Orasanu & Salas, 1991; Rasmussen et al., 1991; Salas, Bowers, & Edens, 2001; Salas & Fiore, 2004; Smith, McCoy, & Layton, 1997). Such distributed systems can now be found in the military, in medicine, in aviation and a variety of other application areas.

Sycara and Lewis (2004) suggested that:

There are three possible functions that software agents might have within human teams:

1. Support the individual team members in completion of their own tasks
2. Assume the role of a (more or less) equal team member by performing the reasoning and tasks of a human teammate
3. Support the team as a whole (p. 204)

and that "teams could be aided along four general dimensions: accessing information, communicating, monitoring, and planning" (p. 227). In considering the design and use of DSSs for such applications, all of the more traditional factors must be considered. In addition, questions regarding the impact of the technology on team situation awareness (Endsley, 2003), rapport, trust, and communication become paramount. It is also important to think of a broader set of design parameters, as distributed approaches to work open up the potential to use a variety of different architectures for distributing the work in terms of the locus of control or responsibility and the distribution of knowledge, data, and information-processing capabilities (Griffith, Sawyer, & Neale, 2003; Sheridan, 1997; Sheridan, 2002; P. J. Smith, McCoy, Orasanu, et al., 1997; P. J. Smith, Billings, et al., 2000; P. J. Smith, Beatty, Spencer, & Billings, 2003). Furthermore, DSSs may be necessary to make effective certain architectures for distributing the work.

Organizational Failures

The discussions above focus on design-induced errors made by the system operators. It is equally important for the design team to recognize that part of the design process is ensuring that the management of the organization into which the DSS is introduced will provide an effective a safety net. This means that the design team needs to ensure that the organization has established effective procedures to detect significant problems associated with the introduction and use of the DSS, and that such problems are communicated to the levels of management where responsibility can and will be taken to respond effectively to these problems (Horton & Lewis, 1991).

CASE STUDIES

Below, case studies are presented focusing on the designs of three quite different DSSs. Within these case studies, a number of the issues raised earlier are discussed in more detail within the context of real-world applications, along with presentations of design solutions that provide concrete illustrations of how to deal with those issues.

Case Study A—Distributed Work in the National Airspace System: Ground Delay Programs

Earlier in this chapter, we indicated that a distributed-work system can be characterized in terms of the assignment of control or decision-making authority in relationship to the distribution of knowledge and data. The Federal Aviation Administration's (FAA's) Enhanced Ground Delay Program provides a very informative example for illustrating the significance of these parameters when designing a distributed system.

The Application Area

The FAA's Enhanced Ground Delay Program (GDP) has been in operation since 1998. Its goal is to help match the arrival demand at an airport with the available capacity, where demand could exceed capacity because of a weather constraint at that airport, a closed runway, and so on.

The strategy underlying the use of GDPs is to delay departures to the constrained airport, holding these departures on the ground when that airport's arrivals are predicted to exceed capacity. Such a strategy has potential benefits for the FAA in terms of reducing air traffic controller workload (reducing the need for the use of airborne holding or some other air traffic control tactic). It also has potential benefits for the NAS Users, as it can be more economical to delay departures, holding these flights on the ground (reducing fuel consumption due to airborne holding as well as reducing the potential for diversions).

The Original Approach to Distributing Work in a GDP

In principle, there are a variety of different ways to implement GDPs. Prior to 1998, the FAA used GDPs, but did so using an approach that was much less effective than the currently used procedure. Under this original paradigm, traffic managers at the FAA's Air Traffic Control Systems Command Center (ATCSCC) would set the airport arrival rate (AAR) for an airport, and then a DST would automatically assign an arrival slot to each specific flight, which would then be held on the ground until it was time to depart to make its assigned arrival slot. NAS Users had one opportunity to swap flights in a batch mode (e.g., requesting all of the swaps at one time). This limited their ability to monitor the development of the situation over time and adapt as new information became available. In addition, although NAS Users were asked to cooperate in order to make this program work, there was actually a disincentive for them to provide the FAA with information about cancellations or potential delays. If the cancellation of a flight was reported to the FAA, the NAS User lost its assigned arrival slot; if it was reported that a flight was going to miss its departure time (because of a mechanical problem, for instance), that flight could be assigned an even greater delay. As a result, NAS Users often chose to withhold this information from the FAA, resulting in unused arrival slots (wasted capacity).

An Alternative Approach to the Design of GDPs

Given these limitations, a joint industry-FAA program evolved to identify and implement a different approach to GDPs. This Collaborative Decision Making program explored methods for still achieving the FAA's primary goal (matching arrival rates to airport capacity), while giving NAS Users the flexibility to better achieve their business goals.

One approach would have been to have NAS Users provide the FAA with all of the knowledge and data about their business concerns relevant to decisions about which flights to assign to which arrival slots. In principle, the FAA traffic managers could have then determined "good" slot allocations for each NAS User, and assigned flights to slots accordingly. Such a decision would have had to somehow take into consideration the competitive nature of this situation and through some decision process determine an "equitable" solution.

It was decided that such an approach would produce unrealistic demands in terms of cognitive complexity and mental workload for these traffic managers, as they would have had to master the relevant expertise of the dispatchers working for each NAS User and to somehow use that knowledge to generate equitable slot assignments. (Dispatchers work for airlines and other general aviation operations that are required to, or choose to, abide by Part 121 regulations of the U.S. Federal Air Regulations, sharing responsibility for each flight with the pilots. They complete the preflight planning—determining the flight's route, departure time, fuel load, and so on—and are responsible for following that flight while it is airborne.)

Strategy for distributing work. Instead, the CDM program developed a new architecture for distributing the work associated with GDPs. This architecture uses a classic systems engineering approach for dealing with complexity, decomposing the overall plan for a GDP into a set of nearly independent subplans that can be developed by different individuals, and that, when completed by each of these individuals, achieves the desired goals (assuming the plan is successfully implemented). The significance of the qualifier "nearly" should not be overlooked in terms of its importance, however. The reality is that there are cases where interactions among these different individuals are required, so the system design has to include mechanisms that support such interactions in order to ensure that the appropriate perspectives are considered in setting up a GDP, and that exceptions can be made for individual flights when that is required to deal with some overriding consideration.

Like the original GDP, this Enhanced GDP starts with the assumption that a "neutral resource broker" is needed because, in situations where a GDP is called for, the NAS Users are competing for a limited resource (arrival slots at the constrained airport). The critical difference, however, is the application of the following principle: "The "referee" (the FAA) should only control the process at the level of detail necessary to achieve its goals concerned with safety and throughput, thus allowing NAS Users the flexibility to try to achieve their own goals to the extent possible subject to this higher-level constraint." Thus, the critical change is the level of abstraction at which ATCSCC (the neutral referee) controls the process. Instead of determining which specific flights are allocated to specific arrival slots, ATCSCC determines the start and stop times for the GDP, the arrival rate for the constrained airport, and the departure airports to be included in the GDP, and then assigns arrival slots to specific NAS Users (such as a specific airline) rather than to specific flights (using a software tool implemented by Metron Aviation called the Flight Schedule Monitor, or FSM, shown in Fig. 13.1).

The NAS User can then decide for itself which of its flights to assign to one of its arrival slots. An airline might, for instance, have flights from Dallas, Chicago, and Miami all scheduled to fly to LaGuardia in New York, all scheduled to arrive between 1800–1815Z, and all included in a GDP for LaGuardia. If that airline was given "ownership" of an arrival slot at 1855Z in a GDP for LaGuardia, it could assign that slot to any one of these three flights based on their relative importance to that airline's business concerns—considering factors such as passenger connec-

tions, aircraft requirements for later flights out of LaGuardia, crew time restrictions, and aircraft maintenance schedules.

Ration-by-schedule (RBS). In order to accomplish this new process, there are several requirements. First, a traffic manager at ATCSCC has to set the start and end times and the airport arrival rate, and determine which departure airports to include. To do this, ATCSCC solicits input from the relevant FAA facilities and the NAS Users. This interaction takes the form of a teleconference that is held every two hours, during which ATCSCC can propose the use of a GDP and provide an opportunity for input from each of the participants. This decision-making process is supported by a simulation or "what-if" function in FSM that allows the user to input a set of parameters (proposed arrival rate, start and end times, and included departure airports) and view the predicted impact in terms of average and maximum delays, and the potential for a spike in demand at the airport after the end time for the GDP. Window 3 in Fig. 13.1 shows such a demand pattern for SFO in one-hour

FSM Live Data Monitor Mode

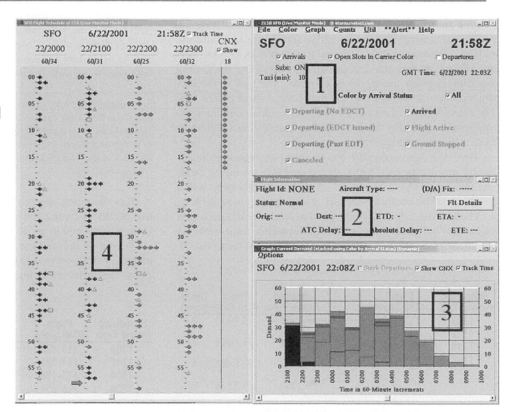

The FSM display consists of four main windows. Clockwise, starting from the upper right, they are:

1. The Control window
2. The Flight Information window
3. The Demand Graph
4. The Timeline

FIGURE 13.1. Screen display from FSM. (Window 1 controls the display of information in FSM; Window 2 provides access to information about specific flights included in the GDP; Window 3 provides an indication of the predicted arrival rate, with the display changing based on the parameters set for the GDP, thus supporting "what-iffing"; Window 4 indicates the predicted arrival times for individual flights.) (Source: Metron Aviation, Inc.)

time bins. The predicted peak demand, for instance, is 45 flights arriving during the 0200Z hour.

After considering these inputs, the ATCSCC traffic manager sets the parameters for the GDP, and lets FSM determine the arrival slot allocations for each NAS User. These allocations are based on a "Ration By Schedule" (RBS) logic. The number of arrival slots each user gets is proportional to its share of the flights that are included in the GDP that were originally scheduled to arrive at the constrained airport. The slots "owned" by that NAS User are initially assigned to its aircraft based on the order of their originally scheduled arrival times (see Fig. 13.2). Thus, if that NAS User originally has 20 flights out of a total of 80 flights that are included in the GDP and are scheduled to arrive at the constrained airport from 1300–1400Z (for a GDP from 1200Z-1900Z), and if the GDP requires this total of 80 to be reduced to 40, then that NAS User gets 10 slots from 1300–1400Z. As a default, the first 10 of its included flights scheduled to arrive after 1300Z are assigned these 10 slots. The arrival times for the remaining 10 flights are pushed back to slots after 1400Z.

Note that, to make this logic work, the GDP needs to cover a period in which, during the GDP or just after its end, there are valleys in the predicted demand that fall below the reduced airport arrival rate (AAR), so that the program doesn't create a huge peak in arrival demand at the end of the GDP. The parameters defining the GDP, along with the slot allocations, are then disseminated to the NAS Users.

Slot swapping. Abstractly, the strategy of controlling the process at a more abstract level (by setting an AAR) and the strategy of using the RBS algorithm to ration slots to NAS Users are critical, as they make it possible to distribute the overall work as relatively independent subtasks. The task of the FAA (the neutral referee) is to set the high-level control parameters. The NAS Users can then work within the resultant constraints to implement plans for their own fleets that are most effective from a business perspective.

The NAS Users accomplish this by swapping flights among assigned arrival slots. Each NAS User can do this by swapping flights among its own slots (see Fig. 13.3), or by making a swap with a slot "owned" by another NAS User, if such a swap is available.

Traditionally, decisions about what to do with a particular flight were handled by the individual dispatcher responsible for that flight. However, this new process involves making global decisions that consider tradeoffs among a number of flights. As a result, instead of requiring some type of potentially time-consuming collaboration among the affected dispatchers in that NAS User's Flight Operations Center, the responsibility for managing participation in a GDP has been assigned to a single senior dispatcher (often a new position referred to as the "Air Traffic Control" or "ATC Coordinator").

Most ATC Coordinators manage the GDP for their fleets using another set of DSSs that allow the ATC Coordinator to:

- Test alternative GDPs (GDPs with different parameters) to see which alternative is best from his fleet's perspective. (This information can be used for discussions during the planning teleconference in which the GDP is proposed, providing input to ATCSCC regarding the GDP parameters that should be used.)

- Swap arrival slots among different flights in the fleet either manually or though an automated process. (It is not unusual to make swaps involving a hundred flights based on passenger connections. This requires a DST that uses an algorithm to estimate the relative importance of the different flights. The ATC Coordinator can, however, constrain or override the solution recommended by the DST.)

- Test for the impact of strategies for canceling flights or delaying their departure times.

Figure 13.3 illustrates swapping flights belonging to the same NAS User. One very important variation of this feature is canceling a flight and then using its slot to swap with another flight. This is important because this provides the incentive for the NAS User to provide the FAA with critical information:

According to the "rules of the game" for GDPs (P. J. Smith et al., 2003), when the NAS User cancels a flight, that User gets to keep its arrival slot and can swap it just like any other slot under its control. In order to make swaps with canceled flights, however, the User must inform the FAA of the cancellation. (accomplished through digital software communications)

This swapping process is further enhanced by a procedure that allows swapping among the different NAS Users. This procedure is referred to as "Slot-Credit Substitution."

Slot-Credit Substitution is best understood in the context of making a decision to cancel a flight. Suppose Airline ABC has flights that arrive every hour on the hour from LGA (LaGuardia

Scheduled Arrival Time	Flight (Call Sign)		Revised Arrival Time	Flight (Call Sign)	Delay (mins)
1300	FLT-123	→	1300	FLT-123	0
1302	FLT-321				
1304	FLT-468	→	1304	FLT-321	2
1306	FLT-654	→	1308	FLT-468	4
		→	1312	FLT-654	6

FIGURE 13.2. Arrival slot allocation using the Ration-By-Schedule algorithm. (Assume that the arrival airport can normally handle one arrival every two minutes. If the GDP reduces the arrival rate by 50% for all arrivals, then the flights are assigned revised arrival slots that keep the original ordering, but show increasing delay over time.)

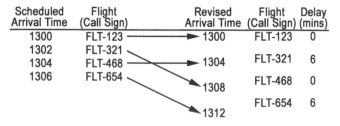

Scheduled Arrival Time	Flight (Call Sign)		Revised Arrival Time	Flight (Call Sign)	Delay (mins)
1300	FLT-123	→	1300	FLT-123	0
1302	FLT-321				
1304	FLT-468	→	1304	FLT-321	6
1306	FLT-654		1308	FLT-468	0
			1312	FLT-654	6

FIGURE 13.3. Arrival slot allocation after swapping FLT-321 and FLT-468. (Note that FLT-468 now has 0 minutes of delay instead of its original 4 minutes of delay.)

Airport) to ORD (Chicago O'Hare). As shown in Fig. 13.4, these flights are scheduled to arrive at ORD at 1000, 1100, 1200, and 1300Z.

Suppose a GDP is issued for ORD that delays these flights, assigning revised arrival times (slots) of 1030, 1200, 1300 and 1400Z, respectively. In principle, each of these flights is still capable of departing on time and therefore arriving at its originally scheduled arrival time if a suitable swap could be found to eliminate its delay due to the GDP. Thus the Earliest Runway Time of Arrival (ERTA) for each of these flights is the same as its originally scheduled arrival time. Each flight could, in principle, therefore be swapped to move its arrival to its ERTA. However, it could not be swapped earlier than its ERTA.

Given this situation, the ATC Coordinator for Airline ABC would be willing to cancel Flight 1 (and put the passengers on the next flight) if this would significantly reduce the delay on later flights. In the case shown in Fig. 13.4, however, even if he cancels Flight 1 the slot released is too early for Flight 2 (a flight with an 1100Z ERTA is too late to make use of a slot with a 1030Z CTA). However, another airline out there wants a slot with a 1030Z CTA, and would gladly give up an 1100Z slot to get it. We know this because the ground delay program software detects that another airline has a flight with an 1100Z CTA and has set its ERTA to 1030Z or earlier.

In this case the steps in the process work as follows:

Airline ABC sends a message saying:

I am willing cancel Flight 1 and give up my 1030Z CTA *only if*

a. I can get a slot at 1100Z and put Flight 2 into it
b. Moving Flight 2 frees up a CTA at 1200Z into which I will put Flight 3
c. Moving Flight 3 frees up a CTA at 1300Z into which I will put Flight 4

The ground delay program software looks for a flight to form the necessary bridge and when it finds the one mentioned above, it approves the entire transaction. (If it could *not* find a suitable flight to form the bridge, it would have rejected the entire transaction and Airline ABC would not have been obligated to cancel Flight 1.)

Note that it is not the setting of the ERTA that kicks off the Slot-Credit Substitution process. (ERTAs are always set as a matter of course, by default set as the originally scheduled or filed arrival time for each flight.) Rather, it is the airline's request for a Slot-Credit Substitution transaction that causes the ground delay program software to search for a suitable bridge.

Overall, then, this process has truly distributed the work by changing the locus of control and the parameters of control of certain decisions. The ATCSCC traffic manager's task is to assign

Flight	Scheduled Arrival Time	ERTA(Z)	CTA(Z)	Delay (mins)
1	1000	1000	1030	30
2	1100	1100	1200	60
3	1200	1200	1300	60
4	1300	1300	1400	60

FIGURE 13.4. Times associated with flights from LGA-ORD with departure delays assigned by a GDP.

the parameters defining the GDP. The ATC Coordinators for each NAS User can then, subject to the constraints imposed by the established GDP, make decisions about which flights to assign to which slots and can even, by setting ERTAs, offer to swap arrival slots with other NAS Users.

Adapting to deal with uncertainty. The GDP process as described thus far basically assumes that the NAS is a deterministic process. If this assumption was true, and if the decision makers had perfect information, then the process would run in a flawless manner, with the right number of flights arriving at the constrained airport at the right time.

However, there are numerous sources of uncertainty in the NAS. The duration and extent of weather constraints are not totally predictable; the estimated time required to clear an aircraft with a blown tire off a runway may be wrong; a flight may miss its assigned departure slot (calculated by working backward from its assigned arrival time and its estimated air time); or a flight may take more or less time than predicted to arrive once airborne. To deal with these uncertainties, certain forms of adaptation have been built into the process.

First, ATCSCC can override the slot allocation for any individual flight, and assign it some other earlier time because of an extenuating circumstance. (In the parlance of the field, such a flight is referred to as a "white hat.") Second, if there are a number of slots before the end of the GDP that have not been filled with active flights, ATCSCC can run a "compression" algorithm that moves later flights up into those slots (subject to the constraints imposed by the ERTAs for those flights and the rule that a flight cannot be moved up to a slot that would require it to depart within the next 30 minutes). Third, ATCSCC can revise the GDP itself because the original predictions or resultant decisions have turned out to be too inaccurate, setting new parameters (such as a new AAR or a new end time), and causing a major shift in arrival slot assignments.

A fourth method for dealing with uncertainty is to pad the original estimate for the AAR. The major reason for doing this is to deal with potential "pop-up" flights. As described earlier, slots are allocated based on the available schedule for arrivals at an airport. However, many general aviation aircraft do not publish a schedule indicating where and when they plan to fly on a given day. They fly on an as-needed basis for their owners. Nevertheless, they still need to be accommodated by the system. To deal with such flights, when setting the AAR using FSM, the ATCSCC traffic manager can indicate the number of pop-ups to plan for, thus leaving some open slots for those flights. When such a flight files its flight plan, it is then assigned the average delay that was assigned to those flights that are now, under the GDP, expected to arrive in the same 15-minute period.

Administrative controls. In addition to these real-time decisions, the overall process is monitored for problems. One such potential problem is the quality of the data submitted by a NAS User. To deal with this issue, three metrics are monitored:

• Time-out cancels. A time-out cancel is a flight that is expected to operate (one that was scheduled, but for which a cancellation message was never submitted), but either never operates, or operates well after its assigned departure slot.

- Cancelled-but-flew flights. A cancelled-but-flew flight is a flight that the participant cancels but that ends up operating.
- Undeclared flights. An undeclared flight is a flight belonging to an NAS User that normally submits a schedule of its expected flights for each day, and that operates without prior notice to the system.

NAS Users must demonstrate acceptable performance in terms of these metrics in order to continue to participate in using the slot-swapping functions provided by GDPs. This helps to ensure sufficient data integrity for the overall process to function as planned.

A second area requiring administrative oversight is a concern over gaming. Suppose, for example, that SFO (San Francisco) has a GDP in effect because of reduced visibility at that airport. A flight that wants to fly to SFO should take a delay based on this GDP. However, if it files for OAK (Oakland, an airport very close to SFO) instead, and then requests an amendment to land instead at SFO once airborne, it has violated the spirit of the GDP program. ATCSCC monitors for such practices, and takes administrative actions to prevent them in the future.

Evaluation of the Implemented Solution

There are two components of the benefits from the Enhanced Ground Delay Program that has been described above in terms of its major features. The first is the benefit to the NAS Users derived from swapping flights in order to reduce delays for their higher-priority flights. These savings are clearly very substantial, but would be very difficult and time consuming to estimate. As a result, to date no study has been conducted to put a dollar figure on that class of benefits.

The second benefit arises from the use of the compression process to fill in unused arrival slots, therefore reducing overall delays. Figure 13.5 shows the frequency of use of the Enhanced GDP process since the introduction of the compression process in 1999. Note that, over the past two years, it has not been un-

common to run as many as 100 GDPs per month. Table 13.1 then provides an estimate of the minutes and dollars saved by the compression process alone, using a conservative average cost per minute of operation for a flight of $42. These estimates suggest that the use of the compression process alone reduced delay by 28,188,715 minutes from March 18, 1999 to June 30, 2005, thus saving over 1 billion dollars.

Case Study A—Conclusions

The Enhanced GDP illustrates a deliberate effort to design a distributed work system. The design strategy is to have one party (ATCSCC as a neutral referee) set a high-level constraint (the airport arrival rate), and to then let the other involved parties (the NAS Users) find solutions that meet their business needs as effectively as possible subject to that constraint (by swapping slots among flights to favor the highest-priority flights). Taking this approach, knowledge and data remain distributed, matching the locus of control for the various decisions that need to be made.

This distribution also serves to reduce the cognitive complexity of the task confronting each individual. The ATCSCC traffic manager doesn't need to know anything about user priorities, and the ATC Coordinator for an NAS User doesn't need to become an expert at setting airport arrival rates. Finally, because exceptions are necessary to override poor decisions when they are proposed by someone due to a slip or due to insufficient access to the relevant knowledge or data, there must be a mechanism to support interactions among the different participants in order to exchange the necessary information and modify those decisions.

To make this process work, however, certain other requirements arise during its design:

- To make it politically acceptable, non-participants (NAS Users who do not choose to participate in slot swapping, and so on) must not be penalized. Such non-participants are initially assigned slots just like everyone else, but their flights are

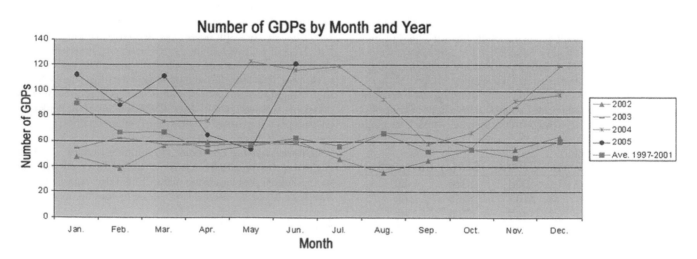

FIGURE 13.5.   Frequency of use of the enhanced GDP by year.

TABLE 13.1. Time and Money Savings from the Use of the Compression Process in GDPs since 1999

- Data Included in Analysis:
  - All compression from Mar. 18, 1999 – June 30, 2005
  - 3,573,796 flights included
  - 28,188,715 total minutes reduced
  - $1,183,926,030 (@$42/min) dollars saved
- Departure Delay Performance
  - Average reduction: 7.7 minutes
- Percentiles:
  - Reductions greater than 15 minutes: 12.3% (440,889 flights)
  - Reductions greater than 30 minutes: 6.1% (218,886 flights)
  - Reductions greater than 45 minutes: 3.8% (137,424 flights)
  - Reductions greater than 60 minutes: 2.6% (93,558 flights)

never moved to a later slot as part of any swapping. (They may actually benefit from compression, however, if they are moved up to an earlier slot that is empty.)

- There must be an incentive for information sharing.
- DSTs must be available to provide:
  - digital communication;
  - simulation of the predicted impact of alternative decisions (what-if capabilities);
  - algorithmic support for identifying, evaluating, and implementing complex decisions involving large numbers of flights (swapping);
  - information displays that allow the user to understand the situation as relevant to that user's responsibilities, and to effectively evaluate the solution recommended by the technology;
  - control by the user to override the DSTs recommendations in cases where the technology exhibits brittleness.
- Organizational changes in roles and responsibilities (e.g., new specialists at ATCSCC with the knowledge and skills to develop GDPs; ATC Coordinators for the NAS Users to look at tradeoffs in delaying alternative flights).
- The overall procedures must support adaptation, so that decisions can be modified as the scenario evolves.

In summary, this case study serves to illustrate the importance of looking at design from a broad systems perspective when dealing with a distributed work environment, considering not just the detailed design issues necessary to implement a given DST, but also considering the alternative strategies for distributing the work and the organizational changes required to match people with certain skills and knowledge with corresponding responsibilities, and to make the system resilient in the face of slips, mistakes, and unanticipated events.

Case Study B—Human Interactions with Autonomous Mobile Robotic Systems

A rapidly emerging technical area is that of autonomous mobile robotic systems. An autonomous mobile robotic system can perform complex tasks for extended periods with no human intervention, and thus represents a large step beyond remote control. But no system is truly autonomous—interaction with its human owners and operators is inevitable. This case study illustrates the issues and needed approaches to provide an effective and responsible relationship between a human supervisor and a team of autonomous mobile robotic systems.

Remote control of mobile robotic vehicle systems has become commonplace over the past two decades. Remote control depends on a human operator to visualize the situation around the remote vehicle and to provide control inputs to maneuver the vehicle in space and to manage its sensors and manipulators to achieve the desired mission. Examples of remote control of robotic vehicles include deep ocean exploration and repairs, bomb disposal operations, and handling of very hazardous materials such as radioactive waste.

While remote control has been successfully used for many situations in which it would be highly undesirable to attempt to send a human, the requirement to maintain continuous information and control flow between the vehicle and the human operator has been limiting. Over the past decade, significant investment has been made in research and development of autonomous mobile robotic systems with promising results for land, air, and undersea vehicles.

The study of human interaction with remotely operated and autonomous systems has a long history. The classic work was by Sheridan and Verplank (1978) because of their study of issues for remote operations of undersea robotic vehicles, and has been recently updated in Parasuraman et al. (2000). This work defines ten levels of autonomy, ranging from fully manual to fully autonomous. Most significantly, the levels are distinguished by defining the nature of the interactions between the human and the machine.

The highest six levels are particularly relevant to this case study:

10. The computer decides everything, acts autonomously, ignoring the human
9. (The computer) informs the human only if it, the computer, decides to
8 (The computer) informs the human only if asked, or
7. (The computer) executes automatically, then necessarily informs the human, and
6. (The computer) allows the human a restricted time to veto before automatic execution, or
5. executes that suggestion if the human approves, or . . . (p. 933)

Thus, like many other issues in human-computer interaction, creating a decision-support system for control of a team of autonomous entities requires careful attention to the cognitive characteristics of humans.

The Application Area

During the mid-1990s, as a result of the earlier successful demonstrations of intelligent systems technology within the military, key technology decision makers with the defense industry research and development community recognized that it was possible to create an unmanned combat air vehicle.

The unmanned combat air vehicle, or UCAV, was envisioned as a platform for performing very dangerous missions or, alternatively, very routine missions. An example of a dangerous mission for modern combat aircraft is *suppression of enemy air defenses*, or SEAD. In the SEAD mission, a team of UCAVs would be assigned the role of finding and attacking enemy surface-to-air missile sites, so that the airspace over the battle area would be safe for manned aircraft.

The development of unmanned reconnaissance aircraft has made steady progress, resulting in the deployment of the Predator and Global Hawk unmanned aircraft systems. While the Predator is still a remote-controlled system, flown by a human pilot at a ground station, the Global Hawk is a true autonomous aircraft. These successes raised confidence that the UCAV concept was feasible as well.

The demands of combat aviation are, however, quite different from reconnaissance and surveillance. Reconnaissance missions normally are single-aircraft missions, planned so that the aircraft avoids potential threats. The typical mission also has well-defined locations where information is sought, and success depends on collecting the information without being detected.

Combat aviation, in contrast, is conducted in an environment of uncertainty and hostility. The enemy actively seeks to deny information about his location and his intent by blocking and disrupting sensors and communications, and by using decoys. The enemy is also using his defensive weapons to find and destroy his opponent's forces. In this environment, a team of combat aircraft must operate successfully with degraded information and limited contact with human supervisors. The team must be able to find and attack the enemy while avoiding damage to noncombatants. At the same time, the team must be able to recognize and remediate system degradations and failures without assistance.

For an unmanned combat aircraft, the difficulty of this environment is higher still. The human supervisor no longer has the direct sensory perceptions available to a pilot flying on the aircraft. His entire understanding of the situation depends on limited sensor data, and his ability to control the responses to the situation is hindered by time lags and dropouts of data due to remote data transmissions. To be successful, the UCAV concept needs to be robust and flexible in the face of a decidedly hostile environment.

The Need for Decision Aiding

While the current generation of unmanned reconnaissance and surveillance aircraft has been successful, this success has largely been dependent on the skill and direct activity of their human operators. Several humans are required to operate the Predator, which is actually remotely flown and not an autonomous system. Similarly, Global Hawk requires a team of humans that continuously monitors a single aircraft throughout its mission, as well as a significant mission-planning effort prior to launch.

If unmanned combat air vehicles are to be useful, a significant reduction in human-operator effort is required. Modern manned aircraft, such as the F-22 Raptor and the newly designed F/A-35 Joint Strike Fighter, are flown by a single human pilot. While training of a military combat pilot is expensive and takes considerable time, UCAVs will not be accepted as an adjunct to manned aircraft unless the mission can be reliably performed and both

(a) the cost of the unmanned aircraft is less than a manned aircraft and (b) the cost of the human operators needed for one unmanned air vehicle is significantly less than the cost of training and supporting a pilot for a manned aircraft.

As a result, the design goal for unmanned combat aircraft is typically stated as a single operator controlling a flight of four or more UCAVs in a combat setting. Given the issues that confront a human operator of an unmanned combat aircraft mission, researchers (Elmore, Dunlap, & Campbell, 2001; Geddes & Lee, 1998) have recognized that decision aiding is needed across a broad front:

- Situation awareness. The human operator must understand the hostile external environment, the offensive and defensive postures of all of the aircraft under his control, and the internal status of all of the aircraft systems for all of the aircraft under his control. This is a massive amount of information to digest, with the potential for sudden and significant changes at any moment.

- Planning and resource commitment. The UCAVs are assigned a mission, and the human operator is responsible for managing the frequently changing tactical behaviors needed to meet the mission goals. Generating alternative short-term tactical plans, evaluating the alternatives, selecting the most effective plans and committing the aircraft resources to the plans are continuous requirements.

- Coordination and acting. As plans are generated and selected, the human operator must also ensure that the plans are coordinated across the flight, free of conflicts, and performed as expected. At any point in time, multiple plans across subsets of the UCAVs in the flight will be active and will require monitoring.

- Understanding and responding to irregularities. Combat aircraft, sensors, and weapons are extremely complex, and, as a result, behavior anomalies are common. In addition, the enemy seeks to exploit vulnerabilities in systems to create irregularities. Recognizing abnormal situations and responding in a timely way are frequent requirements in combat.

In meeting these cognitive needs of the human operator, the designer must be careful not to create a "black box" automation that blocks human involvement (Wiener & Curry, 1980). More recently, a related but distinctly different challenge has emerged—that of "point intelligence" that is not as robust and flexible as it appears (Geddes, 1997b; Hammer & Singletary, 2004). Point intelligence is designed to solve a particular narrow problem, such as assessment or planning, but does not have broad competency. As Hammer and Singletary (2004) pointed out, this leads to behavioral breakdowns between the human and computer.

The Design Solution

From 2002 through 2005, the U.S. Air Force Research Laboratories sponsored the development and evaluation of an approach to supporting human management of autonomous unmanned combat air vehicles (Atkinson, 2003, 2005). This system, known as AUCTIONEER, combines a cognitive model of decision making with a market auctioning system. The result is a system that is capable of seamless operation across Levels 5–9 of Sheridan's classification of levels of autonomy.

As discussed by Atkinson, the functional organization of a complex vehicle, such as a combat aircraft, normally has a number of layers, as shown in Fig. 13.6. The lowest layer acts to control the physical parts of the aircraft, such as its flight control surfaces, engine control surfaces and fuel metering, sensors, and weapons. The next layer contains functions that integrate the behaviors of the lowest layer into more complex behaviors to perform tasks. For example, the navigation system may act to receive position and velocity data from the sensors and direct the flight controls and engines to new settings in order to fly the aircraft toward a desired waypoint in space. This second layer contains many functions that are typically implemented as conventional automation, with limited provisions for human interactions. Even so, a modern manned combat aircraft cockpit may have over 350 controls for a pilot to manipulate to configure his vehicle, sensors, and weapons systems.

The third layer in Fig. 13.6 provides the overall mission organization of the aircraft's behaviors, so that the intended utility of the system can be achieved. For manned aircraft, this layer is mostly performed by a human pilot. The typical training time for a carefully selected person to reach threshold proficiency as a combat aircraft pilot is two years, and expertise normally requires 10 years of continued training and practice.

The fourth layer, the coordination layer, is often performed by humans who are distantly located as part of a command-and-control structure. With the present generation of manned aircraft and command and control systems, little "computer-to-computer" coordination is available. Most coordination is performed human-to-human.

If unmanned vehicles are to meet their economic goals by using significantly fewer human operators, the mission layer and the coordination layer must both be substantially more autonomous. Yet, because an unmanned combat system must also be able to apply lethal weapons, uninterrupted autonomy for any significant periods of the mission may compromise safety and human responsibility. As a result, operations at Level 10 of

Sheridan's hierarchy are strongly undesirable. However, seamless operation across Levels 5–9 of Sheridan's model of autonomy is strongly required.

The AUCTIONEER system is based on the architecture from Elmore et al. (2001), which has evolved directly from the original Pilot's Associate architecture (Lizza & Banks, 1991; Rouse, Geddes, & Hammer, 1990) of the early 1990s. This architecture uses a distributed system of cognitive engines, providing full situation assessment, planning, coordinating, and acting for each entity in the system, both human and UCAV.

To meet the need for situation awareness, the AUCTIONEER system provides for a distributed model of beliefs that is shared across the unmanned members of the flight and the human operator. The belief model, embodied as a concept graph, allows each aircraft to combine information from other aircraft with its own and with that provided by other sources, as well as input from the human. The belief model also provides for active monitoring for specific events that are coupled to the ongoing and planned tasks. When a task is initiated, specific monitors are activated in the belief model. When monitored events are detected, existing tasks may be modified or new tasks activated by the planning functions.

Distributed planning for AUCTIONEER is performed using a market-based auctioning system that has no centralized auction manager service. Each member of the flight can initiate an auction for a task that it determines needs to be performed (because of a monitored event), and any members of the flight, including the auction initiator, can bid on the task. More than one auction can be proceeding simultaneously. The auction mechanism does not assume that all of the members can receive the auction announcement, or that all of the bids can be received in return because of communications dropouts.

The auctioning system uses a carefully structured representation of the tasks that may need to be performed during the mission. The structure is a "Plan Goal Graph" consisting of a directed acyclic graph with alternating plan and goal nodes. The Plan Goal Graph defines the network of causal relationships between activities described by plan nodes and goal nodes that reflect what each activity is expected to accomplish. The Plan Goal Graph has the additional capability to be used as a model of the intentions of groups of active entities (Geddes, 1997a), and is used in AUCTIONEER to both create the tasks to be auctioned and to interpret the actions of the members of the flight and the human operator as the tasks are performed.

Similar studies by Stenz, Dias, and others at Carnegie Mellon (Stenz & Dias, 1999; Zlot & Stenz, 2003) have shown that market-based auctioning on tree-structured tasks has many computational advantages over either the auctioning of unstructured primitive tasks or the use of algorithms that attempt optimal task allocations. However, successful use of an auctioning system must provide mechanisms for the human operator to understand the processes and outcomes of the auctions and to intervene in auctioned tasks without disrupting the benefits of using market-based auctions as a means for resource allocation.

For example, in Fig. 13.7, the user can see the approved paths and proposed paths of vehicles in the flight; review past, future, and proposed tasking for each target or for each vehicle; and edit proposed and active tasks if desired.

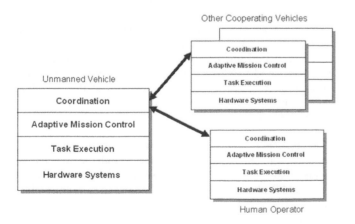

FIGURE 13.6. Behaviors of complex systems are produced in layers.

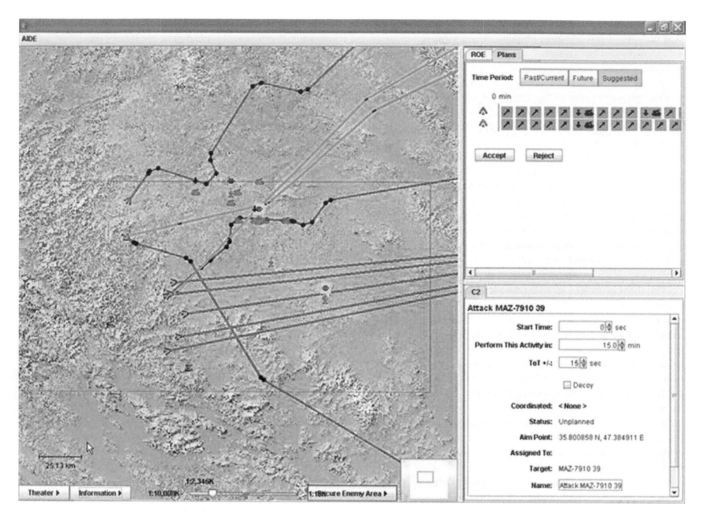

FIGURE 13.7.   The set of UCAVs have begun execution of one coordinated attack while proposing a second coordinated attack against a different target complex. The routes in the lighter grey
(magenta in true color) are for the proposed attack, while the darker routes are the currently
accepted paths. The group of straight paths in the lower right of center are for vehicles that have
not yet won any of the auctioned tasks.

Results

The AUCTIONEER study is particularly useful as a case study
of the effects of opening the "black box" of automation because
the system was studied with both passive and active operator
involvement.

Performance.   In the first study results (Atkinson, 2003),
the auction processes were run without human intervention.
This study established the basis for measurements of effectiveness for the auction environment as a fully autonomous process.
This first study found that the auction process was an effective
method for allocating resources in terms of mission accomplishment. More targets were attacked in a shorter time with
fewer abandoned tasks than what could be achieved by unaided
human control of the vehicles. A second concern was timeliness, because the SEAD mission requires that targets are attacked or suppressed before they can disrupt the friendly forces.

The auction process also resulted in shorter average timelines
to attack targets.

The most significant findings dealt with the stability of tasking. Because the winner of an auction might later find a better
target before finishing the current task, winners could become
auctioneers and "sell" their current tasks to another. Atkinson
found similar results to Zlot and Stenz: that this reselling
process did not result in churning of tasks and in fact led to improved outcomes.

Acceptance.   The second study (Atkinson, 2005) added
the human operator back into the mission, with the ability to intervene in the auction process as described above. In this study,
the overall findings showed that the operator's understanding
and effectiveness were limited by the lack of insight into the
auction process. Even though the underlying decision-support
model used appropriate human cognitive constructs, the system was still a black box to the human because it did not reveal

its internal model at the user interface. Atkinson and her colleagues provided a recommended set of features to open the black box of the auction process.

Communication of beliefs. The operator should be able to view the beliefs held by each member of the flight, and act to modify those beliefs when required. This provides assurance that the members of the flight are acting with a common assessment of the situation and are exploiting all available information.

Communication of intentions in the context of a bid. If the only information available to the operator about an offered bid for a task is its total cost, the operator loses important information about the context of the bid. By allowing the operator to view the chain of intentions that led to the bid, as represented in the Plan Goal Graph, the operator would be able to understand how the bid was composed, and to be assured that the bid was proper.

Communication of direct commands. Operator commands need to be recognized as distinct from task-auction requests. This would allow the operator to act with assurance that his commanded action would be carried out by the aircraft and resource that was directed, rather than incurring an auction response of uncertain outcome. Any additional task reallocations that result from the operator's commands may be automatically handled as auctions.

The findings in this case study underscore two important issues in designing intelligent decision-support systems: (a) the importance of avoiding "black box" automation for high-stakes environments, and (b) the advantages of designing and building decision-support systems that are based on models of human cognitive processes.

While it may seem easier to design and implement approaches that are mathematically elegant, this can result in an automation "black box" that undermines human understanding. Similarly, efforts to reduce the scope of the decision support to solve just one part of the operator's cognitive problem may result in point intelligence that can also degrade performance.

Humans learn to interact with other humans as part of the development of social skills for effective behaviors within their cultures. In cultures based on trust and mutual accomplishment, the ability to communicate about beliefs, goals, and intentions is an important skill. This case study suggests that creating automated behavioral systems that can use explicit models of beliefs, goals, and intentions as the basis for communication with humans is an effective way to open the black box.

Case Study C—Interactive Critiquing as a Form of Decision Support

This case study emphasizes the challenge for interaction designers in dealing with the use of decision-support tools that have the potential for brittle performance due to known or unknown limitations. While there are no perfect solutions to this problem, there are a number of approaches to help reduce its impact.

This case study looks at a tool that incorporates three complementary approaches to the design of an expert system in order to improve overall performance, while reducing the potential impact of brittle performance by the expert system. The first approach to deal with the impact of brittle performance by an expert system is to design a role that encourages the user to be fully engaged in the problem solving, and to apply his knowledge independently without being first influenced by the software. In this case study, the approach explored to achieve this is the use of the computer as a critic, rather than as the initial problem-solver (Fischer, Lemke, Mastaglio, & Morch, 1991; P. Miller, 1986; Silverman, 1992; Tianfield & Wang, 2004). Thus, instead of asking the person to critique the performance of the software, the computer is assigned the role of watching over the person's shoulder. Note that this is more accurately described as having the design team try to anticipate all of the scenarios that can arise, and then, for all of those scenarios, trying to incorporate the knowledge necessary to detect possible slips or mistakes on the part of the user and to provide alerts and assistance in recovering (Pritchett, Vandor, & Edwards, 2002; Wogalter & Mayhorn, 2005).

The second approach to reduce sensitivity to brittleness in the computer's performance is to incorporate metaknowledge into the expert system that can help it to recognize situations in which it may not be fully competent. By doing so, the software may be able to alert the person to be especially careful because the computer recognizes that this is an unusual or difficult case.

The third approach to deal with brittleness is to develop problem-solving strategies that reduce susceptibility to the impact of slips or mistakes. In the software discussed in this case study this is accomplished by incorporating a problem-solving strategy that includes the collection of converging evidence using multiple independent problem-solving strategies and sources of data to arrive at a conclusion.

What follows, then, is a discussion of a specific expert system that incorporates these three strategies for dealing with brittleness. Empirical testing of the system suggests that this approach can significantly enhance performance, even in cases where the software is not fully competent (P. J. Smith & Rudmann, 2005).

The Application Area

The specific problem area considered in this case study is the design of a decision-support tool to assist blood bankers in the identification of compatible blood for a transfusion. One of the difficult tasks that blood bankers must complete as part of this process is the determination of whether the patient has any alloantibodies present in his blood serum, and if so, what particular antibodies are present.

Diagnosis as a generic task. Abstractly, this is a classic example of the generic task of abduction or diagnosis (J. Josephson & S. Josephson, 1994; Psillos, 2002). It involves deciding which tests to run (what data to collect), collecting and interpreting those data, and forming hypotheses, as well as deciding what overall combination of problem-solving strategies to employ. Characteristics of this generic task that make it difficult for people include:

- The occurrence of multiple-solution problems, where more than one "primitive" problem is present at the same time;
- The occurrence of cases where two or more "primitive" problems are present, and where one problem masks the data indicative of the presence of the other;
- The existence of "noisy" data;
- The existence of a large "data space" where data must be collected sequentially, so that the person must decide what data to collect next and when to stop;
- The presence of time stress, in which an answer must be determined quickly (Elstein, Shulman, & Sprafka, 1978; P. J. Smith et al., 1998).

In this application, the "primitive" problems are the individual antibodies that may be present in the patient's blood serum. Time stress can arise when the patient is in an emergency and needs a transfusion quickly.

Sample problem. To illustrate the nature of this diagnosis task, consider the following partial description of an interaction with the decision-support tool AIDA (the Antibody IDentification Assistant) that is the focus of this case study (Guerlain et al., 1999). Initially, the medical technologist needs to determine whether the patient is type A, B, AB, or O, and whether the patient is Rh positive. Then the technologist determines whether the patient shows evidence of any autoantibodies or alloantibodies. As part of this determination, displays like the one shown in Fig. 13.8 are provided by AIDA.

To make visual scanning easier on this data display, some of the rows have been highlighted by the technologist in yellow (shown in light gray in this black-and-white version). In addition, to reduce memory load, the technologist has marked a number of intermediate conclusions, indicating that the patient could have antibodies against the C, E, M, and N antigens (indicated by the technologist in orange in the actual system, but left white in Fig. 13.8), that the f, V, Cw, Lua, Kpa, and Jsa antibodies are unlikely (marked in blue in the actual system, but light gray in Fig. 13.8), and that the other antibodies shown as the labels

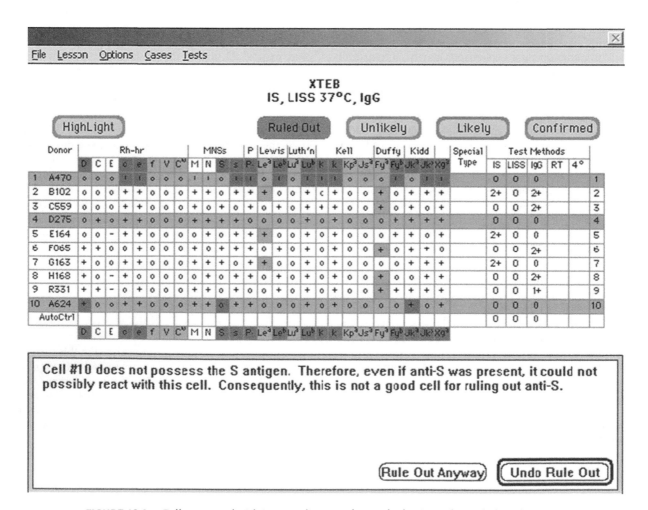

FIGURE 13.8. Full test panel with intermediate results marked using color coded markers provided by AIDA (shown here in black and white). In this case, AIDA has interrupted the user to suggest that an error may have been made in ruling out anti-S.

across the top row can be ruled out (marked in green in the actual system but shown in darker gray on this black-and-white version.) These color-coded intermediate answers are transferred to all other data displays to reduce the memory load for the technologist. Figure 13.8 also provides an example of a critique that AIDA has provided in response to an error made by the technologist in marking anti-S as ruled out. This critique was generated by the expert model (rule-based system) underlying AIDA, which monitors all of the test selections and markings made by the technologist as he solves the case using the interface provided by AIDA.

Thus, this figure serves to illustrate that:

- The technologist is responsible for completing the analysis, deciding which tests to run and which intermediate and final conclusions to reach, and is thus very engaged in the task;

- The computer provides an interface that makes it easy for the technologist to select the tests to run and to view the resultant data (using the highlighting to make visual scanning easier), as well as to remember intermediate conclusions (using color-coded markings to indicate these conclusions);

- Although the primary motivation for the technologist in marking the data forms is to make his task easier, when he does so the computer is provided with a great deal of data regarding what the user is thinking, and can use those data to make inferences based on its expert model about when to provide a critique cautioning the user that he may have made a slip or mistake.

The need for decision aiding.   Initial laboratory and field studies indicated that the task of determining the alloantibodies present in a patient's blood is a difficult one for many technologists. A variety of causes of errors were observed, including slips, perceptual distortions, incorrect or incomplete knowledge (Guerlain et al., 1999), and cognitive biases (Brachman & Levesque, 2004; Bradfield & Wells, 2005; Fraser et al., 1992; Gilovich et al., 2002; Haselton et al., 2005; Kahneman & Tversky, 2000; Plous, 1993; Poulton, 1989).

The Design Solution

It was the judgment of the design team that this task was sufficiently complex that, given the available development resources, it was unlikely that all of the possible scenarios could be anticipated and dealt with by the design team. It was also noted that, for this task, the cost of an erroneous diagnosis was potentially high. Thus it was decided that, instead of automating the task, a decision-support system should be developed that would keep the person very much engaged in the task, and provide other safety nets to reduce the chances of error for those cases in which the computer's knowledge was incomplete. This conclusion was reinforced by a preliminary study regarding the impact of role on diagnostic accuracy in antibody identification. Guerlain et al. (1996) showed that, when the user was asked to critique the computer's performance, instead of having the computer critique the user's performance, the final answer was wrong 31% more often in cases where the computer's knowledge was incomplete when, in that case, the person was assigned the role of critic rather than having the computer critique the person.

Critiquing—additional design considerations.   The literature provided additional guidance in deciding whether and how to develop this software to play the role of a critic. For example, P. Miller (1986) developed a prototype system called ATTENDING that focused on anesthesiology. Based on studies of its use, Miller suggested that critiquing systems are most effective in applications where the user has a task that is frequently performed, but that requires the person to remember and apply a great deal of information in order to complete a case. Miller's conclusion was that, on such tasks, the person is more susceptible to slips and mistakes and would therefore benefit significantly from the decision-support system.

A second consideration was how intrusive the interactions with the knowledge-based system would be for the user. A number of researchers have suggested that an interface that requires that the user spend significant time entering data and informing the computer about what he has concluded is too cumbersome, and will therefore be unlikely to be adopted in actual practice (Berner, Brooks, R. Miller, Masarie, & Jackson, 1989; Harris & Owens, 1986; Miller, 1986; Shortliffe, 1990; Tianfield & Wang, 2004).

A third consideration was concern over the potential for complacency if the person played the role of critic, letting the computer complete an initial assessment and then having the person decide whether to accept this assessment. Parasuraman and Riley (1997) have shown that, in such a role, there is a risk that the person will become overreliant on the computer, and will not adequately apply his knowledge in completing the critique (Metzger & Parasuraman, 2005). (Note, however, that a person could become overreliant even with the roles reversed, as the person might start to get careless and assume the computer will always catch his slips. Administrative controls, based on regular monitoring of the person's performance, might help reduce such complacency, but this is an as-of-yet unexplored aspect of the use of critiquing systems.)

A final consideration was the timeliness of critiques. Waiting for the user to reach a final answer before providing a critique, as has been the case for many critiquing systems, is potentially inefficient and objectionable, as the user may have invested considerable time and effort in arriving at a mistaken answer that the computer could have headed off earlier in the person's problem-solving process. Furthermore, if the critique is given well after the user has made a slip or mistake, it may be difficult for him to remember exactly why he arrived at that intermediate conclusion (thus making it harder to decide whether to accept the computer's critique). This consideration therefore suggests the need for an interface that provides the computer with data about the user's intermediate conclusions rather than just the user's final answer, so that critiquing can be more interactive.

Based on these considerations, AIDA was developed as a critiquing system that supported the user as he completed the antibody identification task. In order to provide immediate, context-sensitive critiques, an interface was developed that encouraged the user to mark intermediate conclusions on the screen.

As suggested earlier, these markings in fact reduced the perceptual and memory loads for the user, thus implicitly encouraging this form of communication and allowing the computer to detect and respond immediately with context sensitive critiques to potential slips and errors made by the person.

Complementary strategies to reduce susceptibility to brittleness.

The considerations outlined above focused on how to keep the person engaged in the diagnosis task, and how to avoid triggering the cognitive biases that can arise if the computer suggests an answer before the person has explored the data himself (Larson & C. Hayes, 2005; P. J. Smith, McCoy, & Layton, 1997). AIDA also incorporated two additional design strategies to reduce the potential for error. One was the incorporation of metaknowledge into the knowledge-based system. The other was to include a problem-solving strategy that was robust even in the face of slips or mistakes by either the person or the computer (e.g., the design team).

Metaknowledge was included to help the computer identify cases in which its knowledge might be incomplete. Such rules were developed by identifying the potential weak points in the computer's problem-solving process. An example was the use of thresholding by AIDA when applying rules such as:

If a test cell's reactions are 0 on a test panel for all three of the testing conditions (IS, LISS and IgG) as they are for the first test cell (Donor A478) shown in Figure 13.8, and if e is present (shown by a + in the column labeled e in the row corresponding to the first test cell) on that test cell but E is not (shown by a 0 in the column labeled E in the row corresponding to the first test cell), then mark e as ruled out.

This heuristic usually produces the correct inference. Since it does not directly reason with some form of probabilistic reasoning (Bochman, 2004; Pearl, 1988; Shafer, 1996; Shafer & Pearl, 1990), however, it can lead to ruling out an antibody that is actually present in the patient's serum with no consideration of the level of uncertainty. This is most likely to happen when the reaction strengths are weak. Thus, AIDA was provided with a rule that monitored for weak reactions on cells, and when this situation was detected and the user was observed trying to complete rule-outs without first enhancing the reactions with some alternative test phase, the system cautioned the user. In this way, AIDA put the user on alert to be especially careful in applying his normal strategies and rules.

A second strategy incorporated into AIDA as protection against brittleness was to monitor for the collection and consideration of converging evidence by the technologist. This problem-solving strategy was observed in use by one of the experts involved in the development of AIDA. She used this strategy to catch her own errors and those of the technologists working under her supervision. The strategy was based on the assumption that any one data source or line of reasoning could be susceptible to error, and that it is therefore wise to accept a conclusion only if independent types of data (test results that are not based on the same underlying data or process) and independent sets of heuristics or problem-solving strategies have been used to test that conclusion. Based on this expert strategy, AIDA monitored the user's problem-solving process to see whether such converging evidence had been collected prior to reaching a final an-

TABLE 13.2. Percentage of Blood Bankers Arriving at the Wrong Final Answer on Four Test Cases

Group	Control Group	Treatmen
Posttest Case 1	33%	0%
Posttest Case 2	50%	19%
Posttest Case 3	38%	0%
Posttest Case 4	63%	0%

swer. If not, AIDA cautioned the user and suggested types of converging evidence that could be collected.

Evaluation of the Implemented Solution

From the standpoint of human-computer interaction, the key question is how effectively this human-machine system performs using this cooperative problem-solving paradigm (Erkens, Andriessen, & Peters, 2003). To gain insights into this, an empirical study was conducted using AIDA.

This study of 37 practitioners at seven different hospitals found that those blood bankers using AIDA with its critiquing functions turned on made significantly fewer errors ($p < .01$) than those that used AIDA as a passive interface. Errors in the final diagnoses (answers) were reduced by 31–63%.

On those cases where AIDA was fully competent (Posttest Cases 1, 2, and 4), errors (% of technologists getting the wrong final answer) were reduced to 0% when the critiquing functions were on (Treatment Group). In the case where AIDA was not fully competent (Posttest Case 2), AIDA still helped reduce errors by 31%.

These empirical results support the potential value of interactive critiquing, supplemented with the metaknowledge and problem-solving strategies that were embedded in the software, as a design strategy for applications in which concerns regarding the potential brittleness of the technology are deemed significant.

CONCLUSION

One goal of this chapter has been to outline different conceptual approaches to support the design of DSSs, including the use of cognitive task analyses and work-domain analyses. A second goal has been to review relevant knowledge about the psychologies of both the designers and users of DSSs, identifying important design questions that need to be considered in the development of such software systems. Finally, the chapter emphasizes the need to take a broader systems perspective when deciding how to integrate DSSs into a work environment, as there are a number of other high-level design parameters (such as deciding how to distribute the work among a team of people and machines) that may offer alternative opportunities for effectively incorporating DSSs into the overall system.

References

Amalberti, R. (1999). Automation in aviation: A human factors perspective. In D. Garland, J. Wise, & V. Hopkins (Eds.), *Handbook of aviation* (pp. 173–192). Mahwah, NJ: Lawrence Erlbaum.

Anderson, J.R. (1993). *Rules of the mind*. Hillsdale, NJ: Lawrence Erlbaum.

Anderson, J. R., Bothell, D., Byrne, N., Douglass, S., Lebiere, C., & Qin, Y. (2004). An integrated theory of the mind. *Psychological Review, 111*(4), 1036–1060.

Andes, R. C., & Rouse, W .B. (1992). Specification of adaptive aiding systems. *Information and Decision Technology, 18*, 195–207.

Annett, J., Duncan, K. D., Stammers, R. B., & Gray, M. J. (1971). *Task analysis*. London: HMSO.

Annett, J., & Stanton, N. (2000). *Task analysis*. London: Taylor & Francis.

Atkinson, M. L. (2003). Contract nets for control of distributed agents in unmanned air vehicles. *Proceedings of 2nd AIAA Unmanned Unlimited Conference*, San Diego, CA.

Atkinson, M. L. (2005). Adding an operator into a task auction paradigm. *Proceedings of the AIAA InfoTech Conference, Workshop and Exhibit,* Arlington, VA, 29–35.

Baddeley, A. (1998). *Human memory: Theory and practice*. Boston: Allyn and Bacon.

Baddeley, A. (1999). *Essentials of human memory*. London: Taylor & Francis Group.

Baecker, R., Grudin, J. Buxton, W., & Greenberg, S. (1995). *Readings in groupware and computer-supported cooperative work: Assisting human-human collaboration*. San Francisco: Morgan Kaufmann.

Bainbridge, L. (1983). Ironies of automation. *Automatica, 19*, 775–779.

Bar-Yam, Y. (2003). *Dynamics of complex systems: Studies in nonlinearity*. Nashville, TN: Westview Press.

Bar-Yam, Y. (2005). *Making things work: Solving complex problems in a complex world*. Cambridge, MA: NECSI Knowledge Press.

Beach, L. R., & Connolly, T. (2005). *The psychology of decision making: People in* organizations (2nd ed.). Thousand Oaks, CA: Sage Publications, Inc.

Bennett, K. B., & Flach, J. M. (1992). Graphical displays: Implications for divided attention, focused attention, and problem solving. *Human Factors, 34*(5), 513–533.

Berner, E., Brooks, C., Miller, R., Masarie, F., & Jackson, J. (1989). Evaluation issues in the development of expert systems in medicine. *Evaluation and the Health Professions, 12*, 270–281.

Billings, C. E. (1997). *Aviation automation: The search for a human-centered approach*. Hillsdale, NJ: Lawrence Erlbaum.

Bisantz, A. M., Roth, E., & Brickman, B. (2003). Integrating cognitive analysis in a large-scale system design process. *International Journal of Human-Computer Studies, 58*, 177–206.

Bittner, K. (2002). *Use case modeling*. Harlow, England: Addison-Wesley Publishing.

Bjork, R. A. (1999). Assessing our own competence: Heuristics and illusions. *Attention and Performance, 17*, 435–459.

Blank, H. (1998). Memory states and memory tasks: An integrative framework for eyewitness memory and suggestibility. *Memory, 6*(5), 481–529.

Blount, S., & Larrick, R. P. (2000). Framing the game: Examining frame choice in bargaining. *Organizational Behavior and Human Decision Processes, 81*(1), 43–71.

Bochman, A. (2004). A causal approach to nonmonotonic reasoning. *Artificial Intelligence, 160*(1–2), 105–143.

Bodker, K., Kensing, F., & Simonsen, J. (2004). *Participatory IT design: Designing for business and workplace realities*. Cambridge, MA: MIT Press.

Bowers, C., Salas, E., & Jentsch, F. (Eds.). (2006). *Creating high-tech teams: Practical guidance on work performance and technology*. Washington, DC: American Psychological Association.

Brachman, R., & Levesque, H. (2004). *Knowledge representation and reasoning*. San Mateo, CA: Morgan Kaufmann.

Bradfield, A., & Wells, G. (2005). Not the same old hindsight bias: Outcome information distorts a broad range of retrospective judgments. *Memory and Cognition, 33*(1), 120–130.

Brehm, S., Kassin, S. M., & Fein, S. (1999). *Social psychology* (4th ed.). Boston: Houghton Mifflin Company.

Burns, C., Bizantz, A., & Roth, E. (2004). Lessons from a comparison of work domain models: Representational choices and their implications. *Human Factors, 46*, 711–727.

Burns, C., & Hajdukiewicz, J. (2004). *Ecological interface design*. Boca Raton, FL: CRC Press.

Burns, C., & Vicente, K. (2001). Model-based approaches for analyzing cognitive work: A comparison of abstraction hierarchy, multi-level flow modeling, and decision ladder modeling. *International Journal of Cognitive Ergonomics, 5*, 357–366.

Byrne, M., & Kirlik, A. (2005). Using computational cognitive modeling to diagnose possible sources of aviation error. *International Journal of Aviation Psychology, 15*(2), 135–155.

Caggiano, D., & Parasuraman, R. (2004). The role of memory representation in the vigilance decrement. *Psychonomic Bulletin and Review, 11*(5), 932–937.

Caldwell, B. (2005). Multi-team dynamics and distributed expertise in mission operations. *Aviation, Space, and Environmental Medicine, 76*(6), 145–153.

Card, S., Moran, T. P., & Newell, A. (1983). *The psychology of human-computer interaction*. Hillsdale, NJ: Lawrence Erlbaum.

Carroll, J. M. (1995). *Scenario-based design: Envisioning work and technology in system development*. New York: John Wiley and Sons.

Carroll, J. M., Neale, D. C., Isenhour, P. L., & McCrickard, D. S. (2003). Notification and awareness: Synchronizing task-oriented collaborative activity. *International Journal of Human-Computer Studies, 58*, 605–632.

Cartwright, D., & Zander, A. (Eds.). (1960). *Group dynamics: Research and theory* (2nd ed.). Evanston, IL: Row and Peterson.

Cary, M., & Carlson, R. (2001). Distributing working memory resources during problem solving. *Journal of Experimental Psychology: Learning, Memory and Cognition, 27*, 836–848.

Chandrasekaran, B. (1988). Generic tasks as building blocks for knowledge based systems: The diagnosis and routine design examples. *Knowledge Engineering Review, 3*(3), 183–210.

Charness, N., Feltovich, P., Hoffman, R., & Ericsson, E. (2006). *The Cambridge handbook of expertise and expert performance*. New York: Cambridge University Press.

Clancey, W. J. (1983). The epistemology of a rule-based expert system— a framework for explanation. *Artificial Intelligence, 20*, 215–251.

Clancey, W. J. (1985). Heuristic classification. *Artificial Intelligence, 27*, 289–350.

Clarke, K., Hardstone, G., Rouncefield, M., & Sommerville, I. (Eds.). (2006). *Trust in technology: A socio-technical perspective*. New York: Springer.

Cockburn, A. (2000). *Writing effective use cases*. Harlow, England: Addison-Wesley.

Crandall, B., Klein, G., & Hoffman, R. (2006). *Working minds: A practitioner's guide to cognitive task analysis*. Cambridge, MA: MIT Press.

Croson, R., & Sundali, J. (2005). The gambler's fallacy and the hot hand: Empirical data from casinos. *Journal of Risk and Uncertainty, 30*(3),195–209.

Dasgupta, P., Chakrabarti, P., & Desarkar, S. (2004). *Multiobjective heuristic search: An introduction to intelligent search methods for multicriteria optimization.* San Mateo, CA: Morgan Kauffmann.

Dechter, R. (2003). *Constraint processing.* San Mateo, CA: Morgan Kaufmann.

Diaper, D., & Stanton, N. (Eds.). (2004). *The handbook of task analysis for human-computer interaction.* Mahwah, NJ: Lawrence Erlbaum.

Eckstein, M. P. (2004). Active vision: The psychology of looking and seeing. *Perception, 33*(8), 1021–1023.

Elmore, W. K., Dunlap, R. D., & Campbell, R. H. (2001). Features of a distributed intelligent architecture for unmanned air vehicle operations. *Proceedings of 2001 Unmanned Systems International, AUVSI,* Baltimore, MD.

Elstein, A. S., Shulman, L. S., & Sprafka, S. A. (1978). *Medical problem solving: An analysis of clinical reasoning.* Cambridge, MA: Harvard University Press.

Endsley, M. (2003). *Designing for situation awareness.* London: Taylor & Francis.

Endsley, M., & Garland, D. (2000). *Situation awareness analysis and measurement.* Mahwah, NJ: Lawrence Erlbaum Associates.

Erkens, G., Andriessen, J., & Peters, N. (2003). Interaction and performance in computer-supported collaborative tasks. In H. Van Oostendorp (Ed.), *Cognition in a digital world* (pp. 225–251). Mahwah, NJ: Lawrence Erlbaum.

Flach, J., Hancock, P., Caird, J., & Vicente, K. (Eds.). (1995). *Global perspectives on the ecology of human-machine systems.* Hillsdale, NJ: Lawrence Erlbaum.

Fischer, G., Lemke, A. C., Mastaglio, T., & Morch, A. I. (1991). The role of critiquing in cooperative problem solving. *ACM Transactions on Information Systems, 9*(3), 123–151.

Forsyth, D. (1998). *Group dynamics.* Florence, KY: Wadsworth Publishing.

Fraser, J. M., Smith, P. J., & Smith, J. W. (1992). A catalog of errors. *International Journal of Man-Machine Systems, 37,* 265–307.

Freeman, F., Mikulka, P., & Scerbo, M. (2004). An evaluation of an adaptive automation system using a cognitive vigilance task. *Biological Psychology, 67*(3), 283–297.

Garb, H. (2005). Clinical judgment and decision making. *Annual Review of Clinical Psychology, 1*(1), 67–89.

Geddes, N. D. (1997a). Large scale models of cooperating and hostile intents. *Proceeding of the 1997 IEEE Conference on Engineering of Computer Based Systems,* Monterey, CA, 94–100.

Geddes, N. D. (1997b). Associate systems: A framework for human-machine cooperation. In Smith, M. J., Salvendy, G., & Koubek, R. J. (Eds.), *Advances in human factors/ergonomics: Vol. 21. Design of computing systems: Social and ergonomic considerations* (pp. 237–242). New York: Elsevier.

Geddes, N., & Lee, J. (1998). Intelligent control for automated vehicles: A decision aiding method for coordination of multiple uninhabited tactical aircraft. *Proceedings of 1998 Unmanned Systems International, AUVSI,* Huntsville, AL, 46–51.

Geddes, N. D., & Lizza, C. S. (1999). *Shared plans and situations as a basis for collaborative decision making in air operations* (SAE Paper 1999-01-5538). Paper presented at the SAE World Aeronautics Conference, Atlanta, GA.

Gentner, D., & Stevens, A. (1983). *Mental models.* Hillsdale, NJ: Lawrence Erlbaum.

Ghallib, M., Nau, D., & Traverso, P. (2004). *Automated planning: Theory and practice.* San Mateo, CA: Morgan Kaufmann Publishers.

Gilovich, T., Griffin, D., & Kahneman, D. (Eds.). (2002). *Heuristics and biases: The psychology of intuitive judgment.* New York: Cambridge University Press.

Gilovich, T., Vallone, R., & Tversky, A. (1985). The hot hand in basketball: On the misperception of random sequences. *Cognitive Psychology, 17,* 295–314.

Gordon, A. (2004). The representation of planning strategies. *Artificial Intelligence, 153*(1/2), 287–305.

Gordon, S., & Gill, R. T. (1992). Knowledge acquisition with question probes and conceptual graph structures. In T. Lauer, E. Peacock, & A. Graesser (Eds.), *Questions and information systems* (pp. 29–46). Hillsdale, NJ: Lawrence Erlbaum.

Gordon, S. E., & Gill, R. T. (1997). Cognitive task analysis. In C. Zsambok & G. Klein (Eds.), *Naturalistic decision making* (pp. 131–140). Hillsdale, NJ: Lawrence Erlbaum.

Grabowski, M., & Sanborn, S. (2003). Human performance and embedded intelligent technology in safety-critical systems. *International Journal of Human-Computer Studies, 58*(6), 637–670.

Grier, R., Warm, J., & Dember, W. (2003). The vigilance decrement reflects limits in effortful attention, not mindlessness. *Human Factors, 45*(3), 349–359.

Griffith, T. L., Sawyer, J. E., & Neale, M. A. (2003). Virtualness and knowledge in teams: Managing the love triangle of organizations, individuals and information technology. *MIS Quarterly, 27,* 265–287.

Grondin, S. (2005). Overloading temporal memory. *Journal of Experimental Psychology: Human Perception and Performance, 31*(5), 869–879.

Guerlain, S., Smith, P. J., Obradovich, J. H., Rudmann, S., Strohm, P., Smith, J. W. et al. (1996). Dealing with brittleness in the design of expert systems for immunohematology. *Immunohematology, 12,* 101–107.

Guerlain, S., Smith, P. J., Obradovich, J. H., Rudmann, S., Strohm, P., Smith, J. W. et al. (1999). Interactive critiquing as a form of decision support: An empirical evaluation. *Human Factors, 41,* 72–89.

Hammer, J. M., & Singletary, B. A. (2004). Common shortcomings in software for autonomous systems. *Proceedings of the AIAA First Intelligent Systems Conference,* Chicago, IL, 19–24.

Handberg, R. (1995). Expert testimony on eyewitness identification: A new pair of glasses for the jury. *American Criminal Law Review, 32*(4), 1013–1064.

Handy, C. (1995). Trust and the virtual organization. *Harvard Business Review, 73,* 40–50.

Harris, S., & Owens, J. (1986). Some critical factors that limit the effectiveness of machine intelligence technology in military systems applications. *Journal of Computer-Based Instruction, 13,* 30–34.

Haselton, M. G., Nettle, D., & Andrews, P. W. (2005). The evolution of cognitive bias. In D. M. Buss (Ed.), *The handbook of evolutionary psychology* (pp. 724–746). Hoboken, NJ: John Wiley and Sons.

Hasling, D. W., Clancey, W. J., & Rennels, G. (1984). Strategic explanations for a diagnostic consultation system. *International Journal of Man-Machine Systems, 20,* 3–19.

Hayes-Roth, B., & Hayes-Roth, F. (1979). A cognitive model of planning. *Cognitive Science, 3*(4), 275–310.

Helander, M., Landauer, T., & Prabhu, P. (Eds.). (1997). *Handbook of human-computer interaction.* Amsterdam: Elsevier.

Hertel, G., Geister, S., & Konradt, U. (2005). Managing virtual teams: A review of current empirical research. *Human Resource Management Review, 15*(1) 69–95.

Hester, R., & Garavan, H. (2005). Working memory and executive function: The influence of content and load on the control of attention. *Memory and Cognition, 33*(2), 221–233.

Hinds, P., & Kiesler, S. (Eds.). (2005). *Distributed work.* Cambridge, MA: MIT Press.

Hinds, P. J., & Mortensen, M. (2005). Understanding conflict in geographically distributed teams: The moderating effects of shared identity, shared context, and spontaneous communication. *Organization Science, 16,* 290–307.

Hoc, J.-M. (2000). From human-machine interaction to human-machine cooperation. *Ergonomics, 43,* 833–843.

Hollnagel, E. (1990). Responsibility issues in intelligent decision support systems. In D. Berry & A. Hart (Eds.), *Expert systems: Human issues,* Cambridge, MA: MIT Press.

Hollnagel, E. (2003). *Handbook of cognitive task design.* Mahwah, NJ: Lawrence Erlbaum.

Hollnagel, E., & Woods, D. (2005). *Joint cognitive systems: Foundations of cognitive systems engineering.* Boca Raton, FL: Taylor & Francis.

Horton, F., & Lewis, D. (Eds.). (1991). *Great information disasters.* London: Association for Information Management.

Hutchins, E. (1990). The technology of team navigation. In J. Galegher, R. Kraut, & C. Egido (Eds.), *Intellectual teamwork: Social and technical bases of collaborative work.* Hillsdale, NJ: Lawrence Erlbaum.

Hutchins, E. (1995). *Cognition in the wild.* Cambridge, MA: MIT Press.

Jagacinski, R. J., & Flach, J. M. (2003). *Control theory for humans: Quantitative approaches to modeling performance.* Hillsdale, NJ: Lawrence Erlbaum.

Janis, I. (1982). *Groupthink* (2nd ed.). Boston: Houghton Mifflin.

Johnson, P., Duran, A., Hassebrock, F., Moller, J., Prietulla, M., Feltovich, P. et al. (1981). Expertise and error in diagnostic reasoning. *Cognitive Science, 5,* 235–283.

Johnston, W. A., & Dark, V. J. (1986). Selective attention. *Annual Review of Psychology, 37,* 43–75.

Jonassen, D., Tessmer, M., & Hannum, W. (1999). *Task analysis methods for instructional design.* Mahwah, NJ: Lawrence Erlbaum.

Jones, P., & Jacobs, J. (2000). Cooperative problem solving in human-machine systems: Theory, models and intelligent associate systems. *IEEE Transactions on Systems, Man and Cybernetics, 30*(4), 397–407.

Jones, E., & Nisbett, R. (1971). The actor and the observer: Divergent perceptions of the causes of behavior. In E. Jones, D. Kanouse, H. Kelley, R. Nisbett, S. Valins, & B. Weiner (Eds.), *Attributions: Perceiving the causes of behavior* (pp. 79–94). Morristown, NJ: General Learning Press.

Jones, P. M., & Mitchell, C. M. (1995). Human-computer cooperative problem solving: Theory, design, and evaluation of an intelligent associate system. *IEEE Transactions on Systems, Man, and Cybernetics, 25,* 1039–1053.

Jones, D. R., & Schkade, D. A. (1995). Choosing and translating between problem representations. *Organizational Behavior and Human Decision Processes, 61*(2), 214–223.

Josephson, J., & Josephson, S. (1994). *Abductive inference.* New York: Cambridge University Press.

Kahneman, D., & Tversky, A. (Eds.). (2000). *Choices, values and frames.* New York: Cambridge University Press.

Katz, N., Lazer, D., Arrow, H., & Contractor, N. (2004). Network theory and small groups. *Small Group Research, 35*(3), 307–332.

Kieras, D. (1991). Towards a practical GOMS model methodology for user interface design. In M. Helander (Ed.), *Handbook of human-computer interaction* (2nd ed.). Amsterdam: Elsevier Science Publishers.

Kieras, D. (2004). GOMS models for task analysis. In D. Diaper & N. Stanton (Eds.), *The handbook of task analysis for human-computer interaction* (pp. 83–116). Mahwah, NJ: Lawrence Erlbaum.

Kieras, D., & Meyer, D. (1997). An overview of the EPIC architecture for cognition and performance with application to human-computer interaction. *Human-Computer Interaction, 12,* 391–438.

Kirwan, B., & Ainsworth, L. (1992). *A guide to task analysis.* London: Taylor & Francis.

Klayman, J., & Ha, Y. (1987). Confirmation, disconfirmation, and information in hypothesis testing. *Psychological Review, 94,* 211–228.

Klein, G. A. (1993). A recognition-primed decision (RPD) model of rapid decision making. In G. Klein, J. Oransanu, R. Calderwood, & C. Zsambok, (Eds.), *Decision making in action: Models and method* (pp. 138–147). Norwood, NJ: Ablex.

Klein, G. (2000). Cognitive task analysis of teams. In J. M. Schraagen, S. Chipman, & V. Shalin (Eds.), *Cognitive task analysis* (417–430). Mahwah, NJ: Lawrence Erlbaum.

Klein, G., Caulderwood, R., & MacGregor, D. (1989). Critical decision method for eliciting knowledge. *IEEE Transactions on Systems, Man and Cybernetics, 19,* 462–472.

Kleinmuntz, D. N., & Schkade, D. A. (1993). Information displays and decision processes. *Psychological Science, 4*(4), 221–227.

Koehler, J. J. (1996). The base rate fallacy reconsidered: Descriptive, normative, and methodological challenges. *Behavioral and Brain Sciences, 9*(1), 1–53.

Kohler, W. (1992). *Gestalt psychology: An introduction to new concepts in modern psychology.* New York: Liveright Publishing Corporation.

Kolodner, J. (1993). *Case-based reasoning.* San Mateo, CA: Morgan Kaufmann Publishers.

Kotovsky, K., Hayes, J. R., & Simon, H. A. (1985). Why are some problems hard? Evidence from Tower of Hanoi. *Cognitive Psychology, 17,* 248–294.

Kulak, D., & Guiney, E. (2003). *Use cases: Requirements in context* (2nd ed.). Harlow, England: Addison-Wesley Publishing.

Kushleyeva, Y., Salvucci, D., Lee, F., & Schunn, C. (2005). Deciding when to switch tasks in time-critical multitasking. *Cognitive Systems Research, 6*(1), 41–49.

Laird, J. E., Newell, A., & Rosenbloom, P. S. (1987). SOAR: An architecture for general intelligence. *Artificial Intelligence, 33,* 1–64.

Larkin, J. H. (1989). Display-based problem solving. In D. Klahr & K. Kotovsky (Eds.), *Complex information processing: The impact of Herbert A. Simon.* Hillsdale, NJ: Lawrence Erlbaum.

Larkin, J. H., & Simon, H. A. (1987). Why a diagram is (sometimes) worth ten thousand words. *Cognitive Science, 11,* 65–99.

Larson, A., & Hayes, C. (2005). An assessment of WEASEL: A decision support system to assist in military planning. *Proceedings of the 2005 Annual Meeting of the Human Factors and Ergonomics Society,* Orlando, 287–291.

Lee, J., & See, K. (2004). Trust in automation: Designing for appropriate reliance. *Human Factors, 46*(1), 50–80.

Lehner, P. E., & Zirk, D. A. (1987). Cognitive factors in user/expert-system interaction. *Human Factors, 29*(1), 97–109.

Lesgold, A., Glaser, R., Rubinson, H., Klopfer, D., Feltovich, P., & Wang, Y. (1988). Expertise in a complex skill: Diagnosing X-ray pictures. In M. Chi, R. Glaser, & M. Farr (Eds.), *The nature of expertise* (pp. 311–342). Hillsdale, NJ: Lawrence Erlbaum.

Levi, D. (2001). *Group dynamics for teams.* London: SAGE Publications.

Lewis, C., & Wharton, C. (1997). Cognitive walkthroughs. In M. Helander, T. Landauer, & P. Prabhu (Eds.), *Handbook of human-computer interaction* (2nd ed.; pp. 717–731). Amsterdam: Elsevier.

Lizza, C. S., & Banks, S. B. (1991, June). Pilot's associate: A cooperative knowledge based system application. *IEEE Expert, 8,* 24–32.

Loftus, E. (1975). Leading questions and the eyewitness report. *Cognitive Psychology, 7,* 560–572.

Logan, G. (2005). Attention, automaticity, and executive control. In A. F. Healy (Ed.), *Experimental cognitive psychology and its applications* (pp. 129–139). Washington, DC: American Psychological Association.

Lowe, R. K. (2003). Animation and learning: Selective processing of information in dynamic graphics. *Learning and Instruction, 13*(2), 157–176.

Luo, Y., Greenwood, P., & Parasuraman, R. (2001). Dynamics of the special scale of visual attention revealed by brain event-related potentials. *Cognitive Brain Research, 12*(3), 371–381.

Lurey, J. S., & Raisinghani, M. S. (2001). An empirical study of best practices in virtual teams. *Information Management, 38,* 523–544.

Mack, R. (1995). Scenarios as engines of design. In J. M. Carroll (Ed.), *Scenario-based design: Envisioning work and technology in system development* (pp. 361–386). New York: John Wiley.

Mackworth, N. (1961). Researches on the measurement of human performance. In H. W. Sinaiko (Ed.), *Selected papers on human factors in the design and use of control systems* (pp. 48–63). New York: Dover Publications. (Original work published 1950).

Madani, O., Condon, A., & Hanks, S. (2003). On the undecidability of probabilistic planning and related stochastic optimization problems. *Artificial Intelligence, 147*(1/2), 5–34.

Majercik, S., & Littman, M. (2003). Contingent planning under uncertainty via stochastic satisfiability. *Artificial Intelligence, 147*(1/2), 119–62.

Marsh, S., & Dibben, M. (2003). The role of trust in information science and technology. *Annual Review of Information Science and Technology, 37*, 465–498.

McCauley, C. (1989). The nature of social influence in groupthink: Compliance and internalization. *Journal of Personality and Social Psychology, 22*, 250–260.

Meister, D., & Enderwick, T. (2002). *Human factors in system design, development and testing.* Mahwah, NJ: Lawrence Erlbaum.

Metzger, U., & Parasuraman, R. (2005). Automation in future air traffic management: Effects of decision aid reliability on controller performance and mental workload. *Human Factors, 47*(1), 35–49.

Meyers, D. (2004). *Psychology* (7th ed.). New York: Worth Publishers.

Michalewicz, Z., & Fogel, D. (2004). *How to solve it: Modern heuristics.* New York: Springer.

Michie, D. (1986). *On machine intelligence* (2nd ed.). Chicester: Ellis Horwood Limited.

Mikulka, P., Scerbo, M., & Freeman, F. (2002). Effects of a biocybernetic system on vigilance performance. *Human Factors, 44*(4), 654–664.

Miller, C., Pelican, M., & Goldman, R. (2000). "Tasking" interfaces to keep the operator in control. *Proceedings of the 5th International Conference on Human Interaction with Complex Systems*, Urbana, IL. 116–122.

Miller, G. A., Galanter, E., & Pribram, K. H. (1960). *Plans and the structure of behavior.* New York: Henry Holt and Company.

Miller, P. (1986). *Expert critiquing systems: Practice-based medical consultation by computer.* New York: Springer-Verlag.

Mital, A., & Pennathur, A. (2004). Advanced technologies and humans in manufacturing workplaces: An interdependent relationship. *International Journal of Industrial Ergonomics, 33*(4), 295–313.

Mitchell, C. M., & Miller, R. A. (1986). A discrete control model of operator function: A methodology for information display design. *IEEE Transactions on Systems. Man and Cybernetics, 16*, 343–357.

Muir, B. (1987). Trust between humans and machines. *International Journal of Man-Machine Studies, 27*, 527–539.

Muir, B., & Moray, N. (1996). Trust in automation 2: Experimental studies of trust and human intervention in a process control simulation. *Ergonomics, 39*(3), 429–460.

Mumford, M., Schultz, R., & Van Doorn, J. (2001). Performance in planning: Processes, requirement, and errors. *Review of General Psychology, 5*(3), 213–240.

Mynatt, C., Doherty, M., & Tweney, R. (1978). Confirmation bias in a simulated research environment: An experimental study of scientific inference. *Quarterly Journal of Experimental Psychology, 30*, 85–95.

Nakatsu, R. T., & Benbasat, I. (2003). Improving the explanatory power of knowledge-based systems: An investigation of content and interface-based enhancements. *IEEE Transactions on Systems Man and Cybernetics Part A-Systems and Humans, 33*(3), 344–357.

Nareyek, A., Fourer, R., & Freuder, E. (2005). Constraints and AI planning. *IEEE Intelligent Systems, 20*(2), 62–72.

Nason, S., Laird, J., & Schunn, C. (2005). SOAR-RL: Integrating reinforcement learning with SOAR. *Cognitive Systems Research, 6*(1), 51–59.

Navon, D. (1978). The importance of being conservative. *British Journal of Mathematical and Statistical Psychology, 31*, 33–48.

Newell, A. (1990). *Unified theories of cognition.* Cambridge, MA: Harvard University Press.

Newell, A., & Simon, H. (1972). *Human problem solving.* Englewood Cliffs, NJ: Prentice Hall, Inc.

Nickerson, R. (1988). Counting, computing, and the representation of numbers. *Human Factors, 30*, 181–199.

Norman, D. A. (1981). Categorization of action slips. *Psychological Review, 88*(1), 1–15.

Norman, D. A. (2002). *The design of everyday things.* New York: Doubleday.

Oliver, D., & Roos, J. (2005). Decision making in high-velocity environments. *Organization Studies, 26*(6), 889–913.

Olson, G. M., & Olson, J. S. (1997). Research on computer supported cooperative work. In M. Helander, T. Landauer, P. Prabhu (Eds.), *Handbook of human-computer interaction* (pp. 1433–1456). Amsterdam: Elsevier.

Olson, G. M., & Olson, J. S. (2003). Groupware and computer-supported cooperative work. In A. Sears & J. Jacko (Eds.), *Handbook of human-computer interaction* (pp. 583–593). Mahwah, NJ: Lawrence Erlbaum.

Orasanu, J., & Salas, E. (1991). Team decision making in complex environments. In G. Klein, R. Calderwood, & C. Zsambok (Eds.), *Decision making in action: Models and methods* (pp. 327–345). New Jersey: Ablex.

Ormrod, J. (2003). *Human learning* (4th ed.). Englewood Cliffs, NJ: Prentice Hall.

Parasuraman, R. (2000). Designing automation for human use: Empirical studies and quantitative models. *Ergonomics, 43*, 931–951.

Parasuraman, R., & Riley, V. (1997). Humans and automation: Use, misuse, disuse and abuse. *Human Factors, 39*, 230–253.

Parasuraman, R., Sheridan, T. B., & Wickens, C. D. (2000). A model for types and levels of human interaction with automation. *IEEE Transactions on Systems Man and Cybernetics, 30*(3), 286–297.

Pashler, H., Johnston, J. C., & Ruthruff, E. (2001). Attention and performance. *Annual Review of Psychology, 52*, 629–651.

Pearl, J. (1988). *Probabilistic reasoning in intelligent systems: Networks of plausible inference.* San Mateo, CA: Morgan Kaufman.

Pennington, N., & Hastie, R. (1992). Explaining the evidence: Tests of the story model for juror decision making. *Journal of Personality and Social Psychology, 62*, 189–206.

Petroski, H. (1994). *Design paradigms: Case histories of error and judgment in engineering.* New York: Cambridge University Press.

Plous, S. (1993). *The psychology of judgment and decision making.* New York: McGraw-Hill.

Poulton, E. C. (1989). *Bias in quantifying judgments.* Hillsdale, NJ: Lawrence Erlbaum.

Preece, J., Rogers, Y., & Sharp, H. (2002). *Interaction design.* New York: John Wiley and Sons.

Preece, J., Rogers, Y., Sharp, H., & Benyon D. (1994). *Human-computer interaction.* Addison-Wesley.

Pritchett, A., Vandor, B., & Edwards, K. (2002). Testing and implementing cockpit alerting systems. *Reliability Engineering and System Safety, 75*(2), 193–206.

Psillos, S. (2002). Simply the best: A case for abduction. *Lecture Notes in Artificial Intelligence, 2408*, 605–625.

Radvansky, G. (2005). *Human memory.* Boston: Allyn and Bacon.

Rasmussen, J. (1983). Skills, rules and knowledge: Signals, signs, symbols and other distinctions in human performance models. *IEEE Transactions on Systems Man and Cybernetics, 13*(3), 257–266.

Rasmussen, J., Brehner, B., & Leplat, J. (Eds.). (1991). *Distributed decision making: Cognitive models for cooperative work.* New York: John Wiley and Sons.

Rasmussen, J., Pejtersen, A., & Goldstein, L. (1994). *Cognitive systems engineering.* New York: John Wiley & Sons.

Rayward-Smith, V., Osman, I., Reeves, C., & Smith, G. (Eds.). (1996). *Modern heuristic search methods.* New York: John Wiley and Sons.

Reason, J. (1991). *Human error.* New York: Cambridge Press.

Reason, J. (1997). *Managing the risks of organizational accidents.* Hampshire, UK: Ashgate.

Reising, D., & Sanderson, P. (2002). Work domain analysis and sensors I: Principles and simple example. *International Journal of Human-Computer Studies, 56*(6), 569–596.

Riegelsberger, J., Sasse, M., & McCarthy, J. (2005). The mechanics of trust: A framework for research and design. *International Journal of Human-Computer Studies, 62*(3), 381–422.

Riesbeck, C. K., & Schank, R. C. (1989). *Inside case-based reasoning.* Hillsdale, NJ: Lawrence Erlbaum.

Roth, E. M., Bennett, K. B., & Woods, D. D. (1987). Human interaction with an 'intelligent' machine. *International Journal of Man-Machine Studies, 27,* 479–525.

Rouse, W.B. (1980). *Systems engineering models of human machine interaction.* New York: Elsevier.

Rouse, W. B., Geddes, N. D., & Hammer, J. M. (1990, March). Computer-aided fighter pilots. *IEEE Spectrum, 8,* 38–41.

Rubin, K. S., Jones, P. M., & Mitchell, C. M. (1988). OFMSpert: Interference of operator intentions in supervisory control using a blackboard structure. *IEEE Transactions on Systems Man and Cybernetics, 18*(4), 618–637.

Russell, S., & Norvig, P. (1995). *Artificial intelligence: A modern approach.* Englewood Cliffs, NJ: Prentice-Hall, Inc.

Sacerdoti, E. D. (1974). Planning in a hierarchy of abstraction spaces. *Artificial Intelligence, 5*(2), 115–135.

Salas, E., Bowers, C., & Edens, E. (Eds.). (2001). *Improving teamwork in organizations: Applications of resource management training.* Mahwah, NJ: Lawrence Erlbaum.

Salas, E., & Fiore, S. (Eds.). (2004). *Team cognition: Understanding the factors that drive process and performance.* Washington, DC: American Psychological Association.

Sarter, N., & Woods, D. (1993). *Cognitive engineering in aerospace applications: Pilot interaction with cockpit automation* (NASA Contractor Rep. 177617). Moffett Field, CA : NASA Ames Research Center.

Scaife, M., & Rogers, Y. (1996). External cognition: How do graphical representations work? *International Journal of Human-Computer Studies, 45,* 185–213.

Schaeken, W. (Ed.). (2005). *Mental models theory of reasoning: Refinements and extensions.* Mahwah, NJ: Lawrence Erlbaum.

Schragen, J. M., Chipman, S., & Shalin, V. (Eds.). (2000). *Cognitive task analysis.* Mahwah, NJ: Lawrence Erlbaum.

Schroeder, R., & Axelsson, A.-S. (Eds.). (2005). *Work and plan in shared virtual environments: Computer supported cooperative work.* New York: Springer.

Schuler, D., & Namioka, A. (Eds.). (1993). *Participatory design: Principles and practices.* Mahwah, NJ: Lawrence Erlbaum.

Scott, R., Roth, E. M., & Deutsch, S. E. (2005). Work-centered support systems: A human-centered approach to intelligent system design. *IEEE Intelligent Systems, 20,* 73–81.

Seamon, J. (2005). *Human memory: Contemporary readings.* Cambridge: Oxford University Press.

Sewell, D. R., & Geddes, N. D. (1990). A plan and goal based method for computer-human system design. *Proceedings of INTERACT 90,* New York, 283–288.

Shafer, G. (1996). *Probabilistic expert systems.* Philadelphia: Society for Industrial and Applied Mathematics.

Shafer, G., & Pearl, J. (Eds.). (1990). *Readings in uncertain reasoning.* San Mateo, CA: Morgan Kaufmann.

Shalin, V. L., Geddes, N. D., Bertram, D., Szczepkowski, M. A., & DuBois, D. (1997). Expertise in dynamic physical task domains. In P. Feltovich, K. Ford, & R. Hoffman (Eds.), *Expertise in context: Human and machine* (pp. 194–217). Cambridge, MA: MIT Press.

Shepherd, A. (2000). *Hierarchical task analysis.* London: Taylor & Francis.

Sheridan, T. B. (1997). Supervisory control. In G. Salvendy (Ed.), *Handbook of human factors* (2nd ed.; pp. 1295–1327). New York: John Wiley and Sons.

Sheridan, T. (2002). *Humans and automation: System design and research issues.* Chidester, England: Wiley.

Sheridan, T. B., & Ferrell, W. R. (1974). *Man-machine studies: Information, control, and decision models of human performance.* Cambridge, MA: MIT Press.

Sheridan, T. B., & Verplank, W. L. (1978). *Human and computer control of undersea teleoperators* (MIT Man Machine Systems Laboratory Tech. Rep 1978.). Cambridge, MA: MIT Man Machine Systems Laboratory.

Shortliffe, E. (1990). Clinical decision support systems. In E. Shortliffe & L. Perreault (Eds.), *Medical informatics: Computer applications in health care* (pp. 466–500). New York: Addison Wesley.

Silverman, B. G. (1992). Survey of expert critiquing systems: Practical and theoretical frontiers. *Communications of the ACM, 35*(4), 106–128.

Skirka, L., Mosier, K., & Burdick, M. (1999). Does automation bias decision making? *International Journal of Human-Computer Systems, 51,* 991–1006.

Smith, P. J., Beatty, R., Spencer, A., & Billings, C. (2003). Dealing with the challenges of distributed planning in a stochastic environment: Coordinated contingency planning. *Proceedings of the 2003 Annual Conference on Digital Avionics Systems,* Chicago, IL., 116–123.

Smith, P. J., Billings, C., Chapman, R., Obradovich, J. H., McCoy, E., & Orasanu, J. (2000). Alternative architectures for distributed cooperative problem-solving in the national airspace system. In M. Benedict (Ed.), *Proceedings of the 5th International Conference on Human Interaction with Complex Systems,* Urbana, IL, 87–93.

Smith, P. J., & Geddes, N. (2003). A cognitive systems engineering approach to the design of decision support systems. In A. Sears & J. Jacko (Eds.), *Handbook of human-computer interaction* (pp. 656–675). Mahwah, NJ: Lawrence Erlbaum.

Smith, P. J., Giffin, W., Rockwell, T., & Thomas, M. (1986). Modeling fault diagnosis as the activation and use of a frame system. *Human Factors, 28*(6), 703–716.

Smith, P. J., McCoy, E., & Layton, C. (1997). Brittleness in the design of cooperative problem-solving systems: The effects on user performance. *IEEE Transactions on Systems, Man, and Cybernetics, 27*(3), 360–371.

Smith, P. J., McCoy, E., Orasanu, J., Billings, C., Denning, R., Rodvold, M. et al. (1997). Control by permission: A case study of cooperative problem-solving in the interactions of airline dispatchers and ATCSCC. *Air Traffic Control Quarterly, 4,* 229–247.

Smith, P. J., McCoy, E., & Orasanu, J. (2000). Distributed cooperative problem-solving in the air traffic management system. In G. Klein and E. Salas (Eds.), *Naturalistic decision making* (pp. 369–384). Mahwah, NJ: Lawrence Erlbaum.

Smith, P. J., McCoy, E., & Orasanu, J. (2001). Distributed cooperative problem-solving in the air traffic management system. In G. Klein and E. Salas (Eds.), *Naturalistic Decision Making,* Mahwah NJ: Lawrence Erlbaum, 369–384.

Smith, P. J., Obradovich, J. H., Guerlain, S., Rudmann, S., Strohm, P., Smith, J., et al. (1998). Successful use of an expert system to teach diagnostic reasoning for antibody identification. *Proceedings of the 4th International Conference on Intelligent Tutoring Systems,* 354–363.

Smith, P. J., & Rudmann, S. (2005). Clinical decision making and diagnosis: Implications for immunohematologic problem-solving. In S. Rudmann (Ed.) *Serologic problem-solving: A systematic approach for improved practice* (pp. 1–16). Bethesda, MD: AABB Press.

Smith, P. J., Stone, R. B., & Spencer, A. (2006). Design as a prediction task: Applying cognitive psychology to system development. In W. Marras & W. Karwowski (Eds.), *Handbook of industrial ergonomics* (2nd ed.). New York: Marcel Dekker, Inc.

St. Amant, R., Freed, A., Ritter, F. & Schunn, C. (2005). Specifying ACT-R models of user interaction with a GOMS language. *Cognitive Systems Research, 6*(1), 71–88.

Stenz, A., & Dias, M. B. (1999). *A free market architecture for coordinating multiple robots* (Tech. Rep. CMU-RI-TR-99-42). Pittsburgh, PA: Robotics Institute, Carnegie Mellon University.

Strauch, B. (2004). *Investigating human error: Incidents, accidents, and complex systems.* Hampshire, UK: Ashgate Publishing.

Suchman, L. (1987). *Plans and situated actions: The problem of human-machine communication.* New York: Cambridge University Press.

Sycara, K., & Lewis, M. (2004). Integrating intelligent agents into human teams. In E. Salas & S. Fiore (Eds.), *Team cognition: Understanding the factors that drive process and performance* (pp. 203–231). Washington, DC: American Psychological Association.

Tetlock, P. (1998). Social psychology and world politics. In D. Gilbert, S. Fiske, & G. Lindzey (Eds.), *The handbook of social psychology: Vol. 2.* (4th ed.; pp. 868–912). New York: McGraw Hill.

Tetlock, P., Peterson, R., McQuire, M., Chang, S., & Feld, P. (1992). Assessing political group dynamics: A test of the groupthink model. *Journal of Personality and Social Psychology, 63,* 403–425.

Tianfield, H., & Wang, R.W. (2004). Critic systems: Towards human-computer collaborative problem solving. *Artificial Intelligence Review, 22*(4), 271–295.

Tobey, D. (2005). *Needs assessment basics.* Alexandria, VA: ASTD Press.

Turban, E., Aronson, J., & Liang, T.-P. (2004). *Decision support systems and intelligent systems* (7th ed.). Englewood Cliffs, NJ: Prentice Hall.

Tufte, E. R. (1983). *The visual display of quantitative information.* Cheshire, CT: Graphics Press.

Tufte, E. R. (1990). *Envisioning information.* Cheshire, CT: Graphics Press.

Tufte, E. R. (1997). *Visual explanations.* Cheshire, CT: Graphics Press.

Tversky, A. (1972). Elimination by aspects: A theory of choice. *Psychological Review, 79,* 281–299.

Tversky, A. (1982). *Judgment under uncertainty: Heuristics and biases.* New York: Cambridge University Press.

Tversky, A., & Kahneman, D. (1974). Judgment under uncertainty: Heuristics and biases. *Science, 185,* 1124–1131.

Vicente, K. (1999). *Cognitive work analysis: Toward safe, productive, and healthy computer-based work.* Mahwah, NJ: Lawrence Erlbaum.

Vicente, K. (2000). Work domain analysis and task analysis: A difference that matters. In J. M. Schraagen, S. Chipman, & V. Shalin (Eds.), *Cognitive task analysis* (pp. 101–118). Mahwah, NJ: Lawrence Erlbaum.

Vicente, K. J. (2002). Ecological interface design: Progress and challenges. *Human Factors, 44*(1), 62–78.

Vicente, K., & Rasmussen, J. (1990). The ecology of human-machine systems II: Mediating "direct perception" in complex work domains. *Ecological Psychology, 2,* 207–249.

Vicente, K., & Rasmussen, J. (1992). Ecological interface design: Theoretical foundations. *IEEE Transactions on Systems, Man and Cybernetics, 22,* 589–606.

Waltz, D. (1975). Understanding line drawings of scenes with shadows. In P. Winston (Ed.), *Psychology of computer vision* (pp. 29–45). Cambridge, MA: MIT Press.

Wason, P. (1960). On the failure to eliminate hypotheses in a conceptual task. *Quarterly Journal of Experimental Psychology, 12,* 129–140.

Watzman, S. (2003). Visual design principles for usable interfaces. In A. Sears & J. Jacko (Eds.), *Handbook of human-computer interaction* (pp. 263–285). Mahwah, NJ: Lawrence Erlbaum.

Wickens, C. D., Lee, J., Liu, Y., & Becker, S. G. (2004). *An introduction to human factors engineering* (2nd ed.). Upper Saddle River, NJ: Pearson/Prentice Hall.

Wiegmann, D., & Shappell, S. (2003). *A human error approach to aviation accident analysis: The human factors analysis and classification system.* Hampshire, UK: Ashgate.

Wiegmann, D., Zhang, H., Von Thaden, T., Sharma, G., & Gibbons, A. (2004). Safety culture: An integrative review. *International Journal of Aviation Psychology, 14*(2), 117–134.

Wiener, E. L., & Curry, R. E. (1980). Flight deck automation: Promises and problems. *Ergonomics, 23*(10). 995–1011.

Wiener, E. L., & Nagel, D. C. (1988). *Human factors in aviation.* New York: Academic Press.

Wilensky, R. (1983). *Planning and understanding: A computational approach to human reasoning.* Reading, MA: Addison-Wesley.

Williams, K. (2005). Computer-aided GOMS: A description and evaluation of a tool that integrates existing research for modeling human-computer interaction. *International Journal of Human-Computer Interaction, 1*(1), 39–58.

Witkin, B. R., & Altshuld, J. S. (1995). *Planning and conducting needs assessments.* Thousand Oaks, CA: Sage Publications.

Wogalter, M. S., & Mayhorn, C. B. (2005). Providing cognitive support with technology-based warning systems. *Ergonomics, 48*(5), 522–533.

Yantis, S. (1998). Control of visual attention. In H. Pashler (Ed.), *Attention* (pp. 223–256). Hillsdale, NJ: Lawrence Erlbaum.

Zhang, J. (1997). The nature of external representations in problem solving. *Cognitive Science, 21*(2), 179–217.

Zhang, J., & Norman, D. A. (1994). Representations in distributed cognitive tasks. *Cognitive Science, 18,* 87–122.

Zlot, R., & Stenz, A. (2003, June 13–14). Multirobot control using task abstraction in a market framework. *Proceedings of the 4th International Conference on Field and Service Robotics.* 64–70.

Zuboff, S. (1988). *In the age of smart machines.* New York: Basic Books.

·14·

ONLINE COMMUNITIES

Panayiotis Zaphiris, Chee Siang Ang, and Andrew Laghos
City University, London

Introduction 276	Log Analysis 281
CMC and Online Communities 276	Content and Textual Analysis 282
Computer Mediated Communication 276	Social Network Analysis (SNA) 282
Online Communities 276	**Case Studies** 284
Examples of CMC and Online Communities 277	Computer Aided Language Learning
Wiki-based communities .278	Communities 284
Online virtual game communities 279	Game Communities and Activity
Analyzing Online Communities:	Theoretical Analysis 285
Frameworks and Methodologies 280	Types of game communities 286
Query-Based Techniques and User Profiles:	Activity theory 286
Interviews, Questionnaires, and Personas 281	Activity theoretical analysis for
Interviews 281	constructionist learning 288
Questionnaires 281	**Discussion and Conclusions** 289
Personas 281	**References** 290

INTRODUCTION

The expansion of the Internet has resulted in an increase in the usefulness of Computer Mediated Communication (CMC) and the popularity of online communities. It is estimated that 25% of Internet users have participated in chat rooms or online discussions (Madden & Rainie, 2003).

This chapter begins with an introduction to the area of CMC and online communities by providing their definitions as well as their advantages and disadvantages. Different types of CMC are then introduced and two special types of online communities are described in more depth. More specifically, we pay special attention to the evolution of Wiki and game-based communities. These two relatively new areas of online sociability create new opportunities and challenges in the way people work, learn, and play online. Wiki-based communities facilitate new modes of social collaboration, where the creation of online content is no longer an individual action but rather it is transformed into a social, collaborative activity. Massively Multi-Player Online Role Playing Games (MMORPGs) have taken the social aspects of computer game playing to a new dimension, where players interact, socialize, and form networks of communities by having fun online.

Online communities are a source of valuable data that, when properly analyzed, can provide us with insights about the social experience people who are part of them have. For this reason, the analysis and evaluation of online communities requires a good understanding of all the available evaluation frameworks and methodologies that exist. We provide a description of the key methods later in this chapter. We then demonstrate the application of some of these methods to two characteristic case studies. Our first case study looks at how learning communities can be analyzed and how results from this analysis can be used for improving the pedagogical value of e-learning. The second case study investigates the use of activity theoretical frameworks in the analysis of computer-game-based communities. We then conclude this chapter with a brief summary and suggestions for new directions in the area of online communities.

CMC AND ONLINE COMMUNITIES

Computer Mediated Communication

It is by now no secret how vital the Internet was, is, and will continue to be in our lives. One of the most important characteristics of this medium is the opportunities it offers for human-human communication through computer networks. As Metcalfe (1992) pointed out, communication is the Internet's most important asset. E-mail is just one of the many modes of communication that can occur by using computers. Jones (1995) pointed out that through communication services, like the Internet, Usenet, and bulletin boards, online communication has for many people supplanted the postal service, telephone, and even the fax machine. All these applications in which the computer is used to mediate communication are called Computer-Mediated Communication (CMC).

December (1997) defined CMC as "the process by which people create, exchange, and perceive information using networked telecommunications systems (or non-networked computers) that facilitate encoding, transmitting, and decoding messages." He emphasized that studies of CMC view this process from different interdisciplinary theoretical perspectives (social, cognitive/psychological, linguistic, cultural, technical, and political) and often draw from fields as diverse as human communication, rhetoric and composition, media studies, human-computer interaction (HCI), journalism, telecommunications, computer science, technical communication, and information studies.

Online Communities

Online communities emerge by using CMC applications. The term *online community* is multidisciplinary in nature, means different things to different people, and is difficult to define (Preece, 2000). To gain a general understanding of what online communities are, Rheingold wrote that "[online] communities are social aggregations that emerge from the Net when enough people carry on those public discussions long enough, with sufficient human feeling, to form webs of personal relationships in cyberspace" (Rheingold, 1993, p. 5).

Online communities are also often referred to as cyber societies, cyber communities, web groups, virtual communities, web communities, virtual social networks, and e-communities, among several others.

Cyberspace is the new frontier in social relationships, and people are using the Internet to make friends, colleagues, lovers, as well as enemies (Suler, 2004). As Korzeny pointed out, even as early as 1978, online communities were formed around interests and not physical proximity (Korzeny, 1978). In general, what brings people together in an online community are common interests such as hobbies, ethnicity, education, and beliefs. As Wallace (1999) pointed out, meeting in online communities eliminates prejudging based on someone's appearance, and thus people with similar attitudes and ideas are attracted to each other.

It is estimated that as of September 2002 there were over 600 million people online (NICA Internet Surveys, 2004). The emergence of the so-called "global village" was predicted years ago (McLuhan, 1964) because of television and satellite technologies. However, it is argued by Fortner (1993) that "global metropolis" is a more representative term (Choi & Danowski, 2002). If one takes into account that the estimated world population of 2002 was 6.2 billion (U.S. Census Bureau, 2004), then the online population is nearly 10% of the world population—a significant percentage which must be taken into account when analyzing online communities. In most online communities, time, distance, and availability are no longer disseminating factors. Given that the same individual may be part of several different or numerous online communities, it is obvious why online communities keep increasing in numbers, size, and popularity.

Preece, Rogers, and Sharp (2002) stated that an online community consists of people, a shared purpose, policies, and computer systems. They identified the following member roles:

• Moderators and mediators who guide discussions/serve as arbiters;

- Professional commentators who give opinions/guide discussions;
- Provocateurs who provoke;
- General participants who contribute to discussions;
- Lurkers who silently observe.

CMC has its benefits as well as its limitations. For instance, CMC discussions are often potentially richer than face-to-face discussions. However, users with poor writing skills may be at a disadvantage when using text-based CMC (ScotCIT, 2003). Table 14.1 summarized the advantages and disadvantages of CMC (ScotCIT, 2003).

Examples of CMC and Online Communities

Examples of CMC include asynchronous communication like e-mail and bulletin boards; synchronous communication like chatting; and information manipulation, retrieval, and storage through computers and electronic databases (Ferris, 1997). Table 14.2 lists the main types of CMC, their mode (synchronous or asynchronous), and the type of media they support (text, graphics, audio, video).

When it comes to website designers, choosing which CMC to employ (for instance, forum or chat room) is not a matter of luck or randomness. Selecting the right CMC tool depends on a lot of factors. For example, in the case of e-learning, the choice of the appropriate mode of CMC will be made by asking and answering questions such as (Bates, 1995; CAP, 2004; Heeren, 1996; Reiser & Gagné, 1983):

- Are the users spread across time zones? Can all participants meet at the same time?
- Do the users have access to the necessary equipment?
- What is the role of CMC in the course?
- Are the users good readers/writers?
- Are the activities time-independent?
- How much control is allowed to the students?

Audio conferencing is a real-time communication mechanism, since the communication happens synchronously. Depending on the application, text chat and graphics may also be supported. Video conferencing, like audio conferencing, offers a useful mode of communication but has the added benefit of allowing participants to see, not just hear, one another.

IRC and chats also support synchronous communication, since they enable the users to carry out conversations using text messaging. MUDs build on chats by providing avatars and graphical environments in which the users can engage in interactive fantasy games (Preece, 2000).

TABLE 14.1. Advantages and Disadvantages of CMC

Advantages
• Time and place independence
• No need to travel
• Time lapse between messages allows for reflection
• Participants have added time to read and compose answers
• Questions can be asked without waiting for a "turn"
• It allows all participants to have a voice without the need to fight for "airtime," as in a face-to-face situation
• The lack of visual cues provides participants with a more equal footing
• Many-to-many interaction may enhance the communication
• Answers to questions can be seen—and argued—by all
• Discussion is potentially richer than in a face-to-face situation
• Messages are archived centrally, providing a database of interactions that can be revisited

Disadvantages
• Communication mainly takes place via written messages, so participants with poor writing skills may be at a disadvantage
• Paralinguistic cues (facial expression, intonation, gesture, body orientation) as to a speaker's intention are not available, except through combinations of keystrokes (emoticons) or the use of typeface emphasis (Italics, bold, capital letters)
• Time gaps within exchanges may affect the pace and rhythm of communications, leading to a possible loss in textual coherence
• The medium is socially opaque; participants may not know which or how many people they may be addressing
• The normal repair strategies of face-to-face communication are not available and misunderstandings may be harder to overcome
• Context and reference of messages may be unclear and misunderstandings may occur

TABLE 14.2. CMC Systems, Their Mode, and the Types of Media that They Support

Type of CMC	Communication Mode	Supports			
		Text	Graphics	Audio	Video
Audio conferencing	Synchronous	Some applications	No	Yes	No
Video conferencing	Synchronous	Yes	Yes	Yes	Yes
IRC	Synchronous	Yes	as attachments	as attachments	as attachments
MUD	Synchronous	Yes	No	No	No
WWW	Sync & Async	Yes	Yes	Yes	Yes
E-mail	Asynchronous	Yes	as attachments	as attachments	as attachments
Newsgroups/BBS	Asynchronous	Yes	No	No	No
Discussion Boards	Asynchronous	Yes	as attachments	as attachments	as attachments
Voice mail	Asynchronous	Some applications	No	Yes	No
Wiki	Asynchronous	Yes	Yes	Yes	Possible
Online Virtual Game Environments	Synchronous	Yes	Yes	Yes	Yes

Websites are usually asynchronous, providing community information and links to other sites, but sometimes also have synchronous software, like chats, embedded in them (Preece, 2000).

E-mail is an asynchronous mode of communication, usually in the form of text. However, the ability to add attachments to e-mail messages makes it possible for audio, video, and graphic files to be used also. Voice mail is an expansion of e-mail whereby users may record themselves speaking out a message and then send that voice file to their contact, instead of typing it. Newsgroups, like e-mail, provide an asynchronous mode of communication, but unlike e-mail where the messages come to the users, it is a "pull" technology meaning the users must go to the UseNet groups themselves (Preece, 2000). Finally, discussion boards, also referred to as "forums" or "bulletin boards," provide an asynchronous mode of communication where the users post messages for others to see and respond to on their own time.

Preece (2000) described in detail some of these types of CMC and their different characteristics. In this chapter, we focus on Wiki and Online Virtual Game Environments that in our view provide a set of new, novel modes of communication and online-community-building systems.

Wiki-based communities. The "Wiki", named for the Hawaiian word "quick," is a new technology that embodies the notion of new media use where everyone is the author. It is a freely expandable collection of hypertexts that can be easily edited by any user with knowledge of a simple markup language. It does not require any specific tools; all you need is a form-capable web-browser client. This simplistic method gives the freedom to everyone reading the page to amend or correct what he or she is reading. This aims to encourage people to contribute to the expansion of the page (Halvorsen, 2005). Every user can be an author by simply clicking the edit button, changing the current text, and submitting the new version, which is then converted automatically into HTML.

Wikis are a system that explicitly supports collaboration and community building as it decentralizes the effort of creating a website from the hands of the few and distributes it to a huge community of Internet users. In a Wiki environment, users are not only editing, they are also encouraged to create their own content and their own pages. A link to existing pages can be made easily, and a new page can be created by making a new link. Thus, apart from contents, the users also codesign the structure of a Wiki site.

The goal of Wiki sites is to become a shared repository of knowledge, with the knowledge base growing over time. Unlike chat rooms, Wiki content is expected to have some degree of seriousness and permanence. In a Wiki, the users create the content in collaboration and over time. Like weblogs, Wikis have been around for some time and are popular among the technology community. However, weblogs can be highly personal while Wikis are intensely collaborative. Whereas weblogs tend to use a modified WYSIWYG editing environment, Wikis use a simple set of formatting commands.

Recently there has been an increasing interest in using Wikis for learning (Wang et al., 2005; Jones, 2003). Although any knowledge-building application that demands the absolute and immutable integrity of the content is not really suitable for a

Wiki, it is useful in situations where communities of people are developing shared ideas, values, or resources. A biology teacher for example could start a Wiki site by posting some material creating a tentative structure of the subject, and uploading some media files. When the students visit, they can expand the contents by modifying or posting more material, or making links to new pages, thus enriching the learning resources. Through shared construction cycles the students feel closer to the learning system as they contribute to its development instead of being passively presented with the information.

A project has already been undertaken to rebuild a web-based learning site for spectroscopy using Wiki technologies with the goal of making the resource more relevant and content-rich, to attract authors from different backgrounds to provide contents in multiple languages to support the international users (Mader, 2004). Perhaps the most famous educational Wiki in existence is Wikipedia (Wikipedia, 2005), a free-content encyclopedia in many languages that anyone can edit. It is the result of a vast collaboration and currently contains over a million entries in multiple languages. The Wikipedia project shows that the model might work, and that groups of people can collaboratively create shared knowledge artifacts.

However, a Wiki site could be highly unstructured because no editorial function examines the contributions or guarantees quality or accuracy of its content. It is the responsibility of the users to ensure correctness and it is their collective responsibility to take care of all aspects of policy, such as appropriate behavior in the community as well as rules (Halvorsen, 2005).

Since anyone can literally contribute anything, vandalism sometimes occurs within Wikis. Therefore, version control is crucial. Each version is saved, so that it is easy to reverse to a former version if necessary (Emigh & Herring, 2005). Studies have confirmed that vandals are often stopped and the vandalized pages are reversed quickly (Viégas, Wattenberg, & Dave, 2004; Aronsson, 2002).

Wiki-based collaboration could be a very practical platform for asynchronous distributed brainstorming; since Wikis allow ideas to be captured quickly, they facilitate elaboration on existing ideas and document the evolution of revisions made to ideas.

With the increasing popularity of Wikis, it is not surprising that recently it has been used to assemble online communities of various interests. The central access for all users has made the Wiki appropriate for project work, document production, the joint development of project concepts, and discussion forums of all types.

The Wiki technology appears to be a suitable approach for supporting cooperative community, especially for knowledge generation. One of the reasons that make it a successful tool for community building is its "anyone-can-contribute" policy, where producing new pages is quick and easy, as is linking them to existing pages.

Moreover, it also provides a simple means of communication through the discussion section associated with each Wiki page as well as the history entry that keeps track of the changes made by every user. In such a way, the users are constructing and discussing the page at the same time within the same space. This results in the rapid appearance of a chaotic network of Wiki pages and the knowledge grows organically. Facilitated by the Wiki technology, users can contribute and integrate knowledge into and obtain knowledge from the existing database created by others.

The particular advantage of the Wiki approach compared to other CMC tools on knowledge generation and exchange systems is the focus on both the process as well as the result of communication (Kittowski, F. F. & Köhler, A., 2002).

Discussion boards focus largely on the process of the communication where opinions are exchanged, conflicts are resolved, and agreements are achieved regarding the documents/ projects that are being worked on. The document/project itself is not being worked/expanded explicitly through the use of discussion boards; instead, only after the consensus is reached will the project/document be worked by extracting ideas from the postings of the discussion.

Content and document management systems are used to organize and facilitate collaborative creation of documents and other content, usually applications for maintaining documents, including version controls, right of access, document protections, and so on. In terms of collaboration, this allows the process of expansion for each document version by several people. Communication processes and discussions that lead to the final product are limited, for example, to annotations.

Wikis, in contrast, permit users to discuss and work on the document simultaneously. Moreover, cooperative production of content becomes very efficient through the realignment of the distinction between author and reader. Thus, the Wikis' features largely fulfill the requirements for a tool to support a cooperative, knowledge-building community.

In fact, many Wikis are collaborative communities. As mentioned previously, Wikis allow anyone to click the "edit button" and change the website. Surprisingly, many Wikis are able to do this successfully without major issues in terms of vandalism. One of the important vandal control features is that the Wiki stores the history of each page. Thus, for each vandal, there is probably a group of people who actually needs the information that was removed through vandalism, and who will reverse the page to its former version. In addition, some Wikis are in fact not completely open as they restrict access, while some even have a democratic error/vandalism reporting system.

Online virtual game communities. With the advent of ubiquitous broadband Internet connection and the increasing graphical processing power of personal computers, a new paradigm of gaming has emerged. Massively Multi-Player Online Role Playing Games (MMORPGs) have changed the game industry dramatically. MMORPGs provide a fictional setting where a large group of users voluntarily immerse themselves in a graphical virtual environment and interact with each other by forming a community of users.

Although the concept of multiplayer gaming is not new, the game world of most local network multiplayer games, as opposed to MMORPG, are simplistic and can accommodate only around 16 concurrent players in a limited space.

An MMORPG enables thousands of players to play in an evolving virtual world simultaneously over the Internet. The game world is usually modeled with highly detailed 3D graphics, al-lowing individuals to interact not only with the gaming environment, but also with other players. Usually this involves the players representing themselves with avatars, the visual representation of the player's identity in the virtual world.

The MMORPG environment is a new paradigm in computer gaming in which players are part of a persistent world that exists independent of the users (Yee, 2005). Unlike other games where the virtual world ceases to exist when players switch off the game, in an MMORPG, the world exists before the user logs on, and continues to exist when the user logs off. More importantly, events and interactions occur in the world even when the user is not logged on, as many other players are constantly interacting, thus transforming the world. To accommodate the large number of users, the worlds in MMORPGs are vast and varied in term of "geographical locations," characters, monsters, items, and so forth. More often than not, new "locations" or items are added by the game developers from time to time according to the demand of the players.

On one hand, an MMORPG like a Role Playing Game (RPG) involves killing monsters, collecting items, developing characters, and so on. However, it contains an extra aspect, which is the internal sociability within the game. Unlike single player games that rely on other external modes of communication (such as mailing lists, discussion forums outside the game) to form the gaming culture, the culture is formed within the MMORPG environment itself.

In such a way, these MMORPG virtual worlds represent the persistent social and material world, which is structured around narrative themes (usually fantasy), where players are engaged in various activities: slaying monsters, attacking castles, scavenging goods, trading merchandise, and so on. On one hand, the game's virtual world represents the escapist fantasy; on another, it supports social realism (Kolbert, 2001).

That means games are no longer meant to be a solitary activity played by a single individual. Instead, the player is expected to join a virtual community that is parallel with the physical world, in which societal, cultural, and economical systems arise. It has been gradually becoming a world that allows players to immerse into experiences that closely match those of the real world: virtual relationships are sought, virtual marriages are held, virtual shops are set up, and so forth.

The MMORPG genre now boasts hundreds of thousands of users and accounts for millions of dollars in revenue each year. The number of people who play the games (and the time they invest in terms of activities within and around the game) is astounding. The MMORPG Lineage (NCsoft, 2005), for example, had more than 2.5 million current subscribers in 2002 (Vaknin, 2002) and within a year, Ultima Online (Electronic Art, 2005) attracted more than 160 million person-hours (Kolbert, 2001).

Such games are ripe for cultural analysis of the social practices around them. Although fundamentally MMORPGs are video games with virtual spaces with which the players interact, they should be regarded not just as a piece of game software; they are a community, a society, and, if you wish, a culture. These games are becoming the most interesting interactive computer-mediated communication and networked activity environment (Taylor, 2002). Thus, understanding the pattern of participation in these game communities is crucial, as such virtual communities function as a major mechanism of socialization of the players to the norms of the game culture that emerges as Squire and Steinkeuhler (in press) noted:

Playing one's character(s) and living in [these virtual worlds] becomes an important part of daily life. Since much of the excitement of the game depends on having personal relationships and being part of [the] community's developing politics and projects, it is hard to participate just a little (Squire, Steinkeuhler, in press)

Recently game designers have tried to stretch the boundaries further by structuring in-game activities to maximize interaction. One of the examples of sociability by design in MMORPG is Star Wars Galaxies (Sony, 2005), which is organized so that players are steered toward certain locations in the game world where social playing is expected to take place (Ducheneaut, Moore, & Nickell, 2004).

Such communities formed around the game can be broadly divided into two categories: in-game and out-of-game communities. Most MMORPGs are created to encourage long-term relationships among the players through the features that support the formation of in-game communities. One of the most evident examples is the concept of guilds. Guilds are a fundamental component of MMORPG culture for people who are natural organizers to run a virtual association that has formalized membership and rank assignments and encourage participation. Sometimes, a player might join a guild and get involved in a guild war in order to fight for the castle. Each guild usually has a leader and several guilds could team up in a war. This involves a complicated leader-subordinate and leader-leader relationship.

In addition, to encourage social interaction, MMORPGs are specially designed in such a way that some game goals are almost impossible to be achieved without forming communities. For example, one player alone could spend a long period collecting all of the items needed to assemble a device. But a guild could ask its members to fan out in small groups and collect all of the necessary components in one day. Complex devices beyond the reach of any individual player could be quickly constructed by the guild. The guild could also accept donations from members and then distribute those contributions to others according to their needs, benefiting everyone through this collaboration (Kelly, 2004).

Apart from relatively long-term relationships such as guild communities, MMORPGs also provide many opportunities for short-term relationship experiences. For example, a player could team up with another player to kill monsters in order to develop the abilities of their avatars ("level up") or some more expert players could help newer players to get through the game.

When trying to win the game, players often need to get information from other resources: guidebooks, discussion forums, other players, and the like. Therefore game playing is generally more concerned with player-player interaction than with player-game interaction. What is at first confined to the game alone soon spills over into the virtual world beyond it (e.g., websites, chat rooms, e-mail) and even life off-screen (e.g., telephone calls, face-to-face meetings).

Apart from these external communities around the game that are mediated through e-mails or online forums (which also exist in many other games), an interesting phenomenon fuses the internal and external game communities. The participation in an external community starts to break the magic circle of the game—that game space is no longer separate from real life—as the out-of-game community trades in-game items for real money.

For example, Norrath, the world of EverQuest, was estimated to have the 77th largest economy in the real world based on buying and selling in online auction houses.

About 12,000 people call it their permanent home, although some 60,000 are present there at any given time. The nominal hourly wage is about USD3.42 per hour, and the labors of the people produce a GNP per capita somewhere between that of Russia and Bulgaria. A unit of Norrath's currency is traded on exchange markets at USD0.0107, higher than the Yen and the Lira. (Castronova, 2001)

Having illustrated the social phenomenon around such playful virtual community, it is believed that it is fruitful to research such communities as we might be able to derive some useful implications on how successful Computer Supported Collaborative Work (CSCW) and Computer Supported Collaborative Learning (CSCL) environments can be designed. For this reason, in the next section we will describe some of the methodologies that can be used in such studies, and later we will present the application of some of these methods to two case studies.

ANALYZING ONLINE COMMUNITIES: FRAMEWORKS AND METHODOLOGIES

There are various aspects and attributes of CMC that can be studied to help us better understand online communities—for instance, the analysis of the frequency of exchanged messages and the formation of social networks, or the analysis of the content of the exchanged messages and the formation of virtual communities. To achieve such an analysis a number of theoretical frameworks have been developed and proposed. For example, Henri (1992) provided an analytical model for cognitive skills that can be used to analyze the process of learning within messages exchanged between students of various online e-learning communities. Mason's work (1991) provided descriptive methodologies using both quantitative and qualitative analysis. Furthermore, five phases of interaction analysis are identified in Gunawardena, Lowe, and Anderson's model (1997):

1. Sharing/Comparing of Information
2. The Discovery and Exploration of Dissonance or Inconsistency among Ideas, Concepts, or Statements
3. Negotiation of Meaning/Co-Construction of Knowledge
4. Testing and Modification of Proposed Synthesis or Co-Construction
5. Agreement Statement(s)/Applications of Newly Constructed Meaning

In this section we provide a description of some of the most commonly used online community evaluation techniques as well as their weaknesses and strengths.

Query-Based Techniques and User Profiles: Interviews, Questionnaires, and Personas

Interviews. An interview can be defined as a type of conversation that is initiated by the interviewer in order to obtain research relevant information. The interview reports have to be carefully targeted and analyzed to make an impact (Usability Net, 2003). Interviews are usually carried out on a one-to-one basis in which the interviewer collects information from the interviewee. Interviews can take place by telephone or face to face (Burge & Roberts, 1993). They can also take place via nonreal-time methods like fax and e-mail, although in these cases they function like questionnaires. Interviews are useful for obtaining information that is difficult to elicit through approaches such as background knowledge and general principles. There are three types of interviews:

1. Structured interviews, which consist of predetermined questions asked in fixed order like a questionnaire;
2. Semi-structured interviews, in which questions are determined in advance but may be reordered, reworded, omitted, and elaborated;
3. Unstructured interviews, which are not based on predetermined questions but instead upon a general area of interest, thus allowing the conversation to develop freely.

Interviews can be used to gain insights about general characteristics of the participants of an online community and their motivations for participating in the community under investigation. The data collected comes straight from the participants of the online communities, whereby they are able to provide feedback based on their own personal experiences, activities, thoughts, and suggestions.

Advantages of interviews (Usability Net, 2003) include (a) what is talked about can address directly the informant's individual concerns; (b) mistakes and misunderstandings can be quickly identified and cleared up; (c) they are more flexible than questionnaires; and (d) they can cover low-probability events.

Disadvantages of interviews include (a) the danger of analyst bias towards his or her own knowledge and beliefs; (b) the accuracy and honesty of responses; and (c) they often must be used with other data-collection techniques to improve quality of data collected.

Questionnaires. A questionnaire is a self-reporting query-based technique. Questionnaires are typically produced on printed paper, but due to recent technologies and in particular the Internet, many researchers engage in the use of online questionnaires, thus saving time and money and eliminating the problem of a participant's geographical distance. Three types of questions can be used with questionnaires: (a) open questions, where the participants are free to respond however they like; (b) closed questions, which provide the participants with several choices for the answer; and (c) scales where the respondents must answer on a predetermined scale.

For online communities, questionnaires can be used to elicit facts about the participants, their behaviors, and their beliefs/attitudes. Like interviews, questionnaires are an important technique for collecting user opinions and experiences they have had using CMC and their overall existence in online communities.

The main advantages of questionnaires are that they are faster to carry out than observational techniques, and they can cover low-probability events. Disadvantages include (a) information is an idealized version of what should, rather than what does, happen; (b) responses may lack accuracy or honesty; (c) there is danger of researcher bias towards the subset of knowledge he or she possesses; and (d) they must be used in conjunction with other techniques for validity.

Personas. Findings from interviews and questionnaires can be further used as a basis for developing user profiles using personas. A persona is a precise description of the user of a system, and of what he or she wishes to accomplish (Cooper, 1999). The specific purpose of a persona is to serve as a tool for software and product design. Although personas are not real people, they represent them throughout the design stage (Blomkvist, 2002) and are best based on real data collected through query-based techniques.

Personas are rich in details, including name, social history, and goals, and are synthesized from findings using query-based techniques with real people (Cooper, 1999). The technique takes into account user characteristics and creates a concrete profile of the typical user (Cooper, 1999).

For online communities, personas can be used to better understand the participants of the community and their backgrounds. Personas can also be used as a supplement to social network analysis (described later in this chapter) to get a greater overview of the characteristics of key participants of a community. Using personas, web developers gain a more complete picture of their prospective and/or current users and are able to design the interfaces and functionality of their systems to be more personalized and suited for the communication of the members of their online communities.

Advantages of personas include (a) they can be used to create user scenarios; (b) they can be anonymous, protecting use privacy; and (c) they represent the user stereotypes and characteristics.

Disadvantages of personas include that if not enough personas are used, users are forced to fall into a certain persona type that might not accurately represent them; and that they are time-consuming.

Log Analysis

A log, also referred to as weblog, server log, or log-file, is in the form of a text file and is used to track the users' interactions with the computer system they are using. The types of interaction recorded include key presses, device movements, and other information about the user activities. The data is collected and analyzed using specialized software tools and the range of data collected depends on the log settings. Logs are also time-stamped and can be used to calculate how long a user spends on a particular task or how long a user lingers in a certain part of the website (Preece, Rogers, & Sharp, 2002). In addition, an analysis of the server logs can help us find out when people visited the site, the areas they navigated, the length of their visits, the frequency of their visits, their navigation patterns, from where they are connected, and details about the computers they are using.

Log analysis is a useful and easy-to-use tool when analyzing online communities. For example, someone can use log analysis to answer more accurately questions like student attendance of an online learning community. Furthermore, logs can identify the web pages users spend more time viewing, and the paths that they used. This helps identify the navigation problems of the website, but also gives a visualization of the users' activities in the virtual communities. For instance, in the case of e-learning communities, the log files will show which students are active in the CMC postings even if they are not active participants (few postings themselves), but just observing the conversations. Preece and Maloney-Krichmar (2003) noted that data logging does not interrupt the community, while at the same time can be used to examine mass interaction.

Advantages of logs (Preece et al., 2002) are that they (a) help evaluators analyze users' behavior; (b) help evaluators understand how users worked on specific tasks; (c) are unobtrusive; and (d) can automatically log large volumes of data.

Disadvantages of logs (Preece et al., 2002) are that (a) powerful tools are needed to explore and analyze the data quantitatively and qualitatively; and (b) there are user privacy issues.

Content and Textual Analysis

Content analysis is an approach to understanding the processes that participants engage in as they exchange messages (McLoughlin, 1996). There have been several frameworks created for studying the content of messages exchanged in online communities. Examples include work from Archer, Garrison, Anderson, and Rourke (2001); Gunawardena, Lowe, and Anderson's (1997) model for examining the social construction of knowledge in computer conferencing; Henri's (1992) content-analysis model and Fahy, Crawford, and Ally's (2001) Transcript Analysis Tool (TAT) which is described in more detail below.

The TAT focuses on the content and interaction patterns at the component level of the transcript (Fahy et al., 2001). After a lengthy experience with other transcript tools and reviews of previous studies, Fahy et al. (2001) chose to adapt Zhu's (1996) analytical model for the TAT. Zhu's model (1996) examined the forms of electronic interaction and discourse, the forms of participation and the direction of participant interaction in computer conferences. The TAT also contains echoes of Vygotskian theory, primarily those dealing with collaborative sense making, social negotiation, and proximal development (Cook & Ralston, 2003). The TAT developers have come up with the following strategic decisions (Fahy, 2003): (a) the sentence is the unit of analysis; (b) the TAT is the method of analysis; and (c) interaction is the criterion for judging conference success and topical progression (types and patterns).

The TAT was designed to permit transcript content to be coded reliably and efficiently (Fahy et al., 2001), while the advantages of TAT are (Fahy, 2003; Cook & Ralston, 2003; Fahy et al., 2001):

It reveals interaction patterns that are useful in assessing different communication styles and online behavioral preferences among participants;

It recognizes the complexity of e-conferences and measures the intensity of interaction;

It enables the processes occurring within the conferences to be noted and recorded;

It probes beyond superficial systems data, which mask the actual patterns of discussion;

It relates usefully to other work in the area;

It discriminates among the types of sentences within the transcript;

It reflects the importance of both social and task-related content and outcomes in transcript-analysis research.

The unit of analysis of the TAT is the sentence. In the case of highly elaborated sentences, the units of analysis can be independent clauses which, punctuated differently, could be sentences (Fahy, 2003). Fahy et al. (2001) have concluded that the selection of message-level units of analysis might partially explain problematic results that numerous researchers have had with previous transcript-analysis work. They also believe that the finer granularity of sentence-level analysis results in several advantages (Fahy, 2003):

Reliability;

Ability to detect and describe the nature of the widely varying social interaction, and differences in networking pattern, in the interactive behavior of an online community, including measures of social network density and intensity;

Confirmation of gender associations in epistolary/expository interaction patterns, and in the use of linguistic qualifiers and intensifiers. Table 14.3 shows the TAT categories (Fahy et al., 2001; Fahy, 2003).

Social Network Analysis (SNA)

Social Network Analysis (SNA) is the mapping and measuring of relationships and flows between people, groups, organizations, computers or other information/knowledge processing entities. The nodes in the network are the people and groups while the links show relationships or flows between the nodes. SNA provides both a visual and a mathematical analysis of human relationships. (Krebs, 2004, p. 1)

Preece (2000) added that SNA provides a philosophy and a set of techniques for understanding how people and groups relate to each other, and has been used extensively by sociologists (Wellman, 1982; Wellman, 1992), communication researchers (Rice, 1994; Rice, Grant, Schmitz, & Torobin, 1990) and others. Analysts use SNA to determine if a network is tightly bounded, diversified, or constricted; to find its density and clustering; and to study how the behavior of network members is affected by their positions and connections (Garton, Haythornthwaite & Wellman, 1997; Hanneman, 1998; Scott, 2000). Network researchers have developed a set of theoretical perspectives of network analysis. Some of these are (Borgatti, 2002),

- Focus on relationships between actors rather than attributes of actors;
- Sense of interdependence: a molecular rather than atomistic view;
- Structure affects substantive outcomes;
- Emergent effects.

"The aim of social network analysis is to describe why people communicate individually or in groups" (Preece, 2000, p. 183), while the goals of SNA are (Dekker, 2002)

- To visualize relationships/communication between people and/or groups using diagrams;

TABLE 14.3. TAT Categories

Category

1: Questioning
The questioning category is further broken down into two types of questions:

1A Vertical Questions
These questions assume a "correct" answer exists, and that they can be answered if the right authority to supply it can be found. Example: "Does anybody know what time the library opens on Saturdays?"

1B Horizontal Questions
For these questions, there may not be only one right answer. These questions invite negotiation. Example: "Do you really think mp3 files should become illegal, or you don't see any harm by them?"

2: Statements
This category consists of two subcategories:

2A Nonreferential Statements
These statements contain little self-revelation and usually do not invite response or dialogue and their main intent is to impart facts or information. Example: "We found that keeping content up-to-date, distribution and PC compatibility issues were causing a huge draw on Ed. Centre time."

2B Referential Statements
Referential statements are direct answers to questions. They can include comments referring to specific preceding statements. Example: "That's right; it's the 1997 issue that you want."

3: Reflections
Reflections are significant personal revelations, where the speaker expresses personal or private thoughts, judgments, opinions, or information. Example: "My personal opinion is that it shouldn't have been a penalty kick."

4: Scaffolding and Engaging
Scaffolding and engaging initiate, continue, or acknowledge interpersonal interaction. They personalize the discussion and can agree with, thank or otherwise recognize someone for their helpfulness and comments. Example: "Thanks Dave; I've been trying to figure that out for ages :)"

5: References/Authorities
Category 5 is compromised of two types:

5A Quotations, references to, paraphrases of other sources.
Example, "You said, 'I'll be out of the city that day.'"

5B Citations, attributions of quotations and paraphrases.
Example: "Mathew, P. (2001). A beginners' guide to mountain climbing."

- To study the factors which influence relationships and the correlations between them;
- To draw out implications of the relational data, including bottlenecks;
- To make recommendations to improve communication and workflow in an organization.

Preece (2002) and Beidernikl and Paier (2003) listed the following as the limitations of SNA:

- More theory that speaks directly to developers of online communities is needed;
- The data collected may be personal or private;
- The analysis of the data is quantitative and specific to the particular network, while common survey data are qualitative and generalize answers on the parent population.

It is also worth pointing out that network analysis is concerned about dyadic attributes between pairs of actors (like kinship, roles, and actions), while social science is concerned with monadic attributes of the actor (like age, sex, and income).

There are two approaches to SNA:

Ego-centered analysis—Focuses on the individual as opposed to the whole network, and only a random sample of network population is normally involved (Zaphiris, Zacharia, & Rajasekaran, 2003). The data collected can be analyzed using standard computer packages for statistical analysis like SAS and SPSS (Garton, Haythornthwaite, & Wellman, 1997).

Whole network analysis—The whole population of the network is surveyed and this facilitates conceptualization of the complete network (Zaphiris et al., 2003). The data collected can be analyzed using microcomputer programs like UCINET and Krackplot (Garton et al., 1997).

The following are important units of analysis and concepts of SNA (Garton et al., 1997; Wellman, 1982; Hanneman, 2001; Zaphiris et al., 2003; Wellman, 1992):

- Nodes: The actors or subjects of study;
- Relations: The strands between actors. They are characterized by content, direction, and strength;
- Ties: Connect a pair of actors by one or more relations;
- Multiplexity: The more relations in a tie, the more multiplex the tie is;
- Composition: This is derived from the social attributes of both participants;
- Range: The size and heterogeneity of the social networks;
- Centrality: Measures who is central (powerful) or isolated in networks;
- Roles: Network roles are suggested by similarities in network members' behavior;
- Density: The number of actual ties in a network compare to the total amount of ties that the network can theoretically support;
- Reachability: In order to be reachable, connections that can be traced from the source to the required actor must exit;
- Distance: The number of actors that information has to pass through to connect one actor with another in the network;
- Cliques: Subsets of actors in a network, who are more closely tied to each other than to the other actors who are not part of the subset.

Social Network Analysis is a very valuable technique when it comes to analyzing online communities, as it can provide a visual presentation of the community and, more importantly, it can provide us with qualitative and quantitative measures of the dynamics of the community. The application of SNA to the analysis of online communities is further demonstrated with a case study later in this chapter.

CASE STUDIES

In this section we present two case studies that demonstrate the use of theoretical and analytical techniques for studying online communities. In the first case study, we demonstrate how the results from an "attitude toward thinking and learning" questionnaire can be combined with social network analysis to describe the dynamics of a computer-aided language-learning (CALL) online community. In the second case study, we present a theoretical activity theory model that can be used for describing interactions in online game communities.

Computer Aided Language Learning Communities

In the first case study we demonstrate a synthetic use of quantitative (SNA) and qualitative (questionnaires) methods for analyzing the interactions that take place in a Computer Aided Language Learning (CALL) course. Data was collected directly from the discussion board of the "Learn Greek Online" (LGO) course (Kypros-Net Inc., 2005).

LGO is a student-centered e-Learning course for learning Modern Greek and was built using a participatory design and distributed constructionism methodology (Zaphiris & Zacharia, 2001). In an ego-centered SNA approach, we have carried out an analysis of the discussion postings of the first 50 actors (in this case the students of the course) of LGO.

To carry out the social network analysis we used an SNA tool called "NetMiner" (Cyram, 2004), which enabled us to obtain centrality measures for our actors. The "in and out degree centrality" was measured by counting the number of interaction partners per each individual in the form of discussion threads (for example, if an individual posts a message to three other actors, then his or her out-degree centrality is three; whereas if an individual receives posts from five other actors then his or her in-degree is five).

Due to the complexity of the interactions in the LGO discussion we had to make several assumptions in our analysis:

- Posts that received zero replies were excluded from the analysis. This was necessary in order to obtain meaningful visualizations of interaction;
- Open posts were assumed to be directed to everyone who replied;
- Replies were directed to all the existing actors of the specific discussion thread unless the reply or post was specifically directed to a particular actor.

In addition to the analysis of the discussion board interactions we also collected subjective data through the form of a survey. More specifically, the students were asked to complete an Attitudes Towards Thinking and Learning Survey (ATTLS). The ATTLS uses 20 Likert scale questions to measure the extent to which a person is a "connected knower" (CK) or a "separate knower" (SK). People with higher CK scores tend to find learning more enjoyable, and are often more cooperative and congenial, and more willing to build on the ideas of others, while those with higher SK scores tend to take a more critical and ar-

gumentative stance to learning (Galotti, Clinchy, Ainsworth, Lavin, & Mansfield, 1999).

The out-degree results of the social network analysis are depicted in Fig. 14.1 in the form of a sociogram, and the in-degree results are depicted in Fig. 14.2. Each node represents one student (to protect the privacy and anonymity of our students, their names have been replaced by a student number). The position of a node in the sociogram is representative of the centrality of that actor (the more central the actor the more active). As can be seen from Fig. 14.1, students S12, S7, S4, and S30 (with out-degree scores ranging from 0.571 to 0.265) are at the centre of the sociogram and possess the highest out-degree. The same students also posses the highest in-degree scores (Fig. 14.2). This is an indication that these students are the most active members of this online learning community, posting and receiving the largest number of postings. In contrast, participants in the outer circle (e.g., S8, S9, S14, etc.) are the least active with the smallest out-degree and in-degree scores (all with 0.02 out-degree scores).

In addition, a clique analysis was carried out (Fig.14. 3) and it showed that 15 different cliques (the majority of which are overlapping) of at least three actors each have been formed in this community.

As part of the ego-centered analysis for this case study we look in more detail at the results for two of our actors: S12, who is the most central actor in our SNA analysis (e.g., with the highest out-degree score), and S9, an actor with the smallest out-degree score. It is worth noting that both members joined the discussion board at around the same time.

First, through a close look at the clique data (Table 14.4) we can see that S12 is a member of 10 out of the 15 cliques, whereas S9 is not a member of any—an indication of the high interactivity of S12 versus the low interactivity of S9. In an attempt to correlate the actors' position in the SNA sociogram with their

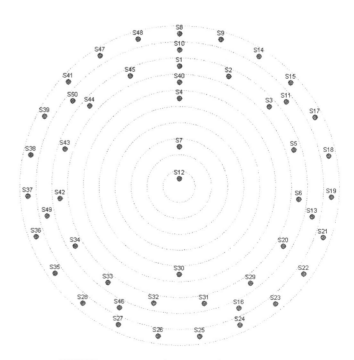

FIGURE 14.1. Out-degree analysis sociogram.

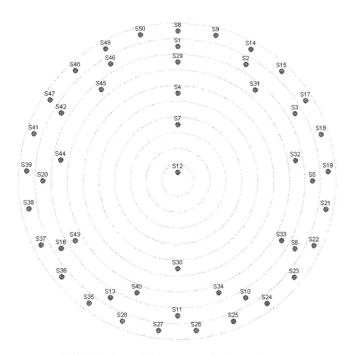

FIGURE 14.2. In-degree analysis sociogram.

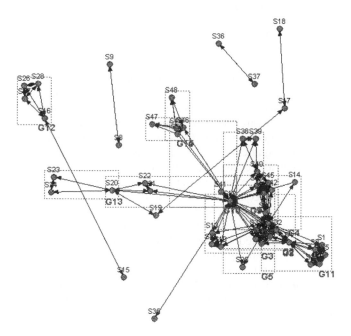

FIGURE 14.3. Clique analysis sociogram.

TABLE 14.4. Clique Analysis of the LGO Discussions

Cliques	Actors
K1	S12, S7, S30, S40, S43, S44, S45
K2	S12, S7, S30, S4
K3	S12, S7, S10, S11, S13
K4	S12, S7, S14
K5	S12, S7, S25
K6	S12, S7, S41
K7	S12, S20, S21, S22
K8	S12, S29, S4, S30, S31, S32, S33, S34
K9	S12, S38, S39, S40
K10	S12, S46, S49, S50
K11	S1, S2, S3, S4, S5, S6, S7
K12	S16, S26, S27, S28
K13	S23, S20, S24
K14	S47, S46, S49, S50
K15	S48, S46, S49, S50

S9 answered them with a score of 1 (strongly disagree). More specifically, S12 strongly agreed that

1. S/He is more likely to try to understand someone else's opinion than to try to evaluate it;
2. S/He often finds herself/himself arguing with the authors of books read, trying to logically figure out why they're wrong;
3. S/He finds that he/she can strengthen his/her own position through arguing with someone who disagrees with them;
4. S/He feels that the best way achieve his/her own identity is to interact with a variety of other people;
5. S/He likes playing devil's advocate, arguing the opposite of what someone is saying.

S9 strongly disagreed with all of the above statements. These are all indications that S12 is a "connected knower" (CK) whereas S9 is a "separate knower" (SK).

This case study showed that the combination of quantitative and qualitative techniques can facilitate a better and deeper understanding of online communities. SNA was found to be a very useful technique for visualizing interactions and quantifying strengths and dynamics in online communities. In combination with the ATTLS, it was possible to identify the key players of the e-Learning community. These members' roles show them to be more powerful and central in the discussions. Identifying their characteristics enables us to reinforce the community by making other participants more active in the discussion board communication. This active learning approach could, in turn, improve the pedagogical value of e-learning within these communities.

Game Communities and Activity Theoretical Analysis

The main motivation of the second case study arises from the more general area of computer game-based learning. Game-based learning has focused mainly on how the game itself can be used to facilitate learning activities but we claim that the educational opportunity in computer games stretches beyond the learning activities in the game per se. Indeed, if you observe

self-reported attitudes towards teaching and learning we looked more closely at the answers these two actors (S12, S9) provided to the ATTLS. Actor S12 answered all 20 questions of the ATTLS with a score of at least 3 (on a 1–5 Likert scale), whereas S9 had answers ranging from 1 to 5. The overall ATTLS score of S12 is 86 whereas that of S9 is 60. A clear dichotomy of opinions occurred on 5 of the 20 questions of the ATTLS. S12 answered all 5 of those questions with a score of 5 (strongly agree) whereas most people playing games, you will likely see them download-ing guidelines from the Internet and participating in online fo-rums to talk about the game and share strategies. In actuality, al-most all game playing could be described as a social experience, and it is rare for a player to play a game alone in any meaning-ful sense (Kuo, 2004). This observation is even more evident in Massively Multiplayer Online Role Playing Games (MMORPGs) which has been discussed earlier in this chapter. For example, the participation in an MMORPG is constituted through language practice within the in-game community (e.g., in-game chatting and joint tasks) and out-of-game community (e.g., the creation of written game-related narratives and fan-sites). The learning is thus not embedded in the game, but it is in the community practice of those who inhabit the game.

Types of game communities. Because of the points mentioned earlier, we believe that the study on computer games should be expanded to include the entire game commu-nity. Computer game communities can be categorized into three classes that we have identified (Fig. 14. 4) (Ang, Zaphiris & Wilson, 2005) as:

Single Game-play Community—This refers to a game com-munity formed around a single-player game. Although play-ers of a single-player game like "The Sims 2" and "Final Fan-tasy VII" play the game individually, they are associated with an out-of-game community that discusses the game either virtually or physically.

Social Game-play Community—This refers to multiplayer games that are played together in the same physical location. It creates game communities at two levels: in-game and out-of-game. Occasionally, these two levels might overlap. The out-of-game interaction might be affected by issues beyond the specific game system; for example, the community starts exchanging information about another game.

Distributed Game-play Community—This is an extension of social game-play community, but it emphasizes the online multiplayer game in which multiple sessions of game are es-tablished in different geographical locations.

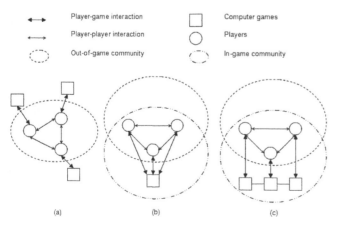

(a) Single game-play community (b) Social game-play community (c) Distributed game-play community

FIGURE 14.4. Types of game communities.

The study of game communities, especially out-of-game communities, from the perspective of education is still very much unexplored. We believe the potential of games in educa-tion is not limited to what is going on in the game. Educators could benefit by studying games as a social community because games are now becoming a culture that permeates the lives of many, especially the younger generation. Black (2004) has in-vestigated the interactions among participants in a virtual com-munity of Japanese comic fans that involve a lot of reading and writing throughout the site. She examines how the community of fans helps each its members with English language writing skills and with cross-cultural understanding. In this section we have pointed out that game communities can emerge from both single-player and multiplayer games. We believe that by fur-ther studying the social interaction in the game community, we will be able to utilize games to learn in a more fruitful way. In the next section, we apply and evaluate one of these models of game communities to a specific scenario in knowledge build-ing using activity theory.

Activity theory. In this case study, we demonstrate how activity theory can be used to analyze an out-of-game commu-nity around a single-player game that is based on construction-ist activities as proposed by Papert (1980). Papert claimed that even for adults, learning remains essentially bound to context, in which knowledge is shaped by the use of external supports. His approach helps us understand how learning is actualized when individual learners construct their own favorite artifacts or objects-to-think-with (Papert, 1980).

Although Papert's theory provided a solid framework for understanding children's, and even adults', ways of learning by designing, it does not offer a systematic framework for analyz-ing the construction activities within a learning community. Analyzing constructionist activities can be useful, as it could help inform constructional design for educational purposes. The most significant analysis includes the learning within a community as well as the development of an individual. We are also interested in finding out how tools such as computers help learners construct artifacts and knowledge. Hence, we would like to draw from the Vygotskian naturalist approach, which emphasizes human activity systems. Lev Vygotsky (1930) formulated a theoretical concept that is very different from the prevailing understanding of psychology that was dominated by behaviorism at that time. This new orientation was a model of tool-mediation and object-orientedness. He proposes the classic triangle model to demonstrate the idea of mediation:

In Fig. 14.5, the subject is the individual engaged in the me-diated action, and the mediating artifact or tool could include physical artifacts and/or prior knowledge of the subject. The ob-ject is the goal/objective of the activity. Although constructionist learning relies very much on computational tools, the concept of mediation is not explicated. Fig. 14.5 shows explicitly that the relationship between the subject and the object is no longer straightforward; instead, it is mediated by the tool. For example, when building a website, the subject is working towards an ob-jective (e.g., to add a table in the web page) using not only the computer (external tools) but also her internal understanding of how websites and computers work (internal tools).

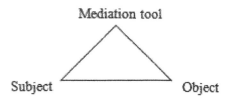

FIGURE 14.5. Vygotsky's mediation.

Leont'ev (1978) extended this notion of activity to differentiate between an individual action and a collective activity by proposing a hierarchy of activity (Table 14.5). Collective activity is connected to the object of the whole community, of which each individual subject is often not consciously aware. An individual action is connected to a conscious goal. Below the collective activity and individual action there is the level of operations that are dependent on the conditions in which the action is performed. Thus, an activity system can be analyzed at three levels: (a) the activity level that is oriented toward the object/objective and carried out by the whole community; (b) the action level that is directed at the individual goal, and (c) the operation level that is elicited by conditions and is performed unconsciously.

This hierarchy is crucial in explaining the learning process in an activity system. We would like to illustrate an example of this hierarchy in learning a foreign language (Table 14.5). The overall objective is to be able to engage in a meaningful conversation. In the beginning, the learner has to work on the grammar and the choice of words at a conscious level. When the learner has reached a higher proficiency level, these actions are transformed into operations. The learner no longer needs to select appropriate words and check grammar rules deliberately, as these have been learned thoroughly and are now operating unconsciously. The consciousness of the learner is now focused on expressing himself properly, depending on the objective of the conversation. Grammatical rules become invisible to the learner and he is only selecting appropriate goals to be achieved. Therefore, it can be inferred that activity theory treats learning as the shift from the higher level (action) to the lower level (operation) in the hierarchy.

Drawing on work by Vygotsky and Leont'ev, Engeström (Engestrom, 2001) viewed all human activities as contextualized within an interdependent activity system. Engeström added collective mediation to Vygotsky's tool mediation and presents the triangle model of activity system (Fig. 14.6).

TABLE 14.5. Hierarchy of Activity

Unit of Analysis	Stimulus	Subject	Language Learning Example
Activity	Object	Community	Engage in a meaningful conversation
Action	Goal	Individual	Sentence construction
Operation	Conditions	Unconscious	Word selections, grammar rules

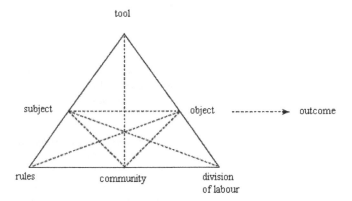

FIGURE 14.6. The triangle activity system diagram.

In the diagram, the subject is the individual or group selected as the point of view of the analysis. The object refers to the raw material or the problem space at which the activity is directed and that is transformed into outcomes with the help of external and internal tools. Tools are the concepts, physical tools, artifacts, or resources that mediate a subject's interactions with an object. The community refers to those with whom the subject shares the same general object. The division of labor (DOL) is the classification of tasks among the members of the community, while the rules are the regulations, norms, and conventions within the activity system.

Constructionist learning can be described and visualized through this activity triangle. Mediated by the tool and the community, the learner externalizes her initial stage of knowledge through object construction. The individual externalization (mediated by the tool) can be broken down into actions and operations. Actions are directed toward a personal goal and are carried out with careful deliberation. For example (Fig. 14.7), in a book-writing activity, the author (an expert word-processor user) will operate (e.g., type) the word processor at the unconscious level, and consciously act on the book (to select appropriate words, and construct meaningful sentences and paragraphs) she is writing. At a certain point, the author encounters a new condition with the word processor that she is not familiar with—for

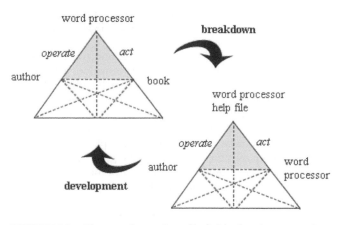

FIGURE 14.7. The transformation of individual action-operation.

example, having to insert a table into the book. Under this new condition, a breakdown is said to have happened. The conscious effort of the author is not any longer placed on the book itself but instead is now placed on the word processor (e.g., to achieve the action (insert tables), the author performs the operation (read help files). Once the author has thoroughly learned about the table insert, she can again act on the book consciously and development is said to have happened.

Activity theoretical analysis for constructionist learning.

We have conducted a study to analyze a Wiki-based game community using activity theory. The goal of this study was to demonstrate the usefulness of activity theory in researching online communities. The collaborative construction activity in a Wiki-supported community devoted to writing a game guidebook is examined. Activity theory is used as an analytical tool in order to investigate this community. In this section, experiences and challenges from the analysis are reported in order to give some insights into how activity theory can be helpful for online community research.

The analysis of online communities can be done through the lenses of various concepts associated with activity theory, such as the levels of activity hierarchy and different perspectives of the triangle model as explained in the previous section.

For instance, we can start our analysis with the most basic aspect of constructionism by simply examining the relationship between subjects and the object. Then, we can analyze Vygotsky's mediation model of activity system consisting of individual actions and tools. The analysis may also be extended to the whole community of this system to include emerging rules as well as division of labor (DOL) that mediate the community. The focus can also be placed mainly on the constructionist concept of externalizing the internal meanings onto a sharable artifact through mediation. More specifically, we can look into these (and other) aspects:

- Subject-object: What are the constructionist actions that act on the object and transform the object into outcome?
- Subject-tool-object: How do actions shift to operations and vice versa? How do tools mediate individual actions and operations? What is the nature of the mediating tools? How do they support knowledge building?
- Subject-rules-object and subject-DOL-object: What is the nature of implicit and explicit rules? How is DOL manifest in the community? How do rules and DOL support knowledge building?

Apart from its focus on both individual and collective development through action-operation transformation, activity theory also helps analyze the tools and capture the rules and the division of labor (DOL) that mediate the actions. These uses must be further explained in order to differentiate individual mediation and collective mediation. Taking the example of our study on game community, individual mediation places its emphasis on how a user uses the tool to write the game guide, without taking into account how other users act in the community. In other words, it is about the affordance of the tool to support what an individual can do.

Collective mediation is about the community, which consists of two major components: rules and DOL. Rules define

what can be done and cannot be done in a community. This should not be confused with the affordance of the tool. The tool might afford certain actions such as writing in abusive language, but the rules might want to ban this action. DOL is self-explained: how the workload is divided among many users in a community.

Activity theory appears to be a promising framework, as it gives an analytical lens on analyzing and interpreting the data. Activity theory provides different perspectives of analysis, as it casts different light on the collected data as researchers can examine it from many perspectives by focusing on different sub-triangles of the activity system diagram. It also helps us examine the learning process—how learning occurs individually and collectively through the transformation of hierarchy of activity from action to operation. Furthermore, both individual and collective aspects are given equal importance. Activity theory informs the development of the whole community as well as the individual's development. It explains how individual development contributes to the community growth, and vice versa.

Based on our study, individual actions help sharpen the mediation tool, while collective actions bring about new rules or refine existing rules that mediate the collective action. In short, activity theory is useful to analyze the community in the following ways:

- It helps understanding of the individual mediation process: subject-tool-object;
- It clearly presents the collective mediation process: subject-community-object;
- It reveals the emerging rules and division of labor in the community.

In a Wiki space, knowledge is socially constructed; it is created individually with tools and negotiated and agreed to within a community based on emerging rules and division of labor (DOL). It starts as a single unit of information (a page in this context) and grows organically and evolves into a complex and well-structured set of knowledge. From our findings, we induce what contributes to the development of the community:

- Users share some historical backgrounds: they already share some of the tools/rules before joining the community, they also share the interest in the same game;
- Users share the same object (goal), which is to build a game guide book;
- A user's individual action: this goal-oriented individual action triggers negotiations that lead to the growth of the space;
- The community's agreement on the object: not only share the same object, the community must be able to negotiate and agree on the object;
- Tools that support these actions and negotiations;
- Emerging rules that coordinate the activity;
- DOL that divide the responsibilities.

The evolution of a knowledge-building community needs more than a group of devoted users who share the same object;

it involves negotiation and agreement among the users on the object. Although every user tends to act toward her or his own goal, it takes the compromise of the entire community to agree on the object.

Activity theory helps online community researchers identify design issues at two aspects: the software's application as tools as well as the social interaction within the community around the tools. It also reveals the development of the community-building process from the individual and collective level, through the shift of activity hierarchy. We thus believe that analyzing online communities with activity theory will yield fruitful results and give insights on online community design.

Finally, we must reiterate the fact that activity theory itself is not limited to what is presented in the triangle diagram as proposed by Engeström. Although his model is tremendously useful, it overlooks several significant concepts of activity theory. One major limitation is the static representation of activity theory. The triangle diagram represents only a snapshot of a particular time, thus making it hard to analyze the activity across time. It is understood that Engeström's model is intended to be open so that it can be used in various domains but this has been proven to pose serious difficulty among the practitioners, as some researchers have started to operationalize it so that it is more practical in day-to-day methodology (Korpela, Soriyan, & Olufokunbi, 2000; Barab, Hay, & Yamagata-Lynch, 2001; Mwanza, 2002). Hopefully with the increasing attention drawn by activity theory, the theory will be expanded and operationalized to fit the purpose of HCI research in general and online community in particular.

DISCUSSION AND CONCLUSIONS

This chapter looked at the definitions of computer mediated communication and online communities and pointed out the multi-disciplinary nature of the definitions and the way online communities are being analyzed and studied. We introduced the different types of CMC and online communities and gave special attention to Wikis and game-based communities. These two relatively new types of communities exhibit new modes of interaction that are worth studying further. We provided an overview of some of the most commonly used techniques and theoretical frameworks for analyzing online communities. Then, we used two case studies to demonstrate the use of those techniques. Although both of our case studies focus on e-learning communities, communities have been used in a multitude of disciplines. Empathic online communities, for example, seem to be gaining a lot of momentum, with people supporting each other. Online communities are widely used for general entertainment and current affairs discussions as well. Professional communities (e.g., business, art, industry-specific) are also being formed within the online environment.

The study of online communities is flourishing, primarily due to the increasing popularity of online services and tools that provide the construction of such networks of users. The study of such complex communities requires the use of a synthesis of methodologies and theoretical foundations. In our first case study, we demonstrated how Social Network Analysis can be used

to model and visualize online community interactions; in the second case study, the theoretical foundations based on activity theory were applied to the domain of game-based communities.

At the beginning of the Internet technology, online communities were solely used for synchronous (often just chat) or asynchronous (most commonly discussion board) textual interaction. This is no longer the case, as people are interacting in online communities using new and more innovative interaction paradigms. Game-based communities, for instance, allow users to represent themselves (using 3D avatars) visually in virtual environments that are often depicted as a fantastic and unrealistic world. They can navigate, change, or even create the virtual world (and thus the context of the community) they interact with. As such, the traditional boundary between author and reader is distorted as the designer (authors) is not the only one who determines what the system is like. Rather, the participants (readers) are coconstructing the virtual world as they are not anymore using the tool to communicate; they are creating and interacting with a virtual environment through which they can meet, socialize, and work with others. Preece (2000) was the first to identify and stress this important social dimension of online communities. Now, this online sociability is becoming more mature and more central to the online communities with which we interact.

This new paradigm of interaction poses new challenges for researchers and practitioners. The importance of appropriate representation of emotions and other social and cultural cues in online communities is now becoming even more important. With textual interfaces there was an attempt to represent these social cues using emoticons. How can this, for example, be transferred to the domain of virtual game communities? Do we want our avatars to behave like us or do we want them to have some alternate and illusionary identities with extraordinary abilities or unusual behaviors? How can we come up with new interfaces and new interaction paradigms that can facilitate this better, in order to cope with the new demands from the users in such online communities?

A second area that is gaining popularity is the research of online communities, or Internet research. Content analysis and query-based techniques were sufficient if what we wanted was a first good impression of the interactions taking place in an online textual community. This is not anymore the case. Studying the sociability of game-based communities, for example, requires the synthesis of more techniques. We might want to immerse ourselves in that community, engage in long-term ethnographic studies of its environment, and get the first-hand experiences of what happens in it. We might have to employ social-cultural theoretical frameworks (e.g., Activity Theory) to get a better understanding of the ways people behave differently from the real world in these inherently social environments. We might have to quantify, using Social Network Analysis for example, the dynamics of the networks and subgroups that evolve around these communities.

The possibility of online communities is unlimited with the emergence of more mature and imaginative virtual worlds. Only by treating them in equality with their physical counterparts that encapsulate the practices of economy, politics, ideology, and everyday life, can we research and study them intellectually.

References

Ang, C. S., Zaphiris, P., & Wilson, S. (2005). Social interaction in game communities and second language learning. *The 19th British HCI Group Annual Conference*, Edinburgh, UK

Archer, W., Garrison, R. D., Anderson, T., & Rourke, L. (2001). A framework for analyzing critical thinking in computer conferences. *European Conference on Computer-Supported Collaborative Learning*, Maastricht, Netherlands.

Aronsson, L. (2002). Operation of a large scale, general purpose Wiki website. Experience from susning.nu's first nine months in service. In J. Á. Carvalho, A. Hübler, & A. A. Baptista (Eds.), *Proceedings of the 6th International ICCC/IFIP Conference on Electronic Publishing* (pp. 27–37). Karlovy Vary, Czech Republic: Verlag für Wissenschaft und Forschung.

Barab, S. A., Hay, K. E., & Yamagata-Lynch, L. C. (2001). Constructing networks of action-relevant episodes: An In Situ research methodology. *The Journal of Learning Sciences, 10*(1 & 2), 63–112.

Bates, A. W. (1995). *Technology, open learning and distance education.* London: Routledge.

Beidernikl, G., & Paier, D. (2003). Network analysis as a tool for assessing employment policy. *Proceedings of the Evidence-Based Policies and Indicator Systems Conference*, London.

Black, R. W. (2004). *Anime-inspired affiliation: An ethnographic inquiry into the literacy and social practices of English language learners writing in the fan-fiction community.* Paper presented at 2004 meeting of American Educational Research Association, San Diego, CA.

Blomkvist, S. (2002). *Persona—an overview, Uppsala University.* Retrieved November 22, 2004, from http://www.it.uu.se/edu/course/homepage/hcinet/ht04/library/docs/Persona-overview.pdf

Borgatti, S. (2000). *What is social network analysis.* Retrieved November 9, 2004, from http://www.analytictech.com/networks/whatis.htm

Burge, E. L. & Roberts, J. M. (1993). *Classrooms with a difference: A practical guide to the use of conferencing technologies.* Ontario: University of Toronto Press.

CAP, University of Warwick (2004). *E-Guide: Using computer mediated communication in learning and teaching.* Retrieved November 8, 2004, from http://www2.warwick.ac.uk/services/cap/resources/eguides/cmc/cmclearning/

Castronova, E. (2001, December). *Virtual worlds: A first-hand account of market and society on the cyberian frontier.* CESifo Working Paper Series No. 618. Fullerton: Center for Economic Studies and Ifo Institute for Economic Research, California State University.

Choi, J. H., & Danowski, J. (2002). Cultural communities on the net— Global village or global metropolis: A network analysis of Usenet newsgroups. *Journal of Computer-Mediated Communication, 7*(3). Retrieved March 8, 2007, from http://JCMC.indiana.edu/vol7/issue3/chai.html

Cook, D., & Ralston, J., (2003). Sharpening the focus: Methodological issues in analyzing on-line conferences. *Technology, Pedagogy and Education, 12*(3), 361–376.

Cooper, A. (1999). *The Inmates are running the asylum.* Indianapolis, IN: SAMS, a division of Macmillan Computer Publishing.

Cyram. (2004). *Netminer for windows.* Retrieved March 8, 2007, from http://netminer.com

December, J. (1997). Notes on defining of computer-mediated communication. *Computer-Mediated Communication Magazine, 3*(1). Retrieved March 8, 2007, from http://www.december.com/cmc/may/1997/jan/december.html

Dekker, A. H. (2002, January–February). A category-theoretic approach to social network analysis. *Proceedings of Computing: The Australasian Theory Symposium (CATS)*. Melbourne, Australia.

Ducheneaut N., Moore R. J., & Nickell, E. (2004). *Designing for sociability in massively multiplayer games: An examination of the "Third Places" of SWG.* Other Players Conference, Denmark.

Electronic Art (2005). *Ultima online.* Retrieved November 8, 2005, from http://www.uo.com/

Emigh, W., & Herring, S. C. (2005). Collaborative authoring on the web: A genre analysis of online encyclopedias. *Proceedings of the 38th Hawai'i International Conference on System Sciences* (HICSS-38). Los Alamitos, CA: IEEE Press.

Engestrom, Y. (2001). Expansive learning at work: Toward an activity theoretical reconceptualisation. *Journal of Education and Work, 14*(1), 133–156.

Fahy, P. J. (2003). Indicators of support in online interaction. *International Review of Research in Open and Distance Learning, 4*(1). Retrieved March 8, 2007, from http://www.irrodl.org/index.php/irrodl/article/view/129/209

Fahy, P. J., Crawford, G., & Ally, M., (2001). Patterns of interaction in a computer conference transcript. *International Review of Research in Open and Distance Learning, 2*(1). Retrieved 8 March 2007, from http://www.irrodl.org/irrodl/article/view/36/73

Ferris, P. (1997). What is CMC? An overview of scholarly definitions. *Computer-Mediated Communication Magazine, 4*(1). Retrieved March 8, 2007, from http:www.december.com/cmc/may/1997/jun/december.html

Fortner, R. S. (1993). *International communication: History, conflict, and control of the global metropolis.* Belmont, CA: Wadsworth.

Galotti, K. M., Clinchy, B. M., Ainsworth, K., Lavin, B., & Mansfield, A. F. (1999). A new way of assessing ways of knowing: The attitudes towards thinking and learning survey (ATTLS). *Sex Roles, 40*(9 & 10), 745–766.

Garton, L., Haythornthwaite, C., & Wellman, B. (1997). Studying On-line Social Networks. In Jones, S. (Ed.), *Doing internet research.* Thousand Oaks, CA: Sage.

Gunawardena, C., Lowe, C., & Anderson, T. (1997). Analysis of a global online debate and the development of an interaction analysis model for examining social construction of knowledge in computer conferencing. *Journal of Educational Computing Research, 17*(4), 397–431.

Halvorsen, R. F. (2005). *TheStateOfWiki.* Retrieved September 22, 2005, from http://heim.ifi.uio.no/~aurilla/web/rune.pdf

Hanneman, R. A. (2001). *Introduction to social network methods.* Retrieved November 9, 2004, from http://faculty.ucr.edu/~hanneman/SOC157/TEXT/TextIndex.html

Heeren, E. (1996). *Technology support for collaborative distance learning.* Doctoral dissertation, University of Twente, Enschede.

Henri, F. (1992). Computer conferencing and content analysis. In A. R. Kaye (Ed.), *Collaborative learning through computer conferencing: The Najaden Papers* (pp. 117–136). Berlin: Springer-Verlag.

Jones G. B. (2003). Blogs and Wikis: Environments for on-line collaboration. *Language Learning & Technology, 7*(2), 12–16.

Jones, S. (1995). Computer-mediated communication and community. *Computer-Mediated Communication Magazine, 2*(3).Retrieved March 8, 2007, from http://www.december.com/cmc/may/pms/mar/jones.html

Kelly, R. V. (2004). *Massively multiplayer online role-playing games: The people, the addiction and the playing experience.* McFarland & Company, U.S.

Kittowski, F. F., & Köhler, A. (2002). Knowledge creating communities in the context of work processes. *ACM SIGGROUP Bulletin, 23*(3), 8–13.

Kolbert, E. (2001, May 28). Pimps and dragons: How an online world survived a social breakdown. *The New Yorker.* Retrieved February 15, 2005, from http://www.newyorker.com/fact/content/?010528fa_FACT

Korpela, M., Soriyan, H. A., & Olufokunbi, K. C. (2000). Activity analysis as a method for information systems development. *Scandinavian Journal of Information Systems, 12*, 191–210.

Korzenny, F. (1978). A theory of electronic propinquity: Mediated communication in organizations. *Communication Research, 5*, 3–23.

Krebs, V. (2004). *An introduction to social network analysis.* Retrieved November 9, 2004, from http://www.orgnet.com/sna.html

Kuo, J. (2004). *Online video games in mental health.* Paper presented at the annual meeting of the American Psychiatry Association, New York.

Kypros-Net Inc. (2005). *The world of Cyprus.* Retrieved March 8, 2007, from http://kypros.org

Leont'ev, A. N. (1978). *Activity, consciousness, and personality.* New York: Prentice-Hall.

Madden, M., & Rainie, L. (2003). Pew Internet & American Life Project Surveys. Pew Internet & American Life Project, Washington, DC.

Mader, S. (2004). OpenSpectrum: A Wiki-based learning tool for Spectroscopy that anyone can edit. *Winter 2005 CONGCHEM: Trends and New Ideas in Chemical Education,* virtual conference. Retrieved March 8, 2007, from http://www.ched-ccce.org/confchem/2005/a/index.html

Mason, R. (1991). Analyzing computer conferencing interactions, *Computers in Adult Education and Training, 2*(3), 161–173.

McLoughlin, C. (1996, January 21–25). A learning conversation: Dynamics, collaboration and learning in computer mediated communication. In C. McBeath & R. Atkinson (Eds.), *Proceedings of the Third International Interactive Multimedia Symposium,* Promaco Conventions, Perth, Western Australia, 267–273.

McLuhan, M. (1964). *Understanding media: The extension of man.* New York: McGraw Hill.

Metcalfe, B. (1992). *Internet fogies to reminisce and argue at Interop Conference* (p. 45). InfoWorld.

Mwanza, D. (2002). *Towards an activity-oriented method for HCI research and practice.* PhD Thesis, Open University.

NCsoft (2005). *Lineage.* Retrieved November 8, 2005, from http://www.lineage.com/

NUA Internet Surveys (2004). Retrieved October 20, 2004, from http://www.nua.ie/surveys/how_many_online/index.html

Papert, S. (1980). *Mindstorms: Children, computers, and powerful ideas.* New York: Basic Books.

Preece, J. (2000). *Online Communities: Designing usability, supporting sociability.* Chichester, UK: John Wiley and Sons.

Preece, J., & Maloney-Krichmar, D. (2003) Online communities: Focusing on sociability and usability. In J. Jacko & A. Sears (Eds.), *Handbook of Human-Computer Interaction* (pp. 596–620). Mawhah, NJ: Lawrence Erlbaum Associates.

Preece, J., Rogers, Y., & Sharp, H. (2002). *Interaction design: Beyond human-computer interaction.* New York: John Wiley & Sons.

Reiser, R. A., & Gagné, R. M. (1983). *Selecting media for instruction.* Englewood Cliffs, NJ: Educational Technology Publications.

Rheingold, H. (1993). *The virtual community: Homesteading on the electonic frontier.* Reading, MA: Addison-Wesley.

Rice, R. (1994). Network analysis and computer mediated communication systems. In S. W. J. Galaskiewkz (Ed.), *Advances in social network analysis.* Newbury Park, CA: Sage.

Rice, R. E., Grant, A. E., Schmitz, J., & Torobin, J. (1990). Individual and network influences on the adoption and perceived outcomes of electronic messaging. *Social Networks, 12*, 17–55.

ScotCIT. (2003). Enabling large-scale institutional implementation of communications and information technology (ELICIT). *Using Computer Mediated Conferencing.* Retrieved November 2, 2004, from http://www.elicit.scotcit.ac.uk/modules/cmc1/welcome.htm

Scott, J. (2000). *Social network analysis: A handbook* (2nd ed.). London: Sage.

Sony (2005). *Star Wars Galaxies.* Retrieved November 8, 2005, from http://starwarsgalaxies.station.sony.com/

Squire, K., & Steinkeuhler, C. (in press): Generating cyberculture/s: The case of Star Wars Galaxies. In D. Gibbs, & K. L. Krause (Eds.), *Cyberlines: Languages and cultures of the Internet* (2nd ed.). Albert Park, Australia: James Nicholas Publishers.

Suler, J. (2004). *The final showdown between in-person and cyberspace relationships.* Retrieved November 3, 2004, from http://www1.rider.edu/~suler/psycyber/showdown.html

Taylor, T. L. (2002). Whose game is this anyway?: Negotiating corporate ownership in a virtual world. *Proceedings of the CGDC Conference,* 227–242. Tampere, Finland

U.S. Census Bureau (2004). *Global population profile 2002.* Retrieved October 20, 2004, from http://www.census.gov/ipc/www/wp02.html

Usability Net. (2003). *UsabilityNet* (2003) Interviews. Retrieved December 3, 2004, from http://www.usabilitynet.org/tools/interviews.html

Vaknin, S. (2002). TrendSiters: Games people play. *Electronic Book Web.* Retrieved February 23, 2002, from http://12.108.175.91/ebookweb

Viégas, F. B., Wattenberg, M., & Dave, K. (2004). Studying cooperation and conflict between authors with history flow visualizations. In E. Dykstra-Erickson & M. Tscheligi (Eds.), *Proceedings of ACM CHI 2004 Conference on Human Factors in Computing Systems,* (pp. 575–582). Vienna, Austria. ACM Press

Vygotsky, L. (1930). *Mind and society.* Cambridge, MA: Harvard University Press.

Wallace, P. (1999). *The psychology of the Internet.* Cambridge, UK: Cambridge University Press.

Wang, H. C., Lu, C. H., Yang, J. Y., Hu, H. W., Chiou, G. F., Chiang, Y. T., et al. (2005). An empirical exploration of using Wiki in an English as a second language course. *Proceedings of 5th IEEE International Conference on Advanced Learning Technologies (ICALT 2005),* Kaohsiung, Taiwan, 155–157.

Wellman, B. (1982). Studying personal communities. In P. M. N. Lin (Ed.), *Social structure and network analysis* (pp. 61–80). Beverly Hills, CA: Sage.

Wellman, B. (1992). Which types of ties and networks give what kinds of social support? *Advances in Group Processes, 9*, 207–235.

Wikipedia (2005). *Wikipedia: The free encyclopedia.* Retrieved October 1, 2005, from http://en.wikipedia.org/wiki/Wikipedia

Yee, N. (2005, May). The demographics and derived experiences of users of massively multi-user online graphical environments. Paper presented at The 55th Internal Communication Association Annual Conference, New York.

Zaphiris, P., & Zacharia, G. (2001). Design methodology of an online Greek language course. *Ext. Abstracts CHI 2001 conference.* ACM Press.

Zaphiris, P., Zacharia, G., & Rajasekaran, M. (2003). Distributed constructionism through participatory design. In Ghaoui, C. (Ed.) *E-education applications: Human factors and innovative approaches* (pp. 164–179). London, UK: Idea Group.

Zhu, E. (1996). Meaning negotiation, knowledge construction, and mentoring in a distance learning course. In *Proceedings of Selected Research and Development Presentations at the 1996 National Convention of the Association for Educational Communications and Technology,* Indianapolis, IN.

•15•

VIRTUAL ENVIRONMENTS

Kay M. Stanney
University of Central Florida

Joseph V. Cohn
Naval Research Laboratory

Introduction 294	**Health and Safety Issues** 301
System Requirements 294	Cybersickness, Adaptation, and Aftereffects 301
Hardware Requirements 295	Social Impact 302
Multimodal I/Os 295	**Virtual Environment Usability Engineering** 302
Tracking systems 296	Usability Techniques 302
Interaction techniques 297	Sense of Presence 302
Augmented cognition techniques 297	Virtual Environment Ergonomics 303
Software Requirements 297	**Application Domains** 303
Modeling 297	VE as a Selection and Training Tool 303
Autonomous agents 298	VE as Entertainment Tool 305
Networks 298	VE as a Medical Tool 306
Design and Implementation Strategies 298	Psychological/cognitive rehabilitation applications 306
Cognitive Aspects 298	Motor rehabilitation applications 306
Multimodal interaction design 298	VE as a System Acquisition Tool 306
Perceptual illusions 299	**Conclusions** 307
Navigation and wayfinding 300	**Acknowledgments** 307
Content Development 300	**References** 307
Products Liability 300	
Usage Protocols 300	

INTRODUCTION

Interactive computing has evolved through the years from cryptic command-based interfaces to a collection of task-based applications to ecologically valid immersive environments, each advance dissolving more of the barrier between users and their desired actions. To many, with virtual environments (VEs) we have reached the panacea in interactive computing. Such environments immerse users in realistic settings, allowing them to engage in an intuitive and intimate manner with their digital universe. Such capability affords the opportunity to "learn by doing," "train like we fight," and "involve me and I understand." While early VEs were low in resolution and sluggish in responsiveness, advances in display and tracking technology have largely resolved these issues such that today VE applications abound (Davies, 2002). Yet, considerable HCI research and development is required to resolve lingering issues, such as cybersickness and usability, if VE technology is to be openly embraced by users. This chapter reviews the current state-of-the-art in VE technology, provides design and implementation strategies, discusses health and safety concerns and potential countermeasures, and presents the latest in VE usability engineering approaches. Current efforts in a number of application domains are reviewed. The chapter should enable readers to better specify design and implementation requirements for VE applications and prepare them to use this advancing technology in a manner that minimizes health and safety concerns.

SYSTEM REQUIREMENTS

Virtual environments provide multimodal computer-generated experiences, which are driven by the hardware and software used to generate a virtual world (see Fig. 15.1). Hardware interfaces consist primarily of the:

- Interface devices used to present multimodal information and sense the VE;
- Tracking devices used to identify head and limb position and orientation;
- Interaction techniques that allow users to navigate through and interact with the virtual world.

Software interfaces include the:

- Modeling software used to generate VEs;
- Autonomous agents that inhabit VEs;
- Communication networks used to support multi-user virtual environments.

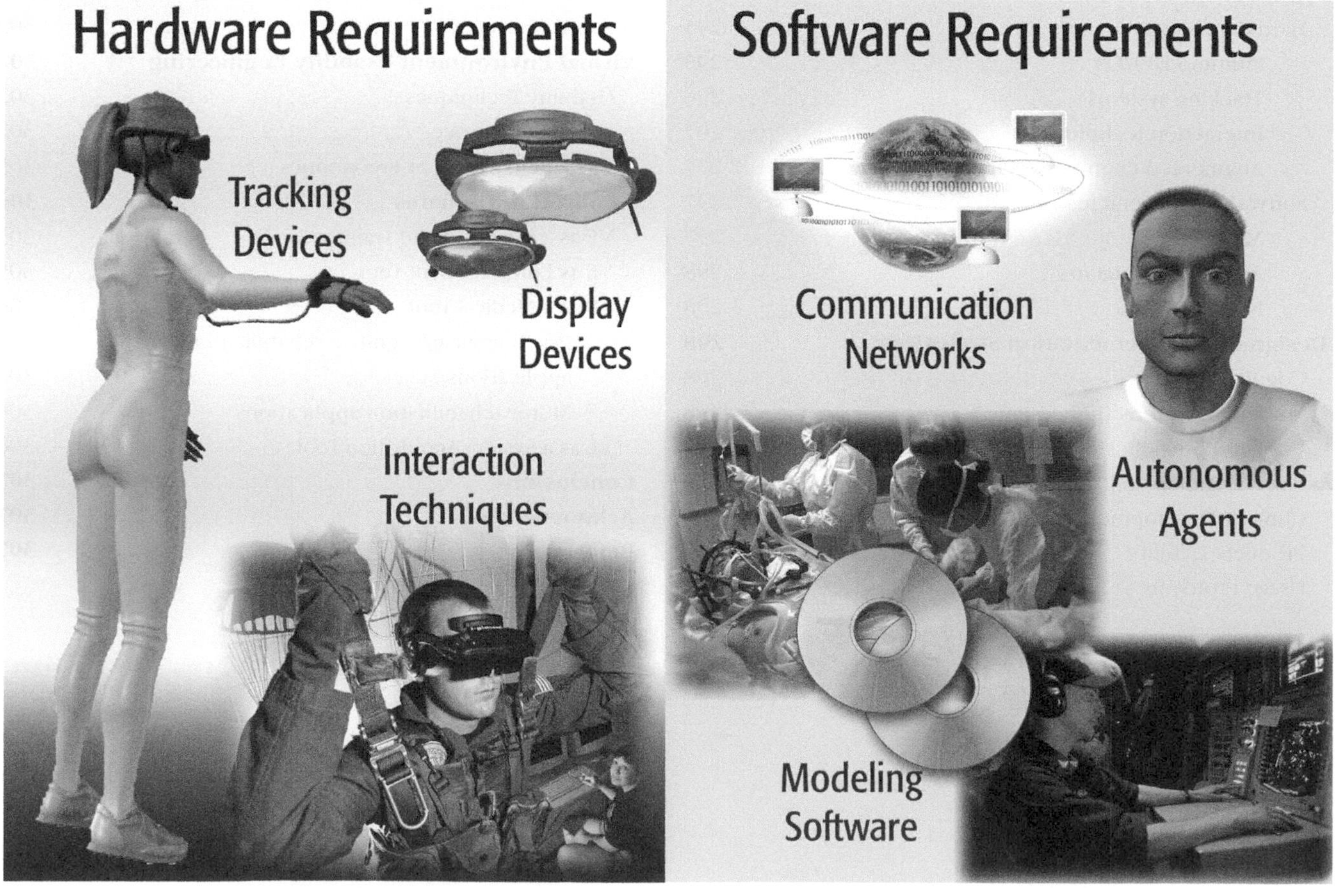

FIGURE 15.1. Hardware and software requirements for virtual environment generation.

Hardware Requirements

Virtual environments require very large physical memories, high-speed processors, high-bandwidth mass-storage capacity, and high-speed interface ports for interaction devices (Durlach & Mavor, 1995). These requirements are easily met by today's high-speed, high-bandwidth computing systems, many of which have surpassed the gigahertz barrier (e.g., AMD's Athlon, Intel's Pentium IV and Itanium, Sun's UltraSPARC IV, IBM's Power5). The future looks even brighter, with promises of massive parallelism in quantum computing, which will allow tomorrow's computing systems to be exponentially faster than their ancestors. With the rapidly advancing ability to generate complex and large-scale virtual worlds, hardware advances in multimodal input/output (I/O) devices, tracking systems, and interaction techniques are needed to support generation of increasingly engaging virtual worlds. In addition, the coupling of augmented cognition and VE technologies can lead to substantial gains in the ability to evaluate their effectiveness.

Multimodal I/Os. To present a multimodal VE multiple, devices are used to present information to VE users. Stanney and Zyda (2002) suggested that the single largest challenge in presenting multimodal VE applications are the advances required in peripheral computer connections. Currently the serial ports used to connect position trackers and other peripheral devices are typically those designed for character input rather than high-speed data transfer. This creates an input port speed problem that must be resolved if multimodal VE systems are to effectively present information to users.

In terms of peripheral devices, the one that has received the greatest attention, both in hype and disdain, is almost certainly the head-mounted display (HMD). One benefit of HMDs is their compact size, as an HMD when coupled with a head tracker can be used to provide a similar visual experience as a multitude of bulky displays associated with Spatially Immersive Displays (SIDs)and desktop solutions. There are three main types of HMDs: monocular (e.g., one display source), biocular (e.g., two displays with separate display and optics paths that show one image), and binocular (e.g., stereoscopic viewing via two image generators) (Vince, 2004). HMDs generally use CRT or LCD image sources, with CRTs typically providing higher resolution, although at the cost of increased weight. When coupled with tracking devices, HMDs can be used to present 3D visual scenes that are updated as a user moves his or her head about a virtual world. Although this often provides an engaging experience, due to poor optics, sensorial mismatches, and slow update rates, these devices are also often associated with adverse effects such as eyestrain and nausea (May & Badcock, 2002). In addition, while HMDs have come down substantially in weight, rendering them more suitable for extended wear, they are still hindered by cumbersome designs, obstructive tethers, suboptimal resolution, and insufficient fields of view. These shortcomings may be the reason behind why, in a recent review of HMD devices, more than a third had been discontinued by their manufacturers (Bungert, 2005). Of the HMDs available, there are several low- to mid-cost models, which are relatively lightweight and provide a horizontal field-of-view and resolution far exceeding predecessor systems.

Low-technology stereo viewing VE display options include anaglyph methods, where a viewer wears glasses with distinct color polarized filters; parallel or cross-eyed methods, in which right and left images are displayed adjacently (parallel or crossed), requiring the viewer to actively fuse the separate images into one stereo image; parallax stereogram methods, in which an image is made by interleaving columns of two images from a left- and right-eye perspective image of a 3D scene; polarization methods, in which the images for the left and right eyes are projected on a plane through two orthogonal linearly polarizing filters (e.g., the right image is polarized horizontally; the left is polarized vertically) and glasses with polarization filters are donned to see the 3D effect; Pulfrich methods, in which an image of a scene moves sideways across the viewer's field of view and one eye is covered by a dark filter so that the darkened image reaches the brain later, causing stereo disparity; and shutter glass methods in which images for the right and left eyes are displayed in quick alternating sequence and special shutter glasses are worn that "close" the right or left eye at the correct time (Halle, 1997; Vince, 2004). All of these low-technology solutions are limited in terms of their resolution, the maximum number of views that they can display, and clunky implementation; they can also be associated with pseudoscopic images (e.g., the depth of an object can appear to flip inside out).

Other options in visual displays include Spatially Immersive Displays (SIDs) (e.g., displays that surround viewers physically with panoramic large field-of-view imagery generally projected via fixed front or rear projection display units; Majumder, 2003), desktop stereo displays, and volumetric displays that fill a volume of space with a "floating" image (Halle, 1997). Examples of SIDs include the Cave Automated Virtual Environment (CAVE) (Cruz-Neira, Sandin, & DeFanti, 1993), ImmersaDesk, PowerWall, Infinity Wall, and VisionDome (Majumder, 1999). Issues with SIDs include a stereo view that is correct for only one or a few viewers, noticeable overlaps between adjacent projections, and image warp on curved screens. Desktop display systems have advantages over SIDs because they are smaller, easier to configure in terms of mounting cameras and microphones, easier to integrate with gesture and haptic devices, and more readily provide access to conventional interaction devices, such as mice, joysticks, and keyboards. Issues with such displays include stereo that is only accurate for one viewer and a limited-display volume. Volumetric displays are advantageous because imagery can be seen from a multitude of viewing angles, generally without the need for goggles. A recent example of this type of display is the Perspecta Spatial 3D, which projects images onto a rotating flat screen inside a glass dome (Favalora et al., 2003). Issues with volumetric displays include low resolution and the tendency for transparent images to lose interposition cues. Also, view-independent shading of objects is not possible with volumetric displays, and current solutions do not exhibit arbitrary occlusion by interposition of objects (Halle, 1997).

The way of the future seems to be wearable computer displays (e.g., Microvision, MicroOptical), which can incorporate miniature LCDs directly into conventional eyeglasses (Tredennick & Shimamoto, 2005). If designed effectively, these devices should eliminate the tethers and awkwardness of current designs, while enlarging the field of view and enhancing resolution.

Stanney and Zyda (2002, p. 3) suggested that "with advances in wireless and laser technologies and miniaturization of LCDs, during the next decade visual display technology should realize the substantial gains necessary to provide high fidelity virtual imagery in a lightweight non-cumbersome manner."

When virtual environments provide audio, the interactive experience is generally greatly enhanced (Shilling & Shinn-Cunningham, 2002). Audio can be presented via spatialized or nonspatialized displays. Just as stereo visual displays are a defining factor for VE systems, so are "interactive" spatialized audio displays (e.g., those with "on-the-fly" positioning of sounds). VRSonic's SoundScape3D (http://www.vrsonic.com/), Aureal's A3D Interactive (http://www.a3d.com/), and AuSIM3D (http://ausim3d.com/) are examples of positional 3D audio technology. Recently there have been promising developments in new sound modeling paradigms (e.g., VRSonic's ViBe technology) and sound design principles that will hopefully lead to a new generation of tools for designing effective spatial-audio environments (Fouad, 2004; Fouad & Ballas, 2000; Hahn, Fouad, Gritz, & Lee, 1998).

Developers must decide if sounds should be presented via headphones or speakers. For nonspatialized audio, most audio characteristics (e.g., timbre, relative volume), are generally considered to be equivalent whether projected via headphones or speakers. This is not so for spatialized audio, in which the information must generally be projected through headphones in order to exploit subtle nuances. However, spatialized audio headphones tend to be more expensive than speakers, and they also tend to require customization to an individual listener for optimum effect. Speaker systems have a larger footprint, are sensitive to room acoustics, and the listener must be placed in the "sweet spot" (e.g., ideal listening position) for the proper 3D audio effect to be realized (Shilling & Shinn-Cunningham, 2002). With true 3D audio, a sound can be placed in any location, right or left, up or down, near or far via the use of a head-related transfer function (HRTF) to represent the manner in which sound sources change as a listener moves his or her head (Begault, 1994; Butler, 1987; Cohen, 1992). Recent advances have allowed for the development of personalized HRTF functions, however, currently, these functions still require a significant amount of calibration time and fail to provide adequate cues for front-to-back or up-down localization (Crystal Rivers Engineering, 1995). Ideally, a more generalized HRTF could be developed that would be applicable to a multitude of users. This may be possible by using a best-fit HRTF selection process in which one finds the nearest matching HRTF in a database of candidate HRTFs by comparing the physiological characteristics of stored HRTFs to those of a target user (Algazi, Duda, Thompson, & Avendano, 2001).

While not as commonly incorporated into VEs as visual and auditory interfaces, haptic devices can be used to enhance aspects of touch and movement of the hand or body segments while interacting with a virtual environment. Specifically, haptic displays can be used to support two-way communications between humans and interactive systems, enabling bidirectional interaction between a user and his or her surroundings (Hale & Stanney, 2004). In general, haptic displays are effective at alerting people to critical tasks (e.g., warning), providing a spatial frame of reference within one's personal space, and supporting hand-eye coordination tasks. Texture cues, such as those conveyed via vibrations or varying pressures, are effective as simple alerts and may speed reaction time and aid performance in degraded visual conditions (Akamatsu, 1994; Biggs & Srinivasan, 2002; Massimino & Sheridan, 1993; Mulgund, Stokes, Turieo, & Devine, 2002). Kinesthetic devices are advantageous when tasks involve hand-eye coordination (e.g., object manipulation), and where haptic sensing and feedback are key to performance. Currently available haptic-interaction devices include static displays (e.g., convey deformability or Braille); vibrotactile, electrotactile, and pneumatic displays (e.g., convey tactile sensations such as surface texture and geometry, surface slip, surface temperature); force feedback systems (e.g., convey object position and movement distances); and exoskeleton systems (e.g., enhance object interaction and weight discrimination) (Hale & Stanney, 2004). Advances in this area are worth pursuing because incorporating haptic feedback in VEs can substantially enhance performance (Burdea, 1996).

The "vestibular system can be exploited to create, prevent, or modify acceleration perceptions" in virtual environments (Lawson, Sides, & Hickinbotham, 2002, p. 137). For example, by simulating acceleration cues, a person can be psychologically transported from his or her veridical location, such as sitting in a chair in front of a computer, to a simulated location, such as the cockpit of a moving airplane. While vestibular cues can be stimulated via many different techniques in VEs, three of the most promising methods are physical motion of the user (e.g., motion platforms), wide field-of-view visual displays that induce vection (e.g., an illusion of self-motion), and locomotion devices that induce illusions of self-motion without physical displacement of the user through space (e.g., walking in place, treadmills, pedaling, foot platforms) (Hettinger, 2002; Hollerbach, 2002; Lawson et al., 2002). Of these options, motion platforms are probably the most advanced. Motion platforms are generally characterized via their range of motion/degrees of freedom and actuator type (Isdale, 2000). In terms of range of motion, motion platforms can move a person in many combinations of translational (e.g., surge-longitudinal motion, sway-lateral motion, heave-vertical motion), and rotational (e.g., roll, pitch, yaw) degrees of freedom (DOF). A single-DOF translational motion system might provide a vibration sensation via of a "seat shaker." A common 6DOF configuration is a hexapod, which consists of a frame with six or more extendable struts (actuators) connecting a fixed base to a movable platform. In terms of actuators, electrical actuators are quiet and relatively maintenance free; however, they are not very responsive and they cannot hold the same load as can hydraulic or pneumatic systems. Hydraulic and pneumatic systems are smoother, stronger, and more accurate; however, they require compressors, which may be noisy. Servos are expensive and difficult to program.

Although other multimodal interactions are possible (e.g., gustatory, olfactory), there has been limited research and development beyond the primary three interaction modes (e.g., visual, auditory, haptic), although efforts have been made to support advances in olfactory interaction (Jones, Bowers, Washburn, Cortes, & Vijaya Satya, 2004).

Tracking systems. Tracking systems allow determination of users' head or limb position and orientation, or the location of handheld devices in order to allow interaction with virtual objects and traversal through 3D computer-generated worlds (Foxlin, 2002). Tracking is what allows the visual scene in a VE to coincide with a user's point of view, thereby providing an egocentric real-time perspective. Tracking systems must be carefully coupled with the visual scene, however, to avoid unacceptable lags (Kalawsky, 1993). Advances in tracking technology have been realized in terms of drift-corrected gyroscopic orientation trackers, outside-in optical tracking for motion capture, and laser scanners (Foxlin, 2002). The future of tracking technology is likely hybrid tracking systems, with an acoustic-inertial hybrid on the market (see http://www.isense.com/products/) and several others in research labs (e.g., magnetic-inertial, optical-inertial, and optical-magnetic). In addition, ultra-wideband radio technology holds promise for an improved method of omnidirectional point-to-point ranging.

Tracking technology also allows for gesture recognition, in which human position and movement are tracked and interpreted to recognize semantically meaningful gestures (Turk, 2002). Tracking devices that are worn (e.g., gloves, bodysuits) are currently more advanced than passive techniques (e.g., camera, sensors), yet the latter hold much promise for the future, as they are more powerful and less obtrusive than those that must be worn.

Interaction techniques. While one may think of joysticks and gloves when considering VE interaction devices, there are many techniques that can be used to support interaction with and traversal through a virtual environment. Interaction devices support traversal, pointing and selection of virtual objects, tool usage (e.g., through force and torque feedback), tactile interaction (e.g., through haptic devices), and environmental stimuli (e.g., temperature, humidity) (Bullinger, Breining, & Braun, 2001).

Supporting traversal throughout a VE, via motion interfaces, is of primary importance (Hollerbach, 2002). Motion interfaces are categorized as either active (e.g., locomotion) or passive (e.g., transportation). Active motion interfaces require self-propulsion to move about a virtual environment (e.g., treadmill, pedaling device, foot platforms). Passive motion interfaces transport users within a VE without significant user exertion (e.g., inertial motion, as in a flight simulator, or noninertial motion, such as in the use of a joystick or gloves). The utility, functionality, cost, and safety of locomotion interfaces beyond traditional options (e.g., joysticks) have yet to be proven. In addition, beyond physical training, concrete applications for active motion interfaces have yet to be clearly delineated.

Another interaction option is speech control. Speaker-independent continuous speech recognition systems are currently commercially available (Huang, 1998). For these systems to provide effective interaction, however, additional advances are needed in acoustic and language-modeling algorithms to improve the accuracy, usability, and efficiency of spoken language understanding.

Gesture interaction allows users to interact through nonverbal commands conveyed via physical movement of the fingers, hands, arms, head, face, or other body limbs (Turk, 2002). Gestures can be used to specify and control objects of interest, direct navigation, manipulate the environment, and issue meaningful commands.

To support natural and intuitive interaction, a variety of interaction techniques can be coupled. Combining speech interaction with nonverbal gestures and motion interfaces can provide a means of interaction that closely captures real-world communications.

Augmented cognition techniques. Augmented cognition is a new computing paradigm in which users and computers are tightly coupled via physiological gauges that measure the cognitive state of users and adapt interaction to optimize human performance (Schmorrow, Stanney, Wilson, & Young, 2005). If incorporated into VE applications, augmented cognition could provide a means of evaluating their educational validity and compelling nature. Neuroscience studies have established that differential aspects of the brain are engaged when learning different types of materials and the areas in the brain that are activated change with increasing competence (Kennedy, Drexler, Jones, Compton, & Ordy, 2005). Thus, if VE users were immersed in an educational experience, augmented cognition technology could be used to gauge if targeted areas of the brain were being activated and dynamically modify the content of a VE learning curriculum if desired activation patterns were not being generated. Physiological measures could also be used to detect the onset of cybersickness (see "Cybersickness, Adaptation, and Aftereffects") and to assess the engagement, awareness, and anxiety of VE users, thereby potentially providing much more robust measures of immersion and presence (see "Sense of Presence"). Such techniques could prove invaluable to entertainment VE applications (cf., Badiqué, Cavazza, Klinker, Mair, Sweeney, Thalmann, et al., 2002) that seek to provide the ultimate experience, military training VE applications (cf., Knerr, Breaux, Goldberg, & Thurman, 2002) that seek to emulate the "violence of action" found during combat, and therapeutic VE applications (cf., North, North, & Coble, 2002; Strickland, Hodges, North, & Weghorst, 1997) that seek to overcome disorders such as fear of heights or flying.

Software Requirements

Software development of VE systems has progressed tremendously, from proprietary and arcane systems, to development kits that run on multiple platforms (e.g., general-purpose operating systems to workstations) (Pountain, 1996). Virtual-environment system components are becoming modular and distributed, thereby allowing VE databases (e.g., editors used to design, build, and maintain virtual worlds) to run independently of visualizers and other multimodal interfaces via network links. Standard APIs (Application Program Interfaces) (e.g., OpenGL, Direct-3D, Mesa) allow multimodal components to be hardware-independent. Virtual environment programming languages are advancing, with APIs, libraries, and particularly scripting languages allowing nonprogrammers to develop virtual worlds (Stanney & Zyda, 2002). Advances are also being made in modeling of autonomous agents and communication networks used to support multiuser virtual environments.

Modeling. A VE consists of a set of geometry, the spatial relationships between the geometry and the user, and the change in geometry invoked by user actions or the passage of time (Kessler, 2002). Generally, modeling starts with building the geometry components (e.g., graphical objects, sensors, viewpoints, animation sequences) (Kalawsky, 1993). These are often converted from CAD data. These components then get imported into the VE modeling environment and rendered when appropriate sensors are triggered. Color, surface textures, and behaviors are applied during rendering. Programmers control the events in a VE by writing task functions, which become associated with the imported components.

A number of 3D modeling languages and toolkits are available that provide intuitive interfaces and run on multiple platforms and renderers (e.g., 3D Studio Max, AC3D Modeler, ZBrush, modo 203, AccuRender, ACIS 3D, Ashlar-vellum Argon/Xenon/Cobalt, Carrara, CINEMA 4D, DX Studio, EON Studio, MultiGen Creator and Vega, RenderWare, solidThinking) (Ultimate 3D Links, 2007). In addition, there are scene management engines (e.g., R3vis Corporation's OpenRM Scene Graph is an API that provides cross-platform scene management and rendering services) that allow programmers to work at a higher level, defining characteristics and behaviors for more holistic concepts (Kershner, 2002; Menzies, 2002). There have also been advances in photorealistic rendering tools (e.g., Electric Image Animation System), which are evolving toward full-featured physics-based global illumination rendering systems (Heirich & Arvo, 1997; Merritt & Bacon, 1997). Taken together, these advances in software modeling allow for the generation of complex and realistic VEs that can run on a variety of platforms, permitting access to VE applications by both small- and large-scale application-development budgets.

Autonomous agents. Autonomous agents are synthetic or virtual human entities that possess some degree of autonomy, social ability, reactivity, and pro-activeness (Allbeck & Badler, 2002). They can have many forms (e.g., human, animal), which are rendered at various levels of detail and style, from cartoonish to physiologically accurate models. Such agents are a key component of many VE applications involving interaction with other entities, such as adversaries, instructors, or partners (Stanney & Zyda, 2002).

There has been significant research and development in modeling embodied autonomous agents. As with object geometry, agents are generally modeled off-line and then rendered during real-time interaction. While the required level of detail varies, modeling of hair and skin adds realism to an agent's appearance (Allbeck & Badler, 2002). There are a few toolkits available to support agent development, with one of the most notable offered by Boston Dynamics, Inc. (BDI) (http://www.bdi .com/), a spin-off from the MIT Artificial Intelligence Laboratory. BDI's products allow VE developers to work directly in a 3D database, interactively specifying agent behaviors, such as paths to traverse and sensor regions. The resulting agents move realistically, respond to simple commands, and travel about a VE as directed. Another option is ArchVision's (http://www.archvision .com/) 3D Rich Photorealistic Content (RPC) People.

Networks. Distributed networks allow multiple users at diverse locations to interact within the same virtual environment. Improvements in communications networks are required to allow realization of such shared experiences in which users, objects, processes, and autonomous agents from diverse locations interactively collaborate (Durlach & Mavor, 1995). Yet the foundation for such collaboration has been built within the Next Generation Internet (NGI). The NGI initiative (http://www.ngi .gov/) aimed to connect a number of universities and national labs at speeds 100 to 1,000 times faster than the 1996 Internet in order to experiment with collaborative-networking technologies, such as high-quality videoconferencing and audio and video streams. The NGI's successor, the Large Scale Networking (LSN) Coordinating Group (http://www.hpcc.gov/iwg/lsn.html), aims to develop technologies and services that enable wireless, optical, mobile, and wireline communications and enhance networking software, with a goal of achieving the terabit per second objective that proved elusive to the NGI.

In addition, Internet2 is using existing networks (e.g., NSF's vBNS—Very high-speed Backbone Network Service) to determine the transport designs necessary to carry real-time multimedia data at high speed (http://apps.internet2.edu/). Distributed VE applications can leverage the special capabilities (e.g., high bandwidth, low latency, low jitter) of these advancing network technologies to provide shared virtual worlds (Singhal & Zyda, 1999).

DESIGN AND IMPLEMENTATION STRATEGIES

While many conventional HCI techniques can be used to design and implement VE systems, there are unique cognitive, content, product liability, and usage protocol considerations that must be addressed (see Fig. 15.2).

Cognitive Aspects

The fundamental objective of VE systems is to provide multimodal interaction or, when sensory modalities are missing, perceptual illusions that support human information processing in pursuit of a VE application's goals, which could range from training to entertainment. Ancillary yet fundamental to this goal is to minimize cognitive obstacles, such as navigational difficulties, that could render a VE application's goals inaccessible.

Multimodal interaction design. Virtual environments are designed to provide users with direct manipulative and intuitive interaction with multisensory stimulation (Bullinger et al., 2001). The number of sensory modalities stimulated and the quality of this multisensory interaction are critical to the realism and potential effectiveness of VE systems (Popescu, Burdea, & Trefftz, 2002). Yet there is currently a limited understanding of how to effectively provide such sensorial parallelism (Burdea, 1996); however, Stanney, Samman, Reeves, Hale, Buff, Bowers et al. (2004) provided a set of preliminary cross-modal integration rules. These rules consider aspects of multimodal interaction, including: (a) temporal and spatial coincidence, (b) working memory capacity, (c) intersensory facilitation effects, (d) congruency, and (e) inverse effectiveness. When multimodal sensory information is provided to users it is essential

VE DESIGN AND IMPLEMENTATIONS STRATEGIES

FIGURE 15.2. Design and implementation considerations for virtual environments.

to consider such rules governing the integration of multiple sources of sensory feedback. VE users have adapted their perception-action systems to "expect" a particular type of information flow in the real world; VEs run the risk of breaking these perception-action couplings if the full range of sensory is not supported or if it is supported in a manner that is not contiguous with real-world expectations. Such pitfalls can be avoided through consideration of the coordination between sensing and user command and the transposition of senses in the feedback loop. Specifically, command coordination considers user input as primarily monomodal and feedback to the user as multimodal. Designers need to consider which input modalities are most appropriate to support execution of a given task within the VE, if there is any need for redundant user input, and whether or not users can effectively handle such parallel input (Stanney, Mourant, & Kennedy, 1998; Stanney et al., 2004).

A limiting factor in supporting multimodal sensory stimulation in VEs is the current state of interface technologies. With the exception of the visual modality, current levels of technology simply cannot even begin to reproduce virtually those sensations, such as haptics and audition, that users expect in the real world. One solution to current technological shortcomings, sensorial transposition, occurs when a user receives feedback through senses other than those expected, that may occur because a command coordination scheme has substituted avail-able sensory feedback for those that cannot be generated within a virtual environment. Sensorial substitution schemes may be one for one (e.g., visual for force) or more complex (e.g., visual for force and auditory; visual and auditory for force). If designed effectively, command coordination and sensory substitution schemes should provide multimodal interaction that allows for better user control of the virtual environment. On the other hand, if designed poorly, these solutions may in fact exacerbate interaction problems.

Perceptual illusions. When sensorial transpositions are used, there is an opportunity for perceptual illusions to occur. With perceptual illusions, certain perceptual qualities perceived by one sensory system are influenced by another sensory system (e.g., "feel" a squeeze when you see your hand "grabbing" a virtual object). Such illusions could simplify and reduce the cost of VE development efforts (Storms, 2002). For example, when attending to a visual image coupled with a low-quality auditory display, auditory-visual cross-modal perception allows for an increase in the perceived quality of the visual image. Thus, in this case if the visual image is the focus of the task, there may be no need to use a high-quality auditory display. Unfortunately, little is known about how to leverage these phenomena to reduce development costs while enhancing one's experience in a virtual environment. Perhaps the one exception is vection (e.g., a compelling illusion of self-motion throughout a virtual world), which is known to be enhanced via a number of display factors, including a wide field of view and high spatial frequency content (Hettinger, 2002). Other such illusions exist (e.g., visual dominance) and could likewise be leveraged if perceptual and cognitive design principles are identified that can be used to trigger and capitalize on these illusory phenomena.

Navigation and wayfinding. Effective multimodal interaction design and use of perceptual illusions can be impeded if navigational complexities arise. Navigation is the aggregate of wayfinding (e.g., cognitive planning of one's route) and the physical movement that allows travel throughout a virtual environment (Darken & Peterson, 2002). A number of tools and techniques have been developed to aid wayfinding in virtual worlds, including maps, landmarks, trails, and direction finding. These tools can be used to display current position, current orientation (e.g., compass), log movements (e.g., "breadcrumb" trails), demonstrate or access the surround (e.g., maps, binoculars), or provide guided movement (e.g., signs, landmarks) (Chen & Stanney, 1999). Darken and Peterson (2002) provided a number of principles concerning how best to use these tools. If effectively applied to VEs, these principles should lead to reduced disorientation and enhanced wayfinding in large-scale virtual environments.

Content Development

Content development is concerned with the design and construction of the virtual objects and synthetic environment that support a VE experience (Isdale, Fencott, Heim, & Daly, 2002). While this medium can leverage existing HCI design principles, it has unique design challenges that arise due to the demands of real-time, multimodal, collaborative interaction. In fact, content designers are just starting to appreciate and determine what it means to create a full sensory experience with user control of both point of view and narrative development. Aesthetics is thought to be a product of agency (e.g., pleasure of being), narrative potential, presence and co-presence (e.g., existing in and sharing the virtual experience), as well as transformation (e.g., assuming another persona) (Church, 1999; Murray, 1997). Content development should be about stimulating perceptions (e.g., sureties, surprises), as well as contemplation over the nature of being (Fencott, 1999; Isdale et al., 2002).

Existing design techniques, for example, from entertainment, video games, and theme parks, can be used to support VE content development. Game development techniques that can be leveraged in VE content development include but are not limited to providing a clear sense of purpose, emotional objectives, perceptual realism, intuitive interfaces, multiple solution paths, challenges, a balance of anxiety and reward, as well as an almost unconscious flow of interaction (Isdale et al., 2002). From theme park design, content development suggestions include: (a) having a story that provides the all-encompassing theme of the VE and thus the "rules" that guide design, (b) providing location and purpose, (c) using cause-and-effect to lead users to their own conclusions, and (d) anchoring users in the familiar (Carson, 2000a, 2000b). While these suggestions provide guidelines for VE content development, considerable creativity is still an essential component of the process. Isdale et al. (2002) suggested that the challenges of VE content development highlight the need for art to complement technology.

Products Liability

Those who implement VE systems must be cognizant of potential product-liability concerns. Exposure to a VE system often produces unwanted side effects that could render users incapable of functioning effectively upon return to the real world. These adverse effects may include nausea and vomiting, postural instability, visual disturbances, and profound drowsiness (Stanney, Salvendy, Deisinger, DiZio, Ellis, Ellison, et al., 1998). As users subsequently take on their normal routines, unaware of these lingering effects, their safety and well-being may be compromised. If a VE product occasions such problems, liability of VE developers or system administrators could range from simple accountability (e.g., reporting what happened) to full legal liability (e.g., paying compensation for damages) (Kennedy & Stanney, 1996; Kennedy, Kennedy, & Bartlett, 2002). In order to minimize their liability, manufacturers and corporate users should design systems and provide usage protocols to minimize risks, warn users about potential aftereffects, monitor users during exposure, assess users' risk, and debrief users after exposure.

Usage Protocols

In order to minimize products liability concerns, VE usage protocols should be carefully designed. A comprehensive VE usage protocol will involve the following activities (see Stanney, Graeber, & Kennedy, in press):

1. Designing VE stimulus to minimize adverse effects:

   - Ensure system lags/latencies are minimized and stable;
   - Ensure frame rates are optimized;
   - Ensure interpupillary distance of visual display is adjustable and set appropriately;
   - Determine if size of field of view is causing excessive vection, such that the spatial frequency content of the visual scene should be reduced;
   - Determine if multimodal feedback is integrated into the VE such that sensory conflicts are minimized.

2. Quantifying stimulus intensity of a VE system using the Simulator Sickness Questionnaire (Kennedy, Lane, Berbaum, & Lilienthal, 1993) or other means and comparing the outcome to other systems (see Stanney et al., 2005). If a given VE system is of high intensity (say the 50th or higher percentile) and is not redesigned to lessen its impact, significant dropouts can be expected.

3. Identifying individual capacity of target user population to resist adverse effects of VE exposure via the Motion History Questionnaire (Kennedy & Graybiel, 1965) or other means.

- For highly susceptible populations redesigning the VE to reduce stimulus intensity or expect high dropout rates;
- Providing warnings for those with severe susceptibility to motion sickness, seizures, migraines, cold, flu, or other ailments.

4. Setting exposure duration and intersession interval to minimize adverse effects by (see Stanney et al., 2005):

- Limiting initial exposures (e.g., 10 minutes or less);
- Setting intersession exposure intervals two to five days apart to enhance individual adaptability;
- Incrementally increasing VE stimulus intensity within one exposure (incremental adaptation) or across multiple exposures (incremental habituation);
- Determining if users are able to complete an adaptive process at each increment of stimulus intensity (e.g., depressing sensitization to preexposure levels or below); if not, lowering stimulus intensity;
- Avoiding repeated exposure intervals occurring less than two hours apart if adverse effects are experienced in an exposure.

5. Educating users regarding potential risks of VE exposure (e.g., inform users they may experience nausea, malaise, disorientation, headache, dizziness, vertigo, eyestrain, drowsiness, fatigue, pallor, sweating, increased salivation, and vomiting).
6. Educating users regarding potential adverse aftereffects of VE exposure (e.g., inform users they may experience disturbed visual functioning, visual flashbacks, and unstable locomotor and postural control for prolonged periods postexposure).
7. Instructing users to terminate VE interaction if they start to feel ill.
8. Providing adequate air flow and comfortable thermal conditions.
9. Adjusting equipment to minimize fatigue.
10. For strong VE stimuli, warning users to avoid extraordinary maneuvers (e.g., flying backward or experiencing high rates of linear or rotational acceleration) during initial interaction.
11. Providing an attendant to monitor users' behavior and ensure their well-being:

- Attendant should terminate exposure if they see red flags (e.g., excessive sweating, verbal frustration, prolonged lack of movement in VE, and less overall movement [e.g., restricting head movement] or if users verbally request termination or complain of excessive symptoms);
- Attendant should assess well being of users' postexposure (e.g., can use battery similar to field sobriety tests; Tharp, Burns, & Moskowitz, 1981).

12. Specifying amount of time postexposure that users must remain on premises before driving or participating in other such high-risk activities. Do not allow individuals who fail postexposure tests or experience adverse aftereffects to conduct high-risk activities until they have recovered (e.g., have someone else drive them home).
13. Calling users the next day or having them call to report any prolonged adverse effects.

Regardless of the strength of the stimulus or the susceptibility of the user, following a systematic usage protocol can minimize the adverse effects associated with VE exposure.

HEALTH AND SAFETY ISSUES

The health and safety risks associated with VE exposure complicate usage protocols and lead to products liability concerns. It is thus essential to understand these issues when utilizing VE technology. There are both physiological and psychological risks associated with VE exposure, the former being related primarily to sickness and aftereffects and the latter primarily being concerned with the social impact.

Cybersickness, Adaptation, and Aftereffects

Motion-sickness-like symptoms and other aftereffects (e.g., balance disturbances, visual stress, altered hand-eye coordination) are unwanted byproducts of VE exposure. The sickness related to VE systems is commonly referred to as "cybersickness" (McCauley & Sharkey, 1992). Some of the most common symptoms exhibited include dizziness, drowsiness, headache, nausea, fatigue, and general malaise (Kennedy et al., 1993). More than 80% of users will experience some level of disturbance, with approximately 12% ceasing exposure prematurely due to this adversity (Stanney, Kingdon, Nahmens, & Kennedy, 2003). Of those who drop out, approximately 10% can be expected to have an emetic response (e.g., vomit), however, only 1–2% of all users will have such a response. These adverse effects are known to increase in incidence and intensity with prolonged exposure duration (Kennedy, Stanney, & Dunlap, 2000). While most users will experience some level of adverse effects, symptoms vary substantially from one individual to another, as well as from one system to another (Kennedy & Fowlkes, 1992). These effects can be assessed via the Simulator Sickness Questionnaire (Kennedy et al., 1993), with values above 20 requiring due caution (e.g., warn and observe users) (Stanney et al., 2005).

To overcome such adverse effects, individuals generally undergo physiological adaptation during VE exposure. This adaptation is the natural and automatic response to an intersensorily imperfect VE and is elicited due to the plasticity of the human nervous system (Welch, 1978). Due to technological flaws (e.g., slow update rate, sluggish trackers), users of VE systems may be confronted with one or more intersensory discordances (e.g., visual lag, a disparity between seen and felt limb position). In order to perform effectively in the VE, they must compensate for these discordances by adapting their psychomotor behavior or visual functioning. Once interaction with a VE is discontinued, these compensations persist for some time after exposure, leading to aftereffects.

Once VE exposure ceases and users return to their natural environment they are likely unaware that interaction with the VE has potentially changed their ability to effectively interact with their normal physical environment (Stanney & Kennedy, 1998). Several different kinds of aftereffects may persist for prolonged periods following VE exposure (Welch, 1997). For example, hand-eye coordination can be degraded via perceptual-motor disturbances (Kennedy, Stanney, Ordy, & Dunlap, 1997; Rolland, Biocca, Barlow, & Kancherla, 1995), postural sway can arise (Kennedy & Stanney, 1996), as can changes in the vestibulo-ocular reflex (VOR), or one's ability to stabilize an image on the retina (Draper, Prothero, & Viirre, 1997). The implications of these aftereffects are that:

1. VE exposure duration may need to be minimized;
2. Highly susceptible individuals or those from clinical populations (e.g., those prone to seizures) may need to avoid or be banned from exposure;
3. Users should be closely monitored during VE exposure;
4. Users' activities should be closely monitored for a considerable period of time postexposure to avoid personal injury or harm.

Social Impact

Virtual environment technology, like its ancestors (e.g., television, computers), has the potential for negative social implications through misuse and abuse (Kallman, 1993). Yet violence in VE is nearly inevitable, as evidenced by the violent content of popular video games. Such animated violence is a known favorite over the portrayal of more benign emotions such as cooperation, friendship, or love (Sheridan, 1993). The concern is that users who engage in what seems like harmless violence in the virtual world may become desensitized to violence and mimic this behavior in the look-alike real world.

Currently, it is not clear whether or not such violent behavior will result from VE exposure, early research, however, is not reassuring. Calvert and Tan (1994) found VE exposure to significantly increase the physiological arousal and aggressive thoughts of young adults (Calvert & Tan, 1994). Perhaps more disconcerting was that neither aggressive thoughts nor hostile feelings were found to decrease due to VE exposure, thus providing no support for catharsis. Such increased negative stimulation may then subsequently be channeled into real-world activities. The ultimate concern is that VE immersion may potentially be a more powerful perceptual experience than past, less interactive technologies, thereby increasing the negative social impact of this technology (Calvert, 2002). A proactive approach is needed that weighs the risks and potential consequences associated with VE exposure against the benefits. Waiting for the onset of harmful social consequences should not be tolerated.

VIRTUAL ENVIRONMENT USABILITY ENGINEERING

Most VE user interfaces are fundamentally different from traditional graphical user interfaces, with unique I/O devices, perspectives, and physiological interactions. Thus, when developers and usability practitioners attempt to apply traditional usability engineering methods to the evaluation of VE systems

they find few if any that are particularly well suited to these environments (for notable exceptions see Gabbard, Hix, & Swan, 1999; Hix & Gabbard, 2002; Stanney, Mollaghasemi, & Reeves, 2000). There is a need to modify and optimize available techniques to meet the needs of VE usability engineering, as well as to better characterize factors unique to VE usability, including sense of presence and VE ergonomics.

Usability Techniques

Assessment of usability for VE systems must go beyond traditional approaches, which are concerned with the determination of effectiveness, efficiency, and user satisfaction. Evaluators must consider whether multimodal input and output is optimally presented and integrated, navigation is supported to allow the VE to be readily traversed, object manipulation is intuitive and simple, content is immersive and engaging, and the system design optimizes comfort while minimizing sickness and aftereffects. The affective elements of interaction become important when evaluating VE systems. It is an impressive task to ensure that all of these criteria are met.

Gabbard, Hix, and Swan (1999) have developed a taxonomy of VE usability characteristics that can serve as a foundation for identifying and evaluating usability criteria particularly relevant to VE systems. Stanney, Mollaghasemi, and Reeves (2000) used this taxonomy as the foundation on which to develop an automated system, *MAUVE* (Multicriteria Assessment of Usability for Virtual Environments), which assesses VE usability in terms of how effectively each of the following are designed: (a) navigation, (b) user movement, (c) object selection and manipulation, (d) visual output, (e) auditory output, (f) haptic output, (g) presence, (h) immersion, (i) comfort, (j) sickness, and (k) aftereffects. *MAUVE* can be used to support expert evaluations of VE systems, similar to the manner in which traditional heuristic evaluations are conducted. Due to such issues as cybersickness and aftereffects, it is essential to use these or other techniques (cf., Modified Concept Book Usability Evaluation Methodology; Swartz, 2003) to ensure the usability of VE systems, not only to avoid rendering them ineffective but also to ensure that they are not hazardous to users.

Sense of Presence

One of the usability criteria unique to VE systems is sense of presence. Virtual environments have the unique advantage of leveraging the imaginative ability of individuals to psychologically "transport" themselves to another place, one that may not exist in reality (Sadowski & Stanney, 2002). To support such transportation, VEs provide physical separation from the real world, by immersing users in the virtual world via an HMD, then imparting sensorial sensations via multimodal feedback that would naturally be present in the alternate environment. Focus on generating such presence is one of the primary characteristics distinguishing VEs from other means of displaying information.

Presence has been defined as the subjective perception of being immersed in and surrounded by a virtual world rather than the physical world one is currently situated in (Stanney, Salvendy, et al., 1998). Virtual environments that engender a high degree of presence are thought to be more enjoyable, effective, and well received by users (Sadowski & Stanney, 2002). To enhance presence, designers of VE systems should spread detail around a scene, let user interaction determine when to reveal important aspects, maintain a natural and realistic, yet simple, appearance, and utilize textures, colors, shapes, sounds, and other features to enhance realism (Kaur, 1999). To generate the feeling of immersion within the environment, designers should isolate users from the physical environment (use of an HMD may be sufficient), provide content that involves users in an enticing situation supported by an encompassing stimulus stream, provide natural modes of interaction and movement control, and utilize design features that enhance vection (Stanney et al., 2000). Presence can be assessed via Witmer and Singer's (1998) Presence Questionnaire or techniques used by Slater and Steed (2000), as well as a number of other means (Sadowski & Stanney, 2002).

Virtual Environment Ergonomics

Ergonomics, which focuses on fitting a product or system to the anthropometric, musculoskeletal, cardiovascular, and psychomotor properties of users, is an essential element of VE system design (McCauley-Bell, 2002). Supporting user comfort while donning cumbersome HMDs or unwieldy peripheral devices is an ergonomics challenge of paramount importance; discomfort could supersede any other sensations (e.g., presence, immersion). If a VE produces discomfort, participants may limit their exposure time or possibly avoid repeat exposure. Overall physical discomfort should thus be minimized, while user safety is maximized.

Ergonomic concerns affecting comfort include visual discomfort resulting from visual displays with improper depth cues, poor contrast and illumination, or improperly set IPDs (Stanney et al., 2000). Physical discomfort can be driven by restrictive tethers; awkward interaction devices; or heavy, awkward, and constraining visual displays. To enhance the ergonomics of VE systems, several factors should be considered, including (McCauley-Bell, 2002):

- Is operator movement inhibited by the location, weight, or window of reach of interaction devices or HMDs?
- Does layout and arrangement of interaction devices and HMDs support efficient and comfortable movement?
- Is any limb overburdened by heavy interaction devices or HMDs?
- Do interaction devices require awkward and prolonged postures?
- If a seat is provided, does it support user movement and is it of the right height with adequate back support?
- If active motion interfaces are provided (e.g., treadmills), are they adjustable to ensure fit to the anthropometrics of users?
- Are the noise and sound levels within ergonomic guidelines and do they support user immersion?

APPLICATION DOMAINS

The wide range of VE types and designs makes them ideal tools for supporting a range of performance-enhancing tasks. Most commonly, VEs are used as training aids; a less common but increasingly well-supported use is as a selection tool, the flip side to the training application. Other applications for VE include their growing potential for providing entertainment, their use as medical rehabilitative tools, and even their implementation within the training system acquisition cycle of the U.S. government.

VE as a Selection and Training Tool

Perhaps the most common application for VEs is as training tools. The U.S. military has focused considerable resources on VE as a means to resolve many of the training deficits that result from the rigors of military life (e.g., sustaining schoolhouse-gleaned skills and knowledge sets during prolonged deployment periods, acquiring new skills and knowledge while away from the schoolhouse, providing large-scale environments for multitrainee exercises, distributing training exercises). As VE training exercises become more closely aligned to real-world conditions, a series of factors, including training transfer effectiveness, cost, logistics and safety, move to the forefront. Further, in terms of applying VE to enhance human performance, training is actually the second stage of a two-stage process. Ideally, one would like to ensure that those individuals for whom training is being provided have a certain degree of "performance capability." In this approach, VE systems can be used as part of a comprehensive performance-enhancement program that focuses on selecting those users with the correct set of Knowledge, Skills, and Abilities (KSAs) and then providing, when needed, training to fine-tune those KSAs.

Given the potentially high cost of developing VEs, this makes sense. Given the broader, yet equally high costs of bringing individuals into an organization and training them to some level of proficiency, it becomes glaringly obvious that having an ability to select individuals who are most likely to benefit from VE training is critical. For example, in the U.S. Air Force the cost of a single individual student pilot failing to complete basic flight school can run to $100,000 (Siem, Carretta, & Mercatante, 1988). Student failure can be attributed to both inadequate selection techniques and deficient training techniques. Clearly, both selection and training play critical roles in producing effective users. The challenge is to develop a program that ensures a smooth union between the two-one, which identifies the best candidates and then provides the optimum training.

Traditional approaches to selection focus on social and psychophysical assessments. For example, aptitude tests, ranging from traditional pen-and-paper-type efforts (Carretta & Ree, 1995) to psychomotor tests (Carretta & Ree, 1993) to computer-based (but not VE!) assessments (Carretta & Ree, 1993) have all been used with varying levels of success. The single most important criticism of each of these approaches is that they are designed to be predictive of future performance, and as such are more often than not abstractions of aspects of the larger task(s) for which the individual is being selected. An alternate approach would be to provide selectees with a method that provides a direct indication of their performance abilities. This distinction, essentially between a test being predictive of performance ability versus indicative of performance ability, has a great impact on selection. A meta-analysis performed within the aviation domain, where much of selection research has focused, found that typical predictive validities (most often reported as either the correlation coefficient, r, or the multiple correlation coefficient, R, and representing the degree to which a given predictor/set of predictors and performance metrics are related) for such assessments range from a low of 0.14 to a "high" of about 0.40 (Martinussen, 1996). Yet, when a simulation component is added to this mix, these values have been shown to improve considerably, pushing correlations towards the 0.60 level (Gress & Willkomm, 1996).

A potential deterrent with using VEs as part of a selection toolkit is the high cost associated with developing both the system as well as the performance measures that need to be integrated with it. However, when considered as one part of a comprehensive performance enhancement program, this concern vanishes. The same effort and cost devoted to developing the system for selection can typically, with minimal additional input, be modified (e.g., via altered scenarios) to meet identified training goals and objectives. It is therefore necessary to explore some of the basic requirements for using VEs as training tools in order to understand how one technology can truly be applied for these multiple purposes.

A constant thread in training research is the notion that in order for training to be effective, the basic skills being taught must show some degree of transfer to real-world performance. Over 100 years ago, Thorndike and Woodworth (1901) laid down the most basic training-transfer principle when they proposed that transfer was determined by the degree of similarity between any two tasks. Applying this heuristic to VE design, one might conclude that the most basic way to ensure perfect transfer is to ensure that the real-world performance elements that are meant to be trained should be replicated perfectly in a virtual environment. This notion of "identical elements" could easily create a serious challenge for system designers even by today's technology standards, as VEs are still not able to perfectly duplicate the wide range of sensorial stimuli encountered during daily interactions with our world (Stoffregen, Bardy, Smart, & Pagulyan, 2003). Countering this somewhat simplistic design approach is Osgood's (1949) principle that greater similarity between any two tasks along certain dimensions will not guarantee wholesale, perfect transfer. The challenge, as noted by Roscoe (1982) is to find the right balance between technical fidelity and training effectiveness.

This suggests that when developing training systems, it is important to take two factors into consideration. The first is how one can measure training transfer, and the second is how one can assess the cost/benefit relationship associated with making the decision to use VEs to provide training. The most basic metric for assessing transfer is the percent of transfer (Wickens & Hollands, 2000) in which the amount of time saved in learning the real world task is expressed in terms of how much time a control group, denied such training, required to learn the task (see Equation 1):

$$\%Transfer = [(Control\ Time - Transfer\ Time)/ \\ Control\ Time]*100 \quad (1)$$

A quick inspection of this metric suggests that it is overly simplistic. Consider a situation in which the percent transfer is high, yet the total amount of time spent training in the transfer platform (e.g., VE) may be greater than that spent by a control group to achieve similar levels of performance. Such an outcome is hardly cost effective. Consequently, in parallel to a basic quantification of transfer, another metric, the Training Effectiveness Ratio (TER; Povenmire & Roscoe, 1973) focusing on training efficiency is needed (see Equation 2).

$$TER = Amount\ of\ time\ saved\ learning\ task \\ by\ Transfer\ Group/Time\ spent\ in\ transfer\ platform \quad (2)$$

All things being equal, these two measures should perfectly quantify VE utility as a training tool. Yet, a third factor, cost, almost always factors in. Indeed, cost is often the single most critical driving factor in determining whether to supplement a training program with VE tools. In terms of training effectiveness, cost can be quantified using the Training Cost Ratio (TCR; Wickens & Hollands, 2000; see Equation 3).

$$TCR = Training\ cost\ in\ real\ platform/ \\ Training\ cost\ in\ transfer\ platform \quad (3)$$

When making the determination to use a VE to provide some (or, less likely, all) of a given set of training, one must consider the interplay between these measures. Specifically, the relationship between training effectiveness, as expressed through the TER, and training cost, as expressed through the TCR, will drive this decision. A good rule of thumb provided by Wickens and Holland (2000) is:

$$TER\ X\ TCR > 1.0,\ VE\ Training\ is\ effective \quad (4)$$
$$TER\ X\ TCR < 1.0,\ VE\ Training\ is\ ineffective \quad (5)$$

A slightly different way of interpreting this was provided by Roscoe (1982) who suggested that, given that the higher the level of technical fidelity the more costly the system, there is a region of design space within which cost/benefit is optimal; move away from this region in either direction, and you run the risk of paying a lot for a system that delivers less than effective (and perhaps even negative) training.

This brief overview should not leave one with the impression that the decision to use—or to not use—VEs for training is simple. Assessing the TER * TCR function is no easy matter. Simply put, the cost of performing the study needed to capture the effectiveness of a training device can far outstrip the perceived immediate impact of incorporating the device into a training curriculum. Often, the type of system being developed requires a significant investment in time, money, and labor in order to be evaluated. Consider a situation in which a new flight simulator is being considered for purchase. In order to determine both the TER and the TCR, one must conduct an actual transfer study, which involves:

1. Removing a set of trainees from their curriculum for experimental purposes;
2. Developing reliable performance metrics;
3. Providing both the experiment group and the control group with adequate time in the real world (e.g., aircraft) environment and the experiment group time in the VE;
4. Dealing with typically low effect sizes, which forces experimenters to use large sample sizes.

Yet, just as the TER * TCR function could guide when to use and not use VEs, it is possible to pull some general principles from the literature. For example, cognitively oriented training, which often features problem solving or decision making as a key training goal, may oftentimes be satisfied by using simple, relatively inexpensive visualization systems (Gopher, Weil, & Bareket, 1994; Morris & Tarr, 2002; Stone, 2002; Figueroa, Bischof, Boulanger, & Hoover, 2005; Milham, Hale, Stanney, Cohn, Darken, & Sullivan, 2004). Contrastingly, motor skill based training, which often requires complex interactions between users and their environments, frequently requires more costly solutions. When considering whether or not to "go virtual," a useful solution would be to first determine the general category of skill to be trained (Cognitive or Motor; cf., Anderson, 1987); estimate the TER and TCR from sources as broadly ranging as basic science studies, published costs of similarly desired systems, and one's own best estimates; and then develop a basic decision matrix to establish a range of TER * TCR values.

VE as Entertainment Tool

In many ways entertainment can be considered a prime technology pusher. Consider the evolution of the VCR, Digital Video Disk, High Definition Television and, more recently, the advent of the all-in-one device (phone, PDA, music/video player) in various forms. Each of these devices captured a unique market niche precisely because they pushed ahead current technology barriers in ways never before considered. Yet, if entertainment provides the technology push necessary to achieve revolutions, the consumers of such technologies provide an equally strong technology pull, placing discrete constraints on this transformation. Often, such constraints are directly at odds with entertainment's evolution. An entertainment developer may create a new game that requires a high-end graphics system, a costly sound system, and perhaps even a tactile (haptic) interface to fully take advantage of the system. The consumers, on the other hand, may have their own satisficing heuristic, requiring that, in addition to providing novel entertainment value, any new technology must cost less than a fixed amount and fit into their back pockets. Unless there is significant demand for a technology to factor these constraints into its development (e.g., a significant consumer base) the resultant product may either be forestalled or, when placed on the market, flounder and vanish overnight.

Until recently, the high cost of the component technologies comprising VEs precluded their becoming anything more than a research test-bed for well-funded labs or an exclusive tool for high-end consumers, like commercial airline companies or the military establishments of entire countries. Yet, after nearly four decades of research and development, VEs have finally begun to realize their potential. This maturation owes much to advancements in the computer industry which, proceeding at a Moore's Law rate, has been able to evolve the simple desktop PC into a powerful graphical rendering system. As technology barriers have been shattered, so too have many of the HCI challenges that were typically associated with earlier technology-limited systems, thus resulting in a number of entertainment-based applications. Current VE entertainment applications include games, online communities, location-based entertainment, theme parks, and other venues (Badiqué et al., 2002). From interactive arcades to cybercafes, the entertainment industry has leveraged the unique characteristics of the VE medium, providing dynamic and exciting experiences. Such applications tend to set the pace, both technological and in terms of content, for advances in VE technology in general. Loosely put, the goal of such entertainment technologies is to provide consumers with experiences that they would not ordinarily encounter in their typical daily activities. Consequently, the goals and objectives of entertainment-based systems are quite broadly defined (and, sometimes, not necessarily defined at all!) and shaped primarily through user feedback vis-à-vis factors such as "ease of use," "engagement level," "interest level," and so forth. These factors are molded to a large extent by popular culture. By attending to these global trends, developers can determine what technologies exist, or need to be created, to support current user demand. Thus, in spite of the optimism generated by such entertainment applications, a cautionary note is warranted. The uses for which typical VEs have been, and continue to be, considered (e.g., primarily as training tools) have a strong scientific basis upon which to define system specifications, and a lengthy history linking these specifications to specific technologies. Consider, for example, the prerequisites for designing a system to train a naval officer to maneuver a ship in a harbor (Hays, Castillo, Seamon, & Bradley, 1997; Nguyen, Cohn, Mead, Helmick, & Patrey, 2001). Since the trainees will, ultimately, be graded on their performances, developers must identify the necessary training goals and objectives (e.g., Osgood's [1949] identical elements), identify the necessary technologies (or develop them if absent), integrate them, test and validate their training efficacy, and ultimately hand them off to be used. Such a developmental process can require tens of millions of dollars, time investments on the order of years, and personnel investments numbering in the dozens. Nevertheless, precisely because training is perceived as (and often is) critical to a given organization's success, training requirements can provide a justification for such large investments. On the other hand, in entertainment, the driving force is delivering customer satisfaction ahead of the competition. As a result, it is quite likely that the vast majority of entertainment developers will assume less risk when developing their products in order to cater to a constantly shifting market interest.

VE as a Medical Tool

Much has been written about applications for VE within the medical arena (cf., Moline, 1995; Satava & Jones, 2002). While some of these applications represent unique approaches to harnessing the power of VE, many other applications, such as simulating actual medical procedures, reflect training applications, and therefore will not be discussed anew here. One area of medical application for which VE is truly coming into its own is medical rehabilitation. In particular, two areas of rehabilitation, psychological/cognitive and motor, show strong promise.

Psychological/cognitive rehabilitation applications. Perhaps the fastest-growing application for VEs in medicine is in the area of phobia treatment (Rizzo, Buckwalter, & van der Zaag, 2002; Satava & Jones, 2002; Thacker, 2003). The reason for this is likely due to the ideal matching between VE's strengths (presenting evolving information with which users can interact in various ways) and such therapy's basic requirements (incremental exposure to the offending environment). Importantly, compared to previous treatment regimens, which oftentimes simply required patients to mentally revisit their fears, VEs offer a significantly more immersive experience. In fact, it is quite likely that many of VE's shortcomings, such as poor visual resolution, inadequate physics modeling underlying environmental cues, and failure to fully capture the wide range of sensorial cues present in the real world, will be ignored by the patient, whose primary focus is on overcoming anxiety engendered by her or his specific phobias. On a practical level, VEs enable patients to simply visit their therapist's office, where they can be provided an individually tailored multimodal treatment experience (Rothbaum, Hodges, Watson, Kessler, & Opdyke, 1996; Emmelkamp, Bruynzeel, & Drost, van der Mast, 2001; Anderson, Rothbaum, & Hodges, 2003).

Motor rehabilitation applications. Many of VE's qualities that make it an ideal tool for providing medical training—such as tactile feedback and detailed visual information (Satava & Jones, 2002)—also make it an ideal candidate for supplementing motor rehabilitation treatment regiments for such conditions as stroke, Parkinson's, and pain treatment (Holden, Todorov, Callahan, & Bizzi, 1999; Gershon, Zimand, Lemos, Rothbaum, & Hodges, 2003). In determining how best to apply VE in these and other treatment regimens, Holden (2005) suggested considering three practical areas in which VE is strongest: repetition, feedback, and motivation. All three elements are critical to both effective learning and regaining of motor function; the application of VE, in each case, provides a powerful method for rehabilitation specialists to maximize the effect of a treatment regimen for a given session, and, because they may reduce the time investment required by therapists (one can simply immerse the patient, initiate the treatment, and then allow the program to execute), to also expand the access of such treatments to a wider population.

Since VE is essentially computer-based, patients can effectively have their attention drawn to a specific set of movement patterns they may need to make to regain function; conducting this in a "loop" provides unlimited ability to repeat a pattern, while using additional visualization aids, such as a rendered cur-

sor or "follow-me" types of cues, to force the patient into moving a particular way (cf., Chua, Crivella, Daly, Hu, Schaaf, Ventura, et al., 2003). As well, it is a relatively simple matter to digitize movement information, store it, and then, based on comparisons to previously stored, desired movement patterns, provide additional feedback to assist the patient. In terms of motivation, treatment scenarios can be tailored to capture specific events that individual patients find most motivating: a baseball fan can practice her movement in a baseball-like scenario; a car enthusiast can do so in a driving environment.

There are certain caveats that must be considered when exploiting VE for rehabilitation purposes, most significantly the potentially rapid loss of motor adaptations following VE exposure. Lackner and DiZio (1994) demonstrated that certain basic patterns of sensorimotor recalibrations learned in a given physical environment can diminish within an hour, post-exposure, although subsequent findings (DiZio & Lackner, 1995) suggest that there are certain transfer benefits that are longer-lasting. Brashers-Krugg, Shadmehr, and Bizzi (1996) provided additional evidence that sensorimotor recalibrations of the type likely to be required for rehabilitation have post-exposure periods in excess of four hours, during which their effects can be extinguished. Most importantly, Cohn, DiZio and Lackner (2000) demonstrated that such recalibrations, when learned in VE, have essentially no transfer to real-world conditions post-exposure. Clearly, more research is needed to understand the conditions under which such transfer effects can be made most effective within the clinical setting.

VE as a System Acquisition Tool

In many areas of research and development, rising costs, with their associated increasing consequences of failure, have all but forced scientists and engineers to focus only on developing those products that have an almost guaranteed chance of succeeding in the marketplace. While this may seem like an ideal situation—after all, who would want to field technologies that won't work? —the net result is an increasing aversion, among companies and their financial backers, to shoulder much risk. Consequently, much-needed breakthrough technologies may be ignored in favor of "sure-thing" technologies, which more often than not do little to push ahead technology barriers.

The United States Department of Defense (DoD), seeking to reverse this trend while maintaining a level of control over risk, developed the process of Simulation Based Acquisition (SBA), a methodology that introduced the use of modeling and simulation (M&S) within the product-development process (Sanders, 1997; Zittel, 2001). Most recently, these M&S solutions have become increasingly integrated into 4D interactive tools that enable human-in-the-loop testing of design and operational principles (Sanders, 1997). As one example that cuts across DoD and industry, consider the development of their newest fighter aircraft, the Joint Strike Fighter (JSF). While more common M&S solutions, such as the ability to simulate individual components digitally before manufacturing them (through the CAD/CAM process), formed a core component of the development process, more advanced M&S solutions, such as VEs, have enabled developers to include users throughout the development process, in order to answer such questions as (a) how users (e.g., pilots) would implement the fighter in combat, using virtual war-gaming techniques; (b) how well individual pilots could handle a given JSF design; and, (c) how the JSF would alter the nature of warfare (Zittel, 2001).

CONCLUSIONS

Clearly, VEs have evolved significantly from Morton Heilig's 1962 Sensorama experience (Burdea & Coiffet, 2003). Yet, despite significant revolutions in component technology, many of the challenges addressed by that primitive system, such as multi-modal sensori-interaction, visual representation, and scenario generation, have yet to be fully resolved more than 40 years later. At the same time, our understanding of the potential such tools have to offer has advanced considerably. No longer simple amusements, these powerful machines can provide educational value, assist in treating core physical and cognitive mal-

adies, and even help design better VE systems. As the uses for which VEs are ideally suited continue to be defined and refined, one can anticipate that current development challenges will be resolved, only to find new ones waiting to take their place.

ACKNOWLEDGMENTS

The author would like to thank Branka Wedell for her contribution of Figs. 1 and 2. This material is based upon work supported in part by the National Science Foundation (NSF) under Grant No. IRI-9624968, the Office of Naval Research (ONR) under Grant No. N00014-98-1-0642, the Naval Air Warfare Center Training Systems Division (NAWC TSD) under contract No. N61339-99-C-0098, and the National Aeronautics and Space Administration (NASA) under Grant No. NAS9-19453. Any opinions, findings, and conclusions or recommendations expressed in this material are those of the authors and do not necessarily reflect the views or the endorsement of the NSF, ONR, NAWC TSD, or NASA.

References

Akamatsu, M. (1994, July). Touch with a mouse. A mouse type interface device with tactile and force display. *Proceedings of 3rd IEEE International Workshop on Robot and Human Communication, Nagoya, Japan,* 140–144.

Algazi, V. R., Duda, R. O., Thompson, D. M., & Avendano, C. (2001, October). The CIPIC HRTF Database. *Proceedings of the 2001 IEEE Workshop on Applications of Signal Processing to Audio and Electroacoustics,* New Paltz, NY, 99–102.

Allbeck, J. M., & Badler, N. I. (2002).Embodied autonomous agents. In K. M. Stanney (Ed.), *Handbook of virtual environments: Design, implementation, and applications* (pp. 313–332). Mahwah, NJ: Lawrence Erlbaum Associates, Inc.

Anderson, J. R. (1987). Skill acquisition: Compilation of weak-method problem solutions. *Psychological Review, 94,* 192–210.

Anderson, P., Rothbaum, B. O., & Hodges, L. F. (2003). Virtual reality exposure in the treatment of social anxiety. *Cognitive and Behavioral Practice, 10*(3), 240–247.

Badiqué, E., Cavazza, M., Klinker, G., Mair, G., Sweeney, T., Thalmann, D., & Thalmann, N. M. (2002). Entertainment applications of virtual environments. In K. M. Stanney (Ed.), *Handbook of virtual environments: Design, implementation, and applications* (pp. 1143–1166). Mahwah, NJ: Lawrence Erlbaum Associates, Inc.

Begault, D. (1994). *3-D Sound for virtual reality and multimedia.* Boston: Academic Press.

Biggs, S. J., & Srinivasan, M. A. (2002). Haptic interfaces. In K. M. Stanney (Ed.), *Handbook of virtual environments: Design, implementation, and applications* (pp. 93–116). Mahwah, NJ: Lawrence Erlbaum Associates, Inc.

Brashers-Krug, T., Shadmehr, R., & Bizzi, E. (1996). Consolidation in human motor memory. *Nature, 382*(6588), 252–255.

Bullinger, H.-J., Breining, R., & Braun, M. (2001). Virtual reality for industrial engineering: Applications for immersive virtual environments. In G. Salvendy (Ed.), *Handbook of industrial engineering: Technology and operations management* (3rd ed.; pp. 2496–2520). New York: Wiley.

Bungert, C. (2005). *HMD/headset/VR-helmet comparison chart.* Retrieved November 1, 2006, from http://www.stereo3d.com/hmd.htm.

Burdea, G. (1996). *Force and touch feedback for virtual reality.* New York: Wiley.

Burdea, G. C., & Coiffet, P. (2003). *Virtual Reality Technology* (2nd ed.). Hoboken, NJ: Wiley.

Butler, R. A. (1987). An analysis of the monaural displacement of sound in space. *Perception & Psychophysics, 41,* 1–7.

Calvert, S. L. (2002). The social impact of virtual environment technology. In K. M. Stanney (Ed.), *Handbook of virtual environments: Design, implementation, and applications* (pp. 663–680). Mahwah, NJ: Lawrence Erlbaum Associates, Inc.

Calvert, S. L., & Tan, S. L. (1994). Impact of virtual reality on young adult's physiological arousal and aggressive thoughts: Interaction versus observation. *Journal of Applied Developmental Psychology, 15,* 125–139.

Carretta, T. R., & Ree, M. J. (1993). Basic attributes test (BAT): Psychometric equating of a computer-based test. *International Journal of Aviation Psychology, 3,* 189–201.

Carretta, T. R., & Ree, M. J. (1995). Air Force Officer Qualifying Test validity for predicting pilot training performance. *Journal of Business and Psychology, 9,* 379–388.

Carson, D. (2000a). Environmental storytelling, part 1: Creating immersive 3D worlds using lessons learned from the theme park industry. Retrieved March 15, 2007, from http://www.gamasutra.com/features/20000301/carson_pfv.htm

Carson, D. (2000b). Environmental storytelling, part 2: Bringing theme park environment design techniques lessons to the virtual world. Retrieved March 15, 2007, from http://www.gamasutra.com/features/20000405/carson_pfv.htm

Chen, J. L., & Stanney, K. M. (1999). A theoretical model of wayfinding in virtual environments: Proposed strategies for navigational aiding. *Presence: Teleoperators and Virtual Environments, 8*(6), 671–685.

Chua, P. T., Crivella, R., Daly, B., Hu, N., Schaaf, R., Ventura, D., et al. (2003). Training for physical tasks in virtual environments:

Tai Chi. *Proceedings of IEEE Virtual Reality Conference 2003* (pp. 87–97). Los Angeles, CA, March 22–26, 2003.

Church, D. (1999, August). Formal abstract design tools. *Games Developer Magazine*, 44–50. Retrieved October 1, 2006, from http://www.gamasutra.com/features/19990716/design_tools_02.htm

Cohen, M. (1992). Integrating graphic and audio windows. *Presence: Teleoperators and Virtual Environments, 1*(4), 468–481.

Cohn, J., DiZio, P., & Lackner J. R. (2000). Reaching during virtual rotation: Context-specific compensation for expected Coriolis forces. *Journal of Neurophysiology, 83*(6), 3230–3240.

Cruz-Neira, C., Sandin, D. J., & DeFanti, T. A. (1993). Surround-screen projection-based virtual reality: The design and implementation of the CAVE. *ACM Computer Graphics, 27*(2), 135–142.

Crystal Rivers Engineering (1995). Snapshot: HRTF Measurement System. Groveland, CA.

Darken, R. P., & Peterson, B. (2002). Spatial orientation, wayfinding, and representation. In K. M. Stanney (Ed.), *Handbook of virtual environments: Design, implementation, and applications* (pp. 493–518). Mahwah, NJ: Lawrence Erlbaum Associates, Inc.

Davies, R. C. (2002). Applications of system design using virtual environments. In K.M. Stanney (Ed.), *Handbook of virtual environments: Design, implementation, and applications* (pp. 1079–1100). Mahwah, NJ: Lawrence Erlbaum Associates, Inc.

DiZio, P., & Lackner, J. R. (1995). Motor adaptation to Coriolis force perturbations of reaching movements: Endpoint but not trajectory adaptation transfers to the non-exposed arm. *Journal of Neurophysiology, 74*(4), 1787–1792.

Draper, M. H., Prothero, J. D., & Viirre, E. S. (1997). Physiological adaptations to virtual interfaces: Results of initial explorations. *Proceedings of the Human Factors & Ergonomics Society 41st Annual Meeting* (pp. 1393). Santa Monica, CA: Human Factors & Ergonomics Society.

Durlach, B. N. I., & Mavor, A. S. (1995). *Virtual Reality: Scientific and technological challenges*. Washington, DC: National Academy Press.

Emmelkamp, P. M. G., Bruynzeel, M., Drost, L., & van der Mast, C. A. P. G., (2001). Virtual reality treatment in acrophobia: A comparison with exposure in vivo. *Cyberpsychology & Behavior, 4*(3), 335–339.

Favalora, G. E., Napoli, J., Hall, D. M., Dorval, R. K., Giovinco, M. G., Richmond, M. J., et al. (2003). 100 million-voxel volumetric display. *SPIE Proceedings of the 16th Annual International Symposium on Aerospace/Defense Sensing, Simulation, and Controls, 4297*, 227–235.

Fencott, C. (1999). Content and creativity in virtual environment design. *Proceedings of Virtual Systems and Multimedia '99*, University of Abertay Dundee, Scotland, 308-317. Retrieved March 15, 2007, from http://www.cs.york.ac.uk/hci/kings_manor_workshops/UCDIVE/fencott.pdf

Figueroa, P., Bischof, W. F., Boulanger, P., & Hoover, H. J. (2005). Efficient comparison of platform alternatives in interactive virtual reality applications. *International Journal of Human-Computer Studies, 62*(1), 73–103.

Fouad, H. (2004). Ambient synthesis with random sound fields. In K. Greenebaum (Ed.), *Audio anecdotes: Tools, tips, and techniques for digital audio*. Natick, MA: A K Peters.

Fouad, H., & Ballas, J. (2000, April). An extensible toolkit for creating virtual sonic environments. Paper presented at the *International Conference on Auditory Displays, ICAD 2000*, Atlanta, GA.

Foxlin, E. (2002). Motion tracking requirements and technologies. In K. M. Stanney (Ed.), *Handbook of virtual environments: Design, implementation, and applications* (pp. 163–210). Mahwah, NJ: Lawrence Erlbaum Associates, Inc.

Gabbard, J. L., Hix, D., & Swan, J. E. II (1999). User-centered design and evaluation of virtual environments. *IEEE Computer Graphics and Applications, 19*(6), 51–59.

Gershon, J., Zimand, E., Lemos, R., Rothbaum, B. O., & Hodges, L. (2003). Use of virtual reality as a distractor for painful procedures in patients with pediatric cancer: a controlled study. *CyberPsychology and Behavior, 6*, 657–661.

Gopher, D., Weil, M., & Bareket, M. (1994) Transfer of skill from a computer game trainer to flight. *Human Factors, 36*(3), 387–405.

Gress, W., & Willkomm, B. (1996).Simulator-based test systems as a measure to improve the prognostic value of aircrew selection. *Selection and Training Advances in Aviation: Advisory Group for Aerospace Reasearch and Development Conference Proceedings, Prague, Czech Republic, 588*, 15-1-15-4.

Hahn, J. K., Fouad, H., Gritz, L., & Lee, J. W. (1998). Integrating sounds and motions in virtual environments. *Presence: Teleoperators and Virtual Environments, 7*(1), 67–77.

Hale, K. S., & Stanney, K. M. (2004). Deriving haptic design guidelines from human physiological, psychophysical, and neurological foundation. *IEEE Computer Graphics and Applications, 24*(2), 33–39.

Halle, M. (1997, May). Autostereoscopic displays and computer graphics. *Computer Graphics, 31*(2), 58–62.

Hays, R. T., Castillo, E. R., Seamon, A. G., & Bradley, S. K. (1997). A virtual environment for submarine ship handling: perceptual and hardware trade-offs. In M. J. Chinni (Ed.), *Proceedings of the 1997 Simulation MultiConference: Military Government, and Aerospace Simulations* (Simulation Series Vol. 29, No. 4, pp. 217–222). San Diego, CA: The Society for Computer Simulation International.

Heirich, A., & Arvo, J. (1997). Scalable Monte Carlo image synthesis [Special issue on Parallel Graphics & Visualization]. *Parallel Computing, 23*(7), 845–859.

Hettinger, L. J. (2002). Illusory self-motion in virtual environments. In K. M. Stanney (Ed.), *Handbook of virtual environments: Design, implementation, and applications* (pp. 471–492). Mahwah, NJ: Lawrence Erlbaum Associates, Inc.

Hix, D., & Gabbard, J. L. (2002).Usability engineering of virtual environments. In K. M. Stanney (Ed.), *Handbook of virtual environments: Design, implementation, and applications* (pp. 681–699). Mahwah, NJ: Lawrence Erlbaum Associates.

Holden, M. (2005). Virtual environments for motor rehabilitation: A review. *CyberPsychology & Behavior, 8*(3), 187–211.

Holden, M., Todorov, E., Callahan, J., & Bizzi, E. (1999). Virtual environment training improves motor performance in two patients with stroke: case report. *Neurology Report, 23*, 57–67.

Hollerbach, J. (2002). Locomotion interfaces. In K. M. Stanney (Ed.), *Handbook of virtual environments: Design, implementation, and applications* (pp. 239–254). Mahwah, NJ: Lawrence Erlbaum Associates, Inc.

Huang, X. D. (1998). Spoken language technology research at Microsoft. *The Journal of the Acoustical Society of America, 103*(5), 2815–2816.

Isdale, J. (2000, April). Motion simulation. *VR News: April Tech Review*. Retrieved (date), from http://vr.isdale.com/vrTechReviews/MotionSim_Links_2000.htm

Isdale, J., Fencott, C., Heim, M., & Daly, L. (2002). Content design for virtual environments. In K. M. Stanney (Ed.), *Handbook of virtual environments: Design, implementation, and applications* (pp. 519–532). Mahwah, NJ: Lawrence Erlbaum Associates, Inc.

Jones, L., Bowers, C. A., Washburn, D., Cortes, A., & Vijaya Satya, R. (2004). The effect of olfaction on immersion into virtual environments. In D. A. Vincenzi, M. Mouloua, & P. A. Hancock (Eds.), *Human performance, situation awareness and automation: Current research and trends* (Vol. 2; pp. 282–285). Mahwah, NJ: Lawrence Erlbaum.

Kalawsky, R. S. (1993). *The science of virtual reality and virtual environments*. Wokingham, England: Addison-Wesley.

Kallman, E. A. (1993). Ethical evaluation: A necessary element in virtual environment research. *Presence: Teleoperators and Virtual Environments, 2*(2), 143–146.

Kaur, K. (1999). *Designing virtual environments for usability*. Unpublished doctoral dissertation, City University, London.

Kennedy, R. S., Drexler, J. M., Jones, M. B., Compton, D. E., & Ordy, J. M. (2005). Quantifying human information processing (QHIP): Can practice effects alleviate bottlenecks? In D. K. McBride & D. Schmorrow (Eds.), *Quantifying human information processing* (pp. 63–122). Lanham, MD : Lexington Books.

Kennedy, R. S., & Fowlkes, J. E. (1992). Simulator sickness is polygenic and polysymptomatic: Implications for research. *International Journal of Aviation Psychology, 2*(1), 23–38.

Kennedy, R. S., & Graybiel, A. (1965). *The Dial test: A standardized procedure for the experimental production of canal sickness symptomatology in a rotating environment* (Rep. No. 113, NSAM 930). Pensacola, FL: Naval School of Aerospace Medicine.

Kennedy, R. S., Kennedy, K. E., & Bartlett, K. M. (2002). Virtual environments and products liability. In K. M. Stanney (Ed.), *Handbook of virtual environments: Design, implementation, and applications* (pp. 543–554). Mahwah, NJ: Lawrence Erlbaum Associates, Inc.

Kennedy, R. S., Lane, N. E., Berbaum, K. S., & Lilienthal, M. G. (1993). Simulator sickness questionnaire: An enhanced method for quantifying simulator sickness. *International Journal of Aviation Psychology, 3*(3), 203–220.

Kennedy, R. S., & Stanney, K. M. (1996). Virtual reality systems and products liability. *Journal of Medicine and Virtual Reality, 1*(2), 60–64.

Kennedy, R. S., Stanney, K. M., & Dunlap, W. P. (2000). Duration and exposure to virtual environments: Sickness curves during and across sessions. *Presence: Teleoperators and Virtual Environments, 9*(5), 466–475.

Kennedy, R. S., Stanney, K. M., Ordy, J. M., & Dunlap, W. P. (1997). Virtual reality effects produced by head-mounted display (HMD) on human eye-hand coordination, postural equilibrium, and symptoms of cybersickness. *Society for Neuroscience Abstracts, 23*, 772.

Kershner, J. (2002). Object-oriented scene management. *GameDev.net*. Retrieved October 15, 2006, from http://www.gamedev.net/reference/articles/article1812.asp

Kessler, G. D. (2002). Virtual environment models. In K. M. Stanney (Ed.), *Handbook of virtual environments: Design, implementation, and applications* (pp. 255–276). Mahwah, NJ: Lawrence Erlbaum Associates, Inc.

Knerr, B. W., Breaux, R., Goldberg, S. L., & Thurman, R. A. (2002). National defense. In K. M. Stanney (Ed.), *Handbook of Virtual environments: Design, implementation, and applications* (pp. 857–872). Mahwah, NJ: Lawrence Erlbaum Associates, Inc.

Lackner, J. R., & DiZio, P. (1994). Rapid adaptation to Coriolis force perturbations of arm trajectory. *Journal of Neurophysiology, 72*(1), 299–313.

Lawson, B. D., Sides, S. A., & Hickinbotham, K. A. (2002). User requirements for perceiving body acceleration. In K. M. Stanney (Ed.), *Handbook of virtual environments: Design, implementation, and applications* (pp. 135–161). Mahwah, NJ: Lawrence Erlbaum Associates, Inc.

Majumder, A. (2003). A practical framework to achieve perceptually seamless multi-projector displays. Unpublished doctoral dissertation, University of North Carolina at Chapel Hill. Retrieved March 15, 2007, from http://www.cs.unc.edu/~welch/media/pdf/dissertation_majumder.pdf

Massimino, M., & Sheridan, T. (1993). Sensory substitution for force feedback in teleoperation. *Presence: Teleoperators and Virtual Environments, 2*(4), 344–352.

Martinussen, M. (1996). Psychological measures as predictors of pilot performance:A meta-analysis. *International Journal of Aviation Psychology, 6*, 1–20.

May, J. G., & Badcock, D. R. (2002). Vision and virtual environments. In K. M. Stanney (Ed.), *Handbook of virtual environments: Design, implementation, and applications* (pp. 29–64). Mahwah, NJ: Lawrence Erlbaum Associates, Inc.

McCauley, M. E., & Sharkey, T. J. (1992). Cybersickness: Perception of self-motion in virtual environments. *Presence: Teleoperators and virtual environments, 1*(3), 311–318.

McCauley-Bell, P. R. (2002). Ergonomics in virtual environments. In K. M. Stanney (Ed.), *Handbook of virtual environments: Design, implementation, and applications* (pp. 807–826). Mahwah, NJ: Lawrence Erlbaum Associates, Inc.

Menzies, D. (2002, July). Scene management for modelled audio objects in interactive worlds. *Proceedings of the 8th International Conference on Auditory Displays*, Kyoto, Japan.

Merritt, E. A., & Bacon, D. J. (1997). Raster3D: Photorealistic molecular graphics. *Methods in Enzymology, 277*, 505–524.

Milham, L. M., Hale, K., Stanney, K., Cohn, J., Darken, R., & Sullivan, J. (2004). When is VE training effective? A framework and two case studies. Poster presented at *The 48th Annual Human Factors and Ergonomics Society Meeting* (pp. 2592–2595). New Orleans, LA, September 20–24, 2004.

Moline, J. (1995). *Virtual Environments for Health Care. White Paper for the Advanced Technology Program (ATP)*. Retrieved September 15, 2006, from the National Institute of Standards and Technology website: http://www.itl.nist.gov/iaui/ovrt/projects/health/vr-envir.htm

Morris, C. S., & Tarr, R. W. (2002, March). Templates for selecting PC-based synthetic environments for application to human performance enhancement and training. *Proceedings of IEEE Virtual Reality 2002 Conference*, Orlando, FL, 109–115.

Mulgund, S., Stokes, J., Turieo, M., & Devine, M. (2002). *Human/machine interface modalities for soldier systems technologies* (Final Report No. 71950-00). Cambridge, MA: TIAX, LLC.

Murray, J. H. (1997). *Hamlet on the Holodeck: The Future of Narrative in Cyberspace*. New York: The Free Press.

Nguyen, L. K., Cohn, J. V., Mead, A., Helmick, J., & Patrey, J. (2001). Realtime virtual environment applications for military maritime training. In M. J. Smith, G. Salvendy, D. Harris, & R. J. Koubek (Eds), *Usability Evaluation and Interface Design: Cognitive Engineering, Intelligent Agents and Virtual Reality* (Vol. 1 of the Proceedings of HCI International 2001) (pp. XXX–YYY). Mahwah, NJ: Lawrence Erlbaum.

North, M. M., North, S. M., & Coble, J. R. (2002). Virtual reality therapy: An effective treatment for psychological disorders. In K. M. Stanney (Ed.), *Handbook of virtual environments: Design, implementation, and applications* (pp. 1065–1078). Mahwah, NJ: Lawrence Erlbaum Associates, Inc.

Osgood, C. E. (1949). The similarity paradox in human learning: A resolution. *Psychological Review, 56*, 132–143

Popescu, G. V., Burdea, G. C., & Trefftz, H. (2002). Multimodal interaction modeling. In K. M. Stanney (Ed.), *Handbook of virtual environments: Design, implementation, and applications* (pp. 435–454). Mahwah, NJ: Lawrence Erlbaum Associates, Inc.

Pountain, D. (1996, July). VR meets reality: Virtual reality strengthens the link between people and computers in mainstream applications. *Byte Magazine*. Retrieved October 20, 2006, from http://www.byte.com/art/9607/sec7/art5.htm.

Povenmire, H. K., & Roscoe, S. N. (1973). Incremental transfer effectiveness of a ground-based general aviation trainer. *Human Factors, 15*(6), 534–542.

Rizzo, A. A., Buckwalter, G. J., & van der Zaag, C. (2002). Virtual environment applications in clinical neuropsychology. In K. M. Stanney (Ed.), *Handbook of virtual environments: Design, implementation, and applications* (pp. 93–116). Mahwah, NJ: Lawrence Erlbaum Associates, Inc.

Rolland, J. P., Biocca, F. A., Barlow, T., & Kancherla, A. (1995). Quantification of adaptation to virtual-eye location in see-thru head-mounted displays. *IEEE Virtual Reality Annual International Symposium '95* (pp. 56–66). Los Alimitos: CA: IEEE Computer Society Press.

Roscoe, S. N. (1982). *Aviation Psychology.* Ames, IA: Iowa State University Press.

Rothbaum, B. O., Hodges, L., Watson, B. A., Kessler, G. D., & Opdyke, D. (1996). Virtual reality exposure therapy in the treatment of fear of flying: A case report. *Behaviour Research & Therapy, 34,* 477–481.

Sadowski, W., & Stanney, K. (2002). Presence in virtual environments. In K. M. Stanney (Ed.), *Handbook of virtual environments: Design, implementation, and applications* (pp. 791–806). Mahwah, NJ: Lawrence Erlbaum Associates, Inc.

Sanders, P. (1997, September/October). Simulation Based Acquisition: An effective, affordable mechanism for fielding complex technologies. *Program Manager,* 72–77.

Satava, R., & Jones, S. B. (2002). Medical applications of virtual environments. In K. M. Stanney (Ed.), *Handbook of virtual environments: Design, implementation, and applications* (pp. 93–116). Mahwah, NJ: Lawrence Erlbaum Associates, Inc.

Schmorrow, D., Stanney, K. M., Wilson, G., & Young, P. (2005). Augmented cognition in human-system interaction. In G. Salvendy (Ed.), *Handbook of human factors and ergonomics* (3rd ed.). New York: John Wiley.

Sheridan, T. B. (1993). My anxieties about virtual environments. *Presence: Teleoperators and Virtual Environments, 2*(2), 141–142.

Shilling, R. D., & Shinn-Cunningham, B. (2002). Virtual auditory displays. In K. M. Stanney (Ed.), *Handbook of virtual environments: Design, implementation, and applications* (pp. 65–92). Mahwah, NJ: Lawrence Erlbaum Associates, Inc.

Siem, F. M., Carretta, T. R., & Mercatante, T. A. (1988). *Personality, attitudes, and pilot training performance: Preliminary analysis* (Tech. Report. N. AFHRL-TP-87-62). Brooks Air Force Base, TX: AFHRL, Manpower and Personnel Division.

Singhal, S., & Zyda, M. (1999). *Networked Virtual Environments— Design and Implementation* (SIGGRAPH Series). New York: ACM Press Books.

Slater, M., & Steed, A. (2000). A virtual presence counter. *Presence: Teleoperators and Virtual Environments, 9*(5), 413–434.

Stanney, K. M., Graeber, D. A., & Kennedy, R. S. (2005). Virtual environment usage protocols. In W. Karwowski (ed), *Handbook of standards and guidelines in ergonomics and human factors* (pp. 381–398). Mahwah, NJ: Lawrence Erlbaum.

Stanney, K. M., & Kennedy, R. S. (1998, October). Aftereffects from virtual environment exposure: How long do they last? *Proceedings of the 42nd Annual Human Factors and Ergonomics Society Meeting,* Chicago, 1476–1480.

Stanney, K. M., Kingdon, K., Nahmens, I., & Kennedy, R. S. (2003). What to expect from immersive virtual environment exposure: Influences of gender, body mass index, and past experience. *Human Factors, 45*(3), 504–522.

Stanney, K. M., Mollaghasemi, M., & Reeves, L. (2000). *Development of MAUVE, the Multi-Criteria Assessment of Usability for Virtual Environments System* (Final Report, Contract No. N61339-99-C-0098). Orlando, FL: Naval Air Warfare Center, Training Systems Division, 8/00.

Stanney, K. M., Mourant, R., & Kennedy, R. S. (1998). Human factors issues in virtual environments: A review of the literature. *Presence: Teleoperators and Virtual Environments, 7*(4), 327–351.

Stanney, K. M., Salvendy, G., Deisigner, J., DiZio, P., Ellis, S., Ellison, E., et al. (1998). *Aftereffects and sense of presence in virtual environments: Formulation of a research and development agenda* (Report sponsored by the Life Sciences Division at NASA Headquarters). *International Journal of Human-Computer Interaction, 10*(2), 135–187.

Stanney, K., Samman, S., Reeves, L., Hale, K., Buff, W., Bowers, C., et al. (2004). A paradigm shift in interactive computing: Deriving multi-modal design principles from behavioral and neurological foundations. *International Journal of Human-Computer Interaction, 17*(2), 229–257.

Stanney, K. M., & Zyda, M. (2002). Virtual environments in the 21st century. In K. M. Stanney (Ed.), *Handbook of virtual environments: Design, implementation, and applications* (pp. 1–14). Mahwah, NJ: Lawrence Erlbaum Associates, Inc.

Stoffregen, T., Bardy, B. G., Smart, L. J., & Pagulayan, R. (2003). On the nature and evaluation of fidelity in virtual environments. In L. J. Hettinger & M. W. Haas (Eds.), *Virtual and adaptive environments: Applications, implications, and human performance issues.* (pp. 111–128). Mahwah, NJ: Lawrence Erlbaum Associates, Publishers.

Stone, R. (2002). Applications of virtual environments: An overview. In K. M. Stanney (Ed.), *Handbook of virtual environments: Design, implementation, and applications.* (pp. 827–856). Mahwah: NJ: Lawrence Erlbaum Associates.

Storms, R. L. (2002). Auditory-visual cross-modality interaction and illusions. In K. M. Stanney (Ed.), *Handbook of virtual environments: Design, implementation, and applications* (pp. 455–470). Mahwah, NJ: Lawrence Erlbaum Associates, Inc.

Strickland, D., Hodges, L., North, M., & Weghorst, S. (1997). Overcoming phobias by virtual exposure. *Communications of the ACM, 40*(8), 34–39.

Swartz, K. O. (2003). *Virtual environment usability assessment methods based on a framework of usability characteristics.* Unpublished master's thesis. Virginia Polytechnic Institute and State University, Blacksburg.

Thacker, P. D. (2003). Fake worlds offer real medicine: Virtual Reality finding a role in treatment and training. *Journal of the American Medical Association, 290*(16), 2107–2112.

Tharp, V., Burns, M., & Moskowitz, H. (1981). *Development and field test of psychophysical tests for DWI arrest* (Department of Transportation Final Report, ODT HS 805 864). Washington, DC: DOT.

Thorndike, E. L., & Woodworth, R. S. (1901). The influence of improvement of one mental function upon the efficiency of the other functions. *Physiological Review, 8*(3), 247–261.

Tredennick, N., & Shimamoto, B. (2005). The vision of Microvision. *Glider Technology Report.* Vol 10, No 10.Retrieved March 15, 2007, from http://www.microvision.com/documents/GTROctober05.pdf

Turk, M. (2002). Gesture recognition. In K.M. Stanney (Ed.), *Handbook of virtual environments: Design, implementation, and applications* (pp. 223–238). Mahwah, NJ: Lawrence Erlbaum Associates, Inc.

Ultimate 3D Links (2007). *Commercial 3D software.* Retrieved March 15, 2007, from http://www.3dlinks.com/links.cfm?categoryid=1&subcategoryid=1#.

Vince, J. (2004). *Introduction to virtual reality* (2nd ed.). Berlin: Springer-Verlag.

Welch, R. B. (1978). *Perceptual modification: Adapting to altered sensory environments.* New York: Academic Press.

Welch, R. B. (1997). The presence of aftereffects. In G. Salvendy, M. Smith, & R. Koubek (Eds.), *Design of computing systems: Cognitive considerations* (pp. 273–276). Amsterdam, Netherlands: Elsevier Science Publishers, San Francisco, CA, August 24–29.

Wickens, C. D., & Hollands, J.G. (2000) *Engineering psychology and human performance* (3rd edition). New Jersey: Prentice Hall.

Witmer, B., & Singer, M. (1998). Measuring presence in virtual environments: A Presence Questionnaire. *Presence: Teleoperators and Virtual Environments, 7*(3), 225–240.

Zittel, R. C. (2001, Summer). The reality of simulation-based acquisition- And an example of U.S. military implementation. *Acquisition Review Quarterly,* 121–132.

·16·

HUMAN-COMPUTER INTERACTION VIEWED FROM THE INTERSECTION OF PRIVACY, SECURITY, AND TRUST

John Karat, Clare-Marie Karat, and Carolyn Brodie
IBM T.J. Watson Research Center

Introduction **312**
HCI and Usability in the Privacy, Security,
and Trust Domains **313**
   Views on Privacy 313
     HCI Studies of Privacy 314
      Anonymity 315
      Personalization 315
      End user views of privacy 315
      Organizational requirements in
      managing privacy 317
     Summary of Findings and Recommendations 317
   Views on Security 318
     Usability of Security Software and Mechanisms 319
     Control Over Security 319
     Key Topic Areas for HCI in Security 320

     Secure design and administration 320
     Authentication mechanisms 320
     Passwords 321
     Biometrics 321
     Other security approaches 322
    Usability Challenges Specific to Security—
    Attack Defense 322
    Summary of Findings and Recommendations 323
   Views on Trust 323
    Factors Influencing Trust 324
    Trust and Risk 325
    Trust and Personalization 325
    Summary of Findings and Recommendations 325
Conclusions and Directions 326
References 327

INTRODUCTION

What do we mean when we say that we trust a computer system? While trust is a complex concept with many different meanings in behavioral science literature, it has a specific meaning when applied to a person's interaction with a technology system. Trust in a system is the extent to which a system behaves as the user expects. Put another way, it is a measure of how much users believe that the system will do what they ask it to do without causing any harm. This view suggests that people can enter an interaction with a system with some ideas about what the interaction will involve, and that their trust in the system can either increase or decrease with experience. Trust becomes important, particularly for people or organizations that rely on people wanting to use their websites or products, because without trust in a system, people will find other ways to conduct business. Parts of the expectations people have about systems are related to the intended function or purpose of the system. They trust that ATMs will correctly distribute money and debit their accounts appropriately. They expect that merchandise ordered through websites will arrive safely at their homes. However, there are also expectations about the handling of information they exchange with the systems. They assume that transactions are secure—that they know with whom they are dealing and that the information reaches only the intended target. They also assume that their privacy is respected—that promises the owners of the system make about how the information will be used or shared are kept. This intersection of trust, security, and privacy is becoming increasingly important, as information technology becomes a more pervasive part of our lives.

As we will discuss below, there can be many aspects and influences on the trust placed by a user in a computer system. We will not try to address them all in this chapter, but will focus on two important contributing factors associated with the risks of harm one might have in interacting with a system. When people interact with systems, they assume that doing so will not cause them any harm. They expect, that is they trust, that the interactions are truthful and without hidden consequences. If they interact with websites, they assume that it represents communication with a specific person or organization. If they disclose information to the websites, they assume that the disclosures will only reach the intended targets and that only appropriate use will be made of the information.

We will consider security as the concept associated with how well a system can protect information it contains. Many factors are associated with security in systems that interact with humans. Perhaps the most important of these is authentication—how a system and a user can be confident of each other's identities. If users and systems have confidence that they know the identities of the others, we generally assume that neither will attempt to do harm to the other. While research in system security is often aimed at minimizing risks associated with malicious attacks on systems, for HCI we will focus on the tradeoffs between rigorous authentication and ease of interaction.

The concept of privacy goes beyond security to examine how well the use of information that the system acquires about a user conforms to the explicit or implicit assumptions regarding that use associated with the personal information. We would like to draw an important distinction when discussing privacy from an HCI view. In general, studies have looked at privacy from two different perspectives. From an end-user perspective, privacy can be considered as restricting access to personal information, or it can be viewed as controlling use of personal information. In the former, the user expresses a wish to be left alone; in the latter, the wish is to use information according to expressed wishes. While much has been said about the end of privacy in the pervasive computing world—meaning that so much is known about each of us that it is futile to worry about privacy—this has really been an attempt to highlight the difficulty individuals face in trying to remain anonymous (the first consideration above). While electronic surveillance is increasingly common, this does not mean that people have to give up their rights to control the use of information collected about them. It is our assumption that collection of more and more data about us is a trend that will continue. However, we also assume that legislation will support people's rights to ensure that appropriate use is made of data collected. In considering privacy, we will generally assume that security in a system is adequate. Thus, we will view data protection failures associated with unauthorized access to information to be security failures or breaches, and those associated with noncompliance to stated privacy policies to be privacy breaches. We will explore the privacy issues in more detail below. For the purposes of this chapter, a simple but useful definition of privacy is "the ability of individuals to control the terms under which their personal information is acquired and used" (Culnan, 2000, p. 21).

In summary, security involves technology to ensure that information is appropriately protected. It involves users in that security features such as passwords for access control or encryption to prevent unintended disclosure often place requirements on users in order to function correctly. Password schemes must be hard to guess but should be easy to remember. Encryption mechanisms must prevent unintended decryption but be transparent (or nearly so) to intended senders and recipients. Privacy involves mechanisms to support compliance with some basic principles (e.g., OECD, 1980; U.S. Fair Credit Reporting Act of 1978, 1978). These suggest that people should be informed about information collection, told in advance what will be done with their information, and given a reasonable opportunity to approve of such use of information. Trust is seen as increasing when security and privacy are provided for. Without trust, it is perceived that people will be less likely to use systems.

As a part of our research, we identified connections between the concepts privacy, security, and trust as they applied to interaction with organizations that collect personal information (PI). This work (Karat, J., Karat, C., Brodie, & Feng, 2005) employed a Contextual Design method that enabled us to identify themes present in data and to understand the relationships among them. One theme involved the relationship of privacy to other concepts. Many of the participants we talked with discussed how privacy related to other concepts such as security, personalization, trust, and education. Fig. 16.1 below shows how each of these concepts relates to privacy. Good security enables and is a building block for privacy, and the interviewees have noted that, as they focus on managing privacy in their organizations, they have found ways to enhance their security as well. The trust that data subjects have in an organization is paramount, and

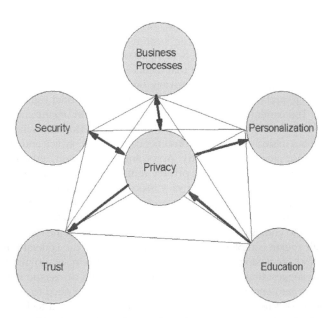

FIGURE 16.1. The relationships between privacy and other concepts.

providing privacy protection is a critical means of ensuring that trust. Privacy education is important for both the organizations and their external users. Many interviewees stated that privacy education of their employees is an ongoing priority now and that the privacy of data subjects' PI cannot be effectively protected until employees (the data users) believe it is important and follow through in their daily actions. Many have initiated training programs and have frequently asked questions (FAQs) available for employees to access to understand how to handle PI correctly. The interviewees in the current research study stated that their customers would only provide PI on the condition that it would be protected and not misused, and this is a critical element in personalization. Privacy is woven into organizations' business processes. Interviewees stated that they are reexamining their business processes in order to protect the personally identifying information (PII) of their employees, customers, constituents, and patients. They are finding redundancies in the collection of PII and are realizing financial benefits from streamlining their processes' privacy management.

The privacy and security functionality of the applications that users experience must also be usable to gain the user's trust. There are unique challenges in making these capabilities usable. We will explore these issues before discussing the research in privacy, security, and trust.

HCI AND USABILITY IN THE PRIVACY, SECURITY, AND TRUST DOMAINS

While the core usability goals of understanding the user, their tasks, and the context of use apply across a wide range of domains, unique aspects of privacy and security present challenges and opportunities when designing security and privacy functionality. First, a key issue to consider is that the use of security

and privacy solutions is not the user's main goal. Users value and want security and privacy functionality, and they are not likely to trust systems that do not provide them, but they regard them as only secondary to completing their primary tasks (e.g., completing an online banking transaction, ordering medications). The central conflict in the minds of many is that security mechanisms are seen as making operations harder, while usability is focused on making operations easier. The user would like the solutions to be as transparent as possible, but users do want to be in control and understand what is occurring. Therein lies the HCI challenge. Thus, the display of information and the interaction methods related to security and privacy solutions need to be accessible when desired by the user.

Second, as more of people's interactions in daily life involve the use of computing technology and sensitive information, disparate types of users must be accommodated. Security solutions in particular have historically been designed with a highly trained technical user in mind. The user community has broadened extensively as organizational business processes have come to include new types of roles and users in the security and privacy area. Many compliance and policy roles in organizations are handled by legal and business process experts who have limited technical skills. Moreover, the end user base includes virtually everyone in the population. The functionality provided by the system for people with different roles must accommodate the skill sets of each. Security and privacy are requirements for doing business as an organization and must be done well, or the organization may lose the user as a customer or worse.

Third, the risk of the negative impact of usability problems is higher for security and privacy applications than for many other types of systems. Although complexity is at the very heart of many security and privacy solutions, from an HCI point of view that complexity is really the enemy of the success of security and privacy. If the system is so complex that the various users groups (e.g., technical users, business users, and end users) cannot understand it, then costly errors will occur. There is a saying in the HCI field that if the user cannot understand functionality, it doesn't exist. In the case of security and privacy, badly designed functionality may put users at more risk than if they used less sophisticated solutions. The increased risk of errors in this domain provides an even greater incentive to include HCI work in system research and development. The user issues in the domain provide unique technical challenges to architects to design the solutions to be simple and effective.

Fourth, users will need to be able to easily update security and privacy solutions to accommodate frequent changes in legislation and regulations. Additionally, different domains (e.g., health care, banking, government) and geographies will have unique requirements. Systems must be designed to enable easy and effective updates to them. While this list is not exhaustive, these challenges provide a unique focus and strong incentive to include HCI in the system life cycle. We discuss the valuable role that HCI can play in more detail below.

Views on Privacy

The rapid advancement of the use of information technology in industry, government, and academia makes it much easier to collect, transfer, and store PI around the world. This raises challenging questions and problems regarding the use and protection of PI (Kobsa, 2002). Questions of who has what rights to information about us for what purposes become more important as we move toward a world in which it is technically possible to know just about anything about just about anyone. As Adams and Sasse (2001, p. 49) stated, "Most invasions of privacy are not intentional but due to designers' inability to anticipate how this data could be used, by whom, and how this might affect users." Deciding how we are to design privacy considerations in technology for the future includes philosophical, legal, and practical dimensions—any or all of which can be considered as within the domain of the field of HCI.

We provide a description of our scope in addressing this area, since privacy can and does mean different things to different people. We are primarily focused on a view of privacy as the right of an individual to control personal information use rather than as the right to individual isolation (National Research Council, 2003; OECD, 1980; U.S. Fair and Accurate Credit Transaction Act, 2003). Organizations commonly provide a description of what kind of information they will collect and how they will use it in privacy policies. In some areas (e.g., the collection and use of health care information in the United States or movement of personal information across national boundaries in Europe), such policies can be required, though the content of the policy is not generally specified in legislation. While there has been considerable consensus around a set of high-level privacy principles for information technology (OECD, 1980), we do not think it is likely that a single privacy policy can be created to address all information privacy needs. For example, there will likely be considerable differences in privacy legislation in different regions of the world (Manny, 2003). Similarly, organizations in different fields (e.g., healthcare, banking, government) need to tailor policies to their domains and needs (Ball, 2003; Baumer, Earp, & Payton, 2000). While we will focus on privacy policy, we acknowledge that privacy is not entirely about "setting rules and enforcing them" (Palen & Dourish, 2002). To implement privacy within an organization, the coordination of people, business processes, and technology is required. Still we do believe that such policies are essential when interacting with technology and organizations in that they enable people to better understand the boundaries between public and private information and technology (Altman, 1975).

It is interesting to note that while privacy policies are not new to most organizations, very little has been done to implement the policies through technology (Smith, 1993). There are emerging standards for privacy policies on websites (Cranor, 2002), but these address machine-readable policy content without specifying how the policy might be created or implemented. The reality is that there is very little capability to have technology actually implement access and disclosure limitations that we might expect from a policy statement like "We will not share your information with a third party without your consent." The emerging focus is on how organizations could create a wide range of policies, and how technology might enable the policies to be enforced and audited for compliance. J. Karat, C. Brodie, C. Karat, and J. Feng (2005) focused on technology to enable usable privacy policy authoring and enforcement, rather than trying to directly address what privacy rights people should have

(e.g., Warren & Brandeis, 1890) or how to deidentify information such as video stored in systems (e.g., Senior, Pankanti, Hampapur, Brown, Tiann, & Ekin, 2004). We think privacy is an important social issue and that technology can enable flexible, reliable and accountable privacy policy (e.g., be privacy enabling) and not just be a force that reduces individual rights.

The current situation with respect to privacy in organizations is described in a Forrester report (Hagen, 2000). This research reveals a mismatch between consumer demands for privacy and enterprise practices in industry. According to this report, although customer concerns about privacy remain high, the majority of executives (58%) believe privacy issues are addressed extremely well by their companies. Most executives do not know whether their customers check the privacy policies or not and few see the need to enhance their privacy practices. Research in the Asia-Pacific region complements these results (Office of the Federal Privacy Commissioner of Australia, 2000). With these results in mind, we suggest that privacy protection must extend across the network into enterprise processes and that there is a need to audit data collection and sharing mechanisms. We agree that technology design generally does reflect concerns of society in general (Ackerman & Mainwaring, 2005), and believe that we are experiencing a shift toward a greater concern for privacy in IT design.

Central to our view of privacy is the notion that the parties involved in information exchanges have implicit or explicit policies with regard to the use of the information. This applies both to the person who the information is about and to the organization collecting and using the information. In the privacy literature on organizations, while there has been some attention given to the generally implicit policies of end users whose PI is being collected and used (often called data-subjects), it is mostly the policies of the organization collecting the information that have been given attention. Smith (1993) described such organizational policies and noted the lack of technology in enforcing the policies. He described the rather unstructured ways in which organizations develop privacy policy—a characterization that has changed little in the nearly 10 years since his research was published in spite of the increased legislation of the past few years. In our work, we try to address both the needs of data subjects and organizations by addressing the gap between policy and practice.

HCI Studies of Privacy

Rapid advances in technology are pushing us closer and closer to Mark Weiser's (1991) vision of a ubiquitous computing world. Technology has permeated every facet of our daily lives—from enabling us to shop online with a few clicks of the mouse, to recommending interest-appropriate movies (Good, et al., 1999), to helping us share documents with colleagues on the team (e.g., Horstmann & Bentley, 1997), to allowing us to estimate when a collaborator may be available for communication (e.g., Begole, Tang, Smith, & Yankelovich, 2002), to reminding us of relevant tasks based on our current context (e.g., Dey & Abowd, 2000), and so on. The benefits of these technological developments promise to improve quality and experience of life by making it more convenient and efficient. In order to effectively provide these benefits, systems need to capture, store, and analyze information that may be deemed sensitive by the individuals concerned. The quantity of information varies from system to system, and the associated sensitivity can vary from individual to individual and context to context. As a result, to achieve their potential, such systems must first overcome various social hurdles. In particular, user perceptions regarding potential violations of privacy have emerged as a key factor that affects technological acceptance and adoption (Herbsleb, Atkins, Boyer, Handel, & Finholt, 2002; Want, Hopper, Falcão, & Gibbons, 1992). For example, the ambitious Total Information Awareness (TIA) initiative proposed by the U.S. government was abandoned before proceeding beyond the planning stages due to its inability to address satisfactorily privacy fears raised by citizens concerned with civil liberties (Associated Press, 2003). As rampant increases in viruses, worms, spam, spyware, and adware threaten to erode user confidence, it has become increasingly important that a system empower users to appropriately manage privacy.

The motivation in studying these issues within the field of HCI lies in the interest in designing technological systems with privacy in mind at the outset. While researchers have acknowledged the existence of privacy concerns in a wide variety of technological domains (e.g., e-mail; Bellotti, 1996; e-commerce; e.g., Ackerman, Cranor, & Reagle, J. 1999; media spaces; e.g., Mantei, Baecker, Sellen, Buxton, Milligan, & Wellman, 1991; data mining; e.g., Thuraisingham, 2002; homes of the future; e.g., Meyer & Rakotonirainy, 2003; and so on), designing effective solutions to address these concerns is challenging. One of the reasons is that relatively few empirical studies have been conducted with the sole aim of studying privacy issues. In addition, empirical investigations of privacy pose numerous methodological challenges. To start with, privacy has proved hard to define due to its highly nuanced and context-dependent nature. Thus, individual differences in perceptions and interpretations of privacy could lead to researcher-introduced bias. For example, in a review of 23 privacy surveys, Harper and Singleton (2001) pointed out that surveys may suffer from manipulative questioning on the one hand, and that unprompted surveys may reveal very little privacy concern on the other hand. It is also possible that certain privacy issues remain undetected because the researchers did not recognize them as such. Moreover, cultural differences in expectations and behaviors regarding these issues tend to be quite profound (Milberg, Burke, Smith, & Kallman, 1995) making it difficult to generalize findings across cultures, or to study settings that involve individuals from multiple cultures. Differences in privacy laws in different countries could make it difficult to isolate actual intention from mere legal compliance. Further, methodologies for studying privacy may themselves be deemed too privacy invasive, causing users to deviate from normal practice and to withhold revealing sensitive aspects. As a result, relying on self-reported attitudes and behavior alone may not provide a valid view of normal practices (Spiekermann, Grossklags, & Berendt, 2001).

With this overview of the privacy and HCI area as context, we will examine some key research topic areas in more depth. These areas will include anonymity, personalization, and end-user views of privacy, including privacy policies, the Platform for Privacy Preferences (P3P), Peer-to-Peer (P2P), ubiquitous computing, and electronic voting. We will conclude this review

with a discussion of the views and requirements of end users within organizations who are responsible for managing the personal data held by the organization.

Anonymity. Anonymity and anonymization in information technology includes the ability of a user to maintain privacy while completing transactions on a network and the ability for a user to keep the data they provide from identifying them personally. In researching anonymizing networks, Dingledine and Mathewson (2005) concluded that, in order for users to be able to preserve their privacy on networks by completing transactions without revealing communication partners, it is critical that the system chosen is usable so that other users can successfully employ the system as well. Anonymizing networks are successful at hiding users among users. An eavesdropper may be able to determine who is using the network but cannot identify who completes a particular transaction, and the larger the group of users on the network, the more anonymous the participants become. The catch phrase in this area is "anonymity loves company" (Reiter & Rubin, 1998).

Malin and Sweeney (2004; Sweeney, 2002) demonstrated that minimal amounts of information believed to be anonymous can be used to personally identify an individual. Since data are aggregated from a myriad of databases in the networked world in which we live, users may falsely believe that they can remain anonymous by providing only minimal bits of information in transactions here and there over time. When the data are aggregated though, the person can be identified easily based on three minimal pieces of information. Users may not think about the types of data from different sources that might be combined and analyzed in order to identify them, and thus there is a false sense of privacy. This research also makes a gray area clear in terms of what data elements might be labeled "PI" versus "PII"— it depends on the other data available in context.

Personalization. J. Karat, C. Karat, and Brodie (2004) stated that the user experience in HCI is different as the systems with which users interact become aware of the context of the interaction. Personalizing interaction involves the use of information about the user to alter the content presented to provide value to the user. In their empirical studies, J. Karat, C. Karat, and Brodie found that the most critical element in the willingness of users to adopt and use personalization systems was user control of personal data. Users want to be in control of their privacy and appreciate the benefits of personalization when they have control over the use of their personal data.

Cranor (2004) complemented this line of research with an analysis of privacy risks associated with personalization and presented a set of design guidelines to reduce privacy risks in personalization systems. Telzrow and Kobsa (2004) completed a meta-analysis of 30 consumer studies of user privacy preferences regarding personalization systems and found that consumer demands and current practice diverge significantly regarding control of personal information. The vast majority of businesses allow neither control over what information is stored nor ability to access it for verification, correction, or updates.

End user views of privacy. Researchers have conducted many studies on end user preferences regarding privacy on the web and the HCI factors that are necessary to satisfy the user's desire for privacy. Jensen and Potts (2004) found that, while most surveys of user concerns about privacy show high rates of behaviors such as reading the privacy policy or taking concrete actions to protect their privacy, informal analysis of log-file data suggests that the actual rates of these user behaviors are much lower. Jensen, Potts, and Jensen (2005) found that users place inappropriate trust in the presence of trust indicators such as the TRUSTe mark or a privacy policy on the website, assuming quality in the presence of these trust indicators rather than understanding that the level of their trust should be dependent on the content of these policies. The users were willing to divulge personal information when it was not warranted. Jensen, Potts, and Jensen recommended that HCI professionals work to increase awareness of the privacy issues related to the policy issues associated with the trust indicators, as unscrupulous online vendors could use the trust indicators to mislead users to accept privacy policies and divulge private information that they would not otherwise willingly do. Buffett, Fleming, Richter, Scott, and Fleming (2004) created a technique for enabling a user to compute the value of the consequences of exchanging personal information on the web. The paper provides a demonstration of the effectiveness of the technique in improving a user's expected utility in a simple privacy negotiation.

Cranor (2005) led the W3C standardization work in the area of the Platform for Privacy Preferences (P3P) policies. P3P policies provide end users with information about the privacy policies of a website before they interact with it. Cranor and a team at AT&T Labs designed and developed the Privacy Bird (Cranor, 2002), a browser-help object that provided summary information about the agreement or lack thereof between the end user's privacy preferences and the privacy policy of the website with which the person was considering interacting. Research continues to improve the usability of the Privacy Bird in communicating to users about website privacy policies and in capturing user privacy preferences.

Peer-To-Peer (P2P) systems enable users to easily share files by downloading simultaneously from multiple sources and by sharing many different file types. Good and Krekelberg (2003) illustrated HCI problems with P2P file sharing systems such as KaZaA. KaZaA, the most popular and widely used in 2003, had an average of 120 million downloads worldwide and 3 million users online at any given time during the course of the study. Good and Krekelberg (2004) conducted a user study of KaZaA and found that a large proportion of users unwillingly shared personal and private files, leaving them at risk of being taken advantage of through unknown exposure of PI. The majority of the users in the study were unable to correctly determine the files they were sharing. Many thought they were sharing no files when they were actually sharing all of the files on their hard drives.

The area of ubiquitous computing presents many new capabilities for users in communicating with the outside world through a number of different mobile, pervasive devices and through many choices in terms of managing their privacy in these new computing contexts. A group of researchers conducted a series of research studies in this domain. In the first study, Lederer, Mankoff, and Dey (2003) conducted a questionnaire-based study on the relative importance of the inquirer and the situation in determining a user's decision regarding the preferred accuracy of PI disclosed through the ubiquitous device. They found that users were more likely to provide the same level of accuracy in the disclosure of PI to the same inquirer in different situations rather than to different inquirers in the same situation. In later research, Hong, Ng, Lederer, and Landay (2004) developed a privacy-risk model for designing privacy-sensitive ubiquitous computing systems and presented two case studies that illustrated the use of the privacy-risk model in the design of applications. The goal was to refine the concepts of privacy to concrete issues that could be addressed by users employing different applications. Hong and Landay (2004) created a toolkit for developers called "Confab" that facilitates the development of privacy-sensitive ubiquitous applications. Confab includes a framework and customizable privacy mechanisms. The tool also includes extensions for managing location privacy. The goal of Confab is to enable application developers to address the range of trust and privacy requirements of end users of ubiquitous computing systems.

There is a growing area of research related to HCI issues and privacy in electronic voting. Hochheiser, Bederson, Johnson, C. Karat, and Lazar (2005) analyzed the current electronic voting systems and determined that increased usability was needed in these systems in three areas. First, voters must have information available to them about the electronic voting machines prior to arriving at the polling place and while there. Second, voters must be able to easily and confidently understand the machine, read the contents of the ballot, and cast their votes. Finally, voters must be able to trust the fairness and accuracy of the electronic voting system. A critical factor in trusting the system is being able to verify their votes without sacrificing their anonymity. Thus, privacy is a critical requirement in the design of successful electronic voting systems. Voter-Verified Paper Trails (VVPTs) can provide auditability of individual votes. It is also crucial to have well-designed audit trails regarding large-scale aspects of the system such as configuration, installation, and voter-count. Shubina and Smith (2004) conducted research on how to design coercion-resistant, voter verifiable electronic voting systems. They stated that it is important that voters can verify that the tally reflects the sum of votes actually cast and that voters can maintain the privacy of their votes and thus not be at risk of coercion from adversaries. They designed and prototyped a system to accomplish these goals. Their design ensured that voters are not subject to coercion from adversaries by providing a voter verification receipt that provides no real information without access to the voting machine or the contents of the cast ballots.

In terms of providing solutions for users to manage their privacy in the future, the European Union is sponsoring research on a system called PRIME (Privacy and Identity Management for Europe). The goal of the PRIME system is to support the user in controlling what PI about him- or herself is revealed to others through interactions with individuals, systems, and organizations (Pettersson et al., 2005). Pettersson et al. (2005) described three alternative user interface paradigms for privacy-enhanced identity management and illustrated how key legal privacy principles from the European Union Directives are associated with these designs. In the report of the first year of research in this effort, the team reported some results of initial usability evaluations of mock-ups of the three paradigms.

Organizational requirements in managing privacy.
HCI research that is underway and focused on the organizational view of privacy includes the SPARCLE Policy Workbench (Karat, J., Karat, C., Brodie, & Feng, 2005) and efforts by Adams and Blanford (2005) in the healthcare domain. J. Karat, C. Karat, Brodie, and Feng employed user-centered design methods with 109 target users to identify organizational privacy requirements, and then they designed and tested a prototype system for users to author privacy policies, implement the policies, and conduct compliance audits of them. The prototype was a Wizard of Oz version of a system, meaning that users were able to have an immersive and dynamic experience with the system capability that seemed real, although there was no functional code behind the screens. Empirical results show that organizational users value highly a set of capabilities that enables policy officers to use natural language to author the policies, perhaps beginning with a template. Users found a visualization of the policies very valuable for communication, review, and reaching consensus across the organization. The implementation capability enabled the experts to approve the nominated mappings between rule elements and database fields and applications in the organization's configuration. The compliance audit capability enabled the users to run general audits to verify that the policy complied with regulations, to verify that the policy was enforced operationally as stated, and to run specific inquiries based on individual requests for information about use of personal information by the organization. The organizational users who were participants in the evaluation study rated the prototyped functionality as being of very high value to them.

C. Karat, J. Karat, Brodie, and Feng (2006) also conducted an empirical study to determine whether the two methods of authoring rules that were prototyped by J. Karat, C. Karat, Brodie, and Feng (2005; natural language with a guide or structured list) enabled users to create higher quality rules than an unguided natural language control method (using a word-processing window). Empirical results demonstrated that both prototyped authoring methods yielded higher quality rules than the control condition. With no training in use of the methods, users were able to author privacy rules, covering about 80% of the required rule elements described in scenarios when using either the natural language with a privacy rule guide tool or a structured list tool, as compared to covering about 40% of the required elements in scenarios using the unguided natural language control condition.

Research and development on the SPARCLE (Server Privacy Architecture and Capability Enablement) policy workbench is continuing. The general rule authoring utility for the privacy domain is functional, working code that has been successfully tested with the privacy policies of banking and finance, healthcare, and government organizations (Brodie, C. Karat, & J. Karat, 2006). Users can create new policies with SPARCLE, can import existing text versions of policies, or can cut and paste sections of policies to form a new policy. Then, SPARCLE employs natural language processing technology to parse and identify the rule elements. Users review and modify the rules, and when they are happy with the policy, the policy is transformed into XACML code (the OASIS international standard for the format of security-access control rules with a privacy profile) for input to the enforcement engine. The team is continuing to design and create

other components in the end-to-end solution for privacy and believes that the policy workbench can be generalized to the security domain and possibly others.

Adams and Blanford (2005) addressed the gap between organizational and user perspectives of security and privacy in the healthcare domain. They conducted ethnographic evaluations of 93 clinical staff, managers, library staff, and IT department members in two hospitals. They found contrasting perspectives on security and privacy within the groups. They employed the concept of communities of practice to identify security and privacy issues within organizations, acknowledging the importance of knowledge gained through a work community in day-to-day work practices as well as formal learning. The community of practice acted as a link between the individual and the organization. In their research, one hospital improved communication and awareness across the organization through a community and user-centered approach to the design and development of an organizational privacy and security application. In the second hospital, a clash between the formal rules and the perspective of the local community of practice as at the heart of an identified security problem. The authors highlighted the importance of designing security and privacy practices within the context of communities of practice.

Summary of Findings and Recommendations

There is a range of HCI research that can inform the usability of the design of privacy mechanisms (see Table 16.1). The review has covered topics related to the networked world in which we live, current privacy legislation, individual differences in defining the meaning of privacy, end user's concern with anonymity while interacting over networks and in terms of the risks of aggregation of data, end user desire for personalization, the impact of some technologies (e.g., pervasive devices, P2P, P3P) on privacy, and end user and organizational views of privacy in a variety of domains. The central theme that emerges is informed user control over the disclosure of personal information. As presented in this section, different views and situational contexts regarding privacy ensure that this is a multifaceted challenge for experts in the HCI and privacy fields to address. In our roles as usability professionals, it is incumbent on us to consider the issues of privacy in the design, review, approval, development, and use of computing systems with which we are involved.

The HCI field will be enriched by considering privacy more fully, and this research will cross-pollinate in several areas. For example, cross-cultural studies of privacy inform the growing interest in HCI in cross-cultural interfaces and coordination. Kumaraguru and Cranor (2005) conducted a study about privacy in India, and the results demonstrated an overall lack of awareness and concern about privacy issues among users as compared to studies of users conducted in the United States. Increased understanding of the diversity and complexity of user preferences, and potential clusters of variables of interest, is providing new impetus for HCI research on individual differences. Moreover, considerations of visualizations and intelligent tutoring systems for privacy in a range of applications and tasks could germinate new emphases in HCI. Privacy in information

TABLE 16.1. Design Factors Impacting Usable Privacy Solutions.

Variable	Impact on HCI and Privacy
Networked World	• There is technology and data collection in most aspects of everyday life in the developed world.
Privacy Legislation	• Requirements for privacy vary by geography and across cultures. Organizations and end users must become knowledgeable.
Individual Differences	• People perceive and define privacy intrusions differently, and privacy solutions must be designed to support varying definitions and choices.
End User Desire for Anonymity	• End users must be informed of the risks of choices involved in different anonymity-enabling situations.
End User Desire for Personalization	• User control of data to be disclosed is critical for using personalization.
Pervasive Devices/Ubiquitous Computing	• Users want control over access to them but want it to be easy to use, fast, and flexible.
Platform for Privacy Preferences (P3P)	• Can inform user of any issues between their own policy and policy of a website they are considering interacting with. • No verification of adherence to policy possible at this time.
Peer-to-Peer (P2P)	• User must be clearly informed about data at risk of disclosure.
Electronic Voting	• User must understand how anonimity is protected and verification is enabled.
Organizational Management of Privacy—Requirements, Approaches, Perspective	• Authors of privacy policies, who have the knowledge of organizational practices, must be able to tie written policies to implementation through technology, with verification through compliance audits. • Critical to understand community of practice issues in an organization in creating privacy polices.

technology is likely to remain a critical issue for the near future. As it directly engages aspects of user control and power, privacy is central to the concerns of HCI. HCI and privacy will remain closely interlocked, and in their intersection are the seeds for successful solutions.

Views on Security

It is broadly recognized that one of the major challenges to the effective deployment of information security systems is getting people to use them correctly (CRA, 2003). As far back as the 1970s, usability with specific reference to security mechanisms was identified as a key principle in the design of secure systems (Saltzer & Schroeder, 1975). Even beyond the domain of electronic information systems, there are many examples of the fact that overly complex security systems actually reduce effective security. For example, Kahn (1967), cited by Anderson (1993), suggested that Russian military disasters of World War II were partly because Russian soldiers abandoned the official army cipher systems because they were too hard to use, and instead reverted to simpler systems that proved easier to crack. Schneiner (2000) summed up the situation: "Security measures that aren't understood by and agreed to by everyone don't work" (p. 373). However, as with many areas in the design of complex systems, recognizing a potential problem does not necessarily create a rush of work to resolve it. Work on making security usable—and balancing the complex relationship between secure and easy to use—is just beginning to become a major topic.

Networked computer systems are increasingly the site of people's work and activity. Millions of ordinary citizens conduct commercial transactions over the Internet, manage their finances, and pay their bills online. Companies increasingly use the Internet to connect different offices and to form virtual teams to tackle mission-critical problems through virtual interaction. For example, interaction between citizens and local and federal government agencies can increasingly be conducted electronically; and the 2004 national elections in Brazil and (to a much more limited extent) the United States saw the introduction of electronic voting, which will no doubt become more widespread.

However, these new opportunities have costs associated with them. Commercial, political, and financial transactions involve disclosing sensitive information. The media regularly carry stories about credit card fraud, identity theft, and hackers breaking into commercial servers. Many people are nervous about committing PI to electronic information infrastructures. Even though modern PCs are powerful enough to offer strong cryptographic guarantees and high levels of security, these concerns remain.

The need for secure systems is broadly recognized, but most discussions of the problem of security focus on the foundational elements of information systems (such as network transmission and information storage) and the mechanisms available to system developers, integrators, and managers to ensure secure operation and management of data. Security, though, is a broader concern and a problem for the end users of information systems as much as for their administrators. Participation in activities such as e-commerce requires that people be able to trust the infrastructures that will deliver these services to them.

This is not quite the same as saying that we need more secure infrastructures. De Paula et al. (2005) suggested that it is important to separate theoretical security (the level of secure communication and computation that is technically feasible) from effective security (the level of security that can practically be achieved in everyday settings). Levels of effective security are usually lower than those of theoretical security. A number of reasons for this disparity have been identified, including poor implementations of key security algorithms (Kelsey et al., 1998), insecure programming techniques (Wagner et al., 2000; Shankar et al., 2001), insecure protocol design (Kemmerer et al., 1994; Schneier & Mudge, 1998), and inadequate operating systems support (Ames et al., 1983; Bernaschi et al., 2000).

One important source of the disparity, though, is problems around the extent to which users can comprehend and make effective use of security mechanisms. Approaches that attempt to make the provision of system security automatic or transparent essentially remove security from the domain of the end user. However, in situations where only the end user can determine the appropriate use of information or the necessary levels of security, this explicit disempowerment becomes problematic.

Perhaps the best consolidation of usability issues for security comes from Sasse and Flechais (2005). They observed that currently users often disclose (or write down) passwords, fail to encrypt confidential messages, switch virus checkers off, and generally engage in behavior counter to good security, and they asked why this is so common. They concluded that most users:

1. Have problems using security tools correctly
2. Do not understand the importance of data, software and systems for their organizations
3. Do not believe that the assets are at risk
4. Do not understand that their behavior puts assets at risk

Whitten and Tygar (1999) identified a "weakest link property" stating that attackers need only to exploit a single weak point in system security. Frequently, the human user proves to be this weakest link—not from malicious intention but from inability to reasonably do the right thing. Sometimes this is a mater of education—people need to be made aware of what to do and why. Some of the blame and burden on making our systems more secure rests in making security systems more usable.

Usability of Security Software and Mechanisms

In a series of studies, researchers at University College, London have explored some of the interactions between usability and security (Adams, Sasse, & Lunt, 1997; Adams & Sasse, 1999). They focused on user-visible elements of security systems, such as passwords. Although many information systems professionals regard users as being uninterested in the security of their systems (and, indeed, likely to circumvent it by choosing poor passwords, etc.), Adams and Sasse's investigations demonstrated that users are certainly motivated to support the security of the system but often are unable to determine the security implications of their actions. The specific problems that they identify with passwords have also led to interesting design alternatives (Brostoff & Sasse, 2000; Dhamija & Perrig, 2000).

In some cases, the complexity of making security work is as much a matter of interface design as anything else. Whitten and Tygar (1999) presented a usability analysis of PGP 5.0, demonstrating the difficulties that users have in completing experimental tasks (in their user study, only 3 out of 12 test subjects successfully completed a standard set of tasks using PGP to encrypt and decrypt e-mail.) The problems that they uncovered were largely problems of interface design, and, in particular, the

poor matching between user needs and the structure of the encryption technology provided to meet these needs.

Zurko and Simon (1996) explored similar concerns in their focus on user-centered security. Their work addressed their perceptions that the inscrutability of conventional security mechanisms makes it less likely that users will employ them effectively. The approach they outlined focused on graphical interfaces and query mechanisms to MAP, an authorization engine. While this approach is clearly helpful, it is limited to a particular area of system security and lacks the real-time feedback.

Control Over Security

One area at the intersection of usability and security that has received some attention is the role of access control in interactive and collaborative systems. For example, Dewan and Shen (1998; Shen & Dewan, 1992) explored the use of access control and meta-access control models as a basis for describing and controlling degrees of information access and management in collaborative systems. This is not simply a technical matter, since the structure and behavior of these internal components can have a significant effect on the forms of interactivity and collaboration they can support (Greenberg & Marwood, 1994).

Many collaborative systems involve privacy issues and need to provide users with control over the disclosure of information. This has spurred a number of researchers to explore the development of privacy control systems that are tailored to the needs of end users. For instance, Dourish (1993) described the relationship between three different security mechanisms for similar multimedia communication systems, each of which reflects assumptions and requirements of the different organizations in which they were developed. Bellotti and Sellen (1993) drew on experiences with multimedia and ubiquitous computing environments to identify the source of a number of potential privacy and security problems. Their primary concepts—disembodiment and dissociation—were visibility problems related to the disconnection between actors and actions that rendered either the actor invisible at the site of action, or actions invisible to the actor.

Based on their investigations of privacy problems in online transactions, Ackerman and colleagues (Ackerman, Cranor, & Reagle, 1999; Ackerman & Cranor, 1999) proposed the idea of privacy critics—semiautonomous agents that monitor online action and inform users about potential privacy threats and available countermeasures. Again, this mechanism turns on the ability to render invisible threats visible.

One important related topic is control over the degree of security available. One of our criticisms of traditional security systems has been their all or nothing approaches. However, some work has attempted to characterize degrees of security provision, as embodied by the idea of "quality of security service" (Irvine & Levin, 1999; Spyropoulou et al., 2000). This builds on earlier work, establishing taxonomy of security service levels (Irvine & Levin, 1999). The fundamental insight is that organizations and applications need to trade off different factors against each other, including security of various forms and degrees, in order to make effective use of available resources (Thomsen & Denz, 1997; Henning, 1999). While this work is directed towards resource management rather than user control, it begins to unpack the security black box and characterizes degrees and qualities of security.

For end users, perceived security can be defined as the level of security that users feel while they are shopping on e-commerce websites. Yenisey, Ozok, and Salvendy (2005) reported a study that aimed to determine items that positively influence this feeling of security by users during shopping and to develop guidelines for perceived security in e-commerce. An experiment allowed users with different security assurances to shop on simulated e-commerce websites. The participants were divided into three groups—shopping for cheap, midrange, and expensive products, respectively. Following the shopping environment, a virtual shopping security questionnaire was presented to the users. Generally, there were no significant differences in item ratings between the groups of different shopping item values. A factor analysis procedure determined two main factors concerning perceived security in e-commerce. The perceived operational factor includes the website's blocking of unauthorized access; emphasis on login name and password authentication; funding and budget spent on security; monitoring of user compliance with security procedures; integration of state-of-the-art systems; distribution of security items within the website; website's encryption strategy; and consolidation with network security vendors. The perceived policy-related factor includes the website's emphasis on network security; top management commitment; effort to make users aware of security procedures; the website's keeping up-to-date with product standards; the website's emphasis on security in file transfers; and issues concerning the Web browser.

Key Topic Areas for HCI in Security

Secure design and administration. Yee (2005) provided guidance for designing and evaluating usable secure software aimed at protecting the interests of the legitimate user. Starting with the work of Saltzer and Schroeder (1975), he divided his guidelines into two general categories—guidelines for authorizing others to access valuable resources and guidelines for communication between the system and the user. For authorization, he suggested associating greater risk to the user with greater effort or less visible operations so that the user's natural tendencies lead to safe operations. For communication, he suggested enabling users to express safe security policies that fit their tasks. The guidelines are not empirically derived, but are gleaned from the experiences of security software designers.

Kandogan and Haber (2005) applied ethnographic approaches to the study of a particular class of users involved in the protection of systems—security administrators. The issues are somewhat different for populations whose job responsibilities include the security of systems used by others when compared to the situation in which users are acting as their own security administrators. However, they represent another important category of user in the area of creating usable security systems. In the ethnographic tradition, they examined their users in typical work situations (e.g., detecting and addressing a security attack) and looked at how well their tools help them to do their job. One important result of their work was the realization that administrators simply have too much to look at in the course of

their daily work and that they specifically need tools to help them understand vast amounts of dynamic data. Their work focused on developing visualization tools to assist in this situation.

Balfanz, Durfee, Smetters, and Grinter, (2004) presented guidelines for usable security systems developed over a course of research that has focused on the intersection of security and usability. The guidelines included advice to focus on designing both security and usability into the system rather than attempting to retrofit either into existing systems. It seems that the lessons learned in HCI work in regards to the need to consider it early and often during design also apply to security (security engineering is an important topic in that field, much as usability engineering is in HCI). Additional guidelines also called for a focus on the user by looking for high-level building blocks that can be used to create user-oriented solutions, rather than assuming users will be skilled at assembling their own tools for effective security. They pointed out that the security community has long held the belief that security is more important than user needs and that users must adapt to requirements in order to assure system security. An empirical study (Balfanz, Durfee, Grinter, Smetters, & Stewart, 2004) showed how hard it is for security researchers to accept the difficulty of using some of their own systems. When they evaluated the time it took to secure a wireless network using a tool they had developed, they were surprised to find that it took users over two hours on average. The authors mentioned that the reaction in the security community was often that the empirical data must be wrong. While these views are changing with the emergence of usable security research groups like this one, we should not expect the difficult trade-offs to be quickly resolved.

Maxion and Reeder (2005) provided a study in which a system specifically designed to assist users in avoiding errors (Salmon) was compared to the standard operating system interface for making file-sharing decisions. Salmon was found to increase successful task completion by 300% (the standard interface used misleading terminology, which caused users to feel that they were sharing files when in fact they were not). Users also spent less time searching for information using Salmon and had a greater proportion of their time on essential task steps rather than security overhead. In their study, they demonstrated that attending to error avoidance in interface design could facilitate usable security.

Authentication mechanisms. Security systems are designed to let authorized people in (the permission problem), and to keep unauthorized people out (the prevention problem). This involves three distinct steps: identification, authentication, and authorization (Renaud, 2005). The identification step asks a person to identify himself—usually by means of a token or an identification string such as an e-mail address or account number. Once the identification token has been provided, the person has to provide some evidence of his or her identity (authentication). This can be done by presenting something he or she knows (e.g., password), something he or she recognizes (e.g., graphical passwords), something he or she holds (e.g., a certificate), or something he or she is (e.g., biometrics). For the first three, authentication depends on the user and the system sharing a secret, which for security purposes should be difficult for someone else to guess, and for usability purposes should be easy for the user to remember. The trade-off between these two purposes can be difficult to resolve, and contributes to discussions of how much security rather than a view of security as something that is either present or not. In the case of biometrics, the system records a digital representation of some aspect of a person's physiology or behavior at enrollment, and this is confirmed at authentication time.

Many authentication scenarios can be strengthened (in a security sense) with public key cryptography. For example, a user can have a smart card that contains a public key and a matching private key. Instead of a password, the user's public key can be placed on file at a remote computer system (authentication server). To authenticate the user, the remote system sends the user a random challenge. The user signs the challenge with his private key and sends the result back to the remote server, which verifies the signature with the public key. In this way, the remote server can verify that the user has possession of the private key without having to receive it. Instead of having the public key on file at the remote system, the smart card can submit both the signed challenge and a public key certificate that has been signed by a third party. In this case, the use of public key technology is called public key infrastructure (PKI).

Whatever methods are used, at each stage of an authentication process, we can ask, "Is it secure?" The real areas of vulnerability are the input mechanism and the user. In the case of knowledge-based authentication, the user must be able to keep the secret and the secret must be hard to discover.

Passwords. Random passwords are currently the most popular user authentication mechanism. As such, they also represent the most common research target for work in the security and usability arena. One might consider them the white rat for HCI work, similar to menu structures or word processor designs of earlier HCI eras. Yan, Anderson, Blackwell, and Grant, (2005) conducted research to examine some of the commonly held beliefs about the security and usability properties of various password generation guidelines. In their work, they confirmed that passwords generated by typical advice (e.g., use of a random sequence of letters and numbers) can be difficult for people to remember, but that more cognitively friendly advice (such as select mnemonic-based passwords) can be much easier to remember without sacrificing the theoretical security levels of random passwords. They also explored additional security screening capabilities—such as screening for weak choices (e.g., obvious dates/places)—and found that such filtering could work well with other advice provided to users. Contrary to beliefs held within security communities, they did not find that theoretical analysis necessarily corresponded to actual system security. For example, encouraging the use of truly random passwords resulted in more frequent writing down and carrying passwords than encouraging the use of mnemonic passwords.

Others have taken up the challenge of providing usable passwords through nontext password schemes. For example, Monrose and Reiter (2005) provided an analysis of graphical passwords. As with text passwords, graphical passwords might be selected by the user (possibly enhancing usability but perhaps decreasing security) or selected by the system (enhancing security at the cost of usability). While there is certainly ample evidence that people exhibit powerful memory for images (e.g.,

Mandler, 1991), it is not completely clear that this translates easily into a superior password mechanism. Results from Monrose and Reiter demonstrated that graphical password schemes suffer drawbacks similar to those of textual schemes. People tended to select memorable graphical passwords enabling them to be more easily attacked, and random graphical figures can be more secure but also more difficult to recall.

Wiedenbeck, Waters, Birget, Brodskiy, and Memon, (2005) experimented with an interesting variant of graphical password. In their PassPoint system, users are provided with an image in which they establish a password by selecting a series of points on the image (the number of points selected and the order in which they are selected are factors in the complexity—and thus the potential security features of the password). Their experimental work indicated that such a system is promising from a usability perspective—people can learn the systems easily and remember their sequences over time. However, performance with such a system was not as good as with a comparison textual password system. It took more time to learn the passwords, and more time to enter them even with practice. The extent to which such performance differences might be due to novelty associated with graphical systems compared to textual systems remains unclear, and is in need of further research before we might expect them to replace the textual passwords with which we have so much experience. As De Angeli, Conventry, Johnson, and Renaud (2005) reported in their work on the graphical authentication systems, successful design of password systems is a complex task and requires considering and weighting a number of factors (such as the trade-off between security and usability).

Some of the trade-offs between security and usability, and some of the tension between the approaches of the two communities can be seen in some recent research lines. Davis, Monrose, and Reiter (2004) provided a study that looked at how user selection of passwords in graphical password schemes might increase the likelihood that passwords can be attacked. For text passwords, user selection is the norm, and there is no evidence that allowing users to select passwords (following some guidelines for length and character type) makes the resulting passwords any easier to brute force guess than the theoretical maximum set of passwords for the set of characters involved. Several new approaches for password selection—including a number that involves allowing the user to select a sequence of images—have been developed to help with password memorability (for a commercially available example, see Real User Corporation, 2002). While there has been research on several of these approaches to indicate that they provide passwords that are more memorable than text passwords, and have similar theoretical protection (entropy), there has not been much research into how effective they are in actual use. The research reported by Davis et al. suggested that people do not select as randomly, and instead form passwords with a limited set of possibilities (perhaps analogous to forming passwords with strings that might be easily associated with an individual such as birthdates). Overall, many questions remain regarding how to optimize both security and usability at the same time in the area of knowledge-based passwords.

Biometrics. As we have discussed, authentication in computer systems has suffered from the limited number of types of mechanisms available for a system to know who it is interacting with. With advances in technology, we have new opportunities (beyond traditional mechanisms such as passwords) for systems to recognize us. Biometrics refers to a means of identification that can be uniquely associated with an individual (e.g., voice patterns, fingerprints, and hand geometry). Biometrics involves the comparison of live anatomical, physiological, or behavioral characteristics to a stored template of a person (Coventry, 2005). Physiological biometrics that have been investigated include those based on fingerprints, hand and/or finger geometry, and the patterns of retinas, veins, irises, and faces. Behavioral biometric techniques that have been advanced include those based on voice, signature, and typing behavior (Peacock, Ke, & Wilkerson, 2005). Other techniques, which are not yet as well developed, include recognition of characteristics such as ear shape, gait, laughter recognition, facial thermograms, and lip shapes.

To begin, it is important to distinguish between the use of biometrics for identification and verification. For identification, the task is to identify an individual out of a population of all possible users. As the population of possible users grows, the demands on identification can also grow, in terms of either performance or accuracy of the recognition system—challenges similar to speech recognition for large vocabulary unconstrained speech. For verification, the task is to verify a particular identity by matching a characteristic to a stored template for the individual. The computational task is much simpler for verification than for identification (where only fingerprints, retinal scanning, and iris scanning have been proven successful for large populations). For the most part, we will talk about the use of biometrics for verification in this chapter.

Coventry (2005) provided an excellent illustration of how biometrics might be used and what some of the current trade-offs are for a common application—automatic teller machines (ATMs). For all techniques, usability issues can be cited as barriers to implementation and acceptance. Fingerprint identification is relatively well developed, but sensors placed in public places are subject to environmental issues (such as dirt or latent prints) as well as user training issues (they can be sensitive to where the finger is placed on the reading device). Retinal scans have attractive potential, but are currently quite invasive and difficult to develop for a full range of users (they require placing the eye close to a sensor location). Speaker verification can be subject to background noise problems and relatively complicated user registration requiring more input than a single word or phrase. Signature verification is attractive because of its long use and association with financial authorization, but current technology is not yet seen as reliable. Typing verification is seen as having some potential, but lacking the reliability necessary for large population verification.

Biometrics researchers have determined that real-life users are the biggest variable in system performance. Ashbourn (2000) suggested several user characteristics for evaluating biometric systems. These included a user's general acceptance of the biometrics concept employed (e.g., Is the idea of having a system read such characteristics acceptable?), general knowledge of the technology (e.g., Does the user understand what is being done?), knowledge of the particular biometric characteristic (e.g., Does the user understand that, if fingerprints are being read, his or her finger should be centrally placed on the

sensing device?), experience with the sensing device, environment of use (e.g., public or private), and transaction criticality.

While biometrics technologies are improving rapidly, inherent performance limitations remain and are extremely difficult to work around, except perhaps by combining multiple technologies or providing for a bypass.

Other security approaches. One other way of authentication by something I have is with smart cards—here referring to a portable device (or card) with processing power and authentication information contained on it. Smart cards essentially add an integrated circuit to familiar plastic credit cards, enabling the use of cryptographic services like random number generation and public key cryptography (Piazzalunga, Salvaneschi, & Coffetti, 2005).

Just (2004) examined the design of challenge question systems, looking for ways to improve the usability/security characteristics of these. Beckles, Welch, and Basney (2005) looked at the usability of security mechanisms in grid computing contexts.

Encryption of data is a well-established mechanism for protecting information. Rather than storing or transmitting data that can be read in the clear (e.g., data that is stored in standard code schemes such as ASCII), information is encrypted using a key and then must be decrypted using a key. While this contributes to achieving some security goals (for example, if an encrypted file containing sensitive information is lost or stolen, it would be a considerably harder task to make use of the information than if an plain data file were lost or stolen), there is a usability cost for this added security—users might have to indicate when and how files are to be encrypted or decrypted. Unfortunately, the complexity of this process has been found to cause user difficulties, variously resulting in inability to use security mechanisms appropriately (Whitten & Tygar, 1999; Caloyannides, 2004; Guttmann, 2003). Work continues, addressing the issues associated with cryptographic approaches to security (e.g., Balfanz, Durfee, & Smetters, 2005).

Usability Challenges Specific to Security—Attack Defense

Much of what we present above should seem familiar to the usability practitioner in the sense that security considerations should be considered as part of the overall system use context. The choice of authentication technique in general should consider the various trade-offs in the techniques available. Biometric techniques can be considered if users are likely to accept them and if they cover the population of expected users. If passwords are employed, they should be easy for users to recall and enter but difficult to guess or steal. Some specific issues arise because of the nature of security as a protection against attack. The fact that it is not just the user, but also that other parties with intention to do harm that must be considered, makes considering usability issues complicated in new ways.

For the majority of usability work, the goal is to make the user's primary task easy. For security considerations, we encounter a different situation—the fact that people might be trying to deceive the user and commit fraud (Conti, Ahamad, & Stasko, 2005). Of considerable recent interest in the security arena is the battle against phishing—the misrepresentation by someone of an identity or website intended to draw the user into interactions that can be harmful to the user. A phishing attack succeeds when a user is tricked into forming an inaccurate model of the interaction and when the user takes actions contrary to their intentions. Attacks often begin with e-mails sent to potential victims, purporting to be from an individual or organization with whom that user has some legitimate reason for interacting.

The issue of phishing has been viewed as a model problem for illustrating usability concerns in security (Dhamija & Tygar, 2005; Miller & Wu, 2005). An Anti-Phishing Working Group (2005) collected and described phishing attacks. Analysis of the reasons why phishing attacks are so often successful points to a number of interesting usability aspects. For example, operating systems and windowing platforms that permit general-purpose graphics also allow attackers to mimic the appearance of legitimate websites. Users tend to habituate on commonly occurring warnings about submitting data over unencrypted connections and can easily fail to notice when they are actually entering information to an insecure website. Because organizations can invest heavily in having their names and logos associated with trust, attackers can take advantage of this association by simply convincing the user that they represent a trusted organization. Because security is generally a secondary goal for users, there is only so much attention we can expect them to pay to attempts at fraud.

End users are not alone in trying to prevent attacks on systems. For people working in organizations with system administrators, the usability problem is partly moved from the end user to the administrator. Here the focus shifts to making tools for users whose primary activities include monitoring the health of the overall system. Security is one of their main tasks. Intrusion detection—the problem of detecting computer attacks in a timely manner—is one of both great difficulty and utmost importance. Finding specific evidence of attack activity in the enormous number of potentially relevant alerts, packets, operating system events, and user actions presents an almost overwhelming task for intrusion detection (ID) analysts. Visualization tools and techniques can be used to increase the effectiveness of intrusion detection analysts by more fully exploiting their visual cognition abilities. Recent work identified promising avenues for research in visualization for intrusion detection (e.g., Goodall, Ozok, Lutters, Rheingens, & Komlodi, 2005). Goodall et al. presented a user-centered visualization based understanding of the work of ID and the needs of analysts derived from the first significant user study of ID. Consistent with good information visualization practice, their tools presented analysts with both at-a-glance understanding of network activity and low-level network link details. Results from preliminary usability testing show that users performed better and found easier those tasks dealing with network state in comparison to network link tasks.

Summary of Findings and Recommendations

We would all like our information to be secure. That is why we would like our systems to be resistant to access or modification from individuals whom we do not authorize to have access to them. This requires that users be aware of security mechanisms even though they would generally prefer not to. The design challenge is to make security appropriately usable—that is,

to make it as easy to use as possible for the intended users, while making it as difficult as possible to circumvent for the potential attacker. This means that security administration and user involve trade-offs that need to be evaluated in the design of a system. Absolute security is a myth—we need to understand that we are looking for appropriate levels of security.

Some of the aspects of security that researchers have investigated are mentioned above and summarized below in Table 16.2. We think that serious consideration of the trade-offs between usability and security just began and that much work remains. While initial approaches involved hiding security controls from users, we believe that this needs to be done in balance with giving the user the ability to control the level of security required.

Views on Trust

Trust is important to every relationship that exists in society, both those of personal and public nature. Trust is related to security and privacy but differs from these concepts in that it is not an objective measure. Trust is based on the perception that the person, organization, or system one is dealing with is reliable. There are many definitions of trust and none of them satisfies every use. These definitions vary across academic domains and with context. When dealing with e-commerce and other computer applications, the user is deciding not only to trust the individual or organization, but also the technology implementation provided by the individual or organization. Kuhlen (1998) defined trust as "allowing us to act as if we have

TABLE 16.2. Design Factors Impacting Usable Security

Automatic Security Versus User-Controlled Security	• Security is not generally the main user task focus for PIM.
	• Risk Management is not a topic that end users explicitly understand.
Authentication Mechanisms	• Access to resources is controlled by knowledge of who the user is.
	• Identification relies on authentication through one (or more) mechanisms.
	• Mechanisms have usability/security trade-offs.
	• Can be "something I know" or "something I have."
Passwords	• Usually textual, but can be other.
	• Users will forget passwords.
	• Users will select passwords that are not optimally secure.
Biometrics	• A form of "something I have" authentication.
	• Not all users might have a characteristic (e.g., fingerprint).
	• Recognition technology is involved, so errors and attacks are possible.
Attack Considerations	• Complete design requires consideration of possible attacks.
	• Anti-phishing approaches.
	• Visualization for system administrator tools.

perfect knowledge about the reliability of that entity even if we do not" (p. 3). Marsh and Dibben (2003) defined trust as concerning "a positive expectation regarding the behavior of someone or something in a situation that entails risk to the trusting party" (p. 466). Jøsang and Presti (2004) defined trust as "the extent to which one party is willing to depend on somebody or something, in a given situation with a feeling of relative security, even though negative consequences are possible" (p. 1). All of these definitions share the concept that trust is about allowing us to feel comfortable and allowing another entity, such as an organization or an organization's computer system, to take one or more actions for us even though there is risk that we will be harmed in some way. In other words, in the domain of e-commerce, the trusting party believes that the trusted organization has implemented privacy and security so that it will protect the PI of the individuals interacting with it.

Earlier in this chapter, we discussed the ongoing research into security and privacy functionality that is intended to protect information that is collected and stored by organizations. Therefore, in this discussion, we will assume that privacy and security protections are in place, and we will concentrate on the factors that affect individuals' perception of trust.

Factors Influencing Trust

Many researchers have studied the factors that influence trust when dealing with online applications. Brodie, C-M. Karat, and J. Karat (2004) showed that greater degrees of transparency and control regarding the use of personal information do increase website visitor trust in the domain of IT equipment. In addition, to control of one's own data and transparency of its use, researchers have found that an individual's familiarity with a website and the perceived credibility and quality of the website also affect trust. Fogg et al. (2001) found that website visitors' perceptions of the credibility of websites were most influenced by the degree to which it was connected to a known organization in the real world and whether it seemed well designed and implemented. Building on this research, Fogg, Soohoo, Danielson, Marable, Stanford, and Tauber (2003) conducted a study with 2,500 participants in order to understand factors that influence the perceived credibility of a website. They found 18 factors that influenced credibility. The biggest factor by far was the design of the appearance of the website. This was followed by the information design or structure, information focus, perceived motivation of the website owner, perceived usefulness of the information, accuracy of the information, name recognition and reputation of the website, the tone of the writing, the identity of the sponsor, functionality on the website, customer service, past experience with the website, clarity of the information, performance on a test, readability, and affiliations (Fogg et al., 2003). They attributed the importance of the design look of the website to the fact that it is very prominent when a visitor looks at a website. This supports Fogg's (2003) theory that credibility is based on noticing a feature and then interpreting its quality to create judgment. This theory is posited as a reason why there are apparent differences in findings relating to the affect of privacy guarantees (e.g., privacy statements and privacy seals) on web-

sites (Palmer, Bailey, & Faraj, 2000; Turow, 2003). Palmer et al. found that privacy seals and statements did contribute to users' trust of websites while Turow found that website visitors do not read privacy statements or understand the meaning of seals and that there is a need to educate the public regarding privacy issues and the web.

Given that familiarity is a factor in establishing trust, Zhang and Ghorbani (2004) identified several factors that influence familiarity, including the individual's knowledge of similar services (prior experience), the number of times he or she has visited the particular website (repeated exposure), the length of each visit to the website (level of processing), and the interval of time between visits (forgetting rate).

Corritore, Kracher, and Wiedenbeck (2003) created a model using similar factors as influencing trust in the online world. In this model, trust is determined by both external factors and the user's perception of the application or website. The external factors include the user's propensity to trust and prior experience in similar situations and the user's perceptions include his or her perception of the credibility of the website, how easy it is to use, and the degree of risk involved (Corritore et al., 2003). Briggs, Simpson, and De Angeli (2004) looked at trust in a narrower domain—advice websites. They developed a model of how people determine whether to trust the advice they receive from websites. The authors developed a 22-point scale designed to break down trust into a set of judgments that can be measured. This list includes whether the user perceives the information to be prepared by an expert or a knowledgeable source, whether comments from other users were available on the website, if the website was owned by a known and respected company, if they had to wait a long time on the website, if different options were suggested by the website, if the website was perceived as hard to use, whether the user felt involved in how the website constructed the advice offered, if the site was perceived to be interactive, if the advice was tailored to the user, if the reasoning was explained to the user, if there was an opportunity to contact a human, if the advice appeared to be impartial, if the advice was perceived to be good and the user trusted it, and if the website behaved in a predictable way (Briggs et al, 2004). This study supported an earlier study that found trust was influenced by the perceived credibility of the source of the information, whether the website was personalized so that information was tailored to the individual, and whether the website was operated in a predictable way (Briggs, Buford, De Angeli, & Lynch, 2002). While many have found the trust affects the use of personalization, Briggs, Simpson, and De Angeli (2004) found that the presence of personalization could contribute to trust if it allowed the visitor to feel that the website is tailored to their needs. Sillence, Briggs, Fishwick, and Harris (2004) had similar results when studying trust of information on health related websites. They proposed a three stage model of trust in which visitors make an initial, rapid assessment of a website based on the design and appearance of the website and then do a more systematic evaluation in which the credibility of the website and the degree to which it appears to be personalized to their situations become more important. The participants were looking for information written by or for people in situations similar to their own. The third stage of this model addressed the maintenance of the trust relationship over time and left future work in this paper (Sillence et al., 2004).

Patrick, Briggs, and Marsh (2005) provided a thorough overview of many issues related to trust, including credibility of information on a website, familiarity with the website, and external factors such as prior knowledge and disposition to trust. They also described how trust is developed slowly over time, but a bad experience or even a processing error can destroy that trust very quickly. They reviewed a number of privacy models that have been operationalized using questionnaires, including Bhattacherjee's (2002) model that described how familiarity leads to trust and to a willingness to do business with a website, and Corritore et al.'s (2003) model that is described above. Patrick et al. (2005) concluded with design guidelines for promoting trust based on the literature that they reviewed.

Trust and Risk

Trust cannot be considered without considering its relationship to risk. As Patrick et al. (2005) pointed out, "Trust is intimately associated with risk" (p. 81), and people evaluate the degree of risk in many situations both online and offline. In the online world, people need different skills to judge the trustworthiness of websites and applications and the organizations that create them correctly. People can become victims of phishing attacks and identity theft if they trust too much. However, Friedman, Kahn, and Howe (2000) pointed out costs to trusting too little, as well as opportunities that can be missed if users trust too little. Friedman et al. discussed several issues that make trust and risk decisions more difficult than face-to-face transactions such as the anonymity of online transactions, the difficulty in assessing the reliability and security of technology, and misleading statements and images. Patrick et al. (2005) referenced work by Chaiken (1980) that described two strategies used depending on the perceived level of risk involved. Where people are not highly involved with the decision, they often make decisions based on appearances, and when they are more deeply involved, they employ strategies that are more systematic (Chaiken, 1980). This suggests that the degree to which users believe they are at risk will determine the strategies they use to assess the credibility of a website or system and the organization behind it. This model is similar to the three-stage model proposed by Sillence et al. (2004) in that, in the first stage of their model, users make quick decisions based on appearance, which is similar to Chaiken's description of when people are not highly involved. In the second stage, Sillence et al. described the use perceptions of the credibility of the information, which is similar to the systematic strategies that Chaiken described when people are more involved.

Trust and Personalization

In the last few years, many e-commerce companies have invested in personalizing the e-commerce websites in order to encourage business. However, personalization is only useful if users are willing to share data with the website, and this has

not always been the case. This requires that users trust that owners of the website will not misuse their data. Hagan (2000) found that website visitors often show their distrust of internet websites by disabling cookies on their computers and entering incorrect information into online forms. Many researchers have studied what aspects of an application or website affect users' willingness to use personalized features of websites. The Office of the Federal Privacy Commissioner of Australia (2000) conducted a large study of the factors that affect whether individuals are willing to share information with a business or other organization. They found that it is important to individuals that they understand how their data will be used and that they have control over who has access to it. Welty and Becerra-Fernandez (2001) examined how information technologies can reduce transaction costs and increase trust by increasing the symmetry of information available to both sides of the transaction in a business-to-business (B2B) setting. Likewise, Brodie, C.M. Karat, and J. Karat (2004) showed that greater degrees of transparency and control regarding the use of PI do increase willingness to share data in order to get access to personalized features of websites in the IT domain. Overall, trust and personalization have a complex relationship in which the presence of personalization functionality that can provide website visitors with information tailored to their situations and needs has been found to lead to a greater degree of trust with a website. At the same time, website visitors must trust the website and the organization behind it enough to share personal information if the functionality is going to be effective. Research has shown that website visitors are willing to share more personal information when they understand how and by whom their data will be used and when they have the ability to control how it is used, which suggests that these are important factors for designers to keep in mind when designing their websites.

Summary of Findings and Recommendations

Trust is necessary in an uncertain world. Given today's world in which the Internet is playing a larger role all of the time while the risks from phishing scams, spyware, and viruses continue to grow, helping users to develop an appropriate level of trust is an important goal. While there has been a great deal of research on why and when people trust in the online world, there is more to do. HCI researchers and practitioners can help by continuing to build an understanding of what leads individuals to trust a website or an application. Table 16.3 contains a summary of the factors identified in the research cited in this chapter that have been found to influence trust in the online world to date. HCI practitioners and website designers need to understand the factors that influence trust and use this knowledge as they design websites and applications. Research to date has tended to focus on a few domains, such as e-commerce and online advice websites, and it is not clear to what extent the findings hold across cultural boundaries. HCI researchers can address these issues by determining to what degree the factors that influence trust generalize across domains and cultures.

TABLE 33.3. Design Factors Impacting Trust.

Perceived Credibility of the Website or Application	• Credibility has been found to be influenced by many factors. • The most common factors cited were website and information design, the perceived usefulness of the data, and reputation of the provider.
User Familiarity	• Familiarity has been found to be influenced by prior experiences with similar sites, repeated exposure to the site, the level of involvement with the site as measured by time spent on each page, and the time between visits to the site.
Predictability of the Behavior of the Website	• Users tended to trust sites that behaved in a predictable fashion.
Perceived Quality of Information Provided	• Judgments of the quality of online advice are related to the perceived credibility of the site.
External Factors	• These include propensity of the user to trust and users' prior experiences in similar situations.
Ease of Use	• Users trusted easy-to-use sites.
Perceived Degree of Risk in Dealing with the Site	• Users were less likely to trust sites where they perceived more was at risk.
Site Personalized for the User	• Does it appear to be tailored for the individual or be created by a person in a situation similar to the individual?
Ability to Control Information Shared with Site	• Users were more willing to trust sites with personal information if they felt they had a control of that information.
Degree of Transparency Regarding Use of Shared Information	• Users were more willing to trust sites when they understood how the data would be used and by whom.
Presence of Privacy/Security Seals and Statements	• Users trust the simple presence of these seals, though it has been shown that these statements and symbols are often not read or understood.

CONCLUSIONS AND DIRECTIONS

The intersection of HCI, privacy, security, and trust is emerging as a critical area for research amid the backdrop of recent world events. Information is being improperly disclosed in ways that cause real harm to people through identity theft. Organizations are concerned about theft of proprietary data and documents. Citizens are becoming increasingly concerned about the use of the vast amounts of data that organizations and governments have about them. The rapid advancement of the use of information technology in industry, government, and academia makes it much easier to collect, transfer, and store sensitive information of a variety of types around the world. We are faced with technical challenges that result from inadequate consideration of security and privacy issues in designing our information systems, which if not addressed will result in a significant deterioration of the trust that people have in IT. While there certainly is considerable research in this area, particularly on the security side, it is becoming increasingly clear that really making our systems secure and enabling appropriate attention to privacy issues will re-

quire more than just a technology focus. While usability has been identified as a major challenge to moving the results of security and privacy research into use in real systems (CRA, 2003), it is not clear that this challenge has reached the interdisciplinary researchers needed to carry out the complex agenda of work.

We see privacy as a complex social issue concerned with individuals' rights to know what information is collected about them and how it might be used. It needs to be concerned with more than just the protection of PII, such as credit card numbers, national identification numbers, and other such identifiers. In isolation, one data element might not identify the individual to whom it relates, however, data aggregation techniques are changing the playing field. Research by Sweeney (2001; Malin & Sweeney, 2004) demonstrated that minimal amounts of information believed to be anonymous can be used to personally identify someone. Questions regarding who has what rights to information about us for what purposes become more important as we move toward a world in which it is technically possible to know just about anything about just about anyone. As Adams and Sasse (2001) stated, "Most invasions of privacy are not intentional but due to designers' inability to anticipate how this data could be used, by whom, and how this might affect users" (p. 50).

Privacy of information held in information systems depends on security—the ability to protect a system's resources from harmful attack or unauthorized access. Security is an essential component in systems that house personal information on employees, clients, constituents, students, customers, and patients. The role of security goes well beyond the protection of PI. For example, it includes corporate and national security information about trade secrets, business practices, payrolls, research, and development; strategy and logistics plans; fuel reserves; and similar things that could wreak havoc if placed in the wrong hands (e.g., by industrial competitors who wish to obtain unfair advantage; terrorists who wish to bring harm to business or government entities, bring down power grids, or destabilize banking and financial systems around the world).

The security functionality in these systems must be usable for the people in the range of user roles with different perspectives and skills who interact with it. If the security aspects of a system are so complex that individuals within the various user roles (e.g., technical users, business users, and end users) cannot understand them, then costly errors will occur. In the HCI field, there is a saying that if the user cannot understand the functionality, it does not exist. In security, the problem may even be worse than the functionality not existing because the organization may think data is secured when it is not. Deciding how we can design usable and effective privacy and security technology capabilities for the future includes philosophical, legal, cultural, business, social policy, human performance, and practical dimensions. Since at its core the goals of HCI and user-centered design are to understand the user, the user's tasks or goals, and the social and physical context in which the user is interacting with the system, all of these dimensions can be considered as within the scope of creating usable and effective privacy and security solutions.

HCI research in privacy and security is a critical area of focus in today's world. Communication and collaboration is occurring between people who have expertise in one or more of the areas of privacy, security, and HCI. This emerging community is beginning to work together to incorporate HCI in the design and development process for usable and effective privacy and security solutions. Some unique challenges to HCI research are related to privacy and security as compared to other domains. First, the use of security and privacy solutions is generally not the user's main goal. Users value and want security and privacy functionality, but they regard this functionality as secondary to completing their primary tasks (e.g., completing a transaction or process). Second, as mentioned above, today's solutions need to be usable for different groups of users with differing skill sets as compared to the past when security solutions in particular were designed with a highly trained and dedicated technical user in mind. Third, the risk of the negative impact of usability problems is higher for security and privacy applications than for many other types of systems. Finally, users must be able to update security and privacy solutions easily to accommodate frequent changes in legislation, regulation, and organizational requirements.

References

Ackerman, M., and Cranor, L. (1999). Privacy critics: Safe-guarding users' personal data. In *Proceedings of the Conference on Human Factors in Computing Systems* (CHI 1999), 258–259. New York, NY: ACM Press.

Ackerman, M., Cranor, L., & Reagle, J. (1999). Privacy in e-commerce: Examining user scenarios and privacy preferences. In *Proceedings of the ACM Conference on Electronic Commerce,* 1–8, New York, NY: ACM Press.

Ackerman, M., & Mainwaring, S. (2005). Privacy issues in human-computer interaction. In L. Cranor & S. Garfinkel (Eds.), *Security and usability: Designing secure systems that people can use* (pp. 381–400). Sebastopol, CA: O'Reilly.

Acquisti, A. (2004). Privacy in electronic commerce and the economics of immediate gratification. *In Proceedings of the ACM Electronic Commerce Conference* (pp. 21–29). New York: ACM Press.

Adams, A., & Blanford, A. (2005). Bridging the gap between organizational and user perspectives of security in the clinical domain. *International Journal of Human Computer Studies, 63,* (1/2), 175–202.

Adams, A., & Sasse, A. (1999). Users are not the enemy: Why users compromise security mechanisms and how to take remedial measures. *Communications of the ACM, 42*(12), 40–46.

Adams, A., & Sasse, A. (2001). Privacy in multimedia communications: protecting uers, not just data. In A. Blandford, J. Vanderdonkt & P. Gray (Eds.), People and computer XV—interaction without frontiers. *Joint Proceedings of HCI2001 and ICM2001,* Springer, Lille, 49–64.

Adams, A., Sasse, M. A., & Lunt, P. (1997). Making passwords secure and usable. Proceedings of HCI on People and Computers XII. Springer, Berlin, 1–19.

Altman, I. (1975). *The environment and social behavior, privacy, personal space, territory and crowding.* Monterey, CA: Brooks/Cole Pub. Co., Inc.

Ames, S., Gasser, M., & Schell, R. (1983). Security kernel design and implementation: An introduction. *IEEE Computer, 16*(7) 14–22.

Anderson, R. (1994). Why cryptosystems fail. *Communications of the ACM, 37*(11), 32–40.

Anti-Phishing Working Group. (2005). *APWG Phishing Archive.* Retrieved 2006, from http://anti-phishing.org/phishing_archive.htm.

Ashbourn, J. (2000). *Biometrics: advanced identity verification,* London: Springer Verlag.

Associated Press. (2003, September 25). Pentagon spy office to close. *Wired News.* Retrieved 2006, from http://www.wired.com/news/print/0,1294,60588,00.html

Balfanz D., Durfee, G., Grinter, R. E., Smetters, D., & Stewart, P. (2004). Network-in-a-box: How to set up a secure wireless network in under a minute. *Proceedings of the 13th USENIX Security Symposium,* 2004, 207–222.

Balfanz, D., Durfee, G., & Smetters, D. K. (2005). Making the impossible easy: Usable PKI. In L. Cranor & S. Garfinkel (Eds.), *Security and usability: Designing secure systems that people can use* (pp. 319–334). Sebastopol, CA: O'Reilly.

Balfanz D., Durfee, G., Smetters, D. K., & Grinter, R. E. (2004, September/October). In search of usable security: Five lessons from the field. *IEEE Security and Privacy, 2*(5), 19–24.

Ball, E. (2003). Patient privacy in electronic prescription transfer. *IEEE Security and Privacy, 1*(2), 77–80.

Baumer, D., Earp, J. B. & Payton, F. C. (2000, December). Privacy in medical records: IT implications of HIPAA. *Computers and Society,* 40–47.

Beckles, B., Welch, V., & Basney, J. (2005). Mechanisms for increasing usability of grid security. *International Journal of Human-Computer Studies, 63*(1/2), 74–101.

Begole, J., Tang, J. C., Smith, R. B., & Yankelovich, N. (2002). Work rhythms: Analyzing visualizations of awareness histories of distributed groups, In *Proceedings of CSCW 2002,* 334–343.

Bellotti, V. (1996). What you don't know can hurt you: Privacy in collaborative computing. In *Proceedings of the HCI Conference on People and Computers XI.*

Bellotti, V., & Sellen, A. (1993). Designing for privacy in ubiquitous environments. *Proceedings of the 1993 European Conference on CSCW,* Kluwer, Milan, 77–92.

Bernaschi, M., Gabrielli, E., & Mancini, L. (2000). Operating system enhancements to prevent the misuse of system calls. *Proceedings of the 7th ACM Conference on Computer and Communications Security,* 174–183. Athens, Greece: ACM Press.

Bhattacherjee, A. (2002). Individual trust in online firms: Scale development and initial trust. *Journal of Management Information Systems, 19*(1), 213–243.

Briggs, P., Buford, B., De Angeli, A., & Lynch, P. (2002). Trust in online advice. *Social Science Computer Review, 20*(3), 321–332.

Briggs, P., Simpson, B., & De Angeli, A. (2004). Personalization and trust: A reciprocal relationship? In C. M. Karat, J. Blom, & J. Karat (Eds.), *Designing personalized user experiences in eCommerce* (pp. 39–56). Dordrecht, Netherlands: Kluwer Academic Publishers.

Brodie, C., Karat, C-M., & Karat, J. (2004). Personalizing interaction. In C.M. Karat, J. Blom, & J. Karat (Eds.), *Designing personalized user experiences in eCommerce* (pp. 185–206). Dordrecht, Netherlands: Kluwer Academic Publishers.

Brodie, C., Karat, C., & Karat, J. (2006). An empirical study of natural language parsing accuracy of privacy policy rules using the SPARCLE Policy Workbench. *Proceedings of the Symposium on Usable Privacy and Security,* 8–19. ACM Digital Library.

Brostoff, S., & Sasse, M. A. (2000). Aare passfaces, more usable than passwords? A field trial investigation. In *Proceedings of HCI 2000,* 405–424. Berlin, Germany: Springer.

Buffet, S., Fleming, M. W., Richter, M. M., Scott, N., & Spencer, B. (2004). Determining Internet users' values for privacy information. *Proceedings of the Second Annual Conference on Privacy, Security, and Trust, 2004,* 79–88.

Caloyannides, M. A. (2004, September/October). Speech privacy technophobes need not apply. *IEEE Security and Privacy, 2*(5), 86–87.

Camenisch, J., & Van Herreweghen, E. (2002). Design and implementation of the Idemix Anonymous Credential System. In *9th ACM Conference on Computer and Communications Security,* 2002, 21–30, New York: ACM Press.

Chaiken, S. (1980). Heuristic versus systematic information processing and the use of source versus message cues in persuasion. *Journal of Personality and Social Psychology, 39,* 752–766.

Conti, G., Ahamad, M., & Stasko, J. (2005). Attacking information visualization system usability: Overloading and deceiving the human. *Proceedings of the Symposium on Usable Privacy and Security,* ACM Digital Library, 89–100.

Corritore, C., Kracher, B., & Wiedenbeck, S. (2003). Online trust: Concepts, evolving themes, a model. *International Journal of Human-Computer Studies, 58*(6), 737–758.

Coventry, L. (2005). Usable biometrics. In L. Cranor & S. Garfinkel (Eds.), *Security and usability: Designing secure systems that people can use* (pp. 175–198). Sebastopol, CA: O'Reilly.

CRA Conference on Grand Research Challenges in Information Security and Assurance. (2003, November). http://www.cra.org/Activities/grand.challenges/security, 16–19.

Cranor, L. (2002). *Web privacy with P3P.* Sebastopol, CA: O'Reilly Media.

Cranor, L. (2004). 'I didn't buy it for myself': Privacy in ecommerce personalization. In C.M. Karat, J. Blom, & J. Karat (Eds.), *Designing personalized user experiences in eCommerce* (pp. 57–74). Dordrecht, Netherlands: Kluwer Academic Publishers.

Cranor, L. (2005). Privacy policies and privacy preferences. In L. Cranor & S. Garfinkel (Eds.), *Security and Usability: Designing Secure Systems That People Can Use* pp. 447–472. Sebastopol, CA: O'Reilly.

Culnan, M. (2000). Protecting privacy online: Is self-regulation working? *Journal of Public Policy and Marketing,* 19, 1, 20–26.

Davis, D., Monrose, F., & Reiter, M. (2004). On user choice in graphical password schemes. *Proceedings of the 13th USENIX Security Symposium, 2004,* 151–164.

De Angeli, A., Coventry, L., Johnson, G., & Renaud, K. (2005). Is a picture really worth a thousand words? Exploring the feasibility of graphical authentication systems. *International Journal of Human Computer Studies, 63*(1/2), 128–152.

de Paula, R., Ding, X., Dourish, P., Nies, K., Pillet, B., Redmiles, D., Ren, J., Rode, J., & Filho, R. (2005). In the eye of the beholder: A visualization-based approach to information systems security. *International Journal of Human-Computer Studies, 63*(1/2), 5–24.

Dewan, P., & Shen, H. (1998). Flexible meta access control for collaborative applications. *Proceedings of the 1998 ACM Conference on CSCW,* 247–256, Seattle: ACM Press.

Dey, A. K., & Abowd, G. D. (2000). CybreMinder: A context-aware system for supporting reminders, In *Proceedings of the International Symposium on Handheld and Ubiquitous Computing, 2000,* 172–186.

Dhamija, R., & Perrif, A. (2000). Deja Vu: A user study using images for authentication. Proceedings of the Ninth USENIX Security Symposium.

Dhamija, R., & Tygar, J. D., (2005). The battle against phishing: Dynamic security skins. In *Proceedings of the Symposium on Usable Privacy and Security,* 77–78. ACM Digital Library.

Dingledine, R., & Matherwson, N. (2005) Anonymity loves company: Usability and the network effect. In L. Cranor & S. Garfinkel (Eds.), *Security and usability: Designing secure systems that people can use,* 547–560. Sebastopol, CA: O'Reilly.

Dourish, P. (1993). Culture and control in a media space. *Proceedings of the 1993 European Conference on CSCW,* 125–137. Milan,: Kluwer.

Fogg, B. J. (2003). Prominence-interpretation theory: Explaining how people assess credibility online. *Proceedings of the Conference on Human Factors in Computing Systems* (CHI 2003), 722–723. New York: ACM Press.

Fogg, B. J., Marshall, J., Laraki, O., Osipovich, A., Varma, C., Fang, N., et al. (2001). What makes websites credible? A report on a large quantitative study. *Proceedings of the Conference on Human Factors in Computing Systems* (CHI 2001) 61–68. New Yor: ACM Press.

Fogg, B. J., Soohoo, C., Danielson, D. R., Marable, L., Stanford, J., & Tauber, E. R. (2003). How do users evaluate the credibility of websites? A study with over 2,500 participants. *Proceedings of the 2003 conference on Designing for User Experiences, 1*–15. New York: ACM Press.

Friedman, B., Kahn, P. H., & Howe, D. C. (2000). Trust online. *Communications of the ACM, 43*(12), 34–40.

Good, N., & Krekelberg, A. (2003). A study of Kazaa P2P file-sharing. *Proceedings of the Conference on Human Factors in Computing Systems—CHI 2003,* 137–144. New York: ACM Press.

Good, N., Schafer, J.B., Konstan, J., Borchers, A., Sarwar, B., Herlocker, J., et al. (1999). Combining collaborative filtering with personal agents for better recommendations. In *Proceedings of the AAAI, 1999,* 439–446.

Goodall, J., Ozok, A., Lutters, W., Rheingens, P., & Komlodi, A. (2005) A user-centered approach to visualizing network traffic for intrusion detection. *CHI '05 extended abstracts on Human factors in computing systems* (pp. 1403–1406). New York: ACM Press.

Greenberg, S., & Marwood, D. (1994). Real time groupware as a distributed system: concurrency control and its effect on the user interface. *Proceedings of the 1994 ACM Conference on CSCW,* 207–217. Chapel Hill, NC: ACM Press.

Guttmann, P. (2003) Plug-and-Play PKI: A PKI your mother can use. *Proceedings of the 12th USENIX Security Symposium,* 45–58.

Hagan, P. R. (2000). Personalization versus privacy. *Forrester Report,* November, 2000.

Harper, J., & Singleton, S. (2001). *With a grain of salt: What consumer privacy surveys don't tell us.* Retrieved 2006 from, http://ssrn.com/abstract=299930.

Henning, R. (1999). Security service level agreements: quantifiable security of the enterprise? *Proceedings of the 1999 Workshop on New Security Paradigms,* 54–60. Caledon Hills, Canada: ACM Press.

Herbsleb, J. D., Atkins, D. L., Boyer, D. G., Handel, M., & Finholt, T. A. (2002). Introducing instant messaging and chat in the workplace. *Proceeding of the Conference on Human Factors in Computing Systems—CHI 2002,* 171–178. New York: ACM Press.

Hochheiser, H., Bederson, B., Johnson, J., Karat, C., & Lazar, J. (2005). The need for usability of electronic voting systems: Questions for voters and policy makers. *National Academy of Sciences, National Research Council Committee on Electronic Voting.*

Hong, J. I., Ng, J. D., Lederer, S., & Landay, J. (2004). Privacy risk models for designing privacy-sensitive ubiquitous computing systems. In *Proceedings of the Conference on Designing Interactive Systems—DIS'04,* 91–100. New York: ACM Press.

Hong, J. I., & Landay, J. (2004). An architecture for privacy-sensitive ubiquitous computing. In *Proceedings of MobiSys'04,* 177–189. New York: ACM Press.

Horstmann, T., & Bentley, R. (1997). Distributed authoring on the web. with the BSCW shared workspace system.*ACM Standards View, 5,* 9–16.

Irvine, C., & Levine, T. (1999). Toward a taxonomy and costing method for security services. *Proceedings of the 15th Computer Security Applications Conference,* 183–188. Phoenix, AZ: IEEE.

Jensen, C., & Potts, C. (2004). Privacy policies as decision-making tools: A usability evaluation of online privacy notices. *Proceedings of the Conference on Human Factors in Computing Systems—CHI 2004* (pp. 471–478). New York: ACM Press.

Jensen, C., Potts, C., & Jensen, C. (2005). Privacy practices of Internet users: Self-reports versus observed behavior. *International Journal of Human-Computer Studies, 63,* 203–227.

Jøsang, A., & Presti, S. L. (2004, April). Analysing the relationship between risk and trust. In T. Dimitrakos (Ed.), *Proceedings of the Second International Conference on Trust Management*, Oxford, 135–145. Berlin: Springer-Verlag.

Just, M. (2004, September/October). Designing and evaluating challenge-question systems. *IEEE Security and Privacy, 2*(5), 32–39.

Kahn, D. (1967). *The codebreakers: The story of secret writing.* MacMillan, New York, NY.

Kandogan, E., & Haber, E. (2005). Security administration tools and practices. In L. Cranor & S. Garfinkel (Eds.), *Security and usability: Designing secure systems that people can use,* 357–378. Sebastopol, CA: O'Reilly.

Karat, C., Brodie, C., & Karat, J. (2005) Usability design and evaluation for privacy and security solutions. In L. Cranor & S. Garfinkel (Eds.), *Security and usability: Designing secure systems that people can use,* 47–74. Sebastopol, CA: O'Reilly.

Karat, C., Karat, J., Brodie, C., & Feng, J. (2006). Evaluating interfaces for privacy policy rule authoring. *Proceedings of the Conference on Human Factors in Computing Systems—CHI 2006.* ACM Press.

Karat, J., Karat, C., & Brodie, C. (2004). Personalizing interaction. In C.-M. Karat, J. Blom, & J. Karat (Eds.), *Designing personalized user experiences in eCommerce* 7–18. Dordrecht, Netherlands: Kluwer Academic Publishers.

Karat, J., Karat, C., Brodie, C., & Feng, J. (2005). Privacy in information technology: Designing to enable privacy policy management in organizations. *International Journal of Human Computer Studies, 63*(1/2), 153–174.

Kelsey, J., Schneier, B., Wagner, D., & Hall, C. (1998). Cryptanalytic attacks on pseudorandom number generators. *Proceedings of the Fifth International Workshop on Fast Software Encryption,* 168–188 Berlin: Springer.

Kemmerer, R., Meadows, C., & Millen, J. (1994). Three systems of cryptographic protocol analysis. *Journal of Cryptography, 7*(2), 79–130.

Kobsa, A. (2002). Personalized hypermedia and international privacy. *Communications of the ACM, 45*(5), 64–67.

Kuhlen, R. (1998, October 1-3). *Trust: A principle for ethics and economics in the global information society.* Paper presented at *The Second UNESCO Congress for Informational Ethics,* Monte Carlo, Monaco.

Kumaraguru, P., & Cranor, L. (2005). Privacy in India: Attitudes and awareness. In *Proceedings of the 2005 Workshop on Privacy Enhancing Technologies (PET2005),* 1–17 30 May.

Lederer, S., Mankoff, J., & Dey, A. K. (2003). Who wants to know what when? Privacy preference determinants in ubiquitous computing. In *Proceedings of the Conference on Human Factors in Computing Systems—CHI–2003,* 724–725. ACM Press.

Malin, B., & Sweeney, L. (2004). How (not) to protect genomic data privacy in a distributed network: Using trail re-identification to evaluation and design anonymity protection systems. *Journal of Biomedical Informatics, 37*(3), 179–192.

Mandler, G. (1991). Your face is familiar but I can't remember your name: A review of dual process theory, relating theory and data. *Journal of Experimental Psychology, Learning, Memory, and Cognition, 8,* 207–225.

Manny, C. H. (2003). European and American privacy: Commerce, rights, and justice. *Computer Law and Security Report, 19*(1), 4–10.

Mantei, M. M., Baecker, R. M., Sellen, A. J., Buxton, W. A. S., Milligan, T., & Wellman, B. (1991). Experiences in the use of a media space. In *Proceedings of the Conference on Human Factors in Computing Systems—CHI'91,* 203–208. New York: ACM Press.

Marsh S., & Dibben M. (2003). The role of trust in information science and technology. In B. Cronin (Ed.). *Annual Review of Information Science and Technology,* 37, 465–498.

Maxion, R., & Reeder, R. W. (2005, July). Improving user interface dependabiity through mitigation of human error. *International Journal of Human-Computer Studies. 63*(1–2), 25–50.

Meyer, S., & Rakotonirainy, A. (2003). A survey of research on context-aware homes. *Proceedings of the Australasian Information Security Workshop Conference on ACSW frontiers 2003, 21,* 159–168.

Milberg, S. J., Burke, S. J., Smith, H. J., & Kallman, E. A. (1995). Values, personal information privacy, and regulatory approaches. *Communications of the ACM, 38*(12), 65–74.

Miller, R., & Wu, M. (2005). Fighting phishing at the user interface. In L. Cranor & S. Garfinkel (Eds.), *Security and usability: Designing secure systems that people can use* (pp. 275–292). Sebastopol, CA: O'Reilly.

Monrose, F., & Reiter, M. (2005). Graphical password systems. In L. Cranor & S. Garfinkel (Eds.), *Security and usability: Designing secure systems that people can use* 157–174. Sebastopol, CA: O'Reilly.

National Research Council. (2003). *Who goes there? Authentication through the lends of privacy.* Washington, DC: National Academies Press.

OECD (1980). *OECD guidelines on the protection of privacy and transborder flows of personal data.* http://www.oecd.org/home/

Office of the Federal Privacy Commissioner of Australia. (2000). *Privacy and Business.* http://www.privacy.gov.au

Office of the Federal Privacy Commissioner (2001, July). *Privacy and the community.* Retrieved 2006, published survey research available at www.privacy.gov.au.

Palen, L., & Dourish, P. (2002). Unpacking 'privacy' for a networked world. *Proceedings of the Human Factors in Computing Systems Conference (CHI 2002),* 129–136. ACM Press.

Palmer, J. W., Bailey, J. P., & Faraj, S., (2000). The role of intermediaries in the development of trust on the WWW: The use and prominence of trusted third parties and privacy statements. *Journal of Computer-Mediated Communication, 5*(3). Retrieved 2006, from http://jcmc.indiana.edu/vol5/issue3/palmer.html.

Patrick, A., Briggs, P., & Marsh, S. (2005). Designing systems that people will trust. In L. Cranor & S. Garfinkel (Eds.), *Security and usability: Designing secure systems that people can use* (pp. 75–100). Sebastopol, CA: O'Reilly.

Peacock, A., Ke, X., & Wilkerson, M. (2005). Identifying users from their typing patterns. In L. Cranor & S. Garfinkel (Eds.), *Security and usability: Designing secure systems that people can use* (pp. 199–220). Sebastopol, CA: O'Reilly.

PET: The workshop on Privacy Enhancing Technologies. (2005). Retrieved in 2006 at http://petworkshop.org/2005/).

Pettersson, J. H., Fischer-Huebner, S., Danielsson, N., Nilsson, J., Bergmann, M., Clauss, S., et al. (2005). Making PRIME usable. *Proceedings of the Symposium on Usable Privacy and Security,* 53–64. ACM Digital Library.

Piazzalunga, U., Salvaneschi, P., & Coffetti, P. (2005). The usability of security devices. In L. Cranor & S. Garfinkel (Eds.), *Security and usability: Designing secure systems that people can use* (pp. 221–244) Sebastopol, CA: O'Reilly.

Real User Corporation. (2002). *The science behind passfaces.* Retrieved 2006, from http://www.realuser.com/published/ScienceBehindPassfaces.pdf

Reiter, M., & Rubin, A. (1998). Crowds: Anonymity for web transactions. *ACM Transactions on Information and System Security,* 1(1), 66–92.

Renaud, K. (2005) Evaluating authentication mechanisms. In L. Cranor & S. Garfinkel (Eds.), *Security and usability: Designing secure systems that people can use* (pp. 103–128). Sebastopol, CA: O'Reilly.

Salzter, J., & Schroeder, M. (1975). The protection of information in computer systems. In *Proceedings of IEEE,* 63(9), 1278–1308.

Sasse, A., & Flechais, I. (2005). Usable security: Why do we need it? How do we get it? In L. Cranor & S. Garfinkel (Eds.), *Security and usability: Designing secure systems that people can use,* 13–30. Sebastopol, CA: O'Reilly.

Schneier, B. (2000). *Secrets and lies: Digital security in a networked world*. New York, NY: Willey Computer Publishing.

Schneier, B., & Mudge, D. (1998). Cryptanalysis of Microsoft's point-to-point tunneling protocol (PPTP). *Proceedings of the 5th ACM Conference on Computer and Communications Security*, 132–141. San Francisco, CA: ACM Press.

Senior, A., Pankanti, S., Hampapur, A., Brown, L., Tian, L., Ekin, A., Connell, J., Shu, C., & Lu, M. (2005). Enabling video privacy through computer vision. *IEEE Security and Privacy, 3*(3), 50–57.

Shankar, U., Talwar, K., Foster, J., & Wagner, D. (2002). Detecting format string vulnerabilities with type qualifiers. In *Proceedings of the 10th USENIX Security Symposium*, Washington, DC, 201–220.

Shen, H., & Dewan, P. (1992). Access control for collaborative environments. *Proceedings of the 1992 ACM Conference on CSCW*, 51–58. Toronto, Canada: ACM Press.

Shubina, A. M., & Smith. S. W. (2004). Design and prototype of a coercion-resistant, voter verifiable electronic voting system. In *Proceedings of the Second Annual Conference on Privacy Security, and Trust*, 2004, 29–39.

Sillence, E., Briggs, P., Fishwick, L., & Harris, P. (2004). Trust and mistrust of online health sites. *Proceedings of the Conference on Human Factors in Computing Systems—CHI 2004*, 663–670. New York: ACM Press.

Smith, J. (1993). Privacy policies and practices: Inside the organizational maze. *Communications of the ACM, 36*(12), 105–122.

Spiekermann, S., Grossklags, J., & Berendt, B. (2001). E-privacy in 2nd generation e-commerce: Privacy preferences versus actual behavior. In *Proceedings of the ACM conference on Electronic Commerce*, 38–47.

Spyropoulou, E., Levin, T., & Irvine, C. (2000). Calculating costs for quality of security service. *Proceedings of the 16th Computer Security Applications Conference*, 334–343. New Orleans, LA: IEEE.

Sweeney, L. (2001). k-Anonymity: A model for protecting privacy. *International Journal on Uncertainty, Fuzziness and Knowledge-based Systems, 10*(5), 557–570.

Teltrow, M., & Kobsa, A. (2004). Impacts of user privacy preferences on personalized systems. In C. M. Karat, J. Blom, & J. Karat (Eds.), *Designing personalized user experiences in eCommerce*, 315–332. Dordrecht, Netherlands: Kluwer Academic Publishers.

Thomsen, D., & Denz, M. (1997) Incremental assurance for multilevel application. *Proceedings of the 13th Computer Security Applications Conference*, 81–89. San Diego, CA: IEEE.

Thuraisingham, B. (2002). Data mining, national security, privacy and civil liberties. *ACM SIGKDD Explorations Newsletter, 4*(2), 1–5.

Turow, J. (2003). *Americans and Online Privacy: The System is broken*. Philadelphia: Annenberg Public Policy Center.

U.S. Fair Credit Reporting Act of 1978. U.S. Code 1681. www.ftc.gov/os/statutes/031224fcra.pdf.

U.S. Fair and Accurate Credit Transaction Act. H.R. 2622, 108th Congress, July 24, 2003.

Wagner, D., Foster, J., Brewer, E., & Aiken, A. (2000). A first step toward automated detection of buffer overrun vulnerabilities. In *Network Distributed Systems Security Symposium*, pp. 1–15, Feb. 2000.

Want, R., Hopper, A., Falcão V., & Gibbons, J. (1992). The active badge location system. *ACM Transactions on Information Systems, 10*(1), 91–102.

Warren, S. A., & Brandeis, L. D. (1890, December). The right to privacy. *Harvard Business Review, 4*, 195.

Welty, B., & Becerra-Fernandez, I. (2001). Managing trust and commitment in collaborative supply chain relationships. *Communications of the ACM, 44*(6), 67–73.

Whitten, A., & Tygar, D. (1999). Why Johnny can't encrypt: A usability evaluation of PGP 5.0. In *Proceedings of the 8th USENIX Security Symposium*, 1999, 169–184.

Wiedenbeck, S., Waters, J., Birget, J., Brodskiy, A., & Memon, N. (2005). PassPoints: Design and longitudinal evaluation of a graphical password system. *International Journal of Human-Computer Studies, 63*, 102–127.

Yan, J., Blackwell, A., Anderson, R., & Grant, A. (2004). Password memorability and security: Empirical results. *IEEE Security and Privacy, 2*(5), 25–31.

Yee, K. (2005). Guidelines and strategies for secure interaction design. In L. Cranor & S. Garfinkel, (Eds.), *Security and usability: Designing secure systems that people can use* (pp. 247–273). Sebastopol, CA: O'Reilly.

Yenisey, M., Ozok, A., & Salvendy, G. (2005) Perceived security determinants in e-commerce among Turkish university students. *Behaviour and Information Technology, 24*(4), 259–274.

Zhang, J., & Ghorbani, A. A. (2004). Familiarity and trust: Measuring familiarity with a website. *Proceedings of the Second Annual Conference on Privacy Security, and Trust*, 2004 23–28.

Zurko, M. E., & Simon, R. T. (1996). *User-centered security*. Paper presented at the New Security Paradigms Workshop.

AUTHOR INDEX

A

Abascal, J., 232
Abbate, J., 219
Abbott, K. R., 223
Aberg, J., 117
Abley, M., 162
Abowd, G. D., 224, 314
Abry, C., 95
Acero, A., 87
Ackerman, M., 173, 314, 315, 319
Ackerman, M. S., 224
Adamczyk, P. D., 111
Adamic, L. A., 222
Adams, A., 314, 317, 319, 326
Adams, E., 167
Adjoudani, A., 89, 92
Agarwala, A., 146
Aggarwal, V., 115, 116
Ahamad, M., 322
Ahlberg, C., 169, 182
Ahrenberg, L., 62, 92
Aiken, A., 318
Ainsworth, K., 284
Ainsworth, L., 247
Aish, R., 146
Aizawa, A., 238
Akamatsu, M., 296
Aldrich, F., 66
Alexander, C. A., 163, 165
Alexandersson, J., 86
Algazi, V. R., 296
Allbeck, J. M., 298
Allgayer, J., 86
Ally, M., 282
Almor, A., 94
Altman, I., 314
Altshuld, J. S., 247
Alty, J. L., 66, 72

Amalberti, R., 246
American Institute of Graphic Arts, 33
Ames, S., 319
Amir, A., 89
Anagnostou, G., 146
Anderson, D., 146
Anderson, J. R., 117, 247, 305
Anderson, K. L., 164
Anderson, P., 306
Anderson, R., 318, 321
Anderson, R. H., 219
Anderson, T., 280, 282
Andes, R. C., 255
Andrews, P. W., 250, 267
Andriessen, J., 268
Ang, C. S., 286
Annett, J., 247
Anti-Phishing Working Group, 323
Apacible, J., 111
Apperley, M., 209
Apperley, M. D., 210
Archer, W., 282
Arens, Y., 71
Argyle, M., 96, 221
Arias, E., 147
Arida, S., 146
Arnold, P., 146
Aronson, J., 246
Aronsson, L., 278
Arrow, H., 219, 255
Arvo, J., 298
Ashbourn, J., 322
Ashman, H., 232, 233, 234, 236, 237, 239
Ashman, H. L., 240
Associated Press, 315
Atkins, D., 224
Atkins, D. E., 219
Atkins, D. L., 315

Atkinson, M. L., 262, 264
Attneave, F., 164
Atwood, M. E., 162
Aung, H. H., 111
Avendano, C., 296
Axelsson, A.-S., 255
Ayers, P., 69
Aykin, N., 33, 34, 35

B
Bacon, D. J., 298
Badcock, D. R., 295
Baddeley, A., 92, 94, 247, 248
Baddeley, A. D., 67
Badiqué, E., 297, 305
Badler, N. I., 298
Baecker, R., 248, 255
Baecker, R. M., 108, 315
Baeg, S. C., 99
Bailey, B. P., 111
Bailey, J. P., 324
Bainbridge, L., 248, 254
Balachandran, A., 221
Balasubramanian, V., 232, 233
Balfanz, D., 320, 322
Balfe, E., 114
Ball, E., 314
Ballas, J., 296
Ballentine, B., 61
Balogh, J., 54
Bälter, O., 107
Bangalore, S., 87
Banks, S. B., 263
Barab, S. A., 289
Bardy, B. G., 304
Bareket, M., 305
Barenfeld, M., 170
Barker, D. T., 56
Barlett, F. C., 168
Barlow, T., 302
Barnard, M., 100
Barnard, P., 67, 75
Barrett, R., 240
Barthelmess, P., 100
Bartlett, K. M., 300
Bartlett, M. S., 123
Bar-Yam, Y., 162, 249
Bar-Yossef, Z., 236
Basney, J., 322
Bass, L., 170
Batchelor, D., 37
Bates, A. W., 277
Battista, G. D., 203
Baumer, D., 314
Baumgartner, V. J., 47, 48
Baus, J., 112, 123
Beach, L. R., 246, 248, 255
Beard, M., 144
Beardsley, P., 146
Beatty, R., 255, 258
Becerra-Fernandez, I., 325
Becker, R. A., 202, 203
Becker, S. G., 246, 247, 284
Becker, T., 86
Beckles, B., 322

Bederson, B., 316
Bederson, B. B., 210, 211
Begault, D., 296
Begole, J., 110, 219, 314
Beidernikl, G., 283
Bellotti, V., 220, 223–224, 315, 319
Benbasat, I., 255
Benford, S., 226
Bengio, S., 100
Ben-Joseph, E., 147, 152, 156
Bennett, K. B., 248, 254
Benoit, C., 86, 87, 89, 92, 95, 98, 99
Bentley, R., 314
Benyon D., 247
Berbaum, K. S., 300, 301
Berdahl, J. L., 219
Berendt, B., 315
Berg, R., 157
Bergman, M. K., 233
Bergmann, M., 316
Bernard, J., 91
Bernardin, K., 100
Bernaschi, M., 319
Berner, E., 267
Berners-Lee, T., 219, 236
Bernsen, N. O., 66, 72, 73
Bernstein, L., 87
Berry, G., 89
Berry, J. D., 10
Bers, J., 87, 98
Bertin, J., 75, 185, 188, 189, 194, 195, 198, 203
Bertram, D., 253
Berzowska, J., 147, 153
Beshers, C., 190, 205
Bessiere, K., 235
Bhalodia, J., 219
Bhattacherjee, A., 325
Bichsel, M., 147
Biderman, A., 147, 152, 156
Bieber, M., 232, 233
Bieger, G. R., 75
Biggs, S. J., 296
Bikson, T.K., 219
Billings, C., 246, 255, 258
Billings, C. E., 254
Billsus, D., 113, 114, 121
Biocca, F. A., 302
Birget, J., 321
Birnbaum, L., 114
Bisantz, A. M., 247
Bischof, W. F., 305
Bittner, K., 246
Bizantz, A., 249
Bizzi, E., 306
Bjork, R. A., 250
Black, R. W., 286
Blackwell, A., 321
Blandford, A., 174
Blanford, A., 317
Blank, H., 250
Blattner, M., 89
Blocher, A., 86
Blomkvist, S., 281
Blount, S., 250
Bly, S., 224

Bly, S. A., 225
Bochman, A., 268
Bodker, K., 247
Boehm, B. W., 174
Boies, S. J., 169
Bolle, R. M., 86, 89
Bolt, R. A., 86, 94, 97, 98
Boneva, B., 221
Bontcheva, K., 116
Booher, H. R., 73
Booth, K. S., 108, 194
Booth, S., 109
Borchers, A., 314
Borgatti, S., 282
Borovoy, R., 157
Bos, N., 225, 226
Bothell, D., 247
Bouas, K. S., 219
Boulanger, P., 305
Bovair, S., 174
Bowers, C., 246, 255, 298, 299
Bowers, C. A., 296
Boxer, S., 37
Boydell, O., 114
Boyer, D. G., 315
Brachman, R., 250, 267
Bradfield, A., 250, 267
Bradley, S. K. 305
Bradner, E., 221
Brailsford, D., 236
Brailsford, T., 241
Brailsford, T. J., 240, 242
Brain, D., 38
Brandeis, L. D., 314
Bransford, J., 168
Brashers-Krug, T., 306
Braun, M., 297, 298
Brave, S., 145, 146, 147, 148, 152
Bray, T., 213
Breaux, R., 297
Bredeweg, B., 117
Breese, J., 125
Bregler, C., 89, 98
Brehm, S., 248, 251
Brehner, B., 246, 255
Breining, R., 297, 298
Breuker, J., 117
Brewer, E., 318
Bricker, P. D., 220
Brickman, B., 247
Briggs, P., 114, 324, 325
Brin, S., 239
Bringhurst, R., 7
Broder, A. Z., 236
Brodie, C., 312, 314, 315, 317, 324, 325
Brodskiy, A., 321
Brombacher, A., 163
Brooke, N. M., 89, 91, 92
Brooks, C., 267
Brostoff, S., 319
Brown, A. B., 165, 167, 175
Brown, C., 226
Brown, G. D. A., 164
Brown, L., 314
Brown, W., 87, 98

Browne, P., 237
Bruce, R., 222
Bruckman, A., 224
Bruckman, A. S., 171
Bruin, J. D., 193
Brusilovsky, P., 113, 240, 241
Bruynzeel, M., 306
Bryant, S. L., 224
Bub, U., 98
Buckwalter, G. J., 306
Budzik, J., 114
Buff, W., 298, 299
Buffet, S., 316
Buford, B., 324
Bull, R., 167
Bullinger, H.-J., 297, 298
Bulthaup, C., 147, 153
Bulthoff, H., 94, 100
Bungert, C., 295
Bunt, A., 108., 116
Burbeck, S., 144
Burchard, P., 211
Burdea, G., 296, 298
Burdea, G. C., 298, 307
Burdick, M., 248, 254
Burge, E. L., 281
Burke, R. D., 114
Burke, S. J., 315
Burnett, G. E., 235
Burns, C., 249, 252
Burns, J., 112
Burns, M., 301
Bush, V., 218
Butler, C. W., 174
Butler, R. A., 296
Buxton, W., 147, 248, 255
Buxton, W. A. S., 156, 157, 315
Byrne, M., 247
Byrne, N., 247
Byun, H. E., 121

C

Cadiz, J., 221
Caggiano, D., 254
Cailliau, R., 233, 234
Caird, J., 253
Cairncross, F., 226
Calder, J., 99
Caldwell, B., 255
Callahan, J., 306
Caloyannides, M. A., 322
Calvert, G., 92, 94, 100
Calvert, S. L., 302
Camp, L. J., 219
Campbell, C., 240
Campbell, R. H., 262
Cano, C., 147, 153
CAP, University of Warwick, 277
Caplan, D., 168
Card, S., 247, 248
Card, S. K., 67, 68, 162, 187, 189, 190, 191, 192, 193, 194, 196, 206, 210, 211, 212, 213, 238, 239
Carenini, G., 116
Carey, K., 221, 224
Carley, K., 219

Carlson, R., 247
Carmel, E., 226
Carpenter, R., 99
Carr, L., 237
Carretta, T. R., 303
Carrithers, C., 169
Carro, R., 242
Carroll, J., 164, 169, 170
Carroll, J. M., 33, 247, 255
Carson, D., 300
Carter, M., 222
Cartwright, D., 255
Caruana, R., 108
Cary, M., 247
Casner, S., 189
Cassell, J., 87, 89
Castellani, S., 173
Castillo, E. R., 305
Castronova, E., 280
Cathiard, M.-A., 95
Caulderwood, R., 249
Cavazza, M., 297, 305
Cawley, T., 237
Cawsey, A., 116, 123
Ceaparu, I., 235
Chai, J., 57
Chaiken, S., 325
Chakrabarti, P., 252
Chan, T., 225
Chandrasekaran, B., 253
Chang, A., 146, 147, 148
Chang, S., 251
Chapman, R., 246, 255
Charness, N., 253
Chase, G., 237
Chavan, A. L., 32
Chelba, C., 87
Chen, F., 95
Chen, J., 241
Chen, J. L., 300
Cheng, K., 235
Chernoff, H., 167
Cheverst, K., 121
Cheyer, A., 87, 99
Cheyer, A. J., 99
Chi, E. H., 191, 212
Chiang, Y. T., 278
Chimani, M., 109
Chiou, G. F., 278
Chipman, S., 247
Chiu, P., 223
Choi, J. H., 276
Choo, S., 147, 153
Choong, Y., 33
Choudhury, T., 89
Christie, B., 67, 221
Chua, P. T., 306
Chuah, M. C., 208
Church, D., 300
Churchill, E., 89
Clancey, W. J., 252, 255
Clark, H., 54
Clark, H. H., 76
Clarke, K., 255
Clarkson, B., 89, 123

Clauer, R., 224
Clausen, C., 29, 37
Clauss, S., 316
Cleveland, W. S., 189
Clifford, C. W. G., 163
Clinchy, B. M., 284
Clore, G. L., 77
Clow, J., 91
Coble, J. R., 297
Cockburn, A., 246
Codella, C., 98
Coetzee, F. M., 236
Coffetti, P., 322
Cohen, J., 146
Cohen, M., 54, 296
Cohen, M. M., 87, 89, 92
Cohen, P., 87, 99
Cohen, P. R., 86, 87, 89, 91, 92, 93, 96, 98, 99, 100
Cohn, J., 306
Cohn, J. V., 305
Coiffet, P., 307
Colella, V., 157
Collins, A., 77
Compton, D. E., 297
Conati, C., 108, 116
Condon, A., 253
Condon, W. S., 87
Connell, I., 174
Connolly, T., 218, 246, 248, 255
Considine, M., 57
Conti, G., 322
Contractor, N., 255
Cook, D., 282
Cook, M., 221
Cooper, A., 281
Corbett, A., 118
Corbett, A. T., 117
Corritore, C., 324, 325
Cortes, A., 296
Coskun, E., 163
Cosley, D., 225
Cottrell, G. W., 164
Coulston, R., 67, 92, 93, 94, 95, 100
Coventry, L., 321, 322
Covi, L. A., 223
Coyle, E., 163
Coyle, M., 114
CRA Conference on Grand Research Challenges in Information
    Security and Assurance, 318, 326
Craig, J., 10
Craig, K. M., 219
Crampton Smith, G., 146
Crandall, B., 247, 249, 251, 253
Cranor, L., 314, 315, 316, 317, 319
Cranor, L. F., 120, 121, 122
Crawford, A., 221
Crawford, E., 107, 124
Crawford, G., 282
Crayne, S., 108
Cristea, A., 242
Crivella, R., 306
Croson, R., 250
Cruz, I. F., 203
Cruz-Neira, C., 295
Crystal Rivers Engineering, 296

Culnan, M., 312
Cummings, A, 219
Cummings, J., 221
Cummings, J. N., 220, 223
Curry, R. E., 262
Cutler, L. D., 146
Cyram, 284
Czerwinski, M., 78

D

Dahlback, N., 62
Dahlbäck, N., 92
Dahley, A., 145, 146, 147, 148, 152
Dair, C., 13, 14
Dalke, A., 87
Dalrymple, M., 86, 87, 96, 98
Dalwood, M., 237
Daly, B., 306
Daly, L., 300
Daly-Jones, O., 221
Danielson, D. R., 324
Danielsson, N., 316
Danis, C., 56
Danis, C. M., 172
Danninger, M., 100
Danowski, J., 276
Dark, V. J., 248
Darken, R., 305
Darken, R. P., 300
Darrell, T., 86, 89
Darves, C., 67
Dasgupta, P., 252
Dave, K., 224, 278
Davies, M., 66
Davies, R. C., 294
Davis, D., 321
Davis, H., 236, 237
Davis, H. C., 236
Davis, M., 3
Davis, S., 237
Dawes, R., 174
Dawkes, H., 205, 207, 208
De Angeli, A., 321, 324
DeAngeli, A., 92, 93, 94, 96
de Angeli, A., 66, 75
de Angelo, T., 75
De Bra, P., 242
December, J., 276
Dechter, R., 252
DeFanti, T. A., 189, 295
Deglise, P., 89
DeGroot, A. D., 170
Deisigner, J., 300, 303
Dekker, A. H., 282
DeKoven, E. A., 109
DelGaldo, E., 33, 48
Delta, H., 89
Dember, W., 254
Demirdjian, D., 100
Denecke, M., 87
Deng, L., 87
Denning, R., 255
Dennis, A. R., 222
den Ouden, E., 163
Dent, M., 239

Denz, M., 319
de Paula, R., 318
De Roure, D., 237
DeSanctis, G., 219
Desarkar, S., 252
DeSurvire, H., 173
Deutsch, S. E., 246
Devine, M., 296
Dewan, P., 319
Dewey, D., 146
Dey, A. K., 314, 316
Dhamija, R., 319, 323
Diaper, D., 247
Dias, M. B., 263
Dibben, M., 255, 324
Dickson, G. W., 219
Diday, E., 239
Dimsdale, B., 203
Ding, X., 318
Dingledine, R., 315
DiZio, P., 300, 303, 306
Doherty, M., 251
Domik, G., 122
Domingos, P., 122
Donis, D. A., 6
Dorval, R. K., 295
Douglass, S., 247
Dourish, P., 223–224, 314, 318, 319
Dowson, S. T., 213
Draper, M. H., 302
Drexler, J. M., 297
Droegemeier, K. K., 219
Drolet, A. L., 225
Drost, L., 306
Drucker, S. M., 225
DuBois, D., 253
Dubrovsky, V., 220
Duchene, D., 87
Ducheneaut, N., 220, 280
Duchnowski, P., 89
Duda, R. O., 296
Dumais, S., 114, 168
Duncan, K. D., 247
Duncan, L., 86, 87, 93, 98
Dunlap, R. D., 262
Dunlap, W. P., 301, 302
Dupont, S., 89
Duran, A., 248
Durfee, G., 320, 322
Durlach, B. N. I., 295, 298

E

Eades, P., 203
Earley, P., 37
Earp, J. B., 314
Eason, K. D., 232
Eccles, R. G., 220
Eckstein, M. P., 248
Eden, H., 147
Edens, E., 255
Edwards, K., 265
Edwards, W. K., 164
Egido, C., 220, 221
Ehrlich, S. F., 223
Eick, S. G., 198, 202, 203, 207

Eisbach, C., 146
Ekenel, H., 100
Ekin, A., 314
Ekman, P., 92
Elashmawi, F. 37
Electronic Art, 279
Ellis, A., 164
Ellis, C., 218
Ellis, C. A., 222
Ellis, J. T., 219
Ellis, S., 300, 303
Ellison, E., 300, 303
Elmore, W. K., 262
Elrod, S., 222
Elstein, A. S., 266
Emigh, W., 278
Emmelkamp, P. M. G., 306
Encarnacao, L. M., 87
Enderwick, T., 246, 254
Endsley, M., 248, 255
Engel, R., 86
Engelbart, D., 218
Engelhardt, Y., 193
Engestrom, Y., 287
English, W., 218
Enns, J. T., 194
Epps, J., 95
Ericsson, E, 253
Erkens, G., 268
Ernst, M., 94, 100
Escudier, P., 87, 96
Esposito, C., 87, 98

F
Fagerjord, A., 241
Fahy, P. J., 282
Fairchild, K. M., 203
Falcão V., 315
Falk, R., 174
Fang, N., 324
Faraday, P., 66
Faraj, S., 324
Farnham, S. D., 225
Farrell, R., 173
Fasel, I., 123
Faust, D., 174
Favalora, G. E., 295
Federal Communications Commission, 220
Federal Trade Commission, 220
Fehling, M., 123
Fein, S., 248, 251
Feiner, S., 190, 205
Feld, P., 251
Feldman, S. I., 219
Felici, J., 7
Felix, C. P., 174
Fell, H., 89
Feltovich, P., 248, 252, 253
Fencott, C., 300
Feng, J., 56, 312, 314, 317
Fenwick, J., 241
Fernandes, T., 33
Fernandez, R., 111
Ferrell, W. R., 247
Ferrier, L., 89

Ferris, P., 277
Fertig, S., 198
Figueroa, P., 305
Filho, R., 318
Findlater, L., 108, 125
Finholt, T., 219, 224, 225
Finholt, T. A., 224, 226, 315
Finn, K., 221, 225
Fiore, S., 255
Fischer, G., 78, 242, 265
Fischer-Huebner, S., 316
Fish, R. S., 224
Fisher, G., 147
Fishkin, K. P., 146
Fishwick, L., 324, 325
Fitton, D., 121
Fitzmaurice, G. W., 156, 157
Fitzpatrick, L., 239
Fjeld, M., 147
Flach, J., 253
Flach, J. M., 248, 253
Flaherty, G., 100
Flake, G. W., 236
Flanagan, J., 87, 89, 99
Flechais, I., 319
Fleming, M. W., 316
Flesch, R. F., 175
Fletcher, H. R., 147, 153
Flickner, M., 86, 89
Fogarty, J., 111
Fogel, D., 252
Fogg, B., 146
Fogg, B. J., 70, 324
Fong, M. W., 87, 92
Forrester Research, Inc., 28, 31
Forsyth, D., 248, 251, 255
Forte, A., 224
Fortner, R. S., 276
Foster, I., 219
Foster, J., 318–319
Fouad, H., 296
Fourer, R., 253
Fowlkes, J. E., 301
Foxlin, E., 297
Fozard, J. L., 164, 167
Francis, A., 57
Francis, W. N., 164
Frank, M. P., 87, 92
Frankel, J. L., 146
Frankowski, D., 114, 225
Franks, J., 168
Fraser, J. M., 250, 267
Fraser, L., 167
Fraser, M., 226
Frazer, J., 146
Frazer, P., 146
Freed, A., 247
Freeman, E., 198, 254
Frei, P., 145, 146, 148
Freitag, D., 108
Freuder, E., 253
Freyne, J., 114
Fridlund, A., 92
Friedman, B., 325
Friesen, W., 92

Fu, X., 221
Fukumoto, M., 98
Fuller, R., 111
Furnas, G. W., 168, 203, 208, 235
Fussell, S. R., 220, 221, 222
Fuster-Duran, A., 96

G

Gabbard, J. L., 302
Gabrielli, E., 319
Gagné, R. M., 277
Galanter, E., 247, 253
Gale, C., 221
Galotti, K. M., 284
Garavan, H., 248
Garb, H., 249
Garcia-Molina, H., 219
Gardiner, M., 67
Gardner, H., 48
Garfield, E., 239
Garg, A., 87, 89, 99, 100
Gargan, R. A., 86, 87, 96, 98
Garland, A., 109
Garland, D., 248
Garrison, R. D., 282
Garton, L., 282, 283
Garzotto, F., 242
Gash, D. C., 224
Gasparetti, F., 114
Gasser, M., 319
Gatica-Perez, D., 100
Gauch, S., 114
Geddes, N., 248, 262
Geddes, N. D., 252, 253, 254,
    255, 262, 263
Geelhoed, E., 221
Geister, S., 255
Gena, C., 123
Gentner, D., 248
George, J. F., 222
Gergle, D., 220, 225
Gershon, J., 306
Gervasio, M. T., 108
Ghallib, M., 253
Ghorbani, A. A., 324
Giaccardi, E., 78
Giangola, J., 54
Gibbons, A., 251
Gibbons, J., 315
Gibbs, S. J., 222
Giffin, W., 250, 251
Giles, J., 224
Gill, R. T., 247, 248, 249
Gilmartin, K. J., 170
Gilovich, T., 250, 267
Giovinco, M. G., 295
Girand, C., 94
Gladwell, M., 172
Glance, N., 222
Glas, D., 146, 149
Glaser, R., 252
Glock, M. D., 75
Glover, E., 236
Goddeau, D., 86, 87
Godefroid, P., 224

Gold, R., 222
Goldberg, A., 144
Goldberg, D., 222
Goldberg, S. L., 297
Golden, E., 111
Goldman, R., 56, 246
Goldstein, L., 246, 248
Gomez, L. M., 168
Gonzalez, V. M., 164
Good, N., 314, 316
Goodall, J., 323
Goodman, J., 87
Gopher, D., 305
Gorbet, M., 146, 147
Gordin, D., 173
Gordon, A., 253
Gordon, S., 249
Gordon, S. E., 247, 248
Goren-Bar, D., 126
Gorniak, P., 100
Gould, J. D., 169, 175
Goulding, J., 239
Grabowski, M., 163, 246
Graeber, D. A., 300, 301
Graham, T., 31
Grant, A., 321
Grant, A. E., 282
Grant, E. C., 92
Grant, K. R., 220
Grasso, A., 173
Grasso, F., 116, 123
Gravier, G., 87, 89, 99, 100
Gray, M. J., 247
Gray, W. D., 162
Graybiel, A., 300
Grayson, D. M., 221
Graziola, I., 126
Green, P., 235
Green, T. G. R., 174, 175
Greenberg, S., 224, 248, 255, 319
Greene, S. L., 171
Greenhalgh, C., 226
Greenwood, P., 254
Greif, I., 218
Gress, W., 304
Grier, R., 254
Griffin, D., 250, 267
Griffith, T. L., 255
Grinter, R. E., 164, 220, 224, 320
Gritz, L., 296
Grondin, S., 248
Grossklags, J., 315
Grudin, J., 218, 221, 223, 248, 255
Gruen, D., 107
Guerlain, S., 266, 267
Guiard-Marigny, T., 92
Guiney, E., 246
Gulli, A., 233, 239
Gumbrecht, M., 222, 224
Gunawardena, C., 280, 282
Gunning, R., 175
Gunther, E., 146
Gupta, A., 100, 221
Gutkauf, B., 122
Guttmann, P., 322

Gutwin, C., 224
Guyomard, M., 86, 87

H

Ha, Y., 251
Haber, E., 320
Hadar, U., 92
Hagan, P. R., 314, 325
Hagerhall, C. M., 163
Hahn, J. K., 296
Hajdukiewicz, J., 249, 252
Hakkani-Tür, D., 112, 123
Halasz, F., 222
Hale, K., 298, 299, 305
Hale, K. S., 296
Hall, C., 318
Hall, D. M., 295
Hall, E. T., 37, 226
Hall, R., 225
Hall, W., 79, 236, 237
Halle, M., 295
Haller, S., 111
Halstead, M. H., 165, 167, 175, 184
Halverson, C., 56
Halverson, C. A., 56
Halvorsen, R. F., 278
Hammer, J. M., 262, 263
Hammond, K., 114
Hammond, K. J., 114
Hammond, N., 119
Hampapur, A., 314
Hancock, P., 253
Handberg, R., 250
Handel, M., 315
Handy, C., 255
Hanks, S., 120, 253
Hanneman, R. A., 282, 283
Hannum, W., 247
Hanson, V. L., 108, 176
Harbusch, K., 86
Hardin, J. B., 219
Hardstone, G., 255
Harman, D., 238, 239
Harper, J., 315
Harris, P., 324, 325
Harris, P. R., 37
Harris, S., 267
Harrison, S., 223
Hart, R. P., 175
Haselton, M. G., 250, 267
Hasling, D. W., 255
Hassebrock, F., 248
Hastie, R., 250
Hauptmann, A., 91
Hauptmann, A. G., 91
Hawley, P., 3
Hay, K. E., 289
Hayes, C., 246, 248, 255, 268
Hayes, J. R., 254
Hayes-Roth, B., 253
Hayes-Roth, F., 253
Hays, R. T., 305
Haythornthwaite, C., 282, 283
Healey, C. G., 194
Healey, J., 111

Hearst, M., 239
Hearst, M. A., 174
Heath, C., 221, 226
Heath, I., 237
Heckerman, D., 125
Heckmann, D., 123
Heeren, E., 277
Hegarty, M., 75, 80
Hegner, S. J., 109
Heim, M., 300
Heirich, A., 298
Helander, M., 246
Helgeson, V., 221
Heller, R. S., 66, 72
Hellerstein, J. L., 165, 167, 175
Helmick, J., 305
Hendrix, A., 28
Hennecke, M. E., 86, 87, 92, 98
Henning, R., 319
Henri, F., 280, 282
Herbsleb, J. D., 224, 225, 315
Herlocker, J., 114, 314
Herlocker, J. L., 225
Hernan, S. V., 219
Hernandez, A., 147, 153
Herring, S. C., 278
Herskovitz, J., 37
Hertel, G., 255
Hester, R., 248
Hettinger, L., 87
Hettinger, L. J., 296, 300
Hickinbotham, K. A., 296
Hilbert, D., 114
Hild, H., 98
Hildebrandt, N., 168
Hill, G., 237
Hindmarsh, J., 221, 226
Hinds, P., 255
Hinds, P. J., 255
Hindus, D., 224
Hines, J., 145, 147, 150
Hirano, M., 147, 151
Hirayama, M., 87, 92
Hix, D., 302
Ho, J., 111
Hoc, J.-M., 246
Hochberg, J., 81
Hochheiser, H., 316
Hodges, L., 297, 306
Hodges, L. F., 306
Hodgins, J., 146
Hofer, E. C., 225
Hoffman, R., 247, 249, 251, 253
Hofstede, G., 37, 38
Holden, M., 306
Hollan, J. D., 66
Holland, J. J., 162, 170
Hollands, J. G., 121, 304
Hollerbach, J., 296, 297
Hollingshead, A. B., 219, 220, 222
Hollnagel, E., 246, 247, 255
Holmback, H., 87, 98
Holmquist, L. E., 146
Holzman, T. G., 90
Hong, J. I., 316

Hong, L., 89
Honold, P., 32
Höök, K., 118, 123
Hoover, H. J., 305
Hopper, A., 315
Horn, D., 56
Horn, D. A., 56
Horstmann, T., 314
Horton, F., 252, 255
Horvitz, E., 99, 108, 111, 114, 125
Houghton, R., 98
Hovel, D., 125
Hovy, E., 71
Howard, M., 220
Howe, D. C., 325
Hu, H. W., 278
Hu, N., 306
Huang, T. S., 87, 89, 98, 99
Huang, W., 221
Huang, X., 87, 100
Huang, X. D., 297
Hudson, S. E., 111
Hürst, W., 89
Hurtle 235
Hutchins 189
Hutchins, E., 246, 248, 255
Hutchins, E. L., 66

I

Ichicawaa, Y., 221
Ihde, S., 86, 89
Ikar, D., 75
Inkson, K., 37
Inselberg, A., 203
Intille, S. S., 111
Iqbal, S. T., 111
Irby, C. H., 144
Irvine, C., 319
Isaacs, E., 220, 221, 224
Isdale, J., 296, 300
Isenhour, P. L., 255
Ishii, H., 142, 145, 146, 147, 148, 149, 150, 151, 152, 153, 156, 157
Itoh, Y., 146
Ivory, M. Y., 174
Iyengar, G., 86

J

Jacko, J. A., 56
Jackson, B. M., 219
Jackson, J., 267
Jacob, R., 146
Jacob, R. J. K., 146, 147
Jacobs, J., 246
Jagacinski, R. J., 248, 253
Jagadeesan, L. J., 224
Jahanian, F., 224
Jain, A., 86, 89, 100
Jalili, R., 98
Jameson, A., 115, 121
Jancke, G., 221
Janis, I., 251
Jansen, B., 238
Jansen, B. J., 237
Janssen, T., 193
Janssen, W., 222

Jebara, T., 89
Jensen, C., 225, 316
Jentsch, F., 246, 255
Jëonsson, A., 92
Ji, Q., 111
Johansen, R., 219
John, B. E., 162, 170
Johnson, G., 321
Johnson, H., 235
Johnson, J., 144, 316
Johnson, M. H., 167
Johnson, P., 248
Johnston, J. C., 248
Johnston, M., 86, 87, 98, 99
Johnston, W. A., 248
Jonassen, D., 247
Jones, C., 174
Jones, C. M., 111
Jones, D. R., 252
Jones, E., 250
Jones, G. B., 278
Jones, L., 171, 296
Jones, M. B., 297
Jones, P., 246
Jones, P. M., 246, 247
Jones, S., 276
Jones, S. B., 306
Jonsson, A., 62
Jonsson, I., 111
Jordan, B., 220
Jøsang, A., 324
Josephson, J., 265
Josephson, S., 265
Jul, S., 235
Julia, L., 99
Just, M., 322
Just, M. A., 75, 80
Juth, S., 99

K

Kahn, D., 318
Kahn, P. H., 325
Kahneman, D., 250, 267
Kaimullah, A., 56
Kaiser, E., 100
Kalawsky, R. S., 297, 298
Kallman, E. A., 302, 315
Kamm, C., 59, 221, 224
Kancherla, A., 302
Kandogan, E., 320
Kaplan, C., 241
Kapoor, A., 226
Karat, C., 56, 312, 314, 315, 316, 317
Karat, C.-M., 324, 325
Karat, J., 56, 312, 314, 315, 317, 324, 325
Karsenty, L., 221
Karshmer, A. I., 89
Kassin, S. M., 248, 251
Kato, H., 146
Katriel, A., 173
Katz, N., 255
Kaur, K., 303
Kawahara, T., 111, 112
Kay, A., 170
Kay, J., 107, 124

Ke, X., 322
Keahey, T. A., 210
Keim, D. A., 198
Keller, A., 165, 167, 175
Kelly, D., 113
Kelly, R. V., 280
Kelsey, J., 318
Kemmerer, R., 319
Kendon, A., 87, 92, 95, 221
Kennedy, K. E., 300
Kennedy, R. S., 297, 299, 300, 301, 302
Kensing, F., 247
Kephart, J. O., 107
Kerr, B., 107
Kershner, J., 298
Kesselman, C., 219
Kessler, G. D., 298, 306
Kieras, D., 175, 247
Kieras, D. E., 174
Kiesler, S., 111, 219, 220, 221, 223, 225
Kingdon, K., 301
Kinshuk, 240
Kirlik, A., 247
Kirwan, B., 247
Kishino, F., 146
Kitamura, Y., 146
Kittowski, F. F., 279
Klayman, J., 251
Klein, G., 247, 249, 251, 253
Klein, G. A., 251, 253
Klein, M. L., 219
Kleinberg, J. M., 238, 239
Kleinmuntz, D. N., 252
Klemmer, S. R., 147
Klinger, A., 163
Klinker, G., 297, 305
Klopfer, D., 252
Knerr, B. W., 297
Knight, J. M., 37
Knight, S., 236
Ko, A. J., 111
Kobayashi, K., 147, 151
Kobayashi, M., 237
Kobsa, A., 86, 111, 113, 123, 314, 315
Koch, P., 111
Koedinger, K. R., 117
Koehler, J. J., 250
Koenemann, J., 238
Koerner, B., 147, 148
Kogan, S., 221, 224
Köhler, A., 279
Kohler, T., 100
Kohler, W., 248
Kohls, L. R., 37
Kolbert, E., 279
Koleva, B., 226
Kollock, P., 225
Kolodner, J., 251, 253
Kolojejchick, J. A., 208
Kolvenbach, S., 223
Komatani, K., 111, 112
Komlodi, A., 323
Konig, Y., 89
Konold, C., 174
Konradt, U., 255

Konstan, J., 225, 314
Konstan, J. A., 120, 225
Koons, D., 87, 89, 94
Kopp, S., 87
Korpela, M., 289
Korzenny, F., 276
Kosbie, D. S., 170
Koschmann, T., 225
Kosslyn, S. M., 194
Kotovsky, K., 254
Koved, L., 98
Kracher, B., 324, 325
Kraemer, K. L., 218, 222
Kramer, K., 157
Krauss, R. M., 220
Kraut, R., 220, 221
Kraut, R. E., 220, 221, 222, 224
Kray, C., 121
Krebs, V., 282
Krekelberg, A., 316
Kricos, P. B., 95
Kriegel, H.-P., 198
Krishnan, M. S., 223
Kristof, R., 76
Krovetz, R., 236
Krueger, H., 147
Krüger, A., 123
Krulwich, B., 121
Kruppa, M., 123
Kucera, H., 164
Kuchinsky, A., 221, 222
Kuhlen, R., 323
Kuhn, A., 29
Kuhn, K., 91, 92, 93, 94, 96, 97
Kulak, D., 246
Kulikowski, C., 99
Kulkarni, Y., 89
Kumar, R., 236
Kumar, S., 99
Kumaraguru, P., 317
Kuo, J., 286
Kurtenbach, G., 223
Kushleyeva, Y., 254
Kuzuoka, H., 221
Kypros-Net Inc., 284

L

Lackner, J. R., 306
Lai, J., 56, 57, 235
Lai, K. Y., 220
Laird, J., 247
Laird, J. E., 175, 253
Lallouache, M.-T., 95
Lallouache, T., 87, 96
Lalmas, M., 238
Lamont, M., 38
Lamping, J., 211
Landauer, T., 246
Landauer, T. K., 168, 173
Landay, J., 316
Landay, J. A., 147
Lane, N. E., 300, 301
Lange, B. M., 223
Langley, P., 123
Lanier, J., 120

Laraki, O., 324
Larkin, J. H., 252, 254
Larrick, R. P., 250
Larson, A., 246, 248, 255, 268
Lathoud, G., 100
Lau, T., 122
Lausen, G., 120
Lavie, T., 75
Lavin, B., 284
Law, S. A., 219
Lawrence, D., 162
Lawrence, S., 236
Lawson, B. D., 296
Layton, C., 246, 248, 252, 255, 268
Lazar, J., 235, 316
Lazer, D., 255
Leatherby, J. H., 91
Lebei, L., 219
Lebiere, C., 247
Lederer, S., 316
Lee, A., 172
Lee, C., 89
Lee, D., 222
Lee, F., 254
Lee, J., 146, 147, 153, 246, 247, 255, 262, 284
Lee, J. W., 296
Lee, R., 147
Lee, Y. S., 32
Lee, Y. V., 87, 92
Le Goff, B., 86, 87, 92
Lehner, P. E., 254
Leichner, R., 221, 222
Leigh, D., 146
Leigh, J., 226
Lemke, A. C., 265
Lemos, R., 306
Lemstrom, K., 163
Lentz, R., 81
Leonhardt, U., 147
Leont'ev, A. N., 287
Leplat, J., 246, 255
Lesgold, A., 252
Lesh, N., 109, 110, 120
Leung, Y. K., 210
Levesque, H., 250, 267
Levi, D., 248, 251, 255
Levie, W. H., 81
Levin, T., 319
Levine, T., 319
Levow, G., 91
Levow, G. A., 61
Lewis, C., 166, 247, 248
Lewis, D., 252, 255
Lewis, J., 98
Lewis, M., 255
Lewis, R., 37
Li, D., 224
Li, F., 219
Li, X., 111
Liang, T.-P., 246
Lieberman, H., 117, 120
Lilienthal, M. G., 300, 301
Lin, M., 56
Lin, X., 112, 199
Linden, G., 114, 120

Ling, D., 98
Lipscomb, J., 98
Lipson, H. F., 219
Liscombe, J., 112, 123
Lisetti, C. L., 111, 123
Litman, D., 111
Litman, D. J., 111
Littlewort, G., 123
Littman, M., 253
Liu, Y., 246, 247, 284
Lizza, C. S., 255, 263
Ljungstrand, P., 146
Lockelt, M., 86
Loftus, E., 250
Logan, G., 248
Longstaff, T. A., 219
Looi, C., 117
Low, R., 92, 94
Lowe, C., 280, 282
Lowe, D., 79
Lowe, K., 92, 94
Lowe, R. K., 248
Lu, C. H., 278
Lu, Y., 163
Luettin, J., 89
Luff, P., 221
Lundmark, V., 220
Lunsford, R., 86, 92, 93, 94, 95, 100
Lunt, P., 319
Luo, Y., 254
Lurey, J. S., 255
Lutters, W., 323
Lynch, C., 118
Lynch, P., 324

M

Macallan distillery, 163
MacDonald, J., 87, 96
MacEachren, A. M., 193, 194, 195
MacGregor, D., 249
Mack, R., 247
Mack, R. L., 170
Mackay, W. E., 219
Mackey, K., 144
Mackinlay, J. D., 187, 189, 190, 191, 192, 193, 194, 195, 196, 198, 210, 211, 212, 238, 239
Mackworth, N., 254
MacLean, K., 92, 94
Madani, O., 253
Madden, M., 276
Mader, S., 278
Madigan, S. A., 164
Maeda, F., 221
Maes, P., 107, 108, 118, 119, 120, 126
Magee, J. S., 221
Maglio, P., 240
Mainwaring, S., 314
Mainwaring, S. D., 224
Mair, G., 297, 305
Majercik, S., 253
Majumder, A., 295
Makhoul, J., 87, 98
Malin, B., 315, 326
Malkin, R., 100
Malone, T., 173

Malone, T. W., 220
Maloney-Krichmar, D., 282
Mancini, L., 319
Mandler, G., 321
Manke, S., 98
Mankoff, J., 316
Mann, W. C., 71
Manny, C. H., 314
Mansfield, A. F., 284
Mantei, M. M., 315
Marable, L., 324
Marchionini, G., 199
Marcus, A., 28, 32, 33, 35, 47, 48
Mark, G., 164
Marks, J., 146
Markus, M. L., 218, 223
Marsh, S., 255, 324, 325
Marshall, J., 324
Marsic, I., 99
Marti, S., 147, 220
Martin, B., 117, 118
Martin, C., 66, 72
Martin, D. L., 99
Martin, F., 157
Martin, J.-C., 86, 87, 89, 99
Martinussen, M., 304
Marwood, D., 319
Marx, M., 61
Masarie, F., 267
Mase, K., 98, 116, 117
Mason, R., 280
Massaro, D. W., 87, 89, 92, 96
Massimino, M., 296
Mastaglio, T., 242, 265
Matchen, P., 171
Matherwson, N., 315
Mathew, P., 283
Matsakis, N. E., 110
Matsushita, Y., 221
Mattis, J., 189, 208
Mavor, A. S., 295, 298
Maxion, R., 320
May, J., 75
May, J. G., 295
May, R. A., 213
Mayhorn, C. B., 265
Maynes-Aminzade, D., 114, 145, 147, 152, 156
Mazalek, A., 146, 153
McCabe, T. J., 174
McCall, K., 222
McCalla, G., 117
McCarthy, J., 255
McCarthy, K., 114
McCauley, C., 251
McCauley, M. E., 301
McCauley-Bell, P. R., 303
McCormick, B. H., 189
McCowan, I., 100
McCoy, E., 246, 248, 252, 255, 268
McCracken, D. D., 235
McCreath, E., 107, 124
McCrickard, D. S., 255
McDaniel, R. G., 170
McDaniel, S. E., 221
McDermott, J., 108

McDonald, D. W., 117, 120
McGee, D., 86, 87, 89, 98, 99, 100
McGee, D. R., 91
McGill, M. E., 189
McGinty, L., 114
McGrath, J. E., 219, 220, 222
McGrath, M., 87
McGrenere, J., 108, 125
McGuire, T. W., 220
McGurk, H., 87, 96
McKevitt, P., 109
McLachlan, P., 92, 94
McLeod, A., 87, 92
McLeod, P. L., 218, 220, 222
McLoughlin, C., 282
McLuhan, M., 276
McMillan, R. D., 219
McMullan, B., 3
McNee, S. M., 120
McNeill, D., 87, 92, 94, 95, 96
McQuire, M., 251
McRoy, S., 111
Mead, A., 305
Meader, D. K., 220, 225
Meadows, C., 319
Medl, A., 99
Meehl, P., 174
Meggs, P. B., 20
Mehandjiev, N., 78
Meier, M., 147
Meier, U., 89, 98
Meister, D., 246, 254
Memon, N., 321
Menzies, D., 298
Mercatante, T. A., 303
Merritt, E. A., 298
Messerschmitt, D. G., 219
Messina, P., 219
Metcalfe, B., 276
Metzger, U., 254, 267
Meyer, D., 247
Meyer, D. E., 175
Meyer, S., 315
Meyers, D., 247
Micarelli, A., 114
Michalewicz, Z., 252
Michalski, R., 239
Michie, D., 252
Mikhak, B., 145, 146, 148
Mikulka, P., 254
Milberg, S. J., 315
Milham, L. M., 305
Millen, D. R., 221, 224
Millen, J., 319
Miller, C., 246
Miller, G. A., 166, 247, 253
Miller, P., 265, 267
Miller, R., 267, 323
Miller, R. A., 254
Miller, S., 87, 98
Milligan, T., 315
Minassian, S., 107
Minneman, S., 223
Mital, A., 246
Mitchell, B. M., 219

Mitchell, C. M., 246, 247, 254
Mitchell, T., 108
Mitrovic, A., 117, 118
Miyake, N., 225
Mobasher, B., 122
Mockus, A., 224
Moffitt, M. D., 108
Moline, J., 306
Mollaghasemi, M., 302, 303
Moller, J., 248
Monk, A., 221
Monrose, F., 321
Moody, P., 107, 220
Moon, B. D., 213
Mooraj, Z., 89
Moore, A., 240, 241
Moore, R. J., 280
Moran, D. B., 86, 87, 96, 98, 99
Moran, R. T., 37
Moran, T. P., 67, 68, 162, 206, 223, 247, 248
Moray, N., 255
Morch, A. I., 265
Morency, L.-P., 89
Morgan, D., 61
Morimoto, C., 86, 89
Morris, C. S., 305
Morris, J., 220
Morris, M. W., 225
Mortensen, M., 255
Morton, J., 167
Mosier, J. N., 223
Mosier, K., 248, 254
Moskowitz, H., 301
Mourant, R., 299
Mousavi, S. Y., 92, 94
Movellan, J. R., 123
Mudge, D., 319
Muir, B., 255
Mukopadhyay, T., 220
Mulgund, S., 296
Müller, C., 112
Muller, M. J., 221, 224
Mullet, K., 75, 76, 81
Multon, F., 86, 87
Mumford, M., 253
Munhall, K. G., 87, 92
Munzner, T., 211
Murphy, R., 31
Murray, J. H., 300
Mwanza, D., 289
Myers, B. A., 170
Mynatt, C., 251

N

Nagel, D. C., 254
Nahmens, I., 301
Nakatsu, R. T., 255
Namioka, A., 247
Napoli, J., 295
Narayanan, N. H., 75, 80
Nardi, B. A., 221, 222, 224
Nareyek, A., 253
Narin, F., 239
Narita, A., 147, 151
Nason, S., 247

Nasoz, F., 111, 123
Nass, C., 75, 76
National Research Council, 226, 314
Nau, D., 253
Naughton, K., 95
Navon, D., 251
NCsoft, 279
Neal, J. G., 86, 87, 98
Neale, D. C., 255
Neale, M. A., 255
Negroponte, N., 86
Nejdl, W., 113
Nelson, T. H., 242
Neti, C., 86, 87, 89, 96, 99, 100
Nettle, D., 250, 267
Newell, A., 67, 68, 162, 175, 206, 247, 248, 251, 252, 253
Newell, B. R., 163
Ng, J. D., 316
Nguyen, L. K., 305
NICA Internet Surveys, 276
Nickell, E., 280
Nickerson, R., 254
Nicolle, C. A., 232
Nie, J., 239
Nielsen, J., 33, 70, 79, 173
Nielson, J., 33, 48, 232, 235
Nies, K., 318
Nijholt, A., 221
Nilsson, J., 316
Nisbett, R., 250
Nisbett, R. E., 37
Noakes, P., 146
Nock, H., 86
Nohria, N., 220
Norman, D. A., 68, 75, 118, 157, 168, 248, 250, 254
North, M., 297
North, M. M., 297
North, S. M., 297
Norvig, P., 109, 252
Not, E., 124
Nunamaker, J. F., 222
Nusbaum, H., 57
Nussbaum, M. A., 32
Nutt, G., 218

O

Oba, H., 146
Obradovich, J. H., 246, 255, 266, 267
O'Conaill, B., 220
O'Connor, K. M., 219, 220, 222
Odobez, J.-M., 100
O'Donnell, R., 173
OECD, 312, 314
Office of the Federal Privacy Commissioner of Australia, 314, 325
O'Hara-Devereaux, M., 219
Ohlsson, S., 118
Oinas-Kukkonen, H., 232, 233
Okada, K., 221
Okuno, H. G., 111, 112
Olfman, L., 219
Olgyay, N., 33
Oliver, D., 253
Oliver, N., 99
Olsen, E., 91, 96
Olson, G. M., 218, 220, 221, 222, 224, 225, 226, 246, 248, 255

Olson, J. S., 218, 220, 221, 222, 223, 224, 225, 226, 246, 248, 255
Olson, M. H., 225
Olufokunbi, K. C., 289
O'Modhrain, S., 146
O'Neill, J., 173
Ono, Y., 37
Opdyke, D., 306
Orasanu, J., 246, 255
Ordy, J. M., 297, 302
Orlikowski, W. J., 224
Ormrod, J., 248
Orth, M., 146
Ortony, A., 77
Oseitutu, K., 56
Osgood, C. E., 304, 305
Osipovich, A., 324
Osman, I., 252
Ostriker, J. P., 219
Otter, M., 235
Oviatt, S., 67, 86, 87, 94, 95, 98, 99, 100
Oviatt, S. L., 86, 87, 89, 90, 91, 92, 93, 94, 95, 96, 97, 99, 100
Owens, J., 267
Oyama, S., 221
Ozmultu, H., 238
Ozok, A., 320, 323
Ozyildiz, E., 87

P

Page, L., 239
Pagulayan, R., 304
Paier, D., 283
Paivio, A., 164
Palen, L., 218, 223, 314
Palmer, J. W., 324
Pan, S., 111
Pangaro, G., 145, 147, 150, 152, 156
Pankanti, S., 86, 89, 314
Pao, C., 86, 87
Papert, S., 68, 149, 170, 286
Parasuraman, R., 246, 248, 254, 261, 267
Paris, C., 116, 123
Park, K. S., 226
Parkes, A. J., 145, 146
Pashler, H., 248
Patel, A., 240
Patera., A., 146
Patrey, J., 305
Patrick, A., 325
Patten, J., 145, 147, 150
Patterson, M., 220
Pausch, R., 91
Pavlovic, V., 87, 89, 98, 99
Pavlovic, V. I., 87
Payton, F. C., 314
Pazzani, M., 113, 121
Pazzani, M. J., 123
Peacock, A., 322
Pearl, J., 268
Pedersen, E., 222
Pejtersen, A., 246, 248
Pelachaud, C., 86, 87, 89, 99
Pelican, M., 246
Pelz, D., 170
Pennathur, A., 246
Pennington, N., 250

Pennock, D. M., 236
Pentland, A., 123
Pentland, S., 89, 100
Pereira, F. C. N., 86, 87, 96, 98
Perrif, A., 319
Pesanti, L. H., 219
Petajan, E. D., 89
Peters, N., 268
Peterson, B., 300
Peterson, R., 89, 251
Petre, M., 174
Petrelli, D., 124
Petroski, H., 251
Pettersson, J. H., 316
Pfleger, N., 87
Phelps-Goodman, E., 147
Phillips, J., 87
Phillips, L. H., 167
Pianesi, F., 126
Piazzalunga, U., 322
Picard, R., 111
Picard, R. W., 111, 123
Pickover, C., 168
Pier, K., 222
Pierce, T., 33
Piernot, P., 146
Pillet, B., 318
Pinski, G., 239
Pinsonneault, A., 218, 222
Piper, B., 147, 152, 153, 156
Pirolli, P., 212
Pittman, J., 86, 87, 98, 99
Pittman, J. A., 99
Platt, J. C., 107
Playfair, W., 189
Plous, S., 248, 250, 267
Poddar, I., 87
Polifroni, J., 86, 87
Pollack, I., 87
Pollack, M. E., 108
Polson, P. G., 174
Poltrock, S. E., 203
Poole, M. S., 219
Popescu, G. V., 298
Porter, J. M., 235
Potamianos, G., 86, 87, 89, 96, 99, 100
Potts, C., 316
Poulton, E. C., 250, 267
Pountain, D., 297
Povel, D., 163
Povel, D. J. L., 163
Povenmire, H. K., 304
Prabhu, P., 246
Prakash, A., 224
Preece, J., 246, 247, 276, 277, 278, 281, 282, 289
Presti, S. L., 324
Pribram, K. H., 247, 253
Price, K., 56
Price, K. J., 56
Prietulla, M., 248
Prinz, W., 223
Pritchett, A., 265
Prothero, J. D., 302
Psillos, S., 265
Putnam, R. D., 220

Q

Qin, Y., 247

R

Rabiner, L. R., 55
Radvansky, G., 247, 248
Raffle, H. S., 145, 146
Rainie, L., 276
Raisinghani, M. S., 255
Rajasekaran, M., 283
Rakotonirainy, A., 315
Ralston, J., 282
Rao, R., 211, 220
Rasmussen, C., 224
Rasmussen, J., 246, 248, 249, 253, 255
Ratti, C., 147, 152, 153, 156
Rauterberg, M., 147, 174
Raven, M. E., 221, 224
Rayward-Smith, V., 252
Reagle, J., 315, 319
Real User Corporation, 321
Reason, J., 248, 251
Recht, B., 147, 150
Reddig, C., 86
Redmiles, D., 318
Redström, J., 146
Ree, M. J., 303
Reeder, R. W., 320
Reeves, B., 75, 76, 242
Reeves, C., 252
Reeves, L., 298, 299, 302, 303
Reicher, G. M., 168
Reilly, J., 114
Rein, G. L., 222
Reiser, R. A., 277
Reising, D., 249
Reisner, P., 174
Reiter, M., 315, 321
Reithinger, N., 86
Rekimoto, J., 146
Remondeau, C., 86, 87
Ren, J., 318
Ren, S., 147
Renaud, K., 320, 321
Rennels, G., 255
Renshaw, E., 107
Resner, B., 147, 148
Resnick, M., 146, 157
Resnick, P., 220, 225
Resnikoff, H. L., 187, 189
Rex, D. B., 213
Reynard, G., 226
Rheingens, P., 323
Rheingold, H., 276
Rhoades, J. A., 219
Rhodes, B., 123, 219
Rhodes, B. J., 114
Riccardi, G., 112, 123
Rice, R., 282
Rice, R. E., 224, 282
Rich, C., 109, 110
Richards, J. T., 176
Richmond, M. J., 295
Richter, M. M., 316
Riedl, J., 225

Riedl, J. T., 191
Riegelsberger, J., 255
Rieman, J., 242
Riesbeck, C. K., 253
Riley, V., 248, 254, 267
Risch, J. S., 213
Ritter, F., 247
Rizzo, A. A., 306
Robert-Ribes, J., 87, 96
Roberts, J. M., 281
Roberts, T. L., 144
Robertson, G., 86, 89
Robertson, G. G., 190, 210, 211, 213, 238, 239
Robertson, G. R., 78
Robin, H., 188, 190, 202
Robinson, E., 221
Robinson, J., 235
Rocco, E., 225
Rockwell, T., 250, 251
Rode, J., 318
Rodvold, M., 255
Roe, D. B., 55
Rogers, Y., 66, 68, 246, 247, 248, 276, 281, 282
Rogozan, A., 89
Rohall, S. L., 107
Rolland, J. P., 302
Rommelse, K., 125
Roos, J., 253
Root, R. W., 224
Roscoe, S. N., 304
Rose, C. F., 92
Rosenblitt, D. A., 220
Rosenbloom, P. S., 175, 247, 253
Ross, A., 89, 100
Roth, E., 247, 249
Roth, E. M., 246, 248, 254
Roth, S. F., 189, 208
Rothbaum, B. O., 306
Roulland, F., 173
Rouncefield, M., 255
Rourke, L., 282
Rouse, W. B., 253, 255, 263
Roy, D., 100
Ruben, H., 168
Rubin, A., 315
Rubin, B., 173
Rubin, D. C., 164
Rubin, K. S., 247
Rubin, P., 87, 98
Rubinson, H., 252
Rudmann, S., 265, 266, 267
Rudnicky, A., 61, 91
Ruhleder, K., 220
Russell, D. M., 212
Russell, M. J., 89, 91, 92
Russell, S., 252
Ruthruff, E., 248
Ruthven, I., 238
Ryall, K., 146
Ryokai, K., 147
Ryu, Y. S., 32

S

Sacerdoti, E. D., 253
Sadowski, W., 302, 303

Saka, C., 92, 94
Salas, E., 246, 255
Salingaros, N. A., 163
Salton, G., 238
Saltzer, J., 318, 320
Salvaneschi, P., 322
Salvendy, G., 33, 300, 303, 320
Salvucci, D., 254
Samman, S., 298, 299
Sanborn, S., 246
Sanders, P., 306
Sanderson, P., 249
Sandin, D. J., 295
Sandry, D., 67
Sandry, D. L., 92, 96
Sano, D., 75, 76, 81
Sanocki, E., 221
Saracevic, T., 237
Sarin, R., 111
Sarin, S. K., 223
Sarter, N., 254
Sarwar, B., 314
Sas, C., 121
Sasse, A., 314, 319, 326
Sasse, M., 255
Sasse, M. A., 319
Satava, R., 306
Satran, A., 76
Satzinger, J., 219
Sawyer, J. E., 255
Scaife, M., 66, 68, 248
Scanlon, T., 75
Scerbo, M., 254
Scha, R., 193
Schaaf, R., 306
Schaeken, W., 248
Schafer, J. B., 114, 225, 314
Schank, R. C., 253
Schatz, B. R., 219
Schell, R., 319
Scherlis, W., 220
Schiano, D. J., 221, 224
Schiele, B., 123
Schkade, D. A., 252
Schlosser, A., 219
Schmandt, C., 61, 220
Schmauks, D., 86
Schmitz, J., 282
Schmorrow, D., 297
Schneider, M., 111
Schneier, B., 318, 319
Schomaker, L., 86, 87, 89, 99
Schraefel, M., 113, 118, 122
Schraefel, M. C., 240
Schragen, J. M., 247
Schroeder, M., 318, 320
Schroeder, R., 255
Schroeder, W., 75
Schuler, D., 247
Schulten, K., 87
Schultz, R., 253
Schulze, K., 118
Schunn, C., 247, 254
Schwartz, J.-L., 87, 96
Schwarz, H., 221, 222

Schwarzkopf, E., 121
Sciarrone, F., 114
Sclabassi, R., 221, 222
ScotCIT, 277
Scott, J., 282
Scott, N., 316
Scott, R., 246
Seamon, A. G., 305
Seamon, J., 247, 248
Sears, A., 56, 125
See, K., 255
Segal, R. B., 107
Sekiyama, K., 96
Selker, T., 240
Sellen, A., 221, 225, 319
Sellen, A. J., 315
Sen, S., 114
Seneff, S., 86, 87
Senior, A., 86, 314
Sethi, Y., 87
Sewell, D. R., 252
Shadmehr, R., 306
Shafer, G., 268
Shahmehri, N., 117
Shaikh, A., 99
Shalin, V., 247
Shalin, V. L., 253
Shallice, T., 68
Shalom Schwarts, S., 37
Shankar, U., 318–319
Shapiro, J. A., 118
Shapiro, S. C., 86, 87, 98
Shappell, S., 250
Sharkey, T. J., 301
Sharma, G., 251
Sharma, R., 87, 98
Sharp, H., 246, 247, 276, 281, 282
Shearin, S., 120
Shelby, R., 118
Shen, H., 319
Shepherd, A., 247
Sheridan, T., 248, 255, 296
Sheridan, T. B., 246, 247, 255, 261, 302
Shilling, R. D., 296
Shimamoto, B., 295
Shin, D. J., 32
Shinn-Cunningham, B., 296
Shmulevich, I., 163
Shneiderman, B., 120, 125, 169, 182, 187, 189, 190, 191, 192, 193, 194, 196, 202, 212, 235
Short, J., 221
Shortliffe, E., 267
Shoval-Katz, A., 75
Shriver, S., 94, 95
Shubina, A. M., 316
Shulman, L. S., 266
Sides, S. A., 296
Sidner, C., 89, 219
Sidner, C. L., 107, 109, 110
Siegel, J., 220, 221, 222
Siem, F. M., 303
Signorini, A., 233, 239
Sillence, E., 324, 325
Silsbee, P. L., 89
Silverman, B., 157

Silverman, B. G., 265
Simmel, D., 219
Simon, H., 252
Simon, H. A., 170, 171, 254
Simon, R. T., 319
Simonsen, J., 247
Simpson, B., 324
Simske, S., 112
Singer, M., 303
Singhal, S., 226, 298
Singletary, B. A., 262
Singleton, S., 315
Siroux, J., 86, 87
Skirka, L., 248, 254
Slagter, R., 221
Slater, M., 303
Slavin, R. E., 225
Smart, L. J., 304
Smeaton, A., 237
Smetters, D., 320
Smetters, D. K., 320, 322
Smith, B., 114
Smith, D., 142, 144
Smith, D. C., 144
Smith, G., 252
Smith, H. J., 315
Smith, I., 86, 87, 98, 99, 220
Smith, J., 266, 314
Smith, J. W., 250, 266, 267
Smith, M. A., 164
Smith, P. J., 246, 248, 250, 251, 252, 255, 258, 265, 266, 267, 268
Smith, R. B., 314
Smith. S. W, 316
Smith-Jackson, T. L., 32
Smits, D., 242
Smyth, B., 114, 115
Snyder, C., T., 75
SOC Data Team, 226
Soergel, D., 199
Sokol, S., 168
Soller, A., 116
Sommerville, I., 255
Sonnemanns, P., 163
Sony, 280
Soohoo, C., 324
Soon, A., 37
Soon, C., 37
Soriyan, H. A., 289
Sparck Jones, K., 238
Sparrell, C., 87, 94
Spehar, B., 163
Spence, C., 92, 94, 100
Spence, R., 191, 205, 207, 208, 209
Spencer, A., 248, 255, 258
Spencer, B., 316
Spencer, H., 9
Spiekermann, S., 315
Spilka, G., 3
Spindle, B., 37
Spink, A., 237, 238
Spool, J. M., 75
Sprafka, S. A., 266
Sproull, L., 219, 225
Spyropoulou, E., 319
Squire, K., 280

Srinivasan, M. A., 296
Stachel, B. B., 107
St. Amant, R., 247
Stammers, R. B., 247
Stanford, J., 324
Stanney, K., 298, 299, 302, 303, 305
Stanney, K. M., 295, 296, 297, 299, 300, 301, 302, 303, 398
Stanton, N., 247
Starner, T., 123, 219
Starr, B., 224
Stash, N., 242
Stasko, J., 322
Steed, A., 303
Steffen, J. L., 198, 207
Stefik, M. J., 212
Stein, B. E., 92, 94, 100
Steiner, T. J., 92
Steinkeuhler, C., 280
Stenz, A., 263
Stephanidis, C., 28
Stepp, R., 239
Stevens, A., 248
Stewart, C., 241
Stewart, C. D., 240
Stewart, P., 320
Stifelman, L. J., 147
Stille, A., 38
Stoffregen, T., 304
Stokes, J., 296
Stone, R., 305
Stone, R. B., 248
Stork, D. G., 86, 87, 92, 96, 98
Storms, R. L., 299
Storrøsten, M., 222
Strauch, B., 250
Straus, S. G., 220
Strickland, D., 297
Strohm, P., 266, 267
Stuart, R., 162
Su, H., 205, 207, 208
Su, Q., 89
Su, V., 145, 146, 147, 148
Subramani, M., 111
Suchman, L., 253
Suenaga, Y., 98
Suhm, B., 86, 87, 89, 91, 93, 99
Suler, J., 276
Sullivan, E., 146
Sullivan, J., 89, 305
Sullivan, J.W., 86, 87, 96, 98
Sumby, W. H., 87
Sumi, Y. 116, 117
Summerfield, A. Q., 87
Summerfield, Q., 87, 92
Summerskill, S. J., 235
Sumner, E. E., 198, 207
Sundali, J., 250
Sunder, S., 220
Suraweera, P., 117, 118
Surendran, A. C., 107
Sutcliffe, A. G., 66, 70–71, 73, 75, 78, 80
Suthers, D. D., 225
Suzuki, H., 146
Swan, J. E. II, 302
Swartz, K. O., 302

Swartz, L., 222, 224
Sweeney, L., 315, 326
Sweeney, T., 297, 305
Sweller, J., 69, 92, 94
Sycara, K., 255
Szczepkowski, M. A., 253

T
Takeda, K., 237
Talwar, K., 318–319
Tamassia, R., 203
Tamioka, K., 32
Tammaro, S. G., 223
Tan, D. S., 78
Tan, S. L., 302
Tanenbaum, A. S., 219
Tang, A., 92, 94
Tang, J., 222
Tang, J. C., 110, 219, 220, 314
Tang, K. P., 111
Tarr, R. W., 305
Tauber, E. R., 324
Taylor, J. M., 108
Taylor, L., 118
Taylor, R. P., 163
Taylor, T. L., 279
Teasley, S., 223, 224
Teasley, S. D., 223
Tees, R. C., 162
Teevan, J., 113, 114
Telang, R., 220
Teltrow, M., 315
Tepper, P., 87
Terveen, L., 117, 120, 225
Terzopoulos, D., 87, 92
Tessmer, M., 247
Tetlock, P., 251
Thacker, P. D., 306
Thalmann, D., 297, 305
Thalmann, N. M., 297, 305
Tharp, V., 301
Thies, S., 122
Thomas, D., 37
Thomas, J., 173
Thomas, J. C., 164, 167, 168, 171, 172, 173, 175, 176
Thomas, M., 250, 251
Thompson, D. M., 296
Thompson, S. A., 71
Thomsen, D., 319
Thomsen, M., 147
Thorisson, K., 87, 94
Thorndike, E. L., 304
Thuraisingham, B., 315
Thurman, R. A., 297
Tian, L., 314
Tianfield, H., 265, 267
Tobey, D., 247
Todorov, E., 306
Tohkura, Y., 96
Tollis, I. G., 203
Tolmie, P., 173
Tomkins, A., 236
Tomlinson, M. J., 89, 91, 92
Torobin, J., 282
Torreano, L., 168

Tractinsky, N., 75
Tracy, W., 8
Traverso, P., 253
Tredennick, N., 295
Trefftz, H., 298
Treisman, A., 67, 81
Trewin, S., 108, 109, 122
Trompenaars, F., 37
Truran, M., 239
Tsandilas, T., 113, 118, 122
Tsimhoni, O., 235
Tufte, E. R., 75, 189, 190, 248, 252
Tukey, J. W., 189
Turban, E., 246
Turieo, M., 296
Turk, M., 86, 89, 297
Turner, C. H., 37
Turow, J., 324
Tversky, A., 250, 252, 267
Tweedie, L. A., 205, 207, 208
Tweney, R., 251
Tygar, D., 319, 322
Tygar, J. D., 323

U
Ueno, S., 111, 112
Ullmer, B., 142, 145, 146, 147, 149, 150, 156
Ultimate 3D Links, 298
Underkoffler, J., 142, 147, 150, 156
Underwood, P., 172
Uribe, T. E., 108
Usability Net, 281
U.S. Census Bureau, 276
U.S. Fair and Accurate Credit Transaction Act, 314
U.S. Fair Credit Reporting Act of 1978, 312

V
Vaknin, S., 279
Valacich, J. S., 222
Valleau, M., 89
Vallone, R., 250
van der Mast, C. A. P. G., 306
van der Veer, G., 221
van der Zaag, C., 306
Van Doorn, J., 253
Vandor, B., 265
van Gent, R., 91
van Kleek, M., 219
VanLehn, K., 118
van Melle, W., 223
Van Mulken, S., 71
Varian, H. R., 225
Varma, C., 324
Vatikiotis-Bateson, E., 87, 92, 98
Veinott, E., 221
Ventura, D., 306
Verbyla, J., 237
Verbyla, J. L. M., 236, 237
Vergo, J., 86, 87, 93
Verplank, W., 144
Verplank, W. L., 261
Vertegaal, R., 221
Vicente, K., 247, 249, 253
Vicente, K. J., 248, 249
Vickers, B., 31
Victor, D., 37

Vidulich, M., 67, 92, 96
Viegas, F. B., 224
Viégas, F. B., 278
Viirre, E. S., 302
Vijaya Satya, R., 296
Villar, N., 121
Vince, J., 295
Vintage Cocktails, 163
Vitali, F., 232, 233
Vivacqua, A., 117
Vo, M. T., 87, 98, 99
Vogel, D. R., J. F., 222
Von Thaden, T., 251
Vygotsky, L., 286

W
Wadge, W. W., 240
Wagner, D., 318–319
Wahlster, W., 86, 87, 111
Waibel, A., 87, 98
Wainer, H., 122
Walendowski, A., 221, 224
Wallace, P., 276
Walston, C. E., 174
Walters, T. B., 213
Waltz, D., 252
Wang, H. C., 278
Wang, J., 98
Wang, M., 99
Wang, M. Q., 91
Wang, R.W., 265, 267
Wang, X., 147, 148
Wang, Y., 147, 152, 156, 252
Want, R., 315
Ware, C., 191
Warm, J., 254
Warren, S. A., , 314
Washburn, D., 296
Wasinger, R., 123
Wason, P., 251
Waters, J., 321
Waters, K., 225
Watson, B. A., 306
Wattenberg, M., 202,
    224, 278
Watts, L., 221
Watzman, S., 248
Waugh, N. C., 164, 167
Weerasinghe, A., 117, 118
Weghorst, S., 297
Weibelzahl, S., 123
Weil, M., 305
Weinberg, G. M., 170
Weinshenk, S., 56
Weiser, M., 157
Weitzman, L., 66
Welch, B., 222
Welch, R. B., 301
Welch, V., 322
Weld, D. S., 122
Wellman, B., 282, 283, 315
Wellner, P., 147, 150
Wells, G., 250, 267
Welty, B., 325
Wen, J., 239

Wen, Z., 116
Wendt, K., 219
Werker, J. F., 162
Wesson, R., 94, 95, 100
Wexelblat, A., 118, 119
Weymouth, T., 224
Wharton, C., 247, 248
White, M., 164
Whittaker, S., 219, 220, 221,
    222, 224, 229
Whitten, A., 319, 322
Wickens, C. D., 67, 92, 96, 121, 246,
    247, 261, 284, 304
Wiedenbeck, S., 321, 324, 325
Wiegmann, D., 250, 251
Wiener, E. L., 254, 262
Wikipedia, 278
Wilbur, S., 220, 221, 225
Wilde, G. J., 111
Wilensky, R., 253
Wilensky, R. L., 109
Wilkerson, M., 322
Wilkins, R., 237
Wilks, A. R., 202, 203
Wilks, Y., 116
Williams, E., 220, 221
Williams, K., 247
Williamson, C., 169
Willkomm, B., 304
Wills, G. J., 203
Wilpon, J. G., 55
Wilson, G., 297
Wilson, S., 286
Winograd, T., 220
Wise, J. A., 163
Wisneski, C., 147, 153
Withgott, M., 146
Witkin, B. R., 247
Witmer, B., 303
Wittig, F., 112
Wogalter, M. S., 265
Wolf, C., 56, 240
Wolfe, R. J., 235
Wolfman, S. A., 122
Wood, A., 153
Wood, C., 98, 99
Wood, D., 57
Woods, D., 246, 254
Woods, D. D., 248, 254
Woodworth, R. S., 304
Wright, M. H., 219
Wu, L., 86, 87, 93, 99
Wu, M., 323

X
Xiao, B., 94, 95, 100
Xue, P., 87, 98

Y
Yacoub, S., 112
Yamagata-Lynch, L. C.,
    289
Yan, J., 321
Yang, C. S., 238
Yang, H., 226

Yang, J., 87, 98
Yang, J. Y., 278
Yankelovich, N., 59, 61,
  219, 314
Yantis, S., 248
Yardley, 29, 37
Yarin, P., 147
Ye, Y., 78
Yedidia, J. S., 146
Yee, K., 320
Yee, N., 279
Yenisey, M., 320
Yli-Harja, O., 163
York, J., 114
York, W., 213
Young, B. C., 114
Young, P., 297
Yu, C. T., 238
Yuille, J. C., 164

Z

Zabowski, D., 108
Zacharia, G., 283, 284

Zakaria, M., 241
Zakaria, M. R., 240, 242
Zancanaro, M., 126
Zander, A., 255
Zaphiris, P., 283, 284, 286
Zhai, S., 86, 89
Zhang, D., 100
Zhang, H., 239, 251
Zhang, J., 225, 254, 324
Zheng, X. S., 111
Zhou, M. X., 115, 116
Zhu, E., 282
Ziegler, C., 120
Zimand, E., 306
Zirk, D. A., 254
Zittel, R. C., 306, 307
Zloof, M., 175
Zlot, R., 263
Zuboff, S., 253
Zuckerman, O., 146
Zukerman, I., 111
Zurko, M. E., 319
Zyda, M., 226, 295, 296, 297, 298

SUBJECT INDEX

A

ACM/SIGCHI, *see* Association for Computing Machinery's Special
    Interest Group on Human-Computer Interaction
Activity theory, 289
Actuated Workbench, 147, 152
Adaptive hypermedia (AH), 240
Adaptive interfaces and agents, 105–130
    adaptable systems, 106
    AgentSalon, 116, 117
    automatic processing, 119
    collaborative assistance, 109
    collaborative filtering, 114
    collaborative learning, 116
    concepts, 106–107
    empirical methods, special considerations concerning,
        123–126
      checking usability under realistic conditions, 126
      comparisons with work of human designers, 125
      early studies of usage scenarios and user requirements, 124
      experimental comparisons of adaptive and nonadaptive
        systems, 125–126
      individual differences, 126
      use of data collected with nonadaptive system,
        123–124
      Wizard of Oz studies, 124–125
    explicit query, 114
    future of user-adaptive systems, 126–127
      advances in techniques for learning, inference,
        and decision, 127
      attention to empirical methods, 127
      diversity of users and contexts of use, 126
      growing need for user-adaptivity, 126–127
      increasing feasibility of successful adaptation, 127
      number and complexity of interactive systems, 126
      scope of information to be dealt with, 126–127
      ways of acquiring information about users, 127
    FXPAL Bar, 114
    intelligent agents, 106

interpretation of naturally occurring actions, 122
I-Spy system, 114
keyhole recognition, 109
key repeat delay interval, 108
Lilsys prototype, 110
Lumière intelligent help system, 125
Microsoft Word 2000, 108
natural language dialogue, 111
obtaining information about users, 121–123
    explicit self-reports and -assessments, 121–122
    low-level indices of psychological states, 123
    naturally occurring actions, 122
    nonexplicit input, 122
    previously stored information, 122–123
    responses to test items, 122
    self-assessments of interests and knowledge, 121–122
    self-reports about objective personal characteristics, 121
    self-reports on specific evaluations, 122
    signals concerning current surroundings, 123
personalization, 106, 112
personalized search, 114
POPFile, 107
privacy issues, 126
prototype "intelligent electronic mail sorter," 107
sensing devices, categories of, 123
Smart Menus, 108, 125
SQL SELECT statement, 117
SQL-Tutor, 117, 121
subtasks, 119
supporting information acquisition, 112–118
    helping users to find information, 112–114
    recommending products, 114–115
    spontaneous provision of information, 114
    support for browsing, 113
    supporting collaboration, 116–117
    supporting learning, 117–118
    support for query-based search or filtering, 114
    tailoring information presentation, 115–116

Adaptive interfaces (*continued*)
  supporting system use, 107–112
    adapting of interface, 108–109
    dialogue control, 111–112
    helping with system use, 109–110
    mediating interaction with real world,
      110–111
    taking over parts of routine tasks, 107–108
  SwiftFile, 107
  transparency, 119
  usability challenges, 118–121
    breadth of experience, 120
    dealing with trade-offs, 120–121
    obtrusiveness, 120
    threats to controllability, 119
    threats to predictability and comprehensibility, 119
    threats to privacy, 120
  user model application, 106
  user modeling servers, 123
  Web Adaptation Technology, 109
Agent communication languages, 99
AgentSalon, 116, 117
AH, *see* Adaptive hypermedia
AIDA, 266, 267, 268
Amazon.com, 106, 112
AmbientROOM, 147
American Anthropologists' Association, 49
"Answer Garden," 173
Articulated reading visualizations, 195
ASR, *see* Automatic Speech Recognition
Association for Computing Machinery's Special Interest Group
    on Human-Computer Interaction (ACM/SIGCHI), 32, 49
ATMs, *see* Automatic teller machines
Attitudes Towards Thinking and Learning Survey (ATTLS), 284
ATTLS, *see* Attitudes Towards Thinking and Learning Survey
Attribute Explorer, 208
AUCTIONEER system, 262, 263
Audiopad, 150–151
Automatic Speech Recognition (ASR), 55
  errors occurring in, 56
  operation, 55
  system accuracy rates, 56
Automatic teller machines (ATMs), 322

B

Backchannel communication, 220
Biased assimilation, 251
Bifocal lens, 209
Big AI game, 78
Biometrics, 321–322
Blogging, 138, 222, 243
Bluetooth, 132
Body syntonic learning, 149
Bookmarks, 79
Bulletin boards, 278
"Buttonification" of the human being, 134

C

CALL course, *see* Computer Aided Language Learning course
Cathode-ray tube (CRT) devices, 29
CAVE, *see* Cave Automated Virtual Environment
Cave Automated Virtual Environment (CAVE), 295
Cell phones
  e-mail done from, 219
  TUI and, 144

ChickClick website, 43, 44
Citation based metrics, 239
CK, *see* Connected knower
Classroom 2000, 224
Clique analysis sociogram, 285
CMC, *see* Computer-mediated communication
COCOMO, 174
Cognitive biases, 250
Cognitive complexity, measures to reduce, *see* Psychological
    simplicity, achievement of
Cognitive maps, 235
Cognitive task analyses (CTAs), 247, 248
Collaborative Decision Making program, 256
Collaborative filtering, 114
Collaborative learning, 116
Collectivism, culture and, 40
Color(s), 19–20
  additive primaries, 19
  basic principles, 19
  coding, MPEG and, 67
  color choices, 20
  color logic, 19–20
  context, 20
  contrast, 20
  decisions regarding color in typography, 20
  how to use color, 19–20
  less is more, 19
  palette of compatible colors, 20
  quantity affects perception, 20
  sacred, 37
  subtractive primaries, 19
  use of color as redundant cue, 20
  use of complementary colors with extreme caution, 20
  viewing medium, 20
Complex Adaptive Systems, psychological simplicity and, 162
Composed Visual Structures, 195
Computer Aided Language Learning (CALL) course, 284
Computer-mediated communication (CMC), 220, 276–280
  advantages and disadvantages of CMC, 277
  computer mediated communication, 276
  examples of CMC and online communities, 277–280
  online communities, 276–277
  online virtual game communities, 279–280
  systems, 277
  wiki-based communities, 278–279
Computer-Supported Cooperative Learning (CSCL), 225
Computer-Supported Cooperative Work (CSCW), 218, *see also*
    Groupware and Computer-Supported Cooperative Work
Concatenated synthesis, 56
Concept demonstrators, 70
Confab, 316
Confirmation bias, 251
Connected knower (CK), 284, 285
Contextual complexity, 169
Continuous plastic TUI, 147
Conversational speech interfaces and technologies, 53–63
  Automatic Speech Recognition, 55
  concatenated synthesis, 56
  crafting of spoken interaction, 58–61
    conversational systems, 59
    correction strategies, 61
    dialogue styles, 59
    directed dialogue, 59
    handling errors, 60–61
    inherent challenges of speech, 59

insertion errors, 61
prompt design, 59–60
providing help, 60
rejection errors, 61
substitution errors, 61
user-initiated, 59
diphones, 56
explicit prompts, 60
formant synthesis, 56
Hidden Markov Models, 55
human-computer interaction, 55
implicit prompts, 60
incremental prompts, 60
interactive voice response systems, 54, 56, 57
invisibility of speech, 59
"in vocabulary" words, 56
natural dialogue study, 59
Natural Language Understanding, 57
prototypical multimodal conversational system, 58
selection of speech technology, 55–58
current capabilities and limitations of NLU systems, 58
current capabilities and limitations of speech-synthesis
software, 57
how ASR works, 55–56
how natural language processing and natural language
understanding work, 57–58
how text-to-speech works, 56–57
speech synthesis types, 56
spoken interface life cycle, 55
system initiated dialogue style, 59
tapered prompts, 60
task model, 55
testing and iterating, 61–62
text-to-speech, 55
UNIX commands, 57
variable prompts, 60
Wizard of Oz study, 62
Cooperative work, *see* Groupware and Computer-Supported
Cooperative Work
Cosmographies, 136
Credit assignment issues, 169–170
CRT devices, *see* Cathode-ray tube devices
CSCL, *see* Computer-Supported Cooperative Learning
CSCW, *see* Computer-Supported Cooperative Work
CTAs, *see* Cognitive task analyses
CUBRICON system, 87
Cultural references, differing, 30
Curlybot, 148
Cyber societies, 276

D

Data encryption, 322
Data-subjects, 314
Decision-support system (DSS), 246
Decision-support systems, human-centered design of,
245–274
AIDA, 266, 267, 268
approaches to design, 247–249
cognitive task analyses and cognitive walkthroughs, 247–249
work-domain analyses, 249
associative activation errors, 250
AUCTIONEER system, 262, 263
availability heuristic, 250
biased assimilation, 251
black box automation, 265

capture errors, 250
case studies, 255–268
distributed work in national airspace system (ground delay
programs), 256–261
human interactions with autonomous mobile robotic systems,
261–265
interactive critiquing as form of decision support, 265–268
cognitive biases, 250
cognitive task analyses, 247, 248
Collaborative Decision Making program, 256
confirmation bias, 251
data-driven errors, 250
description errors, 250
design as prediction task, 254
distributed work in national airspace system (ground delay
programs), 256–261
adapting to deal with uncertainty, 259
administrative controls, 259–260
alternative approach to design of GDPs, 256–260
application area, 256
conclusions, 260–261
evaluation of implemented solution, 260
original approach to distributing work, 256
ration-by-schedule, 257–258
slot swapping, 258–259
strategy for distributing work, 256–257
domain ontology, 252
gambler's fallacy, 250
GOMS model, 247
human interactions with autonomous mobile robotic systems,
261–265
acceptance, 264–265
application area, 261–262
communication of beliefs, 265
communication of direct commands, 265
communication of intentions in context of bid, 265
design solution, 262–263
need for decision aiding, 262
performance, 264
results, 264–265
human performance on decision-making tasks, 249–255
active biasing of user's cognitive processes, 255
additional cognitive engineering considerations, 254–255
cognitive biases, 250–251
complacency and overreliance, 254
descriptive models, 252–253
designer error, 251
distributed work and alternative roles, 255
errors and cognitive biases, 250–252
errors and cognitive biases, implications for design, 251–252
excessive mental workload, 254
human expertise, 252–254
human expertise, implications for design, 253–254
human operator as monitor, 254
lack of awareness or understanding, 254–255
lack of trust and user acceptance, 255
mistakes, 250
normative optimal models, 253
organizational failures, 255
slips, 250
systems approaches to error, 251
hypothesis fixation, 251
interactive critiquing as form of decision support, 265–268
additional design considerations, 267–268
application area, 265–267

Decision-support systems (*continued*)
  complementary strategies to reduce susceptibility
      to brittleness, 268
    design solution, 267–268
    diagnosis as generic task, 265–266
    evaluation of implemented solution, 268
    need for decision aiding, 267
    sample problem, 266–267
  loss-of-activation errors, 250
  mode errors, 250
  NAS User, 256, 258, 260
  overreliance, 255
  point intelligence, 262
  ration-by-schedule, 257, 258
  Slot-Credit Substitution, 258, 259
  time-out cancels, 259
  vigilance performance, 254
  white hat, 259
Decision tree, media resources classification, 73
Degree-of-interest (DOI), 208, 209
Design for attention, multimedia, 80
Dialogue styles, 59
  conversational systems, 59
  directed dialogue, 59
  user-initiated, 59
DiamondHelp, 109, 110
DICTION program, 175
Digital Desk, 147, 150
Digital shadow, 146
Diphones, 56
Direct reading visualizations, 195
Division of labor (DOL), 287, 288
Document clustering, 239
Document vectors, 191
DOI, *see* Degree-of-interest
DOL, *see* Division of labor
Domain ontology, 252
DSS, *see* Decision-support system
Dynamic media, 75, 79

E

Ecosystems, *see* Experience ecosystems, mobile interaction design and
E-mail, 219, 276, 278
EPIC modeling, 175, 247
Excite website, 43
Experience ecosystems, mobile interaction design and, 131–140
  "buttonification" of the human being, 134
  calligraphy, 134
  cell phones, 133, 137
  convergence, 132, 138
  cosmographies, 136
  deconstruction of content, 138
  Graphical User Interface, 133
  handheld devices, 134, 137
  interaction models, 132
  Lomography, 133
  mental models, 136
  mobile experience, 132–133
    content, communication, control, 132
    from interaction to experience, 132–133
    interacting with networked ecosystems, 132
  mobile wireless media, 138
  networked media, 137
  networking, 132
  Palm PDA, 135

podcasting, 138
RFID tags, 137
smart tags, 137
sphere, communities, 137–138
sphere, content, 138–140
  adapting experiences to evolving behaviors, 139
  deconstructed content, 138
  human-human rather than human-machine interaction, 139–140
  narrowcast, 138–139
  seamless experience ecosystems, 139–140
  shaping experience without controlling full ecosystem, 139
sphere, context, 136–137
  physical context, 136–137
  smart objects and active ambient, 137
sphere, of humans and devices, 133–135
  cars and motorcycles, 133
  form follows gesture, 134
  product species, 135
  ritual power of hands, 133–134
  sense of touch, 134–135
sphere, mental models, 136
spheres of interaction, 133–140
techno-cultural phenomena, 132
text messaging, 138
"triple point" interaction, 138
Explicit prompts, 60
Eye, saccadic movement, 9

F

FAQS, *see* Frequently asked questions
FilmFinder, 183, 192, 195
Flesch Formula, 175
Formant synthesis, 56
Fourth-generation hypermedia, 232
Frequently asked questions (FAQs), 313
FXPAL Bar, 114

G

Game communities, 286, 289
GDP, *see* Ground Delay Program
Gender roles, global/intercultural user interface design, 41
Global/intercultural user interface design, 27–52
  case study, 35–36
  ChickClick, 43
  comparison of English–language user community conventions, 31
  cross-cultural theory, 48
  culture dimensions, 36–48
    culture as additional issue for globally oriented UI designers, 37–38
    design issues, 46–48
    Hofstede's dimensions of culture, 3
    individualism versus collectivism, 40–41
    long- versus short-term time orientation, 45–46
    masculinity versus femininity, 41–43
    power distance, 38
    uncertainty avoidance, 43–45
  dimensions, 48
  economic success, 38
  examples of differing cultural references, 30
  Excite website, 43
  future research issues, 48–49
  gender roles, 41
  globalization, 29–35
    advantages and disadvantages, 31
    aesthetics, 35

appearance, 33–35
    color, 35
    critical aspects, 32–33
    definitions, 29–31
    development process, 31–32
    example of specific guidelines, 33–35
    icons, symbols, and graphics, 33–34
    interaction, 33
    language and verbal style, 35
    layout and orientation, 33
    mental models, 33
    metaphors, 32
    navigation, 33
    technology, 32
    typography, 34–35
    user demographics, 32
  legal issues, 32
  localization, insufficient attention to, 28
  mental models, 28
  metaphors, 28
  Planet Sabre, 35–36
  Siemens website, 46
  TeamWare Finnish screen patterns, 37
  time orientation, 45
Globalization, 29–35
  advantages of, 31
  aesthetics, 35
  appearance, 33–35
  case study, 35–36
  color, 35
  critical aspects, 32–33
  definitions, 29–31
  development process, 31–32
  disadvantages of, 31
  example of specific guidelines, 33–35
  icons, symbols, and graphics, 33–34
  interaction, 33
  layout and orientation, 33
  mental models, 33
  metaphors, 32
  navigation, 33
  technology, 32
  typography, 34–35
  user demographics, 32
Global Positioning System (GPS) technology, 123
Global village, 276
Golden mean, 6
Golden rectangle, 6
GOMS model, 247
Google Gmail, 220
GPS technology, see Global Positioning System technology
Graphical User Interface (GUI), 66, 86, 133, 142
Graphic standards, 20–22
  audit, 22
  implementation, 22
  system development, 22
  what system covers, 22
Ground Delay Program (GDP), 256
Groupware and Computer-Supported Cooperative Work, 217–230
  adopting groupware in context, 218
  backchannel communication, 220
  blogs, 222
  buddies, 224
  Classroom 2000, 224
  communication tools, 219–222

blogs, 222
    conferencing tools, 220–222
    e-mail, 219–220
    text, 220–222
    video, 220–222
    voice, 220–222
  computer-supported cooperative learning, 225
  coordination support, 222–224
    awareness, 223–224
    group calendars, 223
    meeting support, 222–223
    workflow, 223
  e-mail, 219
  information repositories, 224
    capture and replay, 224
    repositories of shared knowledge, 224
    wikis, 224
  instant messaging, 224
  integrated systems, 225–226
    collaborative virtual environments, 226
    collaboratories, 226
    media spaces, 225–226
  middleware, 219
  multi-user domains, 226
  paralinguistic cues, 220
  Portholes, 224
  Proxemics, 226
  reinvention of e-mail, 220
  social computing, 225
    social filtering, recommender systems, 225
    trust of people via technology, 225
  technical infrastructure, 219
GUI, see Graphical User Interface
Gunning Fog index, 175

H

HCI, see Human-computer interaction
HCII, see Human-Computer Interaction International
Head-mounted display (HMD), 295
Head-related transfer function (HRTF), 296
Hidden Markov Models (HMMs), 55, 99
HMD, see Head-mounted display
HMMs, see Hidden Markov Models
HRTF, see Head-related transfer function
HTML links, 236
Human-centric Word Processor, 87
Human Computer Interaction (HCI), see also Privacy, security, and trust; Web, HCI and
  modality, 66
  online communities, 276
  psychological simplicity and, 162, 170
  speech interfaces and, 55
  TUI and, 142, 157
Human-Computer Interaction International (HCII), 32
Human image, effectiveness of for attraction, 76

I

IDP, see Information-design process
IM, see Instant messaging
Implicit prompts, 60
Incremental prompts, 60
In-degree analysis sociogram, 285
Individualism, culture and, 40
Information-consuming habits, 239
Information-design process (IDP), 4, 5

Information visualization, 181–215
  applications, 213
  articulated reading visualizations, 195
  bifocal lens, 209
  classing, 192
  composed visual structures, 195, 203–206
    double-axis composition, 204–205
    mark composition and case composition, 205
    recursive composition, 205–206
    single-axis composition, 203–204
  composition types, 204
  data classes, 193
  Data Transformations, 191
  definition, 187
  degree-of-interest, 208, 209
  direct reading visualizations, 195
  document vectors, 191
  dynamic queries, 190
  examples, 182–187
    finding videos with FilmFinder, 182–184
    monitoring stocks with TreeMaps, 184–185
    sensemaking with permutation matrices, 185–187
  exploratory data analysis, 189
  external cognition, 187
  extraction, 212
  FilmFinder, 183, 192, 195
  focus + context attention-reactive abstractions,
        208–211
    alternate geometries, 211
    data-based methods, 208–209
    filtering, 208–209
    highlighting, 209
    micro-macro readings, 209
    perspective distortion, 210–211
    selective aggregation, 209
    view-based methods, 209–211
    visual transfer functions, 209–210
  historical origins, 188–11
  information perceptualization, 182
  interaction techniques, 207
  interactive visual structures, 206–208
    Attribute Explorer, 208
    dynamic queries, 206
    extraction and comparison, 208
    linking and brushing, 206–208
    magic lens (movable filter), 206
    overview + detail, 206
  Kahonen diagram, 199
  knowledge crystallization, 212
  levels of use, 213
  macro-micro reading, 206, 209
  mark composition, 198
  network methods, 202
  object chart, 198
  permutation matrix, 188, 203
  Quantitative Spatial, 192
  recursive composition, 205
  retinal information topographies, 201
  retinal scattergraph, 198
  scientific visualization, 187, 189
  sensemaking with visualization, 211–214
    acquiring information, 212
    acting on it, 212
    creating something new, 212
    knowledge crystallization, 211–212

    levels for applying information visualization, 212–214
    making sense of it, 212
  simple visual structures, 196–203
    connection, 199–202
    enclosure, 202
    networks, 202–203
    trees, 199–202
    1-variable, 197–198
    2-variables, 198
    3-variables, 198–199
    n-variables, 199
  taxonomy, 197
  thresholding, 203
  TreeMap, 185, 186
  View Transformations, 191
  view-value distinction, 191
  visualization reference model, 191–196
    connection and enclosure, 194
    data structures, 191–192
    expressiveness and effectiveness, 195
    mapping data to visual form, 191
    marks, 193
    retinal properties, 194
    spatial substrate, 192–193
    taxonomy of information visualizations, 195–196
    temporal encoding, 194
    visual structures, 192–194
  visual knowledge tools, 214
  visually enhanced objects, 214
  why visualization works, 187
Information Visualizer, 190, 239
Instant messaging (IM), 224
Intelligent agents, 106
Intelligent electronic mail sorter, prototype, 107
Interaction design, *see* Experience ecosystems, mobile interaction
        design and
Interactive prototypes, 70
Interactive voice response (IVR) systems, 54, 56, 57
Intercultural user interface design, *see* Global/intercultural user
        interface design
International Conference on the Internationalization of Products
        and Services (IWIPS), 32, 49
Internationalization, definition of, 29
International Standards Organization (ISO), 28
Internet
  communities, 242
  killer browser, 233
  mobile access of, 234
  original purpose, 289
  security on, 219
inTouch, 143, 148–149
IP Network Design Workbench, 150, 151–152
iPod, 135
ISO, *see* International Standards Organization
ISO standard 14915 on Multimedia User Interface Design, 66
I-Spy system, 114
IVR systems, *see* Interactive voice response systems
iWeaver, 240
IWIPS, *see* International Conference on the Internationalization
        of Products and Services

K
Kahonen diagram, 199
Keyhole recognition, 109
Knowledge crystallization, 211, 212

L

Large Scale Networking (LSN) Coordinating Group, 298
Learner modeling, 117
Learn Greek Online (LGO) course, 284
Learning, multimedia, 68
LGO course, *see* Learn Greek Online course
Lilsys prototype, 110
Linguistic media, 81
Link sorting, 241
Localization
  definition of, 29
  insufficient attention to, 28
Lomography, 133
LSN Coordinating Group, *see* Large Scale Networking Coordinating
    Group
Lumière intelligent help system, 125

M

Macro-micro reading, 206, 209
Magic lenses, 206
Marble Answering Machine, 146
Mark composition, 198
Masculinity (MAS) index, 42
MAS index, *see* Masculinity index
Massively Multi-Player Online Role Playing Games (MMORPGs),
    276, 280, 286
MAUVE system, 302
MDS, *see* Multidimensional scaling
MediaBlocks, 149–150
Media Spaces, 225
Members-Teams-Committee (MTC), 99
Mental models, 136
Microsoft Word 2000, 108
Middleware, 219
Mini-maps, 79
MMORPGs, *see* Massively Multi-Player Online Role Playing Games
Mobile interaction design, *see* Experience ecosystems, mobile
    interaction design and
Mobile wireless media, 138
Model(s)
  authoring, 242
  Document Object, 176
  domain, multimedia user interface design, 70
  GOMS, 247
  GUI, 144
  Hidden Markov, 55, 99
  Human Processor, 67
  mathematical, psychological simplicity and, 174
  mental, 136
  meta-access control, 319
  multimodal interaction, 89
  overlay, 242
  portal user interface, 242
  privacy-risk, 316
  speech recognition, 55
  stereotype, 241
  task, 55, 71
  text to speech synthesis, 57
  trust, 324
  TUI, 145
Mosaic browser, 233
Moving image media, 81
Moving Pictures Expert Group (MPEG), 67
MPEG, *see* Moving Pictures Expert Group
MTC, *see* Members-Teams-Committee

MUDs, *see* Multi-user domains
Multidimensional scaling (MDS), 203
Multimedia user interface design, 65–83
  analog versus discrete media, 66
  audio links, 79
  Big Al game, 78
  cognitive background, 67–69
    emotion and arousal, 68
    learning and memorization, 68–69
    perception and comprehension, 67
    selective attention, 67–68
  color coding, 67
  communication goal, 72
  Concept demonstrators, 70
  contention problems, 68
  decision tree, 73
  definitions and terminology, 66–67
  design for attention, 80
  design for motivation, 82
  design process, 69–82
    aesthetic design, 75–76
    affective effects, 76–77
    design for attention, 79–82
    engagement and attractiveness, 75
    image and identity, 76
    information architecture, 70–72
    interaction and navigation, 77–79
    linguistic media (text and speech), 81–82
    media selection and combination, 72–77
    metaphors and interaction design, 77–78
    moving image media, 81
    navigation, 78–79
    still image media, 81
    users, requirements, and domains, 70
  distinguishing characteristics, 66
  dynamic media, 75, 79
  human image, 76
  interactive prototypes, 70
  ISO standard 14915 on Multimedia User Interface Design, 66
  media selection example, 74
  mini-maps, 79
  modality, 66
  Model Human Processor, 67, 68
  moving images, 79
  Moving Pictures Expert Group, 67
  natural sounds, 75
  navigation controls, 79
  physical media storage, usability implications, 67
  static media, 79
  storyboards, 70
  structural metaphors, 70
  task-driven applications, information needs in, 70
  text media, 79
  thematic congruence, 69
  thematic map, 71
  time-to-market pressure, 66
  usability components, 66
  Virtual Reality, 67
  "within media" design, 66
Multimodal hypertiming, 95
Multimodal interfaces, 85–104
  adaptive temporal thresholds, 95
  advantages, 89–92
  agent communication languages, 99
  basic architectures, 97–99

Multimodal interfaces (*continued*)
  cognitive science underpinnings, 93–97
    complementarity or redundancy, 96
    differences in multimodal interaction, 95–96
    integration and synchronization characteristics of users' input, 94–95
    multimodal interaction, 93–94
    primary features of multimodal language, 96–97
  commercialized, 89
  corpus-collection efforts, 97
  CUBRICON system, 87
  cultural differences, 95
  current status, 86–89
  definition, 86
  differences in graphical user interfaces, 97
  feature-fusion architecture, 98
  future directions, 99–100
  gaze patterns, 96
  goals, 89–92
  graphical user interfaces, 86, 97
  Hidden Markov Models, 99
  high-fidelity simulation testing, 92
  history, 86–89
  hybrid architectures, 99
  LOC-S-V-O word order, 97
  Members-Teams-Committee, 99
  methods, 92–93
  multi-agent architectures, 99
  multimodal hypertiming, 95
  mutual disambiguation of input signals, 91
  myths, 93
  Neural Networks, 99
  Open Agent Architecture, 99
  processing techniques, 97–99
  purpose, 86
  "Put That There" system, 94, 97
  QuickSet, 87
  semantic processing, 86
  sensory perception, 100
  Shoptalk, 87
  temporally cascaded interfaces, 89
  terminology, 90
  theory of working memory, 94
  types, 86–89
  user cognitive load, 92
  user efficiency, 91
Multitasking, 164
Multi-user domains (MUDs), 226
MusicBottles, 153–154
MVIEWS, 87

N

Narrowcast, 138
Natural dialogue study, 59
Natural Interaction Systems, 89
Natural language dialogue, 111
Natural language processing (NLP) technologies, 56, 57
Natural Language Understanding (NLU), 57
Natural sounds, multimedia and, 75
NetMiner, 284
Netscape, 233
Neural Networks (NNs), 99
Next Generation Internet (NGI), 298
NGI, *see* Next Generation Internet
NLP technologies, *see* Natural language processing technologies

NLU, *see* Natural Language Understanding
NNs, *see* Neural Networks

O

Object chart, 198
Object-oriented programming, 171
Online communities, 275–291
  activity theory, 289
  analysis (frameworks and methodologies), 280–283
    content and textual analysis, 282
    interviews, 281
    log analysis, 281–282
    personas, 281
    query-based techniques and user profiles, 281
    questionnaires, 281
    social network analysis, 282–283
  "anyone-can-contribute" policy, 278
  bulletin boards, 278
  case studies, 284–289
    activity theoretical analysis for constructionist learning, 288–289
    activity theory, 286–288
    computer aided language learning communities, 284–285
    game communities and activity theoretical analysis, 285–289
    types of game communities, 286
  clique analysis sociogram, 285
  CMC, 276–280
    advantages and disadvantages of CMC, 277
    computer mediated communication, 276
    examples of CMC and online communities, 277–280
    online communities, 276–277
    online virtual game communities, 279–280
    systems, 277
    wiki-based communities, 278–279
  connected knower, 284, 285
  cyber societies, 276
  discussion, 289
  discussion boards, 279
  division of labor, 287, 288
  document management systems, 279
  EverQuest, 280
  forums, 278
  game-based communities, 289
  global village, 276
  hierarchy of activity, 287
  human-computer interaction, 276
  in-degree analysis sociogram, 285
  mediation model of activity system, 288
  NetMiner, 284
  pull technology, 278
  separate knower, 284, 285
  time-stamped logs, 281
  triangle activity system diagram, 287
  virtual worlds, 279
  Wikipedia, 278
  WYSIWYG, 278
Open Agent Architecture, 99
Overlay model, 242

P

Page design principles, 13–16
  abstracting, 15
  building of page design, 13–15
  chunking, 13
  field of vision, 15
  filtering, 13

gray page or screen, 13
grid, 15
illusion of depth, 16
mixing modes, 13
other page design techniques, 15–16
proximity, 15
queuing, 13
white space, 15
Palm PDA, 135
PassPoint system, 321
Password(s)
  lifetime, 162
  random, 321
PCs, *see* Personal computers
PD, *see* Power distance
PDAs, *see* Personal digital assistants
Permutation matrices, 203
Personal computers (PCs), 54
  desktop metaphor, 136
  GUIs and, 142
  Sony VAIO notebook, 114
Personal digital assistants (PDAs), 54
  AgentSalon, 117
  e-mail done from, 219
  Internet access, 234
  Palm, 135
  tracking with, 224
  TUIs and, 144
  Wukong prototype, 32
Phishing, 322, 325
Pinwheels, 154–155
PKI, *see* Public key infrastructure
Planet Sabre, 35
Platform for Privacy Preferences, 315, 316
Podcasting, 138
POPFile, 107
Portable Voice Assistant, 87
Portals, 240
Portholes, 224
Power distance (PD), 38
PRIME system, 316
Privacy, security, and trust, 311–330
  absolute security, myth of, 323
  automatic teller machines, 322
  Confab, 316
  credibility theory, 324
  data encryption, 322
  data-subjects, 314
  directions, 326–327
  e-commerce websites, 320
  familiarity, 324
  frequently asked questions, 313
  meta-access control models, 319
  PassPoint system, 321
  passwords, 320
  Peer-to-Peer policies, 315
  perceived security, 320
  personalization, 324
  phishing, 322, 325
  Platform for Privacy Preferences, 315, 316
  PRIME system, 316
  privacy, 313–318
    anonymity, 315
    design factors, 318
    end user views, 315–316

HCI studies, 314–317
    organizational requirements in managing privacy, 317
    personalization, 315
    -risk model, 316
    summary of findings and recommendations, 317
  public key infrastructure, 321
  relationships between privacy and other concepts, 313
  security, 318–323
    administrators, 320
    authentication mechanisms, 320–321
    biometrics, 321–322
    control over, 319–320
    design factors, 323
    key topic areas for HCI, 320–322
    passwords, 321
    secure design and administration, 320
    security, other approaches, 322
    summary of findings and recommendations, 323
    usability challenges specific to security (attack defense), 322–323
    usability of security software and mechanisms, 319
  signature verification, 322
  smart cards, 321, 322
  SPARCLE policy workbench, 317
  spyware, 325
  Total Information Awareness initiative, 315
  trust, 323–326
    design factors, 326
    factors influencing, 324–325
    model, 324
    personalization and, 325
    risk and, 325
    summary of findings and recommendations, 325
  user-centered security, 319
  weakest link property, 319
  website design, 324
  Wizard of Oz system, 317
Proxemics, 226
Psychological simplicity, achievement of, 161–179
  aesthetic variables, 163
  "Answer Garden," 173
  case studies, 175–177
    query by example, 175–176
    web accessibility technology, 176–177
  COCOMO, 174
  Complex Adaptive Systems, 162
  complexity and related concepts, 165–170
    complexity, feedback, and interactivity, 169–170
    complexity of system vs. contextual complexity, 169
    complexity of system vs. task complexity, 168–169
    relationship of complexity and distribution, 167
    relationship of complexity and ease of use, 165
    relationship of complexity and naming scheme, 167–168
    relationship of complexity and nature of elements, 167
    relationship of complexity and nonlinearity, 167
    relationship of complexity and number, 166–167
    relationship of complexity and obscurity, 168
    relationship of complexity and structural framework, 168
    relationship of complexity and uncertainty, 166
    relationship of simplicity and complexity, 165–166
  contextual complexity, 169
  credit assignment issues, 169–170
  DICTION program, 175
  Document Object Model, 176
  Flesch Formula, 175
  Gunning Fog index, 175

Psychological simplicity (*continued*)
  in-process metric, 175
  linear regression model, 164
  log of word frequency, 164
  measurement of complexity, 174–175
    a priori mathematical models, 174
    iterative design and testing, 175
    linear regression, 174
    subjective measures, 174
    textual analysis of documentation, 175
  memory load, 167
  multitasking, 164
  nature of psychological complexity as variable, 162–165
  object-oriented programming, 171
  password lifetime, 162
  possible future approaches, 175
  psychological complexity, 164
  query accuracy, 176
  reason for studying psychological complexity, 162
  sources of complexity in development process, 171–174
    deployment, 173
    design, 172
    development, 172–173
    maintenance, 174
    problem finding, 171
    problem formulation and requirements, 171–172
    radical iteration in field, 171
    service, 173–174
    testing, 173
  sources of difficulty for HCI, 170–171
    making computation efficient and effective, 170
    making system understandable and maintainable, 170–171
    making tacit knowledge explicit, 170
    understanding syntax and semantics of communicating with computer, 170
  speech-flake pattern, 168
  Stroop task, 167
  subjective uncertainty, 166
  symmetrized dot pattern, 169
  tangled problems, 170
  task difficulty, 164
  user's input number, 167
  workload, 164
Public key infrastructure (PKI), 321
"Put That There" system, 94, 97

Q
Quantitative Spatial, 192
Query By Example, 175–176
QuickDoc, 87
QuickSet, 87

R
Random passwords, 321
Ration-by-schedule (RBS), 257, 258
RBS, *see* Ration-by-schedule
Relevance feedback (RF), 238
Remote-collaboration interface, 152
Renaissance publisher, 6
Rendered cursor, 306
Retinal information topographies, 201
Retinal properties, information visualization and, 194
Retinal scattergraph, 198
Return-on-investment (ROI) analysis, 31
RF, *see* Relevance feedback

RFID tags, 137
ROI analysis, *see* Return-on-investment analysis

S
SandScape, 152–153
SBA, *see* Simulation Based Acquisition
Scientific visualization, 187, 189
Security, human-computer interaction and, *see* Privacy, security, and trust
Selective attention, 67
SENSAI, 239
Sensetable, 150–151
Separate knower (SK), 284, 285
Shoptalk, 87
SIDs, *see* Spatially Immersive Displays
Siemens website, 46
Simulation Based Acquisition (SBA), 306
SK, *see* Separate knower
Slot-Credit Substitution transaction, 259
Smart cards, 321, 322
Smart Menus, 108, 125
Smart objects, 137
Smart tags, 137
SNA, *see* Social Network Analysis
SOAR modeling, 175, 247
Social computing, 225
Social Network Analysis (SNA), 282, 283
Soft 404s, 236
Sony VAIO notebook PC, 114
Space Physics and Aeronomy Research Collaboratory (SPARC), 221
SPARC, *see* Space Physics and Aeronomy Research Collaboratory
SPARCLE policy workbench, 317
Spatially Immersive Displays (SIDs), 295
Speech, *see* Conversational speech interfaces and technologies
Spyware, 325
SQL SELECT statement, 117
SQL-Tutor, 117, 121
Squint test, visual language, 24
Static media, 79
Stereotype models, 241
Still image media, 81
Storyboards, 70
Stroop task, 167
Student modeling, 117
Subjective uncertainty, 166
SwiftFile, 107

T
Tabletop TUI, 147
Tangible User Interface (TUI), 141–159
  abacus, 156, 157
  Actuated Workbench, 147
  applications, genres of, 146–148
    ambient media, 147–148
    augmented everyday objects, 147
    constructive assembly, 146
    continuous plastic TUI, 147
    interactive surfaces (tabletop TUI), 147
    tangibles with kinetic memory, 146
    tangible telepresence, 146
    tokens and constraints, 146–147
  basic model, 144
  BBS, 146
  body syntonic learning, 149
  clock tool, 142

contributions, 155–156
   coincidence of input and output spaces, 155–156
   double interactions loop, 155
   persistency of tangibles, 155
   space-multiplexed input, 156
   special purpose vs. general purpose, 156
desktop metaphor, 144
Digital Desk, 147
digital shadow, 146
example, 142–144
from GUI to TUI, 142
GUI, 144
instances, 147–155
   Actuated Workbench, 152
   Audiopad, 150–151
   Curlybot, 148–149
   Digital Desk, 150
   InTouch, 148
   IP Network Design Workbench, 151–152
   mediaBlocks, 149–150
   musicBottles, 153–154
   Pinwheels, 154–155
   SandScape, 152–153
   Sensetable, 150–151
   Topobo, 149
intangible representation, 145
inTouch, 143
IP Network Design Workbench, 150, 151
key properties, 145–146
   intangible representations, 146
   tangible representations, 145–146
Marble Answering Machine, 146
model, 145
remote-collaboration interface, 152
success, 145
tabletop TUI, 147
tangible representation as control, 144–145
Ubiquitous Computing, 142
Urban Planning Workbench, 142–144
WYSIWYG, 144
Tangled problems, 170
Tapered prompts, 60
Task difficulty, 164
Task model, 55
TAT, see Transcript Analysis Tool
TeamWare, Finnish screen patterns, 37
Text media, 79
Text-to-speech (TTS), 55
Thematic map, 71
TIA initiative, see Total Information Awareness initiative
Time-out cancels, 259
Topobo, 149
Total Information Awareness (TIA) initiative, 315
Transcript Analysis Tool (TAT), 282
TreeMap, 185, 186
Trust, human-computer interaction and, see Privacy, security, and trust
TTS, see Text-to-speech
TUI, see Tangible User Interface
Typography, 7–9, 13
  Bell Centennial, 8, 9
  contrasts, 14
  definition, 7
  effective, 7
  globalization and, 34
  guidelines, 11–13

   combining typefaces, 11
   contrast in weight (boldness), 11
   decorative typefaces, 12
   highlighting with type, 12
   justified vs. ragged right (flush left), 12
   letter spacing and word spacing, 12
   line length/column width, 12
   line spacing/leading, 12
   output device and viewing environment, 11–12
   positive and negative type, 12–13
  hqx illustration, 10, 13
  legibility, 8
  letter anatomy, 10
  Pixel font, 9
  readability, 8
  type classification, 11
  typeface choice, 8
  typeface size and selection, 9–11
  type family, 7, 10

U

UA, see Uncertainty avoidance
UARC, see Upper Atmospheric Research Collaboratory
Ubiquitous Computing, TUI and, 142
UIs, see User interfaces
Uncertainty avoidance (UA), 43
UNIX commands, 57
UPA, see Usability Professionals Association
Upper Atmospheric Research Collaboratory (UARC), 221
Urban Planning Workbench, 142–144
Usability Professionals Association (UPA), 32, 49
UseNet groups, 278
User interfaces (UIs), 28, 66
User modeling servers, 123

V

Variable prompts, 60
VEs, see Virtual environments
Virtual environments (VEs), 293–310
  application domains, 303–307
   entertainment tool, 305
   medical tool, 306
   motor rehabilitation applications, 306
   psychological/cognitive rehabilitation applications, 306
   selection and training tool, 303–305
   system acquisition tool, 306–307
  design and implementation considerations, 299
  design and implementation strategies, 298–301
   cognitive aspects, 298–300
   content development, 300
   multimodal interaction design, 298–299
   navigation and wayfinding, 300
   perceptual illusions, 299–300
   products liability, 300
   usage protocols, 300–301
  gesture interaction, 297
  hardware and software requirements, 294
  health and safety issues, 301–302
   cybersickness, adaptation, and aftereffects, 301–302
   social impact, 302
  MAUVE system, 302
  rendered cursor, 306
  seat shaker, 296
  Spatially Immersive Displays, 295
  sweet spot, 296

Virtual environments (VEs) (*continued*)
 system requirements, 294–298
  augmented cognition techniques, 297
  autonomous agents, 298
  hardware requirements, 295–297
  interaction techniques, 297
  modeling, 297–298
  multimodal I/Os, 295–296
  networks, 298
  software requirements, 297–298
  tracking systems, 296–297
 training system development, 304
 usability engineering, 302–303
  sense of presence, 302–303
  usability techniques, 302
  virtual environment ergonomics, 303
 volumetric displays, 295
 VRSonic SoundScape3D, 296
Virtual Reality (VR), 67
Visit lists, 79
Visual cues, 16
Visual design principles for usable interfaces, 1–25
 balance, 6
 challenges and opportunities, 24–25
 charts, diagrams, graphics, and icons, 16–19
  building of library, 19
  creation of consistent visual language, 18
  function and style, 18
  guidelines, 18–19
  icons and visual cues, 16
  illustrations and photographs, 17–18
  quality vs. quantity, 18
  reinforce shared meaning, 19
  tables, charts, diagrams, 16
  visuals should reinforce message, 18
  work with professional, 18–19
 color, 19–20
  additive primaries, 19
  basic principles, 19
  color choices, 20
  color logic, 19–20
  context, 20
  contrast, 20
  decisions regarding color in typography, 20
  global world, 20
  how to use color, 19–20
  less is more, 19
  palette of compatible colors, 20
  quantity affects perception, 20
  subtractive primaries, 19
  use of color as redundant cue, 20
  use of complementary colors with extreme caution, 20
  viewing medium, 20
 creating your own guidelines, 24–25
 defining visual design, 3
 design criteria, 7
 designing of experience, 22–23
  effective and appropriate use of medium, 23
  element of time, 23
 design process, 3–4
 fatigue factor, 24
 golden mean, 6
 graphic standards, 20–22
  audit, 22
  implementation, 22

 system development, 22
 what system covers, 22
harmony, 6
holistic process, 5
how the human eye sees, and then reads, 9
informed design process, 4–6
 audit, 5
 design development, 5
 implementation and monitoring, 5–6
interacting design loop, 4
page design, 13–16
 abstracting, 15
 building of page design, 13–15
 chunking, 13
 field of vision, 15
 filtering, 13
 gray page or screen, 13
 grid, 15
 illusion of depth, 16
 mixing modes, 13
 other page design techniques, 15–16
 proximity, 15
 queuing, 13
 white space, 15
process of good design, 4
renaissance publisher, 6
role of designer, 4
saccadic eye movement, 9
simplicity, 6
squint test, 24
typeface size and selection, 9–11
 families of type, 10
 serif and sans serif, 9–10
typographic guidelines, 11–13
 combining typefaces, 11
 contrast in weight (boldness), 11
 decorative typefaces, 12
 highlighting with type, 12
 justified vs. ragged right (flush left), 12
 letter spacing and word spacing, 12
 line length/column width, 12
 line spacing/leading, 12
 output device and viewing environment, 11–12
 positive and negative type, 12–13
universal principles, 6
user's experience, 3
variations in letterforms, 11
 serifs, 11
 stress, 11
 thick and thin, 11
visual design principles, 6
visual design tools and techniques, 6–9
 criteria for good design, 7
 typography, 7–9
 visual design principles, 7
visual language, consistent and appropriate, 23–24
 avoid overuse of saturated colors, 24
 color, 23
 different users' levels of skill, 24
 fatigue factor, 24
 graphics/icons, 23
 guidelines, 24
 legibility, 24
 metaphor, 23
 most difficult common denominator, 24

navigational aids, 23
    other differences to consider, 24
    poster analogy, 24
    readability, 24
    use of "squint test" to check design, 24
  visual mapping, 13
Visual harmony, 6
Visual knowledge tools, 214
Visually enhanced objects, 214
Visual mapping, 13
VR, *see* Virtual Reality
VRSonic SoundScape3D, 296

W

Web
  accessibility, 176–177, 232
  "Answer Garden," 173
  blogs, 222, 243
  browsers, 233
  globalization and, 31
  groupware and, 219
  page(s)
    Document Object Model, 176
    static, 240
  usability problems, 232
Web, HCI and, 231–244
  adaptive and adaptable systems, 240
  adaptive hypermedia, 240
  adaptive hypertext, 240–241
  authoring, 242
  blogs, 243
  browsing and linking, 235–237
    broken and misdirected links, 236
    personalizing links, 236–237
  citation based metrics, 239
  cognitive maps, 235
  communities, 242–243
  correction of links, 236
  difficulty of web use, 232–235
    browsing and linking, 232–233
    context of use, 234–235
    finding things (relevance), 234
    finding things (search and query), 233
    navigation issues, 235
    user interface issues, 234–235
  e-commerce, 234
  feedback, 238
  fourth-generation hypermedia, 232
  glossary links, 237
  HTML links, 236
  information-consuming habits, 239

  Information Visualizer, 239
  intranet systems, 242
  late binding of links, 236
  lexical ambiguity, 239
  link sorting, 241
  methods and models of adaptation, 241
  need vs. desire, 241
  Netscape, 233
  one-size-fits-all approach, 234
  overlay model, 242
  personalization, 234, 240–242
  portals, 240, 242
  query, 233, 237
  real-life education, 241
  reweighting, 238
  searching and querying, 237–239
    community-based ranking algorithms, 238–239
    document clustering, 239
    improved visual interfaces, 239
    query formulation, 237–238
    relevance feedback, 238
    results list, 238–239
  SENSAI, 239
  soft 404s, 236
  stereotype models, 241
  term frequency, 238
  user modeling, 241–242
  user pull of information, 237
  WHURLE, 240
  Wikipedia, 243
  WYSIWYG, 242
Website(s)
  advice, 324
  ChickClick, 43, 44
  design, trust and, 324
  e-commerce, 320
  Excite, 43
  recommendation, 137
  Siemens, 46
WHURLE, 240
Wikipedia, 224, 243, 278
"Within media" design, 66
Wizard of Oz studies, 62, 124
Workflow systems, groupware and, 223
Working memory, theory of, 94
Wukong prototype PDA, 32
WYSIWYG, 144, 242, 278

X

Xerox Parc, 147